THE NIGHT SKY IN SPRING

Latitude of chart is 34°N, but it is practical throughout the continental United States.

To use: Hold chart vertically and turn it so the direction you are facing shows at the bottom.

Chart time (local standard time):
Mid–March....... 11 pm
Mid–April.......... 9 pm
Mid–May........... 8 pm

Star Chart from *GRIFFITH OBSERVER*, Griffith Observatory, Los Angeles

Universe

William J. Kaufmann, III

W. H. Freeman and Company
New York

To my nephew Mark Karwoski with love

Cover image: The inner region of the 30 Doradus Nebula
30 Doradus is the largest nebula in the local group of galaxies. It's diameter is 1000 light years and it is located in a nearby galaxy called the Large Magellanic Cloud, about 160,000 light years from Earth. The tight knot of stars at the center of the photograph is 50 million times more luminous than the Sun. This extremely brilliant central object is either the most massive star known, or it is a dense cluster of extremely massive stars. This photograph was taken by John Wood with the 4-meter telescope at the Cerro Tololo Inter-American Observatory in Chile. (J. S. Mathis, B. D. Savage, and J. P. Cassinelli)

Book design: Valerie Pettis and Sylvia Woodard
Cover design: Valerie Pettis

Library of Congress Cataloguing in Publication Data
Kaufmann, William J.
    Universe.

    Includes bibliographies and index.
    1. Astronomy.   2. Cosmology   I. Title.
QB43.2.K38   1985   523   84-13830
ISBN 0-7167-1673-9

Illustration credits are listed on page 584

A huge bubble of gas (called NGC 2359) surrounds a hot, young star (called HD 56925) which has a very powerful stellar wind. This wind consists of atoms ejected from the star's surface and accelerated to nearly one percent of the speed of light by radiation pressure. Stars of this type are usually called Wolf–Rayet stars, named after two nineteenth century French astronomers who first discovered several examples. The material escaping from HD 56925 has blown a bubble in the surrounding gas and dust left over from the star's birth. This picture was produced by David F. Malin using the 3.9-meter Anglo-Australian telescope. Three black and white photographs taken through red, blue, and green filters were combined to produce this view.

Six different views of NGC 2359 from which astronomers can deduce details of this object. The three photographs labeled "hydrogen," "oxygen," and "nitrogen" were taken through filters transparent only to light emitted by those atoms. The view labeled "H_α/H_β" compares the intensity of radiation from hydrogen atoms at two wavelengths and is sensitive to the amount of interstellar dust (blue indicates a higher density of dust). The view labeled "NII/OIII" compares light emitted by nitrogen and oxygen atoms and is sensitive to the degree of ionization (blue represents high ionization and indicates where atoms have been stripped several electrons). All of these optical views were obtained by M. H. Schneps and E. L. Wright while working at the Kitt Peak National Observatory. The lower-right view is a map showing radio emission and was produced by Schneps, Haschick, Wright, and Barrett using the VLA in New Mexico. Based on these views, Schneps and Wright have inferred details of NGC 2359 shown in the diagram below.

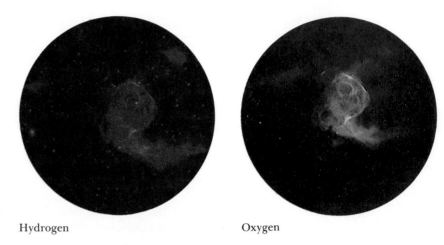

Hydrogen   Oxygen

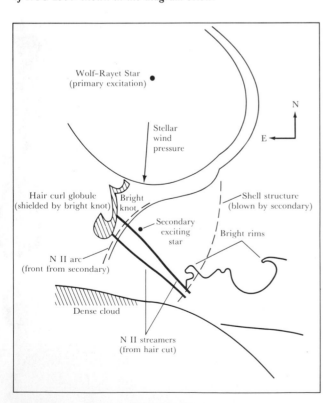

Wolf–Rayet Star
(primary excitation)

N
E

Stellar
wind
pressure

Hair curl globule
(shielded by bright knot)

Bright
knot

Shell structure
(blown by secondary)

Secondary
exciting
star

Bright rims

N II arc
(front from secondary)

Dense cloud

N II streamers
(from hair cut)

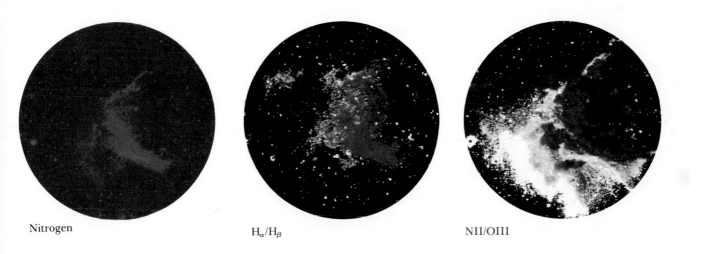

Nitrogen H_α/H_β NII/OIII

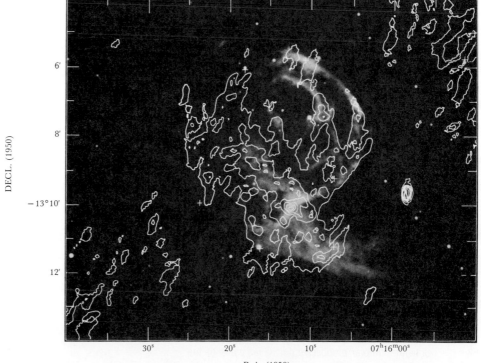

Radio emission

Table of contents

Preface

The popularity of astronomy as a general science course is due in large measure to the inherent intrigue of its subject matter. Since earliest times people have been fascinated by topics that astronomers explore today: the creation of the universe, the formation of the Earth and other planets, the motions of the stars, the structure of space and time. Armed with the powers of observation, the laws of physics and the resourcefulness of the human mind, astronomers survey alien worlds, follow the life cycles of stars, and probe the dim and distant reaches of the cosmos.

Astronomy raises our consciousness because it investigates phenomena and explores realms far removed from our daily experience. Many of the objects that astronomers study are far too vast, distant, or intangible to ever sample directly; many of the phenomena that they observe today occurred very long ago. The dimensions of space and time take on new meaning when viewed in the context of the evolution of the universe.

In writing this book, I have tried to convey more than the intriguing nature of our physical universe, however. I also describe how astronomers have come to know what they know. To show students how scientists reason must be an objective of any first science course. By studying the methods that astronomers have used in exploring the universe and discovering the patterns in the observations they make, we can learn something about the nature of scientific inquiry.

This goal is manifest in the "traditional" organizational scheme of the text, in which the first celestial objects to be examined are those that were observed by the ancient astronomers. This allows us to see how our understanding of the universe developed and to share in the excitement of astronomical discovery. As we move outward from the planets to the stars and galaxies, older earth-based observations and the questions that they provoke are augmented or even supplanted by newer observations, including those outside the visible range and those made from space. These new observations in turn raise a host of new questions, which draw the reader on to the outer limits of our universe and our understanding.

This text is designed to meet the needs of instructors of either a one- or two-term course. The book's twenty-nine chapters are easily divided into two nearly equal parts, the first half dealing with early models of the universe and planetary astronomy and the second half treating stars, galaxies, and cosmology. Instructors may emphasize either half of the book according to preference, covering more or less of the detail as time and the preparation of the students permit.

The first five chapters introduce the foundations of astronomy, including descriptions of such naked-eye observations as eclipses and planetary motions and such basic tools as Kepler's laws and the optics of telescopes. A discussion of the formation of the solar system in Chapter 6 prepares the reader for the next ten chapters, which cover the planets in outward order from the Sun. One of these chapters deals with the Galilean satellites, which are terrestrial worlds in their own right.

Chapter 17 leads into stellar astronomy with a detailed discussion of the properties of light, and Chapter 18, on the Sun, introduces the reader to the general nature of a star. In Chapters 19 through 24, stellar evolution is described chronologically from birth to death. Molecular clouds, star clusters, nebulae, neutron stars, black holes, and various other phenomena are presented in the sequence in which they naturally occur in the life of a star, thus unifying the wide variety of objects that astronomers find scattered about the heavens.

A survey of the Milky Way introduces galactic astronomy in Chapter 25; this is followed by two chapters on galaxies and quasars. The final two chapters, on cosmology, emphasize exciting recent developments in our understanding of the physics of the early universe.

A distinguishing feature of this book is the use of color in its illustrations. Astronomy is founded on observation. Many of our observations have been recorded on film; others are plotted on graphs or sketched in diagrams. Pictures afford a convenient means of summarizing data, and pictures in color indisputably convey more information than do those in black and white. For example, one glance at a color photograph of a planet's cloudtops, or the glowing gases of a nebula, reveals significant details about the object that cannot be gleaned from a black-and-white view alone. In recent years, computer processing of data at nonvisible wavelengths has produced extraordinary false-color views of the X ray, infrared, and radio sky. However, most astronomy texts contain only a sampling of color photographs, usually segregated from the corresponding narrative. In this text, color photographs are integral to the text and are incorporated throughout.

Essays are another special feature of this book; they offer a forum for the personal expression of six renowned astronomers having distinct interests and views. Whether it be to recommend the further exploration of the solar system, the refinement of astronomical instruments, or the consideration of a new idea, each essayist gives the reader a broader understanding of our universe and of what it is to be an astronomer.

Boxed inserts are used to set aside special material from the main body of the text. Some bring together key ideas for ease in learning and for ready reference (for example, Box 1-3 on distances and units of length). Others contain technical and reference information, such as the orbital and physical data for each planet. Still others present arguments, the details of which can be omitted without loss of understanding (an example is Box 27-1 on the relativistic redshift). This segregation permits greater flexibility in an instructor's use of the text and allows the student to easily locate and review topics to which frequent references are made.

The student's ease in understanding astronomy has been a major objective of mine in writing this text. Each chapter begins with a brief, one-paragraph abstract that gives the reader a clear idea of the chapter's contents. The chapter headings are given in the form of declarative sentences to highlight main concepts, and a formal summary outlines the essential facts addressed in each chapter. A series of questions concludes each chapter. These questions are grouped by difficulty and content into three categories: review, advanced, and discussion. Answers to questions that require computation (marked by an asterisk) appear at the end of the book. Care has been taken to formulate questions whose answers require reasoning rather than just memorization. Roger Culver of Colorado State University deserves special recognition for his many contributions in this regard.

I am deeply grateful to many other people who have participated in the preparation of this book. Foremost among them is my developmental editor, Carol Pritchard-Martinez, who has stuck with me through thick and thin. I also thank my editor, Peter Renz, and Neil Patterson for their support and encouragement of this project. Georgia Lee Hadler, who coordinated the production of the book, and Jill Feldheim, who managed the art program, deserve special thanks for their unfailing concern for quality. I acknowledge the fine efforts of my copyeditor, Larry McCombs, who has left his mark on the manuscript, the drawings by Vantage Art, and the marvelous airbrush artistry of George Kelvin.

Many of my colleagues have contributed valuable suggestions and criticisms to the manuscript. Special thanks go to

John K. Lawrence	California State University at Northridge
Richard L. Sears	University of Michigan
David B. Slavsky	Loyola University of Chicago
Joseph S. Tenn	Sonoma State University

each of whom reviewed the manuscript for accuracy and coverage. Selected portions of the manuscript were reviewed in depth by

Robert Allen	University of Wisconsin
John M. Burns	University of Arizona
David S. Evans	University of Texas
Owen Gingerich	Harvard University
J. Richard Gott, III	Princeton University
Paul Hodge	University of Washington
Dimitri Mihalas	High Altitude Observatory
L. D. Opplinger	Western Michigan University
John R. Percy	University of Toronto
Richard Saenz	California Polytechnic State University
Nicholas Wheeler	Reed College
Donat G. Wentzel	University of Maryland
Raymond E. White	University of Arizona

Finally, I thank Thomas F. Scanlon of Grossmont College, who both read several drafts of the manuscript and conducted classroom testing of it, and Andrew Fraknoi of the Astronomical Society of the Pacific, who prepared *Universe in the Classroom: A Resource Guide for Teaching Astronomy*.

In spite of the care exercised in preparing this text, errors may have crept in. I would appreciate hearing from anyone who finds an error or who wishes to comment on the text. You may write to me in care of the publisher. I will respond personally to all correspondence.

William J. Kaufmann, III

Department of Physics
San Diego State University

1 Astronomy and the universe

The Horsehead Nebula
New stars are forming in the clouds of interstellar gas and dust shown in this photograph, which covers an area of the sky approximately $1\frac{1}{3}° \times 2°$. The gases glow because of the radiation emitted by newborn, massive stars. Dark regions (such as the horsehead) are caused by dust grains that block light from the background nebulosity. The very bright star to the left of center is ζ Orionis, the easternmost star in the "belt" of Orion. Most of the stars and nebulosity in this photograph are 1600 light years from Earth. (Royal Observatory, Edinburgh)

Astronomy is the study of the universe. A brief preview here of following chapters provides an outline of the scope and content of astronomy. We introduce important tools such as powers-of-ten notation and angular measure. We learn just a little about the solar system, stars, nebulae, and galaxies—enough to get a sense of where we will be going in this book. Above all, we learn that the universe is indeed comprehensible. Although some questions remain unanswered, we find no reason to think that any aspect of the physical universe is arbitrary or unexplainable.

Modern city dwellers pay little attention to the night sky. If we do glance away from neon lights and television screens toward the heavens, we are likely to see little more than the Moon and a few of the brightest stars.

For those many generations who have lived without electric lights and smog, however, the breathtaking panorama of the night sky has been one of the central experiences of life. Thousands of stars are scattered from horizon to horizon, with the delicate mist of the Milky Way tracing a faerie path through the patterns of brighter stars. The Moon and the planets shift their positions from night to night against this glorious background of "fixed" stars, while the entire spectacle swings slowly overhead from east to west as the night progresses. Our ancestors learned to tell time and directions from these changing patterns in the sky. They mapped the stars into picture outlines that were associated with the most important legends and ideas of their cultures.

Like these earlier people, we find our thoughts turning to profound questions as we gaze at the stars. How was the universe created? Where did the Earth, Moon, and Sun come from? What are the planets and stars made of? And how do we fit in? What are our place and role in the cosmic scope of space and time?

Speculation about the nature of the universe is one of the most ancient human endeavors. The study of the stars transcends all boundaries of culture, geography, and politics. The modern science of astronomy carries on an ancient tradition of observation and speculation, using the newest tools of technology and mathematics. In the most literal sense, astronomy is a universal subject—its subject is indeed the universe.

Modern astronomy has a rich heritage that dates back to antiquity

In some societies, myths and legends provided sufficient explanation. The heavens were thought to be populated with demons and monsters, heroes, gods, and goddesses. Astronomical phenomena were explained as the result of supernatural forces and divine intervention.

We begin by examining some of the astronomical observations and ideas of our ancestors in Chapters 2 and 3. We shall see that the course of civilization has been dramatically affected by the realization that *the universe is comprehensible.* This first glimpse of the power and potential of the human mind is one of the great gifts to come to us from ancient Greece. By observing the heavens and carefully thinking about what we see, we can figure out how the universe operates. For example, we shall see that the ancient Greeks measured the size of the Earth and understood and predicted eclipses.

We shall also see that astronomers have inherited many useful concepts from antiquity. For example, ancient mathematicians invented angles and a system of angular measure that is still used to denote the positions and apparent sizes of objects in the sky. A brief discussion of this useful topic appears in Box 1-1.

There are people who think that astronomy deals with faraway places of no possible significance to life here on Earth. Nothing could be farther from the truth. For example, in Chapter 4 we learn that the seventeenth-century scientist Isaac Newton succeeded in describing how the planets orbit the Sun. In the motions of the planets, we see some of the most fundamental laws of nature revealed in their simplest form, unhampered by air resistance or friction. From Newton's work we obtained our first complete, coherent description of the behavior of the physical universe. The resulting body of knowledge, called **Newtonian mechanics,** speaks in very concrete terms about force, mass, acceleration, momentum, and energy. This understanding had immediate practical application in the construction of machines, factories, buildings, and bridges. It is no coincidence that the Industrial Revolution followed hard on the heels of these advances. Indeed, Newtonian mechanics provided the theoretical and mathematical basis for the Industrial Revolution.

We complete our introduction to astronomy in Chapter 5 with a discussion of the astronomer's most important tool, the telescope. Until recently, everything we knew about the distant universe was based on visible light. Astronomers would peer through telescopes, take photographs of what they saw, and analyze the starlight. Toward the end of the nineteenth century, however, scientists began discovering nonvisible forms of light such as X rays and gamma rays, radio waves and microwaves, and ultraviolet and infrared radiation.

Astronomers recently have constructed telescopes that detect nonvisible forms of light. In orbit, far above the obscuring effects of Earth's atmosphere, these astronomical instruments give us views of the universe vastly different from what our eyes see. This new information

Box 1-1 Angular measure and the small-angle formula

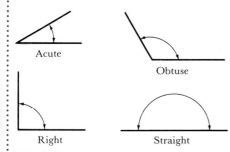

Acute

Obtuse

Right

Straight

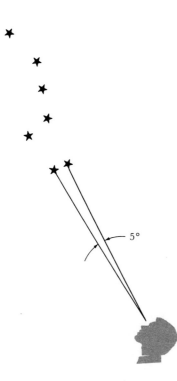

5°

An **angle** is the opening between two lines that meet at a point. Various kinds of angles are shown in the first diagram. **Angular measure** provides a more exact description of the shape or "size" of an angle.

In astronomy, we use angular measure in a wide range of situations. For example, we can most easily use an angle to describe how big an object appears in the sky.

The basic unit of angular measure is the **degree,** designated by the symbol °. A full circle is divided into 360°. A right angle measures 90°, and the angle between the two "pointer stars" in the Big Dipper is about 5°, as shown in the second illustration (bottom left).

Imagine looking up at the full moon. The angle covered by the Moon is nearly $\frac{1}{2}$°. We therefore say that the **angular diameter,** or **angular size,** of the moon is $\frac{1}{2}$°. Alternatively, astronomers say that the Moon **subtends** an angle of $\frac{1}{2}$°. Note that ten full moons could fit side by side between the two pointer stars in the Big Dipper.

To talk about smaller angles, we subdivide the degree into 60 minutes of arc (abbreviated 60 arc min or 60′). A minute of arc is further subdivided into 60 seconds of arc (abbreviated 60 arc sec or 60″). Thus

$$1° = 60 \text{ arc min} = 60'$$

$$1' = 60 \text{ arc sec} = 60''$$

In the *Astronomical Almanac* for 1981, for example, we read that Jupiter had an equatorial diameter of 35.72 seconds of arc on July 3. That is a very convenient and precise statement of how big the planet appeared in Earth's sky on that date.

The angular size of an object can be converted into a linear size (in kilometers or miles, for example) if we know the distance to the object. Astronomers usually are concerned with objects that subtend tiny angles in the sky, and this conversion is accomplished with the **small-angle formula.** Specifically, suppose an object subtends an angle α (measured in seconds of arc) and is at a distance D from the observer, as sketched in the diagram below.

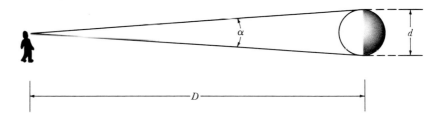

The small-angle formula tells us that the linear size (d) of the object is given by the expression

$$d = \frac{\alpha D}{206,265}$$

For example, on July 3, 1981, Jupiter was at a distance of 824.7 million kilometers from Earth. We have mentioned that Jupiter's angular diameter on that date was 35.72 seconds of arc, so we can calculate the planet's diameter:

$$d = \frac{35.72 \times 824,700,000}{206,265} = 142,800 \text{ kilometers}$$

Incidentally, the number 206,265 is not simply arbitrary. It equals $360 \times 60 \times 60/2\pi$, which is the total number of arc seconds in 360° divided by the circumference of a circle whose radius equals 1.

is crucial to our understanding of familiar objects, such as the Sun, and it gives us important clues about such exotic objects as neutron stars, pulsars, quasars, and black holes.

The remainder of this book presents the substance of modern astronomy in three segments, corresponding to three major steps out into the universe: the planets, the stars, and the galaxies.

By exploring the planets, astronomers uncover clues about the formation of the solar system

In Chapters 6 through 10, we explore the solar system, beginning with sun-scorched Mercury and moving outward toward the frigid depths of space where comets spend most of their time. Throughout this journey, we shall find that our discoveries are relevant to the quality of human life here on Earth. Until recently, our knowledge of such subjects as geology, geophysics, weather, and climate was based on only one planet, Earth. With the advent of space exploration, however, we have a range of other planets with which we can compare and contrast our own. As a result, we are making important strides in understanding the creation and evolution of the Earth and the entire solar system. These investigations give us significant insight into the origin and extent of all our natural resources.

As we examine alien environments, we find an astonishing range of conditions from the oppressive, acid-drenched clouds of Venus to the near-perfect vacuum of interplanetary space. To describe these conditions accurately, we shall need a wide range of numbers, both large and small. To avoid such confusing terms as "a million billion billion," astronomers use the *powers-of-ten* notation described in Box 1-2. You should be sure that you are familiar with this convenient shorthand notation.

As we turn toward the stars in the second half of this book, we shall find that some of our Earth-based traditions become cumbersome. For example, it is fine to use kilometers (or miles) to give the diameters of craters on the Moon or the heights of volcanoes on Mars. It is very awkward, however, to use kilometers to express distances to stars or galaxies. That would be even more absurd than talking about the distance from

Figure 1-1 An astronaut on the Moon
Humanity has taken its first small step out into the universe. As we explore distant worlds, we gain a new perspective on our own planet and a broadened understanding of our relationship to the cosmos. This photograph shows Apollo 11 astronaut Edwin Aldrin at the first lunar landing site on July 21, 1969. (NASA)

Figure 1-2 The Space Shuttle
The Space Shuttle will play an important role in astronomy during the 1980s and 1990s. It will be used to transport a telescope and other astronomical equipment into orbit, far above the obscuring effects of the Earth's atmosphere. Space probes to distant planets and comets also can be launched from the Space Shuttle. This photograph shows the first Space Shuttle rising majestically from Cape Canaveral on its maiden voyage April 12, 1981. (NASA)

New York to San Francisco in inches or millimeters. Astronomers have therefore invented new units of measure such as the *parsec* and the *light year*, as described in Box 1-3. To the annoyance of some people, astronomers do not restrict themselves to one system of measure, but rather use whatever yardsticks seem best suited for the issue at hand. Thus, for example, an astronomer might say that "the supergiant star called Antares has a diameter of 860 million kilometers and is located at a distance of 150 parsecs from Earth." Astronomers are not likely to change their habits in the near future, so you should learn to live with the astronomer's use of different units of measure in different situations.

Box 1-2 Numbers and powers of ten

Astronomy is a subject of extremes. When we speak of the diameter of a galaxy, the temperature inside a star, or the age of the universe, we deal with very large numbers. To avoid such confusing terms as "a million billion billion," we adopt a standard shorthand system. All of the cumbersome zeros that accompany a large number are consolidated into one term consisting of 10 followed by a superscript, or **exponent.** The exponent indicates how many zeros you would need to write out the long form of the number. Thus,

$10^0 = 1$

$10^1 = 10$

$10^2 = 100$

$10^3 = 1000$

$10^4 = 10,000$

and so forth

The exponent tells you how many factors of ten must be multiplied together to give the desired number. For example, ten thousand can be written as 10^4 (read "ten to the fourth") because

$10^4 = 10 \times 10 \times 10 \times 10 = 10,000$

With this notation, numbers are written as a figure between 1 and 10 multiplied by the appropriate power of 10. For example, the distance between the Earth and the Sun can be written as

1.5×10^8 km

After you get used to it, you will find this notation more convenient than "150,000,000 kilometers" or "one hundred and fifty million kilometers."

This shorthand system can be extended to numbers less than one by using a minus sign in front of the exponent. A negative exponent tells you the location of the decimal point as follows:

$10^0 = 1$

$10^{-1} = 0.1$

$10^{-2} = 0.01$

$10^{-3} = 0.001$

$10^{-4} = 0.0001$

and so forth.

(continued)

(Box 1-2, continued)

For example, the diameter of a hydogen atom is

1.1×10^{-8} cm

That is more convenient than saying "0.000000011 centimeters" or "eleven billionths of a centimeter."

Notice that, when the exponent of ten is positive, the decimal point shifts to the right, producing a number greater than one. Conversely, when the exponent is negative, the decimal point is moved toward the left, producing a number less than one.

Using this notation, familiar numerical terms are written as follows:

$$\text{one thousand} = 10^3 \quad = 1000$$
$$\text{one million} = 10^6 \quad = 1,000,000$$
$$\text{one billion} = 10^9 \quad = 1,000,000,000$$
$$\text{one trillion} = 10^{12} \quad = 1,000,000,000,000$$

and also

$$\text{one thousandth} = 10^{-3} \quad = 0.001$$
$$\text{one millionth} = 10^{-6} \quad = 0.000001$$
$$\text{one billionth} = 10^{-9} \quad = 0.000000001$$
$$\text{one trillionth} = 10^{-12} = 0.000000000001$$

Powers-of-ten notation bypasses all those awkward zeros, so that a wide range of circumstances like those in the accompanying illustration can be described in a very convenient fashion. Furthermore, using powers-of-ten notation, it is easy to multiply numbers; you simply add up the exponents:

$$10^2 \times 10^3 = 10^5$$

Thus, for example, "a million billion billion" can be easily written as $10^{6+9+9} = 10^{24}$ (read "ten to the twenty-fourth").

   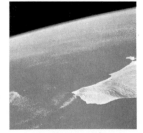 

A sampling of the dimensions of the known universe explored on the journey through the powers of ten is shown here. At center, the Taj Mahal adorns the 10 meter world in reach of our unaided senses. Dimensions grow smaller to the left: first, crystalline skeletons of single-celled diatoms at 10^{-4} meter (.1 millimeter) and at far left, tungsten atoms, 10^{-10} meter in diameter, brought into visibility by a field-emission microscope. At right, looking across the Indian Ocean toward the South Pole through the window of Gemini II, the curvature of the Earth, 10^7 meters in diameter. At far right, a galaxy (M104 in constellation Virgo) measures roughly 10^{18} meters (100,000 light years) in diameter.

Figure 1-3 Jupiter, Io, and Europa
Jupiter is orbited by many moons, four of which are so large that they could qualify as planets in their own right. Two of these giant satellites are shown in this view: ruddy Io on the left, ice-bound Europa on the right. Close-up examination of these worlds has dramatically broadened our understanding of Earth-like planets. This photograph was taken by Voyager 1 in 1979 when the spacecraft was 20 million kilometers (12 million miles) from Jupiter's colorful cloud tops. (NASA)

Box 1-3   Distances and units of length

To understand and appreciate the universe, we shall discuss a wide range of objects, both large and small, from galaxies to atoms. As we go from one subject to another, we shall use the units of length that are best suited to the topic at hand. For example, the distances to the stars are conveniently expressed in light years, whereas the diameters of the planets are more comfortably presented in kilometers. The metric system is very convenient and furthermore it is internationally accepted, so we shall base all our units of length on it.

When discussing objects of more-or-less human dimensions, we shall express sizes and distances in millimeters (abbreviated mm), centimeters (cm), meters (m), and kilometers (km). These units of length are related to each other as follows:

1 millimeter = 0.1 centimeter

1 meter = 100 centimeters

1 kilometer = 1000 meters = 10^5 cm

To make conversions from the old-fashioned English system of inches, feet, and miles, it is helpful to know that

1 inch = 2.54 cm

1 foot = 30.48 cm

1 mile = 1.609 km

Conversions are accomplished by using these equalities to cancel out the unwanted units while introducing the desired units. For example, suppose a NASA publication tells you that "the Saturn V rocket used to send astronauts to the Moon stands about 363 feet tall." You can convert this to meters as follows:

$$363 \text{ ft} \times \frac{30.48 \text{ cm}}{1 \text{ ft}} \times \frac{1 \text{ m}}{100 \text{ cm}} = 111 \text{ m}$$

For the convenience of the reader whose brain is still fettered to the archaic English system, we often give measurements in both systems. Thus, for example, we note that "the diameter of Mars is 6794 km (4222 miles)."

When discussing distances across the solar system, astronomers like to use a unit of length called the **astronomical unit** (abbreviated AU), which is the average distance between the Earth and the Sun:

1 AU = 1.496×10^8 km

(continued)

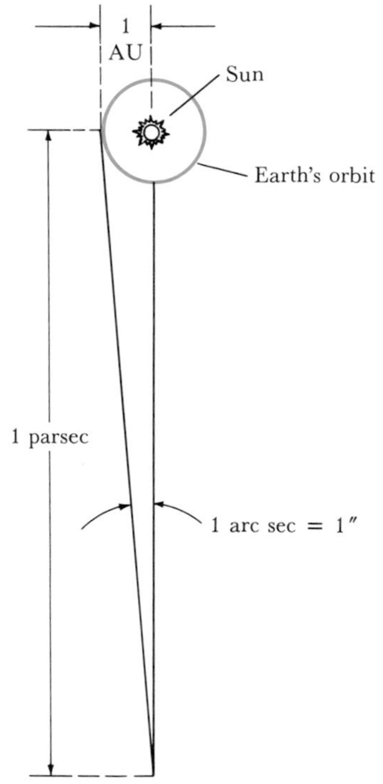

(Box 1-3, continued)

Thus, for example, the distance between the Sun and Jupiter is conveniently stated as 5.2 AU.

When talking about distances to the stars, astronomers choose between two different units of length. First is the **light year** (abbreviated ly), which is the distance light travels in one year:

$$1 \text{ ly} = 9.46 \times 10^{12} \text{ km}$$

or alternatively

$$1 \text{ ly} = 63{,}240 \text{ AU}$$

which is roughly equal to 6 trillion miles. For example, the nearest star (Proxima Centauri) is 4.3 light years from Earth.

The second commonly used unit of length is the **parsec** (abbreviated pc). Imagine taking a journey far into space, beyond the orbit of Pluto. As you look back toward the Sun, the Earth's orbit will subtend a small angle in the sky. One parsec is defined as the distance at which 1 AU subtends an angle of one second of arc, as shown in the diagram. The parsec turns out to be longer than the light year. Specifically,

$$1 \text{ pc} = 3.09 \times 10^{13} \text{ km}$$
$$= 3.26 \text{ ly}$$

Thus, the distance to the nearest star can also be stated as 1.3 pc.

Whether one uses light years or parsecs is a matter of personal taste.

For even larger distances, astronomers commonly use **kiloparsecs** and **megaparsecs** (abbreviated kpc and Mpc) where the prefixes simply mean "thousand" and "million," respectively:

$$1 \text{ kpc} = 10^3 \text{ pc}$$
$$1 \text{ Mpc} = 10^6 \text{ pc}$$

Thus, for example, the distance from Earth to the center of our Milky Way Galaxy is 9 kpc, whereas the distance to a rich cluster of galaxies in the constellation of Virgo is 20 Mpc.

Some astronomers prefer to talk about thousands or millions of light years rather than kiloparsecs and megaparsecs. Once again, the choice is a matter of personal taste.

Diagram labels: 1 AU · Sun · Earth's orbit · 1 parsec · 1 arc sec = 1″

By studying stars and nebulae, astronomers discover how stars are born, grow old, and eventually die

We begin our study of the stars in Chapter 19 with a close-up examination of our own star, the Sun. Once again, we see the surprising impact of astronomy on the course of civilization. In the 1920s and 1930s, physicists figured out how the Sun shines. At the Sun's center, thermonuclear reactions convert hydrogen into helium. This violent process releases a vast amount of energy that eventually makes its way to the Sun's surface and escapes as sunlight. By 1950, physicists had learned how to reproduce this thermonuclear reaction here on Earth. Hydrogen bombs operate on the same basic principles as the energy production at the Sun's center. Thermonuclear weapons stockpiled around the world have a profound effect on international politics and could dramatically influence the future of life on our planet.

As we look deeper into space, we find star clusters and clouds of glowing gas, called **nebulae,** scattered across the sky. In Chapters 20 through 24, we find that these beautiful objects tell us much about the

Figure 1-4   A thermonuclear explosion
Understanding the Sun's source of energy has given humanity the terrifying ability to build thermonuclear weapons. The hydrogen bomb and the thermonuclear reactions at the Sun's center operate under the same basic physical principle: the conversion of matter into energy. This thermonuclear detonation occurred on October 31, 1952, and had an energy output or "yield" equivalent to 10.4 million tons of TNT. (Defense Nuclear Agency)

Figure 1-5 [left]   The Orion Nebula
This beautiful nebula (also called M42 or NGC 1976) is a fine example of a stellar "nursery" where stars are born. Intense radiation from the newly formed stars causes the surrounding gases to glow. Many of the stars embedded in this nebula are less than a million years old. The Orion Nebula is 1600 light years (490 pc) from Earth, and the distance across the nebula is about 23 light years (7 pc). (U.S. Naval Observatory)

Figure 1-6 [right]   The Crab Nebula
This nebula (also called M1 or NGC 1952) is a fine example of a supernova remnant. A dying star exploded, and this beautiful funeral shroud is caused by the gases that were blasted violently into space. In fact, these gases are still moving outward at about 1000 km/sec (roughly 2 million miles per hour). The Crab Nebula is 3600 light years (1.1 kpc) from Earth, and the distance across the nebula is about 6 light years (2 pc). (Lick Observatory)

lives of stars. We discover that stars are born in huge clouds of interstellar gas and dust such as the Orion Nebula shown in Figure 1-5. After billions of years, stars eventually die. Some stars end their lives with a spectacular detonation called a **supernova** that blows the star apart. The Crab Nebula seen in Figure 1-6 is a striking example.

During their death throes, stars return gas to interstellar space. We shall learn that this gas contains many heavy elements created by thermonuclear reactions in the stars' interiors. Interstellar space thus becomes enriched with chemicals that did not exist in earlier times. The situation is analogous to a forest in which decaying leaves and logs enrich the soil for future generations of trees. The Sun and its planets were formed from enriched interstellar material. We therefore arrive at the surprising realization that virtually everything we touch, including the atoms in our bodies, was created deep inside ancient, now-dead stars.

In Chapters 23 and 24 we find that dying stars can produce some of the strangest objects in the sky. Some dead stars become **pulsars** that

emit pulses of radio waves or **bursters** that emit powerful bursts of X rays. Very massive dead stars become **black holes,** surrounded by incredibly powerful gravity from which nothing (not even light) can escape. Many of these bizarre stellar corpses have been discovered in recent years with Earth-orbiting telescopes that detect nonvisible light.

By observing galaxies, astronomers learn about the creation and fate of the universe

Stars are not spread uniformly across the universe but are grouped together in huge assemblages called **galaxies.** Galaxies are the largest individual objects in the universe. A typical large galaxy, like our own Milky Way, contains several hundred billion stars.

We begin the final segment of this book in Chapter 25 with a tour of the Milky Way Galaxy. We discover that our galaxy has beautiful, arching spiral arms (like those of M83 in Figure 1-7) that are active sites of star formation. We are surprised to learn that the center of our galaxy is emitting vast quantities of energy.

Figure 1-7   The galaxy M83
This spectacular galaxy (also called NGC 5236) contains about 200 billion stars. The galaxy's spiral arms are outlined by numerous nebulae that are the sites of active star formation. This galaxy has a diameter of about 35,000 light years (10 kpc) and is at a distance of 12 million light years (4.7 Mpc) from Earth. (Courtesy of R. J. Dufour)

We then move on to explore other galaxies, and we find that they come in a wide range of shapes and sizes. Some galaxies are quite small and contain only a few hundred million stars. Others are veritable monstrosities, huge galaxies that devour neighboring galaxies in a process called "galactic cannibalism."

Some of the most intriguing galaxies appear to be in the throes of violent convulsions. The centers of these strange, distorted galaxies are often powerful sources of X rays and radio waves. In many cases, it looks as though the entire galaxy is being blown apart.

Even more dramatic sources of energy are found still deeper in space. As described in Chapter 27, at distances of billions of light years from Earth we find the mysterious **quasars.** Although quasars look like stars (see Figure 1-8), they are probably the most distant and most luminous objects in the sky. A typical quasar shines with the brilliance of a hundred galaxies. We shall examine data that suggest quasars draw their awesome energy from enormous black holes.

Figure 1-8   The quasar 3C48
Quasars are the most distant and most lumi-
nous objects that astronomers have ever seen.
At first glance, a quasar is easily mistaken for
a faint star. This quasar is thought to be at a
distance of 4.8 billion light years (1800 Mpc)
from Earth. (Palomar Observatory)

Finally, in Chapters 28 and 29, we turn to the most fundamental questions about the creation and fate of the universe. We shall see how the motions of the galaxies reveal that we live in an expanding universe. Extrapolating backwards into the past, we learn that the universe must have been born from an infinitely dense state nearly 20 billion years ago.

Most astronomers believe that the universe began with a cosmic explosion, called the Big Bang, that occurred throughout all space at the beginning of time. During the Big Bang, events happened that dictated the present nature of the universe. We shall learn how astronomers are making significant progress in understanding these cosmic events. Indeed, we may be about to discover the origin of some of the most basic properties of the universe. At the end of the book, we shall see how the motions of the most distant galaxies tell us the ultimate fate of the universe: whether it will expand forever or will someday stop and collapse back on itself.

An underlying theme of this book is the idea that reality is rational. The universe is not a hodgepodge of unrelated things behaving in unpredictable ways. Rather, we find strong evidence for the existence of fundamental principles (usually called the "laws of physics") that govern the nature and behavior of everything in the universe. This is a powerful unifying concept that enables us to explore realms far removed from our earthly experience. Thus a scientist can do experiments in a laboratory to determine the properties of light or the behavior of atoms and then use this knowledge to discover the life cycles of stars and the structure of the universe.

These discoveries have a direct and profound influence on humanity. The past four centuries of civilization clearly show that major scientific advances sooner or later make their way into our lives. The world around us is filled with examples of the impact of science in technology, commerce, medicine, entertainment, and transportation. In the near future we can look forward to the benefits of space technology. Weightlessness and the near-perfect vacuum of space will enable us to manufacture

Figure 1-9   An Earth-orbiting industrial
complex
A new generation of high-technology materials
could easily be manufactured in space. Exotic
alloys, foam metals, ultrapure semiconducting
crystals, and rare vaccines are among the obvi-
ous practical applications of zero-gravity in-
dustry. This artist's conception shows an indus-
trial space station under construction. (NASA)

a wide range of exceptional substances, from exotic alloys to ultrapure medicines.

The dreams of Jules Verne and H. G. Wells pale in comparison to the reality of today. Ours is an age of exploration and discovery more profound than any since Columbus and Magellan set sail across uncharted seas. We have walked on the moon; we have dug in the Martian soil. We have probed the poisonous clouds of Venus and seen the craters on Mercury. We have discovered active volcanoes and barren ice fields on the satellites of Jupiter. We have visited the shimmering rings of Saturn. Never before has so much been revealed in such a short time.

Some people mistakenly believe that astronomy is a sad and depressing subject, telling us that we are insignificant creatures living a brief and meaningless existence on a tiny rock we call the Earth. As you turn the pages of this book, you will come to realize that one of the great lessons of modern astronomy is the awesome power of the human mind to reach out, to explore, to observe, and to comprehend, thereby transcending the limitations of our bodies and the brevity of human life.

Summary

. Important contributions to astronomical knowledge were made by many individuals in many cultures over the centuries.

The universe is comprehensible.

Observation of the heavens has led to discovery of some of the fundamental laws of nature.

. Study of the planets provides information about the Earth's history and resources.

. Study of the stars and nebulae provides information about the origin and history of the Sun.

. Study of the galaxies provides information about the origin and history of the universe.

. Angular measure and powers-of-ten notation are important tools for the study of astronomy.

. A variety of distance units, including the parsec and the light year, are used by astronomers.

. The concepts covered in all three Boxes will be especially helpful in your further study of astronomy.

Review questions

1 What is the advantage to the astronomer of using the light year as a unit of distance?

*2 The average distance to the Moon is 384,000 km and the Moon subtends an angle of $\frac{1}{2}°$. Calculate the diameter of the Moon.

*3 The speed of light is 3×10^{10} cm/sec. How long does it take light to get from the Sun to the Earth?

4 The mass of the Earth is 5.98×10^{27} g and its volume is 1.09×10^{27} cm. Divide the mass by the volume to find the average density of the Earth in g/cm^3.

*5 A hydrogen atom has a radius of about 5×10^{-9} cm. The radius of the observable universe is about 20 billion light years. How many times larger than a hydrogen atom is the universe?

*6 A certain fast food chain advertises that they have sold over 20 billion hamburgers. If this typical hamburger has a diameter of 15 cm, how far out into space would all of these hamburgers reach if placed side by side in a straight line? Would they reach any celestial object?

Advanced questions

*7 At what distance would a person hold a quarter (diameter equal to about 2.5 cm) in order for the quarter to subtend an angle of **(a)** 1 degree? **(b)** 1 arc minute? **(c)** 1 arc second?

*8 The diameter of the Sun is 1.4×10^{11} cm and the distance to the nearest star, Proxima Centauri, is 4.3 light years. If the Sun were reduced to the size of a basketball (about 30 cm in diameter), at what distance would Proxima Centauri be from the Sun on this reduced scale?

*9 Suppose your telescope can resolve objects and features that subtend angles of at least 2 arc seconds. What is the diameter of the smallest crater you can see on the Moon?

Discussion questions

10 How do astronomical observations and experiments differ from those of other sciences?

11 Discuss the meaning and justification of the assumption that "reality is rational."

For further reading

Asimov, I. *The Measure of the Universe.* Harper & Row, 1983. *An extensive journey through the cosmos in half powers of ten.*
Chaisson, E. *Cosmic Dawn.* Little, Brown, 1981.
Jastrow, R. *Red Giants and White Dwarfs, 2nd ed.* Norton, 1979.
King, I. "Man in the Universe." *Mercury,* Nov./Dec. 1976, p. 7.
Morrison, Philip, Morrison, Phylis, and Eames, The Office of Charles and Ray. *Powers of Ten.* Scientific American Books, 1982. *A tour of the universe where each step corresponds to a power of ten.*
Seielstad, G. *Cosmic Ecology.* U. of Calif. Press, 1983.
————, "Cosmic Ecology: A View from the Outside In." *Mercury,* Nov./Dec. 1978, p. 119.

E. *Margaret Burbidge*

Adventure into space*

E. Margaret Burbidge is Professor of Astronomy and Director of the Center for Astrophysics at the University of California, San Diego. She is an observational astronomer; her primary research interests are quasars and active galaxies.

After receiving her B.Sc. and Ph.D. from the University of London, Dr. Burbidge held positions at the University of Chicago, Yerkes Observatory, and the California Institute of Technology. In 1972 and 1973, she was the Director of the Royal Greenwich Observatory in England.

Dr. Burbidge served as President of the American Astronomical Society from 1976 to 1978. During 1982 she was President of the American Association for the Advancement of Science. She has received numerous honors and awards, including the Bruce gold medal of the Astronomical Society of the Pacific, and the Warner Prize (with her astrophysicist husband Geoffrey Burbidge) of the American Astronomical Society. Dr. Burbidge has published extensively and was recently awarded the Russell Lectureship of the American Astronomical Society, its highest honor.

Astronomy is the oldest of the sciences, with roots far back in antiquity. Yet this once-quiet research field is now in many respects the youngest, bubbling with new concepts, new ideas, and above all, new observations.

Entry into the "space age" a quarter century ago began a new era of high technology and launched a voyage of discovery from the solar system to the far universe. The adventure into space encompasses far more, however, than the direct exploration of our own local part of the universe, the solar system. Telescopes and instruments launched into Earth orbit have imaged the sky in X-ray, gamma-ray, ultraviolet, and infrared radiation, producing new discoveries and also raising new questions about star birth, star death, black holes, the center of our Milky Way, active galaxies, clusters of galaxies, and the far universe.

Although my own research concerns the distant universe (galaxies, quasars, radio galaxies), to be described in later chapters in this book, I want to express some thoughts about direct exploration of the solar system. There is a warning to be borne in mind, however; the same technological advances that have made the adventure into space possible have given us the capability of extinguishing life on Earth through nuclear disaster. The first pictures of the Earth as seen from outside were obtained from the Apollo manned space flights, and they brought home, as nothing else could, the fact that we inhabit a bounded surface, in an environment which is home to an ever-growing human population whose ever-increasing waste products it must house.

Exploring the Solar System

After circumnavigating the Earth from poles to equator, the next frontier for exploration is above the Earth's surface—the frontier with the rest of the solar system. And here I have to point out that this exploration is an expensive operation. Nevertheless, the budgets for the exciting ventures of the past two decades, and those planned for the remainder of this century, are but a tiny fraction of the military budget. Also, the expenditures during the past twenty years have pushed technology in many useful fields, especially the development and

*Excerpted from Burbidge, E. M., Science 221, 421–427, July 29, 1983. Copyright © 1983 by the AAAS.

miniaturization of computers and instruments, advances in the science of communications, and the flow of data from Landsat and weather satellites. The most expensive ventures are those involving manned space flight. Yet who will forget the thrill of watching the Apollo 11 landing on the moon and seeing those first steps, by astronauts Aldin and Armstrong! For myself, I cannot look at the Moon without thinking: "We have been there," and without a passing regretful thought that, had I been born a century later, I might have been able to step in those footprints.

Further than the Moon, however, is beyond today's goal for manned space flight, and must depend on unmanned spacecraft. These missions have been spectacularly successful. U.S.S.R. Venera landers on Venus and the U.S. Mariner and Pioneer missions have shown that the hot, enormously dense and dynamic atmosphere of Venus has huge quantities of CO_2, very little water, and is topped by clouds of sulfuric acid. It is indeed as inhospitable a place for manned exploration as Mercury, whose fiery-hot surface, unshielded by any atmosphere, has a surface like the Moon's, cratered by meteoritic bombardment. NASA's future aims include a Venus Radar Mapper, to study Venus's surface topography and its past volcanic activity, and later, a Venus Atmospheric Probe which will plunge into and analyze that dense, turgid atmosphere.

The least inhospitable of our neighboring planets is Mars, and NASA's Viking lander showed unweathered craters produced by past meteoritic bombardment, like the Moon and Mercury, yet also showed dry channels suggesting the flow of liquid in the past. More detailed study is planned for the 1990s by orbiters, and, eventually, by long-lived remotely operated scientific stations on Mars's surface.

Perhaps the most spectacular planetary studies have been those made by Voyagers 1 and 2, flying around and past Jupiter and Saturn and their satellites and Saturn's fantastic ring system. These have been triumphs of planning and technological skill, in the orbital accuracy of the Voyagers and the superb operation of their communications systems. The variety displayed by Jupiter's and Saturn's satelllites again gives tantalizing hints about the early history of the solar system.

The spacecraft Galileo, to be launched by NASA in 1986, will make a more extensive exploration of the whole Jovian system, and send a probe into the swirling, turbulent Jovian atmosphere.

The planets and their satellites by no means comprise all the objects of interest in the solar system. The smaller components—asteroids and comets—are candidates for intensive study by spacecraft that actually make rendezvous with them. The asteroids—solar system debris—are conceivably useful future sources of raw materials.

Beyond discovery and exploration, the ultimate goal of solar system studies is to unravel the secrets of its origin and early history, and to ask ourselves whether planetary systems around Sun-like stars are common or rare.

Other Planetary Systems

Since our Sun is a very average star, one of billions like it in the Milky Way, there may be others in the solar neighborhood with planetary systems like ours. As advances in microwave and infrared technology reveal the existence of more and more complex molecules in dense star-forming interstellar clouds, we can envision a primordial organic mixture out of which life might have evolved elsewhere than on Earth.

But the speculation that radio signals from extraterrestrial intelligence might be detected has led to the question: If life is common throughout our Milky Way, why haven't we already detected it? And that leads to the further question: For what length of time could living creatures in another solar system exist, once they had the necessary technology to send out signals? Perhaps the development of highly organized living creatures is always accompanied by competition, struggle for scarce resources, and warfare. Acquiring the technology to send out strong enough radio or microwave transmissions would presuppose an advanced knowledge of physics, including nuclear physics and the ability to build nuclear bombs. Perhaps any such civilization has but 50 years or so (on our timescale) before it annihilates itself. Should we ask, rather, is the development of civilization as contentious as ours as unavoidable consequence of the possession of "intelligence," and, if so, how many such civilizations have had the wisdom to pass safely through the dangerous warfaring phase to an era of cooperation? This is a task for all of us, and especially scientists, to address, and we should do so in the belief that we shall succeed.

Some astronomical observations, such as the division of the night sky into constellations, were made before history began. Modern astronomers still use constellations described by ancient Babylonians and Greeks. In this chapter, we learn to find our way around the sky, which is conveniently described as the celestial sphere. We find that the seasons are related to the tilt of the Earth's axis of rotation, and that the axis itself is slowly changing its orientation. We learn a little about the calendar and various ways of measuring time. We continue to accumulate terms and tools of measurement that will be useful in our later, more detailed study of astronomy.

The roots of astronomy reach back to the dawn of civilization. This ancient heritage is most apparent in the **constellations.** Perhaps it was shepherds tending their flocks at night, or families sitting by the fire, or priests studying the starry heavens who first imagined pictures among groupings of stars.

You may already be familiar with some of these patterns in the sky such as the Big Dipper, which is actually part of a large constellation called Ursa Major (the Great Bear). Many of these constellations, such as Orion in Figure 2-1, have names from ancient myths and legends. Although some star groupings vaguely resemble the figures they are supposed to represent, most do not.

On modern star charts, the entire sky is divided into 88 constellations. Some constellations cover very large areas (Ursa Major is one of the biggest), others small. The constellations are used to specify certain regions of the sky. Thus, for example, we might speak of "the galaxy M31 in

Andromeda" much as we would refer to "the Ural Mountains in the Soviet Union."

People who spend time outdoors at night are very familiar with the apparent motions of the constellations. If you are a city dweller, you should take the time to observe the basic facts of astronomy yourself. Go outdoors soon after dark, find a spot away from bright lights, and note the patterns of stars in the sky. A few hours later, check again. You will find that the entire pattern of stars (including the Moon, if it is visible) has shifted its position. New constellations have risen above the eastern horizon, while some have disappeared below the western horizon. If you can check again just before dawn, you will find low in the western sky the stars that were just rising when the night began.

The constellations that you can see in the sky change slowly from one night to the next. This shift occurs as the Earth orbits the Sun, as shown in Figure 2-2. The Earth takes a full year to go once around the Sun,

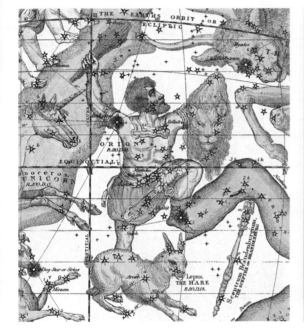

Figure 2-1 Orion

Orion is a prominent winter constellation. From the United States, Orion is easily seen high above the southern horizon from December through March. The photograph was a 4-minute exposure on Kodak Ektachrome film. Because of the time exposure, the colors of the stars are very noticeable. The fanciful drawing of Orion is from an 1835 star atlas. (Courtesy of Robert Mitchell and Janus Publications)

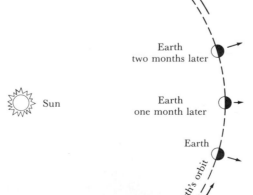

Figure 2-2 Our changing view of the night sky

As we orbit the Sun, the nighttime sky of the Earth gradually turns toward different parts of the sky. Thus, the constellations that we can see change slowly from one night to the next.

and thus the darkened, nighttime side of the Earth is gradually turned toward different parts of the heavens. Specifically, if you follow a particular star on successive evenings, you find that it rises approximately 4 minutes earlier each night. A set of star charts for the evening hours of selected months of the year is included on the endpapers of this book.

Like constellations, many stars have ancient names. Star names and catalogues are discussed in Box 2-1.

It is often convenient to imagine that the stars are located on the celestial sphere

As you gaze at the heavens on a clear dark night, you might think that you can see millions of stars. Actually, the unaided human eye can detect only about 6000 stars over the entire sky. At any one time, you can see roughly 3000 stars because only half of the sky is above the horizon. Of course, the Earth rotates once every 24 hours (that is why we have day and night), and hence the stars rise in the east and set in the west, as do the Sun and Moon. This daily, or **diurnal,** motion of the stars is very apparent in time exposure photographs such as Figure 2-3.

Many ancient societies believed that the Earth is at the center of the universe. They also imagined that the stars are attached to the surface of a huge sphere centered on the Earth. This imaginary sphere, called the **celestial sphere,** is still a useful concept.

Of course, the stars actually are scattered at various distances from Earth. Many of the brightest stars you can see in the sky are 10 to 1000 light years away. These distances are so immense, however, that all the stars appear to be equally remote, fixed to a spherical backdrop. We can use this spherical backdrop as a reference to specify the directions to objects in the sky.

As shown in Figure 2-4, the Earth is at the center of the celestial sphere. We can project key geographical features out into space to establish directions and bearings on the celestial sphere. For example, if the Earth's equator is projected onto the celestial sphere, we obtain the

Figure 2-3 Star trails around the north celestial pole
The Earth rotates once every 24 hours, and hence the stars appear to move across the sky. This time-exposure photograph is centered on the north celestial pole, which is slightly less than 1° from the moderately bright star Polaris. The north celestial pole is directly over the Earth's north pole and is a point about which the heavens appear to revolve. (U.S. Naval Observatory)

Figure 2-4 Celestial coordinates
Astronomers denote positions of objects in the sky by right ascension and declination, a system similar to longitude and latitude on the Earth. Right ascension is measured in units of time (hours, minutes, seconds) eastward along the celestial equator starting from the vernal equinox. Declination is the angle (in degrees, minutes of arc, seconds of arc) north or south of the celestial equator.

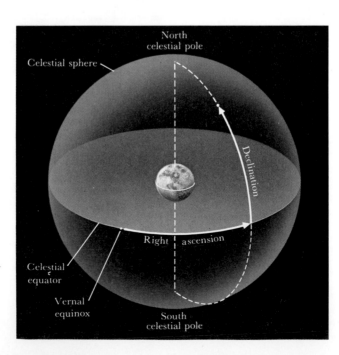

Box 2-1 Star names and catalogues

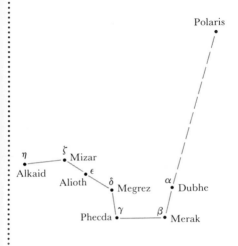

Customs and traditions for naming stars have changed over the centuries. Many of the brightest stars in the sky have Arabic names assigned in medieval times, when astronomy reached its height among Islamic nations. The diagram shows the Big Dipper with the Arabic names for its seven brightest stars. The North Star received its formal name, Polaris, later when Latin was the language used by European astronomers.

As you might suspect, memorizing exotic star names is an unwelcome burden for many astronomers. A simpler system, invented by Johann Bayer in 1603, uses the letters of the Greek alphabet and the constellation names.

The 24 lowercase letters of the Greek alphabet are as follows:

.	α	alpha	. ν	nu
.	β	beta	. ξ	xi
.	γ	gamma	. o	omicron
.	δ	delta	. π	pi
.	ϵ	epsilon	. ρ	rho
.	ζ	zeta	. σ	sigma
.	η	eta	. τ	tau
.	θ	theta	. υ	upsilon
.	ι	iota	. ϕ	phi
.	κ	kappa	. χ	chi
.	λ	lambda	. ψ	psi
.	μ	mu	. ω	omega

In constructing a star name, a Greek letter is used along with the Latin possessive form of the name of the constellation in which the star is located. As examples of the Latin possessive, the twelve constellations of the zodiac are as follows:

Constellation	Possessive
. Aries	. Arietis
. Taurus	. Tauri
. Gemini	. Geminorum
. Cancer	. Cancri
. Leo	. Leonis
. Virgo	. Virginis
. Libra	. Librae
. Scorpius	. Scorpii
. Sagittarius	. Sagittarii
. Capricornus	. Capricorni
. Aquarius	. Aquarii
. Pisces	. Piscium

In most cases, the brightest star in the constellation is α, the second brightest is β, the third is γ, and so on. For example, the brightest star in Libra ("the scales") is called α Librae. This name is more convenient and informative than the Arabic name, Zubenelgenubi.

The 24 letters in the Greek alphabet enable Bayer's system to cover only the brightest two dozen stars in any constellation. Astronomers, however, are often interested in very faint stars, many of which are too dim to be seen with the naked eye. In referring to these fainter stars, astronomers make use of designations from famous star catalogues.

One of the first major star catalogues was produced in Germany in the

(continued)

(Box 2-1, continued)
mid-1800s by F. W. Argelander at the Bonn Observatory. This catalogue lists the positions of 320,000 stars and is called the **Bonner Durchmusterung.** Using this catalogue, astronomers refer to stars by their "BD numbers." Thus, for example, you might be interested in observing the star BD + 5°1668 which happens to be in the constellation of Monoceros ("the unicorn").

Another commonly used catalogue is the **Henry Draper Catalogue,** which was compiled in the United States around 1920. (It is named after a physician whose widow financed the project.) Stars are listed by their "HD numbers" from this catalogue. For example, HD 87901 is α Leonis (also called Regulus), the brightest star in Leo ("the lion").

All across the sky there are faint nonstellar objects such as galaxies, nebulae, and star clusters. As with the faint stars, astronomers refer to these objects by their catalogue designations.

Roughly a hundred of the brightest nonstellar objects are listed in a famous catalogue by the eighteenth century French astronomer Charles Messier. Astronomers often refer to these objects by their "M numbers." For example, the Orion Nebula (see Figure 1-5) is the forty-second object on Messier's list and is therefore called M42.

During the nineteenth century, William Herschel and his son John Herschel observed and catalogued nearly 5000 faint nonstellar objects. This list was enlarged by J. L. E. Dreyer and published in 1888 as the **New General Catalogue.** Further additions were made during the next twenty years so that this catalogue and its supplements ultimately listed nearly 15,000 objects. Astronomers frequently refer to galaxies and nebulae by their "NGC numbers." For example, the Crab Nebula (see Figure 1-6) is called NGC 1952.

celestial equator. The celestial equator divides the sky into northern and southern hemispheres, just as the Earth's equator divides the Earth into two hemispheres.

We can also imagine extending the Earth's north and south poles out into space along the Earth's axis of rotation. This gives us the **north celestial pole** and the **south celestial pole,** also shown in Figure 2-4. The celestial equator and poles are extremely useful in setting up a coordinate system that covers the sky.

To denote positions of objects in the sky, astronomers use a system of "right ascension" and "declination" that is very similar to longitude and latitude. Declination corresponds to latitude. The **declination** of an object is its angular distance north or south of the celestial equator, measured along a circle passing through both celestial poles, as shown in Figure 2-4.

Right ascension corresponds to longitude. Astronomers measure right ascension from a specific point on the celestial equator. This point, called the **vernal equinox,** is one of two locations where the Sun crosses the celestial equator during its apparent annual motion against the background of the celestial sphere (discussed in the next section). In the Earth's northern hemisphere, the season of spring is said to begin when the Sun crosses the vernal equinox in late March. The **right ascension** of an object in the sky is the angular distance from the vernal equinox eastward along the celestial equator to the circle used in measuring its declination (see Figure 2-4). Following traditional practice, astronomers measure this angular distance in time units (hours, minutes, seconds) corresponding to the time required for the celestial sphere to rotate through this angle. One hour of right ascension is equivalent to an angular distance of 15°.

In catalogues of faint stars, galaxies, and nebulae, the position of objects are given by their right ascension and declination. This is valuable information because it tells the astronomer precisely where to point a telescope.

The seasons are caused by the tilt of the Earth's axis of rotation

Figure 2-5 The seasons
The Earth's axis of rotation is inclined 23½° away from the perpendicular to the plane of the Earth's orbit. The Earth maintains this orientation (with its north pole aimed at the celestial north pole near the star called Polaris) throughout the year as the Earth orbits the Sun. Consequently, the amount of solar illumination and the number of daylight hours at any location on Earth varies in a regular fashion throughout the year.

In addition to rotating on its axis every 24 hours, the Earth revolves around the Sun every 365¼ days. The seasonal changes we experience on Earth during a year result from the way the Earth's axis of rotation is tilted with respect to the plane of the Earth's orbit around the Sun.

The Earth's axis of rotation is not perpendicular to the Earth's orbit. Instead, the Earth's axis is tilted by 23½° away from the perpendicular, as shown in Figure 2-5. The Earth constantly maintains this tilted orientation as it orbits the Sun. Thus, during one part of the year, we find the northern hemisphere tilted toward the Sun and the southern hemisphere tilted away. This produces summer in the north and winter in the south. Half a year later, the situation is reversed, with winter in the northern hemisphere (now tilted away from the Sun) while the southern hemisphere experiences summer. As shown in Figure 2-5, the seasons of spring and fall occur between winter and summer, when both hemispheres are receiving roughly equal amounts of illumination from the Sun.

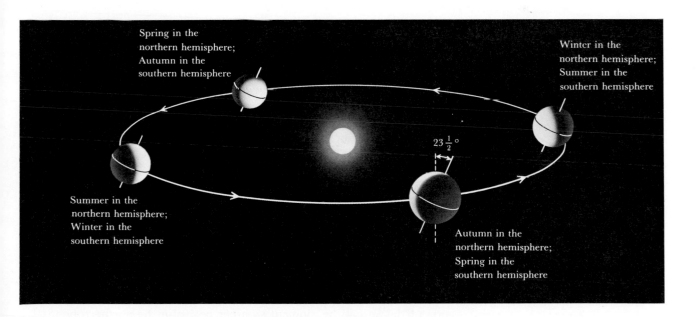

Spring in the
northern hemisphere;
Autumn in the
southern hemisphere

Winter in the
northern hemisphere;
Summer in the
southern hemisphere

$23\frac{1}{2}°$

Summer in the
northern hemisphere;
Winter in the
southern hemisphere

Autumn in the
northern hemisphere;
Spring in the
southern hemisphere

We can represent this seasonal phenomenon on the celestial sphere by examining the Sun's apparent motion against the background constellations. As the Earth moves along its orbit, the Sun appears to shift its position gradually from day to day, as viewed against the background stars. The Sun therefore traces out a path in the sky during the course of a year. This path is called the **ecliptic.** Because of the tilt of the Earth's axis of rotation, the ecliptic and the celestial equator are inclined to each other by 23½°, as shown in Figure 2-6. There are 365 or 366 days in a year and 360° in a circle, so the Sun appears to move along the ecliptic at a rate of approximately 1° per day.

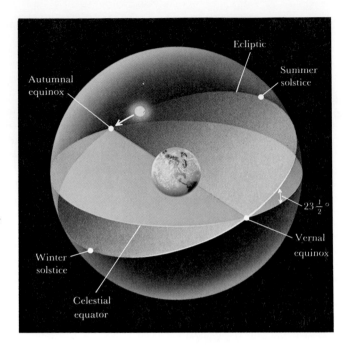

Figure 2-6 Equinoxes and solstice
The ecliptic is the apparent annual path of the Sun on the celestial sphere. The ecliptic is inclined to the celestial equator by 23½° because of the tilt of the Earth's axis of rotation. The ecliptic and the celestial equator intersect at two points called the equinoxes. The northernmost point on the ecliptic is called the summer solstice. The corresponding southernmost point is called the winter solstice.

Figure 2-7 The Sun's daily path
On the first day of spring and the first day of fall, the Sun rises precisely in the east and sets precisely in the west. During summer, the Sun rises in the northeast and sets in the northwest. The maximum northerly excursion of the Sun occurs at the summer solstice. In the winter, the Sun rises in the southeast and sets in the southwest, with its maximum southerly excursion occurring at the winter solstice.

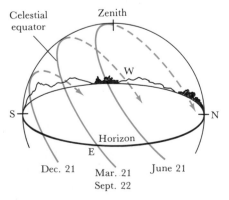

As shown in Figure 2-6, the ecliptic and the celestial equator intersect at only two points, which are exactly opposite each other on the celestial sphere. Both points are called **equinoxes** (from the Latin word meaning "equal night") because, when the Sun appears at either of these points, daytime and nighttime are each 12 hours long at all locations on Earth.

As mentioned earlier, the **vernal equinox** marks the beginning of spring in the northern hemisphere, as the Sun moves northward across the celestial equator about March 21. The **autumnal equinox** marks the moment when fall begins in the northern hemisphere (about September 22) as the Sun moves southward across the celestial equator.

Incidentally, we should always remember that the northern and southern hemispheres experience opposite seasons at any point in time. For example, March 21 marks the beginning of autumn for people in Australia. Obviously, these astronomical terms come from a time when virtually all astronomers lived north of the equator. In keeping with astronomical usage, this book will perpetuate this hemispheric chauvinism.

Between the vernal and autumnal equinoxes, there are two other significant locations along the ecliptic. The point on the ecliptic farthest north of the celestial equator is called the **summer solstice.** It is the location of the Sun at the moment summer begins in the northern hemisphere (about June 21). At the beginning of winter, the Sun is farthest south of the celestial equator at a point called the **winter solstice** (about December 21).

Seasonal changes in the Sun's daily path across the sky are diagrammed in Figure 2-7. On the first day of spring or the first day of fall (when the Sun is at one of the equinoxes), the Sun rises directly in the east and sets directly in the west. Daytime and nighttime are of equal duration.

During the summer months, when the northern hemisphere is tilted toward the Sun, sunrise occurs in the northeast and sunset occurs in the northwest. The Sun spends more than twelve hours above the horizon, and it passes high in the sky at noontime. Incidentally, the point in the sky directly overhead is called the **zenith,** as shown in Figure 2-7. At the summer solstice, the Sun is as far north as it gets, giving the greatest number of daylight hours. Indeed, there are certain locations on Earth (north of the Arctic Circle, as discussed in Box 2-2) where the Sun does not set during summer nights.

During the winter months, when the northern hemisphere is tilted away from the Sun, sunrise occurs in the southeast. Daylight lasts for less than 12 hours as the Sun skims low over the southern horizon and sets in the southwest. Night is longest when the Sun is at the winter solstice. In fact, north of the Arctic Circle, a winter night lasts for 24 hours, as described in Box 2-2.

Box 2-2   Tropics and circles

The inclination of the ecliptic to the celestial equator produces some noteworthy locations on Earth. For example, on any date during the year, the Sun appears directly overhead at "high noon" along a band of locations encircling the Earth. The northernmost place where this happens is called the **Tropic of Cancer,** which circles the Earth at a latitude of $23\frac{1}{2}°$ north of the equator. On the date of the summer solstice, the Sun is at the zenith at high noon as seen from anywhere along the Tropic of Cancer.

The corresponding southern location is called the **Tropic of Capricorn,** exactly $23\frac{1}{2}°$ south of the equator. On the date of the winter solstice, when the Sun has reached its southernmost declination, the Sun passes directly overhead at high noon as viewed from any location along this tropic.

There are also regions on the Earth near the two poles where the Sun spends many consecutive days either above or below the horizon. For example, as seen by someone standing at the north pole, the Sun rises on the day of the vernal equinox and stays above the horizon for six months, setting on the day of the autumnal equinox. During the next six months, the arctic explorer is subjected to six continuous months of night because the Sun is too far south in the sky to be seen from the north pole.

The region around the north pole where you can see the Sun for twenty-four continuous hours on at least one day of the year is bounded by the **Arctic Circle.** The Arctic Circle lies at $23\frac{1}{2}°$ south of the north pole (that is, at a latitude of $90° - 23\frac{1}{2}° = 66\frac{1}{2}°$ N).

The corresponding region around the south pole is bounded by the **Antarctic Circle,** as shown in the diagram below. While antarctic explorers enjoy the "midnight sun" near the time of the winter solstice, the Earth's north polar regions are being subjected to continuous night.

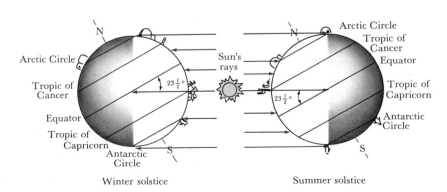

Winter solstice   Summer solstice

(continued)

(Box 2-2, continued)

The Earth's inclination and your own latitude conspire to keep some constellations permanently above your horizon while others are forever hidden from your view. For example, as viewed from the United States, the north celestial pole is always above the horizon. In fact, the elevation of the north celestial pole above the horizon is exactly equal to your latitude on Earth. As the Earth turns, constellations near the north celestial pole revolve about the pole, never rising or setting. Similarly, as shown in the diagram below, there are constellations around the south celestial pole that are too far south to be seen from northern latitudes.

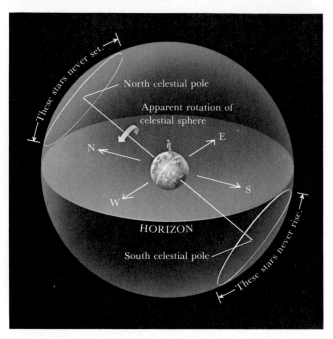

Precession is a slow, conical motion of the Earth's axis of rotation

Ancient astronomers realized that the Moon orbits the Earth. The Moon takes roughly four weeks to go once around the Earth. The word "month" comes from the same Old English root as the word "moon."

As seen from the Earth, the Moon is never far from the ecliptic. In other words, the Moon's path among the constellations is close to the Sun's path. The Moon's path remains within a band called the **zodiac** that extends about 8° on either side of the ecliptic. Twelve famous constellations (listed in Box 2-1) lie along the zodiac, and the Moon is generally found in one of these twelve constellations. As the Moon moves along its orbit, it appears north of the celestial equator for about two weeks, and then it appears south of the celestial equator for about the next two weeks.

Both the Sun and the Moon exert a gravitational pull on the Earth. We will discuss gravity in greater detail in Chapter 4. For now, it is sufficient to realize that gravity is the universal attraction of matter for other matter.

The gravitational pull of the Sun and Moon affect the Earth's rotation because the Earth is not a perfect sphere. Our planet is slightly fatter across the equator than it is from pole to pole. In fact, the equatorial diameter of the Earth is 43 kilometers (27 miles) larger than the diameter

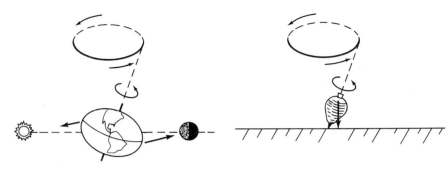

Figure 2-8   Precession
The gravitational pulls of the Moon and the Sun on the Earth's equatorial bulge cause the Earth to precess. As the Earth precesses, its axis of rotation slowly traces out a circle in the sky. The situation is analogous to a spinning top. As the top spins, the Earth's gravitational pull causes the top's axis of rotation to move in a circle.

measured from pole to pole. The Earth is therefore said to have an "equatorial bulge." The gravitational pull of the Moon and the Sun on this equatorial bulge gradually changes the orientation of the Earth's axis of rotation, producing a phenomenon called precession.

The situation is analogous to a spinning toy top, as illustrated in Figure 2-8. If the top were not spinning, gravity would pull the top over on its side. But when the top is spinning, the combined actions of gravity and rotation cause the top's axis of rotation to trace a circle, a motion called **precession.**

As the Sun and Moon move along the zodiac, each spends half the time north of the Earth's equatorial bulge and half the time south of it. The gravitational pull of the Sun and Moon tugging on the equatorial bulge tries to "straighten up" the Earth. In other words, as sketched in Figure 2-8, the gravity of the Sun and Moon tries to pull the Earth's axis of rotation toward a position perpendicular to the plane of the ecliptic. But the Earth is spinning. As with the toy top, the combined actions of gravity and rotation cause the Earth's axis to trace out a circle in the sky while remaining tilted about $23\frac{1}{2}°$ to the perpendicular.

The Earth's rate of precession is fairly slow. At the present time, the Earth's axis of rotation points within $\frac{1}{2}°$ of the star Polaris. In 3000 BC, it was pointing near the star Thuban in the constellation of Draco ("the dragon"). In AD 14,000, the "pole star" will be Vega in Lyra ("the harp"). It takes 26,000 years for the north celestial pole to complete one full precessional circle around the sky (see Figure 2-9).

Of course, the south celestial pole executes a similar circle in the southern sky. As the Earth's axis of rotation precesses, the Earth's equatorial plane also moves. Because the Earth's equatorial plane defines the location of the celestial equator in the sky, the celestial equator also precesses. The intersections of the celestial equator and the ecliptic define the equinoxes, so these key locations in the sky also shift slowly from year to year. In fact, the entire phenomenon is often called the **precession of the equinoxes.** Today the vernal equinox is located in the constellation Pisces ("the fishes"). Two thousand years ago, it was in Aries ("the ram"). Around the year AD 3000, the vernal equinox will move into Aquarius ("the water bearer").

The astronomer's system of right ascension and declination is tied to the positions of the celestial equator and the vernal equinox. Thus, because of precession, the coordinates of stars in the sky are constantly changing. These changes are very small and very gradual but, over the years, they add up. To cope with this difficulty, astronomers always make note of the date (called the **epoch**) for which a particular set of coordinates is precisely correct. Star catalogues and star charts are updated periodically. Most new catalogues and star charts are now being

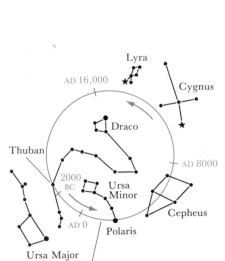

Figure 2-9   The path of the north celestial pole
As the Earth precesses, the north celestial pole slowly traces out a circle among the northern constellations in the sky. At the present time, the north celestial pole is near the moderately bright star Polaris, which serves as the "pole star."

prepared for the epoch 2000. The coordinates in these reference books are precisely correct for January 1, 2000, and they will require only very small corrections over the next few decades.

Traditionally, astronomers are responsible for keeping track of time and the calendar

Astronomers have traditionally been responsible for telling time. The need for accurate time-keeping is universal. The Pharaoh wanted to know when the Nile would flood. The proper seasonal time of religious events is extremely important in most cultures. Even today, we have many reasons to keep the calendar consistent with the cycle of seasons. For example, many of our holiday traditions involve seasonal observances.

In measuring shorter time intervals, we want our system of time-keeping to reflect the position of the Sun in the sky. The Sun's position determines whether we are awake or asleep, whether it is time for breakfast or for dinner. Thousands of years ago, the sundial was invented to keep track of **apparent solar time.** An apparent solar day is the time it takes the Sun to go from one high noon to the next.

For more formal measurements, astronomers use the **meridian,** which is a circle on the celestial sphere that passes through the **zenith** (the point directly overhead) and both celestial poles, as shown in Figure 2-10. Local noon occurs when the Sun crosses the meridian above the horizon; local midnight occurs when it makes the opposite crossing below the horizon, although this crossing cannot be observed directly. The crossing of the meridian by any object in the sky is called a **meridian transit.** If the crossing occurs above the horizon, it is an upper meridian transit. An **apparent solar day** is formally defined as the interval between two successive upper meridian transits of the Sun, observed from any fixed spot on the Earth.

Early observers soon realized that the Sun is *not* a good timekeeper. The length of the apparent solar day (as measured by a device such as a hourglass) varies from one season to another. Similarly, the speed of the Sun's slow eastward progress against the background stars varies over the course of a year.

There are two main reasons for the Sun's nonuniform motion. First, the Earth's orbit is not a perfect circle. The orbit is actually an ellipse, as shown in exaggerated form in Figure 2-11a. As we shall learn in Chapter 4, the Earth moves more rapidly along its orbit when it is near the

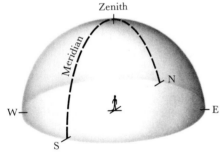

Figure 2-10   The meridian
The meridian is a circle that passes through the observer's zenith and the two celestial poles. Transits of celestial objects across the meridian can be used to measure time.

Figure 2-11   Why the Sun is a poor time keeper
There are two main reasons why the Sun is a poor time keeper. (a) The Earth's speed along its orbit varies throughout the year as shown at left. Hence, the apparent speed of the Sun along the ecliptic is not constant. (b) The projection of the Sun's daily progress along the ecliptic onto the celestial equator (as shown at right) varies throughout the year. Hence, the Sun's net daily eastward progress against the stars is not constant.

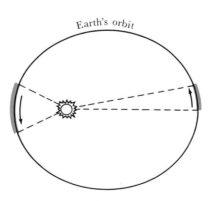

a   A month's motion of the Earth along its orbit

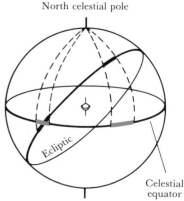

b   A day's motion of the Sun along the ecliptic

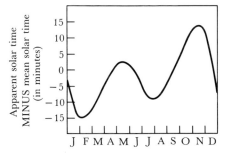

Figure 2-12 The equation of time
The equation of time equals apparent solar time minus mean solar time. It is the amount by which a sundial is in error because the Sun's eastward motion against the background stars is not constant throughout the year.

Sun than it does when it is farther away. This means that the speed of the Sun along the ecliptic varies through the year. The Sun appears to move more than 1° per day along the ecliptic in January when the Earth is nearest the Sun. And conversely, the Sun appears to move less than 1° per day along the ecliptic in July when the Earth is farthest from the Sun.

Even if the Sun moved along the ecliptic at a uniform rate, it would not be a good timekeeper. This is because the ecliptic is inclined by $23\frac{1}{2}°$ to the celestial equator. Near the equinoxes, a significant part of the Sun's motion is in a north–south direction (see Figure 2-11*b*). The net daily eastward progress in the sky is therefore somewhat foreshortened. But near the solstices, the Sun's motion is nearly parallel to the celestial equator, and hence there is no comparable foreshortening at the beginning of summer or winter.

To cope with these difficulties, astronomers invented an imaginary object called the **mean sun** that moves along the celestial equator at a precisely uniform rate. In this context, "mean" is a synonym for "average." Sometimes the mean sun is slightly ahead of the real Sun in the sky and sometimes it is behind. The mean sun moves at a constant rate and serves as a fine timekeeper.

A **mean solar day** is the interval between successive upper meridian transits of the mean sun and is exactly 24 hours long, equivalent to the average length of an apparent solar day. Our clocks and wristwatches are related to mean solar time.

Apparent solar time and mean solar time can differ by as much as a quarter of an hour at certain seasons. The difference between apparent and mean solar time is called the **equation of time** and is graphed in Figure 2-12. This graph tells you how much to add or subtract from your sundial reading to get the mean solar time at your location on Earth.

Time zones were invented for convenience in commerce, transportation, and communication. In a **time zone,** everyone agrees to set all the clocks and watches to the mean solar time for a meridian of longitude that runs approximately through the center of the zone. Time zones around the world are centered about meridians of longitude at 15° intervals. Therefore, in going from one time zone to the next, you must change the time on your wristwatch by exactly one hour. The time zones for North America are shown in Figure 2-13.

Although it is natural to want our clocks and method of time-keeping related to the Sun, astronomers are usually interested in observing the stars. Astronomers therefore find it convenient to use **sidereal time,** which is based on the stars rather than the Sun. Sidereal time is often used when aiming a telescope, and thus most observatories are equipped with a sidereal clock. Box 2-3 discusses sidereal time.

Box 2-3 Sidereal time

If you want to observe a particular star or galaxy, obviously it must be above the horizon. To minimize the distorting effects of the Earth's atmosphere that occur near the horizon, you get the best view when the star or galaxy is high in the sky. Ideally, you would like the object to be on or near the meridian.

To tell which objects are crossing the upper meridian, most observatories are equipped with a **sidereal clock** that tells **sidereal time.** *Sidereal time is*

(continued)

(Box 2-3, continued)

simply the right ascension of any object on the meridian. For example, suppose you want to observe the bright star Regulus in the constellation of Leo ("the lion"). From reference books (such as **The Astronomical Almanac**), you find that coordinates of this star are

R.A. = $10^h\ 07^m\ 28.0^s$

Decl. = $+12°\ 03'\ 03''$

Thus, Regulus will be on the meridian when the sidereal time is 10:07. A sidereal clock is useful because it tells you the right ascension of those objects currently best suited for observation.

From this example, we also see why astronomers measure right ascension in units of time. Because the Earth rotates once a day, it takes nearly 24 hours for all the stars to pass across the meridian. By measuring right ascension in units of time (hours, minutes, seconds), astronomers have a convenient relation between time and star positions. In contrast, declination is measured in units of angle (degrees, minutes of arc, seconds of arc) with a plus sign for north of the celestial equator and a minus sign for south.

Incidentally, it is sometimes convenient to be able to convert between angular and sidereal time measure. For example, you might want to know how many degrees correspond to one hour of right ascension. Because 24 hours of right ascension take you all the way around the celestial equator, we have the relationship that $24^h = 360°$. Therefore, the following equalities hold:

$1^h = 15°$

$1^m = 15'$

$1^s = 15''$

$1° = 4^m$

$1' = 4^s$

$1'' = 0.067^s$

Sidereal time is useful only to people such as astronomers and navigators who deal with the stars. It is different from time on your wristwatch. In fact, a sidereal clock and an ordinary clock tick at different rates.

Ordinary clocks are related to the position of the Sun. Sidereal time is tied to the position of the vernal equinox, because that is the location from which right ascension is measured.

The sidereal day starts when the vernal equinox crosses the meridian, regardless of where the Sun is. A **sidereal day** is the time between two successive upper meridian passages of the vernal equinox. In contrast, we have seen that an **apparent solar day** is the time between two successive upper meridian crossings of the Sun. These two kinds of days are *not* equal for reasons illustrated in the diagram. Because the Earth moves along its orbit around the Sun, the Earth must rotate through nearly 361° to get from one local noon to the next. This extra 1° of rotation corresponds to four minutes of time and is the amount by which a solar day exceeds a sidereal day. To be precise,

1 sidereal day = $23^h\ 56^m\ 4.091^s$

where the hours, minutes, and seconds are in mean solar time. Of course, one day according to your wristwatch is one **mean solar day** and is exactly 24 hours along.

The sidereal clock measures sidereal hours, minutes, and seconds—dividing the sidereal day into exactly 24 sidereal hours. This is why the sidereal clock "ticks" at a slightly different rate than your wristwatch does. Measurements of right ascension are always expressed in sidereal hours, minutes, and seconds. All other time measurements given in this book are expressed in mean solar time unless otherwise stated.

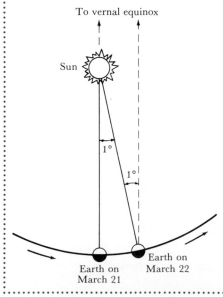

To vernal equinox

Sun

1°

1°

Earth on March 21

Earth on March 22

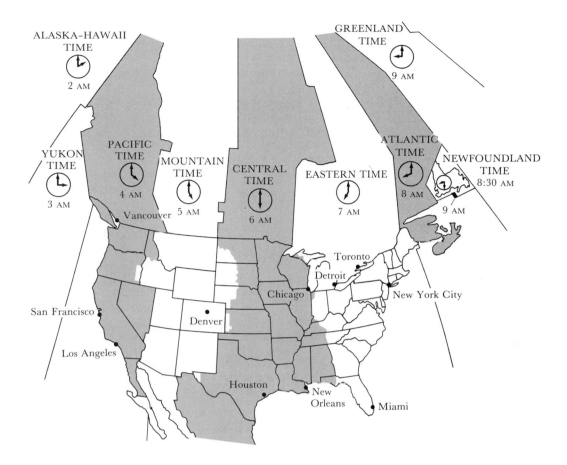

Figure 2-13 Time zones in North America
For convenience, the Earth is divided into 24 time zones centered about 15° intervals of longitude around the globe. Consequently, there are four time zones across the continental United States, giving a three-hour time difference between New York and California.

Just as the day is a natural unit of time based on the Earth's rotation, the year is a natural unit of time based on the Earth's revolution about the Sun. Unfortunately, nature has not arranged things in a convenient fashion. The year does not divide into 365 whole days. Ancient astronomers realized that the length of a year is approximately $365\frac{1}{4}$ days, and Julius Caesar was responsible for establishing the system of "leap years" to account for this extra one-quarter of a day. By adding an extra day to the calendar every four years, Julius Caesar hoped to ensure that seasonal astronomical events (such as the beginning of spring) would occur on the same date year after year.

Caesar's system would be perfect if the year were exactly $365\frac{1}{4}$ days long. Unfortunately, it is not.

Astronomers recognize several different kinds of years. First of all, you could measure the year as the time required for the Sun to return to the same position with respect to the stars in the sky. This is called the **sidereal year** and is equal to 365.2564 mean solar days, or $365^{\text{d}}6^{\text{h}}9^{\text{m}}10^{\text{s}}$.

Although the sidereal year is the true orbital period of the Earth with respect to the stars, it is *not* the kind of year on which we base the calendar. This is because, like Julius Caesar, most people prefer astronomical events to happen on the same date. For example, we want the first day of spring to occur on March 21. But spring begins when the Sun is at the vernal equinox, and the vernal equinox moves slowly against the background stars because of precession. Therefore, to set up a calendar, we want to use a year equal to the time needed for the Sun to return to the vernal equinox. This is called a **tropical year** and is equal to

365.2422 mean solar days, or $365^{d}5^{h}48^{m}46^{s}$. Because of precession, the tropical year is 20 minutes and 24 seconds shorter than the sidereal year.

This discrepancy was known to the ancient Greeks. In the second century BC, Hipparchus calculated the length of the tropical year and obtained a value only about six minutes different from the currently accepted value. (Hipparchus also was the first to detect the precession of the equinoxes by comparing his own observations with those of Babylonian astronomers three centuries earlier.) Julius Caesar's assumption that the tropical year equals $365\frac{1}{4}$ years was off by 11 minutes and 14 seconds. That tiny error adds up to about three days every four centuries. Although Julius Caesar's astronomical advisers were aware of the discrepancy, they felt that it was too small to matter. However, by the sixteenth century, the first day of spring was occurring on March 11.

To straighten things out, Pope Gregory XIII instituted a calendar reform in 1582. He began by dropping ten days (October 5, 1582 was proclaimed to be October 15, 1582), which brought the first day of spring back to March 21. Next, he modified Julius Caesar's system of leap years.

Julius Caesar's system added February 29 to every calendar year that is evenly divisible by four. Thus, for example, 1980, 1984, and 1988 are all leap years with 366 days. But we have seen that this produces an error of about three days every four centuries. Thus Pope Gregory decreed that only century years evenly divisible by 400 should be leap years. For example, the years 1700, 1800, and 1900 (which would have been leap years according to Caesar) are not leap years in this improved Gregorian system. But the years 1600 and 2000 (which can be divided evenly by 400) remain leap years.

The Gregorian system is the system that we use today. It assumes that the year is 365.2425 mean solar days, which is very close to the true length of the tropical year. In fact, the error is only one day every 3300 years. That won't cause any problems for a long time.

Summary

. It is convenient to imagine the stars fixed to the celestial sphere with the Earth at its center.

   The surface of the celestial sphere is divided into 88 regions, or constellations.

   The celestial sphere appears to rotate around the Earth once in each day and night; in fact, of course, it is the Earth that is rotating.

   The poles and equator of the celestial sphere are determined by extending the axis of rotation and the equatorial plane of the Earth to the celestial sphere.

   Positions of objects on the celestial sphere are described by specifying the right ascension (in time units) and the declination (as an angle).

. The Earth's axis of rotation is tilted at an angle of $23\frac{1}{2}°$ from the perpendicular to the plane of the Earth's orbit.

   The seasons are caused by this tilt, which tips one hemisphere or the other toward the Sun at certain times of the year.

   Equinoxes and solstices are significant points in the Earth's orbit, determined by the relationship between the Sun's path on the celestial sphere (the ecliptic) and the celestial equator.

- The Earth's axis of rotation moves slowly in a conical fashion, a phenomenon called precession.

  Precession is caused by the gravitational pull of the Sun and Moon on the Earth's equatorial bulge.

  Precession of the Earth's axis causes precession of the equinoxes along the ecliptic.

  Because the system of right ascension and declination is tied to the position of the vernal equinox, the date (or epoch) of observation must be specified when giving the position of an object in the sky.

- Time-keeping is an important concern of astronomers.

  Apparent solar time is based upon the apparent motion of the Sun across the celestial sphere; this motion varies with the seasons.

  Mean solar time is based upon the motion of an imaginary mean sun along the celestial equator, producing a uniform mean solar day of 24 hours.

  Sidereal time is based upon the apparent motion of the celestial sphere.

  The sidereal year is the actual orbital period of the Earth; the tropical year measures the period between vernal equinoxes.

  Leap-year corrections in the calendar are needed because the tropical year is not exactly 365 days.

- The concepts covered in Box 2-3 will be especially helpful in your further study of astronomy.

Review questions

1 Why is the ancient idea of the celestial sphere still a useful concept today?

*2 On December 1 at 10:00 p.m. you look toward the eastern horizon and see the bright star Procyon rising. At approximately what time will Procyon rise two weeks later, on December 15?

3 At what location on Earth is the north celestial pole on the horizon?

4 Where do you have to be on the Earth in order to see the Sun at the zenith? If you stay at that location for a full year, on how many days does the Sun pass through the zenith?

5 Where do you have to be on Earth in order to see the south celestial pole directly overhead? What is the maximum possible elevation of the Sun above the horizon at that location? On what date is this maximum elevation observed?

6 At what point on the horizon does the vernal equinox rise?

*7 What is the sidereal time when the vernal equinox rises?

*8 On what date is the sidereal time nearly equal to the solar time?

Advanced questions

9 Consult a star map of the southern hemisphere and determine which, if any, bright southern stars could some day become south celestial "pole" stars.

10 Go to a library that has recent issues of *The Astronomical Almanac*. In the section of these reference books entitled "Bright Stars," look up the coordinates of a particular star (of your choice) on two successive years. By how much did the right ascension and declination of that star change?

11 Using a star map determine which, if any, bright stars could some day mark the location of the vernal equinox. Give the approximate years when that would happen.

Discussion questions

12 Examine a list of the 88 constellations. Are there any constellations that obviously date from modern times? Where are these constellations located in the sky? Why do you suppose they do not have archaic names?

13 Describe the seasons if the Earth's axis of rotation were tilted 0° and 90° to its orbital plane.

For further reading

Useful guides for observing the constellations and the sky:
Chartrand, M. *Skyguide*. Western Pub., 1982.
Kals, W. *The Stargazer's Bible*. Doubleday, 1980.
Menzel, D., and Pasachoff, J. *A Field Guide to the Stars and Planets, 2nd ed.* Houghton-Mifflin, 1983.
Whitney, C. *Whitney's Star Finder*. A. Knopf, 1981.
On the history of constellations, star names, and navigation:
Allen, R. *Star Names: Their Lore and Meaning*. 1899. Dover reprint, 1963.
Brown, H. *Man and the Stars*. Oxford U. Press, 1978.
Gallant, R. *The Constellations: How They Came to Be*. Four Winds Press, 1979.
Gingerich, O. "Notes on the Gregorian Calendar Reform." *Sky & Telescope*, Dec. 1982, p. 530.
————, "The Origin of the Zodiac." *Sky & Telescope*, March 1984, p. 218.
Kunitzsch, P. "How We Got Our Arabic Star Names." *Sky & Telescope*, Jan. 1983, p. 20.
Kyselka, W., and Lanterman, R. *North Star to Southern Cross*. U. Press of Hawaii, 1976.

3 Eclipses and the astronomy of antiquity

The Caracol at Chichen-Itza
This ancient Mayan observatory in the Yucatan was built around AD 1000. Its architecture is based on alignments with important celestial events. Mayan astronomers developed a very accurate calendar and measured the motions of celestial bodies with great precision. Special significance was associate with the planet Venus which inspired sacrificial rites and other ceremonies. (E. C. Krupp)

Ancient cultures made many important astronomical observations and devised many ideas and systems that are still used by astronomers. In this chapter, we learn about the phases of the Moon and their relationship to the Moon's motion about the Earth and the Earth's motion about the Sun. We see how ancient astronomers attempted to measure the size of the Earth and the distances from Earth to the Sun and the Moon. Eclipses of the Sun and the Moon played important roles in many of these early theories and measurements. We learn about the conditions under which eclipses occur and about their physical details. We even learn how to predict the occurrence of total solar eclipses. Finally, we learn how to describe the apparent brightness of objects in the sky, using a system introduced by a Greek astronomer in the second century BC.

The beauty of the star-filled night sky or the drama of an eclipse would suffice to make astronomy fascinating. But there are practical reasons for an interest in the universe as well. Even the ancient Greeks knew the connection between the seasons and the relative orientation of the Sun and Earth. Many early seafaring cultures were aware that the tides are influenced by the position of the Moon.

Ancient civilizations placed great emphasis on careful astronomical observation. Hundreds of impressive monuments, such as Stonehenge (Figure 3-1), that dot the British Isles provide evidence of this astronomical preoccupation. Alignments of the stones point to the rising and setting locations of the Sun and Moon at key times during the year such as solstices. Similar monuments are found in the United States—for

Figure 3-1 Stonehenge
This astronomical monument was constructed nearly 4000 years ago on Salisbury Plain in southern England. Originally, the monument consisted of thirty blocks of grey sandstone, each standing 4 meters (13½ feet) high, set in a circle 30 meters (97 feet) in diameter. These stones were topped with a continuous circle of smaller stones. Inside the circle are geometrical arrangements of other stones, most notably a horseshoe-shaped set of larger stones opening toward the northeast. (Courtesy of the British government)

example, the Medicine Wheel in Wyoming. Constructed high on top of a windswept plateau by the plains Indians, the Medicine Wheel is a circular ring of stones. Certain stones and markers are aligned with the rising points of several bright stars, as well as that of the Sun at the summer solstice.

Aztec, Mayan, and Incan architects in Central and South America designed astronomically oriented buildings. For example, at the ruined city of Tihuanaco in Bolivia, the Temple of the Sun was built with walls aligned north–south and east–west with an accuracy better than one degree. A similar precision was exercised 5000 years ago by the builders of the great Egyptian pyramids which also have sides oriented north–south and east–west.

In addition to temples and tombs, the ancients apparently constructed astronomical observatories. One of the best examples is the Caracol in the Mayan city of Chichen-Itza on the Yucatan Peninsula. Built nearly a thousand years ago, the Caracol's cylindrical tower contains windows aligned with the northernmost and southernmost rising and setting points of both the Sun and the planet Venus. A similar four-story adobe building, probably constructed during the fourteenth century, is located at the Casa Grande site in Arizona. All of these structures bear witness to careful and patient astronomical observations by the people of many ancient cultures and civilizations.

Lunar phases are due to the Moon's orbital motion

Ancient astronomers knew that the Moon shines by reflected sunlight. They also understood that the Moon orbits the Earth. This knowledge was deduced from observations of the Moon's changing **phases** as varying amounts of its illuminated hemisphere are exposed to observers on the Earth.

Figure 3-2 is an extraordinary photograph of both the Earth and Moon from space. The Moon orbits the Earth every 27.3 days, and so it takes approximately four weeks for the Moon to complete its cycle of phases, as shown in Figure 3-3.

Figure 3-2   The Earth and the Moon
The Moon circles the Earth every 27.3 days at an average distance of 384,400 kilometers (238,900 miles). This picture was taken in 1977 from the Voyager 1 spacecraft shortly after it was launched toward Jupiter and Saturn. (NASA)

Figure 3-3   The phases of the Moon
Light from the Sun illuminates one half of the Moon, while the other half is dark. As the Moon orbits the Earth, we see varying amounts of the Moon's illuminated hemisphere. It takes 29½ days for the Moon to go through all its phases.

The phase called **new moon** occurs when the dark hemisphere of the Moon faces the Earth. The Moon is not visible at this phase because it is in the same part of the sky as the Sun.

During the next seven days, Earth-based observers see a phase called **waxing crescent moon,** in which progressively more and more of the illuminated hemisphere is exposed to our view. At **first quarter moon,** the angle between the Sun, Moon, and Earth is 90°, and we see one-half of the illuminated hemisphere and one-half of the dark hemisphere.

During the next week, still more of the illuminated hemisphere can be seen from Earth, giving us a phase called **waxing gibbous moon.** When the Moon stands opposite the Sun in the sky, we see the fully illuminated hemisphere, producing the phase called **full moon.** Moonrise occurs at sunset during full moon.

Over the subsequent two weeks, we see less and less of the illuminated hemisphere as the Moon continues along its orbit. This produces the phases called **waning gibbous moon, last quarter moon,** and **waning crescent moon,** as diagrammed in Figure 3-3.

Because the position of the Sun in the sky determines the local time, it is possible to correlate the Moon's phase and location with the time of the day. For example, during first quarter moon, the Moon is approximately 90° east of the Sun in the sky, and hence moonrise occurs approximately at noon. Figure 3-4 is a series of photographs showing the various phases of the Moon.

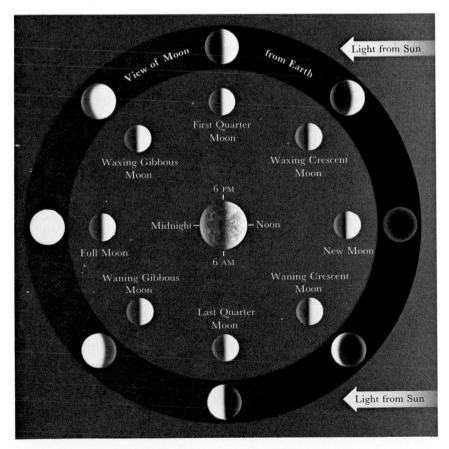

Waxing cresent
(age: 4 days)

First quarter
(age: 7 days)

Waxing gibbous
(age: 10 days)

Full Moon
(age: 14 days)

Figure 3-4   The Moon's appearance
The Moon always keeps the same side facing the Earth. Earth-based observers therefore always see the same craters and lunar mountains, regardless of the phase. (Lick Observatory)

Figure 3-5   The sidereal and synodic months
The sidereal month is the time it takes the Moon to complete one revolution with respect to the background stars. However, because the Earth is constantly moving along its orbit about the Sun, the Moon must travel through slightly more than 360° to get from one new moon to the next. Thus the synodic month is slightly longer than the sidereal month.

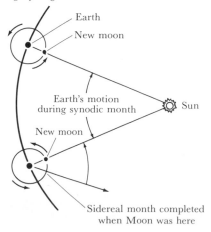

It takes about one month for the Moon to complete one orbit around the Earth. However, astronomers are careful to distinguish between two types of months, depending on whether the Moon's motion is measured relative to the stars or to the Sun. Neither corresponds exactly to the months of our calendar.

The **sidereal month** is the time it takes for the Moon to complete one full orbit of the Earth, measured *with respect to the stars*. This is the Moons' true orbital period and is equal to 27.3 days.

The **synodic month** is the time it takes for the Moon to complete one cycle of phases. For example, it is the time from one new moon to the next. Consequently, the synodic month is measured *with respect to the Sun* and is equal to about 29.5 days.

Of course, the Earth is orbiting the Sun while the Moon goes through its phases. Thus, to get from one new moon to the next, the Moon must travel more than 360° along its orbit, as shown in Figure 3-5. The synodic month thus is approximately two days longer than the sidereal month.

The Moon stays in orbit about the Earth because of the gravitational attraction between these two bodies. Indeed, the Earth's gravity is a primary factor controlling the Moon's orbit. (We shall discuss orbits and gravity in greater detail later in Box 3-1 and in Chapter 4.) In addition to the Earth, however, the Sun also pulls on the Moon. The Sun's gravitational pull causes the Moon sometimes to speed up or slow down slightly in its orbit. Thus the Moon's orbital path is constantly being altered slightly by the Sun's gravity. The size of these minor variations depends on the relative configuration of the Sun, Moon, and Earth.

The final result is that both the sidereal and synodic months are variable. The sidereal month (average length = $27^d7^h43^m11^s$) can vary by as much as seven hours. The synodic month (average length = $29^d12^h44^m3^s$) can vary by as much as half a day.

Waning gibbous
(age: 20 days)

Last quarter
(age: 22 days)

Waning crescent
(age: 26 days)

Ancient astronomers measured the size of the Earth and attempted to determine distances to the Sun and Moon

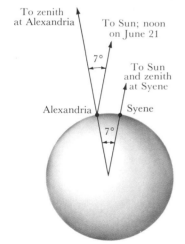

Figure 3-6   Erastosthenes' method of determining the Earth's size
Erastosthenes noticed that the Sun is about 7° south of the zenith at Alexandria when it is directly overhead at Syene. The angle is about one-fiftieth of a circle, so the distance between Alexandria and Syene must be about one-fiftieth of the Earth's circumference.

More than two thousand years ago, Greek astronomers were fully aware of the Earth's spherical shape. Eclipses of the Moon provided the convincing observations. During a **lunar eclipse,** the Moon passes through the Earth's shadow. Ancient astronomers noticed that the edge of the Earth's shadow is always circular. A sphere is the only shape that casts a circular shadow from any angle, so the ancient astronomers concluded that the Earth is spherical.

Around 200 BC, the Greek astronomer Eratosthenes devised a way to measure the circumference of the Earth. He was intrigued by reliable reports from the town of Syene in Egypt (near the modern Aswan) that the Sun shone directly down vertical wells on the first day of summer. Eratosthenes knew that the Sun never appears at the zenith from his home in Alexandria, which is on the Mediterranean Sea, almost due north of Syene. He measured the position of the Sun at local noon on the summer solstice in Alexandria and found that it was about 7° south of the zenith, as sketched in Figure 3-6. Actually, Eratosthenes measured the angle as one-fiftieth of a complete circle (we would say $7\frac{1}{5}°$), and so he concluded that the distance from Alexandria to Syene is one-fiftieth of the Earth's circumference.

In Eratosthenes' day, the distance from Alexandria to Syene was said to be 5000 stades. Therefore, Eratosthenes found the Earth's circumference to be

$$50 \times 5000 \text{ stades} = 250,000 \text{ stades}$$

Unfortunately, no one today is sure of the exact length of the Greek unit called the stade. One guess is that the stade was about $\frac{1}{6}$ kilometer, which would mean that Eratosthenes obtained a circumference of about 42,000 kilometers—within 5 percent of the modern value of about 40,000 kilometers.

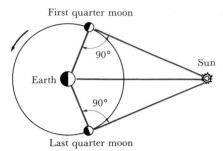

Figure 3-7 *Aristarchus' method of determining distances to the Sun and Moon*
Aristarchus argued that the Moon should take longer to go from first quarter to last quarter than it does from last to first. By measuring this time difference, he determined the angles in the triangles formed by the Sun, Earth and Moon at first and last quarter phases. He was then able to calculate the relative lengths of the sides of these triangles and thereby to obtain the distances to the Sun and Moon. The size of the Moon's orbit is greatly exaggerated in this diagram.

Eratosthenes was one of several brilliant astronomers to emerge from the so-called Alexandrian school, which by his time already had a distinguished tradition. For example, one of the first Alexandrian astronomers, Aristarchus of Samos, devised a method of determining the relative distances to the Sun and Moon, perhaps as long ago as 280 BC.

Aristarchus reasoned that the angle between the Sun, Moon, and Earth is exactly 90° at the moment of first quarter or last quarter moon, as diagrammed in Figure 3-7. Assuming that the Moon moves along its orbit at a uniform rate, Aristarchus pointed out that the Moon should take longer to go from first quarter to last quarter than it does to go from last quarter around to first quarter. The difference between these two travel times can be used to determine the rest of the angles in the triangles in Figure 3-7. Aristarchus apparently believed that the Moon takes one day more to go from first to last quarter than it does to go from last to first. Using geometrical arguments, Aristarchus then deduced the relative lengths of the sides of the triangles in Figure 3-7.

Aristarchus concluded that the Sun is only 20 times farther from us than the Moon is. We now know that the average distance to the Sun is about 390 times larger than the average distance to the Moon. One of Aristarchus's main errors arose from the fact that it is impossible visually to determine the exact instant of first or last quarter. Nevertheless, it is impressive that people were logically trying to measure distances across the solar system more than two thousand years ago.

Aristarchus also used lunar eclipses in an equally bold attempt to determine the relative sizes of the Earth, Moon, and Sun. Observing how long it takes for the Moon to move through the Earth's shadow, Aristarchus estimated that the diameter of the Earth is about three times larger than the diameter of the Moon. To determine the diameter of the Sun, Aristarchus simply pointed out that the Sun and the Moon have the same angular size in the sky. Therefore their diameters must be in the same proportion as their distances. In other words, because Aristarchus believed that the Sun is 20 times farther from the Earth than the Moon, he concluded that the Sun must be 20 times larger than the Moon.

You can now appreciate the significance of Erastosthenes' measurement of the Earth's circumference. The Greeks knew that the Earth's diameter is equal to its circumference divided by the constant called π (pi). Knowing the Earth's diameter, astronomers of the Alexandrian school could calculate the diameters of the Sun and Moon as well as their distances from Earth.

For ease of comparison, Table 3-1 summarizes some ancient and modern measurements of the sizes of the Earth, Moon, and Sun as well as the distances between them. Although some of these ancient measurements are far from the modern values, our ancestors' achievements stand as an impressive exercise in observation and reasoning.

TABLE 3-1 *A comparison of ancient and modern measurements*

	Ancient measure (in kilometers)	Modern measure (in kilometers)
Earth's diameter	13,000	12,756
Moon's diameter	4,300	3,476
Sun's diameter	9×10^4	1.39×10^6
Earth–Moon distance	4×10^5	3.84×10^5
Earth–Sun distance	10^7	1.50×10^8

Eclipses occur only when the Sun and Moon are both on the line of nodes

A **lunar eclipse** occurs when the Moon passes through the Earth's shadow. This can happen only when the Sun, Earth, and Moon are in a straight line at full moon.

A **solar eclipse** occurs when the Earth passes through the Moon's shadow. As seen from Earth, the Moon moves in front of the Sun. Once again, this can happen only when the Sun, Moon, and Earth are aligned. However, for a solar eclipse, the Moon must be between the Earth and the Sun. Therefore a solar eclipse must occur at new moon.

Both new moon and full moon occur at intervals of $29\frac{1}{2}$ days, but solar and lunar eclipses happen much less frequently. This is because the Moon's orbit is tilted slightly out of the plane of the Earth's orbit, as shown in Figure 3-8. The angle between the plane of the Earth's orbit and the plane of the Moon's orbit is about 5°. Because of this tilt, new moon and full moon usually occur when the Moon is either above or below the plane of the Earth's orbit. When the Moon is away from the plane of the Earth's orbit, a perfect alignment between the Sun, Moon, and Earth is not possible, and an eclipse cannot occur.

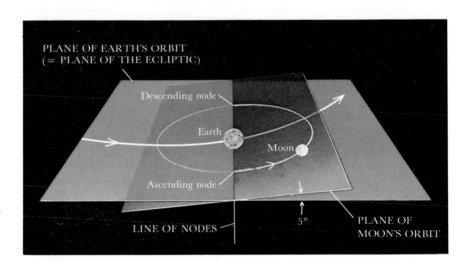

Figure 3-8 The line of nodes
The plane of the Moon's orbit is tilted slightly with respect to the plane of the Earth's orbit. These two planes intersect along a line called the line of nodes.

The plane of the Earth's orbit and the plane of the Moon's orbit intersect along a line called the **line of nodes.** The line of nodes passes through the Earth and is pointed in a particular direction in space, as shown in Figure 3-9. Eclipses can occur only when both the Sun and Moon are on or very near the line of nodes, because only then do the Sun, Earth, and Moon lie along a straight line.

Knowing the orientation of the line of nodes is clearly important to anyone who wants to predict upcoming eclipses. However, predicting eclipses is complicated by the fact that the line of nodes gradually changes its direction in space. The constant gravitational pull of the Sun on the Moon causes the Moon's orbit gradually to shift its orientation in space as described in Box 3-1. The resulting slow westward movement of the line of nodes is one of several details that astronomers must include in their calculations of the dates and times of upcoming eclipses.

It is possible to show that there are at least two but not more than five solar eclipses each year. The last year in which five solar eclipses occurred was 1935. Lunar eclipses occur just about as frequently as solar eclipses. However, the maximum number of eclipses (both solar and lunar) in a year is seven.

Box 3-1 Some details of the moon's orbit

The Moon's orbit about the Earth is an ellipse, as sketched in the accompanying diagram. The Moon is said to be at **perigee** when it is nearest the Earth and at **apogee** when it is farthest from Earth. The line connecting the points of perigee and apogee passes through the Earth and is called the **line of apsides** (see diagram *a*).

The center-to-center Earth–Moon distance varies from a minimum of 356,410 kilometers (221,463 miles) at perigee to a maximum of 406,697 kilometers (252,710 miles) at apogee. Consequently, the apparent size of the Moon as seen from Earth varies over the course of a month. At perigee, the Moon has an angular diameter of 33′ 31″, whereas at apogee the Moon's angular diameter is only 29′ 22″. The average apparent size of the Moon is 31′ 5″, which corresponds to an average Earth–Moon distance of 384,400 kilometers (238,860 miles).

The average speed of the Moon along its orbit is 1.02 kilometers per second (2290 miles per hour). As seen from Earth, the Moon appears to move eastward among the constellations from one day to the next. The Moon's daily eastward progress averages 13.2° (which is 360° divided by the 27.3 days in the sidereal month). In one hour, the Moon moves slightly more than $\frac{1}{2}$°, which is slightly more than its own diameter. This rate of motion means that the time of moonrise is retarded by an average of about 50 minutes from one day to the next.

The mutual gravitational attraction between the Earth and Moon is the primary agent that determines the shape and size of the Moon's orbit. However, the Sun is also constantly tugging on the Moon. The Sun's influence on the lunar orbit is weaker than the Earth's, primarily because the Sun is so much farther away from the Moon than the Earth is. Nevertheless, the Sun's gravity does produce small changes (called "gravitational perturbations") in the lunar orbit.

The Sun's gravitational tugging on the Moon causes both the line of nodes and the line of apsides to shift slowly over the years. The line of nodes gradually moves westward, as sketched in the diagram *b*. This movement is called the **regression of the line of nodes.** It takes 18.61 years for the line of nodes to complete one full rotation.

While the line of nodes is moving westward among the constellations, the Sun's gravity causes the line of apsides to move eastward. It takes 8.85 years for the line of apsides to complete one full rotation. This effect, sketched in the diagram *c*, is simply called the **rotation of the Moon's orbit**.

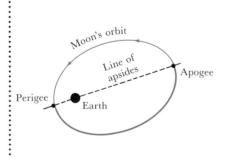

a

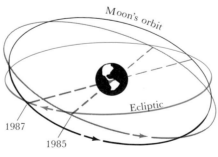

b

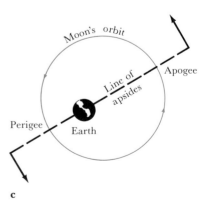

c

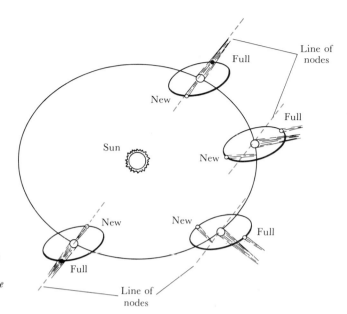

Figure 3-9   Conditions for eclipses
A solar eclipse occurs only if the Moon is very near the line of nodes at new moon. A lunar eclipse occurs only if the Moon is very near the line of nodes at full moon.

Solar and lunar eclipses can be either partial or total, depending on the alignment of the Sun, Earth, and Moon

A lunar eclipse occurs when the Moon, at full phase, moves through the Earth's shadow. However, the Earth's shadow has two distinct parts, as diagrammed in Figure 3-10. The **umbra** is the darkest part of the shadow, from which no portion of the Sun's surface can be seen. In the **penumbra,** only part of the Sun's surface is blocked out.

There are three kinds of lunar eclipses, depending on exactly how the Moon travels through the Earth's shadow. First, the Moon might pass only through the Earth's penumbra. This is called a **penumbral eclipse.** During a penumbral eclipse, none of the lunar surface is completely shaded by the Earth because in the penumbra only part of the Sun is covered by the Earth. At mid-eclipse, the Moon merely looks a little dimmer than usual from Earth. There is no "bite" taken out of the Moon by the Earth's umbra. The Moon still looks full, and thus it is easy to miss a penumbral eclipse.

Figure 3-10   The geometry of a lunar eclipse
People on the nighttime side of the Earth see a lunar eclipse when the Moon moves through the Earth's shadow. The umbra is the darkest part of the shadow. In the penumbra, only part of the Sun is covered by the Earth.

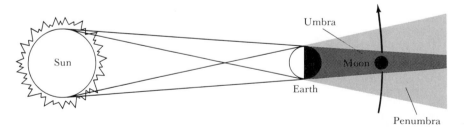

Most people notice a lunar eclipse only if the Moon passes into the Earth's umbra. As the umbral phase of the eclipse begins, a "bite" seems to be taken out of the Moon as part of the lunar surface becomes immersed in the darkest portions of the Earth's shadow. If the Moon's orbit is oriented so that only part of the lunar surface passes through the umbra, then we have a **partial eclipse.** A **total eclipse** of the Moon occurs when the Moon travels completely into the umbra, as sketched in Figure 3-11. The maximum duration of totality occurs when the Moon travels directly through the center of the umbra. The Moon's speed through the Earth's shadow is roughly 1 kilometer per second (2300

Figure 3-11   Various lunar eclipses
This diagram shows the Earth's umbra and penumbra at the distance of the Moon's orbit. Different kinds of lunar eclipses are seen, depending on the Moon's path through the Earth's shadow.

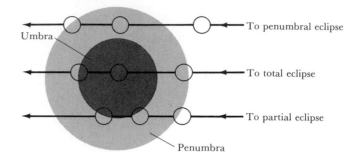

Figure 3-12   A total eclipse of the Moon
This photograph was taken by an amateur astronomer during the lunar eclipse of September 6, 1979. Notice the distinctly reddish color of the Moon. (Courtesy of Mike Harms)

miles per hour), which means that totality can last for as much as 1 hour 42 minutes.

As an example of the frequency of lunar eclipses, Table 3-2 lists all the total and partial eclipses during the 1980s. Penumbral eclipses are not included in this listing.

TABLE 3-2   Lunar eclipses during the 1980s

Date		Percentage eclipsed (100% = total)	Duration of totality
17 July	1981	58	——
9 January	1982	100	1ʰ 24ᵐ
6 July	1982	100	1 42
30 December	1982	100	1 6
25 June	1983	34	——
4 May	1985	100	1 10
28 October	1985	100	0 42
24 April	1986	100	1 8
17 October	1986	100	1 14
7 October	1987	1	——
27 August	1988	30	——
20 February	1989	100	1 16
17 August	1989	100	1 38

Even during totality, the Moon does not completely disappear from the sky. A small amount of sunlight passing through the Earth's atmosphere is deflected into the Earth's umbra. Most of this deflected light is red, and thus the darkened Moon glows faintly in reddish hues during totality, as shown in Figure 3-12.

It is a fortunate coincidence of nature that, as seen from Earth, the apparent diameter of the Moon is almost exactly the same as the apparent diameter of the Sun. Both the Sun and Moon have angular diameters of about $\frac{1}{2}°$. For this reason, the Moon "fits" over the Sun during a **total solar eclipse,** blocking out the dazzling solar disk and not much else. This is important to astronomers because the hot gases (called the **solar corona**) that surround the Sun can be photographed and studied in detail during the few precious moments when the eclipse is total (see Figure 3-13).

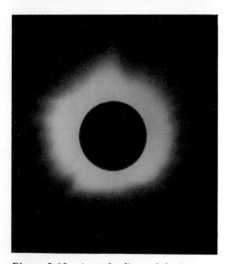

Figure 3-13   A total eclipse of the Sun
During a total solar eclipse, the Moon completely covers the Sun's disk, and the solar corona can be seen. This halo of hot gases extends for thousands upon thousands of kilometers into space. Only the brightest, inner portions of the solar corona are seen in this photograph of the solar eclipse of March 7, 1970. (NASA)

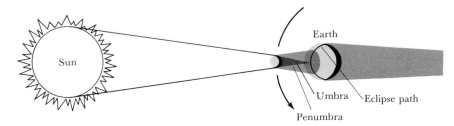

Figure 3-14   The geometry of a total solar eclipse
During a total solar eclipse, the tip of the Moon's umbra traces an eclipse path across the Earth's surface. People inside the eclipse path see a total eclipse, whereas people inside the penumbra see only a partial eclipse of the Sun.

As with the Earth's shadow, the darkest part of the Moon's shadow is called the **umbra.** You must be inside the Moon's umbra in order to see a total solar eclipse, because that is the only region from which the Moon completely covers the Sun. Because the Sun and Moon have nearly the same angular diameter, only the tip of the Moon's umbra reaches the Earth's surface, as sketched in Figure 3-14. As the Earth turns, the tip of the umbra traces an **eclipse path** across the Earth's surface. Only those people standing inside the eclipse path are treated to the spectacle of a total solar eclipse. Figure 3-15 shows the dark spot produced by the Moon's umbra on the Earth's surface.

Immediately surrounding the Moon's umbra is the region of partial shadow called the **penumbra.** From this region, the Sun's surface appears only partially covered by the Moon. During a solar eclipse, the Moon's penumbra covers a large portion of the Earth's surface. Anyone standing inside the penumbra sees a **partial eclipse** of the Sun.

All of the details of a solar eclipse are calculated well in advance and are published in reference books such as *The Astronomical Almanac.* Figure 3-16 shows a typical eclipse map, which displays those areas of the Earth covered by the Moon's umbra and penumbra. During a total eclipse, the Earth's rotation and the orbital motion of the Moon cause the umbra to race along the eclipse path at speeds in excess of 1700 kilometers per hour. Because of the high speed of the umbra, most people along the eclipse path observe totality for only a few moments. Totality never lasts for more than $7\frac{1}{2}$ minutes at any one location on the eclipse path. In a typical total solar eclipse, the angle of the Sun and the Earth–Moon distance produce a duration of totality much less than the maximum $7\frac{1}{2}$ minutes.

The width of the eclipse path depends primarily on the Earth–Moon distance during an eclipse. The eclipse path is widest if the Moon happens to be at the point in its orbit nearest the Earth. This can produce an eclipse path up to 270 kilometers (170 miles) wide. Usually it is much narrower. In fact, sometimes the Moon's umbra does not reach down to

Figure 3-15   The Moon's shadow on the Earth
This photograph was taken from an Earth-orbiting satellite during the total eclipse of March 7, 1970. The Moon's umbra appears as a dark spot on the eastern coast of the United States. (NASA)

the Earth's surface. This happens because the Moon's orbit is slightly elliptical (see Box 3-1). If the alignment for a solar eclipse occurs when the Moon is farthest from the Earth, then the Moon's umbra falls short of the Earth, and no one sees a truly total eclipse. From the Earth's surface, the Moon appears too small to completely cover the Sun, and a thin ring of light is seen around the edge of the Moon at mid-eclipse. An eclipse of this type is called an **annular eclipse.** The length of the Moon's umbra is nearly 5000 kilometers shorter than the average distance between the Moon and the Earth's surface. Thus, the Moon's shadow often fails to reach the Earth, and annular eclipses are slightly more common than total eclipses.

As an example of the frequency of solar eclipses, Table 3-3 lists all the total, annular, and partial eclipses during the 1980s.

Ancient astronomers achieved limited ability to predict eclipses

A total solar eclipse is a dramatic event. The sky begins to darken, the air temperature falls, and the winds increase as the Moon's umbra races toward you. All nature responds: birds go to roost, flowers close their petals, and crickets begin to chirp as if evening had arrived. As totality approaches, the landscape around you is bathed in shimmering bands of

TABLE 3-3   Solar eclipses during the 1980s

Date	Area	Type	Notes
1980 February 16	Atlantic, Kenya, India, China	Total	Max. length 4m 8s
1980 August 10	S. Pacific, Brazil	Annular	
1981 February 4	Pacific, S. Australia and New Zealand	Annular	
1981 July 31	Russia, N. Pacific	Total	Max. length 2m 3s
1982 January 25	Antarctic	Partial	57% eclipsed
1982 June 21	Antarctic	Partial	62% eclipsed
1982 July 20	Arctic	Partial	46% eclipsed
1982 December 15	Arctic	Partial	74% eclipsed
1983 June 11	Indian Ocean, E. Indies, Pacific	Total	Max. length 5m 11s
1983 December 4	Atlantic, Equatorial Africa	Annular	
1984 May 30	Pacific, Mexico, U.S.A., Atlantic, N. Africa	Annular	
1984 November 22/3	E. Indies, S. Pacific	Total	Max. length 1m 59s
1985 May 19	Arctic	Partial	84% eclipsed
1985 November 12	S. Pacific, Antarctica	Total	Max. Length 1m 55s
1986 April 9	Antarctic	Partial	82% eclipsed
1986 October 3	N. Atlantic	Total	Max. length 0m 1s Annular along most of track
1987 March 29	Argentina, Atlantic, Central Africa, Indian Ocean	Total	Max. length of 56s, in Atlantic. Annular along most of track
1987 September 23	Russia, China, Pacific	Annular	
1988 March 18	Indian Ocean, E. Indies, Pacific	Total	Max. length 3m 46s
1989 March 7	Arctic	Partial	83% eclipsed
1989 August 31	Antarctic	Partial	63% eclipsed

light and dark as the last few rays of sunlight peek out from behind the edge of the Moon. And finally the corona blazes forth in a star-studded midday sky. It is an awesome sight.

In ancient times, the ability to predict eclipses must have been very desirable. Archeological evidence suggests that astronomers in many civilizations struggled to predict eclipses, with various degrees of success. The number and placement of certain holes in the ground around Stonehenge suggest some ability to predict eclipses more than 4000 years ago. One of three priceless manuscripts to survive the devastating Spanish conquests shows that the Mayan astronomers of Mexico and Guatemala had a fairly reliable method of eclipse prediction. There are also numerous apocryphal stories such as that of the great Greek astronomer Thales of Miletus. Thales is said to have predicted the famous eclipse of 585 BC, which occurred during the middle of a war. The sight was so awesome and unexpected that the soldiers put down their arms, and peace was declared.

In retrospect, it seems that ancient astronomers actually produced eclipse "warnings" of various degrees of reliability rather than true predictions. Working with historical records, these astronomers generally searched for cycles and regularities from which future eclipses might be anticipated. Indeed, from what you already know, you could produce eclipse warnings.

Figure 3-16   An eclipse map
Details of upcoming eclipses are published in astronomical reference books. Maps such as the one shown here, show the regions of the Earth that will experience a total or a partial eclipse. The Moon's shadow travels generally eastward across the Earth's surface. (Reproduced from The Astronomical Almanac.)

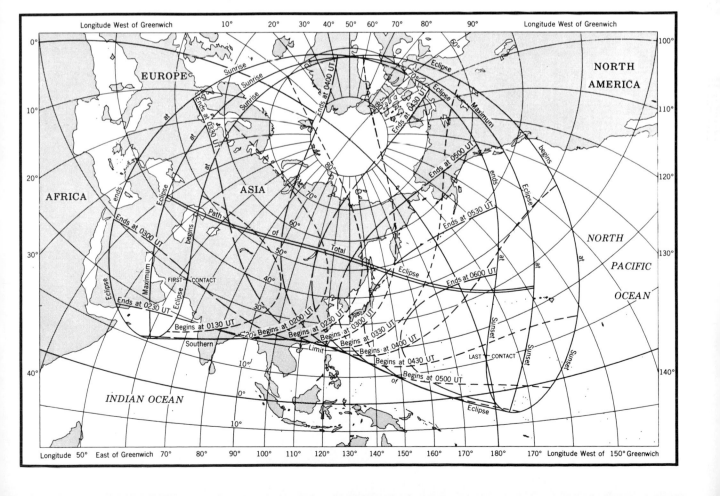

Suppose you observe a solar eclipse in your home town and you want to figure out when you and your neighbors might see another solar eclipse. How would you begin?

First of all, you must remember that a solar eclipse can occur only if the line of nodes points toward the Sun at new moon (recall Figure 3-9). Second, you must know that it takes 29.53 days to go from one new moon to the next. This is called the **lunar month,** or synodic month. Because solar eclipses occur only at the time of the new moon, you must wait several whole lunar months for the proper alignment to occur again.

However, there is a complication: the line of nodes gradually shifts its position with respect to the background stars, as described in Box 3-1. It takes 346.6 days to go from one alignment of the line of nodes pointing toward the Sun to the next identical alignment. This is called the **eclipse year.**

To predict when you will see another solar eclipse, you need to know how many whole lunar months equal some whole number of eclipse years. This will tell you how long you have to wait between virtually identical alignments of the Sun, the Moon, and the line of nodes. The answer is

223 lunar months = 19 eclipse years

because

$$223 \times 29.53 = 19 \times 346.6 = 6585 \text{ days}$$

to within a few hours. This interval is called the **saros.** A more accurate treatment gives the length of the saros as 6585.3 days (that is the same as 18 years $11\frac{1}{3}$ days—give or take a day, depending on the number of leap years included). Eclipses separated by the saros interval are said to form "families."

You might think that you and your neighbors would simply have to wait one full saros interval to get from one solar eclipse to the next. However, because of the extra one-third day, the Earth has rotated by

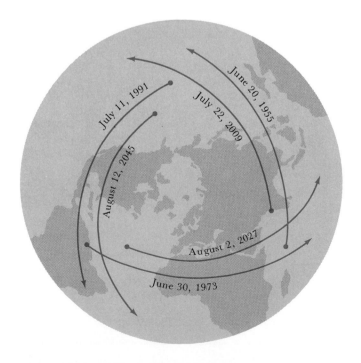

Figure 3-17   A family of eclipses
Eclipses separated by the saros interval (18 years 11.3 days) are said to form "families" because geometrical details, such as the orientation of the line of nodes, are nearly identical. Even the duration of totality is nearly the same for all members of a particular eclipse family.

an extra 120° when the next solar eclipse of a particular family occurs. That puts the eclipse path one-third of the way around the world from you. You therefore must wait three full saros intervals before the eclipse path is back around to your part of the Earth. Figure 3-17 shows a family of solar eclipse paths, each separated from the next by one saros interval. Notice that three saros intervals (54 years 34 days) elapse before the eclipse path again falls on roughly the same part of the Earth.

There is evidence that ancient astronomers knew about intervals such as the saros. It is quite possible that the discovery of such intervals came from lunar eclipses rather than solar eclipses. If you are far from the eclipse path, yet still within the Moon's penumbra, there is a good chance that you would fail to notice a solar eclipse. Even if one-half of the Sun is covered by the Moon, the remaining solar surface provides enough sunlight that the outdoor illumination is not greatly diminished. It is much easier to notice an eclipse of the Moon. When the Moon enters the Earth's shadow, everyone on the nighttime side of the Earth can see the eclipse (unless clouds block the view).

Ancient astronomers established traditions and invented systems that are still used

Ancient astronomers, particularly those of Greece, gave humanity a new and powerful way of thinking about the world. They gave the first clear demonstration that the tools of logic, reason, and mathematics can be used to discover and understand the workings of the universe. This general approach to reality underlies all modern science. Aristarchus even went so far as to suggest that the motions of the planets in the sky can be simply explained if all of the planets as well as the Earth orbit the Sun. As we shall see in the next chapter, this man was nearly 2000 years ahead of his time.

In addition to setting the stage for modern science, ancient Greek astronomers also gave us ideas and established traditions that are still in use. For example, around the year 160 BC, Hipparchus built an observatory on the island of Rhodes. Over a period of several years, he compiled the first comprehensive star catalogue, which listed the coordinates and brightnesses of 850 stars. During the course of this pioneering work, Hipparchus compared his star positions with earlier records dating back to Aristarchus's time. Hipparchus soon noted systematic differences that led him to conclude that the north celestial pole had shifted slightly over the preceding century. He had discovered precession.

While compiling his star catalogue, Hipparchus established a system to denote the brightnesses of stars. Hipparchus's system is the basis of the **magnitude scale** used by astronomers today. Quite simply, Hipparchus said that the brightest stars in the sky are "first magnitude." The dimmest visible stars he called "sixth magnitude." Stars of intermediate brightness he assigned intermediate numbers on this scale of 1 to 6.

With only a few refinements, the same system is used today. As shown in Figure 3-18, the magnitude scale is extended to include very faint stars. For example, with a good pair of binoculars, you can see stars as faint as tenth magnitude. Through some of the largest telescopes in the world, it is possible to see stars as dim as magnitude 20. Photography with long exposure time reveals even dimmer stars.

Modern astronomers use negative numbers to extend Hipparchus's scale to include very bright objects. For example, Sirius is the brightest star in the sky and has a magnitude of $-1\frac{1}{2}$. At its brightest, the planet Venus shines with a magnitude of -4. Of course, the Sun is the brightest object in the sky. Its magnitude is $-26\frac{1}{2}$.

Later in this book, we shall discuss many details about the brightness

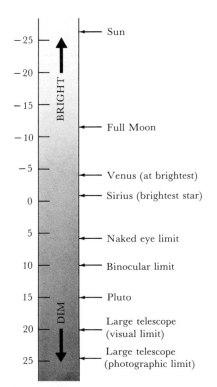

Figure 3-18 The magnitude scale
Astronomers denote the brightness of an object in the sky by its "apparent magnitude." Most stars have apparent magnitudes between 1 and 6. Photography through the largest telescopes reveals stars as faint as magnitude 24.

of stars and galaxies. We shall see how careful observation and logical thinking can extend our knowledge and understanding to realms far beyond our limited daily experiences. Although they have been refined and elaborated, many concepts (such as the magnitude scale) come to us from the people who first discovered how the human mind can unlock the secrets of the universe.

Summary

. Thousands of years ago, many cultures were gathering useful observations of the apparent positions of the Sun, Moon, and planets against the background of the stars.

. The phases of the Moon are caused by the shifting relative positions of the Earth, the Moon, and the Sun.

The Moon completes one orbit around the Earth with respect to the stars in a sidereal month averaging 27.3 days.

The Moon completes one cycle of phases (one orbit around the Earth with respect to the Sun) in a synodic month, or lunar month, averaging 29.5 days.

The lengths of the sidereal and synodic month vary slightly because of the Sun's gravitational pull.

. Ancient astronomers made great progress in determining the sizes and relative distances of the Earth, Moon, and Sun.

Around 200 BC, Erathosthenes measured the Earth's size by comparing the position of the Sun in the sky at the same moment in two different locations.

Soon after 300 BC, Aristarchus of Samos attempted to measure the distances from the Earth to the Sun and the Moon, using the Moon's phases. He also estimated the relative sizes of the Earth, Sun, and Moon.

. A lunar eclipse occurs when the Sun and Moon are both on the line of nodes at full moon. A solar eclipse occurs when the Sun and Moon are both on the line of nodes at new moon. The line of nodes is the line where the planes of the Earth's orbit and the Moon's orbit intersect.

The gravitational pull of the Sun causes the line of nodes gradually to shift its orientation with respect to the stars.

A fairly detailed knowledge of the Moon's orbit around the Earth is needed for an understanding of the intervals between eclipses.

. Depending on the exact relative positions of the Sun, Moon, and Earth, lunar eclipses may be penumbral, partial, or total, and solar eclipses may be partial, annular, or total.

The shadow of an object has two parts: the umbra where the light source is completely blocked, and the penumbra where the light source is only partially blocked.

As the Earth turns, the Moon's umbra races along an eclipse path over the Earth's surface during a total solar eclipse.

. Using an interval called the saros, it is possible to group total solar eclipses into families and to predict the time of the next eclipse in each family.

. Astronomers still use a system for denoting apparent brightness of objects in the sky that was invented by Hipparchus around 160 BC.

Very bright objects have negative values of magnitude (or brightness); very dim objects have large positive values of magnitude.

Review questions

1 Sketch a diagram of the relative positions of the Earth, Moon, Sun, and Voyager 1 when the photograph in Figure 3-2 was taken.

*2 What is the phase of the Moon if it **(a)** rises at 3 AM? **(b)** is crossing the upper meridian at midnight? **(c)** sets at 9 PM? At what time does **(d)** the full moon set? **(e)** the first quarter moon rise? **(f)** the last quarter moon cross the upper meridian?

3 How many more sidereal months than synodic months are there in a year? Why?

4 Which type of eclipse, lunar or solar, do you think most people have seen? Why?

5 Is it possible for a total eclipse of the Sun to be followed three months later by a lunar eclipse? Why?

6 Can one ever observe an annular eclipse of the Moon? Why?

7 In his novel *King Solomon's Mines* author H. Rider Haggard describes a total solar eclipse that can be seen in both South Africa and in the British Isles. Is such an eclipse possible? Why or why not?

Advanced questions

*8 During a lunar eclipse, does the Moon enter the Earth's shadow from the east or from the west? Explain why.

9 If the Moon revolved about the Earth in the same orbit but in the opposite direction, would the synodic month be longer or shorter than the sidereal month? Explain why.

*10 During an occultation or "covering up" of Jupiter by the Moon, an astronomer notices that it takes the Moon's edge 90 seconds to cover Jupiter's disk completely. If the Moon's motion is assumed to be uniform and the occultation was "central" (i.e., center over center), find the angular diameter of Jupiter.

Discussion questions

11 Suppose that you were an ancient astronomer given the task of building a Stonehenge-type observatory-temple to be aligned with the apparent positions and phases of the Moon. Which directions and alignments might you consider important? What sort of observations would be needed to determine these directions? How long would it take to collect the necessary data?

12 Describe the cycle of lunar phases that would be observed if the Moon moved about the Earth in an orbit perpendicular to the plane of the Earth's orbit. Is it possible for solar and lunar eclipses to occur under these circumstances?

For further reading

Cornell, J. *The First Stargazers*. Scribner's, 1981.

Gingerich, O. "Aristarchus of Samos: A Report on a Symposium." *Sky & Telescope,* Nov. 1980, p. 376.

Krupp, E. *Echoes of the Ancient Skies*. Harper & Row, 1983.

Kundu, M. "Observing the Sun during Eclipses." *Mercury,* July/Aug. 1981, p. 108.

Menzel, D., and Pasachoff, J. "Solar Eclipse: Nature's Superspectacular." *National Geographic,* Aug. 1970.

Norton, O. "Repeating Eratosthenes' Observations." *Mercury,* May/June 1974, p. 14.

Sagan, C. "The Shores of the Cosmic Ocean." In Sagan, C. *Cosmos*. Random House, 1980. *An excellent recounting of Eratosthenes' experiment to measure the size of the Earth.*

Whitaker, E. "Why Is the Brightest Lunar Crater Named Aristarchus?" *Sky & Telescope*, Nov. 1978, p. 380.

4 Gravitation and the motions of the planets

Apollo 11 leaving the Moon
The lunar module Eagle returns from the Moon after completing the first successful manned lunar landing. This photograph was taken from the command module Columbia in which the astronauts returned to Earth. All of the orbital maneuvers to dock the Eagle and the Columbia and then set course for Earth were based on Newtonian mechanics and Newton's law of gravity. These same physical principles are used by astronomers to understand a wide range of phenomena from the motions of double stars to the rotation of the entire galaxy. (NASA)

Attempts to explain the motions of the planets led astronomers to an understanding of gravitation. Ancient astronomers believed that the heavens rotate around a stationary Earth. We begin this chapter with a look at the complicated assumptions needed to explain the detailed motions of the planets in such an Earth-centered view of the universe. We then discuss Copernicus's Sun-centered explanation of the universe, which provides simpler general explanations but does not help in explaining the detailed motions of the planets. We learn how Kepler's noncircular orbits for the planets improved and simplified the Sun-centered explanation, and how Galileo's telescopic observations supported the idea that the Earth is one of the planets that all move around the Sun. Next we see how these descriptive successes led to Newton's laws of gravity and motion, which provide a very compact explanation and description of the motions of planets, comets, and satellites. Finally, we take a brief look at the new ideas about space, time, and gravitation introduced in this century by Einstein.

It is not obvious that the Earth moves around the Sun. Indeed, our daily experience strongly suggests that the opposite is true. The daily rising and setting of the Sun, Moon, and stars could lead us to believe that the entire cosmos revolves about the Earth at the center of the universe. That was just what most observers did believe for thousands of years. 51

Ancient astronomers invented geocentric cosmology to explain planetary motions

Ancient Greek astronomers were among the first to leave a written record of their attempts to explain just how the universe works. Their universe consisted of the Earth and those objects that can be seen with the naked eye: the Sun, the Moon, the stars, and five planets. Most Greeks also assumed that the universe revolves about the Earth, thus developing a **geocentric cosmology.** ("Geocentric" means earth-centered. A "cosmology" is a theory of the universe.)

The Greeks (and other ancient cultures) knew of five planets: Mercury, Venus, Mars, Jupiter, and Saturn. The planets are very obvious in the night sky because they slowly shift their positions from night to night with respect to the background of the "fixed" stars in the constellations. Indeed, the word "planet" comes from a Greek term meaning "wanderer." Furthermore, some of the planets are very bright objects in the night sky. At its maximum brilliancy, Venus is 16 times brighter than the brightest star.

Simple observations of the planet's positions against the stars from night to night soon make it clear that the planets do not revolve uniformly about the Earth in concentric circles. Explaining the motions of the five planets was one of the main challenges facing the astronomers of antiquity. It was not easy to develop a comprehensive geocentric theory of the universe.

As seen from Earth, the planets wander primarily across the twelve constellations of the **zodiac.** As mentioned in Chapter 2, these constellations circle the sky in a continuous band centered on the ecliptic. If you follow a planet as it travels across the zodiac from night to night, you find that the planet usually moves slowly eastward against the background stars. This is called **direct motion.** Occasionally, however, the planet seems to stop and then back up for several weeks or months. This occasional westward movement is called **retrograde motion.** These motions are much slower than the daily rotation of the sky caused by the Earth's rotation. Both direct and retrograde motions are best detected by mapping the position of a planet against the background stars from night to night over a long period of observation. An example is the path of Mars from Autumn 1977 through Spring 1978 shown in Figure 4-1.

Figure 4-1 The path of Mars, 1977–1978
From Fall 1977 through Spring 1978, Mars moved across the constellations of Gemini and Cancer. From mid-December through the end of February, Mars's motion was retrograde.

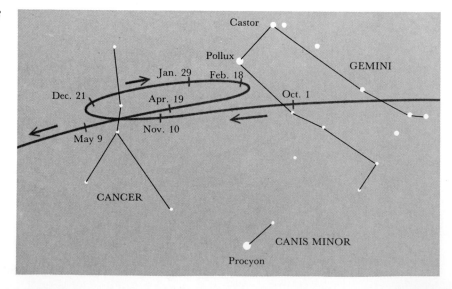

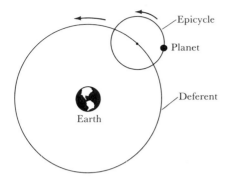

Figure 4-2 A geocentric explanation of planetary motion
Each planet revolves about an epicycle, which in turn revolves about a deferent centered approximately on the Earth. As seen from Earth, the speed of the planet on the epicycle alternately adds to or subtracts from the speed of the epicycle on the deferent thus producing periods of direct and retrograde motion.

The Greeks tried different theories to account for retrograde motion and the exact loops that the planets trace out against the background stars. One of the most successful ideas was proposed by Hipparchus and elaborated by the last of the great Greek astronomers, Ptolemy, who lived in Alexandria during the second century AD. The basic concept is sketched in Figure 4-2. Each planet is assumed to revolve about a small circle called an **epicycle,** which in turn revolves about a larger circle called a **deferent,** that is approximately centered on the Earth. As viewed from Earth, the epicycle moves eastward along the deferent, and both circles rotate in the same direction (counterclockwise in Figure 4-2).

Most of the time, the motion of the planet on its epicycle adds to the eastward motion of the epicycle on the deferent. Thus the planet is seen to be in direct (eastward) motion against the background stars throughout most of the year. However, when the planet is on the part of its epicycle nearest the Earth, the motion of the planet along the epicycle subtracts from the motion of the epicycle along the deferent. This makes the planet appear to slow and then to halt its usual eastward movement among the constellations, even going backward for a few weeks or months. This concept of epicycles and deferents does provide a general explanation of the retrograde loops that the planets execute.

Using the wealth of astronomical data in the library at Alexandria, including records of planetary positions covering hundreds of years, Ptolemy deduced the sizes of the epicycles and deferents and the rates of rotation needed to produce the recorded paths of the planets. After years of laborious work, Ptolemy assembled his calculations in thirteen volumes collectively called the *Almagest.* The positions and paths of the Sun, Moon, and planets were described and predicted with unprecedented accuracy. The *Almagest* became the astronomer's bible. For over a thousand years, Ptolemy's cosmology endured as a useful description of the workings of the heavens.

Eventually, however, things began going awry. Tiny errors and inaccuracies that were unnoticeable in Ptolemy's day compounded and multiplied over the years, especially with regard to precession. Fifteenth century astronomers made some cosmetic adjustments to the Ptolemaic system. However, the system became less and less satisfying as more complicated and arbitrary details were added to keep it consistent with the observed motions of the planets.

Nicolaus Copernicus devised the first comprehensive heliocentric cosmology

Imagine driving on a freeway at high speed. As you pass a slowly moving car, it appears to move backward even though it is traveling in the same direction as your car. This sort of observation inspired the ancient Greek astronomer Aristarchus to suggest a more straightforward explanation of retrograde motion—one in which all planets, including the Earth, revolve about the Sun. The retrograde motion of Mars, for example, occurs when the Earth overtakes and passes Mars, as shown in Figure 4-3. In Aristarchus's day, however, the idea of a moving Earth seemed incompatible with explanations of other phenomena at the Earth's surface. Almost 2000 years elapsed before someone had the insight and determination to work out the details of a **heliocentric** (Sun-centered) **cosmology.** That person was a Polish lawyer, physician, economist, monk, and artist named Nicolaus Copernicus. Especially gifted in mathematics, Copernicus turned his attention to astronomy in the early 1500s.

Copernicus realized that, using a heliocentric perspective, he could determine which planets are closer to the Sun than the Earth and which are farther away. Because Mercury and Venus are always observed fairly

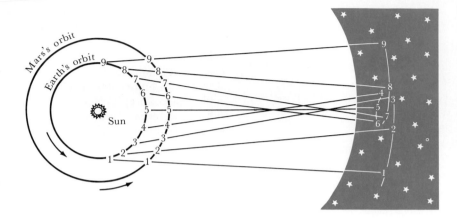

Figure 4-3   A heliocentric explanation of planetary motion
The Earth travels around the Sun more rapidly than Mars. Consequently, as the Earth overtakes and passes this slower-moving planet, Mars appears to move backward for a few months.

near the Sun in the sky, Copernicus concluded that their orbits are smaller than Earth's. The other visible planets—Mars, Jupiter, and Saturn—can be seen in the middle of the night, when the Sun is far below the horizon, which can occur only if the Earth comes between the Sun and a planet. Copernicus concluded that the orbits of Mars, Jupiter, and Saturn are larger than the Earth's orbit.

Astronomers say that Mercury and Venus are **inferior planets** because their orbits are smaller than the Earth's. Mars, Jupiter, and Saturn are called **superior planets** because their orbits are bigger than the Earth's. Other, dimmer superior planets—Uranus, Neptune, and Pluto—were discovered after the telescope came into use.

It is often useful to specify various points on a planet's orbit as sketched in Figure 4-4. These points help us identify certain geometrical

Figure 4-4   Planetary configurations
It is often useful to specify key points along a planet's orbit as shown in these diagrams. These points identify specific geometrical arrangements between the Earth, a planet, and the Sun.

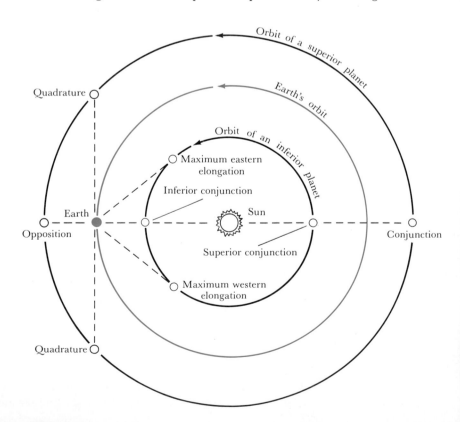

arrangements or **configurations** between the Earth, another planet, and the Sun. For example, when Mercury or Venus is between the Earth and the Sun, we say the planet is at **inferior conjunction.** When Mercury or Venus is on the opposite side of the Sun, we say that it is at **superior conjunction.**

The angle between the Sun and a planet as viewed from the Earth is called the planet's **elongation.** At **maximum eastern elongation,** an inferior planet is as far east of the Sun as it can be. At such times, the planet appears above the western horizon after sunset and is often called an "evening star." Similarly, at **maximum western elongation,** Mercury or Venus is as far west of the Sun as it can possibly be and rises before the Sun, gracing the predawn sky as a "morning star."

When a superior planet is behind the Sun, the planet is at **conjunction** (with elongation of 0°). When it is exactly opposite the Sun in the sky, the planet is at **opposition** (with elongation of 180°). Finally, when the planet's elongation is 90°, it is at **quadrature.**

It is not difficult to determine when a planet happens to be located at one of the key positions sketched in Figure 4-4. For example, when Mars is at opposition, it crosses the upper meridian at midnight. Although it is easy to follow a planet as it moves from one configuration to another, these observations do not immediately provide relevant data about the planet's actual orbit around the Sun. The Earth, from which we make the observations, is also moving. Realizing this, Copernicus was careful to distinguish between two characteristic time intervals, or **periods,** of each planet. The **synodic period** is the time that elapses between two successive identical configurations (as seen from the Earth)—from one opposition to the next, for example, or from one conjunction to the next. The **sidereal period** is the true orbital period of a planet, the time it takes the planet to complete one orbit of the Sun.

Whereas the synodic period of a planet can be determined by observing the sky, the sidereal period must be calculated. Copernicus figured out how to do this (see Box 4-1) and obtained the results shown in Table 4-1.

Box 4-1 Sidereal and synodic periods

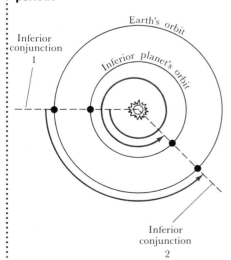

Earth's orbit

Inferior conjunction 1

Inferior planet's orbit

Inferior conjunction 2

Consider an inferior planet orbiting the Sun as shown in the diagram. Let P be the sidereal period of the planet. Let S be the synodic period of the planet. Let E be the sidereal period of the Earth (which Copernicus knew to be equal to $365\frac{1}{4}$ days).

The rate at which the Earth moves around its orbit is $360°/E$ (nearly 1° per day). Similarly, the rate at which the inferior planet moves along its orbit is $360°/P$.

During the inferior planet's synodic period, the Earth covers an angular distance of $S(360°/E)$ around its orbit. In that same time, the inferior planet has covered an angular distance of $S(360°/P)$. Note, however, that the inferior planet has gained one full lap on the Earth. One lap corresponds to 360°, and thus

$$S\left(\frac{360°}{E}\right) + 360° = S\left(\frac{360°}{P}\right)$$

which can be rewritten as

$$\frac{1}{P} = \frac{1}{E} + \frac{1}{S}$$

(continued)

(Box 4-1, continued)

A similar analysis for a superior planet yeilds

$$\frac{1}{P} = \frac{1}{E} - \frac{1}{S}$$

For example, consider Jupiter, whose synodic period is 398.88 days, or 1.092 years. (When astronomers express a time interval in years, they mean Earth years of $365\frac{1}{4}$ days). Of course, $E = 1$ year, exactly. Thus

$$\frac{1}{P} = 1 - \frac{1}{1.092} = 0.084 = \frac{1}{11.9}$$

and we see that it takes 11.9 years for Jupiter to complete one full orbit of the Sun.

TABLE 4-1 *The synodic and sidereal periods of the planets*

Planet	Synodic period	Sidereal period
Mercury	116 days	88 days
Venus	584 days	225 days
Earth	——	1.0 year
Mars	780 days	1.9 years
Jupiter	399 days	11.9 years
Saturn	378 days	29.5 years

Knowing the sidereal periods of the planets, Copernicus devised a straightforward geometrical method of determining the distances of the planets from the Sun. (The mathematical details, which involve some trigonometry, are described in Box 4-2.) His answers turned out remarkably close to the modern values, as shown in Table 4-2.

Box 4-2 Copernicus's method of determining the sizes of orbits

To determine the size of an inferior planet's orbit, Copernicus measured the angle (α) between the Sun and the planet at greatest elongation. As shown in diagram *a*, the triangle formed by the Earth, the inferior planet, and the Sun then contains a right angle. The hypotenuse of the triangle has a length of one astronomical unit (1 AU), and hence the radius of the inferior planet's orbit is equal to $\sin \alpha$, also measured in AU.

Determining the size of a superior planet's orbit is slightly more complicated. First, note the date on which the planet is at opposition. (In diagram *b*, the planets are in positions 1.) After a few months, note the date on which the planet appears at quadrature. (The planets have now moved to positions 2.) Using the number of days that have elapsed, determine the angle β, which is the distance the Earth (moving at roughly 1° per day) has traveled. Knowing the sidereal period of the planet (see Table 4-1), you can determine the angle γ in the same way.

The triangle formed by the Sun, the Earth, and the planet at quadrature contains a right angle. In addition, the short side of the triangle is exactly 1 AU long, and the angle ($\beta - \gamma$) is now known. The hypotenuse of the triangle, which is the radius of the superior planet's orbit, therefore has a length of $1/\cos (\beta - \gamma)$, also measured in AU.

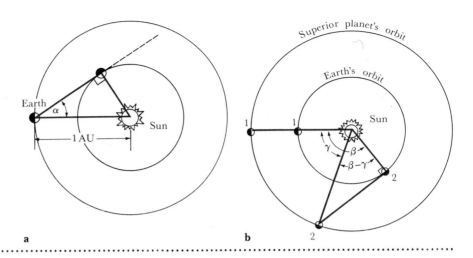

<table>
<tr><td rowspan="2">***TABLE 4-2*** *Average distances of the planets from the Sun (in astronomical units)*</td></tr>
</table>

TABLE 4-2 Average distances of the planets from the Sun (in astronomical units)

Planet	Copernicus	Modern
Mercury	0.38	0.39
Venus	0.72	0.72
Earth	1.00	1.00
Mars	1.52	1.52
Jupiter	5.22	5.20
Saturn	9.07	9.54

From Tables 4-1 and 4-2, it is apparent that, the farther a planet is from the Sun, the longer it takes to travel around its orbit.

Copernicus compiled his ideas and calculations into a book entitled *De Revolutionibus Orbium Celestium* ("On the Revolutions of the Celestial Spheres") that was published in 1543, the year of his death. Although he assumed that the Earth travels around the Sun along a circular path, he found that perfectly circular orbits cannot accurately describe the paths of the other planets. Copernicus had to add an epicycle to each planet to account for the slight variation in speed along its orbit. Thus, according to Copernicus, each planet revolves around a small epicycle, which in turn orbits the Sun along a circular path.

The placement of the Sun at the center of the universe was such a revolutionary proposal that one Renaissance astronomer, Tycho Brahe, tried to test Copernicus's ideas with detailed observations of the sky.

We know that, when we walk from one place to another, nearby objects appear to shift their positions against the background of more distant objects. Tycho Brahe argued that, if Copernicus was correct, nearby stars should shift slightly against the background stars as the Earth orbits the Sun.

Tycho Brahe spent his lifetime making accurate observations of the positions of the stars and planets, achieving an unprecedented level of precision over the years. Yet, in spite of this high degree of accuracy, he could not detect any shifting of star positions. He therefore concluded that Copernicus was wrong.

Actually, the stars are so far away that naked-eye observations could not possibly detect the tiny shifting of star positions that has now been confirmed with telescopic observations. Nevertheless, Tycho Brahe's astronomical records were destined to play an important role in the development of a heliocentric cosmology. Upon his death in 1601, many of Tycho Brahe's charts and books fell into the hands of his gifted assistant, Johannes Kepler.

Johannes Kepler proposed noncircular paths of the planets about the Sun

Until Kepler's time, astronomers had assumed that heavenly objects move in circles. Circles were considered the most perfect and harmonious of all geometric shapes. God was assumed to be in heaven along with the stars and planets and, since God is perfect, he would use only circles to control the motions of the planets. Kepler doubted such arguments and his first major contribution to astronomy was the suggestion that noncircular curves might fit planetary orbits.

Kepler turned from circles to ovals. For years he tried in vain to fit ovals to the orbits of planets about the Sun. Then he began working with a slightly different curve called an **ellipse.**

An ellipse can be constructed with a loop of string, two thumbtacks, and a pencil, as shown in Figure 4-5. An ellipse has two **foci:** each thumbtack is at a **focus.** The longest diameter across an ellipse passes through both foci and is called the **major axis.** Half of that distance is called the **semimajor axis,** whose length is usually designated by the letter a.

To Kepler's delight, the ellipse turned out to be the curve he had been searching for. He published this discovery in 1609 along with other material in a book known today as *New Astronomy*. This important discovery is now called **Kepler's first law** and is stated as follows:

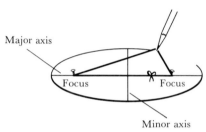

Major axis

Focus   Focus

Minor axis

Figure 4-5 The construction of an ellipse
An ellipse can be drawn with a pencil, a loop of string, and two thumb tacks as shown in this diagram. If the string is kept taut, the pencil traces out an ellipse. The two thumb tacks are located at the two foci of the ellipse.

The orbit of a planet about the Sun is an ellipse with the Sun at one focus.

Kepler also realized that planets do not move at a uniform speed along their orbits. A planet moves most rapidly when it is nearest the Sun, at a point on its orbit called **perihelion.** A planet moves most slowly when it is farthest from the Sun, at a point called **aphelion.**

After much trial and error, Kepler discovered a way to describe how fast a planet moves along its orbit. This discovery, now called the **law of equal areas** or **Kepler's second law,** is illustrated in Figure 4-6. For instance, suppose it takes 30 days for a planet to go from point A to point B. During that time, the line joining the Sun and the planet sweeps out a nearly triangular area. Kepler discovered that the line joining the Sun and the planet sweeps out an equal area during any other 30-day interval. In other words, if the planet also takes a month to go from point C to point D, then the two shaded segments in Figure 4-6 are equal in area. Kepler's second law can be stated as follows:

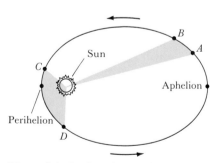

B

A

Sun

C

Aphelion

Perihelion

D

Figure 4-6 Kepler's first and second laws
According to Kepler's first two laws, every planet travels around the Sun along an elliptical orbit with the Sun at one focus in such a way that the line joining the planet and the Sun sweeps out equal areas in equal intervals of time.

A line joining a planet and the Sun sweeps out equal areas in equal intervals of time.

This discovery was also published in *New Astronomy* in 1609.

Kepler was fascinated by the many harmonious relationships in the motions of the planets. His writing is filled with intriguing speculations, including musical scores intended to represent celestial music that the planets make as they travel along their orbits.

One of Kepler's later discoveries stands out because of its impact on future developments. Now called the **harmonic law** or **Kepler's third law,** it gives a relationship between the sidereal period of a planet and the length of its semimajor axis:

The squares of the sidereal periods of the planets are proportional to the cubes of their semimajor axes.

If a planet's sidereal period (P) is measured in years and the length of its semimajor axis (a) is measured in astronomical units, then Kepler's third law is simply stated as

$$P^2 = a^3$$

The length of the semimajor axis is actually the average distance between a planet and the Sun. Using data from Tables 4-1 and 4-2, we can demonstrate Kepler's third law as shown in Table 4-3. This relationship can also be displayed on a graph as in Figure 4-7.

Figure 4-7 Kepler's third law
On this graph, the squares of the periods of the planets (P^2) are plotted against the cubes of their semimajor axes (a^3) The fact that the points fall along a straight line is verification of Kepler's discovery that $P^2 = a^3$.

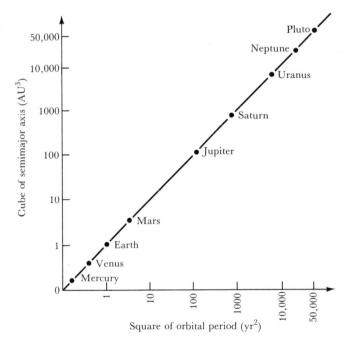

TABLE 4-3 A demonstration of Kepler's third law

Planet	Sidereal period P (in years)	Semimajor axis a (in AU)	P^2	a^3
Mercury	0.24	0.39	0.06	0.06
Venus	0.61	0.72	0.37	0.37
Earth	1.00	1.00	1.00	1.00
Mars	1.88	1.52	3.53	3.51
Jupiter	11.86	5.20	140.7	140.6
Saturn	29.46	9.54	867.9	868.3

This third law was published in 1619. It is testimony to Kepler's genius that his three laws are rigorously obeyed in any situation where two objects orbit each other under the influence of their mutual gravitational attraction. Throughout this book, we shall see that Kepler's laws have a wide range of practical applications. Kepler's laws are obeyed not only by planets circling the Sun, but also by artificial satellites orbiting the Earth and by two stars revolving about each other in a double star system.

Galileo's telescopic discoveries strongly supported a heliocentric cosmology

While Kepler was making rapid progress in northern Europe, an Italian physicist was making equally dramatic observational discoveries in southern Europe. Galileo Galilei did not invent the telescope, but he was the first person to point one of the new devices toward the sky and publish his observations. He saw things that no one had ever dreamed of. He saw mountains on the Moon and sunspots on the Sun. He also discovered that Venus exhibits phases (see Figure 4-8).

After only a few months of observation, Galileo noticed that the apparent size of Venus as seen through his telescope was related to the planet's phase. The planet appears smallest at gibbous phase and largest at crescent phase. There is also a correlation between the phases of Venus and the planet's angular distance from the Sun. These relationships clearly support the conclusion that Venus goes around the Sun (see Figure 4-9).

Galileo also discovered four moons orbiting Jupiter. Figure 4-10 shows a portion of his astronomical records. Astronomers soon realized that these four moons obey Kepler's third law: the cube of a moon's distance from Jupiter is proportional to the square of its orbital period about the planet.

These telescopic observations constituted the first fresh influx of fundamentally new astronomical data in almost 2000 years. In contradiction to prevailing opinions, these discoveries strongly suggested a heliocentric view of the universe. The Roman Catholic church attacked his ideas because they were not reconcilable with certain passages in the Bible or with the writings of Aristotle and Plato. Nevertheless, there was no turning back. Although Galileo was condemned to spend his latter years under house arrest "for vehement suspicion of heresy," his revolutionary ideas would soon inspire a sickly English boy who was born on Christmas day of 1642, less than one year after Galileo died. The boy's name was Isaac Newton.

Figure 4-8 [above]   Crescent Venus
As seen from Earth, Venus exhibits phases that are correlated with the planet's angular size. The planet appears largest at crescent phase (maximum angular diameter: 65.2 arc sec), as shown here. Venus appears considerably smaller at gibbous phase (minimum angular diameter: 9.5 arc sec). (Palomar Observatory)

Figure 4-9 [right]   The changing appearance of Venus
The phases of Venus are correlated with the planet's angular size and its angular distance from the Sun as sketched in this diagram. These observations clearly support the idea that Venus orbits the Sun.

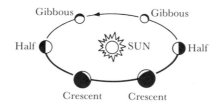

Isaac Newton formulated a description of gravity that accounts for Kepler's laws and almost explains the motions of the planets

Until the mid-seventeenth century, virtually all mathematical astronomy was entirely empirical, characterized by trial and error. From Ptolemy to Kepler, essentially the same approach was used. Astronomers would work directly from data and observations, adjusting ideas and calculations until they finally came out with the right answers.

Isaac Newton introduced a new approach. He made three assumptions, now called Newton's **laws of motion,** about the nature of reality. Newton showed that Kepler's three laws follow logically from the laws of motion and a formula for the force of gravity that he derived from observations and his laws. He used this formula to describe almost exactly the observed orbits of the Moon, comets, and other objects in the solar system.

The first assumption, known as **Newton's first law,** states that

A body remains at rest, or moves in a straight line at a constant speed, unless acted upon by an outside force.

This idea tells us that a force must be acting on the planets. If there were no "outside" force acting on planets, they would leave their curved orbits and move away from the Sun along straight-line paths at constant speeds. This does not happen, and so Newton concluded that the continuous action of this force confines the planets to their elliptical orbits.

Isaac Newton did not invent the idea of gravity. The educated seventeenth-century person had a vague appreciation of the fact that some force pulls things down to the ground. Newton, however, gave us a precise description of the action of gravity. Using his first law, Newton mathematically proved that the force acting on each of the planets is aimed directly at the Sun. This discovery led Newton to suspect that the force pulling a falling apple straight down to the ground is the same as the force on the planets that is always aimed straight at the Sun.

Newton's second assumption, called **Newton's second law,** describes what a force does. Imagine an object floating in space. If you push on the object, it will begin to move. If you continue to push, the object's speed will continue to increase—in other words, it will **accelerate.**

Acceleration is the rate at which a velocity changes. For example, consider an apple falling from a tree (ignoring the effects of air friction). Initially, at the moment the apple's stem breaks, the apple's speed is zero. After one second, its downward speed is 32 feet per second. After two seconds, the apple's speed is 64 feet per second. After three seconds, the speed is 96 feet per second. Because the apple's speed increases by 32 feet per second for each second of free fall, the acceleration is 32 feet per second per second, or 32 ft/sec². In other words, the Earth's gravity produces a constant acceleration of 32 ft/sec² (9.8 m/sec²) downward, toward the center of the Earth.

Newton's second law says that the acceleration of an object is proportional to the force acting on the object. In other words, the harder you push on an object, the greater is the resulting acceleration.

This law is succinctly stated as an equation. Consider an object whose **mass** is m. If a force F acts on the object, it will experience an acceleration a such that

$$F = ma$$

It is important not to confuse the concepts of mass and weight. The **mass** of an object is a measure of the total amount of material in the object and is measured in grams or kilograms. For example, the mass of

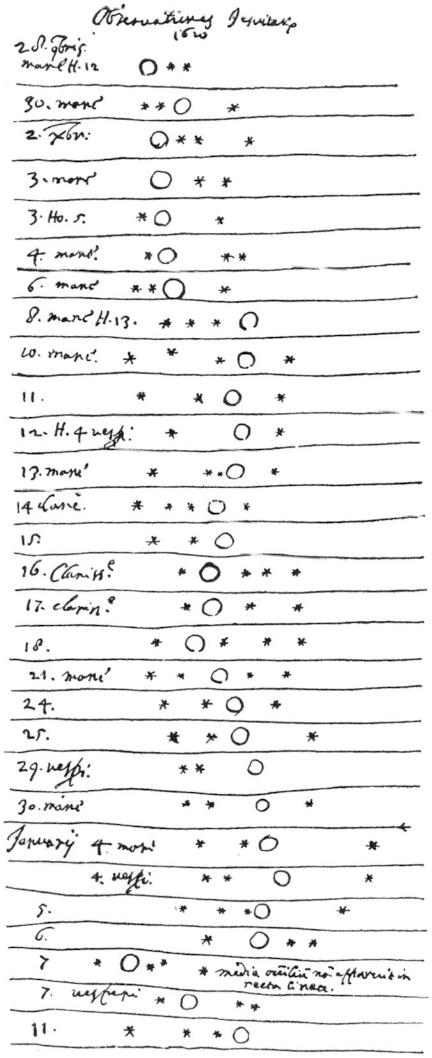

Figure 4-10 Galileo's observations of Jupiter's moons
In 1610, Galileo discovered four "stars" that move back and forth across Jupiter from one night to the next. Galileo concluded that these are four moons that orbit Jupiter much as our Moon orbits the Earth. (Yerkes Observatory)

the Sun is 2×10^{33} g. The mass of a hydrogen atom is 1.7×10^{-24} g. The mass of the author of this book is 75 kg. The Sun, a hydrogen atom, and the author have these masses regardless of where they happen to be in the universe.

In contrast, **weight** is the force with which an object presses down on the ground and is measured in pounds, newtons, or dynes. We can use Newton's second law to relate mass and weight. We have seen that the acceleration due to the Earth's gravity is 32 ft/sec^2, or 9.8 m/sec^2. From the second law ($F = ma$), the force with which the author presses down on the ground is

75 kg $\times$ 9.8 m/sec^2 = 735 newtons = 165 pounds

But that answer is correct only when the author is standing on the Earth. On the Moon he would weigh less. On Jupiter he would weigh more. In space, he would have no weight at all; he would be "weightless." Nevertheless, under all these circumstances, he would always have exactly the same mass. Thus we see that mass is an inherent property of matter unaffected by details of the environment. Whenever we describe the properties of planets, stars, or galaxies, we speak of their masses, never of their weights.

Newton's final assumption, called **Newton's third law,** is the famous statement of action and reaction:

Whenever one body exerts a force on a second body, the second body exerts an equal and opposite force on the first body.

For example, if you weigh 165 pounds, you are pressing down on the floor with a force of 165 pounds. Newton's third law tells us that the floor is pushing up against your feet with a force of 165 pounds. (If it were not, you would fall through the floor.) In the same way, Newton realized that, because the Sun is exerting a force on each planet to keep it in orbit, each planet must be exerting an equal and opposite force on the Sun.

Figure 4-11   Major contributors to modern gravitational theory
Each of these people is responsible for a major contribution or breakthrough in our understanding of gravity. Because of their discoveries and insights, we now send spacecraft to the planets and probe the geometry of the universe.

Copernicus
(1473–1543)

Kepler
(1571–1630)

Galileo
(1564–1642)

Using his three laws and Kepler's three laws, Newton succeeded in formulating a very general statement describing the nature of the force that keeps the planets in their orbits. The force is called **gravity** and Newton's **universal law of gravitation** is stated as follows:

Two bodies attract each other with a force that is directly proportional to the product of their masses and inversely proportional to the square of the distance between them.

In other words, if two objects in space have masses m_1 and m_2 and are separated by a distance r, then the gravitational force F between these two masses is

$$F = G\frac{m_1 m_2}{r^2}$$

If the masses are measured in kilograms and the distance in meters, then the force is measured in newtons. If the masses are in grams and the distance in centimeters, then the force is in dynes. In this formula, G is a number called the **universal constant of gravitation.** From laboratory experiments, G is determined to have the value

$$G = 6.67 \times 10^{-8} \text{ dyne cm}^2/\text{g}^2$$
$$= 6.67 \times 10^{-11} \text{ newton m}^2/\text{kg}^2$$

Using his law of gravity, Newton found that he could mathematically prove the validity of Kepler's three laws. For example, whereas Kepler discovered by trial and error that $P^2 = a^3$, Newton demonstrated mathematically that this equation follows logically from his law of gravity. (Some details are presented in Box 4-3.) Newton also discovered new features of orbits around the Sun. For example, his equations soon led him to conclude that the orbit of an object around the Sun could be any one of a family of curves called **conic sections.**

Newton
(1643–1727)

Einstein
(1879–1955)

Box 4-3   Ellipses and orbits

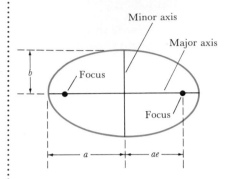

Using his universal law of gravitation and a lot of mathematics, Newton discovered that he could calculate many details of the planets' orbits around the Sun. Here we present some of these details for future reference.

As shown in the diagram, the largest diameter across an ellipse is called the **major axis** and has a length of $2a$. The shortest diameter through the center of the ellipse is called the **minor axis** and has a length of $2b$. The minor axis is the perpendicular bisector of the major axis.

The shape of an ellipse is determined by its **eccentricity,** e. The distance from the center of an ellipse to one focus is ae. Notice that, if $e = 0$, we have a circle.

Kepler's second law tells us that the orbital velocity of a planet varies as it circles the Sun. Newtonian mechanics can be used to calculate the speed of the planet at various points along its orbit. If P is the planet's sidereal period, then the orbital speed (v) at perihelion and aphelion are given by

$$v = \frac{2\pi a}{P}\left(\frac{1 + e}{1 - e}\right)^{1/2} \text{ (at perihelion)}$$

and

$$v = \frac{2\pi a}{P}\left(\frac{1 - e}{1 + e}\right)^{1/2} \text{ (at aphelion)}$$

For example, we have the following data for the Earth:

$a = 1 \text{ AU} = 1.496 \times 10^8 \text{ km}$
$P = 1 \text{ yr} = 3.156 \times 10^7 \text{ sec}$
$e = 0.0167$

where the eccentricity of the Earth's orbit is determined from the annual variation of the Earth–Sun distance. Consequently, the Earth's orbital speed varies from 30.3 km/sec at perihelion to 29.3 km/sec at aphelion.

Although it is commonly said that "the Moon goes about the Earth" or that "the planets revolve around the Sun," such statements are not entirely accurate. When two objects orbit each other, they are actually revolving about a common, stationary point called the **center of mass**. To appreciate this concept, imagine placing a mass (m_1) at one end of a seesaw or teeter-totter and a second mass (m_2) at the other end. The center of mass is the point where you would put the fulcrum to balance the seesaw.

Consider the masses (m_1 and m_2) orbiting each other as shown in the second diagram. The location of their center of mass is given by

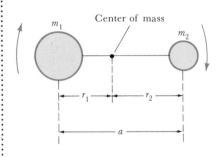

$$m_1 r_1 = m_2 r_2$$

where $r_1 + r_2 = a$ = the distance between the two masses.

The gravitational force between these two masses is given by Newton's law:

$$F = \frac{Gm_1m_2}{a^2}$$

where G is the universal constant of gravitation. From these equations, it is possible to derive the following relationship between the total separation of the two masses (a) and the orbital period (P) with which they revolve about each other:

$$P^2 = \left[\frac{4\pi^2}{G(m_1 + m_2)}\right]a^3$$

This is actually the most general and complete statement of Kepler's third law. In the case of the solar system, the mass of the Sun is thousands of times greater than the masses of any of the planets. Consequently, for all practical purposes, $(m_1 + m_2)$ is equal to the mass of the Sun, $M_\odot$, which is 2×10^{33} g. Thus, Kepler's third law becomes

$$P^2 = \left(\frac{4\pi^2}{GM_\odot} \right) a^3$$

If P is measured in years and a is measured in astronomical units, then the quantity in the parentheses equals 1, and we have $P^2 = a^3$.

A **conic section** is any curve that you get by cutting a cone with a plane, as shown in Figure 4-12. You can get **circles** and **ellipses** by slicing all the way through the cone. You can also get two "open" curves called **parabolas** and **hyperbolas.** For example, comets hurtling toward the Sun from the depths of space sometimes follow parabolic orbits.

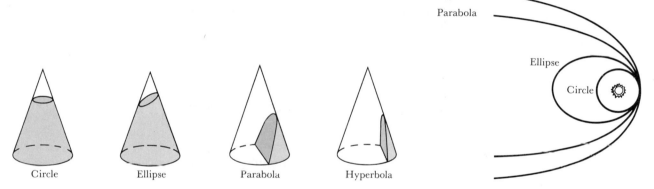

Figure 4-12   Conic sections
A conic section is any one of a family of curves obtained by slicing a cone with a plane, as shown in this diagram. The orbit of one body about another can be any one of these curves: a circle, an ellipse, a parabola, or a hyperbola.

Newton's ideas and methods turned out to be incredibly successful in a wide range of situations. The orbits of the planets and their satellites could be calculated with unprecedented precision. Using Newton's laws, mathematicians proved that the Earth's axis of rotation must precess because of the gravitational pull of the Moon and the Sun on the Earth's equatorial bulge (recall Figure 2-8). Similarly, all the details of the Moon's orbit (recall Box 3-1) could be demonstrated mathematically.

In addition to providing detailed explanations of a variety of known phenomena, Newton's laws and mathematical techniques, collectively called **Newtonian mechanics,** were capable of predicting new phenomena. For example, one of Newton's friends, Edmund Halley, was

Figure 4-13   Halley's comet
This comet orbits the Sun with an average period of about 76 years. During the twentieth century, two appearances of the comet were predicted—one in 1910 and another in 1986. This photograph shows how the comet looked in 1910. (Lick Observatory)

 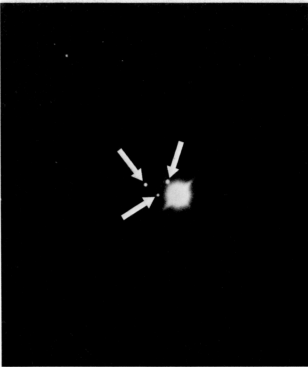

Figure 4-14   Uranus and Neptune
The existence of Neptune (with one moon) was deduced from deviations in the predicted orbit of Uranus (with three moons). This discovery was a major triumph for Newtonian mechanics. Both planets are shown along with some of their satellites. (Lick Observatory)

intrigued by historical records of a comet that was sighted about every 76 years. Using Newton's methods, Halley worked out the details of the comet's orbit and predicted its return in 1758. It was first sighted on Christmas night of 1757, and to this day the comet bears Halley's name.

The perhaps most dramatic success of Newton's ideas involved the discovery of the eighth planet from the Sun. The seventh planet, Uranus, was discovered accidentally by William Herschel in 1781 during a telescopic survey of the sky. Fifty years later, however, it was clear that Uranus was not following its predicted orbit. Two mathematicians, John Couch Adams in England and U. J. Leverrier in France, independently calculated that the deviations of Uranus from its orbit could be explained by the gravitational pull of a yet unknown, more distant planet. They each predicted that the planet would be found at a certain location in the constellation of Aquarius. A brief telescopic search on September 23, 1846, revealed Neptune less than 1° from the calculated position. Although sighted with a telescope, Neptune was really discovered with pencil and paper.

Over the years, Newton's ideas were used in successful predictions and explanations of many phenomena, and thus Newtonian mechanics became the cornerstone of modern physical science. Even today, as we send astronauts to the Moon and spacecraft to the outer planets, Newton's equations are used to calculate orbits and trajectories.

There was, however, one instance in which Newtonian mechanics was not quite in agreement with observations. During the mid-1800s, Leverrier pointed out that Mercury was not following its predicted orbit. As the planet moves along its elliptical orbit, the orbit itself rotates (or precesses) as shown in Figure 4-15. Specifically, Mercury's perihelion shifts by an amount that cannot be explained with Newtonian mechanics. Although the effect is very small (the unexplained excess rotation of

Mercury's major axis is only 43 arc sec per century), all attempts to account for this phenomenon met with failure. Nevertheless, Newton's laws proved so eminently successful in all other situations that most physicists and astronomers simply ignored this tiny discrepancy.

It is a testament to Newton's genius that his three laws were precisely the three basic ideas needed for a full understanding of the motions of the planets. In this way, Isaac Newton brought a new dimension of elegance and sophistication to our understanding of the workings of the universe. That is probably where things would have remained had it not been for the insight, vision, and genius of Albert Einstein.

Albert Einstein's theory states that gravity affects the shape of space and the flow of time

In the cosmologies of most ancient civilizations, the Earth and its inhabitants occupied a special place at the center of the universe. Copernicus's cosmology made the Earth just one of a number of planets orbiting the Sun. This new approach proved very fruitful. Within less than a century, Kepler's accurate description of planetary orbits led directly to Newton's universal law of gravitation. Astronomers and philosophers began to explore the possibility that even the Sun occupies no special location in the universe—that our Sun is merely one of the many stars scattered through space.

Albert Einstein introduced an even more powerful perspective. Einstein believed that the fundamental laws of the universe depend neither on our location in space nor on our motion through space. In other words, the laws of physics should be the same, whether we happen to be sitting on the Earth or moving through space at 95 percent of the speed of light.

Einstein began his work soon after 1900 with a new explanation of the phenomena of electricity and magnetism. The basic properties of electricity and magnetism had been summarized in four equations formulated in 1865 by the great Scottish physicist James Clerk Maxwell. Maxwell's equations are the basis of **electromagnetic theory,** which today has a wide range of practical applications from television sets to microwave ovens. Maxwell's electromagnetic theory predicts differing effects depending upon the motion of electric charges and magnets. For example, a moving electric charge creates a magnetic "field," whereas a stationary electric charge does not. Such predictions imply some absolute or fixed framework of space, within which an object is either stationary or moving. Einstein's goal was to eliminate this assumption of absolute space from electromagnetic theory. He wanted to obtain equations that would depend only upon the *relative* motions of the observer and the electric charges or magnets. He succeeded in 1905 and published his results in a famous paper entitled "On the Electrodynamics of Moving Bodies."

Although Einstein's electromagnetic theory was very intriguing to many physicists, they were disturbed because it introduced some revolutionary concepts in physics. To achieve his goal, Einstein had to abandon the idea of absolute time as well as the idea of absolute space. Whether two events are simultaneous depends upon the position and motion of the observer, said Einstein. Although event A occurs before event B for one observer, a different observer moving at a very high speed relative to the first observer might find that event B occurs before event A. Many people found this concept hard to understand and harder still to accept.

Einstein's theory led to other surprising conclusions: the length of an object depends upon its speed relative to the observer who measures it, and the rate at which a clock ticks similarly depends upon its speed

Figure 4-15 The advance of Mercury's perihelion
As Mercury moves along its orbit, the orbit rotates. Consequently, Mercury traces out a rosette figure about the Sun. This effect is greatly exaggerated in this diagram. The excess rotation of Mercury's orbit (beyond that predicted by Newtonian mechanics) amounts to only 43 arc sec per century.

relative to the observer who times it. The body of knowledge that describes all these effects is called the **special theory of relativity.**

The relativity of distances and time intervals is easily noticeable only at speeds close to the speed of light. For most phenomena, Newtonian mechanics provides accurate predictions and explanations. However, for extremely high speeds (such as those of subatomic particles in particle accelerators), physicists found that special relativity is needed to explain and understand the results of experiments. Using very accurate "atomic clocks," physicists have even confirmed that a clock in a moving jet plane seems to an Earth-based observer to tick slightly more slowly than an identical clock mounted on the Earth. Einstein's theory has been supported by every experiment designed to test it.

Details of special relativity need not concern us here (they will be discussed in Box 24-1), but the newly discovered relativity of space and time was very important in Einstein's desire to understand gravity. Einstein found Newton's description of gravity disappointing because it assumes that space and time are absolute.

In 1916, Albert Einstein formulated a new theory of gravity called the **general theory of relativity.** The basic idea in general relativity is that the mass of an object alters the properties of space and time around the object. Einstein eliminated the idea of a "force of gravity" from his theory. According to the general theory of relativity, gravity causes space to become curved and time to slow down.

These concepts are even more difficult to understand than the concepts of special relativity. One useful analogy is to imagine that space near a massive object (such as the Sun) becomes curved like the surface in Figure 4-16. Imagine a steel ball rolled along this surface. Far from the "well" that represents the Sun, it would move in a straight line along the flat surface. If it passes near the well, however, it will curve toward the well. If it is moving at an appropriate speed with respect to the well, it may move in an orbit around the sides of the well. The curvature of the surface makes the ball follow a curved path toward the well.

It is hard to imagine the three-dimensional curvature of space, but its effects are much the same. Far from the Sun, planets and comets should travel along nearly straight-line paths. Near the Sun, planets and comets travel along curved paths because space itself is curved. The general theory of relativity predicts that only very strong gravitational fields have easily detectable effects upon space and time in their vicinity.

One of the first things Einstein did with his new theory was to calculate the orbits of the planets. With only one tiny exception, general relativity gave almost exactly the same answers as Newtonian theory. Mercury is the only planet to pass so close to the massive Sun (a strong gravitational field) that the curvature of space produces an effect significantly different from that predicted by Newtonian mechanics. Indeed, throughout most of the solar system, the curvature of space is so slight that Einstein's equations are essentially equivalent to Newtonian calculations. Einstein's theory, however, unlike Newtonian mechanics, did predict that Mercury's orbit should be a precessing ellipse, thereby explaining a phenomenon that had frustrated astronomers for half a century.

Einstein also predicted new phenomena to test the validity of his theory. He calculated that light rays passing near the surface of the Sun should appear to be deflected from their straight-line paths because the space through which they are moving is curved. In other words, gravity should bend light rays, an effect not predicted by Newtonian mechanics because light has no mass.

Figure 4-16   The gravitational curvature of space
According to Einstein's general theory of relativity, space becomes curved near a massive object such as the Sun. This diagram shows a two-dimensional analogy of the shape of space around a massive object.

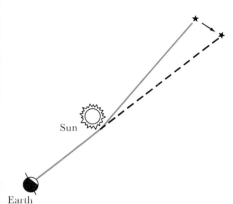

Figure 4-17 The gravitational deflection of light
Light rays passing near the Sun are deflected from their straight-line paths. By relativistic standards, the Sun's gravity is very weak; the maximum angle of deflection is only 1.75 arc sec for a light ray grazing the Sun's surface.

Figure 4-17 shows a beam of light from a star passing by the Sun and continuing down to the Earth. Because the light ray is bent, the star appears shifted from its usual location. At most, the position of a star is shifted by 1.75 arc sec for a light ray grazing the Sun's surface.

This prediction was first tested during a total solar eclipse in 1919. During the precious moments of totality, when the Moon blocked out the blinding solar disk, astronomers succeeded in photographing the stars around the Sun. Careful measurements revealed that the stars were shifted from their usual positions by an amount consistent with Einstein's theory. General relativity had passed another important test.

Einstein made a third prediction. Because gravity causes time to slow down, he stated that clocks on the first floor of a building would tick slightly more slowly than clocks in the attic that are farther from the Earth, as sketched in Figure 4-18. This prediction was tested by analyzing light from very compact stars (whose surface gravity is very strong) and by using extremely accurate clocks first developed in 1960. Again Einstein was proven correct.

During the past several decades, these and similar tests of general relativity have been repeated over and over. In every case, general relativity has been confirmed. This remarkable theory now stands as our most precise and complete description of gravity.

It is important to emphasize that Einstein did not prove Newton wrong. Rather, Einstein demonstrated that Newtonian mechanics is accurate only when applied to low speeds and weak gravity. If extremely high speeds or powerful gravitational fields (such as those of neutron stars and black holes) are involved, only a relativistic calculation will give correct answers.

Einstein's general theory of relativity evoked a sensation when it was proposed. Newtonian gravitation had been overthrown by a radically different approach that worked better.

After the initial excitement, however, interest in general relativity rapidly waned. No one could imagine places where the curvature of space might have a major effect. It was generally believed that Newtonian mechanics would suffice in virtually all circumstances; relativistic theory offered only a little more accuracy, as in the case of Mercury's orbit. The complex mathematics of relativity seemed burdensome, and Newton's simpler approach gave reasonable precision in almost all conceivable circumstances.

In recent years, there has been a reawakening of interest in general relativity, primarily because of dramatic advances in our understanding of the evolution of stars. As we shall see in later chapters, we now have a reasonably complete picture of how stars are born, what happens to them as they mature, and where they go when they die. In particular, it appears that the most massive dying stars are doomed to collapse completely upon themselves, thereby producing some of the most bizarre objects in the universe, black holes. Gravity around one of these massive stellar corpses is so strong that it punches a hole in the fabric of space. Because of this intense gravitational effect, black holes can be described and discussed only in terms of general relativity.

As we set our sights on understanding of the universe as a whole, we again must turn to general relativity. All of the matter in all of the stars and galaxies is responsible for the overall curvature or shape of space. From the viewpoint of general relativity, it is therefore reasonable to ask about the actual shape of the universe. We shall see that the answer suggests the ultimate fate of the cosmos.

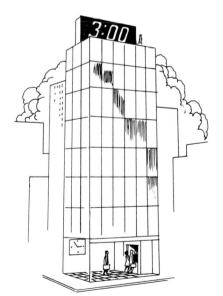

Figure 4-18 The gravitational slowing of time
A clock on the ground floor of a building is closer to the Earth than a clock at a higher elevation. According to general relativity, the clock on the ground floor should tick more slowly than the clock on the roof.

Summary

- Ancient astronomers believed that the Earth is at the center of the universe and invented a complex system of epicycles and deferents to explain direct and retrograde motions of the planets.

- A heliocentric (Sun-centered) theory simplifies the general explanation of planetary motions.

  In a heliocentric system, the Earth is one of the planets. The other planets are divided into inferior planets (with orbits smaller than that of the Earth) and superior planets (with orbits larger than that of the Earth).

  The sidereal period of a planet (measured with respect to the stars) is its true orbital period; its synodic period is measured with respect to the Earth and Sun (for example, from one opposition to the next).

- Ellipses describe the paths of the planets around the Sun much more accurately than circles. Kepler's three laws give important details about elliptical orbits.

- The invention of the telescope led to new discoveries that supported a heliocentric view of the universe.

- Newton based his explanation of the universe on three assumptions or laws of motion. Kepler's laws and extremely accurate descriptions of planetary motions can be deduced from Newton's laws and his universal law of gravitation.

  The mass of an object is a measure of the amount of matter in the object; its weight is a measure of the force with which the gravity of some other object pulls on it.

  In general, the path of one object about another (such as a comet about the Sun) is one of the curves called conic sections: a circle, an ellipse, a parabola, or a hyperbola.

- Although Newtonian mechanics accurately describes and predicts numerous phenomena, Einstein's relativistic theories are more accurate where extremely high speeds or intense gravitational fields are involved.

  The special theory of relativity was designed to eliminate the notion of absolute space and time. The theory leads to the conclusion that measurements of distance and time intervals are affected by the motion of the observer.

  The general theory of relativity explains that gravity causes space to be curved and time to slow down.

  Both the special and general theories have been supported by the results of every experiment designed to test them.

Review questions

1 At what configuration (superior conjunction, greatest eastern elongation, etc.) would it be best to observe Mercury or Venus with an Earth-based telescope? At what configuration would it be best to observe Mars, Jupiter, or Saturn? Explain your answers.

2 Is it possible for an object in the solar system to have a synodic period of exactly one year? Explain your answer.

*3 A line joining the Sun and an asteroid is found to sweep out 5.2 square AUs of space in 1983. How much area is swept out in 1984? In five years?

*4 A comet moves in a highly elongated orbit about the Sun with a period of 1000 years. What is the comet's average distance from the Sun? What is the farthest it can get from the Sun?

*5 Suppose you discovered an alien solar system in which a planet circles a star once every 2 years at an average distance of 4 AU. How does the mass of this star compare with that of our Sun?

6 Give an everyday example of each of Newton's three laws.

*7 Suppose that the Earth were moved to a distance of 10 AU from the Sun. How much stronger (or weaker) would the Sun's gravitational pull be on the Earth?

Advanced questions

*8 Is it possible for a planet's sidereal period to equal its synodic period? Would this planet be closer to the Sun than the Earth is or farther away? Is there a planet that nearly fits this description?

*9 A satellite is said to be in a "synchronous" orbit if it appears to always remain over the exact same spot on Earth. At what distance from the center must such a satellite be placed into orbit? Explain why the orbit must be in the plane of the Earth's equator.

10 Look up the dates of various greatest eastern and western elongations for Mercury in a year of your choice. Does it take longer to go from eastern to western elongation or vice versa? Why do you suppose this is the case?

11 Look up orbital information for the four largest moons of Jupiter. Demonstrate that these data obey Kepler's third law.

Discussion questions

12 Which planet would you expect to exhibit the greatest variation in apparent brightness as seen from Earth? Explain your answer.

13 Use two thumb tacks, a loop of string, and a pencil to draw several ellipses. Describe how the shape of the ellipse varies as you change the distance between the thumb tacks.

For further reading

On renaissance astronomy:
Banville, J. *Kepler: A Novel.* Godine, 1981.
Christianson, G. *This Wild Abyss.* Free Press, 1978.
Cohen, O. "Newton's Discovery of Gravity." *Scientific American,* March 1981.
Durham, F., and Purrington, R. *Frame of the Universe.* Columbia U. Press, 1983.
Gingerich, O. "Copernicus and Tycho." *Scientific American,* Dec. 1973.
Kuhn, T. *The Copernican Revolution.* Harvard U. Press, 1957.
Rogers, E. *Astronomy for the Inquiring Mind.* Princeton U. Press, 1960.
Wilson, C. "How Did Kepler Discover His First Two Laws?" *Scientific American,* Mar. 1972.

On relativity and Einstein:
Baker, A. *Modern Physics and Antiphysics.* Addison-Wesley, 1970.
Gardner, M. *The Relativity Explosion.* Vintage, 1976.
Hoffman, B., and Dukas, H. *Albert Einstein: Creator and Rebel.* NAL Plume Books, 1972.
Kaufmann, W. *Relativity and Cosmology.* Harper & Row, 1977.

5 Light, optics, and telescopes

Cerro Tololo Inter-American Observatory
Astronomers prefer to build observatories on isolated mountaintops far from city lights, where the air is dry, stable, and cloudfree. This aerial view shows the Cerro Tololo Inter-American Observatory in Chile, about 400 km (250 miles) north of Santiago. The large dome houses the 4-meter CTIO telescope shown in Figure 5-13. Several smaller telescopes also share this mountaintop which is at an elevation of 2200 meters (7200 feet) above sea level. (CTIO)

Until very recently, our knowledge about the universe was based almost entirely on the visible light gathered by telescopes. In this chapter, we learn that light is a form of electromagnetic radiation, and that many other forms of such radiation exist. We discuss the two major types of optical telescopes: refractors and reflectors. Then we learn about more exotic telescopes that detect radio waves and other electromagnetic radiation arriving at the Earth. Finally, we take a quick look at the orbiting observatories that are expected to produce a wide range of new information about the universe in the coming decades.

The telescope has been the single most important tool of astronomy. Using a telescope, we can see extremely faint objects in space far more clearly than we can with the naked eye. Telescopes have played a major role in revealing the universe since Galileo first saw craters on the Moon four centuries ago.

Traditionally, telescopes detect visible light. Light from a distant object is brought by either lenses or mirrors to a focus where the resulting image is viewed or photographed. Recently, however, astronomers have built telescopes that detect nonvisible forms of light such as X rays and radio waves. To appreciate these developments and to understand how telescopes work, we must first learn something about the basic properties of light.

Light is electromagnetic radiation and is characterized by its wavelength

Galileo and Newton made important contributions to our modern understanding of light as well as to our theories of gravity and mechanics. Galileo made one of the first attempts to measure the speed of light. Whatever may be the nature of light, it does seem to travel somehow from a source to our eyes, and it travels very swiftly. We see a distant event before we hear the accompanying sound. Does light move instantly from one place to another, or does it have some measureable speed of travel? In the early 1600s, Galileo performed an experiment to measure the speed of light. He stood on one hilltop at night while an assistant stood on another hilltop at a known distance; each of them held a shuttered lantern. First Galileo opened the shutter of his lantern. As soon as the assistant saw the flash of light, he opened his own lantern. Using his pulsebeat, Galileo attempted to measure the time between opening his shutter and seeing the light from the assistant's lantern. From the known distance and time, he could then compute the speed at which light traveled to the distant hilltop and back again.

Galileo found that the measured time did not increase noticeably, no matter how distant the assistant was stationed. He concluded that light travels so rapidly that slow human reactions make it impossible to measure the speed in this fashion. The first reliable measurement was made in 1675 by a Danish astronomer, Olaus Roemer, who studied the motion of Jupiter's moons around that planet (details are discussed in Box 17-1). Much more accurate measurements were made in the mid-1800s, most of them using very elaborate laboratory equipment. From these experiments, we now know that the speed of light in a vacuum is about 3×10^8 m/sec (186,000 miles per second).

A pioneering breakthrough in understanding light came from a simple experiment performed by Isaac Newton in the late 1600s. Newton was familiar with what he called "the celebrated Phenomenon of Colours": a beam of sunlight passing through a glass prism is spread out into the colors of the rainbow (see Figure 5-2). This rainbow, called a **spectrum,** suggested to Newton that white light is actually a mixture of all colors. Passage of white light through a prism separates the light into its component colors. (Most earlier observers had assumed that the colors are somehow added to the light by the prism.) Newton elegantly confirmed his explanation by showing that a second prism can be positioned in such a way as to recombine the colors into white light. From his many experiments with light, Newton concluded that light is composed on indetectably tiny particles. A rival explanation was proposed by the Dutch astronomer Christian Huygens, who suggested that light travels in the form of waves rather than particles.

The English physicist Thomas Young confirmed the wave nature of light in 1801. Young demonstrated that the shadows of objects in light of

Figure 5-1   Palomar Observatory
This dome on Palomar Mountain in southern California houses the world's second largest optical telescope. The telescope's mirror, which collects and focuses starlight, is 5.1 m (200 in.) in diameter. Most major observatories are located far from city lights on mountain tops renowned for stable air and cloud-free skies. (Palomar Observatory)

Figure 5-2   A prism and a spectrum
When a beam of white light passes through a glass prism, the light is broken into a rainbow-colored band called a spectrum. The numbers on the right side of the spectrum indicate wavelengths as described in the text.

7000 Å
6000 Å
5000 Å
4000 Å

a single color are not crisp and sharp. Instead, the boundary between illuminated and shaded areas is overlaid with patterns of closely spaced dark and light bands. These patterns are similar to the patterns produced by water waves passing the edge of a reef or barrier in the ocean. Although some of these results could be explained by the particle theory of light, experimentalists soon obtained an overwhelming range of evidence for the wavelike behavior of light.

Further insight into the wave character of light came from calculations by the Scottish physicist James Clerk Maxwell in the 1860s. As mentioned briefly in Chapter 4, Maxwell had succeeded in describing all the basic properties of electricity and magnetism in four equations. By combining these equations, Maxwell demonstrated that electrical and magnetic effects should travel through space in the form of waves. Furthermore, he proved that these waves should travel with a speed of about 3×10^8 m/sec. Therefore, Maxwell suggested that these waves do exist and are observed as light. His suggestion was soon confirmed by a variety of experiments. Light consists of perpendicular, oscillating electrical and magnetic fields, as shown in Figure 5-3. Because of its electrical and magnetic properties, light is called **electromagnetic radiation.** The distance between two successive wave crests is called the **wavelength** of the light (see Figure 5-3).

More than a century elapsed between Newton's experiments with a prism and the confirmation of the wave nature of light. A primary reason for this delay is that the wavelength of visible light is extremely short, less than a thousandth of a millimeter. To express these tiny distances conveniently, scientists use a unit of length called the angstrom (abbreviated Å and named after the Swedish physicist A. J. Ångström), where $1 \text{ Å} = 10^{-8}$ cm. Experiments demonstrated that visible light has wavelengths covering the range from about 4000 Å for violet light to

Figure 5-3   Electromagnetic radiation
All forms of light consist of oscillating electrical and magnetic fields that move through empty space at a speed of 3×10^{10} cm/sec. The distance between two successive crests is called the wavelength of the light and is usually designated by the lower case Greek letter λ (lambda).

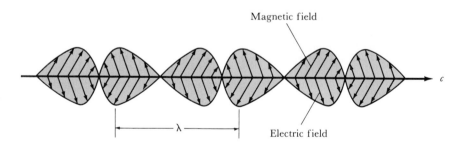

about 7000 Å for red light. Intermediate colors of the rainbow have intermediate wavelengths (see Figure 5-2).

Maxwell's equations, however, placed no restrictions on the wavelengths of electromagnetic radiation. In other words, elecromagnetic waves should exist with wavelengths both longer and shorter than the 4000–7000 Å range of visible light. Researchers began to look for invisible forms of light, forms of light to which the cells of the human retina do not respond.

The British astronomer William Herschel discovered radiation just beyond the red end of the visible spectrum. In 1888, the German physicist Heinrich Hertz succeeded in producing light with wavelengths of a few centimeters, now known as **radio waves.** In 1895, Wilhelm Röntgen invented a machine that produces light with a wavelength shorter than 100 Å, now known as **X rays.** Modern versions of Röntgen's machine are today found in medical and dental offices.

Figure 5-4   The electromagnetic spectrum
The full array of all types of electromagnetic radiation is called the electromagnetic spectrum. It extends from the shortest-wavelength gamma rays to the longest-wavelength radio waves. Visible light forms only a tiny portion of the full electromagnetic spectrum.

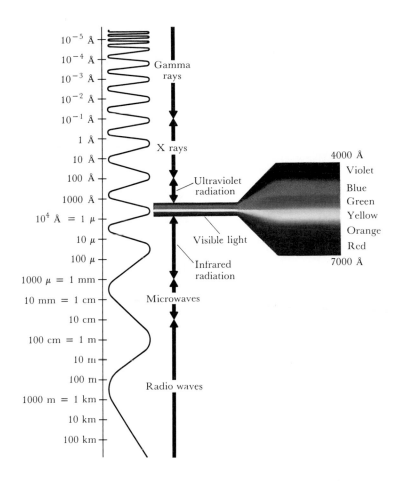

Over the years, radiation in many other wavelength ranges was discovered. Visible light is only a tiny fraction of the full extent of possible wavelengths collectively called the **electromagnetic spectrum.** As shown in Figure 5-4, the electromagnetic spectrum stretches from the longest-wavelength radio waves to the shortest-wavelength gamma rays. For example, at wavelengths slightly longer than visible light, **infrared radiation,** covers the range from about 7000 Å to 1 mm. From roughly 1 mm to 10 cm is the range of **microwaves,** beyond which is the domain of **radio waves.**

At wavelengths shorter than that of visible light, **ultraviolet radiation** extends from about 4000 Å down to 100 Å. X rays have wavelengths between about 100 Å and 0.1 Å, beyond which is the domain of **gamma rays.**

It should be noted that these arbitrary divisions of the electromagnetic spectrum are simply rough boundaries that allow us to identify certain broad sections of the electromagnetic spectrum. (Scientists working with various kinds of electromagnetic radiation have adopted slightly different ways of denoting wavelengths, as described in Box 5-1).

Although these various types of electromagnetic radiation share many basic properties (for example, they all travel at the speed of light), they interact very differently with matter. Your body is transparent to X rays but not to visible light; your eyes respond to visible light but not to gamma rays; your radio detects radio waves but not ultraviolet light. Consequently, astronomers use fundamentally different kinds of telescopes in these various wavelength ranges. For example, a radio telescope that detects radio waves from space is very different from either an X-ray telescope or an ordinary optical telescope. Because optical telescopes are the most common and familiar astronomical tool, we shall discuss them in detail before turning to the exotic instruments that reveal the nonvisible sky.

Box 5-1   Wavelength, frequency, and energy

The visible portion of the electromagnetic spectrum extends from roughly 4000 to 7000 Å, and astronomers working with radiation in that range prefer to express wavelengths in angstroms:

$$1 \text{ Å} = 10^{-8} \text{ cm} = 10^{-10} \text{ m}$$

At slightly longer wavelengths, astronomers working in the infrared usually prefer to express wavelengths in microns (μ). A *micron* is a millionth of a meter and thus

$$1 \mu = 10^{-4} \text{ cm} = 10^{-6} \text{ m}$$
$$1 \mu = 10^{4} \text{ Å}$$

The micron is also called the micrometer (μm).

The microwave portion of the electromagnetic spectrum extends from roughly 1 mm to 10 cm, and astronomers commonly use millimeters or centimeters when discussing wavelengths in this range.

Many of the astronomers who built the first radio telescopes had training and backgrounds in electrical engineering and electronics, and astronomers using radio waves often speak of frequency rather than wavelength. The **frequency** of light is simply the number of wave crests passing by a given point in one second. Because light travels at the speed $c = 3 \times 10^{10}$ cm/sec, the frequency (ν) of light is related to its wavelength (λ) by

$$\lambda = \frac{c}{\nu}$$

For example, hydrogen gas in space emits radio waves with a wavelength of 21.12 cm. The frequency of this radiation is therefore

$$\nu = \frac{c}{\lambda} = \frac{3 \times 10^{10}}{21.12} = 1.420 \times 10^{9} \text{ sec}^{-1}$$

$$= 1{,}420{,}000{,}000 \text{ cycles per second}$$

One cycle per second is also called a hertz (abbreviated Hz) in honor of the physicist who discovered radio waves. Often it is convenient to use the prefix mega (meaning million, abbreviated M) or kilo (meaning thousand, abbreviated k). Thus the frequency of the hydrogen emission is 1420 MHz, or 1420 megahertz.

Throughout the ultraviolet region on the short-wavelength side of visible light, astronomers continue to use angstroms to express wavelengths. However, for X rays and gamma rays, astronomers prefer to speak of the energy that the radiation carries rather than its wavelength.

A beam of light can be regarded as a stream of tiny packets of energy called **photons.** (Twentieth-century physicists have revived the particle theory of light; actually, light is now regarded as having both wave and particle properties.) The energy (E) that a photon carries is directly related to its frequency by the simple equation

$$E = h\nu$$

or equivalently

$$E = \frac{hc}{\lambda}$$

where h is Planck's constant, named after the German physicist who discovered this relationship. Laboratory experiments reveal that $h = 6.625 \times 10^{-27}$ erg sec. The erg is a unit of energy.

The energy of X-ray and gamma-ray photons is most conveniently expressed in a different unit of energy called the **electron volt** (eV):

$$1 \text{ eV} = 1.6 \times 10^{-12} \text{ erg}$$

Thus, Planck's constant may be written as

$$h = 4.135 \times 10^{-15} \text{ eV sec}$$

Consider X rays of wavelength 10 Å. The energy carried by such an X-ray photon is

$$E = \frac{hc}{\lambda} = \frac{(4.135 \times 10^{-15})(3 \times 10^{10})}{10 \times 10^{-8}} = 1.24 \times 10^3 \text{ eV}$$

$$= 1.24 \text{ keV}$$

where the prefix kilo (k) is used to indicate thousands of electron volts.

The X-ray domain of the electromagnetic spectrum, which goes from roughly 100 Å down to 0.1 Å, corresponds to a photon energy range of about 124 eV to 124 keV. Photons with energies higher than about 124 keV are called gamma rays.

A refracting telescope uses a lens to concentrate incoming starlight at a focus

Although light travels at about 3×10^{10} cm/sec in a vacuum, it travels at a slower speed through a dense substance such as glass. The abrupt slowing of light entering a piece of glass is analogous to a person walking from a boardwalk onto a sandy beach: her pace suddenly slows as she steps from the smooth pavement into the sand. Upon exiting a piece of glass, light resumes its original speed, just as a person stepping back onto the boardwalk easily resumes her original pace.

Furthermore, a light ray is bent as it passes from one transparent medium into another at an oblique angle to the surface between the media. This phenomenon is called **refraction** and is caused by the change in the speed of light. Imagine driving a car from a smooth pavement onto a sandy beach. If the car approaches the beach at an angle, one of the front wheels is slowed by the sand before the other, causing the car to veer from its original direction.

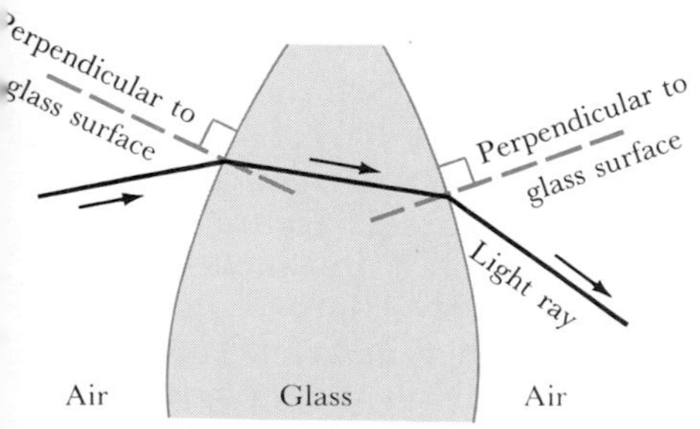

To describe the refraction of a light ray entering a piece of glass, imagine drawing a perpendicular to the surface of the glass at the point where the light strikes the glass, as shown in Figure 5-5. As a light ray goes from a less-dense medium (such as air or a vacuum) into a more-dense medium (such as glass), the light is always bent toward the perpendicular direction. It is bent toward the perpendicular just as a car driving obliquely onto the sand veers toward the direction perpendicular to the pavement–beach boundary. Upon emerging from the other side of a piece of glass, light resumes its original high speed, and the light ray is bent away from the perpendicular direction. The exact amount of refraction depends on the speed of light in the glass, which in turn

Figure 5-5 Refraction
A light ray entering a piece of glass is bent toward the perpendicular. A light ray leaving a piece of glass is bent away from the perpendicular.

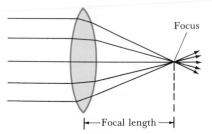

Figure 5-6   A convex lens
A convex lens causes parallel light rays to converge to a focus. The distance from the lens to the focus is the focal length of the lens.

depends on the chemical composition of the glass. Different kinds of glass produce slightly different amounts of refraction.

Because of the refracting property of glass, a convex lens (that is, a lens that is fatter in the middle than at the edges) causes incoming light rays to converge to a point called the **focus,** as shown in Figure 5-6. If the source of light is extremely far away, then the incoming light rays are parallel, and they come to a focus at a specific distance from the lens called the **focal length** of the lens.

Stars and planets are so far away that light rays from these objects are essentially parallel. Consequently, a lens always focuses light from an astronomical object as shown in Figure 5-6. An image of the astronomical object is formed at the focus, and a second lens can be used to magnify and examine this image. Such an arrangement of two lenses is called a **refracting telescope,** or **refractor** (see Figure 5-7). The large-diameter, long-focal-length lens at the front of the telescope is called the **objective lens.** The smaller, short-focal-length lens at the rear of the telescope is called the **eyepiece lens.** Galileo used a small refracting telescope for astronomical observations very soon after the device had been invented in Holland for viewing distant objects on the Earth.

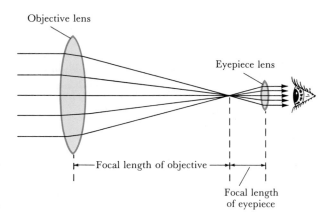

Figure 5-7   A refracting telescope
A refracting telescope consists of a large, long-focal-length objective lens and a small, short-focal-length eyepiece lens that magnifies the image formed at the focus of the objective lens.

The **magnification,** or **magnifying power,** of a refracting telescope is equal to the focal length of the objective lens divided by the focal length of the eyepiece lens. For example, if the objective of a telescope has a focal length of 100 cm, and the eyepiece has a focal length of $\frac{1}{2}$ cm, then the magnifying power of the telescope is 200 (usually written as 200×).

If you build a telescope using only the instructions given so far, you will probably be disappointed with the results. You will see stars surrounded by fuzzy, rainbow-colored halos. This optical defect is called **chromatic aberration** and exists because a lens bends different colors of light through different angles, just as a prism does (recall Figure 5-2).

By adding small amounts of various chemicals to a vat of molten glass, an optician can manufacture different kinds of glass. The speed of light varies slightly from one kind of glass to another, and opticians use this fact to correct for chromatic aberration. Specifically, a thin lens can be mounted just behind the main objective lens of a telescope as diagrammed in Figure 5-8. By carefully choosing two different kinds of glass for these two lenses, the optician can ensure that different colors of light come to a focus at the same point.

Chromatic aberration is the most severe of a host of optical problems

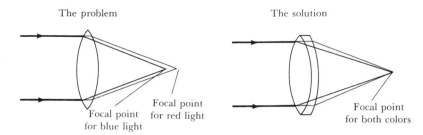

The problem

The solution

Focal point
for blue light

Focal point
for red light

Focal point
for both colors

Figure 5-8   Chromatic aberration
A single lens suffers from a defect called chromatic aberration in which different colors of light have slightly different focal lengths. This problem is corrected by adding a second lens made of a different kind of glass than that of the first lens.

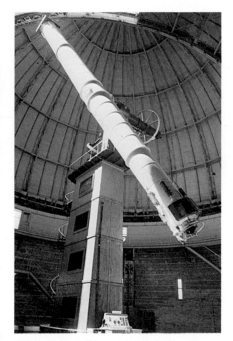

Figure 5-9   A large refracting telescope
This giant refracting telescope was built in the late 1800s and is housed at Yerkes Observatory near Chicago. The objective lens is 102 cm (40 in.) in diameter, and the telescope tube is $19\frac{1}{2}$ m ($63\frac{1}{2}$ ft) long. (Yerkes Observatory)

that must be solved in designing a high-quality refracting telescope. During the nineteenth century, master opticians devoted their lives to overcoming these problems, and several magnificent refractors were constructed in the late 1800s. The largest refracting telescope, completed in 1897, is located at the Yerkes Observatory, not far from Chicago (see Figure 5-9). The objective lens has a diameter of 102 cm (40 in.) and was built under the expert guidance of Alvan Clark. This was not Clark's first attempt at a large telescope. In 1888, he completed a similar instrument for Lick Observatory near San Jose, California. This refractor has an objective lens whose diameter is 91 cm (36 in.), the second largest in the world. Other major refractors are listed in Box 5-2. They all have extremely long focal lengths. For example, the Yerkes refractor has a focal length of 19.35 m ($63\frac{1}{2}$ ft).

Few major new refracting telescopes have been constructed in the twentieth century. There are many reasons for the modern astronomer's lack of interest in this type of telescope. First, because faint light must pass readily through the objective lens, the glass from which the lens is made must be totally free of defects such as bubbles that frequently form when molten glass is poured into a mold. Consequently, the glass for the lens is extremely expensive. Second, glass is opaque to certain kinds of light. Even visible light is dimmed substantially in passing through the thick slab of glass at the front of a refractor, and ultraviolet radiation is largely absorbed by the glass lens. Third, it is impossible to produce such a large lens that is completely corrected to eliminate chromatic aberration. Fourth, it is difficult to support the heavy lens without blocking the path of light into the telescope. All of these problems can be avoided by using mirrors instead of lenses.

Box 5-2   Major refracting telescopes

There are fourteen refractors around the world with objective lenses larger than 65 cm (26 in.) in diameter. They are listed in the table with information such as the name and location of the observatory.

Year completed	Observatory	Location	Objective diameter (cm)	Focal length (cm)
1897	Yerkes Observatory	Williams Bay, Wisconsin	102	1936
1888	Lick Observatory	Mt. Hamilton, California	90	1763
1889	Observatoire de Paris	Meudon, France	83	1616

(continued)

(Box 5-2, continued)

1899	Zentralinstitut für Astrophysik	Potsdam, East Germany	80	1200
1914	Allegheny Observatory	Pittsburgh, Pennsylvania	76	1412
1886	Observatoire de Nice	Mont Gros, France	74	1790
1894	Old Royal Observatory	Greenwich, England	71	848
1896	Archenhold-Sternwarte	Berlin, East Germany	68	2100
1880	Institut für Astronomie	Vienna, Austria	67	1050
1925	Republic Observatory	Johannesburg, South Africa	67	1092
1883	Leander McCormick Observatory	Charlottesville, Virginia	67	991
1873	United States Naval Observatory	Washington, D.C.	66	987
1899	Royal Greenwich Observatory	Herstmonceux, England	66	686
1956*	Mount Stromlo Observatory	Canberra, Australia	66	1080

First used at Johannesburg, South Africa in 1925.

A reflecting telescope uses a mirror to concentrate incoming starlight at a focus

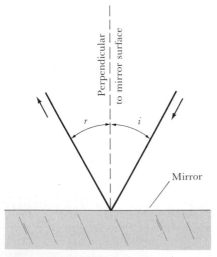

Figure 5-10   Reflection
The angle with which a beam of light approaches a mirror (called the angle of incidence) is always equal to the angle with which the beam is reflected from the mirror (called the angle of reflection). Reflection is accurately described by the equation i = r.

Reflection can be described very simply. To understand reflection, imagine drawing a perpendicular to a mirror's surface at the point where a light ray strikes the mirror, as shown in Figure 5-10. The angle between an arriving (incident) light ray and the perpendicular is always equal to the angle between the reflected ray and the perpendicular. Knowing this, Isaac Newton realized that a concave mirror will cause parallel light rays to converge to a focus as shown in Figure 5-11. The distance between the reflecting surface and the focus is called the **focal length** of the mirror.

An image of a distant object is formed at the focus of a concave mirror. In order to view the image, Newton simply placed a small, flat mirror at a 45° angle in front of the focal point as sketched in Figure 5-12*b*. This secondary mirror deflects the light rays to one side of the **reflecting telescope,** or **reflector,** where the astronomer can place an eyepiece lens to magnify the image. A telescope having this optical design is appropriately called a **Newtonian reflector.** The magnifying power of such a reflecting telescope is calculated in the same way as for a refractor: the focal length of the primary mirror is divided by the focal length of the eyepiece.

Useful modifications of Newton's original design have since been invented. The primary mirrors of many major reflectors are so large that the astronomer can actually sit at the undeflected focal point, directly in front of the primary mirror. The "observing cage" in which the astronomer rides blocks only a small fraction of the incoming starlight. This arrangement is called a **prime focus** (see Figure 5-12*a*).

Another popular optical design, called a **Cassegrain focus,** has the advantage of placing the focal point at a convenient and accessible location. A hole is drilled directly through the center of the primary mirror. A convex secondary mirror placed in front of the original focal point is used to reflect the light rays back through the hole (see Figure 5-12*c*). Alternatively, a series of mirrors can be used to channel the light rays away from the telescope to a remote focal point. Heavy optical equip-

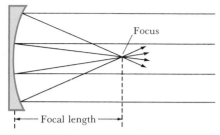

Figure 5-11 A concave mirror
A concave mirror causes parallel light rays to converge to a focus. The distance between the mirror and the focus is the focal length of the mirror.

ment that could not be mounted directly on the telescope is located at the resulting **coudé focus** (see Figure 5-12*d*).

To make a reflector, an optician grinds and polishes a large slab of glass into the appropriate concave shape. The glass is then coated with silver or aluminum or a similar, highly reflective substance. Defects inside the glass such as bubbles or flecks of dirt do not detract from the telescope's effectiveness, as they would in the objective lens of a refracting telescope.

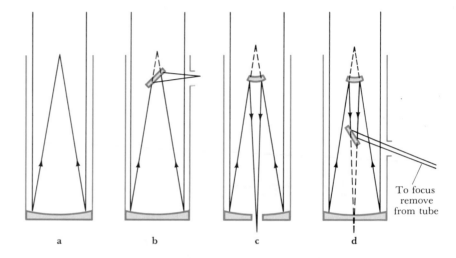

Figure 5-12 Reflecting telescopes
Four of the most popular optical designs for reflecting telescopes are sketched: **(a)** *prime focus,* **(b)** *Newtonian focus,* **(c)** *Cassegrain focus, and* **(d)** *coudé focus.*

Furthermore, reflection is not affected by the wavelength of the light. Therefore, light of all wavelengths is reflected to the same focus without loss of any wavelengths through absorption. (Some problems with absorption and chromatic aberration may arise with the smaller lens used to magnify the image, but the major difficulties caused by the large objective lens in the refractor do not exist in the reflector.) Finally, the mirror can be fully supported by braces on its back, so that a large and heavy mirror can be mounted without much danger of breakage or shape distortion.

Ten reflectors exist with primary mirrors measuring at least 3 m in diameter (see Box 5-3). The largest is located in the Soviet Union, and the second largest is at the Palomar Observatory in southern California.

Box 5-3 Major reflecting telescopes

There are ten reflectors around the world with primary mirrors equal to or larger than 3 m (9.8 ft) in diameter. They are listed in the table with information such as the name and location of the observatory.

Year completed	Observatory	Location	Mirror diameter (meters)
1976	Special Astrophysical Observatory	Zelenchukskaya, U.S.S.R.	6.0
1948	Palomar Observatory	Palomar Mountain, California	5.1
1974	Cerro Tololo Inter-American Observatory	Cerro Tololo, Chile	4.1

(continued)

The Multiple-Mirror Telescope on Mt. Hopkins in Arizona should actually rank third on this list. Its six 1.8-m mirrors have a total area equal to one 4.5-m mirror. In addition, a 4.2-m reflector is under construction at the La Palma Observatory in the Canary Islands.

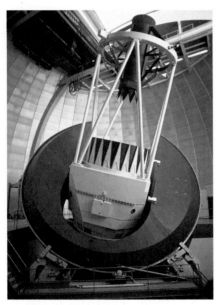

Figure 5-13 The 4-meter telescope at Cerro Tololo
This telescope is located on a mountain top near Santiago, Chile. Its twin is at the Kitt Peak Observatory in Arizona. Both telescopes have been in operation since the early 1970s. (Cerro Tololo Inter-American Observatory)

In the early 1970s, a matching pair of telescopes was built in Arizona and Chile. These two telescopes (see Figure 5-13) allow astronomers to observe the entire sky with essentially the same instrument.

Astronomers strongly prefer large telescopes. A large mirror intercepts and focuses more starlight than does a small mirror (see Figure 5-14). A large mirror therefore produces brighter images and detects fainter stars than does a small mirror. The **light-gathering power** of a telescope is directly related to the area of the telescope's primary mirror. For example, the 200-inch mirror at Palomar Observatory has four times the area of the 100-inch mirror at Mt. Wilson Observatory. Therefore, the Palomar telescope has four times the light-gathering power of the Mt. Wilson telescope.

A large telescope also increases the sharpness of the image and the extent to which fine details can be distinguished. This property is called **resolving power.** With low resolving power, star images are fuzzy and blurred together. With high resolving power, images are sharp and crisp.

The resolving power of a telescope is measured as the angular distance between two adjacent stars whose images can just barely be distinguished under ideal observing conditions. Large modern telescopes—such as those at Palomar, Kitt Peak, and Cerro Tololo—are calculated to have resolving powers better than 0.1 arc sec. In practice, however, this exceptionally high resolving power is never achieved. Turbulence and impurities in the air cause star images to jiggle around, or twinkle. Even

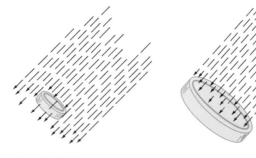

Figure 5-14 Light-gathering power
A large mirror intercepts more starlight than does a small mirror. Large mirrors therefore produce brighter images than do small mirrors.

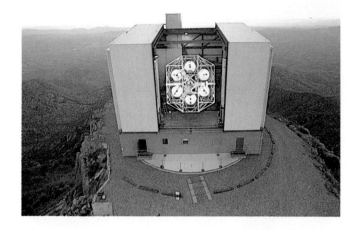

Figure 5-15   The Multiple-Mirror Telescope
This aerial photograph shows the six 1.8-m mirrors that together constitute the first multiple-mirror telescope. The total area of the six mirrors is equal to one 4.5-m mirror. (Multiple Mirror Telescope)

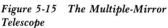
a The problem

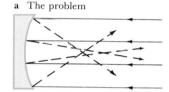
b A solution

One focal point for *all* light rays

c Another solution

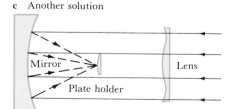

Mirror

Lens

Plate holder

Figure 5-16   Spherical aberration
(a) *Different parts of a spherically concave mirror reflect light to slightly different focal points. This difficulty can be corrected either* **(b)** *by using a parabolic mirror, or* **(c)** *by using a "correcting lens" in front of the mirror.*

through the largest telescopes, a star looks like a tiny circular blob rather than a pinpoint of light.

The angular diameter of a star's image is called the **seeing disk** and is a realistic measure of the best possible resolution that can be achieved. The size of the seeing disk varies from one observatory site to another. At Palomar and Kitt Peak, the seeing disk is roughly 1 arc sec. The best conditions in the world (seeing disk = $\frac{1}{4}$ arc sec) have been reported at the observatory on top of Mauna Kea, a 14,000-foot volcano on the island of Hawaii. Only in the vacuum of outer space could the theoretical resolving power of a large telescope be achieved.

Significant engineering problems are associated with building large reflectors. Very large mirrors (more than about 4 m in diameter) are slabs of glass so heavy that the mirror's shape changes slightly as the telescope is turned toward different parts of the sky. The mirror actually sags under its own weight, thereby detracting from the sharpness of the focus and the quality of the resulting image. New techniques for building thin, light-weight mirrors should help to alleviate this problem.

Another approach is to mount several smaller mirrors together, aimed at the same focal point. The Multiple Mirror Telescope atop Mt. Hopkins in Arizona has six mirrors, each measuring 1.8 m (6 ft) in diameter, mounted together as shown in Figure 5-15. The total light gathering power of this arrangement is equivalent to one $4\frac{1}{2}$-m mirror. The MMT, as it is called, therefore really ranks as the third largest telescope in the world. It has been in operation since 1979, and the design has proven so successful that astronomers around the world are now planning even larger multiple-mirror telescopes.

Just as chromatic aberration plagues refracting telescopes, a defect called **spherical aberration** must be minimized when reflecting telescopes are constructed. At issue is the precise shape of a mirror's concave surface. A spherical surface is easy to grind and polish, but different parts of a spherical mirror have slightly different focal lengths (see Figure 5-16a) resulting in a fuzzy image.

This problem can be eliminated if the mirror's surface is polished to a parabolic shape. A parabola reflects parallel light rays to a common focus (see Figure 5-16b), so many reflecting telescopes have parabolic mirrors.

This is a fine solution as long as the astronomer is not interested in a wide-angle view. Unlike a spherical mirror, parabolic mirrors suffer from

Figure 5-17 The Schmidt Telescope at Palomar

This is one of the largest Schmidt telescopes in the world. Although its mirror has a diameter of 1.8 m (6 ft), the correcting lens at the front of the telescope measures 1.2 m (4 ft) across. An astronomer is shown guiding the telescope as it takes a wide-angle photograph of the sky. (Palomar Observatory)

a different defect called **coma** whereby star images far from the center of the field of view are elongated and look like tiny teardrops.

A telescope with a high-quality, wide-angle field of view uses a spherical mirror to minimize coma. To eliminate spherical aberration, a thin correcting lens is mounted at the front of the telescope (see Figure 5-16*c*). The unique shape of this lens is specifically designed to ensure that all light rays have the same focal point.

This optical arrangement is the basic idea of the **Schmidt telescope,** named after its inventor, Bernhard Schmidt, who built the first prototype in the 1930s. Today there are more than a dozen large Schmidt telescopes at major observatories around the world. One of the largest (see Figure 5-17) is located on Palomar Mountain, a short walk from the giant 5-m reflector. Whereas the 5-m telescope has a field of view only 2 arc min across, the Schmidt telescope produces photographs covering a field 7° in diameter. Schmidt telescopes are designed as cameras; the astronomer using the telescope does not see the view until the photograph is developed.

During the early 1950s, a team of astronomers spent several years photographing the sky with the Palomar Schmidt telescope. This work culminated in the famous National Geographic Society–Palomar Observatory Sky Survey. The entire northern hemisphere and the southern hemisphere down to a declination of −33° are covered in 879 pairs of photographs. Each 6° × 6° segment of the sky was photographed on both blue-sensitive and red-sensitive photographic plates. The blue-sensitive plates recorded primarily blue light, whereas the red-sensitive plates responded only to red light. The views of celestial objects in these two wavelength ranges were often strikingly different from each other (see Figure 5-18).

The Sky Survey was repeated in the early 1980s so that astronomers could search for changes in the sky over the preceding 30 years. In addition, Schmidt telescopes in Chile and Australia have extended this wide-angle coverage to those portions of the southern sky not accessible from Palomar. These magnificent photographs constitute a permanent record of the sky. Many of them reveal large-scale structures that have been overlooked by astronomers using bigger telescopes that have very small fields of view.

A radio telescope uses a large concave dish to reflect radio waves to a focus

Until recently, all information that astronomers gathered about the universe was based on ordinary visible light. But with the discovery of nonvisible electromagnetic radiation, scientists began to wonder if objects in the universe might also emit radio waves, X rays, and infrared and ultraviolet radiation. Surely, views of the universe at these nonvisible wavelengths would enhance our understanding of the cosmos.

The first evidence of nonvisible radiation from outer space came from the work of a young radio engineer, Karl Jansky of Bell Telephone Laboratories. Using long antennas, Jansky was investigating the sources of radio static that affects short-wavelength radiotelephone communication. By 1932, he realized that one kind of radio noise is strongest when the constellation of Sagittarius is high in the sky. The center of our galaxy is located in the direction of Sagittarius, and Jansky concluded that he was detecting radio waves from elsewhere in the galaxy.

Astronomers were not quick to pursue this line of research. Only one person, Grote Reber (an electronics engineer living in Wheaton, Illinois), pursued the matter. In 1936, Reber built the first radio telescope in his

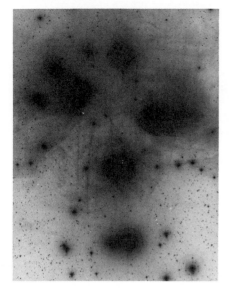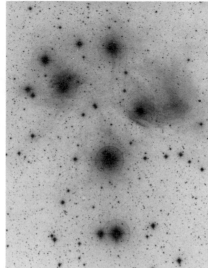

Figure 5-18 The Pleiades on the Sky Survey
Each of these photographs shows a cluster of stars called the Pleiades that is visible to the naked eye in the constellation of Taurus. The blue-sensitive plate is on the left; the red-sensitive plate is in the center. Both pictures are printed exactly as they appear on the original plates: black stars and white sky. Notice that very different features and details are seen in the two contrasting views. For comparison, a full-color photograph is included on the right. The area in each of these views is approximately 1° × 1½°. (Palomar Observatory)

backyard for the purpose of mapping radio emission from the Milky Way. His design was modeled after an ordinary reflecting telescope, with a parabolic "dish" (reflecting antenna) measuring 9.1 m in diameter. The radio receiver at the focal point of the metal dish was tuned to a wavelength of 1.85 m.

By 1944, when Reber completed his map of the Milky Way, astronomers had begun to take notice of these developments. Shortly after World War II, radio telescopes began to spring up around the world. Radio observatories are today as common as major optical observatories.

Like Reber's prototype, the standard radio telescope has a large parabolic dish. A small antenna tuned to the desired frequency is located at the focus, and the incoming signal is relayed to amplifiers and recording instruments typically located in a room at the base of the telescope's pier. Figure 5-19 shows a modern radio telescope.

Figure 5-19 A radio telescope
The dish of this radio telescope is 45.2 m (140 ft) in diameter. It is one of several large instruments at the National Radio Astronomy Observatory near Green Bank, West Virginia. (NRAO)

Astronomers were at first not enthusiastic about detecting radio noise from space in part because of the low resolving power. The resolving power of a telescope decreases with increasing wavelength. In other words, the longer the wavelength, the fuzzier the picture. Because radio radiation has very long wavelengths, astronomers thought that radio telescopes could produce only blurry, indistinct views.

Very large radio telescopes can produce somewhat sharper radio images: the bigger the dish, the better is the resolving power. For this reason, most modern radio telescopes have dishes more than 100 ft in diameter.

There is, however, another clever way to achieve high resolution. Unlike ordinary light, radio signals can be carried over electrical wires. Two radio telescopes separated by many kilometers can therefore be hooked together. This technique is called **interferometry** because the incoming radio signals are made to "interfere" or blend together, so that the combined signal is sharp and clear. The result is very impressive: the effective resolving power is equivalent to that of one gigantic dish with a diameter equal to the distance between the two telescopes.

Interferometry techniques were exploited for the first time in the late 1940s, and astronomers began receiving their first detailed views of radio objects in the sky. Radio telescopes separated by thousands of kilometers were linked together (using radio signals) to give resolving power much higher than that of optical telescopes. This is called **very-long-baseline interferometry (VLBI).** The best possible resolution would be obtained by two telescopes on opposite sides of the Earth. In that case, features as small as 0.00001 arc sec could be distinguished at radio wavelengths—resolution 100,000 times better than the sharpest pictures from ordinary optical telescopes.

One of the finest systems of radio telescopes began operating in 1980 in the desert near Socorro, New Mexico. It is called the *Very Large Array* (VLA) and consists of 27 parabolic dishes, each 26 m (85 ft) in diameter. The 27 telescopes are arranged along the arms of a gigantic Y that covers an area 27 km (17 miles) in diameter. Only a portion of the VLA is shown in Figure 5-20. This system produces radio views of the sky with resolutions comparable to that of the very best optical telescopes.

Figure 5-20   *The Very Large Array*
The twenty-seven radio telescopes of the VLA system are arranged along the arms of a Y in central New Mexico. The north arm of the array is 19 km long; the southwest and southeast arms are each 21 km long. (NRAO)

Telescopes in orbit around the Earth detect radiation that does not penetrate the atmosphere

As the success of radio astronomy began to mount, astronomers started exploring the possibility of making observations at other nonvisible wavelengths. Unfortunately, the Earth's atmosphere is opaque to many wavelengths. Very little radiation other than visible light and radio waves, manages to penetrate the air we breathe.

The transparency of the Earth's atmosphere is graphed in Figure 5-21. Notice the **optical window** through which we see visible light from space and the **radio window** that permits Earth-based radio astronomy. Also notice various transparent regions at infrared wavelengths between 10^4 and 10^5 Å (from 1 to 10 μ). Infrared radiation within these wavelength intervals does penetrate the Earth's atmosphere and can be detected with ground-based equipment. This wavelength range is called the near-infrared because it lies just beyond the red end of the visible spectrum.

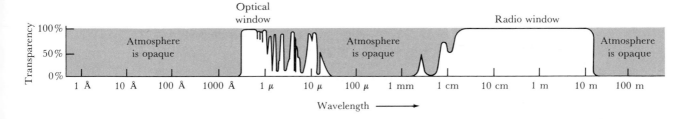

Figure 5-21 Transparency of the Earth's atmosphere
This graph shows the percentage of radiation that can penetrate the Earth's atmosphere as a function of wavelength. Oxygen and nitrogen completely absorb all radiation with wavelengths shorter than 2900 Å. Water vapor and carbon dioxide effectively block out all radiation from about 10μ to 1 cm.

Water vapor is the main absorber of infrared radiation from space. Infrared observatories are therefore located at sites of low humidity. For example, the 14,000-ft summit of Mauna Kea on Hawaii is exceptionally dry, and infrared observations are the primary function of NASA's 3.0-m telescope there.

Another possibility is to take a telescope up in an airplane. That is the basic idea behind the Kuiper Airborne Observatory (KAO) shown in Figure 5-22. The airplane carries a 1-m reflecting telescope to an altitude of 12 km (40,000 ft), placing the observatory above 99 percent of the atmospheric water vapor.

A telescope in Earth orbit offers the best arrangement. On January 25, 1983, the Infrared Astronomical Satellite (IRAS) was launched into a

Figure 5-22 The Kuiper Airborne Observatory
This C-141 jet airplane carries a 1-m reflecting telescope specifically designed for infrared observations. The observing portal (through which the telescope is aimed) can be seen on the fuselage just in front of the wing. (NASA)

Figure 5-23   The Infrared Astronomical Satellite
This satellite contains a reflecting telescope providing complete coverage of the sky at infrared wavelengths that cannot be detected from the ground. The satellite was launched in 1983. (NASA)

900-km-high polar orbit. As shown in Figure 5-23, the satellite is designed around a 57-cm (22½-in.) reflecting telescope that views the entire sky at infrared wavelengths that never penetrate the Earth's atmosphere. With this instrument, astronomers see what the sky looks like in the "far-infrared," the wavelength range from about 100 μ up to 1 mm.

The atmosphere also is transparent to the longest-wavelength ultraviolet light. This wavelength range, extending from about 4000 Å down to 2900 Å, is called the near-ultraviolet because it lies just beyond the violet end of the visible spectrum. Astronomers can easily make ground-based observations in this wavelength range, if they do not use glass lenses in their telescopes. Glass is opaque to the near-ultraviolet (that is the big drawback of refracting telescopes), and all lenses must be made of quartz.

To see the far-ultraviolet, astronomers must make observations from space. During the early 1970s, Apollo and Skylab astronauts carried small ultraviolet telescopes above the Earth's atmosphere to give us some of our first views of the ultraviolet sky. Small rockets have also been used to place ultraviolet cameras briefly above the Earth's atmosphere. A typical view is shown in Figure 5-24, along with a corresponding infrared view from IRAS, a view in visible lights, and a star chart.

Some of the best ultraviolet astronomy has been accomplished by the International Ultraviolet Explorer (IUE), which was launched on January 26, 1978. The satellite (see Figure 5-25) is built around a Cassegrain telescope with a 45-cm (18-in.) mirror and a total focal length of 6.74 m (22 ft). Observations cover the range from 1160 to 3200 Å.

For decades, astronomers have dreamed of having a major observatory in space. Although satellites like the IRAS and IUE give excellent

Figure 5-24   Orion as seen in ultraviolet, infrared, and visible wavelengths
An ultraviolet view (a) of the constellation of Orion was obtained during a brief rocket flight on December 5, 1975. The 100-sec exposure covers the wavelength range 1250–2000 Å. The "false color" view (b) from IRAS displays infrared intensity according to color: red indicates strong 100-μ radiation; green indicates strong 60-μ radiation; and blue shows strong 12-μ radiation. For comparison, an ordinary optical photograph (c) and a star chart (d) are included. (Courtesy of George R. Carruthers, NRL.; NASA; Palomar Observatory)

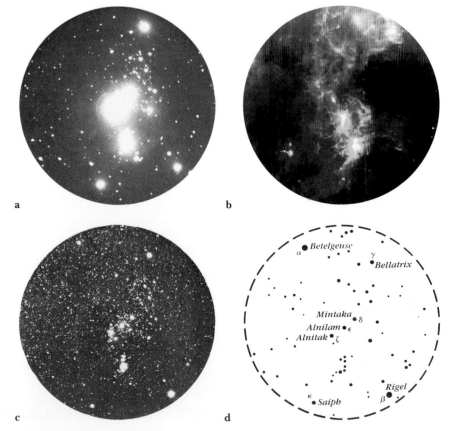

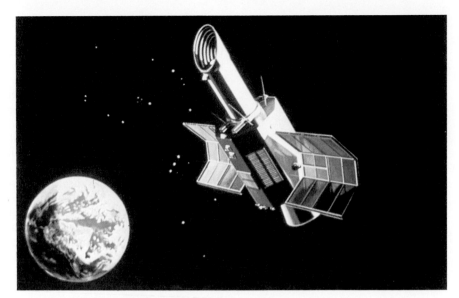

Figure 5-25   The International Ultraviolet Explorer
Since its launch in 1978, this 671-kg satellite has produced superb observations in the far-ultraviolet. The dark blue panels at the mid-section of the satellite are solar-cell arrays that provide electrical power for the radio transmitters and other electronic equipment. (NASA)

views of selected wavelength regions, astronomers are tantalized by the prospects of one very large telescope that could be operated at any wavelength from the infrared through the visible range and out into the far-ultraviolet. This is the mission of the Space Telescope (ST) to be carried aloft by the Space Shuttle about 1986. This instrument will dominate astronomy for the rest of the twentieth century.

The primary mirror of the Space Telescope has a diameter of 2.4 m (94 in.). The resulting light-gathering power is so large and the sky is so dark that the telescope will be able to detect stars as faint as 28th magnitude. For comparison, ambient light scattered by the Earth's atmosphere gives a limit of 24th magnitude for the faintest stars that can be photographed with the 5-m Palomar telescope.

Because the image will not be degraded by atmosphere, the actual resolution of the Space Telescope will be very close to the theoretical value. For example, when operated along with an auxiliary instrument called the faint-object camera (FOC), the telescope will have a resolution of 0.02 arc sec. That is a fiftyfold improvement over the best conditions at observatories such as Palomar, where the seeing disk is 1 arc sec.

Figure 5-26   The Hubble Space Telescope
The Space Shuttle will be used to place this 2.4-m Space Telescope into Earth orbit in 1986. During its anticipated 15-year lifetime, the telescope will be used to study the heavens over a wavelength range from 1100 Å to 1.1μ. (NASA)

The extraordinary light-gathering and resolving powers of the Space Telescope will permit astronomers to make observations that were unthinkable only a few years ago. For example, one observing program will have the Space Telescope search for new planets around other stars. Many familiar nebulae, star clusters, and galaxies whose photographs are scattered throughout this book will be seen with unprecedented clarity and brilliance. Currently unsuspected details and subtleties should be revealed.

Perhaps the greatest surprises will come from the discovery of totally new objects in space. Ever since Galileo turned his telescope toward the skies and saw four moons orbiting Jupiter, each new generation of astronomical instrument has disclosed unimagined objects and processes, often more bizarre than the strangest science fiction. It happened twice in the 1960s, when radio telescopes found quasars and pulsars. It happened again in the 1970s when X-ray telescopes detected bursters. With serendipity so commonplace, many astronomers predict that it will happen again in the 1990s.

Neither X rays nor gamma rays penetrate the Earth's atmosphere, so observations at these extremely short wavelengths also must be done from space. Astronomers got their first quick look at the X-ray sky with brief rocket flights during the late 1940s. Several small satellites launched during the early 1970s viewed the entire X-ray and gamma-ray sky, revealing hundreds of previously unknown sources including at least one black hole.

Although heroic in their day, these preliminary efforts pale in comparison to the detailed views and results from three huge satellites launched between 1977 and 1979. Called High Energy Astrophysical Observatories (HEAO), these satellites each carried an array of X-ray and gamma-ray detectors. Thousands of sources were discovered all across the sky. The second satellite in this series was especially successful in producing high-quality X-ray images of a wide range of exotic objects. Because it was launched near the hundredth anniversary of Albert Einstein's birth, this satellite was called the **Einstein Observatory.** X-ray views from this observatory appear throughout this book, illustrating discussions of the extraordinary objects that produce these high-energy radiations.

Ordinary optical equipment cannot help astronomers who want to observe the X-ray or gamma-ray sky. The energy carried by electromag-

Figure 5-27 [left]   The three HEAOs
Superb views of the X-ray sky and tantalizing glimpses of the gamma-ray sky were obtained from these three Earth-orbiting satellites. Each satellite is roughly the size and mass of a large automobile. In addition to solar panels (for electrical power), the sides of each spacecraft are covered with sensitive X-ray and gamma-ray detectors. (NASA)

Figure 5-28 [right]   The Gamma Ray Observatory
The best views of the high-energy gamma-ray sky will come from this satellite, which is scheduled for launch in the late 1980s. The 8000-kg (19,000-pound) satellite will be placed into Earth orbit by the Space Shuttle. (NASA)

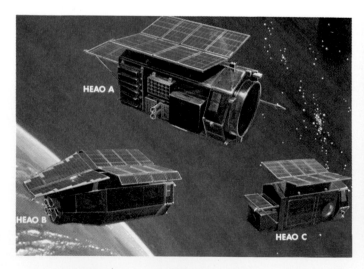

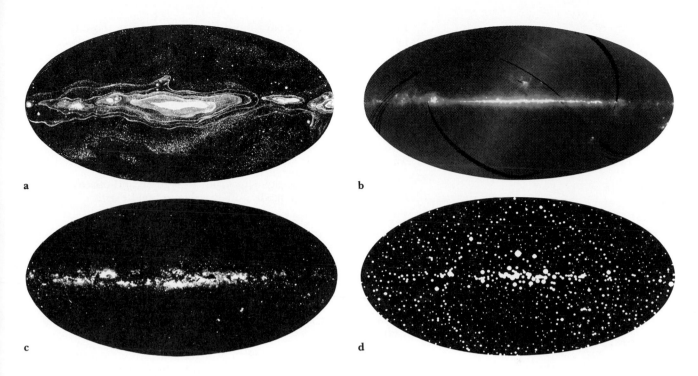

Figure 5-29 *The radio, infrared, visible, and X-ray skies*

These four views show the entire sky at (a) radio, (b) infrared, (c) visible, and (d) X-ray wavelengths. Each illustration is drawn with the Milky Way stretching horizontally across the picture. (Griffith Observatory; Jet Propulsion Laboratory)

netic radiation varies inversely with the wavelength, as mentioned in Box 5-1. In other words, the shorter the wavelength, the higher the energy of a photon. The wavelengths of X rays and gamma rays are so short that these high-energy photons would simply bury themselves in the mirror of an ordinary reflecting telescope. Instead, astronomers use electronic detectors (similar to Geiger counters) that respond to the effects of the high-energy radiation as it passes through sealed containers of gas or electrically charged metal plates carried on satellites.

The shortest-wavelength, highest-energy form of light is gamma radiation. Astronomers have had only their first tantalizing glimpses at this exotic region of the electromagnetic spectrum, and many hopes and expectations lie with the Gamma Ray Observatory (GRO) to be launched in the late 1980s (see Figure 5-28).

The advantages and benefits of these Earth-orbiting observatories cannot be overemphasized. We are no longer limited to the narrow ranges of wavelengths that manage to leak through the shimmering, hazy atmosphere we breathe. For the first time we are really seeing the universe (see Figure 5-29).

Summary

• Light is a wave phenomenon; it consists of oscillating electrical and magnetic fields that travel through space at a speed of about 3×10^{10} cm/sec.

The wavelength of light is associated with its color; wavelengths of visible light range from about 4000 Å for violet light to 7000 Å for red light.

Visible light forms only a small portion of the electromagnetic spectrum. Infrared radiation, microwaves, and radio waves have wavelengths larger than those of visible light. Ultraviolet radiation, X rays, and gamma rays have wavelengths shorter than those of visible light.

We now know that light also has particle-like properties; it can be viewed as a stream of photons. The shorter the wavelength, the greater the energy of the photons.

. Refracting telescopes (refractors) produce images by bending light rays as they pass through glass lenses.

Chromatic aberration is an optical defect whereby light of different wavelengths is bent different amounts by a lens.

Limitations of glass purity, chromatic aberration, and opacity to certain wavelengths have made it inadvisable to build extremely large refractors.

. Reflecting telescopes (reflectors) produce images by reflecting light rays to a focus point from curved mirrors.

Reflectors are not subject to most of the problems that limit the useful size of refractors, although they do have their own problem with spherical aberration, and the weight of the mirror.

. Radio telescopes have large reflecting antennas (dishes) that are used to focus radio waves.

Very large dishes are needed to produce sharp radio images, but arrays of smaller dishes are now used to achieve the same results as those that would be obtained with impossibly large dishes.

. The Earth's atmosphere absorbs much of the radiation that arrives from space.

The atmosphere is transparent chiefly in two wavelength ranges, known as the optical window and the radio window.

For observations at other wavelengths, astronomers depend upon telescopes carried above the atmosphere by high-altitude airplanes, rockets, or satellites.

Satellite-based observatories soon may give us a wealth of new information about the universe, permitting coordinated observation of the sky at all wavelengths.

Review questions

1 Give everyday examples of the phenomena of refractions and reflection.

2 Quite often advertisements appear for telescopes which extoll their magnifying abilities. Is this a good criterion for evaluating telescopes? Explain your answer.

*3 Compare the light-gathering power of the Palomar 200-inch telescope with that of the human eye (pupil diameter = 0.2 in.).

4 Why can radio astronomers observe at any time during the day whereas optical astronomers are mostly limited to nighttime observing?

5 Why do some double stars appear double in blue light but cannot be resolved in red light?

6 What kind of telescope would you use to take a color photograph entirely free of chromatic aberration? Why?

*7 What is the energy in ergs and eV of a photon having a wavelength of 21 cm?

*8 What is the frequency of a photon having a wavelength of 1 mm? How does this compare with the frequencies of the "standard" AM radio broadcast band?

Advanced questions

*9 The four largest moons of Jupiter are roughly the same size as our Moon and are about 528 million km from Earth at opposition. What is the size of the smallest surface features that the Space Telescope (resolution = 0.02 arc sec) will be able to detect? How does this compare with the smallest features that can be seen on the Moon with the unaided human eye (resolution = 1 arc min)?

10 Show by means of a diagram why the image formed by a simple refracting telescope is "upside down."

Discussion questions

11 Discuss the advantages and disadvantages of using a small telescope in Earth-orbit versus a large telescope on a mountain top.

12 If you were in charge of selecting a site for a new observatory, what factors and criteria would you consider important?

For further reading

Bok, B. "The Promise of the Space Telescope." *Mercury,* May/June 1983, p. 60.

Cohen, M. *In Quest of Telescopes.* Sky Pub. and Cambridge U. Press, 1980. *An astronomer's narration of what it is like to observe with the world's largest telescopes.*

Harrington, S. "Selecting Your First Telescope." *Mercury,* July/Aug. 1982, p. 106. *A nice primer on how telescopes work.*

Henbest, N., and Marten, M. *The New Astronomy.* Cambridge U. Press, 1983. *A beautiful album of the universe as seen in all the bands of the electromagnetic spectrum.*

Learner, R. *Astronomy Through the Telescope.* Van Nostrand Reinhold, 1981. *Illustrated history of telescopes and the discoveries they made possible.*

Robinson, L. "The Frigid World of IRAS." *Sky & Telescope,* Apr. 1984, p. 339.

Schorn, R. "Astronomy in the Next Decade." *Sky & Telescope,* Apr. 1982, p. 339.

Sullivan, W. "Radio Astronomy's Golden Anniversary." *Sky & Telescope,* Dec. 1982, p. 544.

Philip Morrison

Windows on the universe

Philip Morrison is an astrophysicist at the Massachusetts Institute of Technology. He did his undergraduate work at Carnegie Institute for Technology before earning his doctorate in theoretical physics at the University of California at Berkeley. He taught physics at San Francisco State College and at the University of Illinois before joining the Manhattan Project in 1943. In 1946 he joined the physics faculty at Cornell University and then, in 1964, he came to MIT, being honored as an Institute Professor in 1973.

Morrison is better known to the public for his advocacy of arms limitation and nuclear disarmament, remarkable contributions as the regular book editor for *Scientific American,* and enthusiasm as a spokesman for science as an intellectual adventure. However, he is equally active in professional science. He has published more than one hundred-forty articles, which are almost evenly divided between those for the general reader and those for specialists.

In the handle of the Big Dipper there is one bright star with a faint one very close by. Seeing both stars—they are called Horse and Rider—has long been held as a test of eyesight, although in fact it is not a very exacting one. In even the smallest telescope or binoculars it is not difficult to see two distinct stars. A somewhat larger telescope shows that the bright star is not truly single, but is itself double, two stars of different brightness, separated by only a fiftieth of the distance that separates Horse and Rider. Looking with a spectroscope at the spectrum of the bright star—thus extending and refining the inborn color sensitivity of every normal eye by splitting the visible color range into a very much finer scale—yields the surprising result that most of the spectral lines of the star are doubled, and that their spacing changes rhythmically in time, repeating itself every few weeks. Each line is contributed by the light from one of two distinct stars, orbiting too close to be seen separately in any telescope. But they emit distinct spectral lines shifted by the Doppler effect of their distinct motions in orbit.

What are the Horse and Rider, then? One fuzzy star . . . to the eye of someone with less than good eyesight; two stars close together . . . to anyone using some magnifying aid; three stars . . . to an observer with a good telescope; four stars . . . to a spectroscopist who watches for a few weeks, and who knows enough about star spectra and orbits to interpret the changing spectral pattern. Some new instrument might show something unexpected there in the field at any time—a fifth object, perhaps.

What we know about the universe is what our instruments have shown us, limited, of course, by our ability to interpret the instrumental findings. The eye is simply the oldest and most common of all astronomical instruments; it evolved over time as our species evolved. Neither the eye nor our instruments is foolproof; all can suffer under illusion. The person with poor eyesight may accept the Horse and Rider as two stars once he gets a view through the telescope. But to recognize the spectroscopic binary star requires another instrument, more knowledge, and more argument.

The better we understand how things work, the more instruments we can build and use, the more we can learn to "see." The quotation marks go on the verb because some of the most valuable instruments that we now use to examine the universe do not literally allow us to see. The eye cannot detect radio energy, but the sky can be mapped with radio-sensitive detectors, and that map presents a radio image of something forever invisible, made entirely recognizable as part of our genetically inherited eyes to the very much wider range of electromagnetic radiation, from radio frequencies 20 octaves lower than the visible, up to extremely high gamma-radiation frequencies, 30 octaves higher than the visible.

Elementary particles also reach us from distant objects in space. To detect neutrinos from the interior of the Sun, experimenters went deep into a mine, for such neutrinos easily penetrate the rock, while the other cosmic-ray particles that might confuse their detectors were there much reduced. We hope soon to detect gravitational waves by the tiny rhythms of shortening and elongating they produce in the ground as they pass by, an effect to be picked up by the timed trembling of mirrors many miles apart, the motions detectable by laser beam.

It is not far-fetched to say that the universe we study is made up of all that our instruments reveal through the best understanding we now have of what all the kinds of signals mean. It is a growing structure of understanding, never certain, and never final, but already much larger, older, more eventful, and more diverse than the wonderful view of the sky we all share any clear night with the ancient astronomers.

The Brookhaven solar neutrino experiment located nearly a mile underground (1480 meters) in the Homestake Gold Mine at Lead, South Dakota. The tank contains 100,000 gallons (380,000 liters) of perchloroethylene. Neutrinos are detected by a reaction that produces radioactive Argon 37. Only about 20 of these radioactive atoms are produced each month. (Brookhaven National Laboratory)

of the universe we seek to understand. To suppose that everything in the vast volume of space and time could be grasped by the human eye would obviously be a provincial and self-centered view.

First of all we try to amplify the faint. Inside the dark bowl of the Big Dipper the big telescopes with their long photographic time exposures show hundreds or even thousands of stars and galaxies. They are as real as the seven stars of the Dipper that the eye picks up directly. Of course at the margin of observability, any instrument, like the eye itself, can be deceptive. Maybe some of the faintest spots are defects in the photographic plate. Perceptions of instruments—like those the eye—demand attentive checking.

Then we magnify the small—like the spacing of the two stars that merge to form the bright star seen single with the eye. We subdivide the color scale yielding spectroscopic analyses of chemical composition and motions; we learn to detect over very brief time intervals, to disclose a pulsar flashing too rapidly for the eye or the photographic plate to see. We extend the red-to-blue color range (about an octave in frequency)

The Caltech 40m gravitational wave detector uses three independent test masses suspended at the corner and ends of an L-shaped vacuum system. A sensitive laser interferometer carefully monitors the separation of the test masses to look for passing gravitational waves from supernova or other astronomical sources. Special foundations help isolate the apparatus from spurious disturbances. This detector is a prototype for much longer and much more sensitive interferometers which may be built in the next few years. (Caltech photo by R. W. P. Drever)

Our solar system

Nebulae in Sagittarius
Planets are probably forming along with new stars in these nebulae (NGC 6559 and IC 1274–5) in Sagittarius. The type of planet to form at a particular distance from a star depends on conditions such as the temperature and substances (rock fragments, ice crystals, gas) at that distance. In the solar system, planets composed primarily of rock formed near the Sun whose heat drove off ices and gas. Far from the Sun, where temperatures are low, planets were able to retain volatile substances resulting in worlds composed primarily of gas. (Royal Observatory, Edinburgh)

Modern instruments and space probes have given us a rich body of information about the solar system. We begin this chapter with a survey of the major physical characteristics of the planets. We find that they fall into two distinct classes, the inner (terrestrial) planets and the outer (Jovian) planets, with Pluto as an oddity that cannot be so easily classified. We then discuss the materials of which the planets are composed and the technique of spectroscopy that provides information about the chemical compositions of distant objects. Finally, we outline current theories about the origin of the elements, the solar system, the planets, and the Sun. Much of the evidence supporting this history will be presented in later chapters.

Looking up at the heavens and wondering about the nature and origin of the Sun, Moon, stars, and planets is a universal human experience. Unlike our ancestors, however, we possess a wealth of information. Especially within the past few decades, telescopic observations and interplanetary spacecraft have given us vast quantities of data from which to draw ideas or to formulate and test theories. The final answers are out there, within our grasp. We know how it all turned out. We see the Sun, planets, moons, asteroids, comets, and meteoroids that make up our tiny niche in the universe. Many of these objects are exceedingly ancient and contain records of the cosmic events that created our solar system.

In addition to gleaning information from the objects that orbit the Sun, we can observe active star formation occurring elsewhere in our galaxy. Stars and planetary systems are now being formed in many beautiful nebulae scattered across the heavens. A general understanding of star creation coupled with knowledge of the Sun and its satellites gives us a fairly comprehensive picture of how the solar system was created. Many details still need to be worked out, but the overall scenario seems remarkably sound and reasonable. For the first time, we can truly appreciate what is unique and what is commonplace about our world. We have begun to fathom our connection with the rest of the cosmos and our place in the universe.

The planets are classified as terrestrial or Jovian by their physical attributes

A brief overview of the solar system distinguishes two classes of planets. First, notice the striking dichotomy in the orbits of the planets as shown in Figure 6-1. The orbits of the inner four planets (Mercury, Venus, Earth, and Mars) are crowded close to the Sun. In contrast, the orbits of the next four planets (Jupiter, Saturn, Uranus, and Neptune) are widely spaced at large distances from the Sun.

As you might expect, the range of surface temperatures that each planet experiences is related to its distance from the Sun. (See Box 6-1 for a discussion of temperature measurements.) The inner four planets are quite warm. For example, noontime temperatures on Mercury climb to 600 K ($= 327°C = 621°F$). At noon in the middle of summer on Mars, it is sometimes as warm as 300 K ($= 27°C = 81°F$). Of course, the outer planets, which receive much less solar radiation, are much cooler. Typical temperatures range from about 150 K ($= -123°C = -189°F$) in Jupiter's cloud tops to 63 K ($= -210°C = -346°F$) on Neptune. Temperature plays a major role in determining whether various substances exist as solids, liquids, or gases, thereby profoundly affecting the appearance of the planets.

Box 6-1 Temperatures and temperature scales

Three temperature scales are commonly used for various purposes. It is useful to be able to convert temperature readings from one scale to another.

Temperatures are expressed throughout most of the world in **degrees Celsius** (°C). The Celsius temperature scale is based on the behavior of water, which freezes at 0°C and boils at 100°C (at sea level on Earth). This scale was once known as the centigrade scale, but it was renamed in honor of the Swedish astronomer Anders Celsius who proposed it in 1742.

For many purposes, scientists prefer to express temperatures in a unit called the **kelvin** (K), named after the British physicist Lord Kelvin (William Thomson), who made many important contributions to our knowledge about heat and temperature. On the Kelvin temperature scale, water freezes at 273 K and boils at 373 K. Because water must be heated through a change of 100 K or 100°C to go from the freezing point to the boiling point, you can see that the "size" of a kelvin is the same as the "size" of a degree Celsius. When considering temperature *changes*, measurements in kelvins and in degrees Celsius are the same. A temperature expressed in kelvins is always equal to the temperature in degrees Celsius *plus* 273. The scientific preference for the kelvin scale arises from a physical interpretation of the meaning of temperature.

All substances are made of **atoms.** These atoms are very tiny (typical atomic diameters are about 10^{-8} cm) and are constantly in motion. The temperature of a substance is directly related to the average speed of its atoms. If something is hot, its atoms are moving at high speeds. If a substance is cold, its atoms are moving much more slowly.
(continued)

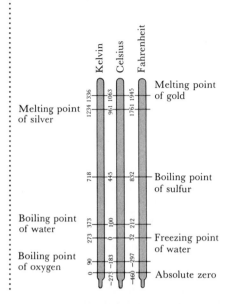

Melting point of silver

Boiling point of water

Boiling point of oxygen

Melting point of gold

Boiling point of sulfur

Freezing point of water

Absolute zero

(Box 6-1, continued)

The coldest possible temperature is the temperature at which the atoms move as slowly as possible (they can never quite stop completely). The minimum possible temperature is called **absolute zero** and is the starting point for the Kelvin scale. Absolute zero is 0 K, or −273°C.

In the United States, many people still use the archaic Fahrenheit scale, expressing temperatures in **degrees Fahrenheit** (°F). When the German physicist Gabriel Fahrenheit introduced this scale in the early 1700s, he intended 0°F to represent the coldest temperature then achievable (with a mixture of ice and saltwater) and 100°F to represent the temperature of a healthy human body. On the Fahrenheit scale, water freezes at 32°F and boils at 212°F. Because there are 180 degrees Fahrenheit between the freezing and boiling points of water, a degree Fahrenheit is only $\frac{5}{9}$ as large as either a degree Celsius or a kelvin.

The following equation is useful to convert from degrees Celsius to degrees Fahrenheit:

$$T_{\mathrm{F}} = \frac{9}{5}T_{\mathrm{C}} + 32$$

where T_{F} is the temperature in °F, and T_{C} is the temperature in °C. To convert in the opposite direction, a simple rearrangement of terms gives the relationship

$$T_{\mathrm{C}} = \frac{5}{9}(T_{\mathrm{F}} - 32)$$

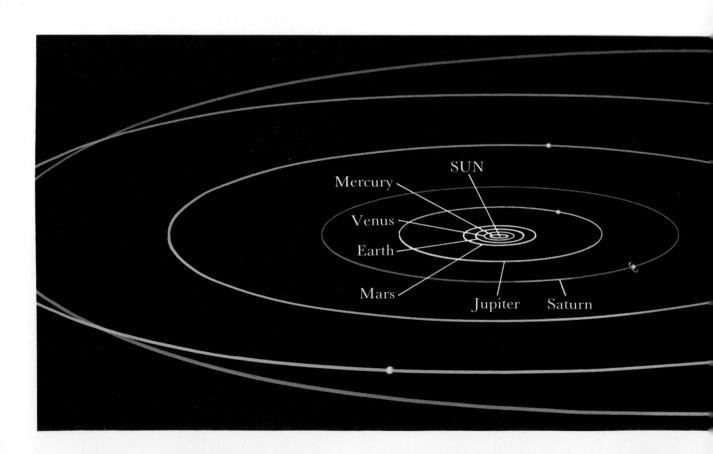

As an example, consider a typical room temperature of 72°F. Using the second equation, we can convert this measurement to the Celsius scale:

$$T_C = \frac{5}{9}(72 - 32) = 22°C$$

To get to the Kelvin scale, we simply add 273 degrees to the value in °C. Thus,

$$72°F = 22°C = 295 \text{ K}$$

The diagram displays the relationships between these three temperature scales.

Notice that most of the planets' orbits are nearly circular. As discussed in Chapter 4, Kepler discovered that these orbits are actually ellipses. Astronomers denote the elongation of an ellipse by its **eccentricity** (review Box 4-3 for details). The eccentricity of a circle is zero, and most planets have orbital eccentricities that are very close to zero. The notable exception is Pluto, with an orbital eccentricity of 0.25. Its highly noncircular orbit sometimes takes Pluto nearer the Sun than its neighbor, Neptune.

The planetary orbits all lie in nearly the same plane. In other words, the orbits of the planets are inclined at only small angles to the plane of the ecliptic. Again, however, Pluto is a notable exception. The plane of Pluto's orbit is tilted at 17° to the plane of the Earth's orbit (see Table 6-1).

Figure 6-1   The solar system
This scale drawing shows the distribution of planetary orbits around the Sun. Four inner planets are crowded close to the Sun, and five outer planets orbit at much larger distances from the Sun.

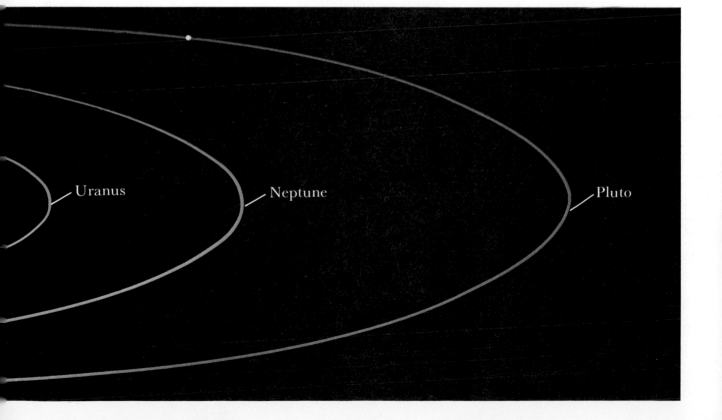

*TABLE 6-1   Orbital characteristics
of the planets*

| | Mean distance from Sun | | Orbital (sidereal) period | Eccentricity | Inclination to the ecliptic (degrees) |
	(AU)	(10^6 km)	(years)		
Mercury	0.39	58	0.24	0.206	7.0
Venus	0.72	108	0.62	0.007	3.4
Earth	1.00	150	1.00	0.017	0.0
Mars	1.52	228	1.88	0.093	1.8
Jupiter	5.20	778	11.86	0.048	1.3
Saturn	9.54	1,427	29.46	0.056	2.5
Uranus	19.18	2,870	84.01	0.047	0.8
Neptune	30.06	4,497	164.79	0.009	1.8
Pluto	39.44	5,900	247.7	0.250	17.2

As we compare the physical properties of the planets, we again find that they fall naturally into two classes—the four inner planets and the four outer planets—with Pluto as an exception. As examples of important properties, let us look next at the sizes, masses, and densities of the planets.

The diameter of a planet can be computed from its apparent angular diameter and distance as described in Box 1-1. For example, at greatest eastern elongation in 1983, Venus was 1.061×10^8 km from Earth and had an angular diameter of $23\frac{1}{2}$ arc sec. Using the small-angle formula, we find that the diameter of Venus is 12,100 km (7520 miles). Similar calculations demonstrate that the inner four planets are quite small. Indeed, the Earth with a diameter of 12,760 km (7930 miles) is the largest. In sharp contrast, the four outer planets are much larger. First place goes to Jupiter, whose equatorial diameter is 143,800 km (89,400 miles). Pluto, despite its position as the outermost planet, is even smaller than the inner planets. Its diameter is only about 3000 km (1900 miles). Figure 6-2 shows the Sun and the planets drawn to the same scale. The diameters of the planets are given in Table 6-2.

Determining the mass of a planet is a more difficult process. It is most easily accomplished if the planet has a satellite. The satellite obeys Kepler's third law, and astronomers can measure the satellite's period and semimajor axis. Therefore, they can calculate the planet's mass from formulas such as those in Box 4-3. If the planet does not have a satellite, astronomers must rely on a comet or spacecraft that passes near the planet. The planet's gravity (which is directly related to the planet's mass) produces a deflection in the path of the comet or spacecraft. By measuring the size of this deflection and using Newtonian mechanics, astronomers can determine the planet's mass. The inner four planets have low masses, whereas the next four planets have substantially larger masses. Again, first place goes to Jupiter, whose mass is 318 times greater than the mass of the Earth. The masses of the planets are given in Table 6-2.

Average density (mass divided by volume) is a physical property that can often be used to deduce important information about the composition of an object. Scientists commonly express average density in grams per cubic centimeter. The inner four planets have very large average

Figure 6-2   The Sun and the planets
*This drawing shows the nine planets in front
of the disk of the Sun, with all ten bodies
drawn to the same scale. The four planets that
have orbits nearest the Sun (Mercury, Venus,
Earth, Mars) are small and are made of rock.
The next four planets from the Sun (Jupiter,
Saturn, Uranus, Neptune) are large and are
composed primarily of gas.*

densities; the average density of the Earth is 5.5 g/cm^3. This may not
seem large until you know that the average density of a typical rock is
about 3 g/cm^3 and the average density of water is 1 g/cm^3. The Earth
must contain a large amount of material that is more dense than rock.
This information provides our first clue that Earth-like planets have iron
cores.

**TABLE 6-2   Physical characteristics
of the planets**

	Diameter		Mass		Average density (g/cm^3)
	(km)	**(Earth = 1)**	**(gm)**	**(Earth = 1)**	
Mercury	4,880	0.38	3.3×10^{26}	0.06	5.4
Venus	12,100	0.95	4.9×10^{27}	0.82	5.2
Earth	12,760	1.00	6.0×10^{27}	1.00	5.5
Mars	6,800	0.53	6.4×10^{26}	0.11	3.9
Jupiter	143,800	11.27	1.9×10^{30}	317.89	1.3
Saturn	120,000	9.44	5.7×10^{29}	95.15	0.7
Uranus	52.300	4.10	8.7×10^{28}	14.54	1.2
Neptune	49,500	3.88	1.0×10^{29}	17.23	1.7
Pluto*	3,000	0.2	10^{25}	0.002	1

*All of the data for Pluto are somewhat uncertain.

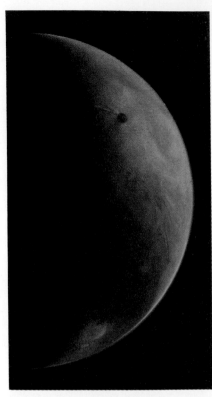

Figure 6-3   A terrestrial planet
Mars is a typical terrestrial planet. Its diameter is only 6800 km, and its average density is 3.9 g/cm³, indicating that the planet is composed of rock. Volcanoes, canyons, and craters are seen in this photograph taken by the Viking 2 spacecraft in August 1976. (NASA)

In sharp contrast, the outer planets have very low densities. Indeed, Saturn has an average density less than 1 g/cm³, less dense than water. This information strongly suggests that the giant outer planets are composed primarily of light elements such as hydrogen and helium. Again, Pluto is an exception. Although Pluto is even smaller than the dense inner planets, its average density seems to be much like those of the giant outer planets. Table 6-2 lists the average densities of the planets.

These differences in mass, size, and density lead us to consider the four inner and the four outer planets as two distinct groups. The inner four planets are called **terrestrial planets** because they resemble the Earth. They are small and dense. Craters, canyons, and volcanos are common on their hard, rocky surfaces (see Figure 6-3). The outer four planets are called **Jovian planets** because they resemble Jupiter (Jove was another name for the Roman god Jupiter). Vast swirling cloud formations dominate the appearance of these enormous gaseous spheres (see Figure 6-4). If these planets have solid cores at all, they are probably not much bigger than the Earth, buried beneath atmospheres that are tens of thousands of kilometers thick.

Although it is usually called a planet, Pluto clearly is an oddity. Its physical properties are not typical of either the terrestrial or the Jovian planets. Some astronomers have suggested that Pluto should be classified with the **asteroids,** tens of thousands of small objects (typical diameter about 40 km) that circle the Sun between the orbits of Mars and Jupiter. The largest asteroid, Ceres, has a diameter of about 750 km.

Another oddity was discovered in 1977 by Charles Kowal of Palomar Observatory—an object, called Chiron, similar to a large asteroid that moves in a highly eccentric orbit about the Sun between the orbits of Saturn and Uranus. Perhaps Pluto and Chiron should be grouped with the asteroids as minor members of the solar system, leaving only eight planets. On the other hand, some astronomers argue that at least seven other objects should be grouped along with the planets as major mem-

Figure 6-4   A Jovian planet
Jupiter is the largest of the Jovian planets. Its equatorial diameter is 143,800 km, and its average density is only 1.3 g/cm³, indicating that the planet is primarily composed of light elements. Two of Jupiter's moons, Io and Europa, are seen in this view taken by the Voyager 1 spacecraft in February 1979. Each of these satellites is approximately the same size as Earth's moon. (NASA)

bers of the solar systems. These seven objects are seven large moons that orbit about four of the planets.

All of the planets except Mercury and Venus, have satellites. More than forty satellites are known; Jupiter and Saturn each have more than a dozen. Certainly dozens of other satellites remain to be discovered. The known satellites fall into two distinct categories. There are seven giant satellites that are roughly comparable to Mercury in size. They are listed in Table 6-3 with relevant data. The other satellites are much smaller (diameters less than 2000 km).

TABLE 6-3   The seven giant satellites

Satellite name	Parent planet	Diameter (km)	Average density (g/cm^3)
Moon	Earth	3,480	3.3
Io	Jupiter	3,630	3.6
Europa	Jupiter	3,130	3.0
Ganymede	Jupiter	5,280	1.9
Callisto	Jupiter	4,820	1.8
Titan	Saturn	5,120	1.9
Triton*	Neptune	4,000	2.0

*The data for Triton are somewhat uncertain.

The recent interplanetary missions of Voyager 1 and 2 have revealed many fascinating characteristics of these giant satellites (see Figure 6-5). For example, Jupiter's satellite Io is one of the most geologically active worlds in the solar system, with numerous volcanos belching sulfur-rich

Figure 6-5   A giant satellite
This view, taken from Voyager 2 in July 1979, shows Callisto, the second largest moon of Jupiter. Its diameter is almost exactly the same as Mercury. Numerous craters pockmark Callisto's ice-bound surface. (NASA)

compounds. Saturn's largest satellite, Titan, has an atmosphere nearly twice as dense as Earth's atmosphere.

Because of their sizes, it is reasonable to include these seven giant satellites in a listing of the planets, even though they happen to orbit other planets instead of orbiting the Sun independently. The hard surfaces (of either rock or ice) of these satellites would place them in the terrestrial category. This expanded definition of a terrestrial planet gives us 10 Earth-like worlds that we can compare to our own planet.

Spectroscopy reveals the chemical composition of the Sun, planets, and stars

Astronomers have learned a great deal about the compositions of the planets. The most accurate determinations come from spacecraft that have landed on planets and made direct chemical analyses of the atmosphere or soil. Unfortunately, we have obtained such direct information for only three worlds: Venus, Moon, and Mars. In all other cases, astronomers must rely on their ability to analyze the sunlight reflected from the distant planets and their satellites. In these circumstances, astronomers bring to bear one of their most powerful tools, *spectroscopy*.

In Chapter 5, we saw that white sunlight can be separated into a spectrum of colors (review Figure 5-2). **Spectroscopy** is the systematic study of such spectra. In the early 1800s, the German physicist Joseph von Fraunhofer reexamined the Sun's spectrum under high magnification and found numerous faint dark lines cutting across the spectrum. In other words, certain very specific wavelengths are missing from sunlight. Figure 6-6 shows a portion of the solar spectrum; including a number of these **spectral lines.**

The significance of Fraunhofer's discovery was revealed by experiments in the mid-nineteenth century: *each chemical produces its own characteristic pattern of spectral lines.* For example, a portion of the spectrum of vaporized iron is shown in Figure 6-6. No other chemical can mimic this particular pattern of lines at these wavelengths. It is iron's own distinctive "fingerprint." The spectral lines of iron appear in the Sun's spectrum, and so some vaporized iron must exist in the Sun's atmosphere.

Spectroscopy is discussed in much greater detail in Chapter 17. For now, we need know only that spectral lines provide extremely reliable evidence about the chemical composition of distant objects. If the spectral lines of a chemical are found in a star's spectrum, it is very safe to conclude that this chemical is present in the star's atmosphere.

Over the years, astronomers have used **spectrographs** (see Box 6-2) to examine and record the spectra of stars and galaxies. As data accumulated, an important fact emerged. No matter where we look—no matter how far we peer into space—we always find the same chemical elements. These naturally occurring elements of which the Earth is made are the same building blocks for all matter everywhere in the universe. The **elements** are fundamental substances because it is impossible to break them down into more-basic chemicals. Each element is composed of only a single kind of atom.

Figure 6-6 Iron on the Sun
The upper spectrum is a portion of the Sun's spectrum from 4200 to 4300 Å. Numerous dark spectral lines are visible. The lower spectrum is a corresponding portion of the spectrum of vaporized iron. Several bright spectral lines are seen against a black background. The fact that the iron lines coincide with some of the solar lines proves that there is some iron (albeit a very tiny amount) in the Sun's atmosphere. (Mount Wilson and Las Campanas Observatories)

Box 6-2 Spectrographs

The spectrograph is one of the astronomer's most important tools, perhaps second only to the telescope itself. In its basic form, a spectrograph consists of a slit, two lenses and a prism arranged (as shown in the diagram *a*) to focus the spectrum of an astronomical object onto a small photographic plate. This optical device typically is mounted at the focal point of a telescope, and the image of the object to be examined is focused on the slit. After the spectrum of a star or galaxy has been photographed, the exposed portion of the photographic plate is covered and light from a known source (usually an iron arc) is focused on the spectrograph slit. This exposes a "comparison spectrum" above and below the spectrum of the astronomical object, as shown in the photograph *b*. The wavelengths of the bright spectral lines of the comparison spectrum are known from laboratory experiments. These bright lines therefore serve as reference markers that can be used to measure the wavelengths of the lines in the spectrum of the star or galaxy.

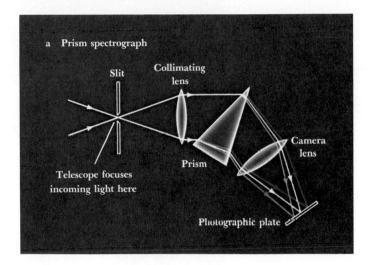

a   Prism spectrograph

There are drawbacks to this old-fashioned spectrograph. A prism does not evenly disperse the colors of the rainbow. The red part of the spectrum is compressed, whereas the blue and violet colors are more spread out. In addition, the blue and violet wavelengths must pass through a fairly great thickness of the prism, which absorbs much of the starlight. Indeed, a glass prism is opaque to the near-ultraviolet. A better device for breaking starlight into the colors of the rainbow is a diffraction grating.

A **diffraction grating** is a piece of glass on which thousands of closely spaced lines are cut. Some of the finest diffraction gratings have as many as 10,000 lines per centimeter. These lines usually are cut by drawing a diamond back and forth across the piece of glass. Spacing of the lines must be very regular. Best results are obtained if the grooves are beveled, as shown in drawing *c*.

A diffraction grating can be used in either of two ways. Incoming starlight can be reflected off the grating (it is then called a **reflection grating**), or the

b

(continued)

(Box 6-2, continued)

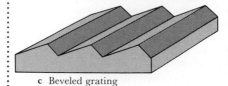

c Beveled grating

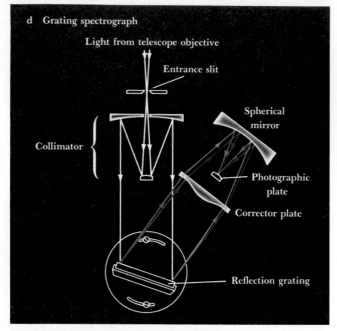

d **Grating spectrograph**

starlight can be passed through the grating (it is then called a **transmission grating**). In either case, light rays leaving various parts of the grating interfere with each other in such a way as to produce a spectrum. Diagram *d* shows the design of a modern grating spectrograph.

In recent years, television-related industries have produced a variety of light-sensitive devices. Many of these devices have proved more useful than photographic film for recording spectra. For example, the film in a standard spectrograph can be replaced with a **photomultiplier** that produces an electric current proportional to the intensity of light entering the device. This current is amplified and used to drive a chart recorder that draws a graph of wavelength versus intensity. The spectral lines appear as peaks or valleys on such a chart.

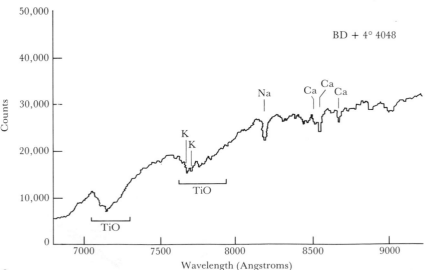

e

Most recently, many observatories have begun to record spectra with a new invention called a charge-coupled device (CCD). A **CCD** is a silicon chip, approximately the size of a postage stamp, that is divided into thousands of tiny squares. For example, the CCD commonly used at the prime focus of the 5-m telescope at Palomar is a centimeter-sized square divided into a 500×500 array of picture elements. When light falls on one of these tiny picture elements (called **pixels**), an electric charge is produced. The charge builds up in proportion to the amount of light that falls on the pixel. When the exposure is finished, electronic equipment is used to measure the amount of charge that has accumulated in each of the pixels. The final result is a graph (counts versus wavelength) on which spectral lines appear as peaks or valleys. An example of a portion of the near-infrared spectrum of the star BD + 4°4048 is shown in diagram *e*. Spectral lines of calcium (Ca), sodium (Na), potassium (K), and titanium oxide (TiO) are identified on the graph.

A list of the chemical elements is most conveniently displayed in the form of a **periodic table** (see Figure 6-7). Each element is assigned a unique **atomic number.** The elements are arranged in the periodic table in sequence of their atomic numbers. With a few exceptions, this sequence also corresponds to increasing average mass of the atoms of the elements. Thus hydrogen (symbol H) has atomic number 1 and is the lightest element. As another example, iron (symbol Fe) has atomic number 26 and is a relatively heavy element. All of the elements that appear in a single vertical column of the periodic table have similar chemical properties. For example, the elements listed in the rightmost column are all gases at Earth-surface conditions of temperature and pressure, and they all tend to be very reluctant to react chemically with other elements.

In addition to nearly 100 naturally occurring elements, Table 6-7 lists several artificially produced elements. All of these elements are heavier than uranium (symbol U) and are highly radioactive, which means that they decay into lighter elements within a very short time after they are created in laboratory experiments.

The **atom** is the smallest possible piece of an element. Typical atomic diameters are about 1 Å. Atoms of all the elements can be broken down into three basic types of subatomic particles: protons, neutrons, and electrons. Box 6-3 discusses some important details of atomic structure.

Figure 6-7 The periodic table of the elements

The periodic table is a convenient listing of the elements, arranged according to their weights and chemical properties.

THE PERIODIC TABLE

1 H																	2 He
3 Li	4 Be											5 B	6 C	7 N	8 O	9 F	10 Ne
11 Na	12 Mg											13 Al	14 Si	15 P	16 S	17 Cl	18 A
19 K	20 Ca	21 Sc	22 Ti	23 V	24 Cr	25 Mn	26 Fe	27 Co	28 Ni	29 Cu	30 Zn	31 Ga	32 Ge	33 As	34 Se	35 Br	36 Kr
37 Rb	38 Sr	39 Y	40 Zr	41 Nb	42 Mo	43 Tc	44 Ru	45 Rh	46 Pd	47 Ag	48 Cd	49 In	50 Sn	51 Sb	52 Te	53 I	54 Xe
55 Cs	56 Ba	57 La	72 Hf	73 Ta	74 W	75 Re	76 Os	77 Ir	78 Pt	79 Au	80 Hg	81 Tl	82 Pb	83 Bi	84 Po	85 At	86 Rn
87 Fr	88 Ra	89 Ac	104	105	106												

58 Ce	59 Pr	60 Nd	61 Pm	62 Sm	63 Eu	64 Gd	65 Tb	66 Dy	67 Ho	68 Er	69 Tm	70 Yb	71 Lu
90 Th	91 Pa	92 U	93 Np	94 Pu	95 Am	96 Cm	97 Bk	98 Cf	99 Es	100 Fm	101 Md	102 No	103 Lr

Box 6-3   Atoms and isotopes

Laboratory experiments during the first part of the twentieth century revealed that the structure of an atom superficially resembles a tiny solar system, as shown schematically in the diagram. Most of the mass of an atom is concentrated in a dense **nucleus** that is composed of **protons** and **neutrons.** Orbiting this massive nucleus like small planets are the **electrons.** Whereas the solar system is held together by gravitational forces, atoms are held together by electrical forces. Each proton carries a positive charge, and each electron carries an equal but opposite negative charge. The attractive electrical forces between the positively charged protons and the negatively charged electrons keep the atom from coming apart.

Although electrons and protons have equal (but opposite) charges, they have very unequal masses. The electron is one of the lightest subatomic particles with a mass of only 9.1×10^{-28} grams. The proton is nearly 2000 times heavier: its mass is 1.7×10^{-24} grams.

A neutron has almost exactly the same mass as a proton. As the name suggests, a neutron has no electric charge—it is electrically neutral. Neutrons serve as buffers between the positively charged protons that are crowded together inside the nucleus. The number of neutrons in a nucleus typically is larger than the number of protons, especially in the case of the heaviest elements.

Under normal circumstances, the number of electrons orbiting an atom is equal to the number of protons in the nucleus, making the atom electrically neutral. Furthermore, the number of protons in an atom's nucleus equals the atomic number for that particular element. A hydrogen nucleus has one proton, a helium nucleus has two protons, and so forth—up to uranium with 92 protons in its nucleus.

The number of protons in the nucleus of an atom determines uniquely what element that atom is. Nevertheless, the same element may have slightly different numbers of neutrons in its nuclei. For example, consider oxygen, which is the eighth element on the periodic table. Its atomic number is 8, and every oxygen nucleus has exactly eight protons, but it can have eight, nine, or ten neutrons. Thus, three slightly different kinds of oxygen, called **isotopes,** exist. The isotope with eight neutrons is by far the most abundant variety and is written as ^{16}O, or oxygen-16. The rarer isotopes with nine and ten neutrons are designated as ^{17}O and ^{18}O, respectively.

The superscript that precedes the chemical symbol for an element is equal to the total number of protons and neutrons in a nucleus of that particular

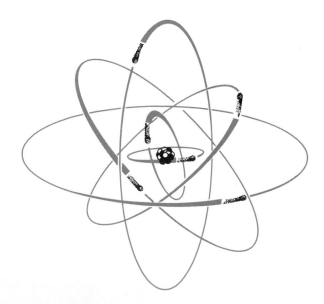

isotope. For example, the common isotope of iron is ^{56}Fe, or iron-56, which means that its nucleus contains a total of 56 protons and neutrons. From the periodic table, however, we see that the atomic number of iron is 26. This means that every iron atom has 26 protons in its nucleus. Therefore, the number of neutrons in a ^{56}Fe nucleus is $56 - 26 = 30$.

It is extremely difficult to distinguish chemically between the various isotopes of a particular element. Ordinary chemical reactions involve only the electrons that orbit the atom, never the neutrons buried in its nucleus.

Although the number of elements is limited, a wide variety of substances exist in the universe because atoms can combine to form various **molecules.** For example, two hydrogen atoms can combine with an atom of oxygen to form a molecule of the substance we call water. Water's chemical formula is H_2O, describing the atoms in its molecule. In a similar way, two oxygen atoms can bond with a carbon atom to produce a molecule of carbon dioxide, whose formula is CO_2. The chemical formula for common table salt is NaCl, which means that salt contains one sodium atom combined with each chlorine atom.

Just as each element can be identified by its spectral lines, molecules also produce unique patterns of spectral lines in the spectra of astronomical objects. For example, Figure 6-8 shows a portion of the spectrum of Venus, with a series of spectral lines caused by carbon dioxide. It is therefore reasonable to conclude that Venus's atmosphere contains this gas. Indeed, direct measurements by Russian and American spacecraft confirm that carbon dioxide gas forms 97 percent of the Venusian atmosphere.

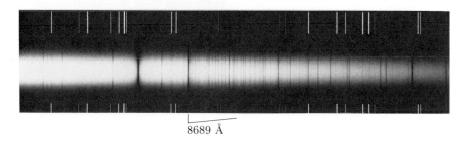

Figure 6-8   Carbon dioxide on Venus
This near-infrared spectrum of Venus shows a series of spectral lines beginning at 8689 Å that is caused by carbon dioxide, the primary constituent of Venus's atmosphere. (Lick Observatory)

8689 Å

The relative abundances of the elements are the result of cosmic processes

Some elements are very common, whereas others are very rare. Hydrogen is by far the most abundant substance in the universe; it makes up nearly three-quarters of the mass of all the stars and galaxies in the universe.

Helium is the second most abundant element. Together hydrogen and helium account for 98 percent of the mass of all the material in the universe. That leaves only 2 percent for *all* the other elements combined.

There is a good reason for this overwhelming abundance of hydrogen and helium. Most astronomers believe that the universe began roughly 20 billion years ago with a violent event called the Big Bang. Only the two lightest elements, hydrogen and helium, emerged from the enormously high temperatures following this cosmic event. All of the other elements on the periodic table were manufactured deep inside stars at later times. In Chapters 21 and 22, we shall explore the processes by which elements are created at a star's center. Indeed, if it were not for the nuclear reactions inside stars, there would be no heavy elements in the universe today.

Near the ends of their lives, stars cast much of their matter out into space. This can be a comparatively gentle process in which a star's outer layers are gradually expelled. Figure 6-9 shows the star HD 65750, which is losing material in this fashion. Alternatively, a star may end its life with a spectacular detonation called a **supernova** explosion, which blows the star apart. Either way, the interstellar gases in the galaxy become enriched with heavy elements dredged up from the dying star's interior, where they have been created. New stars that form from this enriched material thus have an ample supply of heavy elements from which a system of planets, satellites, comets, and asteroids can be formed.

Figure 6-9 The mass-loss star HD 65750 This star is shedding material rapidly. The nebulosity (called IC 2220) around the star is probably being compressed by the out-flow of material from the star. (Anglo-Australian Observatory)

Our solar system is formed of matter created in stars that disappeared billions of years ago. The Sun is a fairly young star, only 5 billion years old. All of the elements other than hydrogen and helium in our solar system were created and cast off by ancient stars during the first 10 billion years of our galaxy's existence. We are literally made of star dust.

Stars create the different elements in different amounts. For example, the elements carbon, oxygen, silicon, and iron are readily produced in a star's interior, whereas gold is created only under special circumstances. Therefore, gold is rare in the solar system, but carbon is not.

A convenient way to express the abundances of the various elements is to say how many atoms of a particular element are found for every trillion (that is, 10^{12}) hydrogen atoms. For example, for every trillion hydrogen atoms in space, there are about 60 billion (6×10^{10}) helium atoms. From chemical analysis of Earth rocks, Moon rocks, and meteorites, scientists have been able to determine relative abundances of the elements in our part of the galaxy. Ten elements are especially abundant; they are listed in Table 6-4.

TABLE 6-4 *Solar abundances of the most common elements*

Atomic number	Symbol	Element	Relative abundance
1	H	Hydrogen	10^{12}
2	He	Helium	6×10^{10}
6	C	Carbon	4×10^{8}
7	N	Nitrogen	9×10^{7}
8	O	Oxygen	7×10^{8}
10	Ne	Neon	4×10^{7}
12	Mg	Magnesium	4×10^{7}
14	Si	Silicon	5×10^{7}
16	S	Sulfur	2×10^{7}
26	Fe	Iron	3×10^{7}

In addition to these ten very common elements, there are five elements that are moderately abundant. They are sodium, aluminum, argon, calcium, and nickel. These elements have abundances in the range of 10^6 to 10^7 relative to the standard trillion hydrogen atoms. All other elements are much rarer. For example, for every trillion hydrogen atoms in the solar system, there are only six atoms of gold.

The planets formed by accumulation of material in the solar nebula during the birth of the Sun

Although hydrogen and helium are the most common elements, they are not plentiful on the inner four terrestrial planets. Warmth from the Sun helped deprive these worlds of the lightest gases. As mentioned in Box 6-1, the higher the temperature of a gas, the greater the speed of its atoms. Temperatures on the inner four planets are comparatively high, and the light-weight atoms of hydrogen and helium move swiftly enough to escape from the relatively weak gravity of these small planets.

A very different situation prevails on the four Jovian planets. Far from the Sun, temperatures are low, and the atoms of hydrogen and helium move slowly. The relatively strong gravity of the massive Jovian planets easily prevents the lightest gases from escaping into space. Indeed, the Jovian planets are composed primarily of hydrogen and helium.

Temperature also must have been an important factor in determining specific conditions inside the vast cloud of gas and dust, called the **solar nebula,** from which the solar system formed. For this reason, it is extremely useful to categorize the abundant elements and their common compounds according to their behaviors at various temperatures.

First, hydrogen and helium are gaseous except at extremely low temperatures and extraordinarily high pressures. Second, rock-forming compounds of iron and silicon are solids except at temperatures exceeding 1000 K. Finally, there is an intermediate class of substances such as water, carbon dioxide, methane, and ammonia. At low temperatures (typically below 200 to 300 K), these common chemicals solidify into solids called ices. At somewhat higher temperatures, they can exist as liquids or gases. In Table 6-5, common planet-forming substances are listed according to these three broad classifications.

TABLE 6-5 Common planet-forming substances

Gas	Ice	Rock
Hydrogen (H)	Water (H_2O)	Iron (Fe)
Helium (He)	Methane (CH_4)	Iron Sulfide (FeS)
Neon (Ne)	Ammonia (NH_3)	Olivine ($(Mg,Fe)SiO_4$)
	Carbon Dioxide (CO_2)	Pyroxene ($CaMgSi_2O_6$)

By observing star formation elsewhere in our galaxy, astronomers can deduce the conditions that led to formation of the solar system. Just before the birth of the Sun, atoms in the solar nebula were so widely spaced that no substance could exist as a liquid. Matter in this vast cloud existed either as gas or as tiny grains of dust and ice. Under these conditions of extremely low pressure, astronomers find it useful to speak of the **condensation temperature** to specify whether a substance is a solid or a gas. Above its condensation temperature, a substance is a gas; below its condensation temperature, the substance solidifies into tiny specks of dust or snowflakes.

Rock-forming substances have very high condensation temperatures, typically in the range 1300–1600 K. The ices have condensation temperatures in the range 100–300 K. The condensation temperatures of hydrogen and helium are so near absolute zero that these common elements always existed, as gases during the creation of the solar system.

Initially, the solar nebula was quite cold. Temperatures throughout the cloud were probably less than 50 K, which is below the condensation temperatures of all common substances except hydrogen and helium. Snowflakes and ice-coated dust grains must have been scattered abundantly across the solar nebula, which had a diameter of at least 100 AU and a mass roughly two or three times the mass of the Sun.

The gravitational pull of the particles on one another caused them to begin a general drift toward the center of the solar nebula. Density and pressure at the center of the solar nebula began to increase, producing a concentration of matter called the **protosun.** Because of gravitational contraction, temperatures deep inside the solar nebula also began to climb. In addition, the solar nebula must have had an overall slight amount of rotation—or **angular momentum,** as it is properly called. Otherwise, everything would have fallen straight into the protosun, leaving nothing behind for the planets. As the solar nebula contracted toward its gravitational center, it was transformed from a shapeless cloud into a rotating flattened disk that was warm at the center and cold at the edge. Astronomers feel confident in describing a flattened, disk-shaped solar nebula because all of the planets today have orbits that are very nearly in the same plane.

Temperatures around the newly created protosun soon climbed to 2000 K. Meanwhile, temperatures in the outermost regions of the solar nebula remained at less than 50 K. Figure 6-10 shows the probable temperature distribution throughout the solar nebula at this preliminary stage in the formation of the solar system. Obviously, all the common icy substances in the inner regions of the solar nebula were vaporized by these high temperatures. Only the rocky substances remained solid. That is why the inner four planets are today composed primarily of dense, rocky material. In contrast, snowflakes and ice-coated dust grains could survive in the cooler, outer portions of the solar nebula. That is why the Jovian planets have such low average densities. Indeed, spectral lines of methane and ammonia have been identified in the spectra of some of

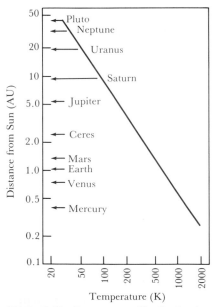

Figure 6-10 Temperature distribution in the solar nebula

This graph shows how the temperature probably varied across the solar nebula as the planets were forming. For the inner planets, temperatures ranged from roughly 1200 K at Mercury to 500 K at the orbit of Mars. Beyond Jupiter, temperatures were everywhere less than 200 K.

the outer planets. In addition, many of the satellites of the Jovian planets are either partially or almost entirely composed of ices.

The terrestrial planets formed from rocky material, whereas the Jovian planets also incorporated vast amounts of ices and gases

The formation of the inner four planets was dominated by the fusing together of solid, rocky particles. Initially, dust grains coalesced and accumulated into objects called **planetesimals,** with diameters of about 100 km. Planetesimal formation took a few million years, as neighboring dust grains and pebbles in the solar nebula collided and were held together by electrostatic or gravitational forces.

During the next stage, gravitational attraction between the planetesimals caused them to collide and coalesce into still larger objects called **protoplanets.** This accumulation of material through the action of gravity is called **accretion.** In recent years, astronomers have used detailed computer simulation to improve our understanding of accretion. The computer is programmed to simulate a large number of planetesimals circling a hypothetical newborn sun along orbits dictated by Newtonian mechanics. Such studies show that accretion continues for roughly 100 million years and typically forms about half a dozen planets.

A particularly successful computer simulation by George W. Wetherill is summarized in Figure 6-11. The calculations begin with 100 planetesimals, each having a mass of 1.2×10^{26} g. This ensures that the total mass (1.2×10^{28} g) equals the mass of the four terrestrial planets (Mercury through Mars) plus their satellites. The initial orbits of these planetesimals are inclined to each other by angles less than 5°, to simulate a thin layer of asteroidlike objects orbiting the protosun.

After an elapsed time of 30 million years, the 100 original planetesimals have coalesced into 22 protoplanets. After 79 million years, 11 larger protoplanets remain. Nearly another 100 million years elapse before the total number of growing protoplanets is reduced to six. Figure 6-11c shows four planets following nearly circular orbits after a total elapsed time of 441 million years. It should be noted, however, that this exercise does not exactly reproduce the inner solar system. In the simulation, the fourth planet from the Sun ends up being the most massive. In fact, however, Earth is nine times more massive than Mars. Nevertheless, agreement with most characteristics of the inner solar system is very striking.

Figure 6-11   Accretion of the terrestrial planets
These three drawings show the results of a computer simulation of the formation of the inner planets. (a) The simulation begins with 100 planetesimals. (b) After 30 million years, these planetesimals have coalesced into 22 protoplanets. (c) This final view is for an elapsed time of 441 million years, but the formation of the inner planets is essentially complete after only 150 million years. (Adapted from "The Formation of the Earth from Planetsimals," by G. W. Wetherill. Copyright © 1981 by Scientific American, Inc.)

a   b   c

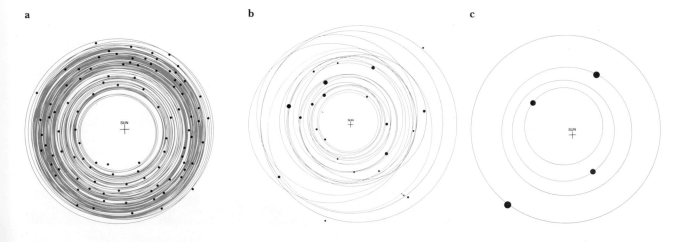

The material from which the inner protoplanets accreted was rich in mineral-forming elements with high condensation temperatures. Iron, silicon, magnesium, and sulfur were particularly abundant, followed closely by aluminum, calcium, and nickel. Energy released by violent impacts of the planetesimals with the growing protoplanets—as well as the decay of radioactive elements—melted all this rocky material. The terrestrial planets therefore began their existence as spheres of molten rock. It was during this time that denser iron-rich minerals sank to the center of the planets, forcing the less-dense silicon-rich minerals to the

Figure 6-12   Formation of the solar system
This series of sketches shows major stages in the birth of the solar system spanning 100 million years. Terrestrial planets accrete from rocky material in the warm, central regions of the solar nebula. Meanwhile, the huge, gaseous Jovian planets form in the cold outer regions. (a) The solar nebula in its initial stages. (b) The early solar system after 50 million years. (c) Planetary formation is nearly complete after 100 million years.

surface. That is why planets such as the Earth have dense iron cores surrounded by less-dense rock.

Of course, the preceding paragraphs present conclusions drawn from many different kinds of evidence, much of which we will discuss in following chapters. Some of the details of this story will probably be revised in coming decades as we gather more evidence from spacecraft and orbiting observatories. However, most of the new discoveries of the past few decades have confirmed the broad outline of this history. Therefore, astronomers are fairly confident that the solar system did form in the fashion we are outlining here.

The formation of the outer planets probably proceeded slightly differently than that of the inner planets. Recall that the original solar nebula was a rotating disk of snowflakes and ice-coated dust particles. Analyses of the behavior of the outer regions of such a disk show that it would probably break up into rings around the developing protosun, separated by regions of space fairly free of matter. The material in each ring then would coalesce to form a large protoplanet. This protoplanet would be largely gaseous because the accumulation of the small particles releases enough energy to vaporize the abundant ices. These huge and massive protoplanets would then sweep up vast amounts of gaseous hydrogen and helium as they move along their orbits. The final result is four huge planets, each with an enormously thick atmosphere surrounding an Earth-sized core of rocky material. Figure 6-12 summarizes this story of the formation of the outer solar system.

During the millions of years while the planets are forming, temperatures and pressures at the center of the contracting protosun continue to climb. Finally temperatures at the center of the protosun reach 8 million degrees, hot enough to ignite thermonuclear reactions, and the Sun is born. As we shall see in Chapter 20, detailed calculations demonstrate that Sun-like stars take approximately 100 million years to form from the prestellar nebula. Therefore, the Sun became a full-fledged star at roughly the same time that the accretion of the inner protoplanets was complete.

A newborn star adjusts fairly violently to the onset of thermonuclear reactions at its core, and often the star's tenuous outermost layers are vigorously expelled into space. This brief burst of mass loss is observed in many young stars across the sky and is called a **T Tauri wind** after the star in Taurus ("the bull") where it was first identified. The T Tauri wind that heralded the birth of the Sun swept the solar system clean of excess gases, thereby preventing further accretion. Many small rocks were left behind to pelt the planets over the next half billion years. For all practical purposes, however, the formation of the solar system was complete when the T Tauri wind began to blow.

The Sun today continues to lose matter gradually in a very mild fashion. This on-going, gentle mass loss is called the **solar wind** and consists of high-speed protons and electrons leaking away from the Sun's outer layers. In later chapters, we shall see how each of the planets carves out its own distinctive cavity in this solar wind.

Summary

. The four inner planets of the solar system share many characteristics and are distinctly different from the four giant outer planets.

The four inner (terrestrial) planets are relatively small (diameters 5000–13,000 km), have high average densities (4–5.5 g/cm^3), and are composed primarily of rock.

The giant outer (Jovian) planets have large diameters (50,000–144,000 km), low densities (1–2 g/cm^3), and are primarily composed of hydrogen and helium.

Pluto, the outermost planet, resembles an icy asteroid.

- Spectroscopy, the study of spectra, provides information about the chemical composition of distant objects.

   Spectral lines serve as distinctive "fingerprints" for elements and chemical compounds in objects from which light comes.

   Each chemical element has a particular kind of atom; atoms of various elements can combine to form molecules of chemical compounds.

   Hydrogen and helium, the lightest elements, were formed shortly after the creation of the universe; the heavier elements were produced much later in the centers of stars and cast into space when the stars died.

   By weight, 98 percent of the matter in the universe is hydrogen and helium.

- Planet-forming substances can be classified as gas, ice, and rock depending on their condensation temperature. The terrestrial planets are composed primarily of rock, whereas the Jovian planets are composed primarily of gas and vaporized ice.

- The solar system formed from a disk-shaped nebula (or cloud) of hydrogen and helium that also contained ice and dust particles.

   The inner planets formed through the accretion of dust particles into planetesimals and then into larger protoplanets.

   The outer planets probably formed through the breakup of the outer nebula into rings of gas and ice-coated dust that coalesced into huge protoplanets.

   The Sun formed by accretion at the center of the nebula; after about 100 million years, temperatures at the protosun's center were high enough to ignite thermonuclear reactions.

   When the protosun became a star, excess gas was vigorously blown away from the Sun thereby ending the process of planet formation.

   The concepts covered in Boxes 6-2 and 6-3 will be especially helpful in your further study of astronomy.

Review questions

1 Why do astronomers almost always use the Kelvin temperature scale in their work rather than the Celsius or Fahrenheit scales?

2 If hydrogen and helium account for 98 percent of the mass of all the material in the universe, why aren't the Earth and Moon composed primarily of these two gases?

3 Why are water (H_2O), methane (CH_4), and ammonia (NH_3) comparatively abundant substances?

4 Suppose you are trying to determine the chemical composition of the atmosphere of a planet by observing its spectrum. Chemicals in the Earth's atmosphere also produce spectral lines (often called *telluric lines*) in the spectra you observe. Can you think of a way of distinguishing between the telluric lines and the spectral lines of the distant planet?

5 Why is it reasonable to suppose that the solar system was formed during a relatively short interval?

6 Would you expect very old stars to possess planetary systems? If so, what types of planets would they have? Explain.

7 Compare and contrast the optics of a prism spectrograph with those of a grating spectrograph (see Box 6-2). What are the advantages and disadvantages?

Advanced questions

*8 Mars has two small satellites, Phobos and Deimos. Phobos circles Mars once every 0.31891 days at an average altitude of 5990 km above the planet's surface. The diameter of Mars is 6790 km. Using this information, calculate the mass and average density of Mars.

*9 At what temperature will the reading on a Celsius thermometer equal that on a Fahrenheit thermometer? At what temperature will the reading on a kelvin thermometer equal that of a Fahrenheit thermometer?

10 Suppose there were a planet having roughly the same mass as the Earth, but located at 50 AU from the Sun. What do you think this planet would be made of? On the basis of this speculation, assume a reasonable density for this planet and calculate its diameter. How many times bigger (or smaller) than the Earth is it?

Discussion questions

11 Propose an explanation for the fact that the Jovian planets are orbited by terrestrial-like satellites.

12 Suppose a planetary system is now forming around some protostar in the sky. In what ways might this planetary system turn out similar to or different from our own solar system?

For further reading

Beatty, J., et al., eds. *The New Solar System, 2nd ed.* Sky Publ. & Cambridge U. Press, 1982. *A fine up-to-date collection of articles by expert planetologists on our modern view of the solar system.*

Chapman, C. *Planets of Rock and Ice.* Scribner's, 1982. *A well-written guide to the terrestrial planets.*

Cohen, B. "Are We Beginning to Understand T Tauri Stars?" *Sky & Telescope,* Oct. 1981, p. 300.

Kaufmann, W. *Planets and Moons.* W. H. Freeman, 1979. *A nontechnical introduction to the solar system.*

Lewis, J. "The Chemistry of the Solar System." *Scientific American,* Sept. 1975. *A special issue devoted to the solar system.*

Murray, B., ed. *The Planets.* W. H. Freeman, 1983. *A collection of articles about the solar system.*

Reeves, H. "The Origin of the Solar System." *Mercury,* Mar./Apr. 1977, p. 7.

Mercury and the Moon

Mercury, like our Moon, has a heavily cratered surface and no atmosphere. Both worlds are shown here to the same scale; Mercury's diameter is 4878 km, whereas the Moon's is 3476 km. For comparison, the distance from New York to Los Angeles is 3944 km (2451 miles). Mercury has a substantial iron-rich core and a magnetic field; the Moon does not. Daytime temperatures at the equator on Mercury reach 430°C (800°F), hot enough to melt lead and tin. One half of Mercury was photographed at close range by a spacecraft in the 1970s. (NASA; Lick Observatory)

Because it orbits near the Sun and is very small, Mercury remained a planet of many mysteries until quite recently. After discussing the difficulties and results of Earth-based visual observations, we turn to the more fruitful studies of the past two decades. We learn that radio and radar observations provided information about the planet's temperature and rotation. Then we look at the dramatic pictures of its surface and the information about its magnetic field and interior that were obtained in 1974 with the spacecraft Mariner 10. We find that Mercury has an Earth-like interior, but that its heavily cratered surface (like the surface of the Moon) retains the scars of events that occurred soon after the planets were formed.

Until 1974, we knew very little about the smallest planet that formed in the warm inner regions of the solar nebula. Today we realize that Mercury has a dual personality: a Moon-like surface but an Earth-like interior. Information about Mercury was difficult to obtain for two simple reasons: Mercury is very small, and it is very near the Sun. Indeed, Mercury is so close to the Sun that most people (including many astronomers) have never seen it.

Earth-based optical observations of Mercury are difficult and often disappointing

Mercury circles the Sun at an average distance of 0.387 AU (57.9 million kilometers = 36.0 million miles). Mercury's orbit is, however, more eccentric than the orbit of any other planet except Pluto. The distance between Mercury and the Sun varies by nearly 24 million kilometers, from 0.306 AU at perihelion to 0.467 AU at aphelion. Figure 7-1 is a scale drawing of the orbits of Mercury and the Earth. (Box 7-1 summarizes data about the planet.)

The best opportunities to see Mercury occur when the planet is positioned as far from the Sun in the sky as it can be—at greatest eastern or western elongation. For a few days near the time of greatest eastern elongation, Mercury appears as an "evening star," hovering low over the western horizon for a short time after sunset. Alternatively, near the time of greatest western elongation, Mercury is glimpsed as a "morning star," heralding the rising Sun in the brightening eastern sky.

Because its orbit is so close to the Sun (see Figure 7-1), Mercury's maximum elongation is only 28°. The celestial sphere rotates at 15° per hour (360° divided by 24 hours), so Mercury never rises more than two hours before sunrise or sets more than two hours after sunset. Unfortunately, the tilt of the Earth's axis and the inclination of Mercury's orbit to the ecliptic often place Mercury much less than 28° from the horizon at the moment of sunset or sunrise. This means that some of the elongations are "favorable" whereas others are "unfavorable," as sketched in Figure 7-2. Mercury's synodic period is 115.9 days (approximately $\frac{1}{3}$ year), and thus a total of six or seven greatest elongations (both eastern and western) occur each year. Typically, only two of these elongations will be favorable for viewing the planet.

Figure 7-1   Mercury's orbit
Mercury circles the Sun every 88 days along an elliptical orbit. The distance between Mercury and the Sun varies from 46 million kilometers at perihelion to 70 million kilometers at aphelion. Earth's orbit is less eccentric than Mercury's, and the average Earth–Sun distance is 150 million kilometers.

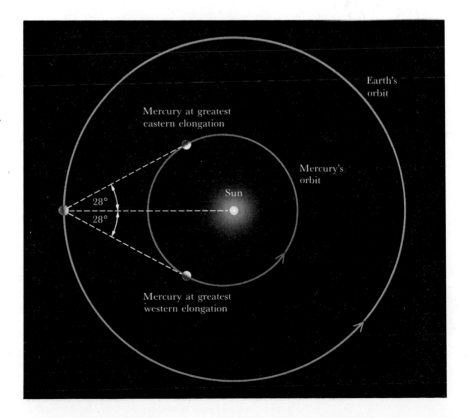

Box 7-1   Mercury data

Mean distance from Sun: 0.387 AU = 5.79×10^7 km
Maximum distance from Sun: 0.467 AU = 6.97×10^7 km
Minimum distance from Sun: 0.306 AU = 4.59×10^7 km
Mean orbital velocity: 47.9 km/sec
Sidereal period: 87.969 days
Rotation period: 58.646 days
Inclination of equator to orbit: 0°
Inclination of orbit to ecliptic: 7° 00′ 16″
Orbital eccentricity: 0.206
Diameter (equatorial): 4878 km
Diameter (Earth = 1): 0.382
Apparent diameter as seen from Earth: maximum = 12.9″
                                                                     minimum = 4.5″

Mass: 3.3×10^{23} kg
Mass (Earth = 1): 0.0558
Mean density: 5.42 g/cm³
Surface gravity (Earth = 1): 0.38
Escape velocity: 4.3 km/sec
Oblateness of planet: 0
Mean surface temperatures: day = +350°C
                                                      night = −170°C

Albedo: 0.1
Brightest magnitude: −1.9
Mean diameter of Sun as seen from Mercury: 1° 22′ 40″

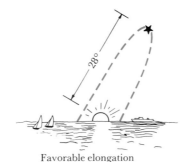 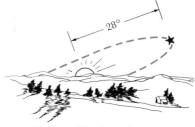

Favorable elongation   Unfavorable elongation

Figue 7-2   Favorable versus unfavorable elongations
Geometrical factors such as the tilt of the Earth's axis, the inclination of Mercury's orbit, and the latitude of the observer on the Earth combine to make an elongation either "favorable" or "unfavorable" for viewing the planet.

Although it is usually inconveniently placed, Mercury is not necessarily dim. Indeed, Mercury is often one of the brightest objects in the sky, shining with a magnitude of −1.9 at greatest brilliancy. That is brighter than the brightest stars.

Like all planets, Mercury shines by reflected sunlight. The fraction of incoming sunlight that a planet reflects is called its **albedo.** Mercury reflects about 10 percent of the sunlight that falls on its rocky surface, and thus its albedo is about 0.1.

Mercury travels around the Sun faster than any other object in the solar system, taking only 88 days to complete a full orbit. Thus Mercury passes through inferior conjunction at least three times a year, and you might expect occasionally to see Mercury silhouetted against the Sun. Such a passage in front of the Sun is called a solar **transit.** In fact, transits of Mercury across the Sun are not very common because Mercury's

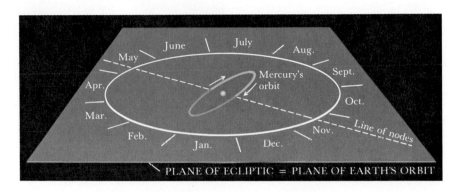

Figure 7-3 The inclination of Mercury's orbit
Mercury's orbit is inclined 7° to the plane of the ecliptic. The angular diameter of the Sun as seen from Earth is only ½°, and so Mercury is usually located several degrees north or south of the Sun at inferior conjunction. Transits are seen only in May and November when both Mercury and the Earth are near the line of nodes of Mercury's orbit.

orbit is tilted 7° to the plane of the ecliptic. As sketched in Figure 7-3, Mercury usually lies well above or below the solar disk at the moment of inferior conjunction.

In order for a transit to be seen, the Sun, Mercury, and the Earth must be in nearly perfect alignment. This occurs only in May and November when the Earth is located near the line of nodes of Mercury's orbit. Only three transits will occur during the remainder of this century:

. November 13, 1986

. November 6, 1993

. November 15, 1999

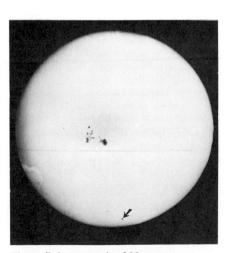

Figure 7-4 A transit of Mercury
Roughly a dozen transits of Mercury occur each century. This photograph shows the tiny planet silhouetted against the Sun during the transit of November 14, 1907. (Yerkes Observatory)

The closer Mercury is to the Sun, the farther it can be from the line between the Earth and the Sun's center and still appear in transit from the Earth. November transits occur when Mercury is near perihelion, and thus November transits are roughly twice as common as May transits. However, the longest transits occur in May when Mercury is near aphelion and therefore is traveling comparatively slowly along its orbit. The maximum duration of a transit is nine hours. Figure 7-4 is a photograph of a transit. Note how tiny the planet seems in comparison to the Sun.

Naked-eye observations of Mercury are best made at dusk or dawn, but the best telescopic views are obtained at midday when the planet is high in the sky, far above the degrading atmospheric effects near the horizon. A yellow filter can eliminate much of the scattered blue light from the sky. The photographs in Figure 7-5, among the finest Earth-based views of Mercury ever recorded on film, were taken at midday.

Because of its small size and nearness to the Sun, you cannot see much surface detail on Mercury through an Earth-based telescope. At best only a few faint, hazy markings can be identified. During the 1880s, the Italian astronomer Giovanni Schiaparelli used these markings in an attempt to produce the first map of Mercury. His telescopic views of Mercury were so vague and indistinct that he made a major error, which

Figure 7-5 Earth-based views of Mercury
These two views are among the finest photographs of Mercury ever produced with an Earth-based telescope. Hazy markings are faintly visible on the tiny planet. (New Mexico State University Observatory)

unfortunately went uncorrected for over half a century. Schiaparelli erroneously concluded that Mercury keeps the same side always facing the Sun.

Radio and radar observations of Mercury revealed its rotation rate

Synchronous rotation, in which the rotation period of an object equals its period of revolution, is a common phenomenon in the solar system. The Moon exhibits synchronous rotation, always keeping the same side exposed to our Earth-based view. The two moons of Mars and many of the satellites of Jupiter and Saturn also keep the same side facing their parent planet. Newtonian mechanics demonstrates that this is a stable situation, usually called a **1-to-1 spin–orbit coupling,** but this is not how Mercury rotates. The first clue about Mercury's true rotation period came in 1962 when astronomers detected radio radiation from the planet.

Every object emits radiation. The wavelength (λ_{max}) at which its most intense radiation is emitted depends on the temperature (T) of the object. Cool objects emit primarily long-wavelength radiation, whereas hotter objects shine with shorter-wavelength radiation. Specifically, if λ_{max} is measured in centimeters and T is in kelvins, then

$$\lambda_{max} = \frac{0.3}{T}$$

This relationship is called **Wien's law** after the German physicist Wilhelm Wien who discovered it in 1893. For example, an object at 100 K emits radiation primarily at a wavelength of 0.003 cm (30 μ). Only at absolute zero does an object emit no radiation at all.

If Schiaparelli had been correct, then one side of Mercury would be in perpetual, frigid darkness, never exposed to the warming rays of the nearby Sun. Thus, radio astronomers at the University of Michigan were surprised in 1962 to discover that the wavelengths of radio radiation coming from Mercury indicated that the temperature on the planet's nighttime side was not nearly as cold as expected. Schiaparelli's belief in synchronous rotation was so well accepted that some astronomers speculated about an atmosphere on Mercury whose winds could carry warmth from the daytime side around to the nighttime side of the planet. As discussed in detail in Box 7-2, however, the temperatures on Mercury are too high and the planet's gravity too weak to retain any substantial atmosphere at all.

Box 7-2 Thermal motion and the retention of an atmosphere

A moving object possesses energy. The faster it moves, the more energy it has. Energy of this type is called **kinetic energy.** If an object of mass m is moving with a speed v, its kinetic energy is given by

$$\frac{1}{2}mv^2$$

This expression for kinetic energy is valid for all objects, both big and small, from atoms and molecules to planets and stars, as long as the speed is small in comparison to the speed of light. If the mass is in grams and the speed in centimeters per second, then the energy is expressed in ergs. For example, a paper airplane with a mass of 2 g drifting through the air at 1 cm/sec has a kinetic energy of 1 erg.

Any substance with temperature above absolute zero possesses **thermal energy.** This thermal energy is the kinetic energy of the moving atoms or molecules in the substance. In a cold substance, the atoms move slowly; in a hot substance, they move much more rapidly.

Consider a gas, such as the atmosphere of a star or planet. If the gas is hot, it probably consists of individual atoms moving at high speeds. For example, the Sun's atmosphere consists primarily of hydrogen and helium atoms. If the gas is cool, the atoms typically combine to form molecules. For example, the Earth's atmosphere consists primarily of nitrogen (N_2) and oxygen (O_2) molecules. The Martian atmosphere is mostly carbon dioxide (CO_2) molecules.

According to the kinetic theory of gases developed during the nineteenth century, the amount of thermal energy per atom or molecule in a gas is given by

$$\frac{3}{2}kT$$

where T is the temperature of the gas in kelvins and k is the Boltzmann constant, which has the value $k = 1.38 \times 10^{-16}$ erg/K.

The thermal energy is just the kinetic energy of the atom or molecule, so we can write the equality

$$\frac{1}{2}mv^2 = \frac{3}{2}kT$$

where v represents the average speed of an atom or molecule in a gas with temperature T. Rearranging this equation, we obtain

$$v = \sqrt{\frac{3kT}{m}}$$

(This value actually is slightly higher than the average speed of the atoms or molecules in the gas, but it is close enough for our purposes here. If you are studying physics, you may know that v is actually the root-mean-square average speed.)

For example, suppose you want to know the average speed of the oxygen molecules that you breathe at a room temperature of 72°F ($= 22$°C $= 295$ K). From a reference book, you can find that the mass of an oxygen atom is 2.66×10^{-23} g. The mass of an oxygen molecule (O_2) is twice the mass of an atom, or $2(2.66 \times 10^{-23}) = 5.3 \times 10^{-23}$ g. Thus the average speed is

$$v = \left(\frac{3(1.38 \times 10^{-16})(295)}{5.3 \times 10^{-23}}\right)^{1/2} = 4.8 \times 10^4 \text{ cm/sec}$$

or almost exactly 1000 miles per hour.

Obviously, atoms and molecules in a gas are moving rapidly, even at

(continued)

(Box 7-2, continued)
moderate temperatures. Indeed, the speeds may be so great that a planet's gravity is not strong enough to retain an atmosphere.

Astronomers find it very useful to speak of the **escape velocity** (or speed) of a planet in trying to decide if an object can permanently leave the planet and escape into interplanetary space. According to Newtonian mechanics, the escape velocity from a planet of mass M and radius R is given by

$$V_{\text{escape}} = \sqrt{\frac{2GM}{R}}$$

where G is the universal constant of gravitation ($G = 6.67 \times 10^{-8}$ dyne cm^2/g^2).

The table gives the escape velocity for various objects in the solar system. For example, astronauts must leave the Earth with a speed greater than 11.2 km/sec (25,100 miles per hour).

	Escape velocity	
	km/sec	**miles/hour**
Mercury	4.3	9,600
Venus	10.3	23,000
Earth	11.2	25,100
Moon	2.4	5,400
Mars	5.0	11,000
Jupiter	59.5	133,000
Saturn	35.6	79,600
Uranus	21.2	47,400
Neptune	23.6	52,800

In a planet's atmosphere, some molecules are moving slowly and others are moving more swiftly. Nevertheless, the *average* speed of a particular kind of molecule can be calculated if you know the planet's atmospheric temperature. A good rule of thumb is this: a planet can retain a gas if the escape velocity is at least six times greater than the average speed of the molecules. In such a case, very few molecules will be moving fast enough to escape from the planet's gravity.

For example, consider the Earth. We saw that the average speed of oxygen molecules is 0.48 km/sec. The escape velocity from the Earth (11.2 km/sec) is much more than six times the average speed of the oxygen molecules, so the Earth has no trouble keeping the oxygen in its atmosphere.

A similar calculation for hydrogen molecules (H_2) gives a different result. At 295 K, the average speed of a hydrogen molecule is 1.9 km/sec. Six times this speed is 11.4 km/sec, which is slightly higher than the escape velocity from the Earth. Thus any hydrogen present in the Earth's atmosphere slowly leaks away into space. In other words, the Earth does not retain hydrogen in its atmosphere.

A breakthrough came in 1965, when Rolf B. Dyce and Gordon H. Pettengill used the giant 1000-foot radio telescope at the Arecibo Observatory in Puerto Rico (see Figure 7-6) to bounce powerful radar pulses off Mercury. The out-going radiation consisted of microwaves of a very

Figure 7-6   The Arecibo radio telescope
This is the largest telescope on Earth. The dish, which is mounted permanently in a bowl-shaped valley in Puerto Rico, measures 305 m (1000 ft) in diameter. The dish is not movable, so this telescope can be used only to examine objects that are near the zenith. (Arecibo Observatory)

specific wavelength. In the reflected signal echoed back from the planet, the microwaves were spread out over a small range of wavelengths. This wavelength shift occurred because of a phenomenon known as the Doppler effect, and it provided detailed information about Mercury's motion.

The **Doppler effect** is a very important physical principle that astronomers use to determine the speed of anything and everything that emits light in the sky. It is described in Box 7-3, set aside here for future reference when we discuss the motions of stars and galaxies. For now, you need know only that the wavelengths of microwaves reflected from the planet's approaching side were shortened, whereas those from the receding side were lengthened. Thus the radar pulse, which went out at one specific wavelength, came back spread over a small wavelength range. From the width of the spread, Dyce and Pettengill deduced Mercury's speed of rotation, and from that they computed its period of rotation. They found that the rotation period almost certainly is between 54 and

---

Box 7-3   The Doppler effect

In 1842, Christian Doppler (a professor of mathematics at Prague) pointed out that wavelength is affected by motion. As shown in the diagram, light waves from an approaching light source are compressed. The circles represent the peaks of waves emitted from various positions as the source moves along. Because each successive wave is emitted at a position slightly closer to you, you see a shorter wavelength than you would if the source were stationary. All the spectral lines in the spectrum of an approaching source are shifted toward the short-wavelength (blue) end of the spectrum. This phenomenon is called a **blueshift.**

Conversely, light waves from a receding source are stretched. You see a longer wavelength than you would if the source were stationary. All the spectral lines in the spectrum of a receding source are shifted toward the longer-wavelength (red) end of the spectrum, producing a **redshift.** In general, the effect of relative motion on wavelength is called the **Doppler effect.**
(continued)

(Box 7-3, continued)

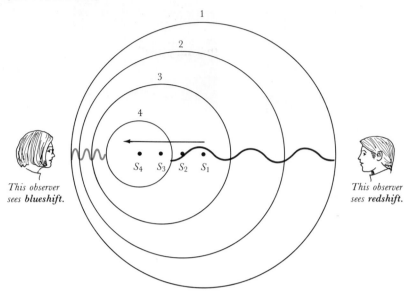

This observer sees blueshift.

This observer sees redshift.

Suppose that λ_0 is the wavelength of a particular spectral line from a source that is not moving. It is the wavelength that you might look up in a reference book or determine in a laboratory experiment for this spectral line. If the source is moving, this particular spectral line is shifted to a different wavelength λ. The size of the wavelength shift is usually written as $\Delta\lambda$, where $\Delta\lambda = \lambda - \lambda_0$. Thus $\Delta\lambda$ is the difference between the wavelength listed in reference books and the wavelength that you actually observe in the spectrum of a star or galaxy.

Christian Doppler proved that the wavelength shift, $\Delta\lambda$, is governed by the following simple equation

$$\frac{\Delta\lambda}{\lambda_0} = \frac{v}{c}$$

where v is the speed of the source measured along the line of sight between the source and the observer. As usual, c is the speed of light (3×10^{10} cm/sec). (This equation is valid only for cases where v is small in comparison to c.)

For example, the spectral lines of hydrogen appear in the spectrum of the bright star Vega (also called α Lyrae). A prominent hydrogen line (called H_α) has a normal wavelength of 6562.85Å, but in Vega's spectrum this line is located at 6562.55Å. The wavelength shift is

$$\Delta\lambda = \lambda - \lambda_0 = 6562.55 - 6562.85$$

$$= -0.30 \text{ Å}$$

and the star is approaching with a speed of

$$v = c\left(\frac{\Delta\lambda}{\lambda_0}\right) = 3 \times 10^5\left(\frac{-0.30}{6562.85}\right) \text{ km/sec}$$

$$= -13.7 \text{ km/sec}$$

(the minus sign indicates motion toward the observer).

A speed determined in this fashion is called a **radial velocity** because v is the component of the star's motion parallel to our line of sight, or along the "radius" drawn from Earth to the star. Of course, a sizeable fraction of a star's motion may be perpendicular to our line of sight. This transverse movement, called **proper motion,** does not affect wavelengths.

64 days. This result is usually written as 59 ± 5 days ("59 plus or minus 5 days"), where 5 days is called the **probable error** in the measurement of 59 days. Note that the probable error is actually a measure of the accuracy (or certainty) of the measurement, not a measure of the error that may really exist in the measurement.

Giuseppe Colombo, an Italian physicist with a long-standing interest in Mercury, found this number very intriguing. Colombo noted that Mercury's sidereal period is 87.969 days, and that

$$\frac{2}{3}\ (87.969 \text{ days}) = 58.65 \text{ days}$$

Colombo therefore boldly speculated that Mercury's true rotation period is exactly 58 days and $15\frac{1}{2}$ hours.

This was not an idle guess. Colombo realized that a rotation period of 58.64 days would mean that Mercury is locked into a **3-to-2 spin–orbit coupling,** meaning that the planet makes 3 complete rotations on its axis for every 2 complete orbits around the Sun. This is a much rarer dynamical stability than the 1-to-1 spin–orbit coupling our Moon and many other satellites exhibit. Both types of spin–orbit coupling are illustrated in Figure 7-7.

Colombo's enlightened guess was dramatically confirmed in the mid-1970s with the flybys of Mariner 10. Mercury does indeed rotate three times about its axis during every two orbits of the Sun. This very slow 58.65-day rotation period is responsible for an unusual phenomenon on Mercury.

Mercury's speed along its orbit varies in accordance with Kepler's second law. The orbital velocity is greatest at perihelion (59 km/sec) and least at aphelion (39 km/sec), as calculated from the formulas in Box 4-3. As seen from Mercury's surface, the Sun rises in the east and sets in the west, just as it does on Earth. When Mercury is near perihelion, however, the planet's rapid motion along its orbit outpaces its leisurely rotation about its axis. The usual east–west movement of the Sun across the sky is interrupted. The Sun actually stops and moves backward (from west to east) for a few Earth days. If you were standing on Mercury watching a sunset near the time of perihelion passage, the Sun would not simply set. It would dip below the western horizon and then come back up, only to set a second time a day or two later.

Figure 7-7   Spin–orbit coupling
(a) The simplest kind of spin–orbit coupling is 1-to-1 coupling, where the rotation period equals the orbital period. The planet always keeps the same side facing the Sun.
(b) Mercury exhibits a 3-to-2 coupling. Mercury rotates three times about its axis during two complete orbits of the Sun.

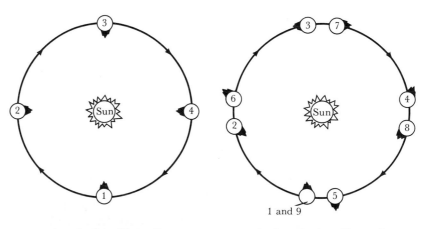

a   1-to-1 spin–orbit coupling   b   3-to-2 spin–orbit coupling

**Photographs from Mariner 10
revealed Mercury's heavily-
cratered, Moon-like surface**

Our first detailed knowledge about Mercury's surface was acquired during the last week of March 1974, when a spacecraft called Mariner 10 coasted within 756 km (470 miles) of the planet's surface. Mariner 10 was launched from Earth on November 3, 1973. On February 5, 1974, the spacecraft flew past Venus at a distance of 5800 km (3600 miles) above the planet's cloudtops. The trajectory of this flyby was designed so that Venus's gravitational pull would redirect the spacecraft toward Mercury. This was the first of several historic "gravity-assisted" missions, in which a close encounter with one planet was used gravitationally to catapult a spacecraft on to another planet.

The Mariner 10 spacecraft is shown in Figure 7-8. It is fairly typical of the spacecraft that were involved in missions to the planets during the 1970s. Two television cameras mounted on a steerable platform obtained pictures of the planet. In addition, a device called an infrared radiometer measured the planet's surface temperature. Two ultraviolet instruments searched for a Mercurian atmosphere, while magnetometers measured Mercury's magnetic field. Throughout the mission, charged particle and plasma detectors examined the solar wind.

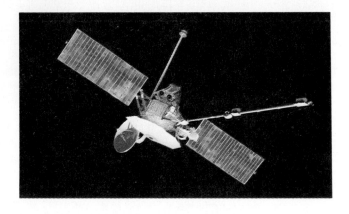

Figure 7-8   The Mariner 10 spacecraft
Mariner 10 weighed 500 kg (1100 pounds) and measured 7.6 m (25 ft) from one end of the extended solar panels to the other. A total of 8122 solar cells provided electricity for the television cameras, magnetometers, spectrometers, and associated equipment. (NASA)

The flight path of Mariner 10 carried the spacecraft through Mercury's shadow, past the darkened, nighttime side of the planet, on March 29, 1974. Naturally, no television pictures could be taken of Mercury's unilluminated hemisphere. Photography was therefore divided into two parts: "incoming views" taken prior to closest approach, and "outgoing views" taken after Mariner 10 emerged from Mercury's shadow. The best incoming and outgoing photomosaics of the entire planet are seen in Figures 7-9 and 7-10.

As Mariner 10 closed in on Mercury, scientists were surprised by the Moon-like pictures appearing on their television monitors. It was obvious that Mercury is a barren, desolate, and heavily cratered world. Figure 7-11 shows a typical closeup view sent back from Mariner 10.

Although the first impression was of a lunar landscape, closer scrutiny of Mercury's surface revealed some significant nonlunar characteristics. For comparison, Figure 7-12 is a photograph of the Moon's southern hemisphere. Notice how the lunar craters are densely packed, one overlapping the next. In sharp contrast, Mercury's surface has extensive **intercrater plains.**

Figure 7-9 [above left]   Mariner 10's incoming view
This view of Mercury was sent back from Mariner 10 as the spacecraft sped toward the planet. Eighteen pictures taken at 42-second intervals were assembled into this photomosaic. The spacecraft was 200,000 km (124,000 miles) from the planet at the time these pictures were taken. (NASA)

Astronomers believe that most of the craters on both Mercury and the Moon were produced during a period of 700 million years immediately after the planets formed. The strongest evidence comes from direct analysis and dating of Moon rocks brought back by the Apollo astronauts. Debris remaining after planet formation rained down on these young worlds, gouging out most of the craters we see today.

Essentially all astronomers agree that the Moon and terrestrial planets must have been completely molten spheres of liquid rock at first. After

Figure 7-10 [above right]   Mariner 10's outgoing view
This view of Mercury was sent back from Mariner 10 as the spacecraft coasted away from the planet, shortly after emerging from Mercury's shadow. Eighteen pictures taken at 42-second intervals were assembled into this photomosaic. These pictures were taken from a distance of 210,000 km (130,000 miles). (NASA)

Figure 7-11 [right]   Mercurian craters and intercrater plains
This view of Mercury's northern hemisphere was taken by Mariner 10 at a range of 55,000 km (34,000 miles) from the planet's surface. Numerous craters and extensive intercrater plains appear in this photograph, which covers an area 480 km (300 miles) wide. (NASA)

only a few hundred million years, their surfaces solidified as the rock cooled. Nevertheless, large meteoroids easily punctured the thin cooling crusts, allowing molten lava to well up from the planets' interiors. Older craters were obliterated as seas of molten rock flooded across portions of the planets' surfaces. As we shall see in Chapter 10, there is clear evidence of extensive lava flooding still visible on the Moon.

The planets did not, however, cool down at the same rate. The volume of a planet is proportional to the cube of its radius, whereas the surface area is proportional to the square of the radius. Consequently, a small planet has a higher ratio of surface area to volume than does a big planet. This means that a small planet can more easily radiate its internal heat into space and can cool off more rapidly than a big planet.

Because Mercury is larger than the Moon, it took longer for a thick protective crust to form on Mercury than it did on the Moon. Throughout its early history, molten rock seeped up through cracks and fissures in Mercury's young, frail crust, and volcanism was probably pervasive. The resulting lava flows certainly inundated many older craters, leaving behind the broad, smooth intercrater plains seen by Mariner 10.

Mariner 10 also revealed numerous long cliffs called **scarps** meandering across Mercury's surface (see Figure 7-13). These scarps probably formed as the planet cooled from its initial molten state. As the interior

Figure 7-12 [above]   Lunar craters
This Earth-based photograph shows a portion of the Moon's southern hemisphere during last quarter moon. Densely packed craters fill the view which covers an area approximately 600 km (370 miles) wide. (Mount Wilson and Las Campanas Observatories)

Figure 7-13 [right]   The Victoria scarp
This view of Mercury's northern limb was taken from a range of 77,800 km (48,300 miles). A long scarp extends southward for several hundred kilometers. This photograph covers an area measuring roughly 550 km (340 miles) across. (NASA)

Figure 7-14 [below]   Unusual, hilly terrain
Jumbled, hilly terrain covers nearly half a million square kilometers of Mercury's "incoming hemisphere." The large, smooth-floored crater near the center of this photograph has a diameter of 170 km (106 miles). (NASA)

solidified, the planet contracted, causing the crust to wrinkle. Most geologists agree that the scarps are the ridges and wrinkles thrust up during this period of compression. The Mercurian scarps are named after famous sailing ships. The Victoria scarp shown in Figure 7-13 takes its name from Magellan's ship, the first to sail around the world. Some scarps rise as much as 3 km (2 miles) above the surrounding Mercurian plains.

While examining the incoming views sent back by Mariner 10, scientists noticed unusual, closely packed, jumbled hills covering a very specific region slightly south of the planet's equator. Figure 7-14 is a wide-angle view of this puzzling terrain. The hills are about 5 to 10 kilometers wide and have elevations between 100 to 1800 meters (300 to 6000 feet). The total area covered by this chaotic terrain is about 500,000 square kilometers (193,000 square miles).

Figure 7-15   The Caloris basin on Mercury
Mariner 10 sent back this view of a huge impact basin on the terminator of Mercury's "outgoing hemisphere." The outer rim of the basin is defined by a ring of mountains up to 2 km (6500 ft) high. The diameter of the basin is 1300 km (810 miles). (NASA)

This peculiar, hilly terrain was explained by what scientists saw in the outgoing views of Mercury. Looking back at the receding planet, Mariner 10 photographed a huge **impact basin** just north of the equator, along the line dividing day from night. (This dividing line is called the **terminator.**) Slightly more than half of the impact basic is hidden on the night side of the terminator in the Mariner 10 views (see Figure 7-15).

This huge impact feature is called the Caloris basin, from the Latin word for "hot," because the Sun is directly over the Caloris Basin during alternating perihelion passages. It is therefore the hottest place on the planet once every 176 days.

The Caloris Basin is 1300 km (810 miles) in diameter. It was formed by the impact of a very large meteoroid near the end of the crater-making period of bombardment that dominated the first billion years of the solar system. We know that the Caloris impact occurred late in this period because there are relatively few fresh craters on the extensive lava flow that filled the basin.

The Caloris Basin is exactly on the opposite side of the planet from the jumbled terrain that Mariner 10 saw on the "incoming hemisphere." The Caloris impact must have been a violent event that shook the planet with earthquake-like disturbances called **seismic waves.** Geologists contend that seismic waves from the Caloris impact were focused as they passed through Mercury (see Figure 7-16). Jumbled hills were pushed up as this concentrated seismic energy reached the far surface of the planet.

Features similar to the Caloris Basin exist on our Moon. The best example is the Orientale Basin, shown in Figure 7-17. Its diameter is 900 km (560 miles). The scarcity of fresh craters attests to its late formation. Furthermore, astronomers' theory about the seismic upthrusting of jumbled hills on Mercury is supported by the fact that similar (although less extensive) chaotic hills are found on the opposite side of the Moon from the Orientale Basin.

Figure 7-16 [left]   The Caloris impact
Seismic waves from the Caloris impact were focused slightly as they traveled through the planet. The focused waves pushed up the numerous jumbled hills on the opposite side of Mercury.

Figure 7-17 [right]   The Orientale basin on the Moon
This is one of several large impact basins on the Moon that resemble the Caloris Basin on Mercury. The outermost ring of peaks, called the Cordillera Mountains, forms a circle 900 km (560 miles) in diameter. This photograph was sent back in 1967 from the Lunar Orbiter IV spacecraft. The black lines across the picture exist because the photograph was transmitted to Earth in thin strips. (NASA)

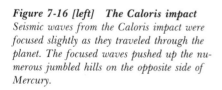

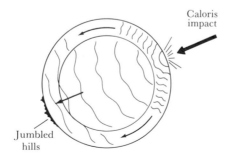

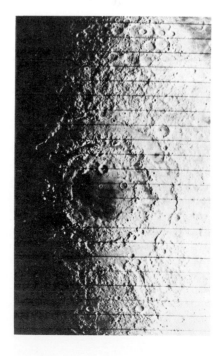

Like the Earth, Mercury has an iron core and a magnetic field

As mentioned in Chapter 6, the average density of the Earth is 5.5 g/cm³, and Mercury's average density is 5.4 g/cm³. Typical rocks from the Earth's surface (as well as from the Moon and Mars) are composed primarily of silicon and other light-weight, mineral-forming elements. These surface rocks have a density of only 3 g/cm³. The higher average density of the Earth comes from abundant quantities of iron that sank toward the center of the Earth while our planet was still entirely molten. Less-dense material was forced toward the surface, dividing our planet naturally into an iron **core** and a surrounding, less-dense, rocky **mantle.**

This process, in which dense elements sink toward a planet's center and force less-dense material toward the surface, is called **chemical differentiation.** It is the result of the action of gravity. Chemical differentiation must have occurred during and immediately after the formation of the solar system, while the terrestrial planets were still entirely molten and internal mass motion could occur on a large scale. By studying how the Earth vibrates during earthquakes, geologists have deduced that our planet's iron core is 6940 km (4310 miles) in diameter. For comparison, the Earth's overall diameter is 12,740 km (7920 miles).

Mercury's average density is very slightly less than the Earth's, so you might suspect that Mercury has proportionally a smaller iron core than Earth's. This is not the case. The Earth is 18 times more massive than Mercury. This larger mass pushing down on the Earth's interior compresses the Earth's core much more than Mercury's core is compressed. Calculations reveal that the uncompressed density of the Earth would be about 4 g/cm³, whereas Mercury's uncompressed density would be 5.3 g/cm³. In fact, Mercury is the most iron-rich planet in the solar system, with iron accounting for 65 to 70 percent of the planet's mass. Mercury's iron core has a diameter of 3600 km (2200 miles). Mercury's overall diameter is 4800 km (3030 miles). Figure 7-18 is a scale drawing of the interior structures of Mercury and Earth.

Independent evidence of Mercury's large iron core came from Mariner 10's magnetometers, which discovered that Mercury has a magnetic field. Iron is the only common element that could account for the magnetic field of a terrestrial planet.

As almost anyone with a compass knows, the Earth has a magnetic field. According to a basic concept in physics, magnetism arises whenever electrically charged particles are in motion. The magnetic field around the Earth is similar to the magnetism that surrounds a coil of wire in which electricity is flowing. Many geologists suspect that the Earth's magnetic field originates with electric currents flowing in the liquid portions of our planet's iron core. These currents are carried around by the Earth's rotation and create the planetwide magnetic field. This process, in which rotation of a planet with an iron core produces a magnetic field, is called the **dynamo effect.**

Scientists were surprised to find that Mercury also has a magnetic field. Mercury rotates much more slowly than the Earth (59 days versus 24 hours), and most scientists believed this leisurely rotation would be too slow to induce a magnetic field by the dynamo effect. Nevertheless, Mercury does have a weak field, but the Earth's field is 100 times stronger.

Mariner 10's magnetometers could not prove that Mercury's field is produced by a dynamo effect as is the Earth's. Mercury's core may be completely solid, and its magnetic field may be a "fossil field" frozen into the core when it solidified. Thus Mercury's magnetism may be similar to that of an iron magnet you can buy in a store.

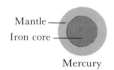

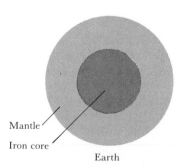

Figure 7-18   The internal structures of Mercury and Earth
Mercury is the most iron-rich planet in the solar system. Mercury's iron core occupies an exceptionally large fraction of the planet's interior.

Mercury's magnetic field shields the planet from the solar wind

Mariner 10's other instruments measured the surface temperature on Mercury and searched for traces of an atmosphere. Temperatures on Mercury vary from 700 K (=427°C = 800°F) at local noon on the equator at perihelion to 100 K (=−173°C = −280°F) at local midnight. This 600 K temperature range is greater than that of any other planet or satellite in the solar system. The only atmosphere Mariner 10 detected was a thin scattering of particles that may have been captured from the solar wind.

As mentioned briefly in the previous chapter, the solar wind is a constant flow of charged particles (mostly protons and electrons) away from the outer layers of the Sun's upper atmosphere. In the vicinity of the Earth, the particles in the solar wind are moving at roughly 400 km/sec, or nearly a million miles per hour.

If a planet possesses a magnetic field, this field will repel and deflect the impinging particles, thereby forming an elongated "cavity" in the solar wind. This cavity is called the planet's **magnetosphere.**

Charged particle detectors on Mariner 10 mapped the structure of Mercury's magnetosphere. The results are shown in Figure 7-19. When the particles in the solar wind first encounter the magnetic field, they are abruptly slowed, producing a bow-shaped **shock wave** that marks the boundary where this sudden decrease in velocity occurs. Still closer to the planet, there is another well-defined boundary where the outward magnetic pressure of the planet's field is exactly counterbalanced by the impinging gas pressure of the solar wind. This boundary is called the **magnetopause.** Between the shock wave and the magnetopause is a turbulent region in which most of the subsonic particles from the solar

Figure 7-19   Mercury's magnetosphere
Mercury's weak magnetic field is just strong enough to carve out a cavity in the solar wind, preventing the impinging particles from directly striking the planet's surface.

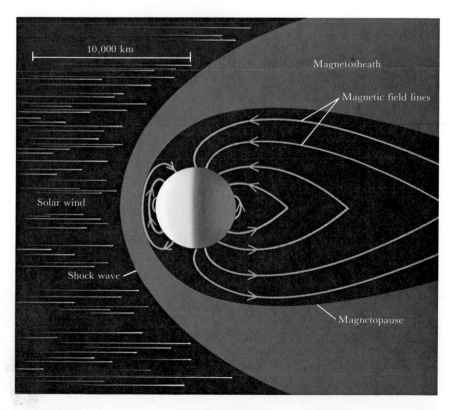

10,000 km

Magnetosheath

Magnetic field lines

Solar wind

Shock wave

Magnetopause

wind are deflected around the planet, just as water is deflected to
either side of the bow of a ship. This turbulent region is called the
magnetosheath. The region inside the magnetopause is the true magnetic domain of the planet.

These features of Mercury's magnetic environment were not entirely
obvious when Mariner first flew past the planet on March 29, 1974. Fortunately, Mariner 10 came back to Mercury for a second and then a
third look. This was possible, thanks to the brilliant suggestion of Giuseppe Colombo. Colombo pointed out that Mariner 10 could easily be
placed in a 176-day orbit about the Sun that would bring the spacecraft
back to Mercury every two Mercurian years (see Figure 7-20). Mariner

Figure 7-20   The orbit of Mariner 10
After coasting past Mercury on March 29,
1974, Mariner 10 went into an orbit whose
period is exactly two Mercurian years.
Consequently, Mercury and the spacecraft pass
close to each other once every 176 days.
However, fuel to stabilize the spacecraft lasted
only for the second and third encounters.

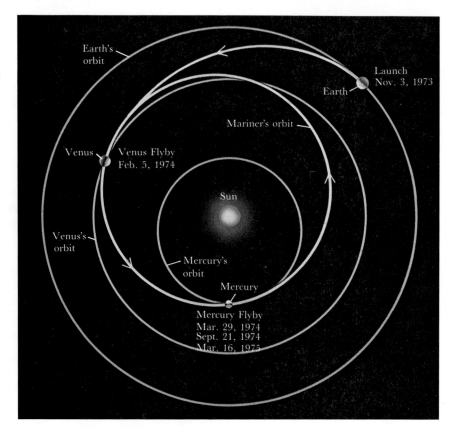

10 came back to Mercury on September 21, 1974, coasting within
48,073 km (29,871 miles) of the planet. The third flyby occurred on
March 16, 1975, when Mariner 10 skimmed over the planet's surface at
the distance of only 327 km (203 miles).

Although important features of Mercury's magnetosphere were determined during these additional flybys, Mariner 10's television cameras did
not see any new portions of Mercury's surface. Because two Mercurian
years equal three Mercurian days, and the orbital period of Mariner 10
was exactly two Mercurian years, the spacecraft was expected to return
to Mercury at the same time of day as on its original visit, with the same
side of Mercury facing the Sun. Indeed, Mariner 10 saw only the familiar craters, so Guiseppe Colombo's theory of a 3-to-2 spin–orbit coupling
was dramatically confirmed.

Mariner 10 still coasts over Mercury's sun-scorched surface every 176 days, but the stabilizing rockets had enough fuel for only the second and third encounters. Today the spacecraft tumbles aimlessly, its eyes blind and its radio voice silent. Astronomers look forward to the possibility of another mission in the 1990s to view the side of Mercury that is forever hidden from Mariner 10.

Summary

. At its greatest eastern or western elongation, Mercury is only 28° from the Sun, so it can be seen only briefly after sunset or before sunrise.

> Solar transits of Mercury occur only in May and November and only about a dozen times per century.

> The poor telescopic views of Mercury's surface led to the mistaken impression that the planet keeps the same side always toward the Sun (1-to-1 spin–orbit coupling).

. Radio observations in 1962 and radar observations in 1965 revealed that Mercury in fact has 3-to-2 spin–orbit coupling; an observer on Mercury would experience $1\frac{1}{2}$ "days" per "year."

. The Mariner 10 spacecraft made three useful passes near Mercury in the mid-1970s, providing pictures of its surface.

> The Mercurian surface is pocked with craters like those of the Moon, but extensive, smooth intercrater plains appear between the craters; these features appear to have formed as the crust of the planet solidified.

> Long and high cliffs called scarps meander across the surface; these scarps probably formed as the planet cooled, solidified, and shrank.

> The impact of a very large object long ago formed the huge Caloris Basin and shoved up jumbled hills on the opposite side of the planet.

. Instruments on Mariner 10 provided information about the magnetic field, temperature, and the very thin atmosphere of Mercury, indicating that the planet has an iron core much like that of the Earth.

> The iron core of Mercury has a radius equal to three-fourths of the planet's radius, whereas the radius of the Earth's core is only slightly more than one-half of the Earth's radius.

> Surface temperatures on Mercury range from 100–700 K, the greatest range of temperature changes known on any of the planets.

> Mercury's atmosphere consists of only a very thin scattering of gas molecules, with a surface pressure about 1/10,000 of the atmospheric pressure at the Earth's surface.

> Mercury's magnetic field produces a magnetosphere surrounding the planet that blocks the solar wind from the surface of the planet.

. The concepts covered in Box 7-3 will be especially helpful in your further study of astronomy.

Review questions

1 Why are naked eye observations of Mercury best made at dusk or dawn, while telescopic observations are best made around noon?

2 Why do astronomers believe that Mercury is "the most iron-rich planet in the solar system"?

3 If the albedo of Mercury were increased, would the planet's surface temperature go up or down? Explain your answer.

4 How can you tell an old crater from a new one?

5 Explain why the Sun is directly over the Caloris Basin on Mercury only during every other perihelion passage.

*6 Find the average kinetic energy of Mercury.

*7 Find the value of λ_{max} for radiation coming from the sunlit side of Mercury.

8 Why is it reasonable to presume that none of the large satellites in the solar system, including our Moon, possess a substantial magnetic field?

Advanced questions

*9 In view of Mercury's 58.6 day rotation period, what difference in wavelength is observed for a spectral line at 5000 Å reflected from either the approaching or receding edge of the planet?

*10 Calculate the minimum molecular weight of a gas that could in theory be retained as an atmosphere by Mercury if the average daytime temperature were 620 K. Are there any abundant gases that meet this minimum criterion? Why doesn't Mercury have an atmosphere of these gases?

Discussion questions

11 If you were planning a return mission to Mercury, in what features and observations would you be particularly interested?

12 What evidence do we have that the surface features on Mercury were not formed during recent geological history?

For further reading

Davies, M., et al., eds. *Atlas of Mercury.* NASA SP-423, 1978. *A full-scale atlas of spacecraft photos and maps.*

Dunne, J., and Burgess, E. *The Voyage of Mariner 10: Mission to Venus & Mercury.* NASA SP-424, 1978.

Hartmann, W. "The Significance of the Planet Mercury." *Sky & Telescope,* May 1976, p. 307.

Murray, B., and Burgess, E. *Flight to Mercury.* Columbia U. Press, 1977.

Murray, B. "Mercury." *Scientific American,* May 1976.

Weaver, K. "Mariner Unveils Venus and Mercury." *National Geographic,* June 1975.

8 Cloud-covered Venus

Venus
Venus and Earth have almost exactly the same size, mass, and surface gravity. However, Venus's thick cloud cover efficiently traps heat from the Sun resulting in a surface temperature (480°C = 900°F) even hotter than on Mercury. Unlike Earth's clouds, which are made of water droplets, Venus's clouds are very dry and contain droplets of concentrated sulfuric acid. This photograph was obtained by the Pioneer Venus spacecraft in 1979. (NASA)

Astronomers have long known that Venus and Earth have many similar properties, such as size, mass, and average density. However, Venus's thick covering of clouds kept Earth-based astronomers from learning much more about the planet for a long time. In this chapter, we see how radio observations from Earth first provided information about the rotation and surface temperature of Venus. Then we turn to the rich and often surprising body of information returned by space probes to Venus. We find that Venus has a dense atmosphere of carbon dioxide with clouds of sulfuric acid droplets, that it is extremely hot and almost entirely lacking water. Venus's surface is surprisingly level and smooth. Finally, we note that our knowledge of Venus's atmosphere has important implications for our theories about the history and the future of our own planet.

At first glance, Venus looks like Earth's twin. The two planets have almost the same mass, the same diameter, the same average density, and the same surface gravity. (Box 8-1 lists basic data about Venus.) However, Venus is closer to the Sun than the Earth is and therefore is exposed to a greater intensity of sunlight, transforming the potentially Earth-like planet into a world that seems extremely hostile to terrestrial organisms. Venus is an inferno whose crushing, poisonous atmosphere is drenched in sulfuric acid.

Box 8-1   Venus data

Mean distance from the Sun: 0.723 AU = 1.082×10^8 km
Maximum distance from the Sun: 0.728 AU = 1.089×10^8 km
Minimum distance from the Sun: 0.718 AU = 1.075×10^8 km
Mean orbital velocity: 35.0 km/sec
Sidereal period: 224.70 days
Rotation period: 243.01 days (retrograde)
Inclination of equator to orbit: 178°
Inclination of orbit to ecliptic: 3° 23′ 40″
Orbital eccentricity: 0.007
Diameter: 12,104 km
Diameter (Earth = 1): 0.949
Apparent diameter as seen from Earth: maximum = 65.2″
                                                        minimum =   9.5″

Mass: 4.87×10^{24} kg
Mass (Earth = 1): 0.8150
Mean density: 5.25 g/cm³
Surface gravity (Earth = 1): 0.903
Escape velocity: 10.3 km/sec
Oblateness of planet: 0
Mean surface temperature: 480°C = 750 K = 900°F
Albedo: 0.76
Brightest magnitude: −4.4
Mean diameter of Sun as seen from Venus: 44′ 15″

The surface of Venus is hidden beneath a very thick, highly reflective cloud cover

Venus's orbit is almost twice as large as Mercury's. Consequently, at its greatest elongation, Venus appears 47° away from the Sun (see Figure 8-1). This is a very comfortable distance for viewing the planet without interference from the Sun's glare. At its greatest eastern elongation, Venus is easily spotted high above the western horizon after sunset (when it is called an evening star). Alternatively, at greatest western elongation, the planet rises nearly three hours before the Sun. With the arrival of dawn, the planet is positioned high in the eastern sky and is often called the morning star.

Venus is easy to identify because it is often one of the brightest objects in the night sky. Venus's cloud cover reflects 76 percent of the sun-

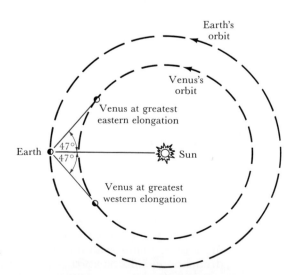

Figure 8-1   Venus's orbit
Venus travels around the Sun along a nearly circular orbit with a period of 224.7 days. The average distance between Venus and the Sun is 108 million kilometers; the average distance between the Earth and Sun is 150 million kilometers. At its greatest eastern elongation, Venus appears as a prominent evening star; at its greatest western elongation, it is a prominent morning star.

light that falls on it (albedo = 0.76), and Venus sometimes reaches a magnitude of −4.4 as seen from the Earth, 16 times brighter than the brightest star. Indeed, only the Sun and Moon outshine Venus at its greatest brilliancy.

Earth-based telescopic views of Venus reveal a thick, nearly feature-less, unbroken layer of clouds (see Figure 8-2). Although this higher re-flective cloud cover makes Venus dazzlingly bright, it is also responsible for our long-standing ignorance of the planet. Because of this perpetual shroud, we did not even know until recently how fast Venus rotates.

Until the twentieth century, very little was gleaned from telescopic observations of the planet. Nineteenth century observers did realize that Venus has an atmosphere. Otherwise, noteworthy viewing of Venus was largely confined to phenomena such as transits, when Venus passes di-rectly in front of the Sun at inferior conjunction.

Transits of Venus are much rarer than those of Mercury. Venus is farther from the Sun than Mercury is, and the required Sun–Venus–Earth alignments occur much less frequently. Indeed, not one transit of Venus will occur during the entire twentieth century. The last Venusian transits occurred in 1874 and 1882. Venusian transits always occur in pairs separated by eight years, and the next pair will occur on June 8, 2004, and June 6, 2012.

Figure 8-2 Earth-based views of Venus
This series of photographs shows how the appearance of Venus changes as it moves along its orbit. The number below each view is the angular diameter of the planet in seconds of arc. Venus reaches greatest brilliancy when its angular diameter is about 40 arc sec. (New Mexico State University Observatory)

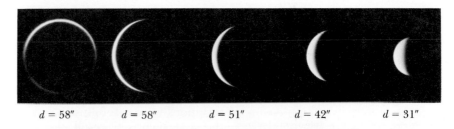

d = 58″   d = 58″   d = 51″   d = 42″   d = 31″

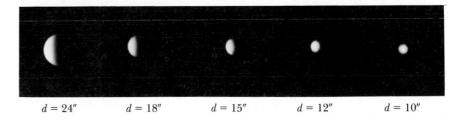

d = 24″   d = 18″   d = 15″   d = 12″   d = 10″

Earth-based optical and radio observations suggested that Venus has retrograde rotation and high surface temperature

In the mid-1950s, astronomers began to realize that Venus has some un-usual characteristics very unlike those of the Earth. Clues about Venus's rotation came in 1956 when Robert S. Richardson of Mount Wilson Ob-servatory measured a small shift in the wavelengths of the spectral lines of reflected sunlight from the planet. As discussed in Box 7-3, the wave-length of a spectral line is affected by relative motion between the source and the observer (the Doppler effect). As Venus rotates, one side of the planet is approaching us, and its spectral lines are blueshifted; the other side of the planet is receding from us and exhibits redshifted spectral lines.

To everyone's surprise, Richardson's observations indicated that Venus's rotation is **retrograde,** or backward. In other words, on the sur-face of Venus, sunrise occurs in the west.

This is surprising because almost all of the other planets rotate about their axes in the same direction that they revolve about the Sun. For example, if you viewed the solar system from a great distance above the Earth's north pole, you would see all the planets going around the Sun counterclockwise. Closer examination would reveal that the vast majority of the planets also rotate counterclockwise on their axes. With only a very few exceptions, even the satellites of the planets move counterclockwise along their orbits. We still do not know how or why Venus's rotation is an exception to this trend.

Another surprising discovery of the 1950s came from astronomers who detected radio waves emitted by Venus's surface. As discussed in Chapter 7, the dominant wavelength of radiation from an object is related to the object's temperature by Wien's law, and at typical planetary temperatures, the dominant radiation from an object is in the infrared range. Measurements of infrared radiation from Venus had indicated a temperature of around 235 K ($= -38°C = -36°F$). However, Venus's thick atmosphere should block infrared radiation just as the Earth's does, so astronomers assumed that this observed radiation came from the top of the cloud layer and provided only a measurement of the temperature in Venus's upper atmosphere.

At any temperature, an object does emit radiation of all wavelengths, but Wien's law refers only to the dominant wavelength. By measuring the relative strengths of radiation at *any* different wavelengths, an astronomer can estimate the temperature of the object emitting the radiation, using more complicated laws of physics. If Venus's atmosphere is anything like Earth's, it should be transparent to radio waves. Therefore, analysis of the relative strengths of radio waves from Venus at different wavelengths should provide a measurement of the planet's surface temperature.

At first, no one could believe the results of such studies, but the observations were carefully repeated again and again, and they always gave the same surprising result: the temperature at Venus's surface is at least 600 K ($= 330°C = 620°F$), the melting point of lead. Recent measurements by spacecraft that landed on the planet indicate a surface temperature of 750 K ($= 480°C = 900°F$). After their initial skepticism, astronomers quickly realized that Venus's high surface temperature might have been expected.

Perhaps you have had the experience of parking your car in the sunshine on a warm summer day. You roll up the windows, lock the car, and go on an errand. After a few hours, you return to your car. You are annoyed to discover that the interior of your automobile has become stiflingly hot, typically at least 20°C warmer than the outside air temperature.

What happened to make your car so warm? First, sunlight entered your car through the windows. The radiation was absorbed by the dashboard, the steering wheel, and the upholstery, raising their temperatures. Every object emits radiation appropriate to its temperature, and Wien's law relates the Kelvin-scale temperature (T) to the dominant wavelength (λ_{max}) at which the most intense radiation is emitted. For example, the Sun's surface temperature is about 6000 K, and thus sunlight has its greatest intensity at wavelength λ_{max}, where

$$\lambda_{max} = \frac{0.3}{T} = \frac{0.3}{6000} = 5 \times 10^{-5} \text{ cm} = 5000 \text{ Å}$$

As you may have expected, 5000 Å is in the middle of the visible spectrum. This sunlight typically raises the temperature of your car's uphol-

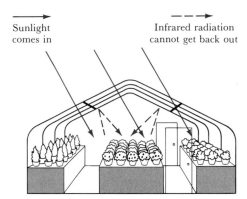

Sunlight comes in

Infrared radiation cannot get back out

Figure 8-3 The greenhouse effect
Incoming sunlight easily penetrates the windows of a greenhouse and is absorbed by objects inside. These objects reradiate this energy at infrared wavelengths to which the glass windows are opaque. The trapped infrared radiation is absorbed by the air and objects, causing the temperature inside the greenhouse to rise. (Recent studies indicate that the physical barrier of glass windows plays a major role in the temperature rise inside a real greenhouse (or car) because the closed windows prevent air mixing. Nevertheless, the greenhouse effect does occur, and is the cause of the high temperature on Venus.)

stery to around 330 K (= 57°C = 135°F). At this temperature, the seats in your car reradiate the energy primarily with

$$\lambda_{max} = \frac{0.3}{T} = \frac{0.3}{330} = 9.1 \times 10^{-4} \text{ cm} = 9.1 \mu$$

Radiation of wavelength 9.1 μ is in the infrared portion of the electromagnetic spectrum. Your car windows are opaque to these wavelengths. This energy therefore is trapped inside your car and absorbed by the air and interior surfaces. As more sunlight comes through the windows and is trapped, the temperature continues to rise.

A similar phenomenon occurs in Venus's atmosphere (and to a lesser extent in Earth's atmosphere). It is commonly called the **greenhouse effect** (see Figure 8-3). Sunlight enters the thick Venusian clouds, where it is absorbed and reradiated at infrared wavelengths that cannot get back out. This trapped radiation produces the high surface temperature of Venus.

Radar observations revealed that Venus's retrograde rotation is coupled to the Earth's motion

In the early 1960s, improved equipment made it possible to send microwave radiation to Venus and detect the waves reflected from that planet. The Venusian clouds are transparent to microwave radiation as well as radio waves. The process of bouncing microwave radiation from an object is usually called **radar,** and the radiation in such a case is sometimes called radar radiation.

Unfortunately, radar observations of Venus are practical only when the planet is near inferior conjunction. As explained in Box 8-2, the intensity of radiation is inversely proportional to the inverse square of the distance from the source. Therefore the intensity of the microwaves sent to Venus decreases as $1/r^2$, where r is the distance traveled. Of course, only a tiny fraction of the outgoing signal is actually reflected back toward the Earth, and this returning radiation again decreases by a factor of $1/r^2$. Consequently, the signal strength of the echo is proportional to $1/r^4$. This is such a drastic reduction that reliable radar observations can be obtained only when the distance between the planets is as small as possible.

The first radar observations of Venus were made in 1961 by a team of scientists headed by William B. Smith at the Lincoln Laboratory of the Massachusetts Institute of Technology. Their goal was to use the Doppler effect (recall Box 7-3) to measure Venus's rotation rate. As with sunlight, radar waves reflected from the approaching side of the planet will be increased in frequency, whereas those from the receding side will be decreased in frequency. The radar waves however, will be reflected from the planet's surface rather than the clouds. Although the microwaves leave Earth at one precise frequency, the echo comes back spread over a small range of frequencies. By measuring the frequency spread, Smith and his colleagues concluded that Venus's rotation is very slow and probably retrograde.

These observations were repeated in 1962 by Roland L. Carpenter and Richard M. Goldstein at the Jet Propulsion Laboratory of the California Institute of Technology. They confirmed a retrograde rotation with a sidereal period of approximately 240 days. Over the next several years, this measurement was refined with NASA's deep space tracking antenna at Goldstone, California (see Figure 8-4). Today we know that

Box 8-2   The inverse-square law and the propagation of light

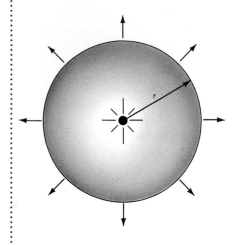

The farther you are from a source of light, the dimmer it appears. But exactly how and why does the brightness of a light depend on the distance to an observer?

Imagine a source of light such as a light bulb or a star. Imagine a sphere of radius r centered on the light source, as shown in the diagram. Suppose that L is the amount of energy emitted by the light source each second. This value L is often called the **luminosity** of the source and is usually measured in ergs per second. For example, the Sun's luminosity is 3.9×10^{33} ergs per second.

As the energy moves away from the source of light, it becomes spread out over larger regions of space. Specifically, the amount of energy passing through each square centimeter of the sphere per second is simply the total luminosity of the source (L) divided by the sphere's surface area ($4\pi r^2$). The resulting quantity is called the **brightness** of the light (b) and is measured in ergs per square centimeter per second:

$$b = \frac{L}{4\pi r^2}$$

For example, the distance from the Earth to the Sun is 1.5×10^8 km (1.5×10^{13} cm), so the brightness of sunlight reaching the Earth is

$$\frac{3.9 \times 10^{33} \text{ erg/sec}}{4\pi(1.5 \times 10^{13} \text{ cm})^2} = 1.4 \times 10^6 \text{ erg/cm}^2 \text{ sec}$$

This quantity is called the **solar constant.**

It is important to remember that the brightness of light that an observer sees or measures is inversely proportional to the square of the observer's distance from the source. This relationship is called the **inverse-square law.** If you double your distance from a source of light, the brightness you see decreases by a factor of four. At triple the distance, the brightness decreases by a factor of nine.

For example, the average distance between Pluto and the Sun is 39.4 AU. Because $(39.4)^2 = 1552$, the brightness of sunlight arriving at Pluto is 1/1552 of that observed from the Earth (at a distance of 1 AU).

the sidereal period of Venus's retrograde rotation is 243.01 days. This is an interesting value. To see why, we must calculate the length of a solar day on Venus and compare it with Venus's synodic period.

First, remember that the sidereal period is measured with respect to the stars (not the Sun). In other words, if you were standing on Venus (and could see through the clouds), you would have to wait 243.01 Earth days to see the same star pass twice overhead. Meanwhile, of course, Venus is moving around the Sun with an orbital period of 224.70 Earth days. Thus Venus's sidereal period is not the same as a solar day, which is the time from one local noon to the next. The length of a solar day on Venus can be calculated from the planet's sidereal rotation and orbital periods by the method given in Box 4-1:

$$\frac{1}{\text{solar day}} = \frac{1}{243.01} - \frac{1}{224.70} = 0.008565 = \frac{1}{116.8}$$

In other words, an astronaut on Venus would have to wait 116.8 days from one local noon to the next.

Figure 8-4   A deep-space tracking antenna
Although this antenna was built by NASA to track interplanetary spacecraft, it can also be used to beam powerful pulses of microwaves toward Venus. The antenna's dish measures 64 m (210 ft) in diameter. (Jet Propulsion Laboratory)

Second, remember that a planet's synodic period is the time interval between successive occurrences of the same configuration, such as the interval from one inferior conjunction to the next. From the data in Table 4-1, we see that Venus's synodic period is 584 days. Note that

$$5 \times 116.8 = 584.0$$

Thus, Venus's synodic period is equal to five Venusian solar days (ignoring the uncertainties in the measurements).

In the previous chapter we saw that Mercury exhibits spin–orbit coupling; the planet's rotation is related to its revolution about the Sun. The case of Venus is more complicated, however, because the synodic period is measured from the Earth. Indeed, Venus's rotation is related to the Earth's revolution about the Sun.

Look at it another way. Earth-based astronomers must wait 584 days between successive inferior conjunctions of Venus. Meanwhile, five solar days elapse on Venus. Thus at each inferior conjunction, it is the same "time of day" on Venus. Consequently, at each inferior conjunction, the same side of Venus is turned toward the Earth.

This unusual relationship between Venus and the Earth almost certainly is the result of the gravitational interaction between the two planets, but no one has been able to explain just how and why this relationship was established.

Space flights to Venus discovered neither a magnetic field nor a magnetosphere

In the 1960s, both the United States and the Soviet Union began sending probes to Venus. The Americans sent fragile, light-weight spacecraft past the planet and studied the Venusian environment with remote sensing devices. The Russians, who possessed more powerful rockets, sent massive vehicles that plunged directly into the clouds.

The first successful mission to Venus was the flight of Mariner 2 in 1962. During this three-month interplanetary voyage, the spacecraft discovered the solar wind. As mentioned in preceding chapters, the solar

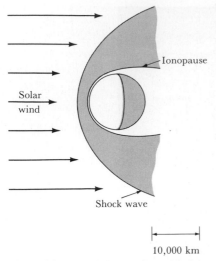

Figure 8-5 *Venus's interaction with the solar wind*
Venus has no magnetic field, so the solar wind strikes the uppermost layers of the planet's atmosphere. Interactions with ions in the upper atmosphere produces a shock wave and ionopause, as shown in this scale drawing.

wind consists of charged particles escaping from the Sun at supersonic speeds. Our first measurement of the speed (roughly 400 km/sec) and density (1 particle/cm^3) of the solar wind came from Mariner 2.

As the probe passed Venus, the magnetometer on Mariner 2 failed to detect any magnetic field whatsoever. This lack of a magnetic field was confirmed during subsequent missions. As discussed in Chapter 7, the Earth's magnetic field is explained by the dynamo effect: currents in the iron core of the rotating Earth produce our planet's magnetism. Venus has nearly the same density and size as Earth, so Venus probably has an iron core very similar to Earth's. Therefore, we conclude that the leisurely rotation rate of Venus is too slow to induce a planetwide magnetic field.

With virtually no magnetic field, Venus is incapable of producing a magnetosphere to protect itself from the solar wind. The solar wind therefore impinges directly on Venus's upper atmosphere, where many of the atoms are stripped of one or more electrons. An atom that is missing one or more electrons is called an **ion.** These ions have positive electric charge because they have lost negatively charged electrons. The electromagnetic interaction between these ions and the supersonic charged particles of the solar wind produces a well-defined shock wave (see Figure 8-5). Along a boundary called the **ionopause** inside the shock wave, the pressure of the ions just counterbalances the pressure from the solar wind. The ionopause is analogous to the magnetopause that surrounds a planet with a magnetic field.

Spacecraft that descended into the clouds provided detailed information about the Venusian atmosphere, which is extremely corrosive

While the Americans were busy landing astronauts on the Moon, Soviet scientists concentrated on building spacecraft that could survive a descent into the Venusian cloud cover. The task proved to be more frustrating than anyone had expected. Finally in 1970, Venera 7 managed to transmit data for a few seconds directly from the Venusian surface. Venera missions during the early 1970s measured a surface temperature of 750 K (480°C = 900°F) and a pressure of 90 atmospheres. One atmosphere (1 atm) is simply the average air pressure at sea level on Earth (10,130 newtons per square meter, or 14.7 pounds per square inch). Thus the Venusian atmosphere weighs down on the planet with a crushing pressure of $\frac{2}{3}$ ton per square inch.

In 1932, Walter S. Adams and Theodore Dunham of Mount Wilson Observatory had identified carbon dioxide (CO_2) in the planet's spectrum (A portion of Venus's spectrum showing carbon dioxide lines appears in Figure 6-8.) Astronomers could not reliably estimate the percentage of this gas in the planet's atmosphere from these early observations, but the lack of other spectral patterns suggested that CO_2 composed most of the Venusian atmosphere. Direct measurements by the Russian probes showed that 96 percent of Venus's dense atmosphere is carbon dioxide.

Because Venus is Earth-like in so many ways, astronomers for a long time assumed that it also would resemble the Earth in having abundant water. The high temperatures would turn water to steam, of course, so the dense clouds were just what would be expected. Science-fiction authors set many stories in the steamy jungles and swamps of Venus before the true surface temperature was known. However, it was not until the late 1960s that Ronald A. Schorn and his colleagues at McDonald Observatory in Texas succeeded in detecting faint water lines in the

Venusian spectrum. Later measurements by Edwin S. Barker, also at McDonald Observatory, showed that water makes up far less than 1 percent of the clouds. If the clouds are dry, what are they made of? Measurements from the Russian spacecraft indicated that nitrogen accounts for most of the 4 percent of the atmosphere that is not CO_2; there are also traces of oxygen, water vapor, and argon. It is not at all obvious why clouds should form in this mixture of gases.

Clues to the complex chemistry of the Venusian clouds came in bits and pieces, beginning in the early 1960s. Earth-based observations as well as measurements during the Mariner 5 and Mariner 10 flybys demonstrated that the clouds efficiently absorb radiation at specific infrared and microwave wavelengths. Godfrey Sill of the University of Arizona and the team of Louise Young and Andrew Young proposed a surprising explanation for this absorption: droplets of sulfuric acid.

When sulfuric acid (H_2SO_4) is mixed with water (H_2O), the ions HSO_4^- and H_3O^+ are produced. These ions absorb infrared and microwave radiation at precisely the wavelengths observed in Venus's clouds. The sulfuric acid would be at least partly responsible for depleting the Venusian atmosphere of water by converting it into H_3O^+. The latest Soviet and American probes leave no doubt that the Venusian clouds are in fact made of droplets of concentrated sulfuric acid.

The sulfuric acid in Venus's atmosphere causes a number of chemical reactions. For example, reactions with fluorides and chlorides in surface rocks give rise to hydrofluoric acid (HF) and hydrochloric acid (HCl). Further reactions produce fluorosulfuric acid (HSO_3F), which is one of the most corrosive substance known to chemists—it can dissolve lead, tin, and most rocks. Indeed, the Venusian clouds are a cauldron of chemical reactions hostile to metals and other solid materials (see Figure 8-6). No

Figure 8-6 Atmospheric chemistry of Venus
A host of chemical reactions occur constantly in the Venusian atmosphere; some are sketched in this diagram. (Adapted from Andrew Young and Louise Young)

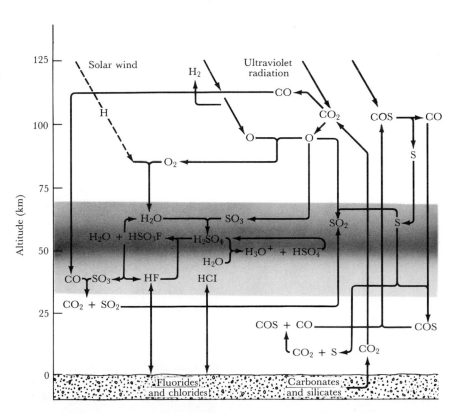

wonder the Russians had so much trouble getting a functioning space-craft down through the clouds to the surface.

While descending through the Venusian clouds, Soviet spacecraft measured atmospheric pressure and temperature. The results are shown in Figures 8-7 and 8-8. Graphs like these are important because they show how pressure and temperature are related to altitude above a planet's surface—fundamental information from which the detailed structure of a planet's atmosphere can be deduced. The pressure and temperature profiles of the Venusian atmosphere are simple: both pressure and temperature decrease smoothly with increasing altitude. As we shall see in Chapter 9, Earth's atmosphere has a much more complicated relationship between temperature and altitude. In the chapters on Jupiter and Saturn, we shall see how pressure and temperature variation with altitude can have a profound effect on a planet's appearance.

Both Soviet and American spacecraft found the top of the Venusian clouds at an altitude of about 68 km (42 miles = 220,000 ft). However, the Russians also discovered the bottom of the clouds at an elevation of about 31 km (19 miles = 100,000 ft) above the ground. Below this elevation, the Venusian atmosphere is remarkably clear.

Figure 8-7 [left]   Temperature in the Venusian atmosphere
The temperature in Venus's atmosphere rises smoothly from a minimum of about 170K (about −100°C, or −150°F) at an altitude of 100 km to a maximum of nearly 750K (about 480°C, or 900°F) on the ground.

Figure 8-8 [right]   Pressure in the Venusian atmosphere
The pressure at the Venusian surface is a crushing 90 atm (1300 pounds per square inch). Above the surface, atmospheric pressure decreases smoothly with increasing altitude.

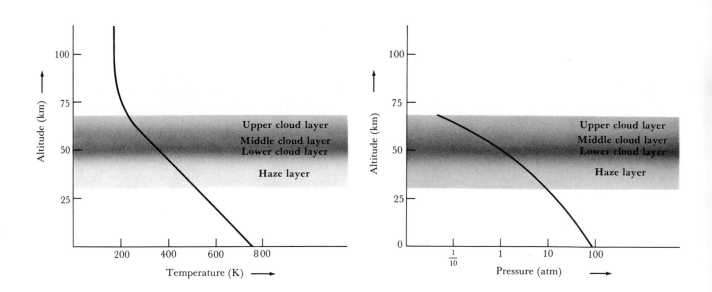

The Pioneer Venus mission discovered layers and convection cells in the Venusian clouds

Although the American space program did not concentrate on Venus, some of the most detailed information about the planet came from the dual Pioneer Venus mission in 1978. This ambitious undertaking began with the launch of the Orbiter Spacecraft shown in Figure 8-9. After one-half year of coasting toward the planet, the craft was slowed by retrorockets and placed in a highly elliptical orbit ranging from about 150 to more than 66,000 km above the Venusian cloud tops.

Meanwhile, the Multiprobe Spacecraft was launched. As shown in Figure 8-10, this second space vehicle consisted of a cylindrical bus and four blunt-nosed probes. Twenty-four days before arrival at Venus, explosive charges separated the large probe from the bus. Four days later, the three smaller probes were similarly deployed. On December 9, 1978, just five days after the arrival of the Orbiter, the instrument-laden bus and its four probes descended into the Venusian clouds. As expected, the bus

Figure 8-9 [left]   Venus Pioneer 1 (the Orbiter spacecraft)
This spacecraft, which is still in orbit about Venus, weighs 590 kg (1300 pounds) and measures 254 cm (8⅓ ft) in diameter. It arrived at Venus five days ahead of the Multiprobe spacecraft shown in Figure 8-10. (NASA)

Figure 8-10 [right]   Venus Pioneer 2 (the Multiprobe spacecraft)
This entire spacecraft (with probes attached) was almost exactly the same size as the Orbiter, but weighed 910 kg (nearly a ton). The large probe measured 142 cm (4⅔ ft) in diameter and weighed 315 kg (693 pounds). Each of the three smaller probes weighed only 90 kg (200 pounds) and had a diameter of 76 cm (2½ ft). (NASA)

burned up in the clouds after sending data about the upper atmosphere back to Earth. The four aerodynamically shaped probes made numerous measurements at lower altitudes, and one of the smaller probes continued to send data from the ground for over an hour.

As the Multiprobe hurtled into the Venusian atmosphere, the Orbiter began a lengthy series of observations and measurements that would keep scientists occupied for several years. For example, the Orbiter sent back numerous photographs of the planet in the ultraviolet wavelengths at which the atmospheric markings stand out best (see Figure 8-11). By

Figure 8-11   Venus from the Orbiter
These four views of Venus were taken in May 1980 at a distance of roughly 50,000 km. All the views show variations of the so-called Y feature caused by the rapid retrograde motion of the clouds around the planet. (Ames Research Center)

following individual cloud markings, scientists determined that Venus's atmosphere rotates in a retrograde direction around the planet in only four days. This rapid atmospheric motion is in sharp contrast to the slow rotation of the solid planet.

Venus appears yellowish or yellow-orange to the human eye, and data from the Multiprobe indicated why: the upper clouds contain substantial amounts of sulfur dust. Over the temperature range of these upper clouds, sulfur is distinctly yellow or yellowish-orange. At lower elevations, large concentrations of sulfur compounds (especially SO_2, OCS, and H_2S) were found along with droplets of sulfuric acid. Because of the tremendous atmospheric pressure, the droplets do not fall as a rain; they are more-or-less permanently suspended in the clouds as an aerosol.

The Multiprobe also discovered four distinct cloud layers. The **upper cloud layer** extends from altitudes of 68 km down to 58 km. A denser and more opaque **middle cloud layer** extends from 58 km down to 52 km. The **lower cloud layer** (from 52 to 48 km) contains the densest and most opaque of the Venusian clouds, even though it is only 4 km thick. Below the cloud layers is a haze layer between 48 and about 31 km. Below this elevation of 100,000 feet, the atmosphere is clear all the way down to the ground.

The Multiprobe also determined the dominant circulation patterns in the Venusian atmosphere. Warmed by the Sun, hot gases in the equatorial regions rise upward and travel in the upper cloud layer toward the cooler polar regions. At these polar latitudes, the cooled gases sink downward to the lower cloud layer, in which they are transported back toward the equator. This process of heat transfer, whereby hot gases rise while cooler gases sink, is called **convection.**

The circulation of Venus's atmosphere is dominated by two huge **convection cells** (one in the northern hemisphere and another in the southern hemisphere) that circulate gases between the equatorial and polar regions of the planet (see Figure 8-12). These convection cells, which are almost entirely contained within main cloud layers, are called **driving cells** because they propel similar circulation cells above and below the main cloud deck somewhat like a meshed set of gears.

Figure 8-12   *The circulation in Venus's atmosphere*

The main convection mechanism in Venus's atmosphere occurs in the clouds. This convection cell drives similar circulation patterns above and below the cloud layer. Circular wind patterns (vortices) in the polar regions force the cloud layer to bulge upward at polar latitudes, producing a polar ring cloud that is clearly visible in many Orbiter photographs. (Adapted from Alvin Seiff)

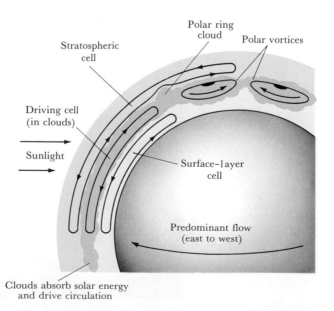

Because the entire atmosphere rotates around the planet in only four days, strong prevailing winds blow from east to west. These winds stretch out the driving cells, producing the characteristic V-shaped, chevron-like patterns that dominate the planet's appearance.

....

Studies of the Venusian atmosphere suggest that volcanoes created the atmospheres of the terrestrial planets

Before the Pioneer Venus mission, there were four main theories, or hypotheses, about the origin of the atmospheres of the terrestrial planets. The first, called the **accretion hypothesis,** held that the "ices" (carbon dioxide, water, and so on) had been trapped in the planets when they were formed and later had been "outgassed" by volcanic activity. The other three hypotheses assumed that the original planets were essentially free of ices, and that the atmosphere came from some outside source. The **solar-nebular hypothesis** held that the ices were gathered by the planets from the remnants of the solar nebula; the **solar-wind hypothesis** argued that the ices were captured from the solar wind; and the **comet-asteroid hypothesis** argued that the ices arrived on the planets in comets, meteoroids, and asteroids that collided with them during the early stages of the solar system when such collisions were relatively common.

The best source of evidence for testing these competing hypotheses was the study of nonreactive (or inert) gases such as neon, argon, and krypton in the planetary atmospheres. These gases are not incorporated into surface rocks because they are chemically unreactive, and they do not escape into space because they are composed of heavy atoms (recall Box 7-2). Any inert gases ever added to an atmosphere should still be there today. Thus the relative abundances of the inert gases in planetary atmospheres should provide some evidence about the sources of the atmospheres.

The atmospheres of Venus, Earth, and Mars contain similar proportions of neon and argon isotopes (especially the ratio of ^{20}Ne to ^{36}Ar), but the Sun's atmosphere has very different ratios of these isotopes. It is reasonable to assume that the Sun's atmosphere has essentially the same composition as the solar wind and as the original solar nebula. We can therefore rule out the solar wind and the solar nebula as likely sources of the planetary atmospheres. Thus, information from the Mars and Venus probes helped limit the likely choice to one between the accretion hypothesis and the comet–asteroid hypothesis.

Venus, Earth, and Mars should have received roughly equal bombardment by comets and asteroids, so they should have similar amounts of inert gases in their atmospheres if the comet–asteroid hypothesis is valid. The Pioneer probes found that argon is 60 times more abundant in Venus's atmosphere than it is in the Earth's. This evidence eliminated the comet–asteroid hypothesis, leaving the accretion hypothesis as the only likely explanation for the atmospheres.

Certainly, planetary atmospheres have been modified over the ages by outside sources of gas. For example, the Venusian atmosphere has a very high ratio of argon to krypton, similar to the ratio that exists in the solar atmosphere. We conclude that the atmosphere of Venus, nearer to the Sun, has been more affected by the solar wind than have the atmospheres of the other terrestrial planets. Nonetheless, the preponderance of evidence suggests that outgassing by volcanoes has been the primary process in formation of atmospheres on the terrestrial planets.

Volcanoes give geologists important information about the gases that

Figure 8-13   Mount St. Helens eruption of May 1980
A preponderance of data suggests that the terrestrial planets obtained their atmospheres from volcanic outgassing. Geologists study this outgassing process in eruptions of volcanoes on Earth. This spectacular recent eruption in the western United States was particularly well studied. (USGS)

probably were spewed into the original atmospheres of Venus, Earth, and Mars. For example, during the Mount St. Helens eruption of 1980 (see Figure 8-13), substantial amounts of sulfuric acid and other sulfur compounds were emitted. As we have seen, these compounds are common in the Venusian atmosphere.

Carbon dioxide and water vapor are among the most abundant gases in volcanic vapors. Venus's atmosphere contains abundant carbon dioxide but very little water. The Earth has abundant water in its oceans and its atmosphere, but there is very little carbon dioxide in the Earth's atmosphere. If Venus and the Earth got their atmospheres mainly by the same outgassing process, then what happened to all the water on Venus? And where is all the carbon dioxide here on the Earth?

In the upper Venusian atmosphere, intense ultraviolet radiation from the Sun breaks water molecules into separate hydrogen and oxygen atoms. The light hydrogen atoms escape into space. Oxygen, which is one of the most chemically active elements, readily combines with other substances in the atmosphere. Thus Venus is left with almost no water.

The carbon dioxide on the Earth is dissolved in the oceans of water and chemically bound into carbonate rocks such as limestone and marble that formed in those oceans. If the Earth became as hot as Venus, so much carbon dioxide would be boiled out of the oceans and baked out of the crust that our planet would soon develop a thick, oppressive carbon dioxide atmosphere much like that of Venus.

Thomas A. Donahue of the University of Michigan has proposed an interesting outline of Venus's early history. Venus and Earth are so similar in size and mass that it is reasonable to suppose that Venusian volcanos outgassed an amount of water vapor roughly comparable to the total content of Earth's oceans. Although some of this water on Venus might originally have collected in oceans, heat from the Sun soon vaporized the liquid to create a thick cover of water-vapor clouds. Calculations demonstrate that this water vapor would have added 300 atm of pressure to the existing 90 atm of carbon dioxide. Thus the early Venusian atmosphere weighed down on the planet's surface with a pressure of three tons per square inch.

This thick, humid atmosphere very efficiently trapped heat from the Sun, creating a greenhouse effect far more extreme than the greenhouse effect that operates today on Venus. Calculations demonstrate that the ground temperature would have increased to 1800 K (2700°F), which is hot enough to melt rock. The Venusian surface was probably molten down to a depth of 450 km (280 miles). Soon, however, dissociation of the water molecules and subsequent loss of hydrogen to space left behind the carbon dioxide atmosphere we find today. Hostile as it may seem, the modern environment on Venus is probably quite mild compared to that of earlier times.

Radar maps of Venus reveal gently rolling hills and two continents

The Pioneer Venus Orbiter mapped 93 percent of the planet's surface, using a radar altimeter that bounced microwaves off the ground directly below the spacecraft. By measuring the time delay of the radar echo, scientists determined the heights and depths of hills and valleys with a probable error of ±200 m. The resulting map is shown in Figure 8-14, which uses a color code to denote altitude.

Venus is remarkably flat. About 60 percent of the planet's surface is covered with gently rolling hills, varying by less than 1 km from the planet's average radius of 6051.4 km. (This value for the planet's radius

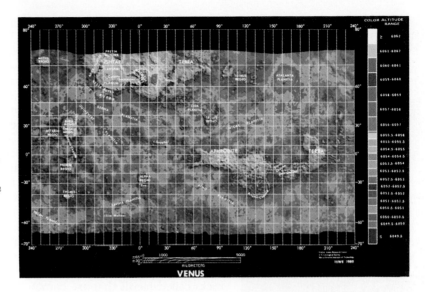

Figure 8-14   A topographic map of Venus's surface
This color-coded map of 93 percent of the Venusian surface is based on data returned from the radar altimeter on the Pioneer Venus Orbiter. The reference elevation (like sea level here on Earth) was chosen as 6051.4 km from the planet's center, the average radius of the planet. Features at this average elevation are shaded blue-green. Yellow and red denote higher elevations, whereas blue and violet indicate lower elevations, as specified on the scale in the figure. (NASA)

is commonly used as a reference level, as we use average sea level here on Earth.)

There are two large "continents" rising well above the generally level surface. In the northern hemisphere, there is **Ishtar Terra,** named after the Babylonian goddess of love. Ishtar is approximately the same size as Australia and consists of a high plateau called Lakshmi Planum, ringed by high mountains. The highest mountain is Maxwell Montes, named after James Clerk Maxwell. The summit of Maxwell is 11 km above the reference level. For comparison, Mount Everest on Earth rises 9 km above sea level. An artist's sketch of Ishtar (with enhanced vertical relief) is shown in Figure 8-15.

The second major Venusian "continent," **Aphrodite Terra** (named after the Greek equivalent of Venus), lies just south of the equator. Aphrodite is slightly bigger than Ishtar and has an area about one-half that of Africa.

The Pioneer Venus Orbiter found evidence of large impact basins. A good example is the enormous circular feature on the southern edge of

Figure 8-15   The continent of Ishtar
This artist's impression shows mainly the Lakshmi Planum region of Ishtar. The highest peak on the planet, Maxwell Montes, appears toward the right side of the drawing. The vertical relief is slightly exaggerated here. (NASA)

Aphrodite. This feature, called Artemis Chasma, is 1800 km in diameter, considerably larger than the largest impact basins on either Mercury or the Moon. A smaller, but very distinct impact crater is located in a formation called **Alpha Regio,** to the west of Aphrodite.

The Pioneer Venus Orbiter also discovered some very large volcanoes in **Beta Regio,** southwest of Ishtar. These volcanoes, Rhea Mons and Theia Mons, rise to altitudes of 6 km, with gently sloping sides that extend over an area 1000 km in diameter. They are among the largest known volcanoes in the entire solar system.

The Russian spacecraft Venera 9 and 10 landed in the vicinity of Beta Regio in 1975. In 1981 Venera 13 and 14 landed about 2000 km south of Beta Regio, near **Phoebe Regio.** Figure 8-16 shows a typical Venera lander. Figure 8-17 is a panoramic view taken by Venera 14. The Venera 14 site seems to be covered with broken, rocky plates. Soviet scientists suggest that this region was covered with a thin layer of lava that fractured upon cooling to create the rounded, interlocking shapes seen in the photograph. This hypothesis agrees with the analysis by Venera's instruments, which indicates that the soil composition is very similar to lava rocks called basalt that are common on Earth and the Moon.

Both Alpha Regio and Beta Regio were well-known Venusian features long before the Venus Pioneer mission. For many years, astronomers had been bouncing radar signals off of Venus in an attempt to map the planet's surface. (As mentioned earlier, Earth-based radar observations are practical only near inferior conjunction, at which time Venus always exposes the same hemisphere to our Earth-based view.) Alpha Regio and Beta Regio were the first two surface features to be identified.

Some of the best Earth-based radar maps of Venus have been produced using the giant 1000-ft dish (see Figure 7-6) at the Arecibo Observatory in Puerto Rico; Figure 8-18 shows a fine example. In these radar images, smooth areas generally appear dark, whereas rough areas appear bright. Notice how prominently Maxwell stands out, as do Rhea Mons and Theia Mons. Comparison with the Orbiter map reveals that the highest features on Venus are, in general, the roughest. A notable exception is Lakshmi Planum, which appears dark in the radar view. This high plateau is therefore remarkably smooth. Much better radar pictures may come from the Venus Radar Mapper (see Figure 8-19) that is to be launched in 1988.

After examining all the data on Venus, most astronomers believe that they are looking at a planet that has evolved geologically in the same way as Earth, but not nearly to the same extent. For example, in Chapter 9 we shall see that the phenomenon of **plate tectonics** dominates Earth's surface. Earth's crust is divided into huge plates that jostle each other, producing earthquakes, volcanoes, chasms, and mountain ranges. On Venus, however, evidence for plate tectonics is found only in isolated

Figure 8-16   A Venera lander
Each of the recent, successful Venera missions involved two spacecraft (a bus and a lander) having a total weight of nearly 10 tons. Upon arrival at Venus, the lander separated from the bus and descended through the Venusian clouds with parachutes. A shock-absorbing ring mounted at the base of the spacecraft cushioned the touchdown. (TASS)

Figure 8-17   The Venera 14 landscape
This historic color view of the Venusian surface has an orange tone because the thick, cloudy atmosphere absorbs the blue component of sunlight. The rocky plates covering the ground may be fractured segments of a thin layer of lava, or they may be crusty layers of sediment that has been cemented together by chemical and wind erosion. (TASS)

Figure 8-18   The Arecibo radar map
By analyzing reflected radar signals from Venus, astronomers are able to piece together a radar map of the planet. Rough areas appear bright, whereas smooth areas appear dark. Only half of the planet has been mapped by radar because Venus exposes the same hemisphere to our Earth-based view at each inferior conjunction. (Arecibo Observatory)

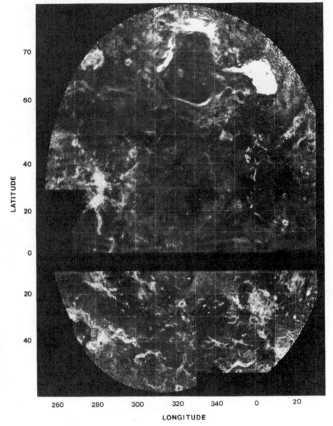

Figure 8-19   The Venus Radar Mapper (VRM)
Space scientists look forward to another probe of Venus in the near future. Using sophisticated radar techniques, this spacecraft will map nearly 90 percent of the planet's surface with much higher resolution than that obtained by the Venus Pioneer Orbiter. Current plans call for a launch in 1988. (NASA)

regions, suggesting that this process is only beginning there. On the other hand, from the small number of impact craters, it is clear that Venus has progressed much further than Mercury along an Earth-like course of evolution.

In a certain sense, Venus is the Earth's twin. Whereas its surface tells us of our geologic past, its atmosphere suggests an awesome future. Five billion years from now, as our Sun swells to become a red giant star, the oceans will boil and shroud the Earth in a thick cloud cover. As the greenhouse effect drives temperatures toward 1000 K, vast amounts of carbon dioxide will be baked out of the rocks. In looking at Venus's atmosphere, perhaps we see distant fate of our own world.

Summary

. Venus is similar to the Earth in its size, mass, average density, and surface gravity, but it is covered by nearly featureless, unbroken, highly reflective clouds that conceal its other features from Earth-based observers.

. Space probes reveal that 96 percent of the Venusian atmosphere is carbon dioxide; most of the balance of the atmosphere is nitrogen, with traces of oxygen, water vapor, and argon.

The clouds are formed of droplets of concentrated sulfuric acid, with substantial amounts of yellowish sulfur dust in the upper layer.

The atmosphere can be divided into layers (whose altitudes are measured here from the average surface elevation): the upper cloud layer (from 68 to 58 km altitude); the denser, more opaque middle cloud layer (58 to 52 km); the densest, most opaque lower cloud layer (52 to 48 km); the haze layer (48 to 31 km); and clear atmosphere (31 to 0 km).

Pressure and temperature decrease smoothly from 90 atm and 750 K at the surface to about $\frac{1}{20}$ atm and 220 K at the top of the upper cloud layer.

Circulation of the Venusian atmosphere is dominated by two huge driving cells in the cloud layers, one cell in the northern hemisphere and one in the southern hemisphere.

The high temperature is caused by the greenhouse effect; the dense CO_2 atmosphere prevents infrared radiation from escaping into space.

- Venus rotates slowly in a retrograde direction with a solar day of 117 Earth days and a sidereal rotation period of 243 Earth days; thus there are slightly less than two Venusian solar days in a Venusian year, and the Sun rises in the west as observed from the surface of Venus.

   The rotation of Venus is coupled to the revolution of the Earth about the Sun; five Venusian solar days elapse between two successive occurrences of any given configuration between the Sun, Venus, and the Earth.

   The Venusian atmosphere rotates rapidly in a retrograde direction around the planet with a period of only about four Earth days.

- Venus has no detectable magnetic field or magnetosphere; it does have a shock wave in the solar wind and an ionopause that is analogous to the magnetopause of a planet with a magnetic field.

- Studies of the Venusian atmosphere support the theory that the atmospheres of the terrestrial planets were formed from "ices" trapped in the original protoplanets and later released in volcanic activity.

   Water was lost from Venus by the action of ultraviolet radiation on the upper atmosphere; carbon dioxide on Earth has been dissolved in the oceans and chemically bound into rocks.

- The surface of Venus is surprisingly flat, mostly covered with gently rolling hills; there are two major continents (regions of higher elevation), some large impact basins, and some large volcanoes.

   The surface of Venus shows little evidence of the motion of large crustal plates that has played a major role in shaping the Earth's surface.

- The concepts covered in Box 8-2 will be especially helpful in your further study of astronomy.

Review questions

1 Venus takes 440 days to move from greatest western elongation to greatest eastern elongation but only 144 days to go from greatest elongation to greatest western elongation. With the aid of a diagram, explain why.

2 As seen from Earth, the magnitude of Venus changes as it moves along its orbit. Describe the main factors that determine Venus's brightness variations as seen from Earth.

3 In earlier astronomy books, Venus is often referred to as the Earth's twin. What physical properties do the two planets have in common? In what physical ways are the two planets dissimilar?

4 Compare the ways in which Mercury and Venus interact with the solar wind.

5 Suppose that Venus had no atmosphere at all. How would the albedo of Venus then compare with that of Mercury or the Moon? Explain your answer.

*6 How much weaker is a radar echo from Venus at inferior conjunction compared with that from Venus at superior conjunction?

Advanced questions

7 Why do you suppose that Venus does not have a magnetic field and Mercury does, even though both planets rotate rather slowly?

*8 How does the solar constant for Venus compare with that for Mercury?

9 Suppose a planet's atmosphere were opaque to visible light but transparent to infrared radiation. How would this affect the planet's surface temperature? Contrast and compare this hypothetical planet's atmosphere with the greenhouse effect in Venus's atmosphere.

Discussion questions

10 Describe the apparent motion of the Sun during a "day" on Venus relative to (a) the horizon and (b) the background stars.

11 If you were designing a space vehicle to land on Venus, what special features would be necessary? In what ways would this mission and landing craft differ from a spacecraft designed for a similar mission to Mercury?

12 Some scientists believe that the Venus Radar Mapper will reveal numerous craters on the Venusian surface. Other scientists disagree. What do you suppose are the "pros" and "cons" in this argument?

For further reading

Beatty, J. "Report from a Torrid Planet." *Sky & Telescope,* May 1982, p. 452. *Venera probe results.*

Beatty, J. "Radar Views of Venus." *Sky & Telescope,* Feb. 1984, p. 110.

Chapman, C. "The Vapors of Venus and Other Gassy Envelopes." *Mercury,* Sept./Oct. 1983.

Dunne, J., and Burgess, E. *The Voyage of Mariner 10: Mission to Venus and Mercury.* NASA SP-424, 1978.

Fimmel, R., et al. *Pioneer Venus.* NASA SP-461, 1983. *A popular-level summary of the mission and its discoveries.*

Pettengill, G., et al. "The Surface of Venus." *Scientific American,* Aug. 1980.

Pollack, J. "The Atmospheres of the Terrestrial Planets." In Beatty, J., et al., eds. *The New Solar System, 2nd ed.* Cambridge U. Press, 1982.

Sagan, C. "Heaven and Hell." In Sagan, C. *Cosmos.* Random House, 1980. *An eloquent chapter comparing Venus and Earth.*

9 Our living Earth

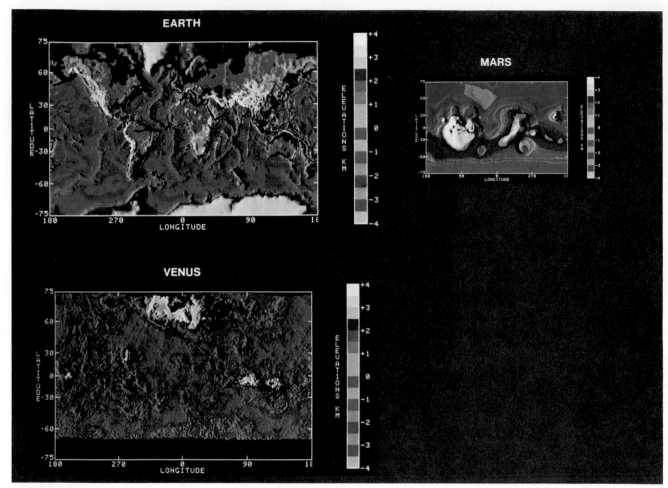

EARTH

MARS

VENUS

The surface topography of Earth, Venus, and Mars

These maps of Earth, Venus, and Mars are reproduced to the same scale. Each map shows surface elevation (Earth's oceans empty) on a color-coded scale. Elevations up to 4 km above the planet's average radius (sea level for Earth) appear in shades of tan, green, and white. Elevations down to 4 km below the planet's average radius appear in various shades of blue. Venus's two continents are prominent, as are Earth's seven continents. The large, high-elevation protrusion on Mars is centered on a major volcanic region. (S. P. Meszaros: NASA)

In this chapter, we look at our own planet as an alien might, viewing it as one of the family of terrestrial planets. We examine the circulation patterns and temperature profile of the Earth's atmosphere and discover that living organisms are largely responsible for our atmosphere's unusual composition. We learn about historical clues found in the rocks of the Earth's crust, and we find that seismic waves give us a surprisingly detailed picture of the Earth's interior. We take a quick look at the Earth's magnetic field and magnetosphere, learning about another unique feature of our planet: the Van Allen belts. Then we discuss the process of plate tectonics that makes the Earth's surface quite different from those of the other terrestrial planets. Finally, we speculate briefly about the possibilities of life elsewhere in the universe.

Of all the planets that orbit the Sun, we are most familiar with the Earth. We walk on its surface, we drink its water, and we breathe its air. Naturally, we know more about the Earth than about any other object in the universe. Sometimes, however, this familiarity makes it difficult for us to view the Earth in its place as a member of the solar system.

Imagine an alien spacecraft approaching the inner solar system. They would see Mars with its thin atmosphere and barren desert landscapes. They would see Venus with its shroud of corrosive clouds hiding a forbiddingly hot surface. Between these two relatively unpromising planets, they would see Earth with an ever-changing ballet of delicate white cloud formations contrasting against the darker browns and blues of the continents and oceans (see Figure 9-1). Perhaps it is only our own bias that leads us to guess that the aliens would focus on the Earth as the most interesting and inviting of the terrestrial planets.

Absorption of sunlight and the Earth's rotation govern the circulation patterns and temperature profile of the nitrogen–oxygen atmosphere

At first, alien visitors might be puzzled to find that Earth's atmosphere is very different from that of its neighbors. The atmospheres of both Venus and Mars consist almost entirely of carbon dioxide, whereas carbon dioxide accounts for only 0.03 percent of Earth's atmosphere. Earth's atmosphere is predominately a 4-to-1 mixture of nitrogen and oxygen, but these two gases are found only in very small amounts on Venus and Mars (see Table 9-1).

TABLE 9-1   The chemical composition of three planetary atmospheres

	Venus	Earth	Mars
Nitrogen	3%	77%	3%
Oxygen	Almost zero	21%	Almost zero
Carbon dioxide	97%	Almost zero	95%
Other gases	Almost zero	2%	2%

Figure 9-1   The Earth
Astronauts often report that Earth is the most invitingly beautiful object visible from their spacecraft. Our blue-and-white world is the largest of the terrestrial planets. This photograph, taken from lunar orbit by Apollo 10 astronauts in 1969, shows the southwest United States, Baja California, and Western Mexico. (NASA)

Why is Earth with its nitrogen–oxygen atmosphere bracketed by two planets possessing atmospheres of nearly pure carbon dioxide? From study of the Earth's geological records, we know that living organisms have been active here for at least the past 3 billion years, and that biological processes such as photosynthesis are largely responsible for the chemical composition of Earth's atmosphere today. As far as we know, large amounts of oxygen in a planetary atmosphere can arise *only* as a direct result of biological activity. So, the aliens might single out the Earth for investigation because its oxygen is an indicator of the possibility of life.

Nearing Earth's surface, aliens would find that our atmosphere has a complicated temperature structure. This is in sharp contrast to the simple temperature profile of the Venusian atmosphere, in which the temperature increases smoothly from about −100°C at an altitude of 100 km to 480°C on the planet's surface (recall Figure 8-7). Earth's temperature profile, with its minima and maxima, is graphed in Figure 9-2. These

Figure 9-2   Temperature profile of Earth's atmosphere
The variation of atmospheric temperature with altitude shows maxima and minima because of interactions between sunlight and various atoms and ions at different elevations.

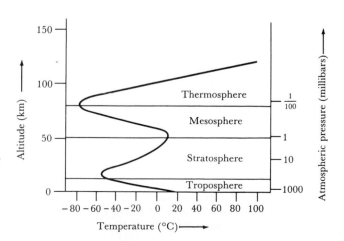

Box 9-1   Earth data

Mean distance from the Sun: $1.000 \text{ AU} = 1.496 \times 10^8$ km
Maximum distance from the Sun: $1.017 \text{ AU} = 1.521 \times 10^8$ km
Minimum distance from the Sun: $0.983 \text{ AU} = 1.471 \times 10^8$ km
Mean orbital velocity: 29.8 km/sec
Sidereal Period: 365.256 days
Rotation Period: 23.9345 hours
Inclination of equator to orbit: 23° 27′
Inclination of orbit to ecliptic: 0°
Orbital eccentricity: 0.0167
Diameter (equatorial): 12,756 km
Mass: 5.976×10^{24} kg
Mean density: 5.52 g/cm^3
Escape velocity: 11.2 km/sec
Oblateness of planet: 0.0034
Temperature range: maximum = 60°C = 140°F
mean = 20°C = 70°F
minimum = −90°C = −130°F
Albedo: 0.39
Mean diameter of Sun as seen from Earth: 0° 32′

temperature variations result from the differing absorption of sunlight at various altitudes.

The average atmospheric pressure at the Earth's surface is 14.7 pounds per square inch, or 1 atm (see Box 9-2). The pressure decreases smoothly with increasing altitude.

Seventy-five percent of the mass of Earth's atmosphere lies below an altitude of 11 km (roughly 7 miles = 36,000 ft) in a region called the **troposphere.** All Earth's weather—clouds, rain, sleet, and snow—occur in this lowest layer. Commercial jets generally fly at the top of the troposphere to minimize buffeting and jostling.

Atmospheric temperature decreases upward to −55°C (= −67°F = 218 K) at the top of the troposphere. Above this level, the region called the **stratosphere** extends from 11 to 50 km above the surface. Ozone (O_3) molecules in the stratosphere efficiently absorb solar ultraviolet rays, heating the air in this layer. The temperature increases upward through the stratosphere to about 0°C (= 32°F = 273 K) at its top. Some scientists are concerned that chemicals we have made and released into the air may destroy much of this ozone layer. If not absorbed by the ozone in the stratosphere, the ultraviolet radiation from the Sun would beat down on the Earth's surface. Because ultraviolet radiation breaks apart most of the delicate molecules that form living tissue, loss of the ozone layer could lead to literal sterilization of the planet. (Recall also the role this radiation has played in eliminating water from the atmosphere and surface of Venus.)

Above the stratosphere, atmospheric temperature declines with increasing altitude in the **mesosphere,** reaching a minimum of about −75°C (= 200 K = −100°F) at an altitude of about 80 km (50 miles). This minimum marks the bottom of the **thermosphere,** above which the Sun's ultraviolet light strips atoms of one or more electrons. Ionized atoms and molecules of oxygen and nitrogen are the most prevalent species. The stripped electrons easily reflect radio waves of a wide range of frequencies. You can tune in a distant AM radio station because its transmissions bounce off of this radioreflective region that surrounds our planet.

Box 9-2   Atmospheric pressure

At sea level on Earth, the air weighs down with a pressure of 14.7 pounds per square inch. By definition, this pressure is called one atmosphere (1 atm). Because we are familiar with this pressure, atmospheric pressure on other planets is usefully expressed in atmospheres also. For example, the atmospheric pressure on Venus is 90 atm. In Chapter 11, we shall learn that the pressure in the sparse Martian atmosphere is only 0.01 atm.

There are, however, other ways of expressing pressure. For example, in the metric system, pressure is measured in dynes per square centimeter rather than pounds per square inch.

$$1 \text{ lb/in.}^2 = 6.89 \times 10^4 \text{ dyne/cm}^2$$
$$1 \text{ dyne/cm}^2 = 1.45 \times 10^{-5} \text{ lb/in.}^2$$
$$1 \text{ atm} = 14.7 \text{ lb/in}^2 = 1.01 \times 10^6 \text{ dyne/cm}^2$$

The fact that 1 atm is equal to almost exactly 1 million dynes per square centimeter has prompted definition of another unit of pressure called a ***bar***, where

$$1 \text{ bar} = 10^6 \text{ dyne/cm}^2 = 0.987 \text{ atm}$$
$$1 \text{ atm} = 1.013 \text{ bar}$$

In other words, a bar is only slightly less than an atmosphere.

In expressing low atmospheric pressures, it is often convenient to use the **millibar**, which is simply 0.001 bar. Thus we say that the atmospheric pressure on Mars is about 10 millibars, approximately the same as the atmospheric pressure at an altitude of 30 km above the Earth's surface (see Figure 9-2).

Finally, there is a way of expressing atmospheric pressure based on the behavior of a barometer. In its simplest form, a barometer can be made by filling a long tube (sealed on one end) with mercury and inverting the tube in a bowl of mercury, as shown in the diagram. Mercury is a very dense liquid and gravity pulls strongly downward on the mercury in the tube. However, atmospheric pressure pushes down on the mercury in the dish and tends to force the mercury back up into the tube, because atmospheric pressure cannot act downward on the top of the column inside the sealed tube. Thus the mercury falls in the tube, leaving a vacuum above the mercury column, until the downward tug of gravity (the weight of mercury in the tube) is just balanced by the atmospheric pressure that pushes the mercury up into the tube.

Atmospheric pressure is commonly expressed as the height of the column of mercury that it supports in a barometer. For example, at sea level on Earth, normal atmospheric pressure supports a column of mercury (Hg) that is 760 mm (29.9 in.) tall. Thus

1 atm = 760 mm Hg = 29.9 in. Hg

Thus, when the weather report says that the barometer reads 30.2 inches, you know that the atmosphere is weighing down on the Earth's surface at sea level with a pressure sufficient to support a column of mercury that is 30.2 inches tall.

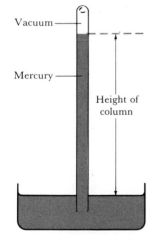

Vacuum

Mercury

Height of column

Earth's atmosphere has intricate circulation patterns. A comparison with Venus is once again enlightening.

Venus's atmosphere exhibits the simplest possible convection pattern: heated gases rise in the equatorial regions, flow in the upper cloud layer to the polar regions where they are cooled, sink to lower altitudes again, and flow back toward the equator in the lower cloud layer. This produces

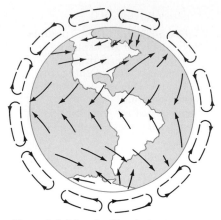

a single equator-to-pole convection cell over each hemisphere (review Figure 8-12). Earth's atmosphere does not exhibit this simple convection pattern because our planet rotates much more rapidly than Venus does. The Earth's relatively fast rotation breaks up convection into three cells per hemisphere (see Figure 9-3).

Were aliens to land on Earth, they would find it radically different from either of its neighbors in yet another way: the Earth is wet, very wet. Nearly 71 percent of the Earth's surface is covered with water. Indeed, an alien space probe to Earth might send back thousands of photographs such as Figure 9-4 representing a random close-up view of Earth's surface. In contrast, Venus and Mars are extremely arid. No place on Earth is as dry as the surfaces of Venus and Mars. By Venusian or Martian standards, our Sahara desert is a veritable swamp.

Figure 9-3 [above] Circulation patterns in Earth's atmosphere
The dominant circulation in the Earth's atmosphere consists of three convection cells in the northern and three cells in the southern hemisphere. The dashed arrows on the Earth's surface indicate the direction of prevailing winds as a result of the circulation in the six cells.

Figure 9-4 [right]   A typical close-up view of Earth's surface
More than two-thirds of Earth's surface is covered with water. In contrast, there is no liquid water on Mercury, Venus, or Mars.

The Earth's crust contains clues about the processes that shape our planet's surface

Rocks that you find beneath your feet are typical specimens of the Earth's outermost layer, or **crust.** Of course, rocks are composed of chemical elements (recall the periodic table in Figure 6-7), but individual chemical elements are rarely found in a pure state. Exceptions include gold nuggets and carbon crystals called diamonds. A less-valuable example, native sulfur, is shown in Figure 9-5.

In describing specimens of the Earth's crust, it is useful to distinguish between minerals and rocks. A **mineral** is a naturally occurring solid composed of a single element of chemical combination of elements, often in the form of crystals. A **rock** is a solid part of the Earth's crust that is composed of one or more minerals.

Quartz, one of the most common rock-forming minerals in the Earth's crust, is one member of an abundant class of minerals called **silicates**

Figure 9-5   Native sulfur
Sulfur is one of the very few chemical elements that is found in a pure state. Most minerals are chemical combinations of various elements.

Figure 9-6 [left]   Three common minerals
A mineral is a specific chemical combination of elements whose atoms are typically arranged in an orderly fashion to produce a crystal. A quartz (SiO_2) crystal is at the left. Feldspar (center) is another silicate. Mica (right) also is a common rock-forming mineral.

Figure 9-7 [right]   Igneous rocks (basalt and granite)
Igneous rocks are created when molten minerals solidify. The sample on the left is basalt, which is a fine-grained mixture of feldspar with ferromagnesian minerals that give the rock its dark color. Granite is on the right. The fine-grained basalt typically forms when molten minerals cool near the Earth's surface; the coarser-grained granite typically forms from slower cooling deeper within the crust.

that are based on the element silicon. A quartz crystal consists of an orderly arrangement of silicon and oxygen atoms.

To identify various minerals, geologists use a number of criteria such as the shape, hardness, and color of the crystals. Three very common rock-forming minerals are shown in Figure 9-6. Each has its own unique, easily identifiable characteristics. These particular minerals are found mixed together in a very common rock called **granite,** which is composed of four minerals: quartz, feldspar, mica, and hornblende.

Geologists cannot classify rocks the same way they classify minerals because a rock contains a mixture of mineral characteristics. For example, the tiny quartz crystals in a piece of granite have very different color and hardness from the shiny flecks of mica. Geologists classify rocks according to how they were created. Specifically, there are three major categories of rocks, corresponding to three rock-forming processes.

Igneous rocks result when minerals cool from a molten state. The formation of igneous rock can be observed directly during a volcanic eruption. Molten rock is called **magma** when it is buried below the surface or **lava** when it flows out upon the surface. Two very common igneous rocks (basalt and granite) are shown in Figure 9-7.

Sedimentary rocks are produced by the action of wind, water, or ice. For example, as winds pile up layer after layer of sand, the grains become cemented together and produce sandstone. Minerals that precipitate out of the oceans can cover the ocean floor with layers of rock such as limestone. These two very common sedimentary rocks are shown in Figure 9-8.

Figure 9-8   Sedimentary rocks (limestone and sandstone)
Sedimentary rocks are produced, often layer by layer, by the action of wind, water, or ice. The sample on the left is limestone, which is primarily calcite with some impurities that give it a dark color. Sandstone is on the right. A sedimentary rock typically collects as loose particles such as soil or sand and then is cemented into a rock later by chemical changes after the rock has been buried a short distance below the Earth's surface.

Sometimes igneous or sedimentary rocks become buried deep beneath the Earth's surface, where they are subjected to enormous pressures and temperatures, changing the rock structure to produce the third variety called **metamorphic rock.** For example, when fine-grained igneous rock is subjected to pressure and heat, it becomes schist. When limestone is metamorphosed, it becomes marble. These two common metamorphic rocks are shown in Figure 9-9.

Figure 9-9   Metamorphic rocks (marble and schist)
When igneous or sedimentary rocks are subjected to high temperatures and pressures deep in the crust, they are changed into metamorphic rock. Marble (left) is produced from limestone; schist (right) is formed from fine-grained igneous rock.

Study of earthquake waves reveals the interior structure of the Earth

Although the specimens shown in Figures 9-7, 9-8, and 9-9 are typical samples of the Earth's crust, they are not representative of our planet's interior. The densities of crustal rocks are typically 3 to 4 g/cm^3, but the average density of the Earth as a whole is 5.5 g/cm^3. The Earth's interior, therefore, must be composed of a substance much denser than the crust.

Iron is a good candidate for this substance because it is the most abundant of the heavier elements (recall Table 6-4), and its presence in the Earth's interior is strongly suggested by the existence of the Earth's magnetic field. Furthermore, iron is common in meteoroids that strike the Earth, suggesting that it was common in the planetesimals from which the Earth formed.

Geologists strongly suspect that the Earth was entirely molten soon after its formation about $4\frac{1}{2}$ billion years ago. Energy released by the violent impacts of numerous meteoroids and asteroids and by the decay of radioactive isotopes melted the solid material collected from the earlier planetesimals. Gravity caused the abundant, dense iron to sink toward the Earth's center, forcing less-dense material to the surface. This process of chemical differentiation produced a layered structure within the Earth: a central core composed of almost pure iron, surrounded by a mantle of dense, iron-rich minerals, which in turn is surrounded by a thin crust of relatively light silicon-rich minerals.

The Earth's interior is as difficult to examine as the most distant galaxies in space. The deepest wells go down only a few kilometers, barely penetrating the surface of our planet. Geologists have, however, deduced basic properties of the Earth's interior by studying earthquakes.

Over the centuries, stresses build up in the Earth's crust. Occasionally, these stresses are relieved with a sudden, vibratory motion called an **earthquake.** The exact origin of an earthquake, called the **focus,** is usually deep within the Earth's crust. The point on the Earth's surface directly above the focus is called the **epicenter.**

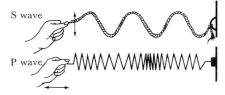

Figure 9-10   Seismic waves
Earthquakes produce two kinds of waves that travel through our planet. S waves are transverse waves, analogous to waves produced by shaking a rope up and down. P waves are longitudinal waves, analogous to those produced by pushing a spring in and out.

Earthquakes produce three different kinds of **seismic waves** that travel around or through the Earth in different ways and at different speeds. Geologists use sensitive **seismographs** to detect and record these vibratory motions. The rolling motion that people feel near an epicenter is due to **L waves,** which travel only over the Earth's surface and are analogous to water waves on the surface of the ocean. The two remaining kinds of waves, called **S waves** and **P waves,** travel through the Earth. **S waves** are said to be **transverse** waves because the vibrations are perpendicular to the direction in which the waves are moving. S waves are analogous to waves produced by a person shaking a rope up and down (see Figure 9-10). In contrast, **P waves** are said to be **longitudinal** waves because the oscillations are parallel to the direction of wave motion, like a spring that is alternately pushed and pulled.

P waves travel almost twice as fast as S waves. Consequently, P waves from an earthquake always arrive at a seismographic station before the S waves do. By measuring the time delay between the arrival of these two kinds of waves, geologists can deduce the distance to the earthquake's epicenter.

Seismic waves do not travel along straight lines through the Earth. Instead, because of varying density and composition of the Earth's interior, both S waves and P waves are bent, or refracted. By studying how these waves are bent, geologists can discover properties of the Earth's interior.

When an earthquake occurs in the Earth's crust, seismographs within a few thousand kilometers of the epicenter record both S waves and P waves. However, on the opposite side of the Earth, only P waves are recorded at seismographic stations. This absence of S waves was first explained in 1906 by the geologist R. D. Oldham, who noted that transverse vibrations such as S waves cannot travel far through liquids. Oldham therefore concluded that our planet has a molten core. Furthermore, there is a region called the **shadow zone** where neither S waves nor P waves from the earthquake are detected (see Figure 9-11). This **shadow zone** results from the specific way in which P waves are refracted at the boundary between the solid mantle and the molten core. By measuring the size of the shadow zone, geologists concluded that the core's diameter is about 7000 km (4300 miles). For comparison, the overall diameter of our planet is 12,700 km (7900 miles).

Figure 9-11   The paths of seismic waves
Seismic waves are refracted as they pass through the Earth. Furthermore, only the P waves can pass through the Earth's liquid outer core. From the paths followed by seismic waves, geologists have deduced the dimensions of the Earth's mantle and core.

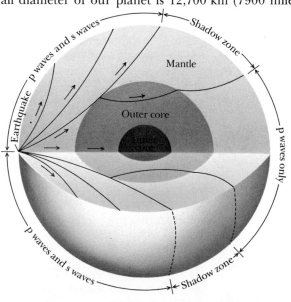

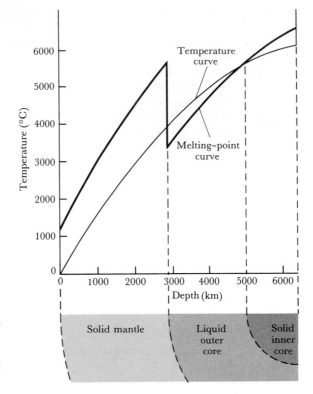

Figure 9-12 Temperature and melting-point curves inside the Earth
The temperature rises steadily from the Earth's surface to its center. By plotting the melting point of rock on this graph, we can deduce which portions of the Earth's interior are solid or liquid.

As the quality and sensitivity of seismographs improved, geologists soon realized that faint traces of P waves could be detected within the shadow zone of an earthquake. This weak refraction of P waves into the shadow zone led seismologist Inge Lehmann in 1936 to conclude that a small, solid **inner core** exists at the center of our planet. The diameter of this inner core is about 2800 km (1700 miles).

The interior of our planet therefore has a curious structure: a liquid core sandwiched between a solid inner core and a solid mantle. To understand why this is so, we must examine the temperature and pressure inside the Earth and their effects on the melting point of rock.

Both temperature and pressure increase with increasing depth below the Earth's surface. The temperature of the Earth's interior rises steadily from about 20°C on the surface to roughly 6100°C at our planet's center (see Figure 9-12).

The Earth's crust is only about 30 km thick and is composed of rocks whose melting points are far greater than typical temperatures in the crust. Hence the crust is solid.

The Earth's **mantle,** which extends to a depth of about 2900 km, is largely composed of minerals rich in iron and magnesium. On the Earth's surface, specimens of these ferromagnesian minerals have melting points slightly over 1000°C. However, the melting point of a substance depends on the pressure to which it is being subjected: the higher the pressure, the higher the melting point. As shown in Figure 9-12, the melting point of the mantle's minerals is everywhere higher than the actual temperature, so the mantle is solid.

At the boundary between the mantle and the outer core, there is an abrupt change in chemical composition from ferromagnesian minerals to almost-pure iron with a small admixture of nickel. This iron–nickel material has a much lower melting point than ferromagnesian minerals, and thus the melting-point curve on Figure 9-12 drops abruptly as it crosses from the mantle to the outer core. Indeed, the melting-point curve remains below the temperature curve down to a depth of about 5000 km. Hence, from depths of about 2900 to 5000 km, the core is liquid.

At depths greater than about 5000 km, the pressure is more than 3 million atmospheres. This pressure is so great that the melting point of the iron–nickel mixture exceeds the actual temperature (see Figure 9-12). Hence the Earth's inner core is solid.

The Earth's magnetic field produces a magnetosphere that captures particles from the solar wind

Although buried deep within the Earth, the liquid outer core has a great influence on the Earth's outermost environment. In Chapter 7, we learned that currents in the molten iron in a planet's interior can give rise to a planet-wide magnetic field through the **dynamo effect.** As a planet rotates, these currents produce a magnetic field, just as a loop of wire carrying an electric current generates a magnetic field. The Earth rotates fast enough to produce a magnetic field that dominates space for tens of thousands of kilometers and dramatically affects Earth's interaction with the solar wind.

In Chapter 7, we saw that Mercury's weak magnetic field carves out a small magnetic cavity in the solar wind. As sketched in Figure 7-19, Mercury's magnetosphere consists of a shock wave, a magnetopause, and a magnetosheath. Earth's magnetosphere also has these three basic features. However, because Earth's magnetic field is about 100 times stronger than Mercury's, our planet's magnetosphere is much bigger than Mercury's.

Figure 9-13 is a scale drawing of Earth's magnetosphere. The bow-shaped **shock wave** occurs where particles of the supersonic solar wind are abruptly slowed to subsonic speeds. The **magnetopause** is the outer boundary of the Earth's magnetic domain. Most of the particles of the solar wind are deflected around the magnetopause through the turbulent region called the **magnetosheath.** To this extent, our magnetosphere is simply a greatly enlarged version of Mercury's. However, deep inside Earth's magnetosphere, our planet's magnetic field is strong enough to trap charged particles that manage to leak through the magnetopause.

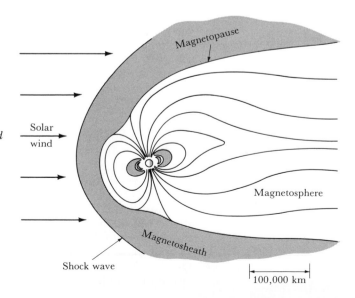

Figure 9-13 Earth's magnetosphere
Earth's magnetic field carves out a cavity in the solar wind. As in the case of Mercury, Earth's magnetosphere consists of a shock wave, a magnetopause, and a magnetosheath. Because of the strength of Earth's magnetic field, our planet is able to trap charged particles in two huge, doughnut-shaped rings called the Van Allen belts.

The particles are trapped in two huge, doughnut-shaped rings called the **Van Allen radiation belts.**

These belts were discovered in 1958 during the flight of the United States' first successful Earth-orbiting satellite. They are named after the physicist who insisted that the satellite should carry a geiger counter to detect charged particles. The inner Van Allen belt extends over altitudes of about 2000 to 5000 km and contains mostly protons. The outer Van Allen belt contains mostly electrons and is about 6000 km thick, centered at an altitude of about 16,000 km above the Earth's surface.

Occasionally, a violent event on the Sun's surface called a **solar flare** sends a burst of protons and electrons toward the Earth. Many of these particles penetrate the magnetopause and overload the Van Allen belts. The excess particles move along the Earth's magnetic field and rain down on the upper atmosphere near the Earth's north and south magnetic poles. As these particles collide with gases in the upper atmosphere, atoms of oxygen and nitrogen are made to fluoresce like the gases in a fluorescent tube. The result is a beautiful, shimmering display called the **northern lights** *(aurora borealis)* or **southern lights** *(aurora australis)*, depending on the hemisphere from which the phenomenon is observed.

Figure 9-14   The northern lights (aurora borealis)
When a deluge of protons and electrons from a solar flare strikes atoms in Earth's upper atmosphere, the gases glow. Aurorae typically occur at an altitude of about 110 km (70 miles) above the Earth's surface. (USNO)

It is remarkable that the Van Allen belts—these two vast features that completely encircle the Earth—were totally unknown until a few decades ago. Such discoveries remind us of how little we truly understand and how much remains to be learned, even about our own planet. Another recent, humbling revelation involves a geological process known as **plate tectonics.**

The process of plate tectonics is responsible for many details of the Earth's surface including mountain ranges, volcanoes, and earthquakes

One of the most important geological discoveries of the twentieth century is the realization that our planet has an active, constantly changing crust. We have learned that the Earth's crust is divided into huge **plates** that constantly jostle each other, producing earthquakes, volcanoes, and oceanic trenches.

The idea of huge, moving plates might occur to anyone who carefully examines a map of the Earth. You can see that South America would fit snugly against Africa, were it not for the Atlantic Ocean. Indeed, the fit between land masses on either side of the Atlantic Ocean is remarkable (see Figure 9-15). This observation inspired people such as Alfred Wegener to propose the idea of "continental drift," suggesting that the conti-

Figure 9-15 [above] Comparing the continents

Africa, Europe, Greenland, and North and South America fit together as though they were once joined. The fit is especially convincing if the edges of the continental shelves (rather than today's shoreline) are used. (Adapted from P. M. Hurley)

nents on either side of the Atlantic Ocean have simply drifted apart. Wegener got the idea of drifting continents around 1910 while looking at a globe of the Earth. After much research, he published in 1924 the theory that there was originally a single, gigantic supercontinent he called Pangaea, which began to break up and drift apart some 200 million years ago. Other geologists refined the theory, arguing that Pangaea first split into two smaller supercontinents they called Laurasia and Gondwanaland, separated by what was called the Tethys Sea. Gondwanaland later split into Africa and South America, while Laurasia divided to become North America and Eurasia. The Mediterranean Sea, according to this theory, is a surviving remnant of the ancient Tethys Sea.

Most geologists, however, initially greeted Wegener's ideas with ridicule and scorn. Although it was generally accepted that the continents do "float" in the denser, somewhat plastic mantle beneath them, few geologists could accept the idea that entire continents could move around the Earth at speeds that must be as great as several centimeters per year. The "continental drifters" could not explain what forces could be shoving the massive continents around.

Then in the mid-1950s, Bruce C. Heezen of Columbia University and his colleagues began discovering long mountain ranges on the ocean floors, such as the Mid-Atlantic Ridge (see Figure 9-16) that stretches all

Figure 9-16 [right] The Mid-Atlantic Ridge

This artist's rendition shows the floor of the North Atlantic Ocean. The unusual mountain range in the middle of the ocean floor is called the Mid-Atlantic Ridge and is caused by lava seeping up from the Earth's interior along a rift that extends from Iceland to Antarctica. (Courtesy of Marie Tharp)

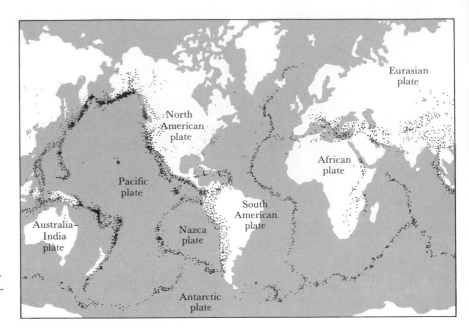

Figure 9-17   The major plates
The boundaries of major plates are the scenes of violent seismic and geologic activity. Most earthquakes occur where plates separate or collide. Plate boundaries therefore are easily identified by plotting earthquakes (each dot is an epicenter) on a map.

Figure 9-18   The mechanism of plate tectonics
Convection currents in the soft upper layer of the mantle (called the asthenosphere) are responsible for pushing around rigid, low-density plates on the crust (called the lithosphere). New crust is formed in oceanic rifts, where lava oozes upward between separating plates. Mountain ranges and deep oceanic trenches are formed where plates collide.

the way from Iceland to Antarctica. During the 1960s, careful examination of the ocean floor revealed that rock from the Earth's mantle is being melted and then forced upward along the Mid-Atlantic Ridge, which is therefore a long chain of underwater volcanoes. This upwelling of new material is forcing the ocean floor to spread. The eastern floor of the Atlantic Ocean is moving eastward, whereas the western floor is moving westward. This **sea-floor spreading** is pushing South America and Africa apart at a speed of roughly 3 cm per year.

Sea-floor spreading, of course, provided just the physical mechanism that had been missing from the theories of continental drift. With some embarrassment, geologists around 1964 began to review the old theories and found a great deal of new evidence to support them. This time, however, the emphasis was upon the motion of large plates of the crust (not simply the continents), and the modern theory of crustal motion came to be known as **plate tectonics.** Geologists today realize that boundaries of Earth's crustal plates can be identified by the seismic activity occurring where these plates are either colliding or separating. The

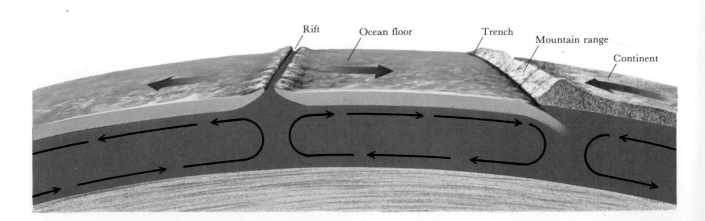

plate boundaries stand out clearly when the epicenters of earthquakes are plotted on a map (see Figure 9-17).

Geologists believe that convective currents in the Earth's mantle are responsible for the movement of the plates. The mantle is not liquid, but it is hot enough to permit an oozing, plastic flow throughout a soft region of the upper mantle called the **asthenosphere.** As sketched in Figure 9-18, hot material (magma) seeps upward along **oceanic rifts** where plates are separating. Cool material sinks back down along **subduction zones** where plates are colliding. Above the plastic asthenosphere is a low-density, rigid layer called the **lithosphere,** divided into plates that simply ride on the convection currents of the asthenosphere. The boundary between crust and mantle (a boundary between different chemical compositions) is within the lithosphere.

The regions where plates meet are the sites of some of the most impressive geological activity on our planet (see Figure 9-19). Great mountain ranges, such as those along the western coasts of North and South America are thrust up by ongoing collisions with the plates of the ocean floor (see also Figure 9-20). Subduction zones, where old crust is pushed back down into the mantle, are typically the locations of deep oceanic trenches such as those off the coasts of Japan and Chile.

Figure 9-19 [left]   The separation of two plates
The plates that carry Egypt and Saudi Arabia are moving apart, leaving the trench that contains the Red Sea. In this view taken by Gemini 12 astronauts in 1966, Saudi Arabia is on the left and Egypt (with the Nile River) is on the right. (NASA)

Figure 9-20 [right]   The collision of two plates
The plates that carry India and China are colliding. The Himalaya Mountains have been thrust upward as a result of this collision. In this photograph taken by Apollo 7 astronauts in 1968, India is on the left, Tibet is on the right, and Mt. Everest is one of the snow-covered peaks near the center. (NASA)

The convection process associated with sea-floor spreading is the primary way in which heat is transported outward from the Earth's interior to the crust. Here again our planet differs radically from its neighbors. One of the surprising results from the radar mapping of the Venusian surface (recall Figure 8-14) is that Venus has no long mountain chains resembling the Mid-Atlantic Ridge. As we shall see in Chapter 11, Mars also lacks tectonic features. Instead, both Venus and Mars have huge volcanoes. Olympus Mons on Mars rises to the altitude of 24 km (15 miles) above the surrounding volcano-studded plains—nearly three times taller than Mt. Everest. Similarly, the summit of Maxwell Montes on Venus is at an altitude of 11 km.

Venus and Mars apparently transport heat outward to their surfaces predominantly by a mechanism called **hot-spot volcanism.** Hot spots deep in the mantles of Venus and Mars squirt molten lava up through the crust. In the absence of any tectonic activity to move the crust around, millions of years of eruptions eventually build the enormous volcanoes that we observe.

Hot-spot volcanism also occurs on Earth. The Hawaiian Islands, which are in the middle of the Pacific plate, are a fine example. There is a hot spot in the Earth's mantle beneath Hawaii that continuously pumps lava up through the crust. However, the Pacific plate is moving northwest at the rate of several centimeters per year. New volcanoes are created over the hot spot as older volcanoes move away from the magma source, become extinct, and eventually erode and disappear beneath the ocean. Thus, the Hawaiian Islands are the most recent additions to a long chain of extinct volcanoes that stretches all the way back to Japan. This chain is a permanent record of the movements of the Pacific Plate over the past 70 million years.

The abundance of life on Earth inspires us to speculate about the possibility of extraterrestrial biology

There is another significant way in which our planet differs radically from its neighbors. Earth is covered with living organisms. Is our world unique, or is it possible that biological phenomena occur elsewhere in the universe? As we look up into the heavens, it is only natural to wonder if we are alone. What sort of information or data might we use to guide our speculation?

On Earth, life managed to get a very early start. Fossil imprints of algaelike organisms are found in sedimentary rocks from Australia that are at least $3\frac{1}{2}$ billion years old (see Figure 9-21). The Earth is only $4\frac{1}{2}$ billion years old and probably spent the first few hundred million years in a largely molten, actively volcanic state. Furthermore, algae are moderately complex organisms that must have evolved from more-primitive life forms. In other words, the Earth's surface must have barely cooled when life emerged! Although it is impossible to draw a general conclusion from only one example, the case of the Earth suggests that, given appropriately hospitable conditions, life forms will appear in a fairly short period of time (say, a few hundred million years).

There are at least two other considerations that suggest life might be a common phenomenon. The first is a classic experiment originally performed by chemists Harold Urey and Stanley Miller at the University of Chicago in 1953. Combinations of four elements available in Earth's primitive atmosphere (hydrogen, nitrogen, carbon, and oxygen) make up 95 percent of living tissue. Knowing this, Urey and Miller placed a mixture of gases thought to be prevalent in the early atmosphere—water vapor, methane (CH_4), ammonia (NH_3), and hydrogen—into a flask and subjected it to electric sparks to simulate lightning flashes that were probably common early in the Earth's history. After a few days, the inside of the flask was coated with a reddish-brown residue rich in large **organic molecules** (the building blocks of living things), including amino acids from which proteins are formed. This experiment thus supports the idea that conditions on a young Earth-like planet would naturally lead to the formation of complex molecules from which living organisms could evolve.

Recent findings by radio astronomers also support this conclusion. Over the past few decades, radio astronomers have found dozens of different organic molecules widely scattered throughout space. Thus the

Figure 9-21   An ancient microfossil
Rocks from Greenland and Australia contain microscopic chainlike fossils believed to be the remnants of microbial organisms that populated the Earth 3 to 3½ billion years ago. The chain of microfossils shown here is about 0.07 mm long. (Courtesy of J. William Schopf)

Figure 9-22   Project Cyclops
Some scientists believe that the existence of intelligent life elsewhere in our Galaxy is sufficiently probable to justify constructing a large system of radio telescopes sensitive enough to detect communications among aliens. This design was never built; it was rejected as too expensive. (NASA)

building blocks of life are in abundant supply in the interstellar clouds from which planets and stars form.

Whereas simple life forms arose early in Earth's history, complex organisms took a long time to evolve. It was only about 400 million years ago that creatures first emerged from the oceans and began slithering across the land. The first primitive mammals made their appearance about 200 million years ago, and the first primates appeared only about 65 million years ago. Advanced life forms, such as human beings, are a very recent phenomenon on our planet.

Is it possible that advanced life forms have evolved on other planets in other star systems? Is it possible that intelligent life exists elsewhere in the universe? Some astronomers think so, and they have made attempts to contact alien civilizations.

One of the most famous attempts involved a three-minute message that was beamed toward a star cluster called M13 in the constellation of Hercules. The message was sent in 1974 with the aid of the giant 305-m radio telescope at the Arecibo Observatory (recall Figure 7-6) and will take 24,000 years to arrive at the cluster. If anything is listening, we might expect an answer in 48,000 years—a rather long time to wait.

If any creatures anywhere in out galaxy are beaming messages towards us with a signal strength at least as powerful as that we used for the M13 message, then our radio telescopes are sensitive enough to detect the communication. For this reason, a few teams of radio astronomers in the United States, The Soviet Union, and elsewhere are involved in a search for extraterrestrial intelligence (SETI). The success of the SETI program depends largely on luck. We would have to be listening at the correct wavelength at exactly the moment that an alien message is arriving.

A more successful program might involve trying to eavesdrop on interalien communications—comparable to an alien detecting the radio and television transmissions that have been leaking away from our planet for the past half-century. Huge systems of radio telescopes be needed to detect these unbeamed signals, because the signals would be extremely weak and diluted when they reach the Earth. Such systems have been conceived. An example called Project Cyclops is shown in Figure 9-22. It would consist of 1000 radio telescopes, each disk measuring 100 m in

diameter. With this system, we could pick up the equivalent of the "Six O'Clock News" from a distance of 1000 light years.

What might an alien civilization say to us? Could we cope with what we might hear? Would we even be able to recognize an alien message if we detected it? If the extraterrestrials are even a few million years more advanced than we are, their attempts to communicate with us might be analogous to a human attempt to beam radio messages at chimpanzees in the wild. As we gaze out into the starfilled night sky, we can only wonder how many other beings may also be looking up at the stars and asking the same questions we ask.

Summary

. The Earth's atmosphere is unlike those of the other terrestrial planets in chemical composition, circulation pattern, and temperature profile.

Earth's atmosphere is 77 percent nitrogen and 21 percent oxygen, with only a small amount of carbon dioxide; this abundance of oxygen is largely due to the biological activities of life forms on the planet.

The atmosphere can be divided into layers called the troposphere, stratosphere, mesosphere, and thermosphere; ozone molecules in the stratosphere absorb ultraviolet light.

The circulation in Earth's atmosphere is complex because of the planet's rapid rotation, with three circulation cells in each hemisphere.

. The outermost layer, or crust, of the Earth is composed of rocks that contain clues about the history of the planet.

There are three major categories of rocks: igneous rocks (cooled from molten material); sedimentary rocks (formed by action of wind, water, or ice); and metamorphic rocks (altered in the solid state by extreme heat and pressure)

Rocks are composed of minerals, naturally occurring elements, or chemical combinations of elements.

. Study of seismic waves shows that the Earth has a small solid inner core surrounded by a liquid outer core; the outer core is surrounded by a dense mantle, which in turn is surrounded by a thin low-density crust.

The inner and outer core are composed of almost-pure iron with some nickel mixed in; the mantle is composed of iron—magnesium (ferromagnesian) minerals; the crust is largely composed of silicate minerals.

There are three types of seismic waves: L waves travel along the planet's surface; transverse S waves and longitudinal P waves travel through the planet's interior.

Both temperature and pressure increase steadily with depth inside the Earth.

. The Earth's magnetosphere is similar to that of Mercury, but larger because of the Earth's stronger magnetic field.

Charged particles from the solar wind are trapped in two huge doughnut-shaped rings called the Van Allen radiation belts.

. The Earth's crust and a small part of the upper mantle form a rigid layer called the lithosphere; this layer is divided into huge plates that move about over the plastic layer called the asthenosphere in the upper mantle.

Movements of the plates (called plate tectonics) are driven by up-welling of molten material at the oceanic rifts, producing sea-floor spreading.

Plate tectonics is responsible for most of the major features of the Earth's surface, but hot-spot volcanism does create some large volcanoes.

. Life apparently appeared on Earth only a few hundred million years after the surface of the planet cooled to a solid state; complex organic molecules are scattered abundantly through the universe; therefore, it seems likely that life will have appeared elsewhere in the universe, wherever a planet exists with favorable conditions.

Review questions

1 Describe the various ways in which the Earth is unique among the planets of the solar system.

2 Why is the Earth's surface not riddled with craters like the surfaces of Mercury, the Moon, and Mars?

3 In what way are the Earth's oceans like an atmosphere? In what way could they be considered part of the crust? How would your answers be different if the Earth were as close to the Sun as Venus is or as far from the Sun as Jupiter is?

*4 What fractions of the Earth's total volume are occupied by the core, the mantle, and the crust?

5 Look up the chemical compositions of quartz, feldspar, mica and hornblende and compare the elements you find with those listed in Table 6-4. Are your findings consistent with the fact that these minerals are very common on the Earth's surface? Explain.

6 Why do you suppose that world-wide television transmissions require the use of relay satellites yet radio programs can be transmitted around the world without the use of such satellites?

*7 As mentioned in the text, Africa and South America are separating at a rate of about 3 centimeters per year. Assuming that this rate has been constant, calculate when these two continents must have been in contact.

Advanced questions

8 The Earth's primordial atmosphere probably contained an abundance of methane (CH_4) and ammonia (NH_3) whose molecules were broken apart by ultraviolet light from the Sun and particles in the solar wind. What happened to the atoms of carbon, hydrogen, and nitrogen that were liberated by this dissociation?

*9 The Earth's atmospheric pressure decreases by a factor of one-half for every $5\frac{1}{2}$ km increase in altitude above sea level. Construct a plot of pressure versus altitude (assume that the sea-level is 1 bar). Discuss the characteristics of your graph. At what altitude is the atmospheric pressure 1 millibar?

Discussion questions

10 The human population on Earth is currently doubling about every 30 years. Describe the various pressures placed on the Earth by uncontrolled human population growth. Can such growth continue indefinitely? If not, what natural and human controls might arise to curb this growth? It has been suggested that over-population problems could be solved by colonizing the Moon or Mars. Do you think that this is a reasonable solution? Explain your answer.

11 A recent scientific study suggests that the continued burning of fossil and organic fuels by humans is releasing enough CO_2 to begin a Venus-like greenhouse effect that will melt the polar ice caps. Estimate the volume of the polar ice caps. Assuming that water and ice have roughly the same density, estimate the amount by which the water level of the world's oceans will rise if the polar caps were to melt completely. What portions of the Earth's surface would be inundated by such a deluge?

12 Imagine that you are an astronomer working at a radio telescope and it becomes apparent that you have detected a message from an advanced alien civilization. Whom would you tell? The military? Politicians? The public? Speculate about the social, political, and economic implications of such an announcement, especially since the alien message might contain a large amount of extremely advanced technical and scientific information. Do you think that astronomers should have an international agreement about what to do in case someone actually does detect such a message?

For further reading

On the Earth:

Akasofu, S. "The Aurora: New Light on an Old Subject." *Sky & Telescope,* Dec. 1982, p. 534.

Calder, N. *The Restless Earth.* Viking Press, 1972. *A good book on our modern view of the Earth by a noted science writer.*

Carrigan, C., and Gubbins, D. "The Source of the Earth's Magnetic Field." *Scientific American,* Feb. 1979.

Hartmann, W. "The Early History of the Planet Earth." *Astronomy,* Aug. 1978, p. 6.

Marvin, U. "The Rediscovery of Earth." In Cornell, J., and Gorenstein, P., eds. *Astronomy from Space.* MIT Press, 1983. *A summary of how we study Earth from space.*

Scientific American magazine had a special issue on the Earth in September 1983.

Wong, C. "Watching the Earth Move from Space." *Sky & Telescope,* Mar. 1978, p. 198. *On measuring continenetal drift.*

On Life Elsewhere:

Discover Magazine, March 1983. *A special section of articles on SETI.*

Goldsmith, D., and Owen, T. *The Search for Life in the Universe.* Benjamin/Cummings, 1980. *An introductory textbook on SETI.*

Papagiannis, M. "The Search for Extraterrestrial Civilizations." *Mercury,* Jan./Feb. 1982, p. 12.

Rood, R., and Trefil, J. *Are We Alone?* Scribner's, 1981. *A popular-level book on SETI.*

Sagan, C., and Drake, F. "The Search for Extraterrestrial Intelligence." *Scientific American,* May 1975.

Tipler, F. "The Most Advanced Civilization in the Galaxy is Ours." *Mercury,* Jan./Feb. 1982, p. 5. *A thought-provoking challenge to the conventional view that intelligent life exists elsewhere in the Galaxy.*

10 Our barren Moon

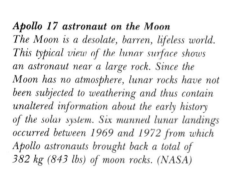

Apollo 17 astronaut on the Moon
The Moon is a desolate, barren, lifeless world. This typical view of the lunar surface shows an astronaut near a large rock. Since the Moon has no atmosphere, lunar rocks have not been subjected to weathering and thus contain unaltered information about the early history of the solar system. Six manned lunar landings occurred between 1969 and 1972 from which Apollo astronauts brought back a total of 382 kg (843 lbs) of moon rocks. (NASA)

The Earth and the Moon form a double-planet system that is unique among the terrestrial planets. In this chapter we learn about the features of the lunar surface that have remained almost unaltered for billions of years. The program of lunar exploration in the 1960s and early 1970s has provided information about the Moon's formation and history. We discuss the Moon's interior and the Moon's lack of an atmosphere or magnetic field. We learn how the Earth and the Moon interact gravitationally to produce tidal effects on both worlds. Finally, we see how radioactive age-dating of rock samples brought back to the Earth by the Apollo astronauts has helped us deduce the early history of the Earth–Moon system.

One of the most conspicuous features of the Earth is its very large satellite, the Moon. Indeed, we are living on one member of a double-planet system. Neither Mercury nor Venus has a natural satellite; Mars does have two moons, but they are very small. The Jovian planets possess several large satellites but, like the moons of Mars, each is very small compared to its parent planet. In all these cases, each planet is thousands of times more massive than any of its satellites. However, the Earth is only 81 times more massive than its Moon. In other words, the masses of the Earth and Moon are roughly comparable. Why did these two worlds, which formed at approximately the same time and place, turn out so vastly different? Whereas the Earth is a dynamic, living planet, our Moon is biologically and geologically dead. Indeed, the only noteworthy events to occur on the lunar surface over the past $3\frac{1}{2}$ billion years are occasional impacts by wayward meteoroids. And, of course, the arrival of humanity.

Box 10-1   Lunar data

Distance from Earth (center to center)
   mean: 384,400 km = 238,860 miles
   closest (perigee): 356,410 km
   farthest (apogee): 406,700 km
Distance from Earth (suface to surface)
   mean: 376,280 km = 233,810 miles
   closest (perigee): 348,290 km
   farthest (apogee): 398,580 km
Sidereal period (fixed stars): 27.321661 days
Synodic period (new moon to new moon): 29.530588 days
Mean orbital velocity: 3680 km/hr
Inclination of lunar equator to ecliptic: 1° 33′
Inclination of lunar equator to orbit: 6° 41′
Inclination of orbit to ecliptic: 5° 09′
Orbital eccentricity: 0.055
Diameter: 3476 km = 2160 miles
Apparent diameter as seen from Earth:
   mean: 31′ 5″
   maximum: 33′ 31″
   minimum: 29′ 22″
Mass: 7.35×10^{22} kg
Mass (Earth = 1): 0.0123
Mean density: 3.34 g/cm^3
Surface gravity (Earth = 1): 0.165
Escape velocity: 2.38 km/sec
Albedo: 0.07
Mean magnitude (at full): −12.7

The Moon's early history can be deduced from the craters, maria, and mountains visible on its surface.

Our satellite consistently provides one of the most dramatic sights in the nighttime sky. The Moon is so large and so near the Earth that some of its surface features are readily visible to the naked eye. Casual observation reveals that the Moon perpetually keeps the same side facing the Earth. As mentioned in Chapter 7, this **synchronous rotation** is a stable situation resulting from the gravitational interaction between Earth and Moon. Nevertheless, with patience, you can see slightly more than one-half of the lunar surface because the Moon wobbles slightly as it moves along its orbit. This wobbling, called **libration,** permits us to view 59 percent of our satellite's surface.

With a small telescope you can see several major different types of lunar terrain (see Figure 10-1). Most prominent are the large, dark, flat areas called **maria** (pronounced MAR-ee-uh). The singular form of this term is **mare** (pronounced MAR-ee); it means "sea" in Latin and was introduced in the seventeenth century when observers using early telescopes thought these were large bodies of water on the Moon. In fact, bodies of liquid water could not possible exist on our airless satellite. Because there is no atmospheric pressure, a lake or ocean would boil furiously and evaporate rapidly into the vacuum of space. Actually, the maria were formed by huge lava flows that inundated low-lying regions

of the lunar surface $3\frac{1}{2}$ billion years ago. Nevertheless, they have retained their fanciful names such as Mare Tranquillitatis (Sea of Tranquillity), Mare Nubium (Sea of Clouds), Mare Nectaris (Sea of Nectar), and Mare Serenitatis (Sea of Serenity).

The largest of the 14 maria is Mare Imbrium (Sea of Showers). It is roughly circular and measures 1100 km (700 miles) in diameter (see Figure 10-2). Although the maria seem quite smooth in telescopic views from the Earth, close-up photographs by the Apollo astronauts reveal small craters and occasional cracks called **rilles** (see Figure 10-3).

Figure 10-1 [left]   The Moon
Our Moon is one of seven large satellites in the solar system. The Moon's diameter (3476 km = 2160 miles) is slightly less than the distance from New York to San Francisco. This photograph is a composite of first-quarter and last-quarter views, so that elongated shadows enhance all surface features. (Lick Observatory)

Figure 10-2 [right]   Mare Imbrium viewed from the Earth
Mare Imbrium (Sea of Showers) is the largest of 14 dark plains that dominate the Earth-facing side of the Moon. Because few craters are seen on the maria, they must have been formed by lava flows occurring late in the Moon's geologic history. (Mount Wilson and Las Campanas Observatories)

Figure 10-3   Details of Mare Tranquillitatis
Close-up views of the lunar surface reveal numerous tiny craters and cracks on the maria. This photograph was taken in 1969 by Apollo 10 astronauts from lunar orbit during a final photographic reconnaissance of potential landing sites for the Apollo 11 team. (NASA)

Figure 10-4   The crater Clavius
*This photograph through the 200-in. Palomar
telescope shows one of the largest craters on the
Moon. Clavius has a diameter of 232 km
(144 miles) and a depth of 4.9 km (16,000
ft) measured from the crater floor to the top of
the surrounding rim. (Palomar Observatory)*

Figure 10-5   Details of a lunar crater
*This photograph, taken from lunar orbit by
Apollo 11 astronauts in 1969, shows a typical
view of the Moon's heavily-cratered far side.
The large crater near the middle of the picture
is approximately 80 km (50 miles) in diameter.
Note the crater's central peak and the numerous tiny craters that pockmark the lunar surface. (NASA)*

Perhaps the most familiar and characteristic features on the Moon are
craters. With an Earth-based telescope, some 30,000 of them are visible,
in size from 1 km to more than 100 km across (see Figure 10-4). Following a tradition established in the seventeenth century, the most prominent craters are named after famous philosophers and scientists.

Craters smaller than about 1 km in diameter cannot be seen from
Earth, simply because of optical limitations such as resolving power of
telescopes. Photographs from lunar orbit reveal millions of smaller
craters that escape the scrutiny of Earth-based observers. Virtually all
craters—both large and small—are the result of bombardment by meteoritic material.

Many of the youngest craters are surrounded by light-colored streaks
called **rays** that were formed by violent ejection of material during impact. Rays emanating from the crater Copernicus, just south of Mare
Imbrium, are visible in the lower portion of Figure 10-2. In addition,
many large craters have a pronounced **central peak** that forms during a
high-speed impact by a sizable meteoroid (see Figure 10-5).

The flat, low-lying, dark maria cover only 15 percent of the lunar surface. The remaining 85 percent of the Moon's surface is light-colored,
heavily cratered terrain at elevations generally higher than those of the
maria. This second, more-common terrain is called the **terrae,** or
highlands. (Terra means "land" in Latin; in this fanciful terminology,
the entire lunar surface is covered by either "land" or "sea").

One of the surprises stemming from the early days of lunar exploration is that there are no maria on the Moon's far side. The lunar far
side consists entirely of heavily cratered highlands (see Figure 10-6). Detailed observations by Apollo astronauts in lunar orbit demonstrated that
the maria on the Moon's Earth-facing side are 2 to 5 km below the aver-

age lunar elevation. In contrast, the cratered terrae on the lunar far side are typically at elevations up to 5 km above the average lunar elevation.

There are other features on the Moon than the flat maria and the cratered terrae. For example, there are **mountain ranges.** Although they take their names from famous terrestrial ranges (for example, the Alps, the Apennines, and the Carpathians), the lunar mountains were not formed in the same way as were the mountains of Earth. Earth's major mountain ranges are created by collisions between lithospheric plates, as discussed in Chapter 9. Lunar mountain ranges, however, were thrust up by the violent impacts of the huge meteoroids that created the maria. Lunar mountains form the rims of vast **basins** that contain the maria. Mountains around the edges of Mare Imbrium can be seen in Figure 10-2; a remarkably circular set of mountains around Orientale Basin is visible in Figure 7-17.

From the features we see on the Moon, we can piece together a probable history of the lunar surface. Like the terrestrial planets, the Moon was molten during the later stages of its formation. The heat energy causing the melting came from the impacts of planetesimals and the decay of radioactive elements. Shortly after the Moon's crust cooled, a period of intense meteoritic bombardment peppered the Moon with numerous craters that form the terrae. Near the end of this cratering period, several large, asteroid-sized objects struck the lunar surface, breaking through the thin crust. Lava welled up, flooding the low-lying areas and producing the maria. Large meteoroids must also have struck the Moon's far side. However, because all the maria are on the Earth-facing side, we theorize that the crust on the Moon's far side was thick enough to prevent penetration to the molten interior. Not much has happened since those ancient days, as indicated by the fact that the mare surfaces have remained almost unchanged since their formation.

Although this scenario seems quite reasonable, we need supporting evidence. When did most of the cratering occur? When did the lava flows flood the mare basins? Are the light-colored highland rocks really older than the darker mare rock, as we would expect from a telescopic examination of the lunar surface? There was only one way to find out:

Figure 10-6 Terrae versus maria
This photograph by Apollo 17 astronauts shows the main difference between the near and far sides of the Moon. The Earth-facing side of the Moon (toward the upper left) is populated with dark, flat maria. The far side of the Moon (toward the lower right) is heavily cratered terrain, with no maria. (NASA)

a journey to the Moon to obtain lunar samples for direct examination and analysis.

Lunar samples would not only help to answer many straightforward questions about the Moon; they would also shed light on the origin of the Earth. Because Earth is a geologically active planet, all traces of Earth's origins have been erased. Typical terrestrial surface rocks are only a few hundred million years old, which is a fraction of the Earth's age. Moon rocks, however, have been undisturbed for billions of years. Lunar exploration could provide a valuable perspective on the creation of the Earth and the birth of the entire solar system.

Lunar exploration was one of the major technological feats of all time

Figure 10-7   *The Crater Alphonsus from Ranger 9 and from Earth*
*Ranger 9 crashed into the floor of the crater Alphonsus on March 24, 1965. **(a)** This photo was taken from an altitude of 415 km (258 miles) only 2 min 50 sec before impact (white circle indicates crash site). Alphonsus has a diameter of 130 km (80 miles), and its central peak rises to an elevation of 1 km (3300 ft) above the surrounding crater floor. **(b)** This view is one of the best Earth-based photographs of the same region. (NASA)*

Human exploration of the Moon's surface is an achievement for which the twentieth century undoubtedly will be remembered. The first step toward lunar exploration was to get a simple probe to the Moon. The Soviet Union achieved this goal first with Luna 2, which sent back data indicating the absence of a lunar magnetic field before it crashed into Mare Serenitatis in September 1959. The American Ranger program launched nine small spacecraft at the Moon. Each Ranger craft carried several television cameras that were designed to return thousands of pictures just before the probe crashed into the lunar surface. The first six probes were failures, but in 1965 the final three Rangers gave scientists their first close-up views of the lunar surface (see Figure 10-7). In the meantime, the Soviet Union had achieved another first by sending a probe around the Moon to obtain pictures of its far side.

The second step was to land a functioning probe on the lunar surface (a soft landing). Again, the USSR was first with Luna 9 in January 1966. The Surveyor program was the American soft-landing effort, and it proved far more reliable than the Ranger program. The five successful Surveyors not only returned thousands of surface photographs to Earth but also dug into the lunar soil to test its physical properties and per-

a
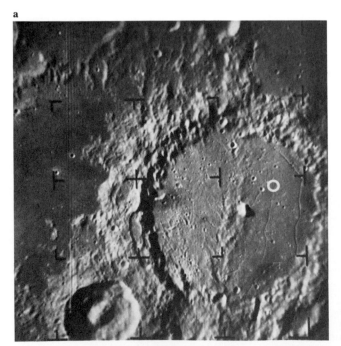

b

Figure 10-8 Surveyor 3 and Apollo 12
Five Surveyor spacecrafts made successful soft landings on the Moon between June 1966 and January 1968. They sent back more than 86,000 pictures of the lunar surface and performed soil analyses to pave the way for the Apollo landings. In this photograph, an Apollo 12 astronaut removes a piece of Surveyor 3 for later examination on Earth. The Apollo 12 spacecraft and radio antenna are visible on the horizon. (NASA)

Figure 10-9 The Alpine Valley
The Lunar Orbiter 5 probe was at an altitude of 247 km above the lunar surface when it took this picture of the Alpine Valley in August 1967. The valley is a cigar-shaped depression 120 km long and up to 10 km wide. It is located in the Alps on the northeastern edge of Mare Imbrium. Notice the long rille that runs the length of the valley. This valley can be seen as a faint, straight-line feature near the right side of the Earth-based photograph in Figure 10-2. (NASA)

formed simple chemical analyses of lunar soil and rocks. (Figure 10-8 shows the Surveyor 3 probe being examined by an astronaut during the later Apollo 12 mission.)

The Ranger probes were primitive versions of the Mariner spacecraft that later explored Mercury, Venus, and Mars (recall Figure 7-8). Many of the design features of the Surveyors were later incorporated in the Viking Landers that today rest on the Martian surface.

While Surveyor and Luna probes were sending back data and pictures from the lunar surface, additional probes were being placed in orbit around the Moon. America's Lunar Orbiter program was designed primarily to make a photographic reconnaissance of the entire lunar surface. The first three Orbiters were placed in low equatorial orbits about the Moon, and the last two were inserted into near-polar orbits. During 1966 and 1967, these five spacecraft returned close-up views of more than $99\frac{1}{2}$ percent of the lunar surface (see Figure 10-9). These detailed photographs confirmed the fact that no maria exist on the Moon's far side (as suggested earlier by lower-quality Soviet photographs).

Careful tracking of the Lunar Orbiters revealed small deviations from the expected orbits, apparently due to gravitational pulls from concentrations of dense material beneath several of the maria on the Earth-facing side of the Moon. The nature of these mass concentrations, or **mascons,** is still being debated. They were first thought to indicate the presence of huge meteorites buried in the great impact basins. (A **meteoroid** is a rock in interplanetary space; a **meteorite** is a rock from space that is found on the ground.) Many geologists, however, now argue that the mass concentrations exist simply because of the lava that flooded the mare basins is more dense than the rock of the surrounding highlands. The largest mascon, producing the greatest anomaly in the Moon's gravitational pull, is located in Mare Imbrium. Mascons are also associated with Mare Serenitatis, Mare Humorum, Mare Nectaris, Mare Crisium, and Mare Orientale.

Photographs from the Orbiters (many with resolutions sharp enough to show features less than 1 m across) revealed many suitable sites for

Figure 10-10   The Saturn 5 launch vehicle
*The mighty Saturn 5 rocket gave the United
States the ability to launch heavy payloads.
Each rocket consisted of three separate stages,
stood 86 m (281 ft) tall, and was topped by
the Apollo spacecraft whose lunar module,
service module, and command module added
another 23 m (82 ft). (NASA)*

human landings on the Moon. Data and pictures from the Surveyors dispelled many worries about the dangers of the lunar surface. For example, some scientists had suggested that the Moon would be covered with a thick layer of dust into which spacecraft would simply vanish.

While the Surveyor and Orbiter programs were being completed, the spacecraft for the Apollo program was being tested (see Figure 10-10). By early 1969, the United States was ready to fulfill President John F. Kennedy's 1961 mandate to send a person to the Moon and safely back to Earth before the end of the decade.

A three-person crew was deemed necessary for a Moon landing so that one astronaut would remain in lunar orbit in the command module while the other two descended to the lunar surface in the lunar module. The two modules would subsequently link up in orbit, the two astronauts would crawl from the lunar module back into the command module along with the rocks they had gathered, the lunar module would then be jettisoned, and all three astronauts would return to Earth in the command module. The historic moment came on July 20, 1969, when people around the world heard Neil Armstrong's announcement: "Tranquility Base here. The Eagle has landed." Humanity had reached the surface of another world.

There were six successful lunar landings in all. The first two, Apollo 11 and Apollo 12, set down in maria. The remaining four landings (Apollo 14 through 17 in 1971 and 1972) were made in progressively more challenging terrain, culminating with the rugged Taurus Mountains, just east of Mare Serenitatis. Major factors in choosing the six landing sites were the astronauts' safety and the exploration of a wide variety of geologically interesting features.

Measurements on the lunar surface show that the Moon has no magnetic field but may have a small, solid core

The lunar modules were packed with an assortment of apparatus and equipment that the astronauts deployed around the landing sites (see Figure 10-11). For example, all of the missions carried magnetometers, which confirmed that the Moon has no magnetic field, suggesting that the Moon today does not have a molten core. Careful magnetic measurements of lunar rocks indicated however, that the Moon did have a weak magnetic field when the rocks solidified billions of years ago. The implication is that the Moon originally had a small, molten, iron-rich core, but that the core solidified as the Moon cooled, so that the lunar magnetic field disappeared. Geologists estimate that, if the Moon does in fact have an iron-rich core, the core is probably less than 700 km (435 miles) in diameter.

Seismometers set up by the astronauts indicated that the Moon exhibits far less seismic activity than does the Earth. Rougly 3000 **moonquakes** were detected per year, whereas a similar seismometer on Earth would record hundreds of thousands of earthquakes per year. Furthermore, typical moonquakes are far weaker than typical earthquakes. A major moonquake measures 0.5 to 1.5 on the Richter scale, whereas a major earthquake is in the range of 6 to 8 on that scale.

Analysis of the feeble moonquakes reveals that most originate at depth of 600 to 800 km below the surface, deeper than the focus of most earthquakes. The depth of the deepest earthquakes is regarded as an indicator of the boundary between the solid lithosphere and the plastic asthenosphere. The lithosphere is brittle enough to fracture and produce seismic waves, whereas the deeper rock of the asthenosphere oozes and flows rather than cracks. Applying the same reasoning to the Moon,

Figure 10-11 The Apollo 15 base
The last three Apollo missions used surface vehicles called Lunar Rovers that greatly enhanced the astronauts' mobility. This photograph shows the Apollo 15 landing site at the foot of the Apennine Mountains near the eastern edge of Mare Imbrium. The hill in the background, called Hadley Delta, is about 5 km behind the lunar module and rises about 3 km above the surrounding plains. (NASA)

we conclude that the Moon's lithosphere is about 800 km thick (see Figure 10-12).

The lunar asthenosphere probably extends down to the iron-rich core at a depth of more than 1400 km below the lunar surface. The asthenosphere and the lower part of the lithosphere presumably are composed of relatively dense ferromagnesian rocks. The upper part of the lithosphere is a less-dense, silicate-rich crust, with an average thickness of about 60 km on the Earth-facing side and up to 100 km on the far side. For comparison, the thickness of the Earth's crust ranges from 5 km under the oceans to about 30 km under major mountain ranges on the continents.

Heat-flow experiments were set up at the Apollo 15 and Apollo 17 bases on the Moon. Sensitive heat detectors measured a heat loss about one-third as great as that measured on the surface of the Earth. The existence of this outward heat flow indicates that the Moon is not totally cold and dead, thus supporting the idea that there is enough heat inside the Moon to create a nonrigid asthenosphere.

Seismometers left on the Moon continued to transmit data for nearly eight years, so geologists were able to monitor meteoritic impacts. Indeed, the Apollo seismometers were sensitive enough to detect a hit by a grapefruit-sized meteoroid anywhere on the lunar surface. Long-term monitoring showed that the Moon is struck by 80 to 150 meteoroids per year with masses in the range of 100 g to 1000 kg (roughly $\frac{1}{4}$ pound to 1 ton).

Figure 10-12 The internal structure of the Moon
Like the Earth, the Moon probably has a crust, a mantle, and a core. The lunar crust has an average thickness of about 60 km on the Earth-facing side but about 100 km on the far side. The crust and solid upper mantle form a lithosphere about 800 km thick. The plastic (nonrigid) asthenosphere probably extends all the way to the base of the mantle. If the Moon has an iron-rich core, it is solid and is less than 700 km in diameter. Although the main features of the Moon's interior are analogous to those of the Earth's interior, the proportions are quite different (compare Figure 9-11).

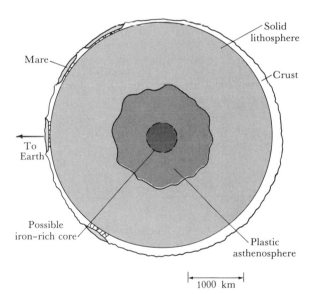

Gravitational interactions produce significant tidal effects in the Earth–Moon system

Long-term monitoring of the seismometers at the Apollo sites on the Moon revealed that the frequency of moonquakes reaches a maximum at new moon and at full moon. Geologists concluded from this pattern that the frequency of moonquakes is influenced by tidal forces.

Anyone who has lived near the ocean is familiar with tides. The shape of the Earth's oceans is distorted by the gravitational pull of the Sun and the Moon (see Figure 10-13). This distortion is greatest when the Sun, Moon, and Earth are aligned (at either new moon or full moon), producing the large tidal shifts in water level that are called **spring tides.** At

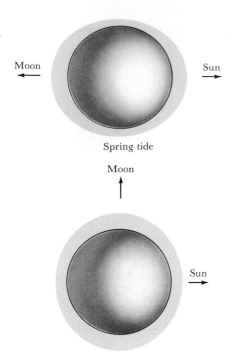

Spring tide

Neap tide

Figure 10-13   Tides on the Earth
*The gravitational forces of the Moon and Sun
deform the oceans. The greatest deformation
(spring tides) occurs when the Sun, Earth, and
Moon are aligned. The least deformation
(neap tides) occurs when the Sun, Earth, and
Moon form a right angle.*

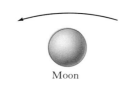

Moon

Earth

Figure 10-14   The Moon's tidal recession
*Because of the Earth's rapid rotation, the tidal
bulge on the oceans is dragged ahead of the
Moon's position in the sky. This bulge on the
leading side of the Earth produces a small forward force on the Moon that causes it slowly
to spiral away from the Earth.*

first quarter and last quarter, when the Sun and Moon are at right angles from the Earth, the tidal distortion of the oceans is least pronounced, producing the smaller tidal shifts called **neap tides.**

It is easy to notice ocean tides: the water level goes up and down. It is
harder to notice, but the solid Earth, too, is deformed and stretched by
the gravitational forces of the Sun and Moon (although to not a much
lesser extent than the oceans).

Just as the gravitational forces of the Sun and Moon deform the
Earth, so also the Sun and Earth deform the Moon. Because the Earth is
81 times as massive as the Moon, the Earth produced very large tidal
deformations of lunar surface. The Moon is squeezed and stretched as it
orbits our planet. The strongest gravitational stresses occur when the
Sun, Moon, and Earth are aligned (at new moon and full moon)—the
time when the Apollo seismometers report the highest frequency of
moonquakes.

Tidal effects also explain why the Moon keeps one side facing the
Earth. Under the Earth's gravitational pull, the Moon is slightly elongated—bulging in much the same way as the Moon's gravity causes the
oceans to bulge out on opposite sides of the Earth (recall Figure 10-13).
Under the gravitational pull of the Earth, the most stable configuration
exists when the tidally induced bulge on the Moon points at the Earth.
Thus, tidal forced induce the common phenomenon of synchronous rotation. We shall see in later chapters that the tiny Martian moons and
many of the satellites of Jovian planets also exhibit synchronous rotation,
keeping the same side always facing the parent planet.

The tidal interactions between the Earth and the Moon have had a
major effect on the relationship between these two bodies over the ages.
The Earth rotates about its axis faster than the Moon revolved around
the Earth. Therefore, as the Moon produces a tidal bulge in the oceans,
the Earth's rotation carries that bulge on around the planet—that is, the
high tide moves on ahead of the Moon rather than staying just below it
(see Figure 10-14). Viewing from the Moon, the bulge on the Earth is
always aimed slightly ahead of the Moon's own position. The gravitational pull of this bulge produces a small but constant force tugging the
Moon ahead in its orbit, speeding up its motion very slightly. As the
Moon's speed of revolution around the Earth increases, it moves gradually into more and more distant orbits around the Earth. In other words,
the Moon is very slowly but relentlessly spiraling away from the Earth at
a rate of about 4 cm per year.

As the Moon moves farther and farther away from the Earth, its sidereal period becomes longer and longer (recall Kepler's third law). Furthermore, constant friction between the oceans and the Earth's surface is
causing the Earth's rate of rotation to decrease. The day is therefore
becoming longer and longer, by approximately 0.002 seconds per century. Indeed, at some point in the very distant future, the Earth will be
rotating so slowly that a solar day will equal a lunar month. At that time,
the Earth's tidal bulge will be aimed directly at the Moon, and therefore
the Moon will stop spiraling away from the Earth. Calculations tell us
that this stable situation will be reached when the solar day and the
lunar month both equal 47 of our present solar days. The Earth will
then keep its same side facing the Moon just as the Moon keeps the
same side facing the Earth.

Geological evidence confirms the long-term tidal recession of the
Moon. Certain marine creatures produce both daily and monthly banded
structures in their shells, thereby allowing biologists to determine the
number of days in a lunar month. An examination of fossils from the

early Cenozoic Era (50 million years ago) shows that the Moon's synodic period was then 29 days instead of the present $29\frac{1}{2}$ days. In addition, ancient corals indicate that the day was only 22 hours long during the Paleozoic Era (400 million years ago). Extrapolation into the distant past suggests that the Moon was nearest the Earth about $4\frac{1}{2}$ billion years ago, at approximately the time when the two bodies formed. At that time, the Earth was probably rotating so rapidly that a day lasted only 5 or 6 hours.

Lunar rocks were formed 3 to $4\frac{1}{2}$ billion years ago

The Apollo astronauts brought back 382 kg (843 pounds) of lunar rocks that proved to be a very important source of information about the early history of our double-planet system. All of the lunar rock samples are igneous rocks, and all of them appear to have formed through cooling of molten lava. The samples are almost completely composed of the same minerals that are found in terrestrial volcanic rocks: pyroxene, olivine, feldspar, and so forth. No sedimentary rocks were found by the the Apollo astronauts, indicating that the Moon has never had an atmosphere or oceans to create such rocks. They found no true metamorphic rocks either, although many of the lunar samples have been modified by meteoritic impacts. Indeed, the entire lunar surface is covered with a layer of find powder and rock fragments produced by $4\frac{1}{2}$ billion years of relentless meteoritic bombardment. This layer, which ranges in thickness from 1 m to 20 m, is called the **regolith** because the term "soil" as used on Earth normally suggests the presence of decayed biological matter.

Astronauts who visited the maria discovered that these dark regions of the Moon are covered with basaltic rock quite similar to the dark-colored rocks formed by lavas from volcanoes on Hawaii and Iceland. The rock of these low-lying lunar plains is called **mare basalt** (see Figure 10-15).

In contrast to the dark maria, the light-colored lunar highlands are covered with a light-colored rock called **anorthosite** (see Figure 10-16). On Earth, anothositic rock is found only in very old mountain ranges such as the Adirondacks in the eastern United States. Anorthosite is rich in calcium and aluminum in comparison to the bare basalts, which have more of the heavier elements such as iron, magnesium, and titanium. Anorthosite therefore has a lower density than basalt. The anorthositic magma apparently floated to the lunar surface when the Moon was molten, solidifying as it cooled to form the lunar crust. The denser mare

Figure 10-15 [left] Mare basalt
This 1531-g (3½- pound) specimen of mare basalt was brought back by Apollo 15 astronauts. This particular sample is called a vesicular basalt because of the tiny holes, or "vesicles," that cover 30 percent of the rock's surface. Gas must have been dissolved under pressure in the lava from which this rock solidified. When the lava reached the airless lunar surface, bubbles formed as the pressure dropped. Some of the bubbles were frozen in place as the rock cooled. (NASA)

Figure 10-16 [right] Anorthosite
The light-colored lunar terrae are covered with a very ancient type of rock called anorthosite. Anorthositic rock is believed to be the material of the original lunar crust. This particular sample, called the "Genesis rock" by the Apollo 15 astronauts who picked it up at the base of the Apennine Mountains, has an age of approximately 4.1 billion years. (NASA)

Figure 10-17   A lunar breccia
Meteoritic impacts can cement rock fragments together to form breccias. This particular impact breccia was collected by Apollo 16 astronauts from the rim of a crater near their landing site. (NASA)

Figure 10-18   A glass-coated breccia
This glass-coated lunar rock was picked up near the rim of a crater by the Apollo 15 astronauts. Natural glass forms whenever molten rock cools very rapidly. Notice the dark color of the shiny glass coating. (NASA)

Figure 10-19   A zap crater
The upper surfaces of many moon rocks are covered with nearly microscopic craters produced by the impacts of high-speed meteoritic dust grains. These glass-lined craters are typically less than 1 mm in diameter. (NASA)

basalts formed later from lava that oozed out from the lunar interior and filled the mare basins.

The Apollo astronauts brought back many specimens of **impact breccias,** which are of various rock fragments that have been cemented together by meteoritic impact (see Figure 10-17). Breccias formed by sedimentary or volcanic action are fairly common on the Earth, but impact breccias are quite rare on the Earth.

Meteoritic bombardment is the only source of "weathering" to which rocks of the lunar surface are subjected. In addition to making breccias and churning up the regolith, meteoritic impacts also melt rocks to produce glass. Many lunar samples are coated with a thin layer of smooth, dark glass created when the surface of the rock was suddenly melted and then rapidly solidified (see Figure 10-18). Furthermore, small black glass beads are common in the lunar regolith. Presumably, these glass spheres were formed from droplets of molten rock hurled skyward by the impact of a meteoroid. Finally, many lunar samples bear the scars of high-speed meteoritic dust. Dust grains traveling at thousands of kilometers per hour produce tiny **zap craters** on the exposed surface of Moon rocks (see Figure 10-19). Because of these tiny, glass-lined craters, Moon rocks often seem to sparkle when held in the sunshine.

Although evidence of meteoritic impacts dominates our impressions of the lunar surface, the rate of this weathering is actually quite slow. Geologists estimate that it takes tens of millions of years to wear away a layer of rock only 1 mm thick. Consequently, features formed 3 billion years ago are well preserved today. Indeed, the astronauts' footprints will remain sharply imprinted on the lunar surface for millions of years to come.

Although lunar rocks bear a strong resemblance to terrestrial rocks, there are some important differences. Whereas every terrestrial rock contains some water, lunar rocks are totally dry. There is absolutely no evidence that water ever existed on the Moon. In the absence of both atmosphere and water, it is not surprising that the astronauts found no traces of life. After the first few Apollo missions, the astronauts were

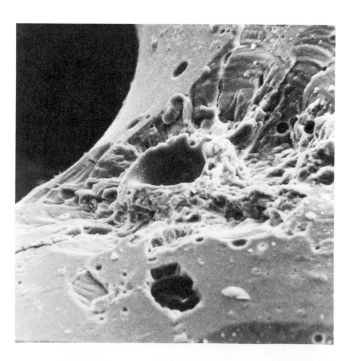

quarantined for a few days to make sure that they had not brought back any lunar microorganisms—germs that might cause some terrible disease. However, in view of the total sterility of the lunar samples, this precaution was soon dropped. The lunar rocks are, however, stored in rooms filled with dry nitrogen. Geologists justifiably fear that the rocks would rust if exposed to the oxygen and water vapor in the Earth's air.

By carefully measuring the abundances of trace amounts of radioactive elements in lunar samples (see Box 10-2), geologists confirmed that anorthosite is more ancient than the mare basalts. This result had been expected because the lunar highlands are densely cratered, whereas the basaltic surfaces of the mare show relatively few craters. Typical anorthositic specimens from the highlands are between 4.0 and 4.3 billion years old; one rock brought back by the Apollo 17 astronauts is nearly 4.6 billion years old. All these extremely ancient specimens representing material from the Moon's original crust. In sharp contrast, all the mare basalts are between 3.1 and 3.8 billion years old. Apparently, the mare basalts solidified from ferromagnesian-rich lavas that gushed up from the Moon's mantle and flooded the mare basins between 3.1 and 3.8 billion years ago—just about the time that the oldest rocks of the Earth's present surface layers were being formed.

Box 10-2 Radioactive age-dating

The Apollo program finally gave geologists the exciting opportunity to get their hands on extremely ancient rocks. By examining these lunar samples, scientists began to piece together a history of important events that happened shortly after the creation of the solar system. To get the story right, however, geologists must accurately determine the ages of the lunar rocks. Fortunately, most rocks contain trace amounts of radioactive elements such as uranium. The relative abundances of various radioactive isotopes and their decay products provide the key that geologists use to determine the ages of rocks.

As we saw in Box 6-3, every atom of a particular element has the same number of protons in its nucleus. However, different isotopes of the same element have different numbers of neutrons in their nuclei. For example, the common isotopes of uranium are ^{235}U and ^{238}U. Each isotope of uranium has 92 protons in its nucleus (correspondingly, uranium is the element 92 on the periodic chart; see Figure 6-7). However, a ^{235}U nucleus contains 143 neutrons, whereas a ^{238}U nucleus has 146 neutrons.

In light-weight elements, the number of protons in the nucleus is typically equal or nearly equal to the number of neutrons. For example, the common isotope of oxygen is ^{16}O, which has eight protons and eight neutrons in its nucleus. In heavy elements, however, the number of neutrons exceeds the number of protons. These extra neutrons serve as buffers between the many positively charged protons crowded together in the nucleus. Without the shielding effect of these extra neutrons, the repulsive electric forces between the closely packed protons would break apart the nucleus. A nucleus is said to be stable if it effectively resists the natural forces that tend to break it apart.

A nucleus is unstable, or radioactive if it contains too many protons or too many neutrons. In either case, the unstable nucleus ejects particles to achieve a stable nucleus. For example, ^{235}U is radioactive and naturally casts off helium (^{4}He) nuclei. This is a common mode of radioactive decay called **alpha decay** that removes two neutrons and two protons from the unstable nucleus. In particular, we find

$$^{235}_{92}\text{U} \rightarrow ^{231}_{90}\text{Th} + ^4_2\text{He}$$

(continued)

(Box 10-2, continued)

where the subscripts refer to the number of protons that each nucleus contains. Because uranium decays into thorium, ^{235}U is called the **parent isotope** and ^{231}Th is called the **daughter isotope.**

Some radioactive isotopes decay rapidly, whereas others decay slowly. Physicists find it convenient to talk about the decay rate in terms of an isotope's half-life. The **half-life** of an isotope is the time interval in which one-half of the parent nuclei decay into daughter nuclei. For example, the half-life of ^{235}U is 710 million years. If you start out with 1 g of ^{235}U, after 710 million years you will have only $\frac{1}{2}$ g of ^{235}U left; the other $\frac{1}{2}$ g of your sample has turned into other elements.

When ^{235}U decays into ^{231}Th, the process doesn't stop there. The thorium isotope is extremely unstable and has a half-life of only 25.6 hours. At a very rapid rate, ^{231}Th decays into an isotope of protactinium, ^{231}Pa. The protactinium in turn decays into actinium. And so on, and so on, until a stable isotope of lead (^{207}Pb), which does not decay, is finally formed. The entire **decay series** of ^{235}U is shown in the first table.

The decay series of ^{235}U

Parent isotope	Half-life	Daughter isotope
^{235}U	710,000,000 years	^{231}Th
^{231}Th	25.6 hours	^{231}Pa
^{231}Pa	34,000 years	^{227}Ac
^{227}Ac	21.6 years	^{227}Th
^{227}Th	18 days	^{223}Ra
^{223}Ra	11.7 hours	^{219}Rn
^{219}Rn	3.9 seconds	^{215}Po
^{215}Po	0.002 second	^{211}Pb
^{211}Pb	36 minutes	^{211}Bi
^{211}Bi	2.2 minutes	^{207}Tl
^{207}Tl	4.8 minutes	^{207}Pb

Notice that only the first step in the decay of ^{235}U takes a long time. Compared to this first step, all the other steps occur very quickly. In other words, we can describe the decay series simply as the conversion of urnaium (^{235}U) into lead (^{207}Pb) with a half-life of 710 million years. If the relative abundances of ^{235}U and ^{207}Pb in a rock are known, then its age can be calculated.

There are several parent–daughter isotope combinations that are useful for "age-dating" rocks. Four are listed in the second table.

Radioisotopes used in rock-dating

Radioactive parent isotope	Half-life (billions of years)	Stable daughter isotope
Potassium (^{40}K)	1.3	Argon (^{40}Ar)
Rubidium (^{87}Rb)	47.0	Strontium (^{87}Sr)
Uranium (^{235}U)	0.7	Lead (^{207}Pb)
Uranium (^{238}U)	4.5	Lead (^{206}Pb)

To see how geologists date rocks, consider the very slow conversion of rubidium (^{87}Rb) into strontium (^{87}Sr). Over the years, the amount of ^{87}Rb in a rock decreases, while the amount of ^{87}Sr experiences a corresponding increase. Dating the rock is not simply a matter of measuring its ratio of rubidium to strontium, however, because the rock already had some strontium in it when it formed. Geologists must determine how much *fresh* strontium came from the decay of rubidium after the rock's formation. To do this, geologists use a third isotope whose concentration has remained constant as a reference. In this case, they use ^{86}Sr, which is stable and is not created by any radioactive decay; it is said to be nonradiogenic. Specifically, dating a rock entails comparing the ratio of radiogenic and nonradiogenic strontium (^{87}Sr/^{86}Sr) to the ratio of radioactive rubidium to nonradiogenic strontium (^{87}Rb/^{86}Sr). Because the half-life for the conversion of ^{87}Rb into ^{87}Sr is known, the rock's age can be calculated from these ratios.

This particular isotope combination turned out to be highly reliable in dating the lunar rocks. The graph shows data for lunar lava sample 10044 brought back from Mare Tranquillitatis by the Apollo 11 astronauts. The rock is a typical mare basalt. At various locations around the sample, geologists D. A. Papanastassiou and G. J. Wasserburg measured the abundance ratios ^{87}Sr/^{86}Sr and ^{87}Rb/^{86}Sr. The data were then plotted on a graph of ^{87}Sr/^{86}Sr versus ^{87}Rb/^{86}Sr. Each dot on the accompanying graph represents a pair of measurements from one tiny part of rock 10044.

Notice that the data points fall along a straight line. This means that, although different amounts of rubidium and strontium were found in various parts of the rock, everywhere the same percentage of ^{87}Rb had decayed into ^{87}Sr. Because the half-life of the ^{87}Rb $\rightarrow$ ^{87}Sr decay is known, lines can be drawn on this graph corresponding to the abundance ratios that would be found in rocks of various ages. These lines, called **isochrons,** are steeper for greater ages because ^{87}Rb/^{86}Sr decreases as the rock ages, while ^{87}Sr/^{86}Sr increases.

The isochron for 4½ billion years, the age of the oldest lunar rocks, is shown on the graph. The data for rock 10044, however, lie along a shallower line that corresponds to an age of 3.7 billion years. That is the age of the rock. We conclude that igneous rock sample 10044 solidified when the Moon was slightly less than a billion years old. This rock is thought to be a sample of lava that flowed out onto the lunar surface after the impact of a large meteoroid that penetrated the young Moon's fragile crust.

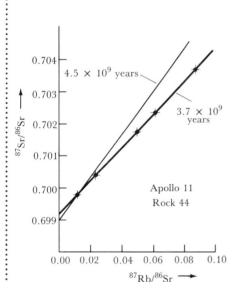

The Moon probably formed by accretion during the final stages of solar-system formation

Volatile elements (such as potassium and sodium) melt and boil at relatively low temperatures, whereas **refractory elements** (such as titanium, calcium, and aluminum) melt and boil at much higher temperatures. Compared to terrestrial rocks, the lunar rocks have slightly greater proportions of refractory elements and slightly lesser proportions of volatile elements. The implication is that the Moon formed from material somewhat hotter than the material from which the Earth was created. Some of the volatile elements boiled away, leaving the young Moon relatively enriched in the refractory elements.

Prior to the Apollo program, there were three different theories about the origin of the Moon. First, The **fission theory** holds that the Moon was pulled out from (or broken off of) the Earth. Second, the **cocreation theory** proposes that the Earth and the Moon were formed at the same time but separately. Third, the **capture theory** suggests that the Moon was formed elsewhere in the solar system and drawn into orbit about the Earth by gravitational forces. Unfortunately, evidence that permits a confident choice of one theory over the other is lacking; the debate continues. The lunar rocks, however, supply some significant clues.

One fact used to support the fission theory is that the Moon's average density (3.3 g/cm^3) is much less than the Earth's average density (5.5 g/cm^3), as would be expected if the Moon was formed from the low-density rock near the Earth's surface. However, most geologists counter this argument by pointing to the significant differences between lunar and terrestrial rocks—particularly the differences in water content and in abundances of volatile and refractory elements. These differences are inconsistent with the theory that the Moon was torn out of the Earth's crust and upper mantle.

Proponents of the capture theory use these differences to argue that the Moon formed elsewhere and later was captured by the Earth. After all, neither Mercury nor Venus nor Mars has a large satellite—why should the Earth have been unique in forming as a double-planet system? Jupiter and Saturn do have satellites even larger than our Moon, but those satellites (and the smaller moons of Uranus) all follow orbits that lie in the equatorial planes of their parent planets. The rotation of each Jovian planet and the revolution of its satellites share the same orientation in space, as would be expected if the planet and satellites were formed in a single process. The plane of the Moon's orbit, on the other hand, is much closer to the plane of the ecliptic than it is to the plane of the Earth's equator. This pattern would be expected if the Moon originally was in orbit about the Sun and later was captured by the Earth.

There are difficulties with the capture theory, however. If the Earth did capture the Moon intact, then a host of conditions must have been met entirely by chance. The Moon must have coasted to within 50,000 km of the Earth at exactly the right speed to leave a solar orbit and adopt an Earth orbit without ever actually hitting our planet. Of course, just because something is highly improbable is not proof that it did not happen. Most geologists agree that the Moon must have formed in the same general part of the solar system as the Earth did. An analysis of meteorites shows that the abundances of the isotopes of oxygen (^{16}O, ^{17}O, and ^{18}O) differ measurably among rocks from different parts of the solar system. The abundances in terrestrial and lunar rocks are sufficiently similar to indicate that both worlds formed at nearly the same distance from the Sun.

Although very special conditions must be met for a planet to capture from solar orbit one giant rock, much less stringent conditions exist for the capture of swarms of tiny rocks. For this reason, most scientists tend to favor some sort of cocreation or coaccretion theory of lunar formation. Great numbers of rock fragments must have swarmed about the protosun during the formation of the solar system. Most of this rocky debris probably orbited our infant star in the plane of the ecliptic along with the protoplanets. Energy radiated from the contracting protosun would have baked the water and volatile elements out of these smaller rock fragments, many of which were soon captured into orbit about the protoearth. Then, just as planetesimals accreted to form the protoearth in orbit about the Sun, the fragments in orbit about the Earth accreted to form the Moon—or so says the most widely accepted theory.

The Moon probably reached roughly its present mass by about 4.6 billion years ago. The heat produced by rock fragments falling into the protomoon and by the decay of such radioactive isotopes as ^{26}Al (aluminum-26) must have melted the lunar surface to a depth of hundreds of kilometers. After a few hundred million years, the rain of rock

Figure 10-20   Eratosthenes
Eratosthenes is a young crater 61 km in diameter on the southern edge of Mare Imbrium. Another young crater, Copernicus, is near the horizon in this photograph taken by the Apollo 17 astronauts in 1972. Both craters also appear near the lower edge of the Earth-based view in Figure 10-2. (NASA)

fragments tapered off, and the most radioactive isotopes had largely decayed into stable isotopes. The Moon gradually cooled and the low-density lava floating at the Moon's surface began to solidify into the anorthositic crust that exists today. The heavy barrage of large rock fragments ended about 4 billion years ago, the final impacts producing the ancient craters that cover the lunar highlands.

Recorded among the final scars at the end of this crater-making era are the impacts of more than a dozen asteroid-sized objects, each measuring at least 100 km across. As these huge rocks rained down on the young Moon, they blasted out the vast mare basins. Meanwhile, heat from the decay of long-lived radioactive elements such as uranium and thorium began to melt the inside of the Moon. Then, over the period from 3.8 to 3.1 billion years ago, great floods of molten rock gushed up from the lunar interior, filling the impact basins and creating the maria we see today.

Very little has happened on the Moon since those ancient times. A few fresh craters have been formed (see Figure 10-20), but the astronauts visited a world that has remained largely unchanged for over 3 billion years. The history of our own planet was just beginning 3 billion years ago, as the first organisms began populating the new oceans.

Many questions and mysteries still remain. The six Apollo and three Luna landings have brought back samples from only nine locations, barely scratching the Moon's surface. We still know very little about the Moon's far side and nothing at all about the Moon's poles. Is the Moon's interior molten? Does the Moon really have an iron core? How old are the youngest lunar rocks? Did lava flows occur over western Oceanus Procellarum only 2 billion years ago, as crater densities suggest? If the cocreation theory is correct, why don't Venus and Mars have large satellites like our Moon? Is the Moon really geologically dead, or does it look dead simply because our examination of the lunar surface has been so cursory?

One of the greatest adventures of all time is over. During a few historic moments, humanity reached out and touched another world. An incredible dream became a reality.

Summary

. The Earth-facing side of the Moon displays light-colored, heavily cratered highlands (terrae) and dark-colored, smooth-surfaced maria; mountain ranges also exist.

. Much of our knowledge about the Moon came from lunar exploration in the 1960s and early 1970s.

Detailed orbits of spacecraft around the Moon revealed the presence of mass concentrations (mascons) corresponding to many of the maria.

The Moon's far side has no maria.

The Moon has no magnetic field today, although it did have a weak magnetic field when the lunar rocks solidified billions of years ago; we suspect that the Moon today does not have a liquid iron core.

Analysis of seismic waves from moonquakes and meteoroid impacts indicates that the Moon's crust is thicker than the Earth's (and thickest on the far side of the Moon), a mantle with a thickness equal to about 75 percent of the Moon's radius, and perhaps a small, solid iron core.

The Moon's lithosphere is far thicker than that of the Earth, and the Moon's asthenosphere probably extends from the base of the lithosphere to the core.

. Gravitational interactions between the Earth and Moon produce tides in the oceans of the Earth and in the solid bodies of both worlds.

Tidal interactions lock the Moon into synchronous rotation with the Earth and cause a very small constant acceleration of the Moon in its orbit, thereby causing it to spiral very slowly outward from the Earth.

. The anorthositic crust exposed in the highlands was formed between 4.0 and 4.3 billion years ago, whereas the mare basalts solidified between 3.1 and 3.8 billion years ago.

Meteoroid impacts have been the only significant "weathering" agent on the Moon; the Moon's regolith ("soil" layer) was formed by meteoritic action, and glasses and breccias of meteoritic origin are very common.

All of the lunar rock samples are igneous rocks formed largely of minerals found in terrestrial rocks; the lunar rocks contain no water and also differ from terrestrial rocks by being relatively enriched in the refractory elements and depleted in the volatile elements.

. The most widely accepted theory of lunar origin holds that the Moon accreted from rock fragments orbiting the Earth very soon after the Earth itself had formed.

The Moon was molten in its early stages, and the anorthositic crust solidified from low-density magma that floated to the lunar surface; the mare basins were created later, by the impact of planetesimals and filled with lava from the lunar interior.

The Moon's surface has undergone very little change in the past 3 billion years.

....

Review questions

1 Why is more lunar detail visible through a telescope when the Moon is near quarter phase than when it is at full phase?

2 Why are nearly all of the lunar maria located on the side of the Moon facing the Earth? Why are there more craters on the far side of the Moon than the near side?

3 How would you prove to someone that the Moon has no atmosphere?

4 Why do you suppose that no Apollo mission landed on the far side of the Moon?

5 How would our theories of the Moon's history have been affected if the Apollo astronauts had discovered sedimentary rock on the Moon?

6 Some people who support the fission theory have proposed that the Pacific Ocean basin is the scar left when the Moon pulled away from the Earth. Explain why this idea is probably wrong.

7 Imagine that you are planning a lunar landing mission. What type of landing site would you select in order to obtain bedrock? Where might you land to search for evidence of recent volcanic activity?

Advanced questions

*8 Estimate the maximum rate at which the Moon is gaining mass from meteoritic impacts. On the basis of your calculation, how long will it be before the Moon doubles its mass? Compare your answer with the age of the universe (about 20 billion years) and explain why your extrapolation into the distant future is probably unreasonable.

*9 Find the average distance between the Earth and the Moon when the length of the day and the lunar month will both be equal 47 of our present days.

10 Can you think of any other weathering processes which might occur on the Moon besides those related to meteoritic impacts?

Discussion questions

11 The idea has been advanced that without the presence of the Moon in our sky, astronomy would have developed far more slowly. Please comment.

12 Compare the advantages and disadvantages of exploring the Moon with astronauts as opposed to mobile unmanned instrument packages.

13 Describe how you would empirically test the idea that human behavior is related to the phases of the Moon. What problems are inherent in such testing?

For further reading

Cadogan, P. "The Moon's Origin." *Mercury,* Mar./Apr. 1983, p. 34.
Cooper, H. *Apollo on the Moon* and *Moon Rocks.* Dial, 1970. *Two accounts by a respected journalist of the Apollo 11 mission and the analysis of the material the astronauts brought back to Earth.*
French, B. *The Moon Book.* Penguin, 1977. *A layperson's introduction to the Moon by a lunar scientist.*
Goldreich, P. "Tides and the Earth-Moon System." *Scientific American,* Apr. 1972.
Hartmann, W. "The Moon's Early History." *Astronomy,* Sept. 1976, p. 6.
Lewis, R. *The Voyages of Apollo: The Exploration of the Moon* Quadrangle, 1975. *The story of the Apollo program.*
Weaver, K. "First Explorers on the Moon: The Incredible Story of Apollo 11." *National Geographic,* Dec. 1969.

Mars
Many of the major features of the Martian surface are seen in this photograph from Viking 2 as the spacecraft approached the dawn side of the planet in August 1976. Clouds flank the western slopes of the huge volcano Olympus Mons near the top of the photograph. In the middle is the vast rift canyon called Valles Marineris. This canyon stretches nearly 4000 km along the planet's equator. At the bottom of the photograph, carbon dioxide snow lines the floor of the Argyre Basin and surrounding craters. (NASA)

Mars has many Earth-like characteristics and scientists have long speculated about the possibility of Martian life. Around 1900, some astronomers claimed to have seen networks of linear features ("canals") on the Martian surface, leading to speculation about an advanced alien civilization there. The Mariner and Viking probes reported many surprising facts about Mars—including the discovery of an enormous volcano and a huge canyon—but they found no canals and no sign of life. Most of this chapter is devoted to a review of the information returned by the Viking missions about the Martian surface, interior, satellites, atmosphere, and lack of biological activity. Many questions about the Martian environment could be answered by another mission to this fascinating world.

Mars is the only planet whose surface features can be seen through Earth-based telescopes. Early telescopic observers reported seasonal variations that seemed to indicate vegetation on the Martian surface, and some reported geometrical patterns that might represent a planetwide system of artificial canals. Scientists speculated about the possibility of life on Mars, and science-fiction writers wove popular stories about invasions of Earth by hostile Martians. Ironically, the first actual invasion came in 1976 when automated spacecraft from the Earth landed on the Martian surface. These probes sent back pictures and data indicating that Mars is a barren and desolate world. The possibility of some form of Martian life has not been completely ruled out, but it now seems quite likely that Mars is as sterile and lifeless as the Moon.

Box 11-1   Mars data

Mean distance from the Sun: 1.524 AU $= 2.279 \times 10^8$ km
Maximum distance from the Sun: 1.667 AU $= 2.491 \times 10^8$ km
Minimum distance from the Sun: 1.381 AU $= 2.067 \times 10^8$ km
Mean orbital velocity: 24.1 km/sec
Sidereal period: 686.98 days = 1.88 years
Rotation period: $24^h\ 37^m\ 23^s$
Inclination of equator to orbit: 23° 59′
Inclination of orbit to ecliptic: 1° 50′ 59″
Orbital eccentricity: 0.093
Diameter: 6796 km
Diameter (Earth = 1): 0.532
Apparent diameter as seen from Earth: maximum = 25.7″
minimum = 3.5″

Mass: 6.42×10^{23} kg
Mass (Earth = 1): 0.107
Mean density: 3.94 g/cm^3
Surface gravity (Earth = 1): 0.380
Escape velocity: 5.0 km/sec
Oblateness of planet: 0.006
Surface temperatures: maximum =   20°C = 293 K =   70°F
minimum = −140°C = 133 K = −220°F

Albedo: 0.16
Brightest magnitude: −2.8
Mean diameter of Sun as seen from Mars: 21′

Earth-based observations suggested that Mars might have some form of extraterrestrial life

The first reliable record of surface features on Mars is found in observations recorded by the Dutch physicist Christian Huygens in November 1659. Using a refracting telescope of his own design, Huygens identified a prominent, dark, triangular feature that we now call Syrtis Major. After observing this feature for several weeks, Huygens concluded that the rotation period of Mars is approximately 24 hours. This was the first in a series of observations that would soon lead to speculations about life on Mars which seemed at first to be remarkably similar to the Earth.

The first accurate measurements of Mars's rotation period were made in 1666 by the Italian astronomer Giovanni Domenico Cassini. A Martian solar day is nearly $37\frac{1}{2}$ minutes longer than an Earth solar day. Cassini also was the first person to see the Martian polar caps, which bear a striking superficial resemblance to the arctic and antarctic polar caps on Earth (see Figure 11-1). More than a century elapsed, however, before

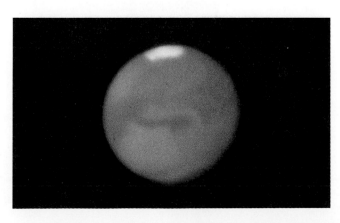

Figure 11-1 Mars viewed from the Earth
This high-quality, Earth-based photograph of Mars was taken a few days after the opposition of 1971 when the Earth–Mars distance was only $56\frac{1}{4}$ million kilometers (35 million miles). At that time, Mars presented a disk nearly 25 arc sec in angular diameter. Circumstances as favorable as these will not be repeated for the rest of the century. This photograph accurately portrays what you might typically see through a moderate-sized telescope under excellent observing conditions. Notice the prominent southern polar cap. (Courtesy of Stephen M. Larson)

the famous German-born, English astronomer William Herschel first suggested that the Martian polar caps are made of ice.

Herschel was also the first to determine the inclination of Mars's axis of rotation. Just as Earth's equatorial plane is tilted $23\frac{1}{2}°$ from the plane of its orbit, Mars's equator makes an angle of nearly 24° with its orbit. This striking coincidence means that Mars experiences Earth-like seasons. Mars takes nearly two years to orbit the Sun, however, so the Martian seasons last nearly twice as long as the seasons on Earth.

The Martian surface exhibits interesting seasonal variations. During spring and summer in a Martian hemisphere, the polar cap shrinks while the dark markings (which often look greenish) become very distinct. Half a Martian year later, with the approach of fall and winter, the dark markings fade while the polar cap grows. The implication of Martian vegetation was so strong that, in 1802, the German mathematician Karl Friedrich Gauss proposed that we signal the Martian inhabitants by drawing huge geometrical patterns in the Siberian snow. His plan was never carried out.

There are favorable and not-so-favorable times to observe Mars. Because Mars's orbit is noticeably elliptical (see Figure 11-2), the best views are obtained when Mars is simultaneously at opposition and near perihelion. This configuration is called a **favorable opposition** because the Earth–Mars distance can be as small as 56 million kilometers (35 million miles). At such times, Mars appears brilliant and red in the southern summer sky, outshining all the stars and gleaming with a magnitude as bright as −2.8. Through a telescope, the ruddy planet presents a disk nearly 26 arc sec in diameter. That is the same angular diameter as a moderate-sized (50-km) lunar crater viewed from the Earth. The Earth-based photograph in Figure 11-1 was taken during the favorable opposition of

Figure 11-2 The orbits of Earth and Mars
Earth catches up with Mars every 780 days (nearly 2⅟ years). These close encounters, properly called oppositions, are the best times to observe the red planet. Becuase Mars's orbit is noticeably elliptical, however, some oppositions afford better views than others. The last favorable opposition occurred on August 10, 1971. The next favorable oppositions will occur in 1986 and 1988, as listed in Box 11-2.

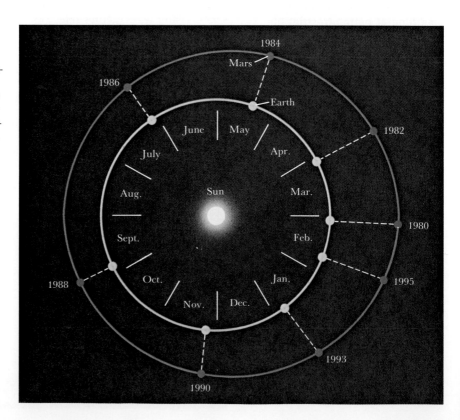

1971. The next favorable oppositions will occur in 1986 and 1988 (see Box 11-2).

At unfavorable oppositions, such as those when Mars is near aphelion, the Earth–Mars distance can be as large as 101 million kilometers (63 million miles). And when Mars is not at opposition, the Earth–Mars distance can be much greater than 100 million kilometers. Mars then appears as a tiny reddish dot, virtually devoid of features even when viewed through a large telescope.

Box 11-2   Martian oppositions: 1980–2000

The table lists the ten Martian oppositions during the last two decades of the twentieth century.

Date of opposition	Earth–Mars distance (millions of kilometers)	Apparent diameter (arc seconds)	Magnitude
Feb. 25, 1980	100	14	−1.0
Mar. 31, 1982	94	15	−1.2
May 11, 1984	79	17	−1.8
July 10, 1986	60	23	−2.4
Sept. 28, 1988	58	24	−2.6
Nov. 27, 1990	77	18	−1.7
Jan. 7, 1993	93	15	−1.2
Feb. 12, 1995	100	14	−1.0
Mar. 17, 1997	97	14	−1.1
Apr. 24, 1999	85	16	−1.5

Speculations about life on Mars were furthered by observations of the Italian astronomer Giovanni Virginio Schiaparelli during the favorable opposition of 1877. Schiaparelli reported seeing forty straight-line features crisscrossing the Martian surface. He called these dark, linear features *canali,* which was soon mistranslated into English as **canals.** The alleged discovery of canals on Mars seemed to imply the existence on Mars of intelligent creatures capable of substantial engineering feats. This speculation led Percival Lowell, who came from a wealthy Boston family, to finance a major new observatory near Flagstaff, Arizona, primarily for additional studies of Mars. By the end of the nineteenth century, Lowell had reported observations of 160 Martian canals (see Figure 11-3).

Not all astronomers saw the canals. In 1894, the American astronomer Edward Emerson Barnard, working at the Lick Observatory, complained that "to save my soul I can't believe in the canals as Schiaparelli draws them." Incidentally, Barnard apparently was the first person to report seeing craters on Mars. However, he did not publish these observations for fear of ridicule.

The skepticism of cautious observers such as Barnard was drowned out by the more fanciful pronouncements of Lowell and his colleagues.

Figure 11-3 Martian canals
During moments of "good seeing," when the air is exceptionally calm and clear, some Earth-based observers reported networks of lines crisscrossing the Martian surface. Some people suggested that these lines, called canals, are irrigation ditches constructed by intelligent creatures. This drawing and corresponding photograph were made during the opposition of 1926. (Lick Observatory)

It soon became fashionable to speculate that the Martian canals form an enormous planet-wide irrigation network designed to transport water from melting polar caps to vegetation near the equator. In view of the planet's reddish, desertlike appearance, Mars was thought to be a dying planet whose inhabitants must go to great lengths to irrigate their farmlands. No doubt the Martians would readily abandon their arid ancestral homeland—invade the Earth with its abundant resources. Hundreds of science fiction stories and dozens of monster movies owe their existence to the *canali* of Schiaparelli.

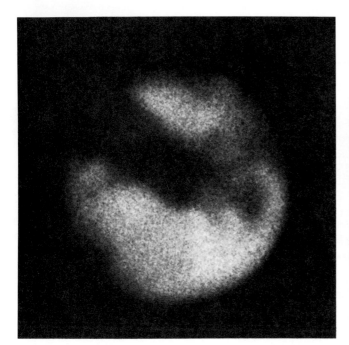 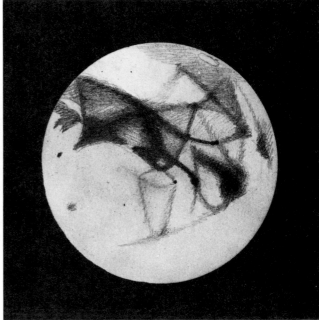

Space probes to Mars found craters, volcanoes, and canyons, but no canals

Fanciful speculation came to abrupt end with the flyby of Mariner 4 in the summer of 1965. As the spacecraft coasted past the planet, it sent back 21 pictures covering less than 1 percent of the Martian surface. Although many of the views were indistinct, they clearly showed numerous flat-bottomed **craters.** From studies of the relatively crater-free maria, scientists knew that the major crater-making era in the solar system ended a few billion years ago. Like the Moon, the Martian surface must therefore be extremely ancient and has apparently experienced few changes over the past 3 billion years. The prospect of abundant Martian life suddenly seemed extremely remote.

Confirming observations of numerous craters were supplied by the flybys of Mariner 6 and Mariner 7 in the summer of 1969. Each spacecraft carried two television cameras that sent back a total of 201 pictures (see Figure 11-4). Not one single canal was observed. Schiaparelli's *canali* had been an optical illusion.

Like Mariner 4, Mariners 6 and 7 concentrated primarily on the heavily cratered southern hemisphere of Mars. Astronomers were quick to point out that the partially eroded, flat-bottomed appearance of the Martian craters could be a result of **dust storms** that rage across the planet's surface. From time to time over the past two centuries, astronomers have

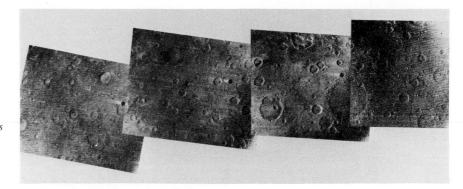

Figure 11-4   Martian craters
Numerous flat-bottomed craters are seen in this mosaic of four views taken by Mariner 6 in 1969. The area shown in these photographs measures roughly 3500 by 700 km (2200 by 400 miles). (NASA)

watched the faint surface markings disappear under a reddish-orange haze as thin Martian winds stirred up finely powdered dust. Some of these storms actually obscure the entire planet. Deposits of dust from these storms in the craters over the ages give the craters their characteristic flat-bottomed shape.

These early Mariner missions also confirmed that the Martian atmosphere is primarily composed of carbon dioxide. In 1947 the American astronomer Gerard P. Kuiper had detected spectral lines of this gas in the spectrum of sunlight reflected from Mars, but data from spacecraft were required to demonstrate that carbon dioxide forms 95 percent of the Martian atmosphere. These measurements also indicated that the Martian atmosphere is extremely thin. Atmospheric pressure on the Martian surface is typically in the range of 5 to 10 millibars—roughly the same as the air pressure at an altitude of 30 km (100,000 ft) above the Earth's surface. This extremely low atmospheric pressure was unexpected, especially in view of the huge dust storms visible from Earth. The material whipped up by the rarified Martian winds must be extremely fine-grained powder. This assumption is consistent with the fact that numerous dust storms over 3 billion years have not completely obliterated the craters.

The flybys of the 1960s gave us only fleeting glimpses of a small fraction of the Martian surface; the best was yet to come. After traveling 400 million kilometers, Mariner 9 went into orbit about Mars on November

Figure 11-5   Mariner 9
This 1-ton spacecraft went into orbit about Mars in November 1971. It operated flawlessly for nearly a full year, sending back a total of 7329 close-up pictures of the Martian surface and the two tiny Martian moons. (NASA)

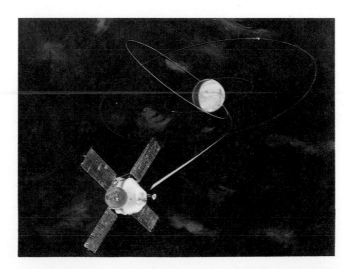

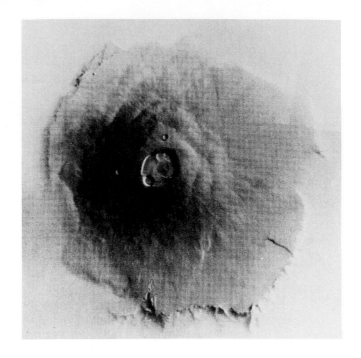

Figure 11-6   Olympus Mons
This computer-enhanced mosaic of four Mariner 9 views looks straight down on the largest volcano on Mars. Olympus Mons is 2¾ times as tall as Mount Everest, and its cliff-ringed base measures nearly 600 km (370 miles) in diameter. (NASA)

13, 1971. To everyone's dismay, Mars was embroiled in a planetwide dust storm that completely obscured the surface. After a few weeks, however as the dust started to subside scientists realized that they were looking down on several **volcanoes** far taller and broader than any comparable protrusions on Earth.

The largest Martian volcano, Olympus Mons, rises 24 km (15 miles) above the surrounding plains—nearly three times as high as Mount Everest (see Figure 11-6) The highest volcano on Earth is Mauna Loa in the Hawaiian Islands, whose summit is only 8 km above the ocean floor. As we saw in Chapter 9, both the Martian volcanoes and the Hawaiian Islands, were formed by **hot-spot volcanism.** Recall, however, that the process of plate tectonics is also at work in the case of the Hawaiian Islands. The huge size of Olympus Mons strongly suggests a lack of plate tectonics on Mars. A hot spot in Mars's mantle kept pumping lava up-

Figure 11-7   The Olympus caldera
This view of the summit of Olympus Mons is based on a mosaic of six pictures taken by one of the Viking orbiters. The caldera consists of overlapping, collapsed volcanic craters and measures roughly 70 km across. The volcano itself is wreathed in midmorning clouds that formed from ice (H₂O) brought upslope by cool air currents. The cloud tops are about 8 km below the volcano's peak. (NASA)

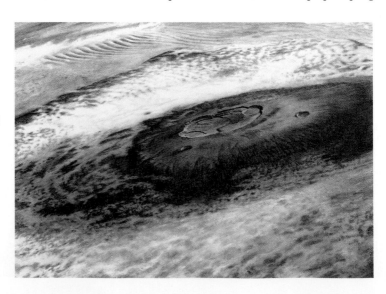

ward through the same vent for millions of years, producing one giant volcano rather than a long chain of smaller volcanoes.

The base of Olympus Mons is ringed with cliffs, has a diameter of nearly 600 km, and covers an area as big as the state of Missouri. At the volcano's summit are several overlapping volcanic craters, forming a **caldera** large enough to contain the state of Rhode Island (see Figure 11-7).

Olympus Mons is one of several very large volcanoes clustered together on a huge bulge called the Tharsis Rise that covers an area 2500 km in diameter and is centered just north of the Martian equator. Ground levels over this vast dome-shaped region are typically 5 to 6 km higher than the average ground level for the rest of the planet. Another major, although less dramatic, grouping of volcanoes is located nearly on the opposite side of the planet from Tharsis in a region called Elysium. None of these volcanoes seem to be active today.

For reasons we do not understand, most of the volcanoes on Mars are in the northern hemisphere, whereas most of the craters are in the southern hemisphere. Between the two hemispheres, Mariner 9 discovered a vast canyon running roughly parallel to the Martian equator (see Figures 11-8 and 11-9). If this canyon were located on Earth, it could stretch all the way from New York to Los Angeles. In honor of the four Mariner spacecraft that revealed so much of the Martian surface, this enormous chasm is called Valles Marineris.

Valles Marineris stretches 4000 km, beginning with heavily fractured terrain in the west and ending with ancient cratered terrain in the east. Many geologists suspect that Valles Marineris is a fracture in the Martian crust caused by internal stresses, not unlike the rift valleys on Earth that result from plate tectonics such as the Red Sea (recall Figure 9-19), which is 3000 km long.

It seems reasonable that plate tetonics could have occurred on Mars only when the planet was very young. In Chapter 7, we learned that small bodies cool much more rapidly than large bodies. Mars's diameter is only half that of the Earth, and its mass is one-tenth of the Earth's mass (only about twice the mass of Mercury). It appears that tectonic processes never really get started on worlds as small as Mercury or the Moon. Although they do begin on a world as large as Mars, they apparently are halted fairly soon as the rapidly cooling lithosphere becomes thicker and thicker.

Figure 11-8 [left] A segment of Valles Marineris
This photograph was taken from an altitude of 2000 km by Mariner 9 and shows an area 300 by 400 km, which is nearly the same size as the state of Pennsylvania. Earth's Grand Canyon is only as big as one of the tributary canyons seen in this photograph. (NASA)

Figure 11-9 [right] Eastern Valles Marineris
This mosaic was constructed from fifteen photographs taken by Viking Orbiter 1 and shows an area 1800 by 2000 km. North is toward the upper left. The eastern end of Valles Marineris merges with chaotic, cratered terrain toward the upper right. (NASA)

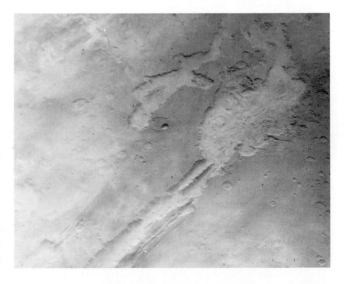

Surface features indicate that water once flowed on Mars

Although Valles Marineris was not formed by water erosion, the Mariner 9 photographs did include many features that look exactly like dried-up riverbeds (see Figure 11-10). These features were totally unexpected because liquid water cannot exist today on Mars. As you know, water is liquid only over a certain range of temperature and pressure: at low temperatures, water becomes ice; at high temperatures, it becomes steam. The atmospheric pressure above a body of water also affects the state of the water. If the pressure is very low, molecules escape from the liquid's surface, causing the water to vaporize. Any liquid water on Mars would boil furiously and rapidly evaporate into the thin Martian air. Data from Mariner 9 furthermore demonstrate that the Martian atmosphere is very dry; it never rains on Mars. If all the water vapor could somehow be squeezed out of the Martian atmosphere, it would not fill one of the five Great Lakes in North America.

The idea that water carved the Martian riverbeds seemed so preposterous that many scientists searched for other possible explanations. Low-viscosity lava flows? Liquid methane or ammonia? None of the hypotheses seemed promising.

In 1976, the Viking orbiters supplied evidence strongly supporting the idea that liquid water once raged across the Martian surface in great torrents. Numerous Viking photographs show erosional features that look like the results of flash floods in the deserts of Arizona and New Mexico. Scientists were faced with a problem that soon became known as the "mystery of the missing water."

Scientists immediately looked to the Martian polar caps, as a probable location of the missing water. It was not clear; however, how much of the polar caps consists of frozen carbon dioxide (dry ice) and how much is water ice.

In 1972, Mariner 9 sent back photographs from orbit as summer came to the northern hemisphere of Mars. During the Martian spring, the polar cap receded very rapidly (see Figure 11-11). This rapid shrink-

Figure 11-10 [left]   An ancient riverbed
This pair of photographs from Mariner 9 shows a 700-km-long riverbed in Mars's heavily-cratered southern hemisphere. Features such as these were unexpected because liquid water cannot now exist on the Martian surface. This particular riverbed is called Ma'adim Vallis and is located about 20° south of the Martian equator. (NASA)

Figure 11-11 [right]   The northern polar cap
This view of the northern polar cap was taken by Mariner 9 in August 1972 from a distance of 13,700 km (8500 miles). By observing the rate of shrinkage with the approach of Martian summer, scientists deduced that the residual polar cap contains a substantial amount of frozen water. (NASA)

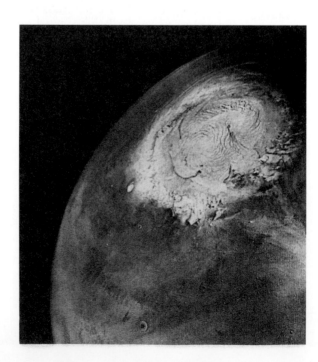

age strongly suggested that a thin layer of carbon dioxide frost was evaporating quickly in the sunlight. However, with the arrival of summer, the rate of recession abruptly slowed, suggested that a thicker layer of water ice had been exposed. Scientists concluded that the **residual polar caps** that survive through the Martian summers contain a large quantity of frozen water. Calculating the volume of water ice in the residual caps is difficult, however, because we do not know how thick the layer of ice is. Scientists therefore are still debating the exact amount of frozen water stored in the Martian polar caps.

Another important clue about water on Mars came from the Viking orbiters in 1976. Many close-up views show flash-flood erosion features where the water appears to have emerged from collapsed, jumbled terrain (see Figure 11-12). These photographs confirmed earlier suspicions that frozen water might form a layer of permafrost under the Martian surface similar to that beneath the tundra in far northern regions on Earth. Apparently, heat from volcanic activity occasionally melts this subsurface ice. The ground then collapses as millions of tons of rock push the water to the surface, thereby producing a brief flash-flood as the water quickly boils away.

Figure 11-12 A flood in the Capri plateau
The melting of subsurface ice and subsequent collapse or downfaulting of the Martian surface easily explain the features seen in the mosaic taken by Viking Orbiter 1 in 1976. A torrent of water liberated by this process apparently flowed eastward (to the left) across the surrounding plains. The area seen here measures 300 by 300 km. (NASA)

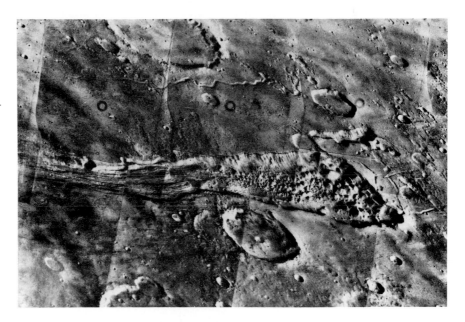

From photographs such as Figure 11-12, scientists estimated the widths and depths of the Martian flash-flood channels. These dimensions suggest peak flood discharges of 10^7 to 19^9 m^3/sec. For comparison, the average discharge of the Amazon is 10^5 m^3/sec. The largest known flash flood on the earth occurred about 2 million years ago in eastern Washington, when a natural dam gave way. The peak discharge is estimated to have been 10^7 m^3/sec. Most of the Martian flash floods therefore must have been greater than anything known on Earth.

As noted in previous chapters, water vapor and carbon dioxide are two of the most common gases in volcanic vapors. Mars's surface gravity, although only one-third as strong as Earth's, is nevertheless capable of keeping these gases from escaping into space. Ultraviolet light from the Sun can dissociate water into hydrogen and oxygen, and these gases are light enough to escape into space. Detailed calculations show, however, that Mars could have lost only a small fraction of its water in this fashion. It is therefore reasonable to suppose that much of the carbon dioxide and water vapor outgassed during Mars's early history still remains on the planet. Some is in the polar caps, and some is in subsurface ice and permafrost. Some water and carbon dioxide also may be chemically bound in the rocks and sand that cover the Martian surface.

The Viking landers sent back close-up views of the Martian surface

The discovery that water once existed on the surface of Mars rekindled speculation about Martian life forms. By the mid-1960s, it was clear that Mars has neither civilizations nor fields of plants, but some sort of microbial life forms still seemed possible. Searching for Martian microbes was one of the main objectives of the ambitious and highly successful Viking missions.

The two Viking spacecraft were launched from Earth during the summer of 1975 and arrived at Mars almost a year later (see Figure 11-13). Each spacecraft consisted of two modules: an orbiter and a lander. The modules were linked together throughout the journey and remained together in orbit about Mars for several weeks. The photographs of the surface sent back by the two orbiting spacecraft were studied carefully to select landing sites; the Viking landers were not designed to survive a landing in a boulder-strewn area.

On July 20, 1976—a month after arriving at Mars—the Viking 1 lander separated from the orbiter and began its descent to the planet's surface (see Figure 11-14). A heat shield, a parachute, and retrorockets slowed the lander from its original speed of 16,000 km/hr (about 10,000 mph) to less than 2 m/sec (about 5 mph) at touchdown. The landing site was a rocky field called Chryse Planitia (the Golden Plains) at latitude 22° north of the Martian equator.

The second Viking lander followed 45 days later and set down in Utopia Planitia (the Utopian Plains) on September 3, 1976. Utopia is at latitude 48° north of the Martian equator, nearly on the opposite side of the planet from Chryse. Viking Lander 2 was therefore much farther

Figure 11-13   Mars from Viking 1
As the Viking 1 spacecraft approached Mars, it sent back this view that predominantly shows the southern hemisphere. Valles Marineris is seen near the top of this picture. The huge Argyre Basin and numerous craters appear just below the center of the photograph. (NASA)

Figure 11-14 [left] The Viking descent
This drawing schematically shows the sequence of events during the descent of one of the Viking spacecrafts on Mars. A heat shield, a parachute, and retrorockets were used to ensure a gentle landing. (NASA)

Figure 11-15 [right] The Viking lander
From the base of the foot pads to the top of the antenna, each Viking lander is about 2 m (6 ft) tall and weighed 576 kg (1270 pounds) without fuel. An extendable arm for retrieving soil samples is visible in the foreground. The two cylinders (one with an American flag) protruding from the top of the lander contain the two television cameras. The dish antenna was used to relay data to Viking orbiter overhead which in turn sent the information back to Earth. (NASA)

north than Viking Lander 1. As shown in Figure 11-15, each of the Landers is roughly the size of a small automobile.

The Viking television cameras were activated immediately after landing. The cameras had been stowed for the long voyage in a downward-looking position, so the first pictures from Mars showed a footpad firmly resting on the Martian regolith. Rocks at both sites appeared to be igneous. In fact, many of the rocks at Utopia were vesicular, indicating that gas had been dissolved under pressure in the Martian lava prior to its eruption onto the surface (see Fig. 11-16).

The Viking Lander 1 site at Chryse is a moderately cratered plain near the mouth of a large flash-flood channel. The upraised rim of a crater is visible on the horizon in Figure 11-17. Such craters are proba-

Figure 11-16 The first views from Mars
The first photographs from Viking Lander 1 at Chryse (top) and from Viking Lander 2 at Utopia (bottom) were transmitted to Earth immediately after the successful landings. The rocks in both views have sizes in the range of 10 to 20 cm (4 to 8 in.). Notice the vesicular appearance of the Utopia rocks, evidence of their igneous origin. (NASA)

Figure 11-17   The Viking Lander 1 site
This view looks southward over the Chryse plains. The horizon is about 3 km (2 miles) from the spacecraft. Part of the upraised rim of a crater on the horizon is visible at the upper left. A few light-colored patches of exposed bedrock also appear in this picture. (NASA)

bly the source of the jagged rocks that litter the scene. Some light-colored bedrock is exposed near the horizon.

Some small sand drifts were seen at the Viking 1 site. Careful scrutiny of pictures taken over several months reveals that these drifts are pushed around slightly by the Martian winds. New rocks are occasionally exposed, while others are covered up, providing what may be the best explanation for the seasonal changes that have been observed from Earth for the past two centuries. Seasonal winds transporting fine red dust alternately expose and then cover dark rocks, thereby mimicking vegetational color changes over the long Martian year.

The Viking Lander 2 site at Utopia is situated 1500 km closer to the north pole than is Chryse. As shown in Figure 11-18, Utopia is remarkably crater-free, although many of the rocks in this view may have been ejected from a large crater about 2000 km north of the spacecraft.

Figure 11-18   A panorama of the Viking Lander 2 site
This mosaic of three views shows about 180° of the Utopia plains. Northwest is at the left; southeast on the right. The flat, featureless horizon is approximately 3 km (2 miles) away from the spacecraft. (NASA)

Instruments on the Viking landers sent back detailed information about the Martian atmosphere

In all of the pictures sent back from Mars, the sky has a distinctly pinkish-orange tint. This coloration is thought to be caused by extremely fine-grained dust suspended in the Martian atmosphere.

Both landers carried equipment to measure meteorological conditions on Mars. These instruments promptly confirmed that the surface atmospheric pressure is in the range of 6 to 8 millibars, as had been expected from previous missions. Atmospheric pressure at sea level on Earth is about 1000 millibars, so the density of the Martian atmosphere is less than 1/100 that of the Earth's atmosphere. Direct chemical analysis also confirmed the high carbon dioxide content: the Viking instruments reported a carbon dioxide abundance of 95 percent, nitrogen 2.7 percent, and argon 1.6 percent. The remaining fraction of a percent is mostly oxygen and carbon monoxide, with a small amount of water vapor.

Above the Martian surface, atmospheric density decreases with increasing altitude, much as it does on Earth. Measurements obtained during the descents of the two landers are displayed in Figure 11-19, with the density profile of Earth's atmosphere for comparison.

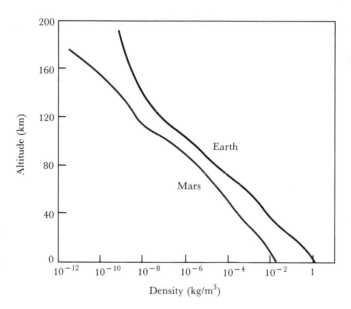

Figure 11-19   Density of the Martian atmosphere
At ground level, the density of Mars's atmosphere is less than one-hundredth the density of Earth's. Up to 100 km altitude, the density of Earth's atmosphere decreases more rapidly than that of Mars. Above 100 km, this relationship is reversed.

While one set of instruments measured atmospheric density, other instruments recorded atmospheric temperature as the spacecraft plunged down through the tenuous Martian atmosphere (see Figure 11-20). As we saw in Chapter 9 (recall Figure 9-2), atmospheric temperature on Earth exhibits a maximum at an altitude of 50 km because of the absorption of ultraviolet radiation by our ozone layer. Mars has no ozone layer, and the temperature profile of the Martian atmosphere exhibits relatively little variation with altitude.

After only a few weeks on the Martian surface, the Viking data showed clearly that the atmospheric pressure at both landing sites was dropping steadily. Mars seemed to be rapidly losing its atmosphere, and some scientists joked that all the air would be gone by Christmas. A straightforward explanation was, however, readily available: winter was coming to the southern hemisphere. At the Martian south pole, it was so cold that large amounts of carbon dioxide were solidifying out of the atmosphere, covering the ground with dry-ice snow.

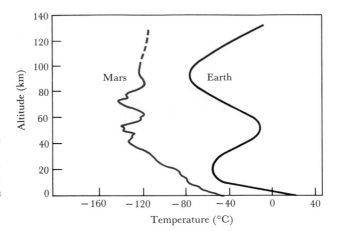

Figure 11-20 Temperature of the Martian atmosphere
Unlike Earth, Mars lacks an ozone layer in its atmosphere. Consequently, the temperature profile of the Martian atmosphere does not show a maximum at 50 km altitude as does the Earth's temperature profile.

Figure 11-21 [left] Winter at Utopia
This picture, taken in May 1979, shows a thin frost layer that lasted for about a hundred days at the Viking Lander 2 site. Freezing carbon dioxide adheres to water-ice crystals and dust grains in the atmosphere, causing them to fall to the ground. The sky is therefore not as pink as it was in the summertime view of Figure 11-18. (NASA)

Figure 11-22 [right] Diurnal temperature variation
The lower curve shows the daily temperature variation at the Viking 1 site. The upper curve shows the daily temperature variation at a desert site in California. The daily temperature range on Mars is about three times greater than that on Earth. The thin, dry Martian air does not retain heat as well as the Earth's atmosphere does.

In early 1977, when spring came to the southern hemisphere, the dry-ice snow rapidly evaporated, and the atmospheric pressure returned to prewinter levels. With the arrival of winter in the northern hemisphere, another decrease in atmospheric pressure was observed as dry-ice snow blanketed the northern latitudes (see Figure 11-21). Thus the Martian surface experiences extreme seasonal variations in temperature and atmospheric pressure.

Aside from an occasional dust storm, the weather on Mars seems boring to someone accustomed to the temperate zones of the Earth. Atmospheric pressure varies with the seasons in a regular and predictable fashion. The atmospheric temperature varies with the time of day, also in a monotonously repetitive way. Actually, the daily temperature variation on Mars is quite similar to the temperature changes observed in a desert on Earth. The warmest time of day occurs about two hours after local noon, and the coldest temperatures are recorded just before sunrise. The thin, dry Martian atmosphere is not capable of retaining much of the day's heat, however, so the range of the daily temperature swings on Mars is larger than that on Earth (see Figure 11-22).

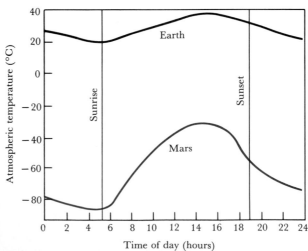

Geological analysis by the Viking landers showed that the Martian regolith contains abundant iron

Just as meteorologists were delighted with the success of the atmospheric measurements, geologists were very pleased with the chemical analysis that the Viking landers performed on the Martian regolith. Some of the most important data came from a device called an X-ray fluorescence spectrometer, which showed that rocks at both lander sites are rich in iron and magnesium. Specifically, iron and silicon comprise about two-thirds of a Martian rock's content. Surprisingly high concentrations of sulfur also were found. Sulfur is 100 times more abundant in Martian rocks than in terrestrial rocks, but potassium and aluminum are significantly less abundant in the Martian rocks than in terrestrial rocks.

Each Viking lander had a scoop at the end of a mechanical arm to obtain regolith and rock samples for analysis (see Figure 11-23). Bits of the regolith were observed to cling to a magnet mounted on the scoop, indicating that a significant fraction of the regolith (3 to 7 percent by weight) is magnetic material, possibly the iron-rich mineral maghemite. In general, the Martian regolith at both lander sites could best be described as an iron-rich clay. The familiar reddish color of Mars may in fact be caused by rust (iron oxides) in the regolith.

The high iron content of the Martian crust is interesting because Mars has particularly a lower average density (3.9 g/cm^3) than the other terrestrial planets (more than 5 g/cm^3 for Mercury, Venus, and Earth). It appears that Mars is relatively lacking in iron and magnesium, and furthermore that these elements are distributed throughout the body of the planet rather than concentrated in a dense core. In other words, Mars apparently did not undergo chemical differentiation. The reasons for this unusual history are poorly understood, and geologists hoped to obtain important information about the interior of Mars from seismometers on the Viking landers. Unfortunately, one of the seismometers failed to deploy. With only one properly functioning seismometer, geologists were unable to pinpoint the locations and depths of marsquakes, so the structure of the Martian interior remains largely unknown.

The absence of an Earth-like iron-rich core seems consistent with the almost total absence of a planetwide magnetic field. Soviet missions demonstrated that Mars's magnetic field is less than 0.004 times as strong as the Earth's field. Mars rotates nearly as fast as the Earth, but an Earth-

Figure 11-23 Digging in the Martian soil
Viking's mechanical arm with its small scoop protrudes from the right side of this view of the Chryse plains. Several small trenches dug by the scoop in the Martian regolith appear near the left side of the picture. (NASA)

like magnetic dynamo definitely is not operating on Mars. If Mars does have an iron-rich core, it is reasonable to assume that the core is not molten.

Mars's magnetic field is so weak that it is largely ineffectual in warding off the solar wind. In fact, the solar wind may actually impinge directly on the outermost layers of the Martian atmosphere, just as it does on Venus (recall Figure 8-5). Unfortunately, our knowledge of this aspect of the Martian environment is woefully incomplete. An alternative possibility is that Mars's weak field just manages to carve out a magnetosphere just barely enclosing the planet's atmosphere.

The two Martian moons resemble asteroids

Two tiny moons move around Mars in orbits that are close the the planet's surface. These satellites are so small that they were not discovered until the favorable opposition of 1877. While Schiaparelli was seeing *canali,* the American astronomer Asaph Hall spotted these two moons, which orbit very close to the planet and almost directly above the Martian equator. He named them Phobos ("fear") and Deimos ("panic"), after the mythical horses that drew the chariot of the Greek god of war, Mars.

Phobos is the nearer and larger of the Martian moons. It circles Mars in only 7 hours 39 minutes, moving over the Martian surface at an average distance of only 6000 km. (Recall that our Moon orbits at an average distance of 376,280 km from the Earth's surface.) This orbital period is much shorter than the Martian day; Phobos rises in the west and gallops across the sky in only 5 hours, as viewed by an observer near the Martian equator. During this time, Phobos appears several times brighter in the Martian sky than Venus does from Earth.

Deimos, which is farther from Mars and somewhat smaller than Phobos, appears about as bright from Mars as Venus does from Earth. Deimos's orbit, 20,000 km above the Martian surface, is at almost the right distance for a synchronous orbit—an orbit in which a satellite seems to hover above a single location on the planet's equator. Consequently, as seen from the Martian surface, Deimos takes about three full days to creep slowly from one horizon to the other.

The Mariner 9 mission treated astronomers to their first close-up views of the Martian satellites (see Figure 11-24). Numerous additional views were provided by the Viking orbiters. Phobos and Deimos were revealed to be jagged, heavily cratered, football-shaped rocks. Phobos was found to be roughly 27 by 21 by 19 km, and Deimos slightly smaller, at roughly 15 by 12 by 11 km. It was further revealed that both Phobos and Deimos rotate synchronously about Mars, always keeping the same side facing the red planet.

The origin of the Martian moons is unknown. Phobos and Deimos might have formed by gravitational accretion of tiny rocks that orbited Mars in the distant past, in much the same way as our own Moon formed, but on a much smaller scale. If this theory is correct, the Martian moons probably are composed of basaltic rock like our own satellite.

A second possibility is that Phobos and Deimos are captured asteroids. Mars is quite near the asteroid belt, where thousands of large rocks orbit the Sun. It is possible that two of these asteroids wandered close enough to Mars to become permanently trapped by the planet's gravitational field. If this is correct, then the Martian moons are probably composed of meteoritelike material, which is generally less dense than basaltic rock. Thus, future accurate measurements of the densities of Phobos and Deimos may help to resolve this debate.

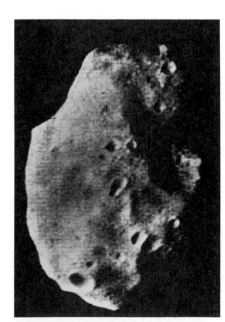

Figure 11-24   Phobos
Mars's larger satellite is shown in this photograph sent back by Mariner 9. The area covered by this photograph is 28 by 20 km (17 by 14 miles). Mars's other moon, Deimos, is somewhat smaller. Both are pockmarked with craters. (NASA)

Biological experiments failed to detect any conclusive evidence of life on Mars

The meteorological and geological equipment on the Viking landers dramatically broadened our understanding of the Martian environment, but most anticipation and excitement centered on the Viking biological experiments. Each lander carried a compact biological laboratory designed to perform three different tests for microorganisms in the Martian soil. In addition, other equipment was expected to shed light on the possibility of Martian life.

For example, the television cameras might reveal some life forms. Biologists joked about "titanium mushrooms" or other strange creatures that the biological tests might fail to detect yet should be classified as living because they grew and multiplied. Careful scrutiny of the thousands of pictures sent back over many months, however, failed to reveal any objects or any changes that could be attributed to biological processes.

A machine impressively called a gas-chromatograph mass spectrometer performed detailed chemical analyses of the Martian soil. At both landing sites, very surprising results were obtained: absolutely no organic compounds were detected in the soil. This is strong evidence that life has never existed on Mars, because this equipment should at least have detected the corpses of Martian microorganisms or their waste products. Furthermore this negative result is surprising because meteorites falling on Mars should have contaminated the planet's surface with an easily detectable amount of organic material. Yet the equipment, which was undoubtedly in perfect working order, failed to find any organic compounds whatsoever. This was our first hint that natural chemical processes on Mars may literally sterilize the planet.

The three biological experiments carried out by the Viking landers were based on the general proposition that living things alter their environment. They eat, they breathe, and they give off waste products. In each of the three experiments, a sample of Martian regolith was placed in a closed container with or without some nutrient substance. The container was then examined for any changes in its contents.

The **gas-exchange experiment** was designed to detect processes that could be broadly classified as respiration. A small sample of Martian regolith was placed in a sealed container with a controlled amount of gas and nutrients. The gases in the container were monitored to see if their chemical composition changed.

The **labeled-release experiment** was designed to detect processes resembling metabolism. A small sample of Martian regolith was moistened with nutrients containing radioactive carbon atoms. Any organisms in the regolith should eat the food and emit gases containing the telltale radioactive carbon.

The **pyrolytic-release experiment** was designed to detect photosynthesis, the biological process by which plants on Earth synthesize organic compounds from carbon dioxide, using sunlight as an energy source. In the Viking experiments, a regolith sample was placed in a container along with radioactive carbon dioxide and exposed to artificial sunlight. If plantlike photosynthesis were to occur, some of the radioactive carbon from the gas should become incorporated into the microorganisms in the regolith.

The first data returned by the Viking biological experiments caused great excitement: in almost every case, rapid and extensive changes were detected inside the sealed containers. Further analysis of the data, however, led to the conclusion that these changes were due solely to nonbiological chemical processes. It appears that the Martian regolith is rich in chemicals that effervesce (fizz) when moistened. A large amount of oxygen apparently is tied up in the regolith in the form of unstable chemicals called peroxides and superoxides that break down in the presence of water to release oxygen gas.

The chemical reactivity of the Martian regolith may be a direct result of the absence of a protective ozone layer in the Martian atmosphere. Ultraviolet radiation from the Sun beats down on the Martian surface, easily dissociating molecules of carbon dioxide (CO_2) and water vapor (H_2O), by knocking off oxygen atoms, which then become loosely attached to chemicals in the regolith. Alternatively, this process can produce ozone (O_3) and hydrogen peroxide (H_2O_2), which also become incorporated in the regolith. In all these cases, the loosely attached oxygen atom makes the regolith extremely reactive in the presence of water.

Here on Earth, hydrogen peroxide is commonly used as an antiseptic. When you pour hydrogen peroxide on a wound, the liquid fizzes and froths as the loosely attached oxygen atoms chemically combine with organic material, thereby destroying germs. Perhaps this is why the Viking landers failed to detect any organic compounds in the Martian regolith. The superoxides and peroxides in the regolith make it literally antiseptic.

More Martian probes are needed to answer important questions about Mars and about the Earth

The next mission to Mars will use a small satellite called the Mars Geoscience/Climatology Orbiter, or MGCO. The launch is scheduled for 1990. Using remote-sensing techniques, MGCO will analyze the mineral composition of the Martian surface and also determine the role of water in the Martian climate. The results could lead to a deeper understanding of the Earth's weather and geology. The modest MGCO project, however, is the only mission to Mars scheduled by the United States during the remainder of the twentieth century.

Although the possibility of Martian organisms seems quite remote, some scientists point out that the polar regions of Mars may have conditions more suitable for life. These regions could be explored by another mission to the Martian surface that could also send back data about the Martian interior. Ideally, this ambitious undertaking would utilize a rover vehicle (see Figure 11-25) that would send back data and pictures

Figure 11-25 The Mars Rover
Future missions to Mars could use a mobile vehicle that would send back data from a variety of sites. Unfortunately, because of budget cutbacks, the United States has no firm plans for such a mission. (NASA)

from a wide range of sites. Another possibility is to land a spacecraft that would scoop up some Martian rocks and regolith, blast off, and bring the samples back to Earth for laboratory analysis. The Soviet Union has been particularly successful with missions of this type to the Moon. Their unmanned Luna 16, 20, and 24 probes all returned samples of Moon rocks to Earth. Indeed, Soviet scientists are vigorously planning a similar assault on Mars, with the ultimate goal of establishing a manned base there early in the twenty-first century. A courageous mission of this magnitude would realize some of the great dreams of science-fiction writers within our lifetimes.

Summary

- Earth-based observers found that the Martian solar day is very slightly longer than that on the Earth, that Mars has polar caps that expand and shrink with the seasons, and that the Martian surface undergoes seasonal color changes; a few observers reported a network of linear features that came to be called "canals"; these observations led to many speculations about Martian life.

    The best Earth-based views of Mars are obtained at favorable oppositions, when Mars is simultaneously at opposition and near perihelion.

- Spacecraft returned close-up views showing that the Martian surface has numerous flat-bottomed craters, several huge volcanoes, a vast canyon, and dried-up riverbeds—but no canals or other planet-spanning linear features.

- The Martian atmosphere is composed mostly of carbon dioxide with only very slight traces of water vapor; the Martian atmospheric pressure is about one-hundredth that of the Earth's atmosphere and shows seasonal variations.

    Mars has no ozone layer and a very weak magnetic field, so its surface is exposed to ultraviolet radiation.

- Liquid water would quickly boil away in Mars's thin atmosphere, but the polar caps do contain a considerable amount of frozen water, and a layer of permafrost may exist beneath the Martian regolith.

    The flash-flood features and dried riverbeds observed on the Martian surface are probably created when volcanic activity melts frozen water just beneath the regolith.

    The winter expansion of the polar caps is due to deposition of a thin layer of frozen carbon dioxide (dry ice).

- Great dust storms sometimes blanket Mars. Very fine-grained dust in the atmosphere gives the Martian sky a pinkish-orange tint.

- Mars apparently never experienced chemical differentiation; the planet might not have an iron-rich core.

- Mars has two very small, football-shaped satellites that move in orbits very close to the surface of the planet; they may be captured asteroids.

- Chemical reactions in the Martian regolith, together with ultraviolet radiation from the Sun, apparently act to sterilize the Martian surface.

- Little is known about the interior of Mars; even the absence of a dense core has not been confirmed by seismic studies.

Review questions

1 Explain why Mars has the longest synodic period of all the planets although its sidereal period is only 687 days.

2 Compare the cratered regions of Mercury, the Moon, and Mars. Assuming that the craters on all three worlds originally had equally sharp rims, what can you conclude about the environmental histories of these worlds?

3 Explain why the Earth's sky is blue, that of Venus is yellow, and that of Mars is pink.

4 How would you tell which craters on Mars were formed by meteoritic impacts and which were formed by volcanic activity?

5 What two types of Martian terrain suggest that Mars may have had a denser atmosphere in the past than it does today? What third type of terrain suggests that this episode of a denser atmosphere must have been comparatively short-lived?

6 With carbon dioxide just about as abundant in the Martian atmosphere as in the Venusian atmosphere, why do you suppose that there is no greenhouse effect on Mars?

7 Why do you suppose that Phobos and Deimos are not round like our Moon?

Advanced questions

*****8** Imagine that you are on Mars, looking back at the Earth. Assuming that the human eye has a resolving power of 50 arc sec, would you see the Earth and Moon as a "double star"?

9 Is it reasonable to suppose that the polar regions of Mars might harbor life forms even though the Martian regolith is sterile at the Viking lander sites?

Discussion Questions

10 Suppose that someone told you that the Viking mission failed to detect life on Mars because the tests were designed for *terrestrial* life forms and not Martian life forms? How would you respond?

11 Compare the scientific opportunities for long-term exploration offered by the Moon and Mars. What difficulties would be inherent in establishing a permanent base or colony on each of these two worlds?

12 Imagine that you are an astronaut living at a base on Mars. Describe what your day might be like, what you would see, the weather, the spacesuit you would wear, etc. Suppose you and your colleagues have a motorized vehicle for exploring the planet. Where would you like to go?

For further reading

Batson, R., et al., eds. *The Atlas of Mars*. NASA SP-438, 1979. *Detailed maps and photomosaics of the red planet.*

Carr, M. "The Surface of Mars: A Post Viking View." *Mercury,* Jan./Feb. 1983, p. 2.

Cooper, H. *The Search for Life on Mars*. Holt, Rinehart, & Winston, 1980. *A profile of the Viking project by a well-known journalist.*

Gore, R. "Sifting for Life in the Sands of Mars." *National Geographic,* Jan. 1977.

Moore, P. "Mars—Then and Now." *Mercury,* Mar./Apr. 1980. *A good article on the history of Mars exploration.*

Sagan, C. "Blues for a Red Planet." In *Cosmos*. Random House, 1980.

Veverka, J., et al. "The Puzzling Moons of Mars." *Sky & Telescope,* Sept. 1978, p. 186.

12 Jupiter: lord of the planets

Jupiter with Io and Europa
Jupiter is the largest and most massive planet in the solar system. Jupiter's diameter equals 11¼ Earth diameters; Jupiter's mass equals 318 Earth masses. However, Jupiter is composed primarily of hydrogen and helium in an abundance very similar to the Sun's chemical composition. This photograph from Voyager 1 in 1979 shows Europa (right) and Io (center). Both satellites are nearly the same size as our Moon. The Great Red Spot, a large persistent storm in the Jovian atmosphere roughly the same size as Earth, is on the left. (NASA)

About 70 percent of the mass of the solar system outside the Sun is concentrated in the single giant planet Jupiter. In this chapter, we learn about the unusual features of this active, vibrant world, whose multicolored and turbulent clouds are surrounded by an enormous magnetosphere. We learn more about atmospheric behavior in general as we discuss the data from space probes sent past Jupiter, and we discover that much of Jupiter's mass is composed of liquid metallic hydrogen. We conclude our discussion by noting the intriguing possiblity that life may exist within the thick Jovian atmosphere. We postpone our discussion of Jupiter's Earth-like satellites until the next chapter.

More than any other single factor, temperatures throughout the young solar nebula dictated the final forms and natures of the planets that orbit the Sun. In the warm, inner regions of this ancient nebula, surviving dust grains consisted primarily of metals, silicates, and oxides. The temperature was too high to allow substantial condensation of volatile substances such as water, methane, and ammonia. The four planets that formed close to the Sun were therefore composed almost entirely of rocky material. Their surface gravities were too low and their surface temperatures too high to retain any of the abundant but lightweight hydrogen and helium gases that made up most of the solar nebula.

The orbit of Jupiter, however, is more than 5 times as large as the orbit of the Earth, and hence sunlight reaching Jupiter is less than $\frac{1}{25}$ as bright as that reaching the Earth. In the past, as now, it was cold at this distance of one-half billion miles from the Sun (recall Figure 6-12). In the young solar nebula, the dust grains at this distance from the protosun were covered with thick, frosty coatings of frozen water, methane, and ammonia. These volatile substances therefore became important constituents of the planets that accreted in the outer reaches of the solar system.

Jupiter, in its large orbit 5.2 AU from the Sun, may have formed in much the same fashion as the terrestrial planets—by accretion of the dust grains (coated with frozen gases in this case) into a great number of planetesimals, which in turn accreted to form a huge protoplanet. Many scientists, however, think that the Jovian planets were formed in a two-step process. First, accretion led fairly quickly to formation of a large protoplanet having at least several times the mass of the Earth. Then, the strong gravitational pull of this protoplanet attracted and retained substantial quantities of the hydrogen and helium existing in the cool outer reaches of the solar nebula. The Canadian–American astrophysicist Alastair Cameron carried out calculations showing that this gathering of lightweight gases would become very efficient and rapid after the protoplanet had grown beyond a certain mass.

The final result was the largest planet in the solar system—a planet composed (by weight) of 82 percent hydrogen, 17 percent helium, and only 1 percent all other elements. This composition closely matches what is believed to have been the original composition of the solar nebula; it is also very similar to the composition of the Sun's atmosphere today (recall Table 6-4). As we shall see later in Chapters 13, Saturn—and to a lesser extent, perhaps Uranus and Neptune—are thought to have formed in much the same way as Jupiter.

Box 12-1   Jupiter data

Mean distance from the Sun: 5.203 AU = 7.783×10^8 km
Maximum distance from the Sun: 5.455 AU = 8.157×10^8 km
Minimum distance from the Sun: 4.951 AU = 7.409×10^8 km
Mean orbital velocity: 13.1 km/sec
Sidereal period: 11.86 years
Rotation period: equatorial = $9^h 50^m 30^s$
               internal   = $9^h 55^m 30^s$
Inclination of equator to orbit: 3° 04′
Inclination of orbit to ecliptic: 1° 18′
Orbital eccentricity: 0.048
Diameter: equatorial = 143,800 km
        polar   = 135, 200 km
Diameter (Earth = 1): equatorial = 11.27
                  polar   = 10.60
Apparent diameter as seen from Earth: maximum = 50.1″
                                 minimum = 30.4″
Mass: 1.90×10^{27} kg
Mass (Earth = 1): 317.89
Mean density: 1.314 g/cm^3
Surface gravity (Earth = 1): 2.64
Escape velocity: 61 km/sec
Oblateness of planet: 0.0637
Mean surface temperature: −110°C at cloud tops
Albedo: 0.51
Brightest magnitude: −2.6
Mean diameter of Sun as seen from Jupiter: 6′ 09″

Huge and massive Jupiter is composed largely of lightweight gases

Jupiter is huge: its mass is 318 times greater than the mass of Earth. Indeed, the mass of Jupiter is $2\frac{1}{2}$ times the combined masses of all the other planets, satellites, asteroids, meteoroids, and comets in the solar system. Jupiter's equatorial diameter is $11\frac{1}{4}$ times as large as the Earth's diameter. Jupiter's volume is about 1430 times larger than the Earth's.

Jupiter's average density can be computed from its mass and size: it is only 1.33 g/cm^3. This low average density is entirely consistent with the picture of a huge sphere of hydrogen and helium, compressed by its own gravity. In fact, back in the 1930s, the Swiss–American astronomer Rupert Wildt first suggested that Jupiter is mostly hydrogen, basing his hypothesis largely upon the calculated average density of the planet. Unfortunately, it is very difficult to detect hydrogen and helium in Jupiter's atmosphere because these molecules do not produce prominent spectral lines at visible wavelengths in the light reflected from the planet. Therefore, evidence confirming the composition of Jupiter was slow in coming.

Wildt did find very prominent spectral lines of methane (CH$_4$) and ammonia (NH$_3$) in the spectrum of sunlight reflected from Jupiter. That each of these molecules contains three or four hydrogen atoms was one bit of evidence that led Wildt to suspect a great abundance of hydrogen in the atmosphere. At the low temperatures in Jupiter's upper atmosphere, however, hydrogen exists only as H$_2$ molecules, which produce very weak spectral lines (they were finally detected in 1960). Conclusive confirmation of helium in the Jovian atmosphere was not obtained until the 1970s, when infrared instruments on space probes passing near the planet detected evidence of collisions between hydrogen molecules and helium atoms. This indirect evidence finally confirmed Wildt's brilliant hypothesis about the composition of Jupiter's atmosphere.

Through an Earth-based telescope, Jupiter appears as a colorful, intricately banded sphere (see Figure 12-1). Figure 12-2 is a close-up view from Voyager 1. The most prominent features are alternating dark and

Figure 12-1 [left] Jupiter from Earth
Many belts and zones are easily identified in this Earth-based view of the largest planet in the solar system. The Great Red Spot appears toward the lower right. (U.S. Naval Observatory)

Figure 12-2 [right] Jupiter from Voyager 1
This view was sent back from Voyager 1 in 1979 when the spacecraft was only 30 million kilometers from the planet. Features as small as 600 km across can be seen in the turbulent cloud tops of this giant planet. Complex cloud motions surround the Great Red Spot. (NASA)

light bands parallel to Jupiter's equator, shaded in subtle tones of red, orange, brown, yellow, and blue. The dark, reddish bands are called **belts;** the light-colored bands are called **zones.** These are not the only conspicuous markings. A large, reddish oval called the **Great Red Spot** is often visible in Jupiter's southern hemisphere. This remarkable feature, which has been observed since the mid-1600s, appears to be a long-lived storm in the planet's dynamic atmosphere. Many careful observers have reported smaller spots and blemishes that last for only a few weeks or months in Jupiter's turbulent clouds.

The best time to observe Jupiter is, of course, when the planet is at opposition. At such times, Jupiter outshines all the stars in the sky (its brightest magnitude is −2.6) and through a telescope presents a disk nearly 50 arc sec in diameter—roughly twice as great as the angular diameter of Mars under the most favorable conditions. Jupiter takes almost a dozen years to orbit the Sun, so the planet appears to meander across the twelve constellations of the zodiac at the rate of approximately one per year. Successive oppositions occur at intervals of about 13 months (see Box 12-2).

Box 12-2   Oppositions of Jupiter

The synodic period of Jupiter is 398.9 days, or approximately a year and a month. This interval between successive oppositions is obvious in the table, which lists all oppositions in the last two decades of the twentieth century.

Date	Diameter (arc sec)	Magnitude
Feb. 24, 1980	44.7	−2.1
Mar. 26, 1981	44.2	−2.0
Apr. 25, 1982	44.4	−2.0
May 27, 1983	45.5	−2.1
June 29, 1984	46.8	−2.2
Aug. 4, 1985	48.5	−2.3
Sept. 10, 1986	49.6	−2.4
Oct. 18, 1987	49.8	−2.5
Nov. 23, 1988	48.7	−2.4
Dec. 27, 1989	47.2	−2.3
Jan. 28, 1991	45.7	−2.1
Feb. 28, 1992	44.6	−2.0
Mar. 30, 1993	44.2	−2.0
Apr. 30, 1994	44.5	−2.0
June 1, 1995	45.6	−2.1
July 4, 1996	47.0	−2.2
Aug. 9, 1997	48.6	−2.4
Sept. 16, 1998	49.7	−2.5
Oct. 23, 1999	49.8	−2.5
Nov. 28, 2000	48.5	−2.4

Details of Jupiter's rotation give clues about the planet's internal structure

Although it is the largest and most massive planet in the solar system, Jupiter has the fastest rate of rotation. At its equatorial latitudes, Jupiter completes a full rotation in only 9 hours 50 minutes 30 seconds. However, Jupiter is not a solid, rigid object. By following features in the belts and zones, we see that the polar regions of the planet rotate a little more slowly than do the equatorial regions. Near the poles, the rotation period is about 9 hours 55 minutes 41 seconds. The first person to notice this **differential rotation** of Jupiter was the Italian astronomer Giovanni Domenico Cassini in 1690. This was the same gifted observer who gave us the first accurate determination of Mars's rotation rate.

Jupiter's colorful cloudtops are the turbulent, uppermost layer of Jupiter's thick atmosphere. Are the observed rotation rates of these clouds anything like the rotation rates of deeper levels or that of a solid central core? Some intriguing clues come from the fact that Jupiter emits radio waves. In particular, radio emissions with wavelengths in the range of 3 to 75 cm vary slightly in intensity with a period of 9 hours 55 minutes 30 seconds. As we shall see later, this radio emission is directly associated with Jupiter's magnetic field, which is anchored deep inside the planet. Thus radio observations reveal Jupiter's **internal rotation period,** which is slightly different from the atmospheric rotation that we observe through a telescope.

Jupiter's rapid rotation profoundly affects the overall shape of the planet. Even a casual glance through a small telescope shows that Jupiter is slightly flattened, or **oblate.** The diameter across Jupiter's equator (143,800 km) is 6.37 percent larger than the diameter from pole to pole (135,200 km). Thus Jupiter is said to have an **oblateness** of 6.37 percent, or 0.0637.

If Jupiter were not rotating, it would be a perfect sphere. A massive, nonrotating object naturally settles into a spherical shape, in which every atom on its surface experiences the same intensity of gravity aimed directly at the object's center. However, Jupiter is rotating, and thus every part of the planet also experiences an outward-directed centrifugal force that is proportional to the distance from the axis of rotation. Equatorial regions are farther from the planet's axis of rotation than are the polar regions, and hence the equatorial diameter is slightly larger than the polar diameter. This centrifugal stretching of the equatorial dimensions gives Jupiter its characteristic oblate shape.

At every point throughout Jupiter, the inward force of gravity is exactly balanced by the outward pressure of the compressed material plus the outward centrifugal effects of the planet's rotation. This balance is called **hydrostatic equilibrium.** The shape and density of a planet adjust themselves to ensure that this balance is maintained.

Jupiter's shape is an excellent indicator of its internal structure. Two planets with the same mass, same average density and same rotation will have slightly different oblateness if one planet has a compact core and the other does not. The less-dense material surrounding a compact core extends to a larger distance from the planet's center and therefore experiences greater deformation by centrifugal forces than does the planet with a uniform distribution of matter. All other things being equal, therefore, a planet with a dense core will be more oblate than will a planet without a central core.

Detailed calculations by William B. Hubbard of the University of Arizona and Soviet scientists V. Zharkov and V. Trubitsyn strongly suggest that 4 percent of Jupiter's mass is concentrated in a dense rocky core. In other words, Jupiter's oblateness is consistent with a rocky core nearly 13 times as massive as the entire Earth. Some of this core was probably the original "seed" around which proto-Jupiter accreted. However, additional amounts of rocky material were added at later times by meteoritic material falling onto the planet.

Although Jupiter's rocky core is 13 times as massive as the Earth, it is probably not much bigger than our planet. The tremendous crushing weight of the remaining 305 Earth masses of Jupiter's bulk compresses the core down to a sphere 20,000 km in diameter (Earth's diameter is 12,800 km). The pressure at Jupiter's center is about 80 million atmospheres, and the rocky material of Jupiter's core is therefore squeezed to a density of about 20 g/cm³. The corresponding temperature at the planet's center is probably about 25,000 K. In contrast, the temperature at Jupiter's cloudtops is only 165 K.

Radio observations revealed Jupiter's magnetic field and metallic hydrogen interior

In the 1950s, astronomers began discovering radio emissions from Jupiter. Most of this radiation is **thermal;** it comes from the planet's surface and is exactly what would be expected from an object emitting heat. Some of the radiation, however, is **nonthermal** and is found in two broad wavelength ranges. At wavelengths of a few meters are sporadic bursts of **decametric ("ten-meter") radiation** that probably is caused by electrical discharges and lightning in Jupiter's clouds. (As we shall see in Chapter 13, these discharges may be the result of complex electromagnetic interactions between Jupiter and its large satellite, Io.) At wavelengths of a few tenths of meters (particularly 3 to 75 cm) is the **decimetric ("tenth-meter") radiation** that revealed the planet's internal rotation period. This decimetric radiation also provided important clues about the interior of Jupiter's enormous bulk between its rocky core and its colorful cloudtops.

The first person to explain the source of Jupiter's decimetric radiation was the Soviet astrophysicist Iosif S. Shklovskii. In the 1950s, Shklovskii argued convincingly that decimetric radiation is characteristic of radio waves emitted by high-speed electrons moving through a magnetic field. Astronomers now realize that this process is very important in a wide range of circumstances throughout the universe—from pulsars and quasars to the radio emissions of entire galaxies. Whenever electrons traveling near the speed of light encounter a magnetic field, they spiral around the magnetic field and emit copious radio waves. Energy produced in this fashion is called **synchrotron radiation.**

If some of Jupiter's radiation is indeed synchrotron radiation, then Jupiter must have a magnetic field. From the intensity of this radiation, it was soon apparent that Jupiter's magnetic field must be very strong—many times stronger than Earth's. Additional calculations showed that a molten iron center inside Jupiter's rocky core could not possibly create such an enormous planetwide magnetic field. What then could be the source of Jupiter's magnetism?

Jupiter is largely composed of hydrogen. A hydrogen atom consists of a single proton orbited by a single electron. Deep inside Jupiter, however, pressures are so great that the electrons are stripped from their protons. The result is a liquidlike mixture of protons and electrons. The electrons, no longer bound to protons, are free to wander around and thereby create electric currents, just as moving electrons in a copper wire constitute an electric current. In other words, the highly compressed hydrogen deep inside Jupiter behaves like a metal; it is therefore called **liquid metallic hydrogen.**

Detailed calculations by American scientists Edwin E. Salpeter and

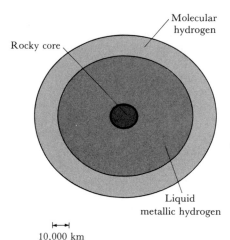

Rocky core

Molecular hydrogen

Liquid metallic hydrogen

10,000 km

Figure 12-3   The structure of Jupiter *Most scientists believe that Jupiter probably has a rocky core approximately 20,000 km in diameter. This core is surrounded by a 40,000-km-thick layer of liquid metallic hydrogen, which is responsible for Jupiter's powerful magnetic field. The outer 20,000-km-thick layer is composed primarily of ordinary hydrogen.*

David J. Stevenson strongly suggest that molecular hydrogen is transformed into liquid metallic hydrogen when the pressure exceeds 3 million atmospheres. This transition occurs at a depth of approximately 20,000 km below Jupiter's cloudtops. Thus the internal structure of Jupiter consists of three distinct regions (see Figure 12-3): the rocky core is surrounded by a 40,000-km-thick layer of liquid metallic hydrogen, which is in turn surrounded by a 20,000-km-thick layer of ordinary hydrogen. All of the colorful cloud patterns that we see through telescopes are located in the outermost 100 km of the outer layer of molecular hydrogen.

It is clear from Figure 12-3 that the vast majority of Jupiter's enormous bulk is electrically conductive liquid metal. Because of Jupiter's rapid rotation, electric currents in this thick layer of liquid metallic hydrogen generate a powerful magnetic field, in much the same way that liquid portions of the Earth's core produce the Earth's magnetic field. The intrinsic strength of Jupiter's magnetic field is 19,000 times greater than that of the Earth's field. The Jovian magnetic field is so much stronger than ours because Jupiter's liquid metallic region is so much larger than the Earth's and because Jupiter rotates so much faster than the Earth does.

Spacecraft mapped details of Jupiter's enormous magnetosphere

Jupiter's powerful magnetic field surrounds the planet with an enormous magnetosphere, large enough to envelop the orbits of many of its moons. But from the Earth, our evidence of this magnetosphere is limited to the faint hiss of radio static. For many years astronomers realized that our understanding and appreciation of this extraordinary planet would be vastly increased if we could send spacecraft past Jupiter to probe its magnetosphere and view its dynamic atmosphere at close range. This program was carried out by four historic missions to Jupiter during the 1970s.

The first Jupiter flyby occurred in December 1973, when Pioneer 10 coasted to within 131,000 km (81,000 miles) of the planet's cloudtops (see Figure 12-4). An identical spacecraft called Pioneer 11 followed in December 1974, passing Jupiter at a distance of only 46,400 km (29,000 miles)

Figure 12-4   The Pioneer spacecraft
This drawing shows one of two identical spacecraft that flew past Jupiter in the 1970s. Instead of using solar panels, the spacecraft drew its power from two electric generators containing radioactive isotopes. Numerous instruments on board the spacecraft made measurements of Jupiter's magnetic environment and sent back pictures of Jupiter's clouds. (NASA)

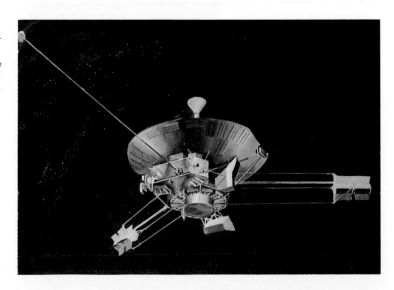

Each Pioneer spacecraft carried an array of instruments to detect charged particles, associated radiation, and magnetic fields. The Pioneer missions therefore gave us our first look at the Jovian magnetosphere and revealed its awesome dimensions. The volume surrounded by the shock wave is nearly 30 million kilometers across. In other words, if you could see Jupiter's magnetosphere from the Earth, it would cover an area in the sky 16 times larger than the full Moon.

Pioneers 10 and 11 crossed Jupiter's magnetopause several times, reporting the boundary at distances ranging from 3 to 7 million kilometers above the planet. Evidently, Jupiter's magnetosphere is sensitive to fluctuations in the solar wind and expands and contracts by a factor of two, very rapidly and frequently. In contrast, dramatic changes in the size of the Earth's magnetosphere are extremely rare, in spite of occasional "gusts" in the solar wind.

The inner regions of our magnetosphere are dominated by two huge Van Allen Belts (recall Figure 9-13) that are filled with charged particles. The same would probably be true of Jupiter if it were not rotating so rapidly. Centrifugal forces due to Jupiter's rapid rotation, however, spew the particles out into a huge electrically charged **current sheet** (see Figure 12-5). This current sheet lies in the plane of Jupiter's magnetic equator. Jupiter's magnetic axis is inclined 11° from the planet's axis of rotation, and the orientation of the magnetic field is the reverse of Earth's (a compass would point toward the south pole on Jupiter).

Figure 12-5 Jupiter's magnetosphere
Like other planetary magnetospheres, Jupiter's has a bow shock wave, a magnetosheath, and a magnetopause. In Jupiter's case, gas pressure from a hot plasma keeps the magnetosphere inflated, thereby holding off the solar wind. Particles trapped inside Jupiter's magnetosphere are spewed out into a vast current sheet by the planet's rapid rotation. Jupiter's axis of rotation is inclined to its magnetic axis by about 11°.

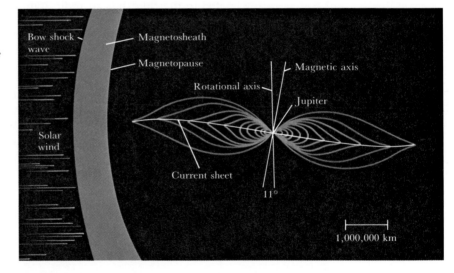

The flybys of Voyagers 1 and 2 in 1979 confirmed the overall picture relayed by the Pioneer spacecraft, but the Voyagers also carried instrumentation designed to provide more data about the Jovian magnetosphere. With planets such as the Earth and Mercury, the magnetopause occurs where the pressure of the planet's magnetic field counterbalances the pressure of the solar wind. The Voyager data indicate that this is not the case with Jupiter.

Instruments carried by the Voyager spacecraft indicated that the inner regions of Jupiter's magnetosphere contain a very hot, gaslike mixture of charged particles called a **plasma.** A plasma is formed when a gas is heated to extremely high temperatures, at which electrons are torn off the atoms of the gas. Thus a plasma is an electrically neutral mixture of positively charged ions and negatively charged electrons. The plasma that envelops Jupiter consists primarily of electrons and protons along with some ions of helium, sulfur, and oxygen. These charged particles are caught up in Jupiter's rapidly rotating magnetic field and accelerated to extremely high speeds.

The Voyager instruments relayed a startling discovery: the plasma surrounding Jupiter has an astounding temperature of 300 million to 400 million kelvins. The Jovian plasma is the hottest place in the solar system! Even the center of the Sun is only 20 million kelvins. However, the Jovian plasma is very thin: the average density in the hottest regions is about one charged particle per 100 cm^3.

Because of its high temperature, Jupiter's huge plasma has a great deal of energy and exerts pressure that holds off the solar wind. The Voyager data suggest that the pressure balance between the solar wind and the hot plasma inside the magnetosphere is precarious. A gust in the solar wind can blow away some of the plasma, and the magnetosphere then deflates rapidly to as little as one-half its original size. Charged-particle detectors carried by the Voyagers recorded several bursts of hot plasma that may have been associated with magnetospheric deflations. However, additional electrons and ions accelerated by Jupiter's rotating magnetic field soon replenish the plasma, and the magnetosphere expands again.

Pictures taken during flybys showed many details in Jupiter's clouds

The most dramatic data to come from the Pioneer and Voyager flybys are the spectacular, colorful, close-up pictures of Jupiter's dynamic atmosphere. The two projects carried very different imaging systems. The Pioneer spacecraft rotated rapidly for stability, and pictures of Jupiter were assembled from a series of thin strips produced as the cameras of the whirling spacecraft scanned across the planet's face. The Voyager spacecraft, designed several years later with more sophisticated technology, employed television systems. As expected, the Voyager pictures were superior to the Pioneer views in resolution, color discrimination, and detail.

Even after allowing for differences in the photographic equipment, it was obvious that Jupiter's atmosphere had undergone some fundamental changes during the four years between the Pioneer and Voyager flybys (see Figure 12-6).

The changes in Jupiter's cloud cover are most apparent in the area surrounding the Great Red Spot. Over the past three centuries, Earth-based observers have reported many long-term variations in the spot's size and color. At its largest, the Great Red Spot measured 40,000 by 14,000 km, so large that three Earths could fit side-by-side across it. At other times (such as in 1976–1977), the spot faded from view. During the Voyager flybys of 1979, the Great Red Spot was only slightly larger than the Earth.

Figure 12-7 shows the two contrasting views of the Great Red Spot. During the Pioneer flybys, the Great Red Spot was embedded in a broad white zone that dominated the planet's southern hemisphere. By the time of the Voyager missions a few years later, however, the cloud structure had changed dramatically. A dark belt had broadened and encroached on the Great Red Spot from the north, and the entire region was apparently embroiled in a much greater degree of turbulence.

Pioneer 10, December 1973

Pioneer 11, December 1974

Voyager 1, March 1979

Voyager 2, July 1979

Figure 12-6   Four views of Jupiter
These four close-up pictures of Jupiter span 5½ years and show major changes in the planet's upper atmosphere. Noticeable changes are apparent even in the Voyager views, which were separated by only four months. All four pictures show the Great Red Spot. The black dot on the Pioneer 10 view is the shadow of Jupiter's moon Io. (NASA)

Careful examination of cloud motions in and around the Great Red Spot reveal that the spot rotates counterclockwise with a period of about six days. Furthermore, winds to the north of the spot are blowing to the west, whereas winds south of the spot are moving toward the east. The circulation around the Great Red Spot is therefore like a wheel spinning between two oppositely moving surfaces (see Figure 12-8). This surprisingly stable wind pattern has survived for at least three centuries.

Infrared observations probed the vertical structure of Jupiter's atmosphere

Infrared data from the Pioneer and Voyager spacecraft confirmed an extraordinary fact that had been deduced from Earth-based observations: Jupiter emits more energy than it receives from the Sun—in fact, nearly twice as much. Many scientists believe that this excess heat escaping from Jupiter is energy left over from the formation of the planet 4½ billion years ago. As gases from the solar nebula fell into the protoplanet, vast amounts of gravitational energy were converted into the thermal energy that still radiates into space.

In order for heat to flow upward through the Jovian atmosphere and radiate out into space, the temperature must be warmer deep inside the atmosphere and cooler at the cloudtops. Infrared measurements confirm that temperature rises with increasing depth in the Jovian cloud cover.

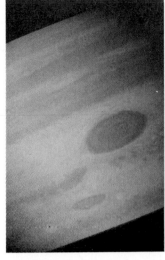
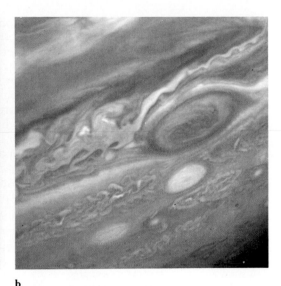

a   b

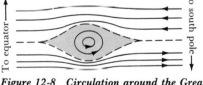

Figure 12-8 Circulation around the Great Red Spot
The Great Red Spot spins counterclockwise, completing a full revolution in about six days. Meanwhile, winds to the north and south of the spot blow in opposite directions. Consequently, circulation associated with the spot resembles a wheel spinning between two oppositely moving surfaces. (Adapted from Andrew P. Ingersoll)

Figure 12-9 Infrared and visible views
(a) *This infrared photograph was taken through the 200-in. Palomar telescope. The brightest parts of the image correspond to holes in the clouds where deeper and warmer regions of the Jovian atmosphere are visible. Dark parts of the image correspond to the cool cloudtops.* **(b)** *This image in visible light was taken by Voyager 1 at almost the same time the infrared picture was taken. Comparisons between the two photographs show that cloud color is correlated with depth in the Jovian atmosphere. The bluish and brown clouds are roughly 100°C warmer and 100 km lower than the red and whitish clouds.* (NASA)

This is analogous to the Earth's atmosphere, which is cool at the highest cloudtops (at the top of the troposphere) but is warmer near the ground.

Over the range of temperatures in the Jovian atmosphere, the gases emit energy primarily as infrared radiation. Figure 12-9 shows nearly simultaneous photographs of Jupiter at infrared and at visible wavelengths. In the infrared picture, the brighter parts of the image correspond to hotter temperatures. There is a striking correlation between brightness in the infrared image and color in the visible-light image. In other words, the differing colors in Jupiter's clouds correspond to differing temperatures and hence to differing depths in the atmosphere. (It is customary to discuss these features in terms of depths measured from the cloudtops rather than altitudes because the exact location of any solid surface on Jupiter is unknown.) Bluish clouds correspond to the brightest parts of the infrared picture, so these clouds must be the warmest and hence the deepest layers that we can see in the Jovian atmosphere. Brown clouds form the next highest layer, followed by whitish clouds, with the red clouds in the highest layer.

The Jovian atmosphere has a minimum temperature of about 110 K (= −160°C = −260°F) at an altitude above the cloudtops where the atmospheric pressure is about 100 millibars. By analogy with the Earth's

a   b

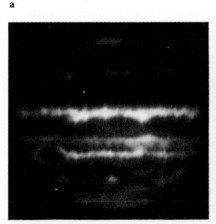

atmosphere (recall Figure 9-2), we can call this level the boundary between the stratosphere and the troposphere. As on the Earth, all of the weather on Jupiter takes place below the stratosphere. At 100 km below the top of the Jovian troposphere, conditions are fairly hospitable by human standards (see Figure 12-10).

From the observed spectral lines and the computed conditions of temperature and pressure, scientists conclude that the three Jovian cloud layers have different chemical compositions. The uppermost cloud layer (which contains the red and whitish clouds) is composed of crystals of frozen ammonia. About 25 km deeper in the troposphere, ammonia (NH_3) and hydrogen sulfide (H_2S) combine to produce ammonium hydrosulfide (NH_4SH) crystals. About 100 km down from the top of the troposphere, the layer of bluish clouds is composed of water droplets and snowflakes of frozen water.

Figure 12-10 The vertical structure of Jupiter's upper atmosphere
This graph displays the temperature and pressure profiles of Jupiter's upper atmosphere, as deduced from measurements at infrared and radio wavelengths. Three major cloud layers are shown, along with the colors that predominate at various depths.

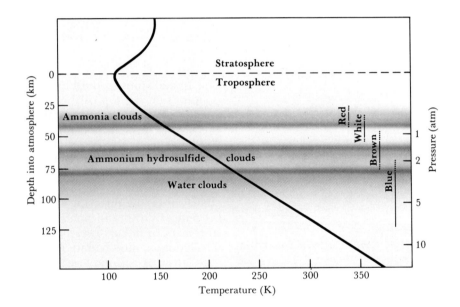

Studies using computers explain some of the phenomena we see in the Jovian clouds

Because the Great Red Spot appears dark in infrared photographs, we know that its cool clouds are located near the top of the Jovian cloud layers. The Great Red Spot is therefore a high pressure system.

Anyone familiar with weather forecasting knows that the Earth's weather is dramatically affected by high-pressure and low-pressure systems. A **high-pressure system** (commonly called a "high") is simply a place where a more-than-usual amount of air happens to be located. This excess air weighs down on the Earth's surface to produce high barometric readings (recall Box 9-2). On the other hand, a **low pressure system** (commonly called a "low") is a place where there is less-than-normal amount of air producing low barometric readings. If we could see air, highs would look like bulges in the atmosphere (where extra amounts of air are piled up), and lows would look like depressions, or troughs.

Gravity has a strong influence on the basic dynamics of the Earth's weather. Gravity causes the air to flow "downhill" from the high-pressure bulges into the low-pressure troughs. But because the Earth is rotating, the windflow from the highs toward the lows is not along straight

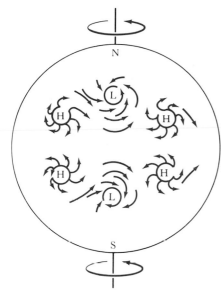

Figure 12-11 Cyclonic and anticyclonic wind flows
Because of a planet's rotation, cyclonic wind flowing into a low-pressure region or anticyclonic winds flowing out of a high-pressure region rotate either clockwise or counterclockwise, depending on the hemisphere in which the weather system is located.

lines. Instead, the Earth's rotation causes a deflection of the winds, producing either clockwise or counterclockwise flow about the high and low pressure regions. This deflection is the result of the **Coriolis effect** which occurs whenever motion is superimposed on rotation.

Imagine riding on a merry-go-round. If you throw a ball across the merry-go-round, you find that the ball is deflected to one side. The Coriolis effect explains that this deflection occurs because the platform from which you observe the ball's motion is rotating. In the same way, winds flowing into a low or out of a high are deflected from straight-line paths, resulting in either clockwise or counterclockwise motion.

Figure 12-11 shows the resulting wind flow about highs and lows. In the northern hemisphere, winds blowing toward a low-pressure region seem to rotate counterclockwise about the low, forming a **cyclone.** Winds blowing away from a high-pressure region seem to rotate clockwise about the high, forming an **anticyclone.** In the southern hemisphere, the directions of rotation are reversed: the cyclonic winds about a low rotate clockwise, and the anticyclonic winds about a high rotate counterclockwise.

This basic pattern of wind flow is the same on Earth, Jupiter, or any other rotating planet. Thus the six-day counterclockwise rotation of the Great Red Spot in Jupiter's southern hemisphere is consistent with infrared observations—the reddish cloudtops are cool and at a high elevation. The Great Red Spot is an anticyclone, a long-lasting high-pressure bulge in Jupiter's southern hemisphere.

The Voyager pictures showed other anticyclones in Jupiter's southern hemisphere. These features appear as **white ovals** (two such ovals are visible in Figure 12-7), indicating that their cloudtops are not quite as high as the Great Red Spot. Nevertheless, wind flow in these ovals clearly is counterclockwise.

Most of the white ovals are observed in Jupiter's southern hemisphere. In contrast, **brown ovals** are more common in Jupiter's northern hemisphere (see Figure 12-12). Whereas the white ovals are the high-

Figure 12-12 A brown oval
Large brown ovals in Jupiter's northern hemisphere are caused by openings in the main cloud layer that reveal warm, dark-colored gases below. Voyager 1 was 4 million kilometers from Jupiter when this picture was taken. The length of this oval is roughly equal to the Earth's diameter. (NASA)

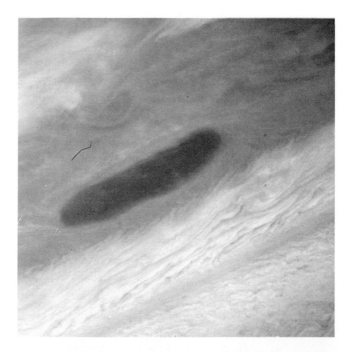

Figure 12-13   The northern and southern hemispheres
Computer processing was used to construct these views that look straight down onto Jupiter's north and south poles. (a) In the northern view, notice that light-colored plumes are evenly spaced around the equatorial regions. Several brown ovals are visible. (b) In the southern view, notice that the three biggest white ovals are separated from each other by almost exactly 90° longitude. In both views, the banded belt–zone structure is absent near the poles. The ragged black spot is an area not photographed by the spacecraft. (NASA)

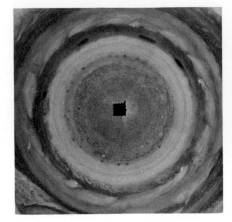

a   North Pole

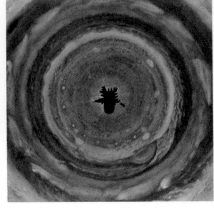

b   South Pole

altitude cloudtops of high-pressure systems, the brown ovals result from holes in Jupiter's cloud cover that permit us to see down into warmer regions of the Jovian atmosphere. Like the Great Red Spot, a white oval is apparently long-lived; Earth-based observers have reported seeing them in the same locations since 1938. However, a brown oval lasts for only a year or two. Computer-generated Figure 12-13 shows how Jupiter would look if you were located directly over the planet's north pole or south pole. Regular spacing of cloud features such as ripples, plumes, and light-colored wisps is also obvious.

Computer processing was also used to "unwrap" Jupiter, and producing a map like views of the planet such as Figure 12-14. Note the changes that occurred during the four months between the flybys of Voyager 1 and Voyager 2. Regular spacing of light-colored plumes is apparent in the equatorial regions of both pictures. The Great Red Spot moved westward whereas the white ovals moved eastward during the interval between the two flybys.

Figure 12-14   A comparison of Voyager 1 and Voyager 2 views
Computer processing was used to produce these two "unwrapped" views of Jupiter from (a) Voyager 1 and (b) Voyager 2. Each view was aligned with respect to Jupiter's magnetic axis so that displacements to the right or left represent real cloud motions. Notice that the Great Red Spot moved westward while the white ovals moved eastward. (NASA)

a   Voyager 1 view

b   Voyager 2 view

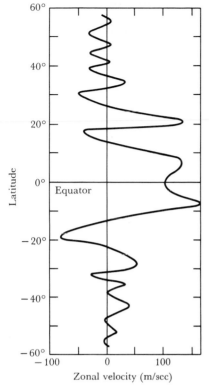

Figure 12-15 Prevailing zonal winds
*Jupiter's atmosphere is dominated by a persist-
ent pattern of counterflowing eastward and
westward winds. This graph shows wind velocities over latitudes ranging from 60° north to
60° south of the Jovian equator.*

These beautiful Voyager photographs might suggest that Jupiter's clouds are the result of incomprehensible turmoil. You may wonder if there is anything constant in the Jovian atmosphere. Surprisingly, there is. Telescopic observations over the past 80 years, along with the Voyager data, demonstrate that the wind speeds in the Jovain atmosphere are remarkably stable. Although Jupiter's colorful bands change quite rapidly, the underlying wind patterns do not.

Jupiter's persistent wind patterns consist of counterflowing eastward and westward winds. The speeds of these **zonal jets** are plotted in Figure 12-15. Although the winds near Jupiter's equator blow eastward with speeds slightly greater than 100 m/sec (224 miles per hour), reversals of wind direction occur in a regular pattern from the equator toward either pole. Many scientists suspect that this stationary pattern of alternating wind flow directions is caused by currents deep inside Jupiter.

Computer simulations involving whirlpools and eddies caught between counterflowing streams help us understand both long-term and short-term features in Jupiter's clouds. Andrew P. Ingersoll and his colleagues at the California Institute of Technology have pioneered these calculations. For example, Figure 12-16 shows the behavior of a small, unstable whirlpool. This whirlpool, technically called a **vortex,** is spinning too slowly to remain intact and so is torn apart by the counterflowing winds. Larger, rapidly rotating vortices do survive, however, in these simulations. The white ovals and the Great Red Spot endure by simply rolling with the wind currents (recall Figure 12-9). Figure 12-17 shows a simulation in which two stable vortices merge to form a larger stable vortex. The long-lived white ovals apparently maintain themselves in this fashion.

**Figure 12-16 [left]   The demise of an
unstable vortex**
*In this computer simulation, a small vortex is
rotating too slowly to remain intact. After
slightly more than one week, the vortex is
pulled apart between counterflowing zonal jets.
(Adapted from Andrew P. Ingersoll)*

**Figure 12-17 [right]   The merging and
maintenance of stable vortices**
*This computer simulation shows the collision
and merger of two rapidly spinning, stable
vortices. The result is a larger vortex, along
with the ejection of some material. Mergers of
this type occur around the Great Red Spot.
(Adapted from Andrew P. Ingersoll)*

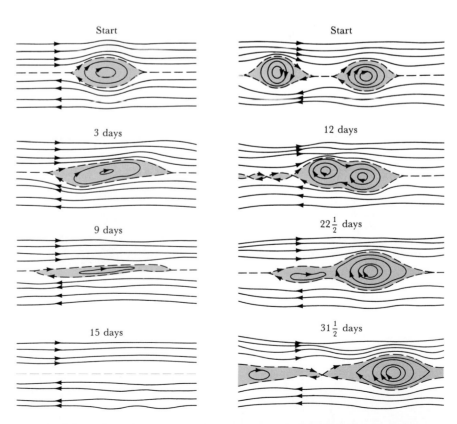

Many aspects of Jupiter's clouds, such as their colors, are still poorly understood

The 1970s saw a dramatic increase in our understanding and appreciation of the largest planet in the solar system, but many mysteries still remain. For example, we still do not know what gives Jupiter's clouds their colors. The crystals of ammonia, ammonium hydrosulfide, and frozen water in the three main cloud layers are all white. The browns, blues, reds, and oranges therefore must be due to other chemicals. Some scientists believe that sulfur, which can assume many different colors depending on its temperature, plays an important role in the clouds. Others think that phosphorus might be involved, especially in the Great Red Spot.

Perhaps the most intriguing proposal is that Jupiter's colors are due to organic compounds, possibly resulting from biological activity. Many of Jupiter's colors are similar to the colors of organic material created in the Urey–Miller experiments. As discussed in Chapter 9, these experiments involved passing electrical discharges (to simulate lightning in Earth's early atmosphere) through a mixture of methane, ammonia, hydrogen, and water vapor. These gases are abundant in Jupiter's atmosphere, as is ample evidence of lightning (see Figure 12-18). Many of the amino acids produced in the Urey–Miller experiments may have been brewing for 4 billion years in Jupiter's clouds. Is it possible that Jupiter's clouds harbor extraterrestrial life?

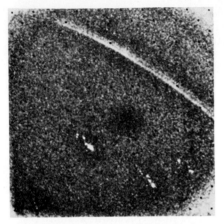

Figure 12-18 Aurorae and lightning in the Jovian night
This photograph of Jupiter's dark side is a 3-minute exposure from Voyager 1. A huge aurora arches across the northern horizon. In the lower half of the picture, flashes from about 20 enormous lightning bolts illuminate the clouds. (NASA)

Figure 12-19   The Galileo mission
The next U.S. mission to Jupiter will involve an atmospheric probe that will make direct measurements and chemical analyses as it descends through the Jovian cloud cover. Unfortunately, development of this sophisticated spacecraft has been severely delayed, by budget cutbacks. (NASA)

To explore these issues, the next United States mission to Jupiter will carry an atmospheric probe. Using a heat shield and a parachute, the probe will descend through the Jovian clouds, making a wide range of measurements and chemical analyses (see Figure 12-19). Just as the Voyager flybys revealed phenomena that stagger the imagination, the Galileo mission undoubtedly will provide a few answers and many new mysteries.

Summary

. Jupiter, with a mass of 318 Earth masses, is composed of 82 percent hydrogen, 17 percent helium, and only 1 percent all other elements. This chemical composition is very similar to that of the Sun.

Jupiter may have a rocky core with a mass of about 13 Earth masses; this core is surrounded by a 40,000-km-thick layer of liquid metallic hydrogen and outer layer of molecular hydrogen gas about 20,000 km thick.

The visible features of Jupiter (belts, zones, the Great Red Spot, ovals, and colored clouds) exist in the outermost 100 km of the molecular hydrogen layer.

The outer layers of the Jovian atmosphere show differential rotation, with the equatorial regions rotating slightly faster than the polar regions; the internal rotation rate determined from study of radio waves is nearly the same as the polar rotation rate.

Because of its rapid rotation, Jupiter is noticeably oblate.

. Jupiter has a very strong magnetic field created by currents in the metallic-hydrogen layer; its magnetosphere is huge and contains a vast current sheet of electrically charged particles.

The particles in the current sheet produce strong radio waves through the process of synchrotron radiation.

The Jovian magnetosphere encloses a plasma of charged particles that is hotter than the center of the Sun but has very low density; the magnetosphere exists in a delicate balance between pressures from the plasma and from the solar wind, and its size fluctuates drastically.

. The colored ovals visible in the Jovian atmosphere represent gigantic storms in the troposphere; some such as the Great Red Spot are very stable and persist for many years.

The ovals are cyclonic or anticyclonic storms created at the boundaries between zonal jets (wind streams) moving in opposite directions around the planet.

. Jupiter emits more heat than it receives from the Sun; presumably the planet is still cooling.

. There are three cloud layers in Jupiter's troposphere. The reasons for the distinctive colors of these different layers are not yet known.

. Some scientists speculate that life may exist in layers of the Jovian atmosphere where pressures and temperatures are not too different from those on the Earth; organic molecules almost certainly exist there.

Review questions

1 Why do the magnitudes of Jupiter at opposition (see Box 12-2) vary so little compared to the oppositions of Mars (see Box 11-2)?

2 Which planet, Mars or Jupiter, is easier to observe with an Earth-based telescope? Explain your answer.

3 If Jupiter does not have any observable solid surface and its atmosphere rotates differentially, how are astronomers able to determine the planet's rotation rate?

4 Why do astronomers believe that Jupiter does not have a large iron-rich core even though the planet possesses a strong magnetic field?

5 Explain the statement "Since the Great Red Spot appears dark in infrared photographs, we know that its cool clouds are located at a high elevation."

6 What data and techniques have been used to determine the internal structure of Jupiter?

7 Compare and contrast Jupiter's magnetosphere with the magnetosphere of a terrestrial planet like Earth. Why is the size of the Jovian magnetosphere variable whereas Earth's is not?

Advanced questions

*8 Find the location of the center of mass of the Sun–Jupiter system. Is it inside or outside the Sun?

9 Estimate the wind velocities in the Great Red Spot, which rotates with a period of about 6 days.

10 What sort of experiment would you design in order to establish whether Jupiter has a rocky core?

Discussion questions

11 Describe some of the semi-permanent features in Jupiter's atmosphere. What factors influence their longevity? Compare and contrast these long-lived features with some of the transient phenomena seen in Jupiter's clouds.

12 Suppose that you were designing a mission to Jupiter that involved an airplane-like vehicle that would spend many days (months?) flying through the Jovian clouds. What observations, measurements, and analyses should this aircraft make? What dangers might the aircraft encounter and what design problems would you have to overcome?

For further reading

Beatty, J. "The Far Out Worlds of Voyager 1." *Sky & Telescope,* May 1979, p. 423; June 1979, p. 516. "Voyager's Encore Performance." *Sky & Telescope,* Sept. 1979, p. 206.

Gore, R. "Voyager Views Jupiter." *National Geographic,* Jan. 1980.

Ingersoll, A. "Jupiter and Saturn." *Scientific American,* Dec. 1981.

Johnson, T., and Yeates, C. "Return to Jupiter: Project Galileo." *Sky & Telescope,* Aug. 1983, p. 99.

Morrison, D., and Samz, J. *Voyage to Jupiter.* NASA SP-439, 1980. *An excellent book on the Voyager mission as reported by one of the participants.*

Washburn, M. *Distant Encounters: The Exploration of Jupiter and Saturn.* Harcourt, Brace, Jovanovich, 1983.

The Galilean satellites of Jupiter

The Galilean satellites with our Moon and Titan

Jupiter's four large satellites are shown here along with our Moon and Saturn's largest moon, Titan. All six worlds are reproduced to the same scale. Io has numerous active volcanoes and Europa has a smooth icy surface; both are roughly the same size as our Moon. Ganymede and Callisto are each covered with a 1000 km thick layer of ice; both are roughly the same size as Mercury. The remaining large satellite of the solar system, Triton, is not shown in this montage; Voyager 2 will attempt to obtain photographs of Triton during a flyby of Neptune in 1989. (S. P. Meszaros; NASA)

Galileo's discovery in the early 1600s of four giant moons circling Jupiter provided the first proof of objects in the universe that definitely do not move around the Earth. As astronomers learned more about the Galilean satellites, they realized that these moons are about the same size as our Moon and Mercury. When the Pioneer and Voyager space probes passed near Jupiter in the 1970s, they sent back a wealth of surprising photographs and data about the Galilean satellites. In this chapter, we discuss the most prominent and interesting features of these terrestrial worlds—explosive volcanic eruptions, "oceans" of ice, and plate tectonics operating in layers of frozen water. We find that the Galilean satellites probably formed around Jupiter in a small-scale version of the process that created the solar system. We also learn more about the origin of Jupiter's radio emissions. The giant moons of Jupiter prove to be important in our understanding of processes on terrestrial planets, largely because they differ from the inner planets in so many ways.

Galileo Galilei called them the "Medicean Stars" to attract the attention of a wealthy Florentine patron of the arts and sciences. Since those fateful nights in 1610, the four giant moons of Jupiter have played an important role in our understanding of the universe. To Galileo, they were observational evidence supporting the heretical Copernican cosmology. For the modern astronomer, the Voyager flybys of 1979 revealed four extraordinary terrestrial worlds, different from anything astronomers had ever seen or even imagined. We call them the **Galilean satellites.** They are named after the mythical lovers and companions of Zeus: Io, Europa, Ganymede, and Callisto.

The Galilean satellites are easily seen with Earth-based telescopes

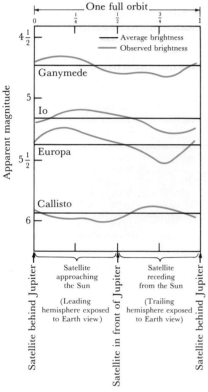

Figure 13-1   Brightnesses of the Galilean satellites
The apparent magnitude of each Galilean satellite varies with a period equal to its orbital period. This means that all four satellites orbit about Jupiter with synchronous rotation. (Based on observations by Dale Cruikshank and David Morrison)

When viewed through an Earth-based telescope, the Galilean satellites look like mere pinpoints of light. With patience, you can follow these four worlds as they orbit Jupiter. Because their orbital periods are fairly short (ranging from 1.8 days for Io to 16.7 days for Callisto), major changes in the positions of the satellites are easily noticed from one night to the next. By consulting a current issue of *The Astronomical Almanac,* you can even schedule your observations to include transits, eclipses, and occultations. When one of the moons passes between Jupiter and the Sun in a transit, the satellite's shadow is seen as a black dot against the planet's colorful cloudtops. In an eclipse, one of the moons seems suddenly to disappear and later reappear as it moves behind Jupiter, passing into and later out of the planet's enormous shadow. In an occultation, a satellite passes in front a star blocking its light from Earth-based view. Because the moons' apparent magnitudes range from about 4.7 for Ganymede to nearly 6 for Callisto, you might think that all four satellites should be visible to the naked eye. A telescope almost always is necessary, however, because of the overwhelming glare of Jupiter.

Although the Galilean satellites appear as tiny stars, Earth-based techniques have been used to measure their diameters. For example, when a satellite emerges from Jupiter's shadow, it does not blink on instantly. Instead, there is a very brief time interval during which the satellite gets brighter and brighter as more and more of its surface is exposed to sunlight. The duration of this interval depends on the orbital speed of the satellite (which is known from Kepler's laws) and the diameter of the satellite. Consequently, by accurately measuring the time it takes for a satellite to move into or out of Jupiter's umbra, we are able to calculate the satellite's diameter.

Stellar occultations have been used in a similar fashion. By measuring the time it takes for one of the satellites to pass in front of a background star we can calculate the satellite's diameter. In addition, Galilean satellites occasionally occult each other. All four satellites orbit Jupiter in the plane of the planet's equator. Every six years, the Earth passes through this equatorial plane; for a few days, mutual occultations of the satellites occur. Once again timing the occultations enables us to calculate the satellite diameters.

The brightness of each of the moons varies slightly as it moves along its orbit. These variations can logically be attributed to dark and light areas on the surface that are alternately exposed to or hidden from our view as the satellite rotates. Careful measurements showed that the apparent magnitude of each satellite varies with a period equal to the satellite's orbital period (see Figure 13-1). In other words, each Galilean satellite rotates exactly once on its axis during each trip around its orbit. Hence each Galilean satellite has synchronous rotation, keeping the same hemisphere perpetually facing Jupiter just as our Moon keeps the same side facing the Earth.

Data from spacecraft greatly improved our knowledge of the Galilean satellites

Very accurate measurements of the diameters of the Galilean satellites came from the Voyager flybys because direct photography revealed measurable disks (see Figure 13-2). These measurements confirmed and refined the data obtained from Earth. The two inner Galilean satellites (Io and Europa) are approximately the same size as our Moon. The two outer satellites (Ganymede and Callisto) are comparable in size to Mercury. The Voyager data (accurate to better than ±10 km) are listed in Box 13-1.

Box 13-1 The Galilean satellites

Of the 17 known satellites that orbit Jupiter, four are large enough to be classified as terrestrial worlds. The table lists basic data about these four worlds; data about Mercury and our Moon are included for comparison.

Name	Mean distance from Jupiter (km)	Sidereal period (days)	Diameter (km)	Mass (kg)	Mass (Moon = 1)	Mean density (g/cm^3)
Io	412,600	1.77	3,632	8.92×10^{22}	1.21	3.55
Europa	670,900	3.55	3,126	4.87×10^{22}	0.66	3.04
Ganymede	1,070,000	7.16	5,276	1.49×10^{23}	2.03	1.94
Callisto	1,880,000	16.69	4,820	1.06×10^{23}	1.44	1.81
Mercury	——	——	4,878	3.30×10^{23}	4.49	5.42
Moon	——	——	3,476	7.35×10^{22}	1.00	3.34

As early as the 1920s, fairly accurate determinations of the masses of the Galilean satellites were made from Earth. The satellites often pass near each other as they orbit Jupiter. During these encounters, gravitational interactions between the satellites cause slight perturbations of their orbits. By measuring these tiny orbital deflections and using Newtonian mechanics, astronomers were able to calculate the masses of the satellites. Europa was discovered to be the least massive (its mass is two-thirds that of our Moon). Ganymede is by far the most massive member of the quartet, with at least twice the mass of our Moon. As you might expect, the Pioneer and Voyager flybys produced the best determinations of the satellite masses. Data calculated from deflections of the spacecraft trajectories are given in Box 13-1.

As soon as reliable mass and diameter measurements were available, it became apparent that the average densities of the satellites are related to their distances from Jupiter. The innermost satellite, Io, has the highest average density (3.55 g/cm^3), which is slightly more dense than our Moon. The next satellite, Europa, has an average density of 3.05 g/cm^3 which is not quite as dense as our Moon. Recalling that typical rocks in

Figure 13-2 Io and Europa
The Galilean satellites look like pinpoints of light when viewed through an Earth-based telescope, but the Earth-orbiting Space Telescope will produce views of Jupiter comparable to this photograph from Voyager 1. Surface features as small as 400 km across are visible. Io (left) and Europa (right) are each approximately the same size as our Moon. The Space Shuttle is scheduled to carry the Space Telescope aloft in 1986. (NASA)

the Earth's crust have densities around 3 g/cm³, we find it reasonable to suppose that both Io and Europa are made primarily of rocky material.

The outer two satellites also exhibit decreasing density with increasing distance from Jupiter. As listed in Box 13-1, both Ganymede and Callisto have average density less than 2 g/cm³, indicating that a sizable fraction of each world is made of something less dense than rock. Figure 13-3 includes views of the four satellites sent back by Voyager 1.

Figure 13-3   The Galilean satellites
Io and Europa have diameters and densities comparable to our Moon and are composed primarily of rocky material. Although Ganymede and Callisto are roughly as big as Mercury, their average densities are low. Each of these outer satellites is covered with a thick layer of water and ice. (NASA)

Io   Europa

Ganymede   Callisto

The formation of the Galilean satellites probably mimicked the formation of the solar system

The arrangement of the Galilean satellites parallels the arrangement of the planetary characteristics when grouped according to distance from the Sun. Moving outward from the Sun, average density declines from more than 5 g/cm³ for Mercury to less than 1 g/cm³ for Saturn (recall Table 6-2). Scientists therefore began to suspect that the same general processes that formed the solar system were at work during the formation of the Galilean satellites, although on a much smaller scale.

The low densities of Ganymede and Callisto suggest that these two satellites are composed of roughly equal amounts of rock and ice. Indeed, in the early 1970s, American astronomer John Lewis pointed out that these low densities are exactly what one would expect for relatively small objects formed by the accretion of ice-covered dust grains. More recently, NASA scientists James Pollack and Fraser Fanale constructed theoretical models for the formation of the Galilean satellites, including in their simulation the fact that Jupiter emits twice as much energy as it receives from the Sun. They calculated that frozen water could be retained and incorporated into satellites at the distances of Ganymede and Callisto, but that only rocky material would condense at the orbits of Io and Europa because of Jupiter's warmth. Thus Jupiter's gravity and heat produced two distinct classes of Galilean satellites, just as warmth from the protosun caused the dichotomy between the small, dense, rocky inner planets and the huge, gaseous, low-density outer planets.

Confirmation of water ice on the Jovian satellites came from Earth-based spectroscopic observations during the early 1970s. Various teams of astronomers measured the intensity of reflected sunlight at wavelengths from 0.3 to 5.3 μ. The spectra of Europa and Ganymede exhibited strong absorptions at 1.5 and 2.0 μ (see Figure 13-4). Scientists recognized these spectral lines as the pattern characteristic of water-ice molecules. Although Callisto also was expected to have an icy surface, the ice spectral lines were much weaker in its spectrum. Indeed, Callisto is the dimmest of the Galilean satellites, reflecting much less light over a wide range of wavelengths than does any of its companions. Scientists therefore began to suspect that Callisto is covered with a dark, dusty coating of meteoritic material that subdues the reflective properties of the underlying layer of ice.

Calculations and observations prior to the Voyager flybys suggested that Io might be volcanically active

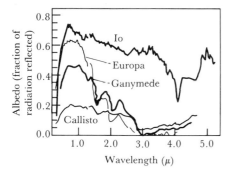

Figure 13-4 Infrared spectra of the Galilean satellites
Reflected sunlight from Europa and Ganymede shows prominent absorption at infrared wavelengths due to water ice. Notice that Io's spectrum differs radically from the other three, strongly suggesting that the innermost Galilean satellite is unusual. The strong absorption at 4.1 μ in Io's spectrum is probably produced by frozen SO_2. The depressed reflectivity of Callisto is due to a dark coating, perhaps meteoritic dust. (Adapted from Roger Clark and Thomas McCord)

The quality of Earth-based observations of the Galilean satellites improved dramatically during the 1970s. Infrared spectra provided the first indication that Io is a very unusual world. In 1974, Robert Brown of the University of Arizona announced his detection of sodium atoms in the vicinity of Io. These atoms were soon discovered to be part of a vast, tenuous cloud of material that envelops the satellite and extends for tens of thousands of kilometers along its orbit. Only three scientists had the insight, however, to recognize the unusual circumstances that profoundly affect Io.

Three days before Voyager 1 flew past Io, Stanton J. Peale of the University of California at Santa Barbara an Patrick M. Cassen and Ray T. Reynolds of NASA's Ames Research Center published calculations indicating that Io is acted upon by tremendous tidal forces. As it orbits Jupiter, Io is repeatedly caught in a gravitational tug-of-war between the huge planet on one side and the other Galilean satellites on the other. This gravitational battle distorts Io's orbit, causing the satellite to vary its distance from Jupiter. As the distance between Io and Jupiter varies, tidal stresses on Io squeeze and flex the satellite. This constant tidal flexing in turn causes frictional heating of Io's interior.

Calculations show that heat pumped into Io in this fashion could be as great as 10^{13} watts, equivalent to 2400 tons of TNT exploding every second. This energy must eventually make its way to the satellite's surface, so Peale, Cassen, and Reynolds predicted that "widespread and recurrent volcanism" should exist on Io. Although the paths of the Voyager spacecraft were designed to give maximal photographic coverage of the surfaces of the Galilean satellites (see Box 13-2), no one expected to obtain photographs of erupting volcanoes on Io. After all, a spacecraft making a single trip past the Earth would be very unlikely to catch a large volcano in the act of eruption.

Box 13-2   The Voyager flybys

From Earth-based observations, all four Galilean satellites were known to keep one side constantly facing Jupiter, just as the same side of our Moon constantly faces Earth. With this fact in mind, the trajectories of the two Voyager spacecraft were planned to maximize pictorial coverage of the satellites' surfaces. For example, as Voyager 1 approached Jupiter on March 5, 1979, it photographed Io's outward-facing and trailing hemisphere. After

(continued)

(Box 13-2, continued)

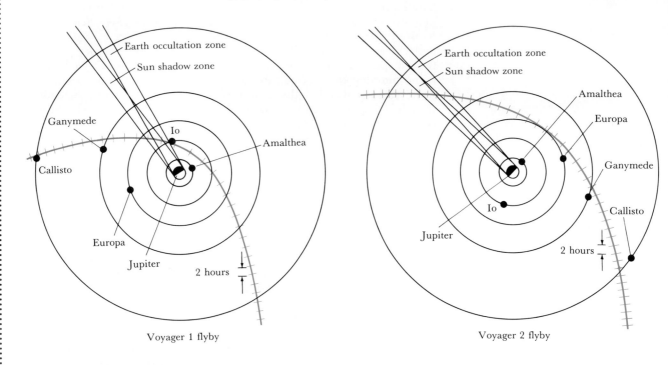

Voyager 1 flyby   Voyager 2 flyby

passing within 277,000 km of Jupiter, the spacecraft went on to view the Jupiter-facing hemispheres of Ganymede and Callisto, flying past them at high latitudes so that their north poles could be examined.

As shown in the diagram, Voyager 2 encountered Callisto and Ganymede before reaching Jupiter on July 9, 1979. The spacecraft was therefore able to photograph the outward-facing hemispheres of these two worlds, primarily from southern latitudes. Consequently, nearly 80 percent of the surfaces of Ganymede and Callisto were photographed with a resolution of 5 km or better.

Europa, whose Jupiter-facing side had been seen from afar by Voyager 1, was photographed at a much closer distance by Voyager 2. One-fourth of Europa's surface was viewed with a resolution of 5 km or better.

Io and Amalthea revolve about Jupiter so rapidly that Voyagers 1 and 2 together photographed both satellites with intermediate resolution (20 km) at all latitudes. Some details of the flybys are given in the table.

Satellite	Closest approach (km)		Best resolution (km)	
	Voyager 1	**Voyager 2**	**Voyager 1**	**Voyager 2**
Amalthea	420,000	558,270	8	11
Io	18,640	1,127,920	1	21
Europa	732,270	204,030	33	4
Ganymede	112,030	59,530	2	1
Callisto	123,950	212,510	3	4

The Voyager probes discovered several small moons and a ring around Jupiter

The first moon to be photographed at close range by Voyager 1 was a tiny reddish satellite called Amalthea, one of Jupiter's many small moons. Amalthea circles the planet in only 11.7 hours along an orbit much closer to Jupiter than that of Io (see diagram in Box 13-2). Amalthea proved to be an irregularly shaped object similar to a large asteroid (see Figure 13-5). It measures about 270 km along its longest axis and 155 km in the shortest direction. Amalthea therefore is about ten times larger than the moons of Mars but ten times smaller than the smallest Galilean satellites. Like the Galilean satellites, Amalthea orbits Jupiter with synchronous rotation, keeping its longest axis pointed toward the planet.

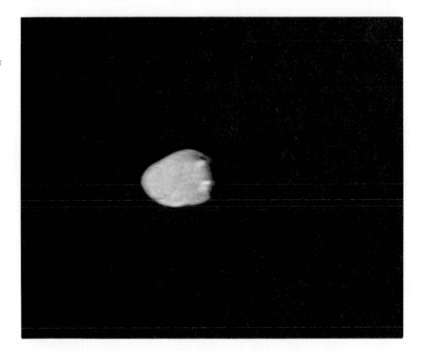

Figure 13-5   Amalthea
Tiny, reddish Amalthea is one of Jupiter's innermost satellites. It is an irregularly shaped, asteroidlike object measuring 270 km along its longest dimension. This photograph was taken by Voyager 1 at a distance of 425,000 km and has a resolution of 8 km. Amalthea was discovered in 1892 by the American astronomer E. E. Barnard at the Lick Observatory. (NASA)

A total of 16 satellites are now known to orbit Jupiter (see Box 13-3); three were discovered during the Voyager flybys. These newly discovered moons along with Amalthea and the Galilean satellites all orbit Jupiter in the plane of the planet's equator. In contrast, the remaining eight moons are all extremely tiny (estimated diameters typically less than 50 km) and circle Jupiter along very large orbits that are inclined at steep angles to the planet's equatorial plane. Furthermore, the outermost four satellites all have retrograde orbits. Because of the oddities of their orbits, it is reasonable to suppose that the eight outer moons may be wayward asteroids captured by Jupiter's powerful gravitational field.

In addition to discovering several moons, the Voyager cameras also revealed a faint ring around Jupiter (see Figure 13-6). This ring, which is probably composed of dust and rock fragments, lies in Jupiter's equatorial plane, even closer to the planet than the orbit of Amalthea. The sharp outer boundary of the ring is only 1.81 Jupiter radii from the planet's center. This ring, the small inner moons, and the Galilean satellites form a system of orbiting material that apparently is a common feature of the Jovian planets. As we shall see in the next two chapters, both Saturn and Uranus also have rings and families of satellites whose orbits lie in their planets' equatorial planes.

Box 13-3 Jupiter's family of moons

A total of 16 satellites are known to orbit Jupiter. Nearest Jupiter are four small moons. Next come the four giant Galilean satellites. The remaining eight moons are all very tiny and are located at large distances from Jupiter. All 16 satellites are listed in the table with data about their orbits.

Name	Average radius of orbit (km)	Average radius of orbit (Jupiter radii)	Period (days)	Year of discovery
Metis	128,000	1.80	0.30	1981
Adrastea	130,000	1.81	0.30	1979
Amalthea	181,000	2.55	0.49	1892
Thebe	222,000	3.11	0.67	1980
Io	422,000	5.95	1.77	1610
Europa	671,000	9.47	3.55	1610
Ganymede	1070,000	15.10	7.15	1610
Callisto	1880,000	26.60	16.70	1610
Leda	11,110,000	156	240	1974
Himalia	11,470,000	161	251	1904
Lysithea	11,710,000	164	260	1938
Elara	11,740,000	165	260	1904
Ananke	20,700,000	291	617R	1951
Carme	22,350,000	314	692R	1938
Pasiphae	23,300,000	327	735R	1908
Sinope	23,700,000	333	758R	1914

NOTE: *The annotation R for the periods of the outermost four satellites indicates that these moons move in a retrograde direction about Jupiter.*

Figure 13-6 Jupiter's ring

A portion of Jupiter's faint ring is seen in this photograph from Voyager 2. The ring is closer to Jupiter than any of the planet's satellites and is probably composed of tiny rock fragments. The brightest portion of the ring is about 6000 km in width. The outer edge of the ring is sharply defined, but the inner edge is somewhat fuzzy. A tenuous sheet of material extends from the ring's inner edge all the way down to the planet's cloudtops (NASA)

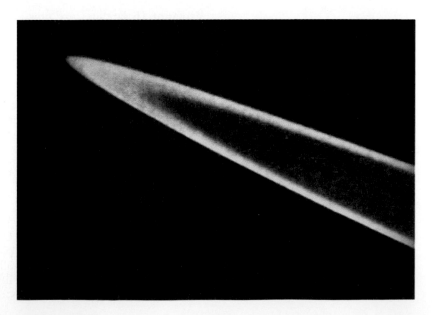

Io is covered with colorful deposits of sulfur compounds ejected from numerous active volcanoes

Within a few hours after Voyager 1 passed Amalthea, Io loomed into its view, and the probe began sending back a series of strange and unexpected pictures such as the one shown in Figure 13-7. Baffled by what they were seeing, scientists at the Jet Propulsion Laboratory in Pasadena (where the pictures were being received and enhanced by computer processing) jokingly compared Io to pizzas and rotten oranges.

A major clue to these puzzling vistas was uncovered several days after the Jupiter flyby when Linda Morabito, a navigation engineer at JPL, noticed a large umbrella-shaped cloud protruding from Io in one photograph. She had discovered an erupting volcano. Careful reexamination of the close-up photographs revealed a total of eight giant eruptions. Thus was the brilliant Peale–Cassen–Reynolds prediction of "widespread and recurrent volcanism" confirmed.

The volcanoes on Io are named after gods and goddesses traditionally associated with fire in Greek, Norse, Hawaiian, and other mythologies. Figure 13-8 shows two views of the symmetric plume of Prometheus.

The plumes and fountains of material spewing from Io's volcanoes rise to astonishing heights of 70 to 280 km above the satellite's surface. To reach these altitudes, the material must emerge from the volcanic vents with speeds between 300 and 1000 m/sec. That is much greater than the ejection speeds from the most violent terrestrial volcanoes. For example, Vesuvius, Krakatoa, and Mount St. Helens have eruption velocities of only around 100 m/sec. Scientists therefore began to suspect that Io's volcanoes operate in a fundamentally different way than do volcanoes here on Earth. Evidence of such differences was found in the Voyager pictures and data.

No impact craters like those on our Moon were seen on Io, indicating that the satellite's surface is extremely young—perhaps less than 100 million years old. Material from the volcanoes apparently obliterates impact craters soon after they are created.

The Voyager cameras did reveal numerous black dots on Io, which apparently are the volcanic vents from which the eruptions occur. These black spots are typically ten to fifty km in diameter, and they cover five percent of Io's surface. Lava flows radiate from many of these black dots (see Figure 13-9), some of which are located at the origins of volcanic plumes.

Evidence supporting the volcanic nature of the black spots came from infrared interferometer spectrometers carried on each of the Voyager

Figure 13-7   Io
This close-up view of Io was taken by Voyager 1 at a range of about 860,000 km. Notice the extraordinary range of colores from white, yellow, and orange to black. Scientists believe that these brilliant colors are due to surface deposits of sulfur ejected from Io's numerous volcanoes. (NASA)

Figure 13-8   Prometheus on Io
These two views from Voyager 1 were taken 2 hours apart and show details of the plume of the volcano called Prometheus. The plume's characteristic umbrella shape is silhouetted against the blackness of space. When viewed against the light background of Io's surface, jets of material give the plume a spiderlike appearance. This plume rises to an altitude of 100 km above Io's surface. (NASA)

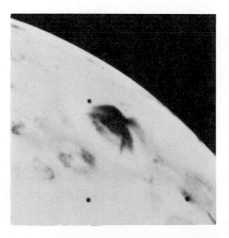

Figure 13-9   A volcanic center on Io
No impact craters are seen in close-up pictures such as this one taken by Voyager 1 at a range of 130,000 km. Long, meandering lava flows radiate from many of the black dots that apparently are the sites of intense volcanic activity. This photograph covers an area measuring 1000 by 800 km, approximately twice the size of California. (NASA)

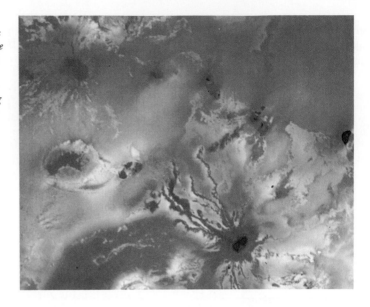

spacecraft. By measuring the intensity of infrared radiation across Io's surface, scientists discovered that some of the black spots have temperatures as high as 20°C, in sharp contrast to the surrounding surface temperature of only − 146°C. In fact, Voyager's cameras captured the actual onset of a minor eruption (see Figure 13-10).

After the discovery of widespread volcanic activity on Io, scientists soon concluded that sulfur ejected from the volcanoes is responsible for the satellite's brilliant colors. Sulfur is normally bright yellow. If heated and then suddenly cooled, however, it can assume a range of colors from orange and red to black.

The infrared spectrometer on Voyager 1 detected sulfur dioxide (SO_2) in the plumes of material erupting from Io's volcanoes. Sulfur dioxide is an acrid gas commonly discharged from fumaroles and volcanic vents here on Earth. When this gas is released into the cold vacuum of space from eruptions on Io, it crystallizes into white snowflakes. It is likely that the whitish deposits on Io (see Figure 13-7) are frozen sulfur dioxide (SO_2 frost or snow).

Figure 13-10   The venting of gases on Io
These two photographs were taken 6 hours apart by Voyager 1. Bright bluish patches appeared over a black crescent-shaped feature. These blue-white spots are probably caused by sulfur dioxide gas escaping from Io's interior. (left) This view was taken at a range of 374,000 km. (right) This clearer photograph was taken at a distance of 130,000 km. (NASA)

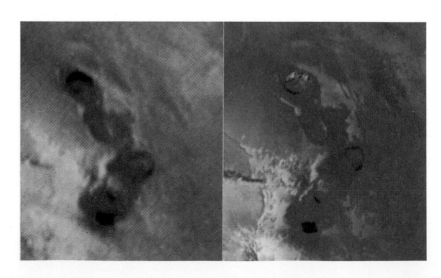

The abundant sulfur and sulfur dioxide suggest an explanation of the mechanism of Io's volcanoes. Indeed, the term "volcano" may be the wrong word altogether.

We have seen that material is ejected from volcanic vents on Io at much higher velocities than those observed in even the most explosive volcanic eruptions here on Earth. Another major difference is what Io's volcanic vents are not located at the tops of volcanic mountains. Many volcanoes on Earth and Mars have an easily recognized conical shape with a caldera at the summit. Few of Io's calderas (the black spots) are associated with any major topographical relief.

In all of these regards, Io's volcanoes are more similar to terrestrial geysers, than to terrestrial volcanoes. In a geyser, such as those in Yellowstone Park in Wyoming, water seeps down to volcanically heated rocks, is suddenly changed to steam, and erupts explosively through a vent. Geologist Susan Kieffer calculates that if Old Faithful geyser were erupting under the low gravity and vacuum that surround Io, the geyser would send a plume of water and ice to an altitude of 40 km.

Both sulfur and sulfur dioxide are molten at depths of only a few kilometers below Io's surface because of the heat generated by the tidal flexing of the satellite. Planetary geologists Eugene Shoemaker and Bradford Smith have pointed out that sulfur dioxide could be the principal propulsive agent driving Io's eruptions. Just as the explosive conversion of water into steam produces a geyser on Earth, the sudden conversion of liquid sulfur dioxide into a high-pressure gas could produce an eruption on Io. Specific calculations by Susan Kieffer indicate that this explosive expansion of sulfur dioxide could result in eruption velocities up to 1000 m/sec.

The material erupting from Io's geyserlike volcanoes is composed primarily of sulfur and sulfur dioxide. Voyager's instruments failed to detect any other abundant gases, such as the water vapor and carbon dioxide that are emitted from terrestrial volcanoes. Apparently, Io has been completely outgassed by volcanic activity extending over hundreds of millions of years. Io's surface gravity is comparable to that of our Moon, so Io has not been able to retain its volatile gases; the satellite has almost no atmosphere. The atmospheric pressure on Io, primarily due to sulfur dioxide, is a scant 10^{-4} millibars.

It is estimated that each volcano on Io ejects roughly 10,000 tons of material per second. Although this material is a hot mixture of molten sulfur and sulfur dioxide gas under high pressure as it gushes from a volcanic vent, the gas–liquid mixture rapidly cools and solidifies in the cold, nearly perfect vacuum around the satellite. It then takes about one-half hour for the fine particles of sulfur dust and sulfur dioxide snow to fall back down onto Io's surface.

Altogether, Io's volcanoes and vents eject an estimated 100 billion tons of matter each year. This represents deposition of a sulfur-rich layer of average thickness 10 m over Io's entire surface each year. Io's surface is therefore constantly changing, and it is probably safe to say that there are no long-lived or even semipermanent features on the satellite. This rate of deposition is easily sufficient to obliterate rapidly any impact craters.

Although nearly 100,000 tons of sulfur and sulfur dioxide erupts from Io each second, not all of this matter returns to the satellite's surface. Apparently a small fraction—perhaps 10 tons per second— manages to escape Io's gravity and become part of Jupiter's magnetosphere. High temperatures in the magnetosphere easily strip one or more electrons from each sulfur or oxygen atom. The result is a huge doughnut-shaped ring called the **Io torus,** circling Jupiter at the distance of Io's orbit, that contains a plasma composed primarily of electrons and ions of sulfur and oxygen. Although these ions are not plentiful, they produce radiation that had been detected by Earth-based astronomers several years before the Voyager fly-bys. The sodium cloud that Robert Brown discovered in 1974 is only a small part of this vast torus.

The Voyager flybys also illuminated Io's role in the bursts of decametric radio radiation (that is, radiation with wavelengths of tens of meters) that come from Jupiter. Jupiter rotates once every ten hours, while Io takes 1.77 days to complete an orbit. Thus Jupiter's magnetic field constantly sweeps past Io at a high speed. Because of this rapid motion through the magnetic field, a complex and powerful electromagnetic interaction is set up between Jupiter and Io. Io develops a strong electric charge, and an electric current flows between the satellite and Jupiter. Although the details are not yet fully understood, this electric current is believed to be responsible for the decametric bursts first observed by Bernard Burke and Kenneth Franklin in 1955.

Europa is covered with a smooth layer of ice, crisscrossed with numerous cracks

Voyager 1 did not pass near Europa, but Voyager 2 captured the excellent view shown in Figure 13-11. Europa is a very smooth world (no mountains and very few craters) crisscrossed with a spectacular series of streaks and cracks. Most of the cracks appear to be filled with dark-colored material, but some are filled with a light-colored substance.

We have seen that Europa's average density is about 10 percent less than the average density of our Moon. Spectroscopic observations from Earth indicated that Europa has frozen water on its surface. These two facts suggest that Europa's surface may be covered with an ice layer 100 km thick. That would be consistent with the remarkable smoothness of Europa's surface. An "ocean" of ice 100 km deep would certainly hide mountain ranges and other topographic features. But what causes the network of cracks? And why are impact craters so rare?

We have learned that tidal squeezing is responsible for volcanism on Io, which in turn gives the satellite many of its extraordinary characteris-

Figure 13-11 Europa
Europa's ice surface is covered by numerous streaks and cracklike features that give the satellite a fractured appearance. The streaks are typically 20 to 40 km wide. This picture was taken by Voyager 2 at a distance of 240,000 km; surface features as small as 5 km across can be seen. (NASA)

tics. Europa is caught in a similar tidal tug-of-war, with Jupiter to one side and the two largest Galilean moons periodically passing on the opposite side. However, the tidal effects of Jupiter on Europa are considerably less than those on Io. Whereas the gravitational force of Jupiter is inversely proportional to the square of the distance from the planet (recall Box 4-3), the corresponding tidal force is inversely proportional to the cube of the distance. In other words, because Europa is about 1.6 times farther from Jupiter than Io is, the tidal effects on Europa are $1/(1.6)^3$, or only one-quarter, as great as those on Io.

NASA scientist Partrick Cassen has suggested that the tidal flexing of Europa is responsible for the network of cracks that cover its surface. Indeed, some of the darkest streaks follow paths along which the tidal stresses are calculated to be strongest. Although the tidal flexing of Europa is far too weak to produce volcanoes, it is thought to supply enough energy to jostle and churn the satellite's icy coating. This would explain why only a few small impact craters have survived to the present time, even though Europa's surface is much older than Io's.

There are arguments against Cassen's theory. Presumably, the streaks on Europa are caused by cracks in the icy coating through which water gushed up and then froze. These cracks are a few tens of kilometers wide. To accommodate all the cracks, Europa's surface area would have increased by 10 to 15 percent since the cracks first began to appear. It seems unreasonable to suppose that Europa is expanding like an inflating balloon. Apparently, more is happening on Europa than meets the eye. Perhaps old surface material is somehow pulled back down into the mushy layer below the ice coating and recycled, just as the Earth's crust is pulled back down into the Earth's mantle in subduction zones (recall Figure 9-18). Thus Europa's surface may represent a water-and-ice version of plate tectonics.

Ganymede and Callisto have heavily cratered, icy surfaces

Whereas only 100 km of ice and water on top of an otherwise rocky world suffice to explain Europa's density, a much thicker layer of water and ice must surround Ganymede and Callisto. To be consistent with average densities slightly less than 2 g/cm^3, the rocky cores of these two outer satellites must be enveloped in mantles of water and ice nearly 1000 km thick. The diagrams in Figure 13-12 show the probable interior structures of all four Galilean satellites.

The two outer Galilean satellites exhibit the kind of ancient, cratered surface normally associated with Moon-like landscapes. Indeed, Ganymede even looks somewhat like our Moon (see Figure 13-13). Of course, the craters on both Ganymede and Callisto are made of ice rather than rock.

Ganymede is the largest satellite in the solar system. Its diameter is 5270 km which is slightly greater than the diameter of Mercury. Although Ganymede looks vaguely like our Moon, there are significant differences. Ganymede has two very different kinds of terrain that are distinguished by both appearance and age (see Figure 13-14). Dark, polygon-shaped regions are presumed to be the oldest surface features because they exhibit a high density of craters. Light-colored, heavily grooved terrain found between the dark angular islands is much less cratered and is therefore younger.

The largest single feature on Ganymede is a vast, dark, circular island of ancient crust called Galileo Regio (see Figure 13-13). It measures 4000 km in diameter and covers nearly one-third of the hemisphere of Ganymede that faces away from Jupiter. It is the only surface feature on the Galilean satellites that can be detected with Earth-based telescopes.

It is easy to distinguish between young and old craters on Ganymede. The youngest craters are surrounded by bright rays of freshly exposed ice (see Figure 13-14). Older craters clearly have been covered with deposits of dark meteoritic dust. The most ancient craters (sometimes called ghost craters or palimpsests—an archaeological term referring to a parchment that was scraped clean and written over again) are barely visible on Galileo Regio and other dark islands of old crust. The degradation and near obliteration of the oldest craters probably also involved the slow plastic flow of Ganymede's icy surface.

The Voyager photographs also show small craters on Ganymede to be better preserved than large ones. Eugene Shoemaker of the U.S. Geological Survey points out that the preservation of craters is linked to the thermal and structural history of the satellite's icy crust. When Ganymede formed $4\frac{1}{2}$ billion years ago, it may have been completely covered with an ocean of liquid water roughly 1000 km deep. During the next 200 million years, the water cooled and a thick coating of ice developed. Today this layer of solid ice is probably about 100 km thick; beneath it lies a 900-km-thick sloshy mantle of water and ice. The dark, angular islands are ancient remnants of the original crust.

Craters also tell us about the history of the younger, light-colored terrain. This terrain is covered with numerous grooves, and cratering varies in density from about the same as on the ancient crust down to one-tenth that amount. We thus suspect that this grooved terrain was formed over a long time period. The process probably began very early in the satellite's history and continued through the period of intense meteoritic

Figure 13-12   Interiors of the Galilean satellites
These cross-section diagrams show the probable internal structures of the four Galilean satellites, based on their average densities and information from the Voyager flybys.

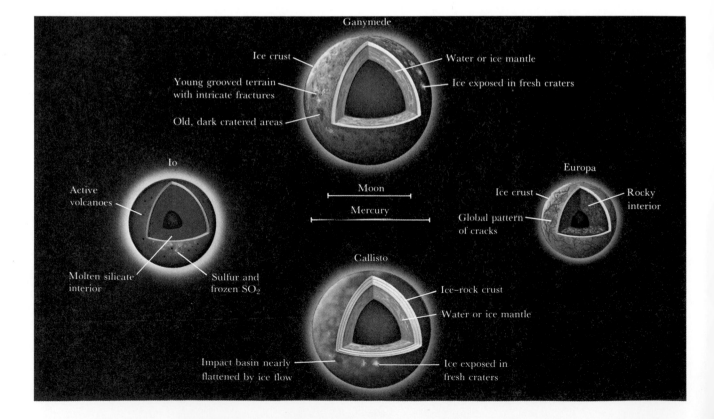

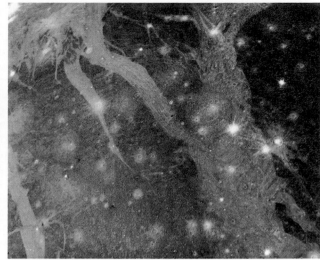

Figure 13-13 [left] Ganymede
This view from Voyager 2 was taken at a distance of 1.2 million kilometers and shows the hemisphere that always faces away from Jupiter. This hemisphere is dominated by a huge, dark circular region called Galileo Regio, which is the largest remnant of Ganymede's ancient crust. (NASA)

Figure 13-14 [right] Young and old terrain on Ganymede
This close-up view of Ganymede was taken by Voyager 2 at a range of only 312,000 km. Features as small as 5 km across can be seen. Dark, angular islands of Ganymede's ancient crust are separated by younger, light-colored, grooved terrain. The southwest edge of Galileo Regio appears at the right side of the picture. (NASA)

bombardment. The age of the grooved terrain therefore ranges from about $4\frac{1}{2}$ to $3\frac{1}{2}$ billion years.

High-resolution photographs such as Figure 13-15 show that this grooved terrain actually consists of parallel mountain ridges up to 1 km high and spaced 10 to 15 km apart. These features suggest that plate tectonics may have dominated Ganymede's early history. Water seeping upward through cracks in the original crust would freeze and force apart fragments of the original crust, producing the jagged, dark islands of old crust separated by bands of younger, light-colored, heavily grooved ice. Thus the cracks play a role in Ganymedean plate tectonics analogous to the role of the oceanic rifts on the Earth. But unlike thinner-crusted Europa where tectonic-like activity perhaps occurs even today, tectonics on Ganymede bogged down 3 billion years ago as the satellite's cooling crust froze to unprecedented depths.

Figure 13-15 Grooved terrain on Ganymede
This picture, taken by Voyager 1 at a range of 145,000 km, shows an area roughly as large as the state of Pennsylvania. The smallest visible features are about 3 km across. Numerous parallel mountain ridges are spaced 10 to 15 km apart and have heights up to 1000 m. (NASA)

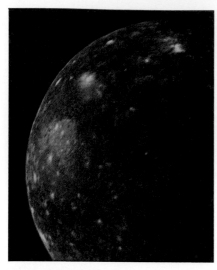

Figure 13-16   Callisto
Callisto, the outermost Galilean satellite, is almost exactly the same size as Mercury. Numerous craters pockmark Callisto's icy surface, as seen in this mosaic of views from Voyager 1 taken at a distance of about 400,000 km. A huge impact basin called Valhalla dominates the Jupiter-facing hemisphere of this frozen, geologically inactive world. A series of concentric rings up to 3000 km in diameter surrounds the impact site. (NASA)

Callisto, Jupiter's outermost Galilean satellite, looks very much like Ganymede: numerous impact craters are scattered over a dark, ancient, icy crust. (see Figure 13-16). There is one major, obvious difference, however: Callisto has no younger, grooved terrain. From this we infer that tectonic activity never began on Callisto. Perhaps because of its greater distance from Jupiter, the ocean that enveloped young Callisto $4\frac{1}{2}$ billion years ago froze more rapidly and to a greater depth than on Ganymede, thereby forever preventing tectonic processes. Indeed, Callisto's icy crust may be several times thicker than Ganymede's, extending to depths of several hundred kilometers. It is bitterly cold on Callisto: Voyager's instruments measured a noontime temperature of $-118°C$ ($-180°F$), and the nighttime temperature plunges to $-193°C$ ($-315°F$).

Voyager 1 photographed a huge impact basin on Callisto. Located on Callisto's Jupiter-facing hemisphere, this Valhalla Basin was produced by the impact of an asteroid-sized object very early in the satellite's history. Valhalla consists of a large number of concentric rings, separated by 50 to 200 km and having diameters ranging up to 3000 km.

The great age of Valhalla is inferred both from the presence of overlying impact craters and from the absence of substantial vertical relief of the concentric rings. The Valhalla impact probably occurred around 4 billion years ago, when the satellite's young, relatively thin crust still permitted plastic flow to reduce the height of the upraised rings in the ice.

The Voyager pictures also showed traces of a Valhalla-like impact on Ganymede's Galileo Regio. Segments of a system of concentric rings cover a large portion of this dark island of ancient crust. However, no obvious impact feature is found at the geometric center of this ring system. Apparently the development of grooved terrain completely obliterated the impact basin.

In looking at the Galilean satellites, we see a neat, orderly progression from a high-density, geologically active world (Io) nearest Jupiter to a low-density, geologically dead world (Callisto) farthest from Jupiter. As the spacecraft sped on toward their rendezvous with Saturn, many scientists expected another orderly system to be neatly laid out for their examination. They were in for a rude surprise.

Summary

. The four Galilean satellites of Jupiter orbit about the planet in the plane of its equator with synchronous rotation; their periods of rotation and revolution about the huge planet are relative short (from 2 to 17 days).

The inner two Galilean moons (Io and Europa) are roughly the size of our Moon and have densities similar to that of the Moon; the outer two Galilean moons (Ganymede and Callisto) are roughly the size of Mercury and are lower in density than the Moon or Mercury.

Several small moons (including Amalthea) and a faint ring of small rock fragments also orbit in the plane of the equator, closer to Jupiter than the Galilean satellites.

Eight small moons move in much larger orbits that are noticeably inclined to the plane of Jupiter's equator; some in retrograde orbits.

. Io is covered with a colorful layer of sulfur compounds deposited by very frequent and explosive eruptions from volcanic vents.

Io's volcanic eruptions resemble terrestrial geysers; the energy heating Io's interior comes from tidal forces that flex the moon as it passes between the planet and the other large moons.

The Io torus is a ring of electrically charged particles circling Jupiter at the distance of Io's orbit; interactions between this ring and Jupiter's magnetic field produce the strong decametric radio emissions associated with the planet.

. Europa is covered with a smooth layer of frozen water that is crisscrossed by an intricate pattern of long cracks, probably produced by tidal flexing of the moon.

. The heavily cratered surface of Ganymede is composed of frozen water; large polygons of dark, ancient surface are separated by regions of heavily grooved, lighter-colored, younger terrain.

Plate tectonics apparently operated during the early history of Ganymede.

. Callisto also has a heavily cratered crust of frozen water, but plate tectonics apparently never operated on this moon, presumably because it quickly developed a very thick solid crust.

The impact basin called Valhalla on Callisto provides evidence of plastic flow in the icy crust of the moon.

. The inner moons, Galilean satellites, and ring of Jupiter probably formed through a process of accretion similar to the process that formed the solar system about the Sun, but on a smaller scale; the outer moons possibly were asteroids captured later by Jupiter's gravity.

. The Galilean satellites help us to understand processes on terrestrial planets because they provide examples of these processes operating under conditions quite different from those that prevail on the inner planets of the solar system.

Review questions

1 How does the Galilean satellite system resemble the solar system? How is it different?

2 With all of its volcanic activity, why doesn't Io possess an atmosphere?

3 How would you account for the existence of the non-Galilean satellites of Jupiter as well as Jupiter's ring of material?

4 Long before the Voyager fly-bys, Earth-based astronomers reported that Io appear brighter than usual for a few hours after emerging from Jupiter's shadow. Based on what we know about the material ejected from Io's volcanoes, explain this brief anomalous brightening of Io.

5 Compare and contrast the surface features of the four Galilean satellites discussing the relative geological activity and evolution of these four satellites.

6 Compare and contrast Valhalla (see Figure 13-16) with Mare Orientale on the Moon and the Caloris Basin on Mercury.

Advanced questions

*7 Pick one of the Galilean satellites and calculate the mass of Jupiter using the data in Box 13-1.

8 Using the diameter of Io (3630 km) as a scale, estimate the height to which the plume of Prometheus rises above the surface of Io in Figure 13-8.

*9 If material is ejected into space from Io's volcanoes at the rate of 10 tons/sec, how long will it be before Io loses ten percent of its mass? How does your answer compare with the age of the solar system?

*10 How long does it take for Ganymede to enter or leave Jupiter's shadow?

Discussion questions

11 Speculate on the possibility that Europa, Ganymede, or Callisto might harbor some sort of marine life.

12 Suppose that you were planning four missions to the moons of Jupiter that would land a spacecraft on each of the Galilean satellites. What kinds of questions would you want these missions to answer and what kinds of data would you want your spacecraft to send back? In view of the different environments on the four satellites, how would the designs of the four spacecraft differ? Be specific about the possible hazards and problems that each spacecraft might encounter in landing on the satellites.

For further reading

Griffin, R. "Barnard and His Observations of Io." *Sky & Telescope,* Nov. 1982, p. 428.

Johnson, T. "The Galilean Satellites." In Beatty, J., et al., eds. *The New Solar System, 2nd ed.* Sky Pub. and Cambridge U. Press, 1982.

Johnson, T., and Soderblom, L. "Io." *Scientific American,* Dec. 1983.

Morrison, D. "Four New Worlds: The Voyager Exploration of Jupiter's Satellites." *Mercury,* May/June 1980, p. 53.

Soderblom, L. "The Galilean Moons of Jupiter." *Scientific American,* Jan. 1980.

14 The spectacular Saturnian system

Jupiter, Saturn, and Earth
This montage shows Jupiter, Saturn, and Earth reproduced to the same scale. Saturn, like Jupiter and the Sun, is composed primarily of hydrogen and helium. Note that the cloud features on Saturn are much less distinct than on Jupiter. Saturn has a lower surface gravity than Jupiter which causes the Saturnian cloud layers to occur at greater depths in Saturn's atmosphere than in Jupiter's. Sunlight reflected from these deeper cloud layers suffers a greater amount of absorption on Saturn thereby giving the planet its faded appearance compared to Jupiter. (S. P. Meszaros; NASA)

Earth-based observers have long been fascinated by Saturn with its system of thin, flat rings circling the planet. The Voyager flybys gave us new insights on this complex system, revealing surface features on many Saturnian satellites and details about the thick atmosphere of Titan, Saturn's largest moon. In this chapter, we discover parallels between the atmospheric structures of Saturn and Jupiter, which suggest differences in how these two planets evolved. We explore the complex interactions between the ring system and the many satellites. We learn about the terrestrial nature of Titan, where methane snow and rain may fall through a nitrogen atmosphere beneath a smog of carbon-hydrogen compounds. Finally, we review Saturn's myriad of moderate-sized moons, whose surface details pose many new puzzles for planetary scientists.

The magnificent rings of Saturn make this planet one of the most spectacular objects visible in the nighttime sky to an amateur astronomer using a small telescope. Saturn is so far away, however, that our Earth-based telescopes reveal only the coarsest, large-scale features. Astronomers were totally unprepared for the breathtaking vistas that unfolded as the Voyager spacecraft sped past Saturn's shimmering cloudtops in the early 1980s.

251

Box 14-1   Saturn data

Mean distance from the Sun: 9.539 AU = 1.427×10^9 km
Maximum distance from the Sun: 10.069 AU = 1.507×10^9 km
Minimum distance from the Sun: 9.009 AU = 1.347×10^9 km
Mean orbital velocity: 9.6 km/sec
Sidereal period: 29.46 years
Rotation period: equatorial = $10^h\ 13^m\ 59^s$
                    internal   = $10^h\ 39^m\ 25^s$
Inclination of equator to orbit: 29°
Inclination of orbit to ecliptic: 2° 29'
Orbital eccentricity: 0.0556
Diameter: equatorial = 120,660 km
          polar   = 108,600 km
Diameter (Earth = 1): equatorial = 9.44
                       polar   = 8.40
Apparent diameter as seen from Earth: maximum = 20.9''
                                      minimum = 15.0''

Mass: 5.69×10^{26} kg
Mass (Earth = 1): 95.15
Mean density: 0.69 g/cm^3
Surface gravity (Earth = 1): 1.16
Escape velocity: 35.6 km/sec
Oblateness of planet: 0.102
Mean surface temperature: −180°C at cloudtops
Albedo: 0.61
Brightest magnitude: −0.3
Mean diameter of Sun as seen from Saturn: 3' 22''

Earth-based observations reveal gaps between Saturn's rings as well as faint markings on the planet

When Galileo focused his telescope on Saturn in the early 1600s, he saw few details, but he did notice two puzzling lumps protruding from opposite edges of the planet's disk. In 1655, the Dutch astronomer Christian Huygens observed Saturn with a better telescope and noted that there were no protrusions. Having faith in Galileo's observations, Huygens suggested that Saturn is surrounded by a flattened ring that just happened to be edge-on as viewed from Earth in 1655 and was therefore temporarily invisible. This brilliant deduction was confirmed ten years

Figure 14-1   Saturn from the Earth
This is one of the best views of Saturn ever produced by an Earth-based observatory. Sixteen original color images taken with the 1.5 m telescope at the Catalina Observatory on March 11, 1974, were combined to make this photograph. Note the prominent Cassini division in the rings and the belts and zones in the Saturnian atmosphere. (NASA)

later when Saturn had moved far enough along its orbit so that the ring could again be seen.

As the quality of telescopes improved, details of the ring and of Saturn's cloud cover became visible. In 1675, G. D. Cassini discovered a dark division in the ring that looks like a gap about 5000 km wide. Astronomers also discovered stripes in Saturn's clouds similar to the belts and zones on Jupiter (see Figure 14-1). The contrast between Saturn's belts and zones is less dramatic than the colorful patterns in Jupiter's atmosphere. Saturn's stripes are not restricted to equatorial regions as are Jupiter's; the alternating dark and light bands extend into the polar regions on Saturn.

After Cassini's discovery of a gap in Saturn's ring, astronomers began to view the ring as a system of rings. The **Cassini division** separates the outer **A ring** from the brighter **B ring** closer to the planet. By the mid-1800s, astronomers using improved telescopes were able to detect a faint **C ring** (or crepe ring) that lies just inside the B ring (see Figure 14-2).

Figure 14-2 Saturn's classic rings
The three broad rings of Saturn are clearly visible in this photograph sent back by Voyager 1. The faint C ring exists in the region of 1.21 to 1.53 Saturn radii from the planet's center. The bright B ring occupies the region from 1.53 to 1.95 Saturn radii. The 5000-km-wide Cassini division lies between the B ring and the A ring, which occupies the region between about 2.03 and 2.26 Saturn radii. (NASA)

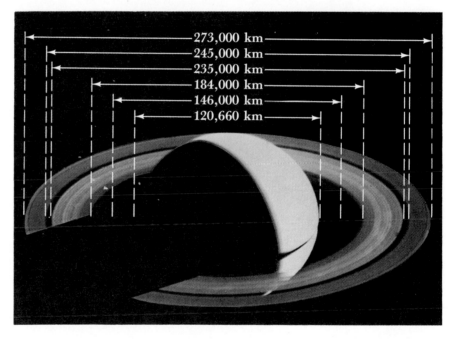

Earth-based views of the Saturnian ring system change dramatically as Saturn orbits slowly about the Sun (a Saturnian year is equal to $29\frac{1}{2}$ Earth years). This change is observed because the rings, which lie in the plane of Saturn's equator, are tilted 29° from the plane of Saturn's orbit. Thus, over the course of a Saturnian year, the rings are viewed from various angles by the Earth-based observer (see Figure 14-3). At one time, the observer looks "down" on the rings; one-half of a Saturnian year later, the "underside" of the rings is exposed to view from Earth. At intermediate times, the rings are seen edge-on, and they then disappear entirely from the view of the Earth-based observer. Thus we conclude that the rings are very thin—less than 2 km in thickness according to recent estimates.

Figure 14-3   The changing appearance of Saturn's rings
Saturn's rings are tilted 27° from the plane of Saturn's orbit. Earth-based observers, therefore, see the rings at various angles as Saturn moves around its orbit. Note that the rings seem to disappear entirely when they are viewed edge-on. (Lowell Observatory)

Saturn's rings are composed of numerous fragments of ice and ice-coated rock

Astronomers have known for more than 100 years that Saturn's rings cannot possibly be solid, rigid, thin sheets of matter. The Scottish physicist James Clerk Maxwell proved mathematically in 1857 that such a broad, thin, rigid sheet would break apart and he therefore concluded that Saturn's rings are composed of "an indefinite number of unconnected particles."

Supporting observational evidence came four decades later when James Keeler at Lick Observatory observed Doppler shifts in sunlight reflected from Saturn's rings. As explained in Box 7-3, relative motion between a source of light and an observer produces a displacement, or shift, of spectral lines. By measuring the Doppler shifts of spectral lines, Keeler determined that the inner portions of Saturn's rings are moving more rapidly about the planet than are the outer portions, as would be expected if the rings are formed of numerous tiny moonlets, each circling Saturn along its own individual orbit. Indeed, the orbital speeds measured for various parts of the rings are in complete agreement with the orbital velocities predicted from Kepler's third law.

Saturn's rings are very bright, with an albedo of 0.8, so the particles that form the rings must be highly reflective. Astronomers had long suspected that the rings are made of ice and ice-coated rocks, but confirming evidence was not obtained until the early 1970s, when American astronomers Gerard Kuiper and Carl Pilcher identified the spectral features of frozen water in the near-infrared spectrum of the rings. Additional measurements from Earth-based observatories and the Voyager spacecraft tells us that the temperature of the rings ranges from $-180°C$ ($-290°F$) in the sunshine to less than $-200°C$ ($-300°F$) in Saturn's shadow. Water ice is in no danger of melting or evaporating at these temperatures.

Voyager observations also showed that the ring particles range in size from snowflakes less than 1 mm in diameter up to icy boulders that measure tens of meters across. It seems reasonable to suppose that all of this material is ancient debris that has failed to accrete into satellites. Indeed, the ring particles are so close to Saturn that they can never accrete into satellites.

Imagine a collection of small fragments of rock orbiting a planet. The gravitational attraction between neighboring rocks tends to pull the rock fragments together to produce a larger satellite. However, the various rock fragments are at differing distances from the parent planet and thus they experience differing gravitational attractions from the planet that tend to keep the fragments separated. At a distance from the planet's center called the **Roche limit,** these attractive and disruptive effects are exactly balanced. Inside the Roche limit, fragments will not accrete (fall together) to form a larger body; instead they will tend to spread out into a ring around the planet. (As you might expect, the Roche limit for Saturn lies just outside the outer edge of the A ring.) All large satellites are found only outside a planet's Roche limit. If any large moon were to come inside a planet's Roche limit, the planet's gravity would cause the satellite to break apart into fragments.

The Roche limit applies only to objects held together by gravity. In a rock fragment, chemical bonds (electromagnetic forces) between atoms and molecules hold the rock together. These chemical forces are much stronger than the disruptive gravitational tidal effects of a nearby planet and thus the rock does not break apart. In the same way, people walking around on the Earth's surface (which is inside the Earth's Roche limit) are in no danger of coming apart because we are held together by comparatively strong intermolecular forces rather than gravity.

The internal structure of Saturn was deduced from measurements of its oblateness

Earth-based observations indicated that the composition and structure of Saturn are similar to those of Jupiter. The average density of Saturn is 0.7 g/cm^3, only about one-half that of Jupiter. Such a low density means that the planet must be composed largely of very light elements. Spectral lines for methane (CH_4) and ammonia (NH_3) in sunlight reflected from Saturn are a strong indication of a high abundance of hydrogen in the planet.

Saturn is even more oblate than Jupiter (examine Figure 14-1). Saturn's equatorial diameter is about 10 percent larger than its polar diameter—that is, its oblateness is about 0.10. As discussed in Chapter 12, this large oblateness is evidence of a dense core surrounded by less-dense

Box 14-2 Theoretical models of Jupiter and Saturn

The oblateness of a planet (sometimes called its ellipticity) is usually represented by the symbol e and is computed as

$$e = \frac{R_e - R_p}{R_e}$$

where R_e is the planet's equatorial radius and R_p is its polar radius. Jupiter's ellipticity is 0.064, whereas Saturn's is 0.102. Saturn therefore is more oblate than Jupiter.

As explained in Chapter 12, the oblateness of a planet is directly related to its speed of rotation and to the degree to which its mass is concentrated near its center. Saturn rotates a little more slowly than Jupiter does: a solar day near Saturn's equator is about 24 minutes longer than a solar day near Jupiter's equator. The greater oblateness of Saturn cannot be due to faster rotation, so it must be due to a greater concentration of mass at Saturn's center. This concentration cannot be due to greater compression because Saturn is smaller and less dense than Jupiter. We conclude, therefore, that Saturn's rocky core must be larger and more massive than Jupiter's. Detailed calculations suggest that about 26 percent of Saturn's mass is contained in its rocky core, whereas Jupiter's core contains only about 4 percent of that planet's mass.

The average density of Saturn (0.69 g/cm^3) is much less than that of Jupiter (1.31 g/cm^3) despite Saturn's more massive core, so Saturn's density must decrease more rapidly than does Jupiter's from the core toward the surface. The table lists of details of a model of the internal structures of Jupiter and Saturn as computed in a study by Andrew Ingersoll of the California Institute of Technology. In this table, the outer layer of molecular hydrogen is listed as "layer 1," and the mantle of liquid metallic hydrogen is listed as "layer 2"—these names are used simply to conserve space in the table. Note the sharp change in density at the boundaries between the various layers, where there is an abrupt change in the composition or the state of the matter making up the planet.

Level	Planet	Distance from center of planet (km)	(radius = 1)	Pressure (bars)	Density (g/cm^3)	Temperature (K)	Percentage of planet's mass below this level
Near top of layer 1	Jupiter	71,000	1.0	1	0.0002	165	100
	Saturn	60,000	1.0	1	0.0002	140	100
Bottom of layer 1	Jupiter	54,000	0.76	3,000,000	1.1	10,000	77
	Saturn	28,000	0.46	3,000,000	1.1	9,000	43
Top of layer 2	Jupiter	54,000	0.76	3,000,000	1.2	10,000	77
	Saturn	28,000	0.46	3,000,000	1.2	9,000	43
Bottom of layer 2	Jupiter	10,000	0.15	42,000,000	4.4	20,000	4
	Saturn	16,000	0.27	8,000,000	1.9	12,000	26
Top of core	Jupiter	10,000	0.15	42,000,000	15	20,000	4
	Saturn	16,000	0.27	8,000,000	7	12,000	26
Center of planet	Jupiter	0	0.00	80,000,000	20	25,000	0
	Saturn	0	0.00	50,000,000	15	20,000	0

NOTE: Layer 1 is the outer layer of molecular hydrogen; layer 2 is the mantle of liquid metallic hydrogen.

material that is distorted into an oblate shape by the planet's rotation. From the information about the average density, the oblateness, and the probable chemical composition, astronomers were able to construct a model of the internal structure of Saturn (see Box 14-2). This model suggests that Saturn's structure resembles that of Jupiter: a solid, rocky core surrounded by a mantle of liquid metallic hydrogen, which in turn is surrounded by a layer of molecular hydrogen. The relative thicknesses of the layers differ, however, in the models for Saturn and Jupiter (see Figure 14-4).

Figure 14-4 Internal structure of Saturn and Jupiter
There are three distinct layers in Saturn's interior, as there are in Jupiter's. The planet's rocky core is surrounded by a layer of liquid metallic hydrogen, which in turn is enveloped in a thick layer of molecular hydrogen. Both diagrams are drawn to the same scale.

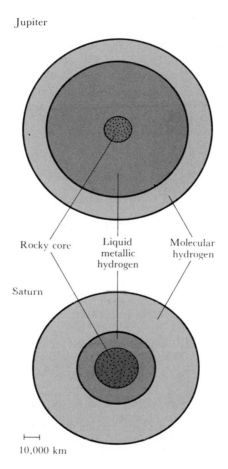

Jupiter

Rocky core   Liquid metallic hydrogen   Molecular hydrogen

Saturn

10,000 km

Saturn has a Jupiter-like magnetosphere with Earth-like radiation belts

Saturn's mantle of liquid metallic hydrogen, like that of Jupiter, is thought to be the source of the planet's magnetic field. Information about Saturn's magnetosphere was sent back during the Pioneer 11 flyby in August 1979 (see Figure 14-5). Saturn's magnetic field is somewhat weaker than Jupiter's, as was expected because of Saturn's slightly slower rotation and much smaller volume of liquid metallic hydrogen (review Figure 14-4).

Pioneer 11 sent back data indicating that Saturn's magnetosphere resembles that of Jupiter, but that it contains radiation belts similar to those of Earth instead of a huge current sheet like that in Jupiter's magnetosphere. Although Saturn's magnetic field is almost as strong as Jupiter's, it contains far fewer charged particles than does Jupiter's. There are two reasons for this deficiency of charged particles in Saturn's magnetosphere. First, Saturn lacks a continuous supply of particles like the volcanoes of Io that dump 1 ton of material into Jupiter's magnetosphere each second. Second, many charged particles are absorbed by the fragments in Saturn's rings, thus depleting the particle concentration in the inner magnetosphere. The charged particles that do exist in Saturn's magnetosphere are concentrated in radiation belts similar to the Van Allen belts of the Earth's magnetosphere.

Like Jupiter, Saturn emits more radiation than it receives from the Sun

Figure 14-5 [left]   The best view of Saturn from Pioneer 11
Pioneer 11 was the first of three spacecraft to visit Saturn. Although Pioneer's instruments were designed to measure physical conditions around the planet, the spacecraft did send back pictures of modest quality. This view was taken at a range of 2½ million kilometers. The satellite to the left of the planet is Rhea, which is 1500 km in diameter—roughly one-half the size of our Moon. (NASA)

Figure 14-6 [right]   A Voyager flyby
After passing by Jupiter, both Voyager spacecraft continued past Saturn. Voyager 1 passed near Saturn in November 1980. Voyager 2 followed 9 months later, coasting past the planet in August 1981. Voyager 1 then headed out of the solar system. Voyager 2 moved on toward a January 1986 Uranus flyby, to be followed by an August 1989 pass near Neptune. (NASA)

Both Jupiter and Saturn have internal sources of energy. Each planet radiates more energy than it receives in the form of sunlight. As we saw in Chapter 12, Jupiter's internal heat is thought to be the remnant of the energy trapped as the planet accreted from the original solar nebula 4½ billion years ago. Jupiter has been slowly cooling off ever since as this energy escapes in the form of infrared radiation.

Saturn is both smaller and less massive than Jupiter. One would thus expect Saturn to cool more rapidly than Jupiter and hence to emit less energy today. But, in fact, Saturn radiates about 2½ times as much energy as it receives from the Sun, whereas Jupiter emits about 1½ times the heat it absorbs from sunlight. What factors might explain why Saturn emits so much more heat than Jupiter?

For some time before the space probes to the Jovian planets, astronomers had suspected that both Jupiter and Saturn have compositions similar to that of the original solar nebula (and to that of the Sun's atmosphere today). Each of these giant planets is massive enough and cool enough to have retained all of the gases that originally accreted from the solar nebula. As we saw in Chapter 12, the Voyager flybys (see Figure 14-6) did confirm that Jupiter's atmosphere has a solar abundance of elements (by weight, 82 percent hydrogen, 17 percent helium, and 1 percent all other elements). Surprisingly, however, the Voyager spacecraft reported that Saturn's atmosphere has less helium than expected. The chemical composition of Saturn's atmosphere, by weight, is 88 percent hydrogen, 11 percent helium, and 1 percent all other elements.

Edwin E. Salpeter of Cornell University and David J. Stevenson of the

California Institute of Technology (Caltech) offer a brilliant hypothesis that links Saturn's apparent deficiency of helium to the excess heat radiated by the planet. According to this theory, Saturn did indeed cool more rapidly than Jupiter. This triggered a process analogous to the development of a rainstorm here on Earth. When the air is cool enough, humidity in the Earth's atmosphere condenses into raindrops that fall to the ground. In the case of Saturn, helium droplets rain downward from the planet's atmosphere toward its core. Helium appears to be deficient in Saturn's upper atmosphere merely because it has fallen farther down into the planet. Furthermore, as the helium droplets descend through the molecular hydrogen, the two gases rub against each other. The resulting friction produces heat that eventually escapes from Saturn's surface.

Precipitation of helium from Saturn's clouds is calculated to have begun 2 billion years ago, and the resulting release of energy adequately accounts for the extra heat radiated by Saturn since that time. Similar calculations for Jupiter indicate that Jupiter is only now reaching the stage where a significant amount of helium precipitation can begin in its outer layers. Saturn has therefore given us important clues about the course of Jupiter's future evolution.

The Voyager flybys sent back information about the Saturnian atmosphere

Saturn's atmosphere, like Jupiter's, contains methane (CH_4), ammonia (NH_3), and water vapor (H_2O). These compounds are the simplest combinations of carbon, nitrogen, and oxygen with hydrogen. Also like Jupiter, Saturn has three distinct cloud layers: an upper layer of frozen ammonia crystals; a middle layer of crystals of ammonia hydrosulfide (NH_4SH), which is a combination of ammonia and hydrogen sulfide (H_2S); and a lower layer of frozen water crystals.

Although their atmospheres have similar structures and compositions, Saturn and Jupiter are not identical in appearance. Saturn's clouds lack the colorful contrast of Jupiter's (see Figure 14-7), although some of the

Figure 14-7   Saturn from Voyager 2
Voyager 2 sent back this picture when the spacecraft was still 2 months and 34 million kilometers away from its closest approach to the planet. Two bright cloud patterns are visible in the northern hemisphere. Two of Saturn's icy moons, Rhea and Dione, appear near the lower-right side of this view. (NASA)

Voyager photographs do show faint hints of belts and zones (see Figure 14-8). After special computer processing to exaggerate greatly the colors in the Voyager photographs, details such as storm systems and ovals did become visible (see Figure 14-9).

The differing appearances of Saturn and Jupiter are related to the different masses of the two planets. Jupiter's strong surface gravity

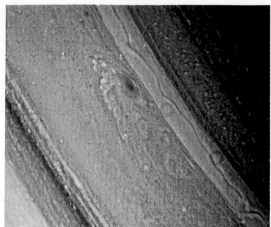

Figure 14-8 [left] Saturn's clouds from Voyager 1
This view of Saturn's cloudtops was taken by Voyager 1 at a range of 1¾ million kilometers. Note that there is substantially less contrast between belts and zones here than there is on Jupiter. The shadow of Dione appears at the bottom of the picture. (NASA)

Figure 14-9 [right] Eddy currents in Saturn's atmosphere
Computer processing exaggerates the colors in this Voyager 2 picture of Saturn's northern midlatitudes. The wavy line in the light blue ribbon is a pattern moving eastward at 150 m/sec (300 miles per hour). The dark oval and two puffy, blue-white spots below it are eddies drifting westward at roughly 20 m/sec (40 miles per hour). (NASA)

compresses its atmosphere, so that the three cloud layers are compressed into a range of 75 km in the upper atmosphere of the planet. Saturn's somewhat weaker surface gravity subjects its atmosphere to less compression, so that the same three cloud layers are spread out over a range of nearly 300 km (see Figure 14-10). The colors of Saturn's clouds are less dramatic because deeper layers are partly obscured by the thick atmosphere above them.

By following features in the Saturnian clouds, scientists have determined wind speeds in the planet's upper atmosphere. There are counterflowing eastward and westward currents in Saturn's upper atmosphere, as there are in Jupiter's. However, Saturn's equatorial jet is much broader and much faster than Jupiter's (see Figure 14-11). In fact, wind speeds near Saturn's equator approach 500 m/sec (1000 miles per hour), which is approximately two-thirds the speed of sound.

Figure 14-10 Temperature profiles of Jupiter and Saturn
The structures of the upper atmospheres of Jupiter and Saturn are displayed on these graphs of temperature versus depth. Note that Saturn's atmosphere is more "spread out" than Jupiter's, which is a direct result of Saturn's weaker surface gravity. (Adapted from Andrew P. Ingersoll)

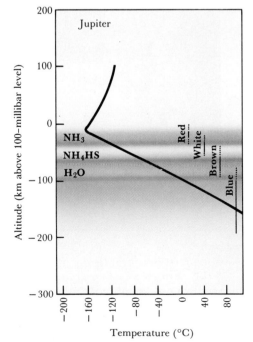

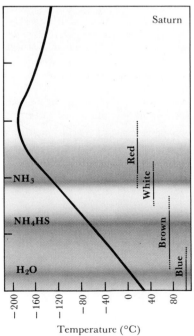

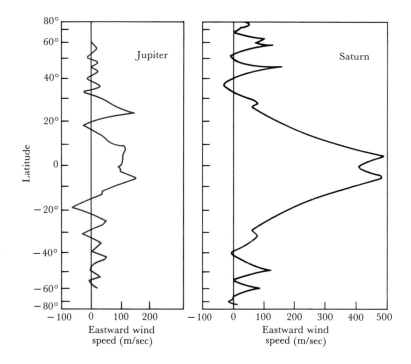

Figure 14-11 Wind speeds on Jupiter and Saturn
Average wind speeds on Jupiter and Saturn are plotted for latitudes ranging from 80° north to 80° south. Positive numbers are eastward velocities; negative numbers are westward velocities. Although both planets exhibit counterflowing currents, Saturn's equatorial zonal jet is much broader and faster then Jupiter's.

Saturn's rings consist of thousands of narrow, closely spaced ringlets

During the 1979 flyby, Pioneer 11 sent back a photograph of modest quality that seemed to show a thin ring just beyond the outer edge of the A ring. The existence of this ring was confirmed by a sudden drop in the level of radiation reported by Pioneer 11's charged-particle detectors as the probe passed through the region where the ring would have been expected to absorb electrons and protons from the magnetosphere. This previously unknown ring is now called the **F ring.**

During the Voyager flybys of 1980 and 1981, the cameras sent back pictures showing unexpected details in the ring structures of Saturn. The broad rings were seen to consist of hundreds upon hundreds of closely spaced bands, or **ringlets,** of particles (see Figure 14-12). Although intriguing suggestions have been proposed, scientists still do not

Figure 14-12 Saturn's rings from Voyager 1
Voyager 1 took this view of Saturn's rings from a distance of about 1½ million kilometers. The C ring scatters light differently than the A or B rings do, so that the C ring has a bluer color. The broad Cassini division is clearly visible, as is the narrow Encke division within the A ring. The very thin F ring is visible just beyond the outer edge of the A ring. (NASA)

understand just why Saturn's A, B, and C rings are divided into thousands of thin ringlets.

The Voyager cameras also sent back the first high-quality pictures of the F ring. It is visible just beyond the outer edge of the A ring in Figure 14-12. Close-up views revealed a startling and mysterious fact: the F ring is kinky and braided; it actually consists of several intertwined strands (see Figure 14-13). One Voyager 2 image shows a total of five strands, each about 10 km across. Voyager scientists were at a complete loss to explain this complex structure. The braids, kinks, knots, and twists in the F ring pose one of the most challenging puzzles in modern astronomy.

Figure 14-13   Details of the F ring
This Voyager 1 photograph of the F ring was taken from a distance of 750,000 km. The total width of the F ring is about 100 km. Within this span are several discontinuous strands, each measuring roughly 10 km across. (NASA)

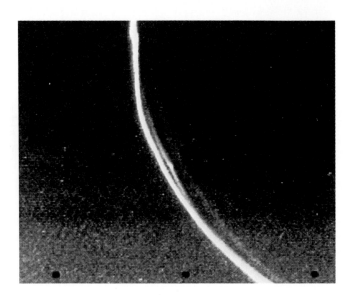

The sizes and densities of particles in the rings were deduced from studies of scattered sunlight

Through Earth-based telescopes, we see only the sunlit side of Saturn's rings. From this perspective, the B ring appears very bright, the A ring appears moderately bright, the C ring appears dim, and the Cassini division appears dark. The proportion of sunlight reflected back toward the Sun (the albedo) is directly related to the density of fragments or particles in the ring. The B ring is bright because it has a high density of ice and rock fragments, whereas the darker Cassini division has a lower density of fragments.

Eight hours after it had crossed from the northern to the southern side of the rings, Voyager 1 took the photograph that appears as Figure 14-14. The Sun was shining down on the northern side of the rings at the time of the Voyager flybys, so Figure 14-14 shows the sunlight that passes *through* the rings. As expected, the B ring looks darkest here because little sunlight gets through its dense concentration of fragments, whereas the Cassini division looks very bright because sunlight passes relatively freely through its low density of fragments. However, the fact that the Cassini division does appear bright in this photograph is clear evidence that it does contain some fragments. If it contained no fragments at all, we would see the black of space through it. The brightness that we see in Figure 14-14 must be sunlight scattered by some fragments in the Cassini division.

Another division is visible in Figure 14-14 in the outer half of the A ring. This division has a width of 270 km and is named the **Encke division** after the German astronomer Johann Franz Encke, who reported seeing it in 1838. Many astronomers have argued, however, that Encke's report was erroneous because his telescope did not have sufficient resolving power to produce an image of this narrow gap in the rings. The first undoubted observation of the Encke division was made in the late 1880s by the American astronomer James Keeler, using the newly constructed 36-in, refractor at the Lick Observatory. For this reason, the Encke division has been called the "Keeler gap" by some astronomers.

The process by which light bounces in various directions from particles or fragments in its path is called **scattering.** The proportion of light scattered in various directions depends upon both the size of the particles and the wavelength of the light. By carefully measuring the albedo at various wavelengths and at various angles to the incident sunlight, scientists can estimate the size of the particles or fragments that are scattering the sunlight. Similar measurements can be made with radiation of other wavelengths to get more accurate estimates of the size of fragments much larger than the wavelengths of visible light.

Voyager scientists carefully measured the brightness of the rings from many angles as the spacecraft flew past Saturn. They also measured how radio waves emitted by the spacecraft transmitters were scattered by the particles in the rings. From these data, G. Leonard Tyler and Ahmed Essam A. Marouf of Stanford University have determined that the largest particles in Saturn's rings are roughly 10 m across. More abundant are snowball-sized particles about 10 cm in diameter. The smallest particles are just a few microns wide—smaller than snowflakes.

The F ring consists primarily of tiny, micron-sized particles. The A and B rings have a mixture of particles of various sizes. Cassini's division and the C ring, which scatter only a small amount of light back toward Earth, contain relatively few tiny particles.

Subtle differences in color from one ring to the next also give important clues about the nature of the particles in the rings. These differences are clearly visible in Figure 14-15, in which the colors have been exaggerated by computer processing. Scientists believe that these color variations correspond to slight differences in the chemical compositions of the rings. Although the main chemical constituent is frozen water,

Figure 14-14 [left]   Details of the A ring
This view of the underside of Saturn's rings was taken from a distance of 740,000 km by Voyager 1. Both the Encke division and the thin F ring are clearly visible toward the right side of the picture. The Cassini division and the outer edge of the B ring are on the left side of the picture. The Cassini division appears bright in this view of the shaded side of the rings. (NASA)

Figure 14-15 [right]   False-color view of ring details
Computer processing severely exaggerates subtle color variations in this view of the sunlight side of the rings from Voyager 2. Note that the C ring and Cassini's division appear bluish. Also note distinct color variations across both the A and B rings. (NASA)

Figure 14-16   Spokes in the B ring
Radial, wedge-shaped spokes are visible in the central, most-opaque portions of the B ring. The micron-sized dustlike particles do not scatter sunlight back toward the Sun. Thus the spokes appear dark in this picture from Voyager 2 that was taken with the Sun behind the camera. (NASA)

tiny trace amounts of other chemicals (perhaps coating the surfaces of the ice particles) are probably the cause of the colors seen in the computer-enhanced view. These trace chemicals have not been identified. Nevertheless, the existence of color variations and their probable persistence over millions of years suggests that the icy particles do not wander around or migrate substantially from one ringlet to the next.

Voyager scientists also noted radial "spokes" in the B ring, particularly in black-and-white photographs such as Figure 14-16. Studies of light scattered from these spokes suggest that they are made of fine dustlike particles that are suspended a few tens of meters above the main plane of the rings. Perhaps the buildup of electric charges on the ring's particles (through the absorption of protons and electrons from Saturn's magnetosphere) produces electrostatic forces that suspend fine dust particles at a distance from the larger ring fragments, thereby producing the spokes that we see.

Saturn's innermost satellites affect the appearance and structure of the rings

Astronomers had realized before the Voyager flybys that one of Saturn's moons, Mimas, has an effect on the ring system. Mimas is a moderate-sized satellite that orbits Saturn every 22.6 hours. According to Kepler's third law, particles in Cassini's division should orbit Saturn every 11.3 hours. Consequently, on every second orbit, particles in Cassini's division line up between Saturn and Mimas. During these repeated alignments, the combined gravitational forces of Saturn and Mimas cause small fragments to deviate from their original orbits. In this way, the 2-to-1 resonance with Mimas depletes Cassini's division of dust that would otherwise scatter sunlight back toward Earth; hence Earth-based astronomers see the division as a dark band.

In addition to revealing new details of the A, B, C, and F rings, the Voyager cameras also discovered three new ring systems: the D, E, and G rings. The **D ring** is Saturn's innermost ring system. It consists of a series of extremely faint ringlets located between the inner edge of the C ring and the Saturnian cloudtops. The **E ring** and the **G ring** both lie quite far from the planet, well beyond the outer edge of the A ring (see Box 14-3 for details). Both of these outer ring systems are extremely faint, fuzzy, and tenuous. Each lacks the ringlet structure that is so prominent in the main ring systems. The E ring lies along the orbit of Enceladus, one of Saturn's icy satellites. Some scientists suspect that

Box 14-3   Saturn's rings

Several broad ring systems surround Saturn. The A, B, and C rings are visible through Earth-based telescopes. Although observers have often reported seeing additional, faint ring systems, the true locations of the D, E, F, and G rings were not established until the Voyager flybys in the early 1980s. Information about each of the seven ring systems is given here in order outward from the planet.

- **The D ring** extends from Saturn's cloudtops out to the inner edge of the C ring at 1.21 Saturn radii (73,000 km) from the planet's center. The D ring was discovered by the Voyager spacecraft and consists of a series of very faint ringlets.

- **The C ring** was discovered in 1850 and can be seen through Earth-based telescopes under good observing conditions. The C ring extends from 1.21 Saturn radii out to the inner edge of the B ring at 1.53 Saturn radii (92,000 km) from the planet's center.

- **The B ring** is the brightest of Saturn's ring systems and extends from 1.53 Saturn radii out to the inner edge of the Cassini division at 1.95 Saturn radii (117,600 km) from the planet's center.

- **The A ring** extends from the outer edge of the Cassini division at approximately 2.03 Saturn radii out to 2.26 Saturn radii (136,300 km) from the planet's center. The sharp outer edge of the A ring is controlled by the "shepherd satellite" called 1980S28.

- **The F ring** is very thin and is located at 2.33 Saturn radii (140,600 km) from the planet's center. The F ring, which was discovered during the Pioneer 11 flyby, is only 100 km wide. The F ring is confined by two "shepherd satellites" called 1980S26 and 1980S27.

- **The G ring** is extremely faint and tenuous. It is located at about 2.8 Saturn radii (169,000 km) from the planet's center and lacks the detailed ringlet structure that dominates the appearance of the inner rings.

- **The E ring** like the G ring, is extremely faint. The E ring extends from about 3.5 to 5.0 Saturn radii and is devoid of fine-scale ringlet structure. The material in the E ring may be associated with the satellite Enceladus, whose orbit is 3.95 Saturn radii from Saturn's center.

water geysers on Enceladus are the source of ice particles in the E ring, in much the same way that Io's volcanoes produce a torus of material along its orbit around Jupiter.

The Voyager cameras also discovered two tiny satellites that follow orbits on either side of the F ring (see Figure 14-17). Peter Goldreich of Caltech and Scott Tremaine at Princeton University pointed out that the gravitational forces of these two satellites keep the F ring particles in place. The outer satellite moves around Saturn at a slightly slower speed than that of the ice particles in the ring. As the ring particles pass by the outer satellite, they experience a tiny gravitational tug that tends to slow them down. Consequently, these particles lose a little energy, causing them to fall into orbits a little closer to Saturn. Meanwhile, the inner satellite orbits the planet a little faster than the F ring particles do. As the satellite moves past the particles, its gravitational pull tends to speed them up, thereby nudging them into a slightly higher orbit. The combined effect of these two satellites therefore focuses the icy particles

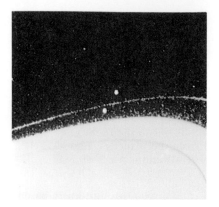

Figure 14-17   The F ring and its two shepherds
Two tiny satellites, each measuring about 50 km across, orbit Saturn on either side of the F ring. The gravitational effects of these two shepherd satellites focus and confine the particles in the F ring to a band about 100 km wide. This Voyager 2 picture was taken from a range of 10½ million kilometers. (NASA)

into a well-defined, narrow band about 100 km wide. Because of their confining influence, these two moons are called **shepherd satellites.**

A shepherd satellite also is responsible for the sharp outer edge of the A ring. This tiny moon circles Saturn just beyond the outer edge of the A ring. As particles near the edge of the A ring pass by the slowly moving shepherd satellite, they feel a gravitational drag that slows them down slightly, preventing them from wandering into orbits farther from Saturn.

Nearly two dozen moons are known to orbit Saturn (see Box 14-4). They can be divided into three categories according to size. First there is planet-sized Titan, one of the terrestrial worlds of the solar system. With a diameter of 5150 km, Titan is intermediate in size between Mercury and Mars. Second, there are six moderate-sized satellites, all of which were discovered before 1800. These six satellites have properties and diameters that lead us to group them in pairs: Mimas and Enceladus (400 to 500 km in diameter), Tethys and Dione (roughly 1000 km in diameter), and Rhea and Iapetus (about 1500 km in diameter). Finally, there are several very tiny moons that may be captured asteroids (as in the case of Phoebe, Saturn's outermost satellite) or jagged fragments of ice produced by impacts and collisions (as in the case of the shepherd satellites).

Titan has a thick, opaque atmosphere rich in methane, nitrogen, and hydrocarbons

Figure 14-18   Titan
This view of Titan was taken by Voyager 2 from a distance of 4½ million kilometers. Very few features are visible in the thick, unbroken haze that surrounds this large satellite. The main haze layer is located nearly 300 km above Titan's surface. (NASA)

Long before the Voyager flybys, astronomers knew that Titan is an extraordinary world. It was discovered in 1655 by Christian Huygens, the same year in which he proposed that Saturn has rings. By the early 1900s, several scientists had begun to suspect that Titan has an atmosphere because it is cool enough and massive enough to retain heavy gases. Confirming evidence came in 1944 when the American astronomer Gerard P. Kuiper discovered spectral lines of methane in the sunlight reflected from Titan. Titan therefore became the only satellite in the solar system known to have an appreciable atmosphere.

Because of its atmosphere, Titan was a primary target for the Voyager missions. To everyone's chagrin, however, the Voyagers spent hour after precious hour sending back featureless images such as Figure 14-18. Titan's cloud cover is so thick that it blocks any view of the surface. Indeed, the haze surrounding Titan is so dense that very little sunlight penetrates down to the ground; the surface of Titan must be a dark gloomy place.

In size, mass, and average density, Titan is quite similar to the largest Jovian satellites. Thus we would expect Titan's internal structure to resemble those of Ganymede and Callisto—a rocky core surrounded by a mantle of frozen water nearly 1000 km thick.

Titan's thick atmosphere, however, distinguished it from all other satellites. The atmospheric pressure at Titan's surface is 1.6 bars, or 60 percent *greater* than the atmospheric pressure at sea level on Earth, although Titan's surface gravity is lower than that of Earth. Thus, considerably more gas must be weighing down on Titan than on Earth. Specifically, about ten times more gas lies above each square centimeter of Titan's surface than lies above 1 cm^2 of Earth's surface.

What factors led to the formation of Titan's atmosphere that did not exist in the case of either Ganymede or Callisto? For one, Titan formed in a much cooler part of the solar nebula. In contrast to the warmer conditions near Jupiter, the ices from which Titan accreted probably

Box 14-4   Saturn's satellites

Nearly two dozen satellites are known to orbit about Saturn. The table lists the most prominent of these, with data about their orbits and physical properties. All of the larger satellites are spherical, and the diameter of the satellite is listed in the size column for each of these satellites. The smaller satellites are nonspherical, and three dimensions (width, length, and height) are given for each of these satellites. The masses (and hence the densities) of the smaller satellites are not yet known.

Satellite 1980S28 is the A-ring shepherd; satellites 1980S27 and 1980S26 are the F-ring shepherds. Satellites 1980S3 and 1980S1 are called coorbital satellites because they move in almost the same orbit. Tethys, 1980S13, and 1980S25 also are coorbital satellites, as are Dione and 1980S6. In the latter two cases, the tiny satellites occupy specific locations along the orbits of the larger moons where a balance exists between the gravitational pulls of Saturn and the larger moon. These locations, called the Lagrangian points, are discussed in Box 16-1.

Name of satellite	Distance from center of Saturn (km)	(Saturn radii)	Orbital period (hours)	Size (km)	Density (g/cm³)	Albedo
1980S28	137,670	2.28	14.45	10 × 10 × 20	——	0.4
1980S27	139,350	2.31	14.71	70 × 50 × 40	——	0.6
1980S26	141,700	2.35	15.08	55 × 45 × 35	——	0.6
1980S3	151,420	2.51	16.66	70 × 60 × 50	——	0.4
1980S1	151,470	2.51	16.67	110 × 100 × 80	——	0.4
Mimas	185,540	3.08	22.62	392	1.4	0.7
Enceladus	238,040	3.95	32.88	500	1.2	1.0
Tethys	294,670	4.88	45.31	1060	1.2	0.8
1980S13	294,670	4.88	45.31	17 × 14 × 13	——	0.6
1980S25	294,670	4.88	45.31	17 × 11 × 11	——	0.8
Dione	377,420	6.26	65.69	1120	1.4	0.6
1980S6	378,060	6.26	65.69	18 × 16 × 15	——	0.5
Rhea	527,100	8.74	108.42	1530	1.3	0.6
Titan	1,221,860	20.25	382.69	5150	1.88	0.2
Hyperion	1,481,000	24.55	510.64	205 × 130 × 110	——	0.2
Iapetus	3,560,800	59.02	1,903.94	1460	1.2	0.5–0.05
Phoebe	12,954,000	214.7	13,210.8	220	——	0.06

contained substantial amounts of frozen methane and ammonia. As Titan's interior became warm (through the decay of naturally occurring radioactive isotopes), these ices vaporized, producing an atmosphere around the young satellite. Methane (CH_4) is stable in the sunlight and remained in the atmosphere, but ammonia (NH_3) is easily broken down into nitrogen and hydrogen by the Sun's ultraviolet radiation. Titan's gravity is too weak to retain hydrogen, so it escapes into space. In fact, the Voyagers instruments detected a huge torus of dilute hydrogen gas surrounding Saturn along Titan's orbit. Even today, hydrogen is escaping from Titan at a substantial rate.

The dissociation of ammonia and subsequent loss of hydrogen leaves Titan with an abundant supply of nitrogen. Voyager data suggest that roughly 90 percent of Titan's atmosphere is nitrogen. The two next-most-abundant gases are argon and methane.

The interaction of sunlight with methane (CH_4) induces chemical reactions that produce a variety of carbon–hydrogen compounds called **hydrocarbons.** Voyager's infrared spectrometers detected small amounts of many hydrocarbons such as ethane (C_2H_6), acetylene (C_2H_2), ethylene (C_2H_4), and propane (C_3H_8) in Titan's atmosphere. Furthermore, nitrogen combines with these hydrocarbons to produce other compounds such as hydrogen cyanide (HCN), some of which are the building blocks of the organic molecules on which life is based. There is little reason to suspect life on Titan; its surface temperature of 95 K ($-178°C = -288°F$) is prohibitively cold. Nevertheless, a more detailed study of the chemistry of Titan may shed light on the origins of life on Earth.

On Earth, the atmospheric pressure and temperature are near the **triple point** of water, so that water is found on Earth in all three phases: liquid, solid, and gas. The atmospheric pressure and temperature at Titan's surface, however, are near the triple point of methane. Methane therefore may play a role on Titan similar to that water plays on Earth. Methane snowflakes may fall onto frozen methane polar caps, while methane raindrops descend into methane rivers, lakes, and seas at warmer locations. Some recent studies have suggested, however, that methane does not exist in all three phases on Titan.

Some molecules are capable of joining together in long, repeating molecular chains, forming substances called **polymers.** Many of the hydrocarbons and carbon–nitrogen compounds in Titan's atmosphere form such polymers. Droplets of some polymers remain suspended in the atmosphere to form the kind of mixture called an **aerosol** (see Figure 14-19), but the heavier polymer particles settle down onto Titan's surface, probably covering it with a thick layer of sticky, tarlike goo. Darrell F. Strobel of the Naval Research Laboratory estimates that the deposits of hydrocarbon sludge on Titan may be $\frac{1}{2}$ km deep.

Figure 14-19   Titan's atmosphere
The Voyager data suggest that Titan's atmosphere has three distinct layers. The uppermost layer absorbs ultraviolet radiation from the Sun. The middle layer is opaque to visible light. The lowest layer is an aerosol of particles suspended in the atmosphere. Methane rainclouds may exist near Titan's surface, which is at a temperature of about 95 K. (Adapted from Tobias Owen)

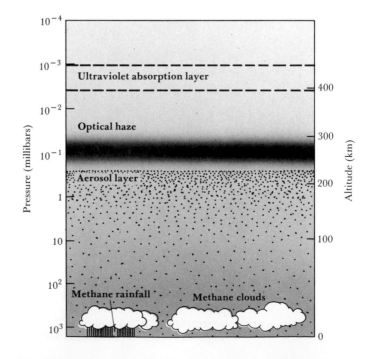

The features of the icy surfaces of Saturn's six moderate-sized moons provide clues to their histories

Saturn's six moderate-sized satellites share several common characteristics. All six circle the planet along **regular orbits,** meaning in the plane of Saturn's equator and in the same direction that the planet itself rotates. All six also exhibit synchronous rotation. Finally, all six have average densities in the range of 1.2 to 1.4 g/cm^3, a value consistent with a composition largely of ice.

Mimas, the innermost of the six middle-sized moons, is heavily cratered. In fact, one of the craters on Mimas (see Figure 14-20) is so large that the impact forming it must have come close to shattering the moon into fragments.

Although the cratered surface of Mimas indicates that it is a geologically dead world, its neighbor and near-twin Enceladus has extensive regions of surface that lack craters (see Figure 14-21). These regions apparently have been "resurfaced" within the past 100 million years by some form of geological activity. Furthermore, Enceladus is the most highly reflective large object in the solar system, with an albedo greater than that of newly fallen snow. Enceladus seems to be covered with extremely pure ice, totally free of the contaminating rock or dust that is thought to explain the lower albedos of the other icy satellites.

One hypothesis suggests that Enceladus, like Jupiter's moon Io, is heated by tidal forces. The satellite Dione orbits Saturn in 65.7 days, almost exactly twice Enceladus's orbital period of 32.9 days. Enceladus therefore is repeatedly caught in a tidal tug-of-war between Saturn and Dione. The tidal flexing thus produced may melt Enceladus's interior, forming water geysers on Enceladus in much the same way that volcanoes are produced in Io. Such geysers might produce deposits of fresh ice and snow on the surface; they might also be the source of the particles that form the thin E ring about Saturn. No direct evidence of such geysers (or of a molten interior) was obtained by the Voyager probes.

Tethys and Dione are the next-largest pair of Saturn's six moderate-sized satellites. Thethys is heavily cratered like Mimas (see Figure 14-22). Also like Mimas, Tethys has one exceptionally huge impact crater; called Odysseus, it is 400 km in diameter. Stretching three-fourths of the way around Tethys is a 2000-km-long complex of valleys called Ithaca

Figure 14-20   Mimas
Mimas is the smallest and innermost of Saturn's six moderate-sized satellites. This view was taken by Voyager 1 at a range of nearly 500,000 km. The huge impact crater, named Arthur after the legendary English king, is 130 km in diameter; Mimas is only 400 km in diameter. (NASA)

Figure 14-21 [left]   Enceladus
This high-resolution image of Enceladus was obtained by Voyager 2 from a distance of 191,000 km. Ice flows and cracks strongly suggest that Enceladus's surface has been subjected to recent geological activity. The youngest, crater-free ice flows are estimated to be less than 100,000 years old. (NASA)

Figure 14-22 [right]   Tethys
Voyager 1 obtained this view of Tethys at a distance of 1.2 million kilometers. Ithaca Chasma is the long canyon near the edge of the sunlit hemisphere. (NASA)

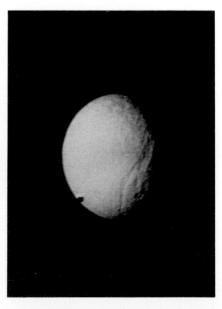

Figure 14-23   Dione
Bright wispy streaks criss-cross Dione's trailing hemisphere. (A similar network of light-colored wisps covers the trailing hemisphere of Rhea.) The nature and cause of these streaks are not known. This view of Dione was obtained by Voyager 1 from a distance of 695,000 km. (NASA)

Chasma that may be a huge crack produced by the impact that created Odysseus.

Dione is slightly larger than Tethys. The surface of its leading hemisphere (the hemisphere pointing toward its direction of orbital motion) and its trailing hemisphere are quite different. Dione's leading hemisphere is heavily cratered like the surface of Tethys, but its trailing hemisphere is covered by a network of strange wispy markings (see Figure 14-23). These wisps may be troughs and valleys in the icy surface, or they may be fresh deposits of frozen water formed by outgassings from Dione's interior.

Rhea and Iapetus are the largest and most distant pair of Saturn's moderate-sized moons. Rhea resembles Dione; its trailing hemisphere is covered by a network of wispy markings, whereas its leading hemisphere is heavily cratered (see Figure 14-24). Whatever process produced the wisps on one side of Dione has apparently also been active on one side of Rhea.

Figure 14-24   Rhea
The leading hemisphere of Rhea is heavily cratered, like the leading hemisphere of Dione. This view was taken by Voyager 1 at a distance of 128,000 km. Surface features as small as 3 km across are visible. (NASA)

Iapetus is the most distant of Saturn's moderate-sized satellites, orbiting $3\frac{1}{2}$ million kilometers from the planet. Iapetus has been known to be an unusual moon ever since its discovery because its albedo varies widely in a regular cycle as it moves about the planet. The Voyager probes confirmed earlier hypotheses about extreme differences between the leading and trailing hemispheres of Iapetus (see Figure 14-25). The leading hemisphere is as black as asphalt, whereas the trailing hemisphere is highly reflective like the surfaces of the other moderate-sized moons.

Most scientists agree that the darkness of Iapetus's leading hemisphere is due to a relatively thin layer of some light-absorbing substance on the surface, but the nature and source of this dark coating are unknown. One possibility is that the coating was formed by substances outgassed from the satellite's interior. Another intriguing hypothesis involves Phoebe, the most distant of Saturn's known satellites.

Phoebe moves in a retrograde direction about an orbit tilted well away from the plane of Saturn's equator. This would lead us to suspect that Phoebe is a captured asteroid, and its very dark surface does indeed resemble the surfaces of a particular class of asteroid called carbonaceous asteroids because of their high carbon content. Perhaps bits and pieces

Figure 14-25 Iapetus
Voyager 2 took this photograph of Iapetus from a distance of 1.1 million kilometers. The leading hemisphere of Iapetus is extremely dark. One theory suggests that the leading hemisphere has been coated with material drifting inward from Saturn's outermost satellite, Phoebe. (NASA)

of this dark charcoal-like material drifting toward Saturn were swept up on the leading hemisphere of Iapetus.

The Voyager spacecraft revealed many new and unsuspected facts about Saturn's complex system of rings and satellites. There is much yet to be learned about the Saturnian system, and this information should bring better understanding of the creation and evolution of other planets, including Earth. The giant moon Titan may even harbor important clues about the origin of life on Earth. Scientists hope for future missions to Saturn that can gather new data, but in the meantime they wonder about the surprises that may be in store as Voyager 2 moves on toward its flybys of Uranus and Neptune.

Summary

• Saturn is circled by a system of thin, broad rings lying in the plane of the planet's equator, which is tilted from the plane of Saturn's orbit. This tilt causes the rings to be seen at various angles by an Earth-based observer over the course of a Saturnian year.

> The faint C ring lies nearest to Saturn; just outside it is the much brighter B ring; outside the B ring is a dark gap called the Cassini division, and beyond that is the moderately bright A ring. Other fainter rings were confirmed by the Voyager probes.

> The rings are composed of numerous fragments of ice and ice-coated rock ranging in size from a few microns to around 10 m.

> The rings exist inside the Roche limit of Saturn, where disruptive tidal forces are stronger then the attractive gravitational forces between the fragments.

> Each of the major rings is composed of a great many narrow ringlets.

> The faint, narrow F ring just outside the A ring has an intricately braided structure of unknown origin.

> Some of the ring boundaries are produced by shepherd satellites, whose gravitational pulls restrict the orbits of the ring fragments.

• Saturn's internal structure and atmosphere are similar to those of Jupiter, but smaller Saturn has a larger rocky core and a thinner mantle of liquid metallic hydrogen, resulting in Saturn's more oblate shape and weaker magnetic field.

Saturn's atmosphere contains less helium than does Jupiter's; this may be due to precipitation of helium downward into the planet, with the resulting friction accounting for Saturn's surprisingly strong heat radiation ($2\frac{1}{2}$ times as great as the incoming solar energy).

The cloud layers in Saturn's atmosphere are spread out over a greater altitude range than those of Jupiter, so that their colors are obscured in a view from outside the atmosphere; Saturn does have belts, zones, zonal jets, and ovals like those of Jupiter.

. Saturn's magnetosphere is similar to Jupiter's except that it has Earth-like radiation belts instead of a current sheet.

. The largest Saturnian satellite, Titan, is a terrestrial world with a very dense nitrogen atmosphere in which methane may play a role similar to that of water on the Earth; a variety of hydrocarbons are formed by interaction of sunlight with methane, forming an aerosol layer in Titan's atmosphere and probably a very thick sludge on its surface.

. Six moderate-sized moons circle Saturn in regular orbits: Mimas, Enceladus, Tethys, Dione, Rhea, and Iapetus. All are probably composed largely of ice, but their surface features and histories vary significantly.

. Saturn has more than dozen much smaller satellites whose tilted orbits and retrograde revolution suggest that they are captured asteroids.

....

Review questions

1 Is there any way we can infer from naked eye observations that Saturn is the most distant of the naked eye planets? Explain.

2 Why do you suppose there are fewer transits, eclipses, and occultations of the Saturnian satellites than of the Galilean satellites?

3 It has been claimed that Saturn would float if one had a large enough bathtub. From the data in Box 14–1, calculate Saturn's average density and comment on this claim.

4 Explain why Saturn is more oblate than Jupiter even though Saturn rotates more slowly than Jupiter.

*5 The Space Telescope will have a resolving power of 0.02 arc sec. Is this sufficient to produce resolved images of any of Saturn's satellites? If so, which ones?

6 Explain how shepherd satellites operate. Is "shepherd satellite" an appropriate term for these objects? Explain.

7 Why do you suppose that Titan is the only satellite in the solar system to possess a substantial atmosphere?

....

Advanced questions

*8 It is well known that Cassini's division involved a 2-to-1 resonance with Mimas. Does the location of the Keeler gap correspond to a resonance with one of the other satellites? If so, which one? (Hint: The Keeler gap is located 133,500 km from Saturn's center.)

*9 Find the escape velocity of Titan. Assuming an average atmospheric temperature comparable to that of Saturn, what is the limiting molecular weight of gases that could be retained by Titan's gravity?

Discussion questions

10 Compare and contrast the internal structures and atmospheres of Jupiter and Saturn. Wherever possible, describe the reasons for differences or similarities in terms of physical parameters such as the mass, chemical composition, and surface gravity.

11 Comment on the suggestion that Titan may harbor life forms.

12 NASA and the Jet Propulsion Laboratory are currently planning a mission to Saturn that will place a spacecraft in orbit about the planet in the late 1990s. In your opinion, what issues should be examined, what data should be collected, and what kinds of questions might such a mission answer?

For further reading

Beatty, J. Reports on the Voyager encounters with Saturn. *Sky & Telescope,* Jan., Oct., and Nov. 1981.

Gore, R. "Saturn: Riddle of the Rings." *National Geographic,* July 1981.

Ingersoll, A. "Jupiter and Saturn." *Scientific American,* Dec. 1981.

Morrison, D. *Voyages to Saturn.* NASA SP-451, 1982. *A good, lavishly illustrated guide to the Voyager discoveries.*

Osterbrock, D., and Cruikshank, D. "Keeler's Gap in Saturn's A Ring." *Sky & Telescope,* Aug. 1982, p. 123.

Owen, T. "Titan." *Scientific American,* Feb. 1982.

Pollack, J., and Cuzzi, J. "Rings in the Solar System." *Scientific American,* Nov. 1981.

David Morrison

David Morrison is acting Vice Chancellor for Research and Graduate Study and Professor of Astronomy at the University of Hawaii. He received his PhD from Harvard University in 1969, where Carl Sagan was his dissertation advisor. His commitment and contributions to astronomy are long-standing. He was recently President of the Astronomical Society of the Pacific and is a Councillor of the American Astronomical Society. He was a member of the Voyager Imaging Science Team and is the author of *Voyage to Jupiter* and *Voyages to Saturn,* two authoritative works on the Voyager missions. He has also served as Assistant Deputy Director of the Office of Space Science at NASA and is presently the chairman of NASA's Solar System Exploration Committee. (Photograph by Andrew Fraknoi)

The future of planetary exploration

The triumphant Voyager flybys of Saturn in 1980 and 1981 represented the culmination of the first golden era of planetary exploration. In the two decades that began in 1962 with the first successful interplanetary spaceflight to Venus, robot probes from Earth visited every one of the planets known to ancient peoples. Spacecraft bearing such names as Pioneer, Venera, Viking, and Voyager set sail on unknown seas and radioed back to our planet images of worlds previously undreamed of. While these tiny robots were providing the initial reconnaissance of the planetary system, and even more ambitious Apollo Project landed 12 astronauts on the surface of the Moon and returned them safely to Earth. Never before in history had our horizons expanded so rapidly. And never before had the results of exploration been so widely shared among the people of the Earth.

For reasons that may be understood by a future generation of historians and sociologists, that initial burst of exploratory energy has, in the 1980's, largely run its course. Now we speak of space as an arena for a "star wars" arms race, and justify the Space Shuttle as a "truck" to launch commercial ventures in low-Earth orbit. The great Apollo Saturn rockets lie rusting on the grass at Cape Kennedy and the NASA Space Center in Houston, while spacecraft meant to transport astronauts to the moon rest instead in museums. Neither the United States nor the U.S.S.R., the two contestants in the space race of the 1960s, has sent a mission of any kind to the Moon in more than a decade.

At the same time, the U.S. program of unmanned planetary exploration has also slowed dramatically, with only two launches scheduled during the entire decade of the 1980s. The Soviets, in contrast, continue their flights to Venus while expanding their targets to include Mars and Comet Halley. The Europeans and Japanese also are initiating plantary programs with flights to Comet Halley in 1986. There is real prospect for the first time in 20 years that the United States will find itself taking a back seat to other nations in space exploration.

Surely, however, this lull in the U.S. space program is but a temporary pause along the road to the planets. The Earth is itself a planet, and the solar system is the proper extended environment for humanity. It is unthinkable that we should refuse to face the challenge of exploring our neighbors in the planetary system.

In spite of the great successes of the planetary program of 1962–1981, we have only begun to explore the worlds of the solar system. Spacecraft have orbited and landed on only the Moon and the two nearest planets. Elsewhere we are dependent on the snapshots returned from flybys. We know Jupiter and Saturn and their spectacular families of satellites and rings no better than the tourist who stops in a city for 24 hours, snaps photos of a couple of prominent sights, chats briefly with a taxi driver, and flys on to the next destination. And we have never been to the outer planets at all, although Voyager 2 will provide an initial glimpse of Uranus in 1986.

One of the most obvious gaps in our current overview of the solar system lies in the area of primitive bodies. All of the larger planets and satellites formed originally from planetesimals composed of the original material of the solar nebula. Subsequently, this material was thermally altered as the planets and satellites heated and evolved geologically. On Earth, which is a very active planet, very few rocks are more than a billion years old, while even the Moon, a much less active world, preserves little information from the first few hundred million years of planetary history. If we wish to unlock the secrets of the birth and early evolution of the solar system, we must look to bodies too small to have been heated and chemically modified. These small remnants of creation are the comets and asteroids. Several spacecraft will speed past Comet Halley in 1986, but this is only a beginning. The continued investigation of comets and asteroids should become a major focus of a renewed program of planetary exploration.

Other broad objectives are easily defined. After the exciting discoveries of the volcanoes of Io and the complex organic chemistry of Titan, how can we fail to return with better instruments and longer stay times? If we wish to understand the atmospheres of the giant planets, we must probe them directly with instrumented entry vehicles. Viking revealed a complex Mars with evidence of plentiful water sometime in its past, yet we have not returned to follow up on these and other discoveries. Even as close to home as our own Moon, no spacecraft has ever photographed the Poles, and our knowledge of the farside remains surprisingly meager.

In the United States the broad scientific objectives of planetary exploration are set by the Space Science Board of the National Academy of Sciences, while specific mission plans to accomplish these goals are the responsibility of NASA and its panels of scientific and engineering experts. The Space Science Board has published a three-volume description of objectives for the exploration of the inner planets, the outer planets, and the small bodies. A specific NASA mission plan now exists for a series of flights through the year 2000. In response to the fiscal austerity imposed on space science by recent administrations, most of the proposed missions are modest in scope, placing great emphasis on economy. Nevertheless, these plans offer good prospects for planetary exploration in the late 1980s.

First there is the much-delayed Galileo orbiter and probe to Jupiter, to arrive at the giant planet in 1988. That same year a Venus Radar Mapper mission is scheduled for launch, to be followed by a Mars orbiter mission in 1990 and a first U.S. comet mission in 1992 or 1993. Other missions planned for the 1990s include a lunar orbiter, a Saturn orbiter, a probe to Titan, and missions to initiate the study of the asteroids.

None of these missions is as complex as Viking, and certainly none comes even close to the level of commitment represented by Apollo. In fact, the entire NASA plan can be supported for less than half the annual funds devoted to planetary exploration throughout most of the 1960s and 1970s. For all of its public and scientific appeal, planetary missions have never represented more than 0.04 percent of the U.S. federal budget. A major planetary mission can be funded from beginning to end for the cost of taking one battleship out of mothballs, and all the planetary scientists in the nation can be happily supported on the overruns associated with the development of even a rather small new weapons system.

Many scientists hope that we can aspire to greater things. Even with our fitful and faltering efforts, we have revolutionized our concept of the planetary system during the past two decades. There is little doubt that history will remember the 1960s and 1970s as the time when humanity began to break the bonds that hold us to our own planet. In comparison to the accomplishments of this exploration program, most other human events seem petty from a historical perspective.

If we could invest but a fraction of the wealth we squander on other pursuits, the planets would be ours. And I refer not just to weapons expenditures. The coins fed into video games in "amusement arcades" would finance a planetary exploration program far beyond today's most optimistic projections. The money spent annually on pizza in the United States would be enough to place humans on Mars in 15 years. For as little as the cost of one movie admission from each citizen of the United States, we could return a pristine piece of a comet for study in terrestrial laboratories.

While I do not propose to eliminate video arcades or pizza or movies from the U.S. scene, I do hope that some of our vast wealth and energy can be directed again toward the stars. It may be that the future wellbeing (or even survival) of humanity will depend on better understanding of the Earth as a planet, or on tapping the resources of the Moon. No one can predict the future with any confidence, but surely it would be the height of folly to ignore our planetary neighbors or to turn back from the space frontier we have begun to explore.

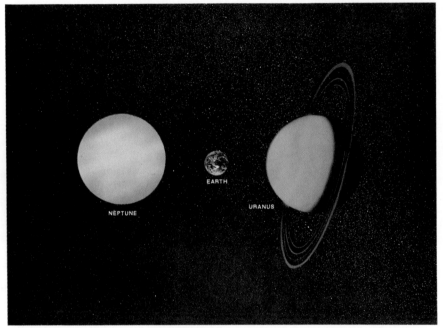

Uranus, Neptune, and Earth
This artist's rendition shows Uranus, Neptune, and Earth to the same scale. Uranus and Neptune are very similar in mass, size, and chemical composition. Voyager 2 will attempt to take the first close up photographs of Uranus in 1986 followed by the first close up pictures of Neptune in 1989. Uranus's rings are dark and narrow, unlike Saturn's, which are broad and bright. Some astronomers suspect that Neptune may also be surrounded by rings. (S. P. Meszaros; NASA)

Until Voyager 2 passes Uranus in 1986 and Neptune in 1989, we are limited in our knowledge of these outer worlds to what can be determined from Earth-based observations. In this chapter we learn that the outer two Jovian planets are distinctly different from the inner two Jovian planets in many characteristics. We learn that Uranus has a unique orientation and an orbiting system of thin rings and five satellites. We find that Neptune has two moons with unusual orbits. Finally, we discuss Pluto and its recently discovered moon Charon, learning that they closely resemble a double asteroid. We begin the chapter, however, with a discussion of the discovery of the outer Jovian planets—a strange story that culminates in the finding of Pluto.

At first glance, Uranus and Neptune appear to be twins. They have nearly the same size, the same mass, the same density, and the same chemical composition. Only a few years ago, however, astronomers regarded the Earth and Venus as near-twins. Our ideas changed rapidly as spacecraft probed the Venusian environment. We were startled to discover how different Venus and Earth are, despite their superficial similarities. In view of the unimagined worlds revealed to us in recent years, we curb our desire to speculate and extrapolate about Uranus and Neptune. Instead, we wait patiently for the Voyager 2 flybys and the launch of the Space Telescope, which will give us more information about these two, remote Jovian planets that today we see only dimly across billions upon billions of kilometers of space.

Uranus was discovered by chance, but Neptune's existence was predicted with Newtonian mechanics

Uranus was discovered by chance on March 13, 1781, by the little-known astronomer William Herschel. Herschel was a German musician who had emigrated to England and then became fascinated with astronomy. Using a telescope that he had constructed, Herschel was systematically surveying the sky when he noticed a faint, fuzzy object that he first thought to be a distant comet. By the end of 1781, however, it had become clear that the object's orbit is planetlike—far beyond Saturn and far beyond regions where comets can normally be seen. Herschel had discovered the seventh planet from the Sun and, in doing so, had doubled the diameter of the known solar system.

Herschel acquired instant fame. He was appointed court astronomer to King George III and was given the financial resources to pursue his scientific interests. During the following decades, Herschel made discoveries that established him as one of the great astronomers of all time.

Although Herschel is credited with discovering Uranus, many other astronomers had sighted the planet in earlier times and mistaken it for a dim star. At opposition, Uranus reaches a magnitude of +5.6, which means that it can be seen faintly with the naked eye under good observing conditions. Uranus is plotted as a dim star on at least 20 star charts drawn between 1690 and 1781.

By the beginning of the nineteenth century, it had become painfully clear that astronomers could not predict the exact orbit of Uranus. The observed positions of Uranus were slightly different from the positions predicted with Newtonian mechanics. By 1830, the discrepancy between observation and prediction had become so large (2 arc min) that some scientists began to suspect a flaw in Newton's laws. The Astronomer Royal of Great Britain, George Airy, was one of those who thought that Newton's law of gravitation might not be accurate at very large distances from the Sun.

In 1843, a 24-year-old student at Cambridge University in England began to explore an earlier suggestion that the gravitational pull of some unknown, distant planet might be causing Uranus to deviate slightly from its predicted orbit. John Couch Adams began trying to calculate the orbit of a planet about the size of Uranus that could account for the unexplained perturbations in Uranus's orbit. He obtained from Airy all of the data available about Uranus's exact observed positions. After two years of hard work, Adams reached the conclusion that Uranus had caught up with and passed a more distant planet during 1822. Uranus accelerated slightly as it approached the unknown planet and then decelerated slightly after 1822 as it receded from the planet. Adams predicted that the unknown planet would be found at a certain position in the constellation of Aquarius.

In October 1845, Adams submitted his calculations and prediction to Airy, but Airy was unconvinced by the work of this lowly student, and the matter was dropped. Meanwhile, in France, the more-established astronomer Urbain Leverrier independently performed the same calculations. In June 1846, Leverrier published his prediction of an unknown planet beyond Uranus. Leverrier's predicted position for the planet differed from Adam's predicted position by less than 1°.

Shocked by Leverrier's agreement with Adams, Airy quickly instructed James Challis, director of the Cambridge Observatory, to begin a thorough search for the unknown planet in the constellation of Aquarius. The search was hampered, however, by the lack of an up-to-date star map of this region at Cambridge Observatory. Many uncharted stars were observed, and each had to be followed over many nights to see whether it showed any planetary motion with respect to the other stars.

While the search proceeded in Cambridge, Leverrier wrote to Johann Gottfried Galle at the Berlin Observatory, urging a search for the unknown planet. The letter was received in Berlin on September 23, 1846, and the planet was sighted that same night. Because of the excellent star charts available in Berlin, Galle was quickly able to locate the only uncharted star of the expected brightness in the predicted location.

Some French astronomers insisted that the new planet should be named Leverrier, but Leverrier himself proposed the name Neptune. After many years of debate between English and French astronomers, credit for the discovery of Neptune eventually came to be divided equally between Adams and Leverrier.

An interesting postscript to this story is that Galileo may have sighted Neptune 233 years before anyone else did. In January 1613, Neptune happened to be located very near Jupiter in the sky. At this time, Galileo was regularly observing the motions of the four large satellites of Jupiter. Galileo's drawings of Jupiter and its satellites show a "star" less than 1 arc min from the predicted location of Neptune during those winter nights. Neptune can be as bright as magnitude +7.7, making it visible through small telescopes. Galileo even notes in his observing log that on one night this star seemed to have moved relative to the other stars. It is therefore quite possible that Galileo had sighted Neptune.

Earth-based observations provide basic information about Uranus and Neptune

Even through a moderately large telescope, both Uranus and Neptune are dim and uninspiring sights. Each planet appears as a small, hazy, featureless disk with a faint greenish-blue tinge (see Figures 15-1 and 15-2).

We know very little about Uranus and Neptune because they are so very far away. Although we have reasonably accurate data on their masses, sizes, and densities (see Boxes 15-1 and 15-2), the rotation rates of Uranus and Neptune are still poorly known. The latest observations suggest that the rotation periods of Uranus and Neptune are about 24 and 18 hours, respectively.

Although Uranus's rotation period is not particularly unusual, the orientation of its axis of rotation is unique in the solar system. Uranus's

Figure 15-1 [left]   Uranus
Nearly 3 billion kilometers from the Sun, Uranus receives only $\frac{1}{400}$ the intensity of sunlight that we experience here on Earth. Uranus is therefore a dim and frigid world that never shines brighter than magnitude +5.7 in our skies. Even though its diameter is about four times Earth's diameter, Uranus always shows us a disk less than 4 arc sec across. (New Mexico State University Observatory)

Figure 15-2 [right]   Neptune
At $4\frac{1}{2}$ billion kilometers from the Sun, Neptune's surface receives only a $\frac{1}{900}$ the intensity of sunlight that we receive on Earth. Neptune is therefore dimmer than Uranus and never shines brighter than magnitude +7.7 in our skies. Although about the same size as Uranus, Neptune looks smaller through Earth-based telescopes because it is farther away. The maximum possible angular diameter of Neptune's disk is 2.2 arc sec. That is roughly the same size as a dime seen from a distance of 1 km. (New Mexico State University Observatory)

Box 15-1   Uranus data

Mean distance from the Sun: 19.182 AU = 2.8696×10^9 km
Maximum distance from the Sun: 20.088 AU = 3.004×10^9 km
Minimum distance from the Sun: 18.275 AU = 2.735×10^9 km
Mean orbital velocity: 6.8 km/sec
Sidereal period: 84.01 years
Rotation period: 23.9 hours
Inclination of equator to orbit: 97.9°
Inclination of orbit to ecliptic: 0.77°
Orbital eccentricity: 0.047
Diameter: 52,290 km
Diameter (Earth = 1): 4.10
Apparent diameter as seen from Earth: maximum = 3.7″
                                     minimum = 3.1″
Mass: 8.66×10^{25} kg
Mass (Earth = 1): 14.54
Mean density: 1.2 g/cm^3
Surface gravity (Earth = 1): 0.79
Escape velocity: 21.2 km/sec
Oblateness of planet: 0.02
Surface temperature (at cloudtops): 57 K
Albedo: 0.35
Brightest magnitude: + 5.6
Mean diameter of Sun as seen from Uranus: 1′ 41″

Box 15-2   Neptune data

Mean distance from the Sun: 30.058 AU = 4.4966×10^9 km
Maximum distance from the Sun: 30.316 AU = 4.537×10^9 km
Minimum distance from the Sun: 29.800 AU = 4.457×10^9 km
Mean orbital velocity: 5.4 km/sec
Sidereal period: 164.79 years
Rotation period: 18 hours
Inclination of equator to orbit: 28.8°
Inclination of orbit to ecliptic: 1.77°
Orbital eccentricity: 0.009
Diameter: 49,500 km
Diameter (Earth = 1): 3.88
Apparent diameter as seen from Earth: maximum = 2.2″
                                     minimum = 2.0″
Mass: 1.03×10^{26} kg
Mass (Earth = 1): 17.23
Mean density: 1.66 g/cm^3
Surface gravity (Earth = 1): 1.12
Escape velocity: 23.6 km/sec
Oblateness of planet: 0.027
Surface temperature (at cloudtops): 57 K
Albedo: 0.35
Brightest magnitude: + 7.7
Mean diameter of Sun as seen from Neptune: 1′ 04″

axis of rotation lies very nearly in the plane of the planet's orbit (see Figure 15-3). Consequently, as Uranus moves along its 84-year orbit, the planet's north and south poles alternately point toward or away from the Sun, producing extremely exaggerated seasonal changes on the planet. During the summertime near Uranus's north pole, the Sun remains above the horizon for nearly 42 Earth years while southern latitudes are

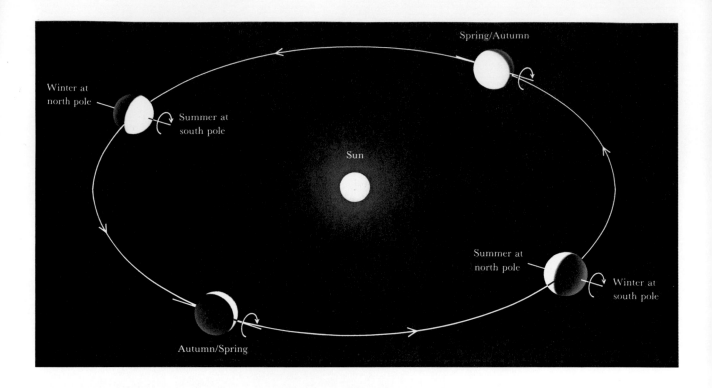

**Figure 15-3   Exaggerated seasons
on Uranus**

*Uranus's axis of rotation is tilted so steeply
that it lies nearly in the plane of the planet's
orbit. Seasonal changes on Uranus therefore
are severely exaggerated. For example, during
midsummer at Uranus's south pole, the Sun
appears nearly overhead for many Earth years,
while the planet's northern regions are subjected to a long, continuous winter night.*

Figure 15-4 [left]   Uranus: the best view
*This picture was produced by an automated
36-in. telescope carried by a balloon to an altitude of 80,000 ft, above the degrading effects
of the Earth's atmosphere. Uranus appears
totally featureless, although continent-sized
markings as small as 2000 km across should
have been visible. (Project Stratoscope II,
Princeton University)*

**Figure 15-5 [right]   Infrared spectra of
Uranus and Neptune**
*The near-infrared spectra of Uranus and
Neptune show broad absorption due to meth-
ane from 9600 to 10,000 Å. These spectra
were obtained with the 4-m telescope at Kitt
Peak by Harold Reitsema, Bradford Smith,
and Stephen Larson.*

subjected to a continuous, frigid winter night. For the next 42 Earth
years, the situation is reversed.

Scientists have known about Uranus's unusual orientation for many
years. Yet no one has ever tackled the problem of calculating the effects
of the exaggerated seasons on the planet's climates. This reluctance
stems in part from the absence of clues about the planet's atmospheric
circulation patterns. Our very best pictures of Uranus show absolutely no
cloud features at all (see Figure 15-4). Astronomers hope that the flyby

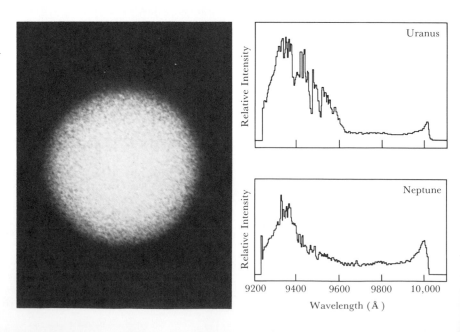

of Voyager 2 in January 1986 will yield views of cloud patterns on Uranus.

The unusual orientation of Uranus's rotation axis should produce an unusual magnetosphere if Uranus has a magnetic field. Evidence suggesting the presence of a magnetosphere came from the Earth-orbiting International Ultraviolet Explorer in 1982. The satellite detected from Uranus strong ultraviolet emission characteristic of aurorae created when charged particles in the solar wind are accelerated down into a planet's upper atmosphere, where they collide with hydrogen atoms and molecules. Examining Uranus's magnetosphere will be another of the major objectives of the Voyager 2 flyby.

Uranus and Neptune have similar interiors that differ from those of Jupiter and Saturn

Using spectroscopy, astronomers have learned about the chemical composition of Uranus and Neptune. Figure 15-5 shows the near-infrared spectra of the two planets. The spectra are remarkably similar, each showing a broad absorption due to methane between 9600 and 10,000 Å. Hydrogen has also been detected, and the presence of helium has been deduced.

The outer two Jovian planets are significantly different from the inner two Jovian planets. The very large spheres of Jupiter and Saturn are composed primarily of hydrogen and helium solar abundances. Uranus and Neptune, however, are distinctively smaller and less massive than Jupiter or Saturn. If Uranus and Neptune also had solar abundances of the elements, their smaller masses would produce less compression and therefore lower average densities than those of Jupiter and Saturn. In fact, however, Uranus and Neptune have average densities comparable to or greater than those of Jupiter or Saturn. We conclude therefore that Uranus and Neptune must contain greater proportions of the heavier elements than the solar abundances. The pioneering work of Rupert Wildt at Yale University in the 1930s suggested that the interiors of Uranus and Neptune contain significant amounts of oxygen, nitrogen, carbon, silicon, and iron in addition to abundant hydrogen and helium.

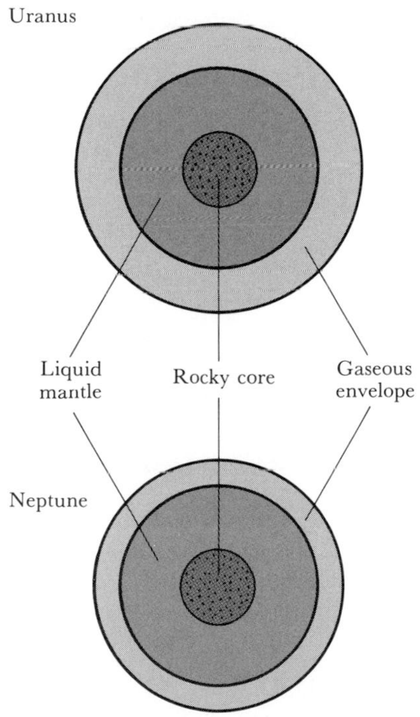

From the known physical properties of Uranus and Neptune, William Hubbard and J. J. MacFarlane have calculated that the interiors of these planets have three-layered structures (see Figure 15-6). Each planet probably has a rocky core composed primarily of iron and silicon. The masses and densities of the planets suggest pressures at their centers on the order of 20 million bars and central temperatures of roughly 7000 K. Surrounding the rocky cores are liquid mantles of water (H_2O), ammonia (NH_3), and methane (CH_4). Thus the mantles contain abundant amounts of oxygen, nitrogen, and carbon in combination with hydrogen. The outer layers of the two planets are predominantly hydrogen and helium in gaseous state at low density.

Although similar to Uranus in internal structure, Neptune emits slightly more radiation than it receives from the Sun, whereas Uranus does not. Although farther from the Sun, Neptune's cloudtops register the same temperature as Uranus's (47 K = $-216°C$ = $-357°F$).

As we saw in preceding chapters, both Jupiter and Saturn have internal energy sources. Jupiter is still releasing energy trapped inside the planet during its formation, whereas Saturn's excess heat probably comes from the precipitation of helium. However, neither of these mechanisms would work on Uranus or on Neptune because these outer planets are too small and contain too little hydrogen and helium. Some astronomers

Uranus

Liquid mantle   Rocky core   Gaseous envelope

Neptune

Figure 15-6 The interiors of Uranus and Neptune
Uranus and Neptune have very similar interior structures. Each planet has a rocky core surrounded by a liquid mantle of water, methane, and ammonia, which in turn is enveloped in a low-density gaseous layer composed largely of hydrogen and helium.

speculate that the planets' atmospheres may somehow regulate the flow of heat energy into space.

Uranus's atmosphere is cold and clear to great depths; there is no haze, and clouds have never been seen. In contrast, Neptune's upper atmosphere seems to contain a substantial haze that may consist of aerosol particles or ice crystals. Some astronomers suspect that differences in atmospheric structure and the intensity of sunlight (Uranus receives $2\frac{1}{2}$ times more sunlight than Neptune) produce an effect that controls how much heat can escape from the interiors of the two planets. The details of this mechanism are poorly understood. We hope that data from Voyager 2 will shed light on this issue.

Uranus is orbited by five moderate-sized, icy satellites and a system of thin rings

Uranus is orbited by five satellites each with its orbit in the planet's equatorial plane (see Figure 15-7). The two largest moons (Titania and Oberon) were discovered by William Herschel in 1789. Another English astronomer, William Lassell, discovered Ariel and Umbriel in 1851. Uranus's smallest and dimmest known moon, Miranda, was first detected by Gerard Kuiper in 1948, using the 2.1-m McDonald Observatory reflector.

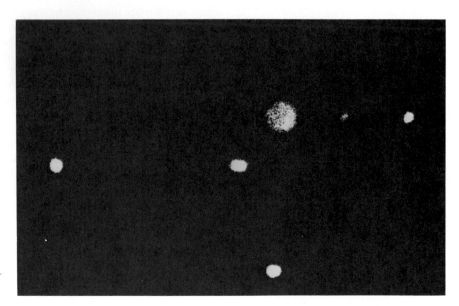

Figure 15-7   Uranus and its five moons
This photograph of Uranus and its satellites was taken in infrared wavelengths by William Sinton at the Mauna Kea Observatory on Hawaii. (Institute for Astronomy, University of Hawaii)

In 1982, astronomers H. R. Brown, Dale Cruikshank, and David Morrison used NASA's 3-m infrared telescope on top of Mauna Kea in Hawaii to measure the diameters and albedos of the four largest Uranian moons. Titania and Oberon have diameters of nearly 1600 km, whereas those of Ariel and Umbriel are 1300 and 1100 km, respectively. Thus all four are comparable in size to Saturn's moderate-sized satellites.

Cruikshank's pioneering observations of water-ice absorption in the infrared spectra of the Uranian moons reveal another similarity to Saturn's satellites. Presumably, ice composes a large fraction of the Uranian moons. Albedo measurements indicate, however, that Uranus's moons are much less reflective than Saturn's. The albedos of Oberon, Titania,

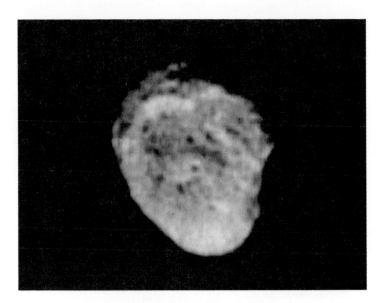

Figure 15-8 An icy, dust-covered satellite
This view of the Saturnian moon Hyperion was obtained by Voyager 2 during its flyby in 1981. The reflectivity of Hyperion's surface is similar to that of the Uranian moon Ariel. Views of the Uranian satellites will be transmitted back to Earth from Voyager 2 in January 1986. (NASA)

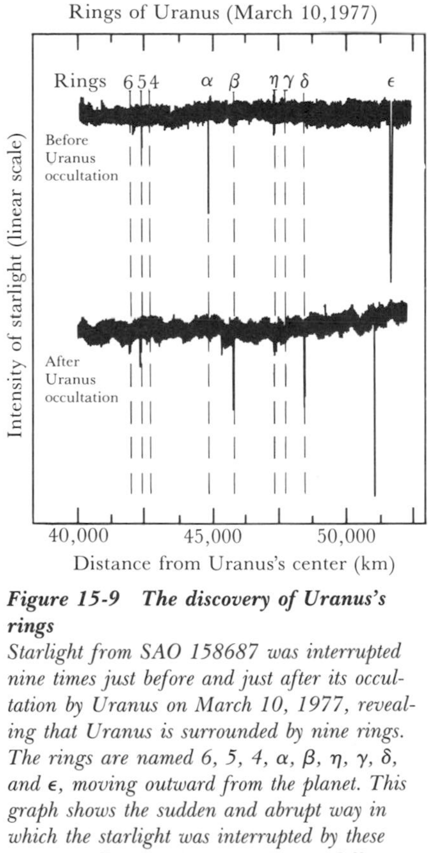

Rings of Uranus (March 10, 1977)

Intensity of starlight (linear scale)

Rings   6 5 4   α β   η γ δ   ε

Before Uranus occultation

After Uranus occultation

40,000   45,000   50,000

Distance from Uranus's center (km)

Figure 15-9 The discovery of Uranus's rings
Starlight from SAO 158687 was interrupted nine times just before and just after its occultation by Uranus on March 10, 1977, revealing that Uranus is surrounded by nine rings. The rings are named 6, 5, 4, α, β, η, γ, δ, and ε, moving outward from the planet. This graph shows the sudden and abrupt way in which the starlight was interrupted by these exceptionally narrow rings. The two different positions for the ε ring are a result of its elliptical shape. (From observations by James L. Elliot and colleagues)

and Umbriel are comparable to that of Callisto, the darkest of the Galilean satellites. The albedo of Ariel is similar to that of Hyperion, which Voyager photographs showed to be a reddish, irregularly shaped body (see Figure 15-8). Presumably, the Uranian satellites are icy worlds covered with substantial coatings of darker material. Very little is known about tiny Miranda.

Another striking similarity between Uranus and Saturn is that each planet is surrounded by a ring system. Uranus's rings were discovered by accident, as was the planet itself. On March 10, 1977, Uranus was scheduled to occult (move in front of) a faint star called SAO 158687, as seen from the Indian Ocean and adjacent areas on Earth. This event afforded an excellent opportunity to measure Uranus's diameter and to study its upper atmosphere. Because Uranus's position and speed along its orbit were already known, astronomers needed only to measure how long the star was hidden to deduce the planet's size. In addition, by carefully measuring how the starlight faded as Uranus passed in front of the star, they planned to deduce important properties of Uranus's upper atmosphere.

With these goals in mind, several astronomers journeyed to the Indian Ocean, including a team headed by James L. Elliot of Cornell University in NASA's Kuiper Airborne Observatory (recall Figure 5-22). To everyone's surprise, the star SAO 158687 briefly blinked on and off several times just before the occultation and then again immediately after the occultation. The astronomers concluded that Uranus is surrounded by a series of very narrow rings. In all, nine rings were counted. They are named by numbers and Greek letters (see Figure 15-9).

Uranus's rings are very different from those of Saturn's. Whereas Saturn's rings are broad and bright, Uranus's are narrow and dark. Most of the Uranian rings are less than 10 km wide. The widest ring, which also happens to be the outermost ring, is called the ε ring and has a width of about 100 km.

Uranus's nine rings are all located between 1.60 and 1.95 Uranian radii from the planet's center, well within the planet's Roche limit. An examination of data from subsequent occultations confirms that the spaces between the rings are free of debris. Some sort of mechanism, perhaps involving shepherd satellites, efficiently confines particles to these nine narrow orbits.

Typical particles in Saturn's rings are chunks of ice with the reflectivity and dimensions of snowballs, but typical particles in Uranus's rings could be compared with lumps of coal. The Uranian rings reflect only a small percentage of the sunlight that falls on them.

Six of the Uranian rings are slightly noncircular. The most eccentric ring is the outer ϵ ring. Scrutiny of data from several occultations suggests that the ϵ ring is lumpy and may consist of several strands. Perhaps it is kinky and braided like Saturn's F ring (recall Figure 14-13), which also happens to have an overall width of about 100 km.

Neptune's two largest moons, Triton and Nereid, follow unusual orbits

Figure 15-10 [left]   Neptune and Triton
Triton, one of the seven planet-sized satellites in the solar system, is seen here near Neptune. Triton circles Neptune in a retrograde orbit once every six days. The size of Triton's orbit is decreasing so rapidly that the satellite will be inside Neptune's Roche limit in less than 100 million years. (Yerkes and McDonald Observatories)

Figure 15-11 [right]   Nereid
Tiny Nereid, Neptune's second largest satellite, was discovered in 1949 and is indicated by the arrow in this photograph. Triton appears as a "lump" just below the overexposed image of Neptune. Nereid's orbit has the greatest eccentricity of any satellite orbit in the solar system. (Yerkes and McDonald Observatories)

It is reasonable to suppose that Neptune, like the other three Jovian planets, is surrounded by rings. Neptune passes in front of a star roughly once a year, and astronomers have searched for the telltale momentary dimming of the star light. Although the observations have been inconclusive, they have revealed the probable existence of a small satellite, less than 200 km in diameter, well inside the orbit of Neptune's largest moon, Triton.

Triton (see Figure 15-10) was discovered by William Lassell in 1846, just five years before he discovered two Uranian satellites. Along with Titan, the Galilean satellites, and our Moon, Triton is one of the seven giant moons of our solar system. Although Triton's size is somewhat uncertain (its diameter is estimated to be 4000 km), it clearly is of planetary dimensions.

Neptune's second largest moon, Nereid (see Figure 15-11), was discovered by Gerard Kuiper in 1949, just one year after he discovered Miranda. Like Miranda, Nereid is quite small; its diameter is estimated to be somewhat less than 1000 km.

Triton and Nereid have unusual orbits. Triton's orbit is retrograde. Nereid's orbit is the most eccentric of any satellite in the solar system. The distance between Nereid and Neptune varies from 1.4 million kilometers to 9.7 million kilometers as this tiny moon moves along its highly

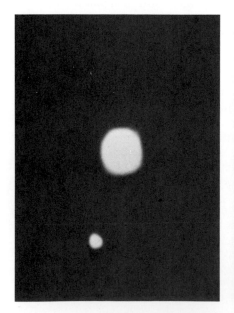

elliptical path. Triton completes its backward orbit in only 6 days, but Nereid takes nearly 360 days to go once around Neptune.

An extraordinary fact has emerged from recent dynamical studies of Triton: its orbit is decaying. That is, Triton is slowing and spiraling in toward Neptune. In about 10 million to 100 million years, Triton will be inside Neptune's Roche limit. The giant satellite then will be torn to pieces by Neptune's tidal forces. Thus, even if Neptune does not already have rings, it will develop a spectacular ring system as rock fragments gradually spread out along the satellite's former orbit.

Spectroscopic studies suggest that Triton's surface is covered with rock rather than ice. Dale Cruikshank's data from NASA's 3-m infrared telescope on Hawaii also indicate the presence of a sea of liquid nitrogen on Triton. Traces of a methane atmosphere have also been detected. Triton will be a fascinating target for Voyager 2 in 1989.

Pluto was discovered during a laborious search of the heavens

Speculations about a ninth planet date back to the late 1800s, when a few astronomers suggested that an unknown planet might be perturbing Neptune's orbit. Such statements were more than a bit premature. Nevertheless, encouraged by the fame of Adams and Leverrier, several people set out to become the discoverer of "Planet X." Two Bostonian gentlemen, William Pickering and Percival Lowell, were prominent leaders in this effort.

Modern calculations show that there were no statistically significant perturbations of Neptune's orbit. It is therefore not surprising that no planet was found at the positions predicted by Pickering, Lowell, and others. Nevertheless, the search continued.

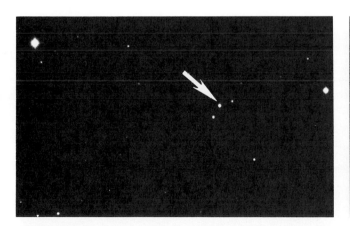

Figure 15-12 Pluto
Pluto was discovered in 1930 by searching for a dim, starlike object that slowly moves in relation to background stars. These two photographs were taken one day apart. Although Pluto does not show a measurable disk, astronomers estimate that the planet is roughly the same size as our Moon. (Lick Observatory)

Before he died in 1916, Percival Lowell urged that a special wide-field camera be constructed to help with the search for Planet X. After many delays, the camera was finished in 1929 and installed at the Lowell Observatory in Arizona where a young astronomer, Clyde W. Tombaugh, had joined the staff to carry on the project. On February 18, 1930, Tombaugh finally discovered the long-sought planet. It was disappointingly faint (15th magnitude) and showed no discernible disk. The planet was named Pluto (the mythological god of the underworld, whose name has Percival Lowell's initials as its first two letters), and the discovery was publicly announced on March 13, the 149th anniversary of the discovery of Uranus. Two typical photographs of Pluto appear as Figure 15-12.

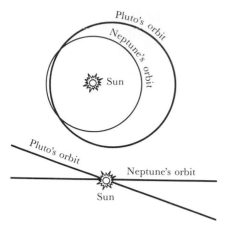

Figure 15-13   Pluto's orbit
Pluto's orbit is more elliptical and more steeply inclined to the plane of the ecliptic than is the orbit of any other planet in the solar system. In fact, the orbit is so eccentric that Pluto is occasionally closer than Neptune is to the Sun.

Although the measurements are uncertain, Pluto seems to be slightly smaller than our Moon. Therefore, even if Pluto is made of solid rock, its mass must be smaller than the Moon's mass. This small mass could not account for the gravitational perturbations that Pickering and Lowell had thought they found in Neptune's orbit. Therefore, Tombaugh kept searching. For the next 13 years, he examined endless pairs of photographs (like the pair in Figure 15-12), looking for the telltale displacement of a faint starlike image. None was ever found. If a trans-Pluto planet exists, it must be fainter than 18th magnitude. If there is a tenth planet, it must be either very small (more like an asteroid than a planet in size) or very far away.

Pluto's orbit about the Sun is more elliptical and more steeply inclined to the plane of the ecliptic than the orbit of any other planet in the solar system (see Figure 15-13). In fact, Pluto's orbit is so eccentric that Pluto is sometimes closer to the Sun than Neptune is. For example, when Pluto is at perihelion in 1989, it will be more than 100 million kilometers closer than Neptune is to the Sun.

There are many things that we do not know with confidence about Pluto. We are not sure of its mass, its size, or its average density (see Box 15-3). But we do know how fast it rotates. A solar day on Pluto lasts 6 days 9 hours 18 minutes. Although we cannot see any surface details on Pluto, its magnitude varies with this period. Apparently there are light or dark features—perhaps a large frozen lake—that are periodically exposed to our Earth-based view, causing the brightness variations that divulge the planet's rotation period.

Pluto and its moon Charon are roughly comparable in size

In 1978, while examining some photographs of Pluto, James W. Christy of the U.S. Naval Observatory noticed that the image of the planet was slightly elongated. Pluto seemed to have a lump on one side (see Figure 15-14). Numerous other photographs of Pluto were promptly examined; some of them showed this lump whereas others did not. It was soon evident that the lump is actually a satellite that is occasionally far enough

Figure 15-14   Pluto and Charon
Pluto's moon Charon appears as a slight elongation or lump on one side of this greatly enlarged image of the planet. Pluto and Charon are separated by only 19,700 km. The Pluto–Charon system deserves to be called a double planet because these two objects resemble each other in mass and size more closely than do any other planet–satellite pair in the solar system. (Courtesy of James Christy and Robert Harrington)

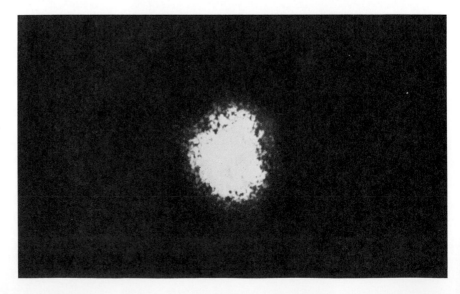

Box 15-3   Pluto data

Mean distance from the Sun: 39.44 AU = 5.900×10^9 km
Maximum distance from the Sun: 49.19 AU = 7.375×10^9 km
Minimum distance from the Sun: 29.69 AU = 4.425×10^9 km
Mean orbital velocity: 4.7 km/sec
Sidereal period: 247.7 years
Rotation period: 24.6 hours
Inclination of equator to orbit: 50° (?)
Inclination of orbit to ecliptic: 17.2°
Orbital eccentricity: 0.250
Diameter: 3000 km (?)
Diameter (Earth = 1): 0.2 (?)
Apparent diameter as seen from Earth: 0.2″
Mass: 10^{22} kg (?)
Mass (Earth = 1): 0.002 (?)
Mean density: 1 g/cm^3 (?)
Surface gravity (Earth = 1): 0.4 (?)
Escape velocity: 5 km/sec (?)
Surface temperature: 50 K (?)
Albedo: 0.4
Brightest magnitude: + 14
Mean diameter of Sun as seen from Pluto: 50″

away from Pluto to show up as a slight elongation in the planet's image. Christy proposed that the newly discovered moon be christened Charon (pronounced like Karen) after the mythical boatman who ferried souls across the River Styx to Hades, the domain ruled by Pluto.

Charon's orbit is quite remarkable. First, the average distance between Charon and Pluto is a scant 19,700 km—less than $\frac{1}{20}$ the distance between the Earth and our Moon. Second, Charon's orbital period is 6.3874 days, the same as the rotational period of Pluto. In other words, Pluto rotates synchronously with the revolution of its satellite. As seen from the satellite-facing side of Pluto, Charon neither rises nor sets but instead seems to hover in the sky, as if perpetually suspended above the horizon.

Charon and Pluto are remarkably comparable in both size and mass. The measurements are rather uncertain, but reasonable estimates place Pluto's diameter at 3000 km and Charon's diameter at 1400 km.

We know both the separation and the orbital period of the Pluto–Charon system, so we can use Kepler's third law (recall Box 4-3) to determine the total mass of Pluto and Charon. For Pluto and Charon together, the mass is 0.0026 Earth masses.

Although Kepler's third law does not give the individual masses, we can make some headway by assuming that Pluto and Charon have the same average density because they are probably made of the same or similar material. This means that their individual masses are proportional to their volumes, which in turn are proportional to the cubes of their radii. We find that Pluto is probably about ten times as massive as Charon. Relating these results to more familiar quantities, we find that Pluto's mass is about $\frac{1}{5}$ that of our Moon, whereas Charon's mass is about $\frac{1}{50}$ that of our Moon. With these masses, the average densities of Pluto and Charon turn out to be roughly 1 g/cm^3, quite similar to those of the icy worlds that orbit Saturn.

Pluto may once have been a satellite of Neptune

In size, mass, and density, Pluto and Charon resemble the satellites of the Jovian planets. Especially in view of Pluto's unusual orbit about the Sun, it is reasonable to wonder whether Pluto might be an escaped satellite that once orbited Neptune. Perhaps some sort of cataclysmic event occurred in the ancient past to reverse the direction of Triton's orbit, fling Nereid into its highly elliptical orbit, and catapult Pluto away from Neptune altogether.

That Pluto is an escaped satellite of Neptune was first proposed by R. A. Lyttleton of Cambridge University in 1936 and vigorously championed by Gerard P. Kuiper, the father of modern planetary astronomy. Problems with this proposal have, however, been long recognized. For one, the present-day orbits of Neptune and Pluto do not intersect. The two planets are never closer than 384 million km, despite the proximity suggested in Figure 15-13. If Pluto were catapulted away from Neptune (for example, by a near collision with Triton and Nereid), why does Pluto's orbit not pass through Neptune's orbit at the site of this catapulting event?

With the discovery of Charon, Thomas Van Flandern and Robert S. Harrington of the U.S. Naval Observatory made the intriguing proposal that an unknown, massive, planetlike object may have been involved in Pluto's escape from Neptune. Perhaps Triton, Nereid, and Pluto all orbited Neptune along well-behaved regular orbits in the ancient past. Then, in a near-collision, the unknown planet severely perturbed the orbits of all three satellites. Triton and Nereid were swung into their present unusual orbits. At the same time, while catapulting Pluto into a heliocentric orbit, tidal forces from the unknown planet tore Pluto into two pieces.

This Flandern–Harrington hypothesis explains why the orbits of Neptune and Pluto do not intersect: the gravitational perturbation by the unknown planet supplied enough energy and momentum to tip Pluto's orbit away from Neptune's. The tidal fragmentation of Pluto would probably have exposed extensive areas of fresh ice, uncontaminated by a darker coating of meteoritic dust and rocky debris. The resulting varied reflectivity across the planet's surface is perhaps the cause of the nonuniform albedo observed during Pluto's rotation. To have had this effect on Neptune's moons, Van Flandern and Harrington estimate that the unknown planet's mass must be about five times the mass of the Earth.

Where might this unknown planet be today? Van Flandern and Harrington suggest that the catastrophic encounter with Neptune's satellites flung the planet into a very large elliptical orbit whose average distance from the Sun is 50 to 100 AU. In such an orbit, the planet would take 350 to 1000 years to circle the Sun and would be so faint that it would be extremely difficult to discover.

Pluto and Charon resemble a double asteroid

Although intriguing, the idea that there might be a massive trans-Pluto planet is by no means commonly accepted by astronomers. One of the main objections concerns the resemblance of the Pluto–Charon system to a pair of asteroids. As we shall see in the next chapter, there is increasing evidence that some asteroids may be double objects—two comparably sized chunks of rock in close proximity to each other. Some astronomers vigorously argue that the ninth planet from the Sun should be demoted to asteroid status.

This viewpoint is supported by the recent discovery of a large asteroid quite far from the Sun. In November 1977, Charles Kowal (working at the Palomar Observatory) noticed a faint streak on a photograph of stars taken with the 48-in. Schmidt telescope. Suspecting that the streak might be the blurred image of a slowly moving asteroid, Kowal followed the object. Its elongated orbit was soon found to lie mostly between the orbits of Saturn and Uranus.

Kowal named the remote asteroid Chiron (pronounced Kiron, and not to be confused with the name of Pluto's moon) after a wise centaur in Greek mythology. Chiron circles the Sun once every 50.68 years along a highly elliptical orbit that extends slightly inside the orbit of Saturn. At aphelion, Chiron is 18.9 AU from the Sun, or quite near Uranus's orbit. Calculations suggest that Chiron will someday either collide with Saturn or Uranus, or during a near-miss will be subjected to gravitational perturbations that will eject the asteroid from the solar system altogether.

Chiron, whose diameter is estimated to be 100 km, is just one of thousands of rock and ice fragments that orbit our Sun. The vast majority of these asteroids are located in the **asteroid belt** between the orbits of Mars and Jupiter—or so we assume. Perhaps an observational **selection effect** inhibits our viewing of extremely remote asteroids because they are so dim. With the discovery of Chiron, some astronomers have begun to speculate that there might be another asteroid belt between the orbits of Saturn and Uranus. Perhaps there really is a lot of unknown interplanetary debris out there, of which Pluto and its moon just happen to be the largest pieces.

Summary

. Uranus was discovered by chance; Neptune was discovered at a location predicted with Newtonian mechanics; Pluto was discovered after a long search; if another planet exists beyond Pluto, it is either very small or very far away.

. Both Uranus and Neptune have three-layered internal structures: a rocky core surrounded by a liquid mantle of water, ammonia, and methane, with an outer gaseous envelope composed predominantly of hydrogen and helium.

. Uranus is unique among the planets and satellites of the solar system in that its axis of rotation lies nearly in the plane of its orbit, producing greatly exaggerated seasonal changes on the planet.

> Uranus has a system of thin, dark rings and five satellites similar to the moderate-sized moons of Saturn.

. Neptune has a terrestrial-sized satellite, Triton, that moves in a retrograde orbit and a small satellite, Nereid, that moves in a very eccentric orbit; Triton's orbit is decaying, and the satellite will be torn apart to form a new ring system within about 100 million years.

> Triton may have a rock surface with liquid nitrogen seas and a very thin methane atmosphere.

. Pluto and its moon Charon have roughly comparable masses and move in a highly elliptical orbit that is steeply inclined to the plane of the ecliptic.

> The Pluto–Charon system may represent an escaped satellite of Neptune or a double asteroid moving in a very large orbit about the Sun.

. Chiron is a recently discovered asteroid that moves in a highly elliptical orbit between the orbits of Saturn and Uranus; its existence suggests that asteroids may be more common in the outer solar system than previously suspected.

Review questions

1 Could astronomers of antiquity see Uranus? If so, why do you suppose it was not recognized as a planet?

2 Why do you suppose that the discovery of Neptune is rated as one of the great triumphs of science while the discoveries of Uranus and Pluto are not?

3 On the basis of Earth-based observations and our current understanding, would you expect the Voyager spacecraft to discover magnetic fields and magnetospheres about Uranus and Neptune? Explain.

4 Why is it reasonable to suppose that Neptune is surrounded by rings? How might you detect such a ring system using an Earth-based telescope?

5 How can astronomers distinguish a faint solar system object like Pluto from background stars in the same field of view?

***6** How much fainter does the Sun appear from Pluto than it does from Earth?

7 Would you expect to see impact craters on Pluto and Charon? Explain.

Advanced questions

***8** If Earth-based telescopes can resolve angles down to 0.25 arc sec, how large could an object be at Pluto's average distance from the Sun and still not present a resolvable disk?

9 Suppose that you were standing on Pluto. Describe the motions of Charon relative to the Sun, the stars, and your horizon. Would you ever be able to see a total eclipse of the Sun? Under what circumstances would you never see Charon?

Discussion questions

10 Compare and contrast Triton and Titan.

11 Discuss the controversies surrounding the nature and origin of Pluto. Enumerate the ways in which Pluto resembles a planet, an asteroid, and a former satellite. What sorts of data or observations from Voyager 2 and the Space Telescope might elucidate Pluto's origin?

12 NASA and the Jet Propulsion Laboratory have tentative plans to place spacecraft in orbit about Uranus and Neptune early in the twenty-first century. What kinds of data should be collected and what questions would you like to see answered by these missions?

For further reading

Belton, M. "Uranus and Neptune." *Astronomy,* Feb. 1977, p. 6.
Bennett, J. "The Discovery of Uranus." *Sky & Telescope,* Mar. 1981, p. 188.
Chaikin, A. "New Light on Cold Worlds." *Sky & Telescope,* July 1983, p. 23.
Elliot, J., et al. "Discovering the Rings of Uranus." *Sky & Telescope,* June 1977, p. 412.
Harrington, R., and B. "The Discovery of Pluto's Moon." *Mercury,* Jan./Feb. 1979, p. 1.
Mulholland, D. "The Ice Planet [Pluto]." *Science 82,* Dec. 1982, p. 64.
Tombaugh, C. "The Search for the Ninth Planet: Pluto." *Mercury,* Jan./Feb. 1979, p. 1

16 **Interplanetary vagabonds**

Comet rendezvous and asteroid flyby
This artist's conception shows the Mariner Mark II spacecraft during a proposed mission to a comet and an asteroid. After close-up examination of one or two asteroids, the spacecraft will rendezvous with a comet for detailed remote-sensing studies. Several comets with short periods and nearly circular orbits in the vicinity of the asteroid belt are likely candidates for this mission. Scientists hope that this spacecraft will be launched in 1990. (JPL)

The planets are not the only objects that move in orbits about the Sun. In this chapter we discuss the asteroids, meteoroids, and comets that are small but significant members of the solar system. We learn that these objects provide major clues about the history of the solar system. We also discuss speculations that meteoroids may have played a role in the origin of life on Earth and in the extinction of more than one-half of the species living on Earth some 63 million years ago.

When William Herschel in 1781 discovered a new planet beyond Saturn, he proposed that it be named Georgium Sidus in honor of King George III of Great Britain. Astronomers in other countries rejected this suggestion, instead adopting the name Uranus that had been proposed by a young German astronomer, Johann Elert Bode. Bode is far more famous, however, for his earlier popularization of a simple rule that describes the distances of the planets from the Sun. Although Bode did not discover this rule, it is usually known today as Bode's law. Most astronomers now regard this "law" as merely a coincidence, but it did lead directly to the discovery of a large number of previously unknown objects that orbit about the Sun.

Bode's law led astronomers to search for a planet between the orbits of Mars and Jupiter

Bode's rule for remembering the distances of the planets from the Sun goes like this:

1 Write down the sequence of numbers 0, 3, 6, 12, 24, 48, 96, ... (Note that each number after the second one is simply twice the preceding number.)

2 Add 4 to each number in the sequence.

3 Divide each of the resulting numbers by 10.

As shown in Table 16-1, the final result is a series of numbers that corresponds remarkably well to the distances (in AU) of the planets from the Sun.

TABLE 16-1 Bode's law

Bode–Titius progression	Planet	Actual distance (AU)
$(0 + 4)/10 = 0.4$	Mercury	0.39
$(3 + 4)/10 = 0.7$	Venus	0.72
$(6 + 4)/10 = 1.0$	Earth	1.00
$(12 + 4)/10 = 1.6$	Mars	1.52
$(24 + 4)/10 = 2.8$	?	
$(48 + 4)/10 = 5.2$	Jupiter	5.20
$(96 + 4)/10 = 10.0$	Saturn	9.54
$(192 + 4)/10 = 19.6$	Uranus	19.18
$(384 + 4)/10 = 38.8$	Neptune	30.06
$(768 + 4)/10 = 77.2$	Pluto	39.44

Astronomers regarded Bode's rule as merely a useful trick for remembering the planetary distances until Herschel's unexpected discovery of Uranus very near the orbit predicted by Bode's scheme for a planet beyond Saturn. Suddenly it seemed far more likely that Bode's rule might actually represent some physical property of the solar system. The numerical sequence soon became known as **Bode's law**—an unfortunate name because it neither is a physical law nor was it invented by Bode. It had first been published in 1766 by Johann Titius, a German physicist and mathematician.

Astronomers now looked with new interest at the "missing planet" in the sequence of Bode's law—the gap between the orbits of Mars and Jupiter. Six German astronomers who jokingly called themselves the Celestial Police organized an international group to begin a careful search for the missing planet. Before their search had gotten underway, however, the anticipated announcement came from Sicily.

The Sicilian astronomer Giuseppe Piazzi was carefully preparing a map of faint stars in the constellation of Taurus. On January 1, 1801, he noticed a dim, previously uncharted star that shifted its position slightly over the next several nights. Suspecting that he might have found the missing planet of Bode's law, Piazzi excitedly wrote to Bode in Berlin.

Unfortunately, Piazzi's letter did not reach Bode until late March. By that time, Piazzi's object was too near the Sun to be seen. To make matters worse, Piazzi had a limited number of observations and the best mathematical techniques of the day could only predict an orbit if the object were traveling along a circle or parabola. An elliptical orbit was suspected, and thus astronomers feared that Piazzi's object may have been lost.

Upon hearing of the plight of the astronomers, the brilliant young mathematician Karl Friedrich Gauss took up the challenge and developed a general method of computing orbits from observations. This method yields an orbit from only three sets of observations, regardless of the conic section along which an object is traveling. In other words, from only three separate measurements of an object's right ascension and declination, its orbit can be calculated.

In November 1801, Gauss had finished his computations and predicted that Piazzi's object would be found in the constellation of Virgo. One of the "Celestial Police," Baron Franz Xavier von Zach, sighted Piazzi's object on December 31, 1801, only a short distance from the position given by Gauss. At Piazzi's request, the object was named Ceres (pronounced SEE-reez) after the protecting goddess of Sicily.

Ceres orbits the Sun once every 4.6 years at an average distance of 2.77 AU. This is in remarkable agreement with the distance "predicted" by Bode's law for the missing planet. But Ceres is very small. At opposition, its magnitude is only +7.4, and its diameter is estimated to be a scant 1000 km. Ceres thus does not qualify as a full-fledged planet and astronomers continued the search.

Numerous small objects circle the Sun between the orbits of Mars and Jupiter

On March 28, 1802, Heinrich Olbers discovered another faint, starlike object that moved against the background stars. He called it Pallas. Like Ceres, Pallas orbits the Sun every 4.6 years at an average distance of 2.77 AU. Pallas is even dimmer and smaller than Ceres, reaching a magnitude of only +8.0 at opposition and having an estimated diameter of only 600 km. Obviously, Pallas is not the missing planet either.

The discovery of two small objects with similar orbits at the distance expected for the missing planet led astronomers to suspect that Bode's missing planet might have somehow broken apart or exploded. The search for other small objects therefore continued. Only two more were found—Juno and Vesta—until the mid-1800s, when telescopic equipment and techniques had improved. Then astronomers began to stumble across many more such objects circling the Sun between the orbits of Mars and Jupiter. The objects are today called **asteroids** or **minor planets.**

The next major breakthrough came in 1891 when the German astronomer Max Wolf began using photographic techniques to search for asteroids. A total of 300 asteroids had been found up to that time—each painstakingly discovered by scrutinizing the skies for faint, uncharted stars that shift their positions slowly from one night to the next. With the advent of astrophotography, however, the floodgates were opened. An astronomer simply aims a camera-equipped telescope at the stars and takes a long exposure. If an asteroid happens to be in the field of view, it leaves a distinctive, blurred trail on the photographic plate because of its movement along its orbit during the long exposure (see Figure 16-1). Using this technique, Wolf alone discovered 228 asteroids during his observations.

Although thousands of asteroids have been sighted, only 3000 have well determined orbits. An additional 6000 asteroids have "passable"

Figure 16-1 Two asteroids
*Asteroids are detected by their blurred trails on
time-exposure photographs of the stars. The
images of two asteroids are seen in this picture. Astronomers sometimes find asteroids
accidentally while photographing various
portions of the sky for other purposes.
(Yerkes Observatory)*

orbits, while the orbits of 20,000 more have never been determined. The
orbits of all officially discovered asteroids are published annually in the
famous Soviet catalogue *Ephemerides of Minor Planets.*

To become the official discoverer of an asteroid, you must do a lot
more than produce one photograph with a blurred trail whose path does
not match any known orbit listed in the *Ephemerides.* You must track the
asteroid long enough to compute an accurate and reliable orbit (important data may be available from your colleagues who inadvertently photographed your asteroid on earlier occasions). Then you must prove the
accuracy of your orbit by locating the asteroid again on at least one succeeding opposition. At that time, an official number is assigned to your
asteroid (Ceres is 1, Pallas is 2, and so forth). You are also given the
privilege of selecting a name for your asteroid.

Ceres is unquestionably the largest asteroid. With a diameter of nearly
1000 km, Ceres accounts for about 30 percent of the mass of all the asteroids combined. Only three asteroids (Ceres, Pallas, and Vesta) have
diameters greater than 300 km. Thirty other asteroids have diameters
between 200 and 300 km, and there are 200 more asteroids bigger than
100 km across.

Astronomers estimate that roughly 100,000 asteroids exist that are
bright enough to appear on Earth-based photographs. The vast majority
are less than 1 km across. Like Ceres, Pallas, and Juno, most asteroids
circle the Sun at distances between 2 and $3\frac{1}{2}$ AU. This region of the solar
system between the orbits of Mars and Jupiter is called the **asteroid belt**
(see Figure 16-2). Asteroids whose orbits lie entirely within this region
are called **belt asteroids.**

The combined matter of all the asteroids (including an estimate for
those not yet officially known) would produce an object barely 1500 km
in diameter, considerably smaller than our Moon. If the asteroids are the
fragments of Bode's missing planet, it cannot have been large enough to

rank with the terrestrial planets. Furthermore, geologists and physicists have never been able to produce a good theory to explain how a planet could fragment or explode. There is little support today for the hypothesis of a shattered planet. It seems more reasonable to suppose that the asteroids are debris left over from the formation of the solar system out of the solar nebula.

Constant gravitational perturbations caused by the enormous mass of Jupiter probably kept planetesimals from ever accreting into larger objects in the region between Mars and Jupiter. The missing planet never had a chance to form. What remains today in the gap between the orbits of Jupiter and Mars appears to be simply a remnant of the scattered debris from the original solar nebula that elsewhere accreted into planets.

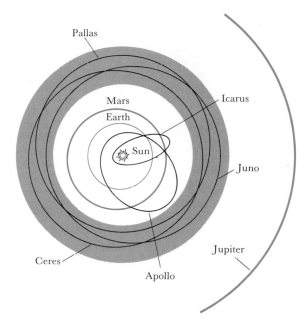

Figure 16-2   The asteroid belt
Most asteroids orbit the Sun in a belt of width $1\frac{1}{2}$ AU between the orbits of Mars and Jupiter. The orbits of Ceres, Pallas, and Juno are indicated. The orbits of two Apollo asteroids (Apollo and Icarus) also are shown.

Jupiter's gravity affects the structure of the asteroid belt and captures asteroids along its orbit

Consider a belt asteroid moving along its orbit between the orbits of Jupiter and Mars. Each time the faster-moving asteroid catches up with and passes massive Jupiter, it experiences a slight gravitational tug toward Jupiter. This tug tends to alter the asteroid's orbit slightly. However, over the ages, these close passes occur at various points along the asteroid's orbit, so their effects cancel one another.

Now imagine an asteroid that circles the Sun once ever 5.93 years, which is exactly half of Jupiter's orbital period. On every second trip around the Sun, the asteroid finds itself lined up between Jupiter and the Sun, again and again, always at the same location and with the same orientation. These gravitational perturbations add up and deflect the asteroid from its original 5.93-year orbit, leaving a gap in the asteroid belt. According to Kepler's third law, a period of 5.93 years corresponds to a semimajor axis of 3.28 AU. Because of Jupiter, there are no asteroids that orbit the Sun at this average distance.

Similarly, we would expect to find a gap corresponding to an orbital period of one-third Jupiter's period, or 3.95 years. Other gaps should exist for other simple relationships between the periods of asteroids and

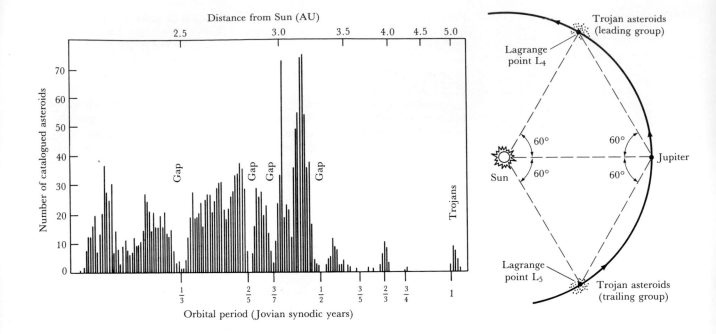

Figure 16-3 [left] *The Kirkwood gaps*
This histogram displays the numbers of aster-oids at various distances from the Sun. Notice that very few asteroids have orbits whose or-bital periods correspond to simple fractions (such as ¹/₂, ²/₇, ²/₅, ¹/₃) of Jupiter's orbital period. Gravitational perturbations due to re-peated alignments with Jupiter have deflected asteroids away from these orbits.

Figure 16-4 [right] *The Trojan asteroids*
Asteroids are trapped at the two Lagrange points along Jupiter's orbit by the combined gravitational forces of Jupiter and the Sun. Asteroids at these locations are named after Homeric heroes of the Trojan War.

Jupiter. The data graphed in Figure 16-3 show that such gaps do exist. They are called **Kirkwood gaps** in honor of the American astronomer Daniel Kirkwood, who first drew attention to them. The divisions in the ring structure of Saturn (recall Chapter 14) may have a similar origin, created by gravitational perturbations of the Saturnian moons on the icy fragments in the rings.

Although Jupiter's gravitational pull depletes certain orbits in the as-teroid belt, it captures asteroids at certain locations much farther from the Sun. As explained in Box 16-1, there are two locations along Jupi-ter's orbit where the gravitational forces of the Sun and Jupiter work together to hold asteroids in orbit. These two locations are called the **Lagrange points** L_4 and L_5. Point L_4 is located one-sixth of the way around Jupiter's orbit ahead of the planet, and point L_5 occupies a simi-lar position behind the planet (see Figure 16-4).

The asteroids trapped at Jupiter's Lagrange points are called **Trojan asteroids,** and they are named individually after heroes of the Trojan War. Nearly two dozen Trojan asteroids are catalogued, and some as-tronomers believe that there may be as many as 700 rock fragments or-biting near each Lagrange point.

Asteroids occasionally collide with each other and with the inner planets

In addition to the belt asteroids and the Trojan asteroids, there are other asteroids distinguished by highly elliptical orbits that bring them into the inner regions of the solar system. This class is usually divided into two groups: asteroids that cross Mars's orbit (called **Amor asteroids** after their prototype, Amor), and asteroids that cross Earth's orbit (called **Apollo asteroids** after their prototype, Apollo).

It is probable that the Amor and Apollo asteroids are somehow re-lated. Edward Anders of the University of Chicago argues that gravita-tional perturbations by Mars deflect Amor asteroids into Earth-crossing orbits, causing Amor asteroids to become Apollo asteroids.

Occasionally an Apollo asteroid passes quite close to Earth. For

Box 16-1 Lagrange points and the restricted three-body problem

In the late 1700s, the great French mathematician Joseph Louis Lagrange (pronounced la-GRAHNZH) tackled the famous three-body problem—the task of predicting the motions of three objects moving freely in space under the mutual influences of their gravitational forces. Lagrange succeeded in solving a restricted form of the problem in which one of the objects is assumed to be small enough that its gravity has no effect on the other two more massive objects. (Mathematicians later proved that there is no way to find a precise solution for the general three-body problem.) Among the practical applications of a solution for the restricted three-body problem are (1) plotting a course for a spaceship from the Earth to the Moon, and (2) calculating the path of an asteroid affected chiefly by the gravitational pulls of Jupiter and the Sun.

Lagrange discovered some interesting facts. The easiest way to display the combined gravitational fields of the two massive objects (we shall call them M_1 and M_2, where $M_1 > M_2$) is to draw equipotential contours. The gravitational field has a constant strength at each point along an equipotential contour. Thus we can think of the equipotential contour map as a sort of topographic map, showing "hills" and "valleys" in the gravitational field (see the top diagram). Any small object in this field will feel a force pulling it in the "downhill" direction. The massive objects M_1 and M_2 (along with the entire gravitational field) rotate about the center of mass of the system, which is nearer M_1 than M_2 because $M_1 > M_2$.

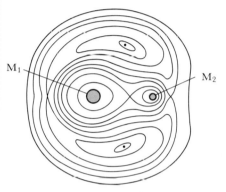

M_1 M_2

The equipotential contours

Note that there are three locations along a line joining M_1 and M_2 where contour lines cross. A tiny object placed at one of these locations would be balanced unsteadily at a local "flat spot" in the gravitational field. These three points are the unstable Lagrange points L_1, L_2, and L_3 (see the lower diagram). They are said to be unstable points because, if the small object moves ever so slightly away from the exact Lagrange point, it will then experience a gravitational pull in the "downhill" direction away from the Lagrange point. An object such as an asteroid therefore cannot be permanently trapped at one of the unstable Lagrange points.

Notice, however, that there are two points at the bottoms of teardrop-shaped "valleys" on either side of the line joining M_1 and M_2. These are the stable Lagrange points L_4 and L_5. If a small object is moved slightly away from one of these points, it will experience a gravitational pull in the "downhill" direction back toward the Lagrange point. Therefore a small object such as an asteroid can be permanently trapped at one of the stable Lagrange points. Indeed, the Trojan asteroids move in small orbits around the two stable Lagrange points in the Sun-Jupiter system.

The Trojan asteroids at L_4 and L_5 along Jupiter's orbit have been known for many years. (The discovery of the first one in 1906 provided the first practical proof of Lagrange's theoretical ideas about these points.) Of course, stable Lagrange points exist at many places in the solar system. While passing Saturn, the Voyager spacecraft discovered tiny satellites at the L_4 and L_5 points of the Saturn–Tethys and the Saturn–Dione systems. Small clouds of dustgrain-sized particles have been observed at the L_4 and L_5 points of the Earth–Moon system. A group called the L_5 Society argues that the L_5 point of the Earth–Moon system would be the ideal location for a huge space station with a permanent human population. Despite careful searches, no asteroids have been found at the stable Lagrange points of the Earth–Sun and the Saturn–Sun systems.

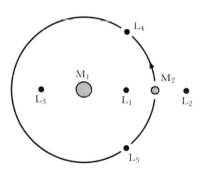

L_4

M_1 M_2

L_3 L_1 L_2

L_5

The five Lagrange points

Figure 16-5   Eros
Like all Apollo asteroids, Eros occasionally passes near the Earth. This photograph was taken in February 1931, when the distance between Eros and the Earth was only 23 million kilometers. The dimensions of Eros are roughly 10 by 20 by 30 km. The asteroid rotates with a period of 5.27 hours. (Yerkes Observatory)

example, Figure 16-5 shows Eros as it passed within 23 million kilometers of our planet in 1931. On June 14, 1968, Icarus passed Earth at a distance of only 6 million kilometers. One of the closest near-misses in recent history occurred on October 30, 1937, when Hermes passed us at a distance of 900,000 km—only a little more than twice the distance to the Moon.

During these close encounters, astronomers can examine the details of the asteroids. For example, an asteroid's magnitude often is observed to vary in a periodic fashion, presumably because of the different surfaces turned toward us as the asteroid rotates. Thus periodic magnitude variations reveal the asteroid's rate of rotation. Typical asteroid rotation periods are in the range of 5 to 20 hours.

Careful scrutiny of an asteroid's magnitude variations can also reveal the asteroid's shape and dimensions. Only the largest asteroids (such as Ceres, Pallas, and Vesta) are spherical, because only they have enough gravity to pull themselves into a spherical shape. Smaller asteroids permanently retain the odd shapes produced by interasteroid collisions. A small asteroid looks dim when seen end-on but appears brighter when seen broadside. Therefore, measurements of an asteroid's magnitude variations tell astronomers a lot about its shape.

There is ample evidence that interasteroid collisions successfully fragment asteroids into small pieces. In 1918, the Japanese astronomer Kiyotsugu Hirayama drew attention to groups of asteroids that share nearly identical orbits. These groupings are called **Hirayama families** and presumably resulted from the fragmentation of parent asteroids. For example, of the three dozen known Amor asteroids, 20 can be grouped into four Hirayama families.

A collision between kilometer-sized asteroids must be an awesome event. Typical collision velocities are estimated at 1 to 5 km/sec (2000 to 11,000 miles per hour), which is more than sufficient to shatter rock. In collisions at the low end of this velocity range, the resulting fragments may not achieve escape velocity from each other and will reassemble because of their mutual gravitational attraction. Alternatively, several large fragments may end up orbiting each other. This is probably what happened to both Pallas and Victoria, which are **binary asteroids,** each consisting of a main asteroid and a large satellite. Only in a high-velocity collision is there enough energy to shatter the asteroid permanently and create a Hirayama family.

Interasteroid collisions produce numerous chunks of rock, many of which eventually rain down on Venus, Earth, and Mars. Fortunately, for us, the vast majority of these asteroid fragments (usually called **meteoroids**) are quite small. On rare occasions, however, a large fragment does collide with our planet. The result is an **impact crater** whose diameter depends on both the mass and speed of the impinging object, as graphed in Figure 16-6.

Figure 16-6   Diameters of impact craters
This graph shows how the diameter of an impact crater is related to the mass and impact velocity of the impinging object.

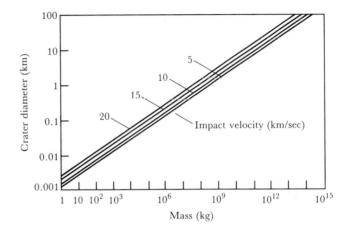

One of the most impressive and best-preserved terrestrial impact craters is the famous Barringer Crater near Winslow, Arizona. The crater measures 1.2 km across and is 200 m deep (see Figure 16-7). The crater was formed 25,000 years ago when an iron-rich object measuring roughly 50 m across struck the ground with a speed estimated at 11 km/sec (25,000 miles per hour). The resulting blast was equal to the detonation of a 20-megaton hydrogen bomb.

Iron is one of the more abundant elements in the universe (recall Table 6-4) as well as one of the most common rock-forming elements

Figure 16-7   The Barringer Crater
An iron meteoroid measuring 50 m across struck the ground in Arizona 25,000 years ago. The result was this beautifully symmetrical impact crater measuring 1.2 km in diameter and 200 m deep at its center. (Meteor Crater Enterprises)

(recall Table 6-5), so it is not surprising that iron is an important constituent of asteroids and the smaller objects called meteoroids. Another element, iridium—one of several elements that geologists call **siderophiles,** or "iron lovers"—is common in iron-rich minerals but is very rare in ordinary rocks. Measurements of iridium in the Earth's crust can therefore tell us about the rate at which meteoritic material has been deposited on the Earth over the ages. Geologist Walter Alvarez and his physicist father Luis Alvarez from the University of California at Berkeley were involved in such measurements in the late 1970s.

Working at a site of exposed marine limestone in the Apennine mountains in Italy, the Alvarez team discovered an exceptionally high abundance of iridium in a dark-colored layer of clay between the limestone strata (see Figure 16-8). Since this discovery in 1979, a comparable layer of iridium-rich material has been uncovered at a variety of sites around the world. In all cases, geological dating reveals that this apparently worldwide layer of iridium-rich clay is about 63 million years old.

Paleontologists were quick to realize the profound significance of this particular age. Around sixty-three million years ago, all of the dinosaurs became extinct. In fact, at that time, a staggering 65 percent of all the species on Earth disappeared within a relatively brief span of time.

The Alvarez discovery suggests a startling explanation for the dramatic extinction of more than half of the life-forms that inhabited our planet at the end of the Mesozoic era. Perhaps an asteroid hit the Earth. An asteroid 10 km in diameter slamming into the Earth would have thrown enough dust into the atmosphere to block out sunlight for several years. As plants died for lack of sunshine, the dinosaurs would have starved to death along with many other creatures in the food chain based on vegetation. Eventually the dust settled, depositing an iridium-rich layer around the world. Tiny, rodentlike creatures who could ferret out seeds and nuts were prominent among the animals who managed to survive the holocaust, setting the stage for the rise of mammals in the Cenozoic era.

Figure 16-8 The iridium-rich layer of clay
This photograph of strata in the Apennine Mountains in Italy shows a dark-colored layer of iridium-rich clay sandwiched between white limestone (below) from the late Mesozoic era and greyish limestone (above) from the early Cenozoic era. This iridium-rich layer may be the result of an asteroid impact that caused the extinction of the dinosaurs. The coin is the size of a U.S. quarter. (Courtesy of Walter Alvarez)

Some geologists and paleontologists are not yet convinced that such a meteoroid impact did produce the "great dying out" at the end of the Mesozoic era, but many scientists agree that this hypothesis fits the available evidence better than other explanations that have been offered.

Meteorites are classified as stones, stony irons, or irons, depending on their composition

A **meteoroid,** like an asteroid, is a chunk of rock in space. There is no official dividing line between meteoroids and asteroids, but the term asteroid is generally applied only to objects larger than a few hundred meters across.

A **meteor** is the brief flash of light (sometimes called a shooting star) that is visible at night when a meteoroid strikes the Earth's atmosphere (see Figure 16-9).

If a piece of rock survives its fiery descent through the atmosphere, the object that reaches the ground is called a **meteorite.** People have been finding specimens for thousands of years, and descriptions of meteorites appear in ancient Chinese, Greek, and Roman literature. Our ancestors placed special significance on these "rocks from heaven." There are numerous examples of **meteorite veneration** such as the sacred black stone of Kaaba enshrined at Mecca.

Figure 16-9   A meteor
A meteor is produced when a piece of interplanetary rock or dust strikes the Earth's atmosphere at a high speed. Exceptionally bright meteors, such as the one shown in this long exposure (notice the star trails), are usually called fireballs. (Courtesy of Ronald A. Oriti)

The extraterrestrial origin of meteorites was hotly debated as late as the eighteenth century. Upon hearing a lecture by two Yale professors, President Thomas Jefferson is said to have remarked, "I could more easily believe that two Yankee professors could lie than that stones could fall from Heaven." Although several **meteorite falls** had been widely witnessed and specimens had been collected (as, for example, in the 1751 fall near Zagreb, Yugoslavia), many scientists were reluctant to accept the idea that rocks fall to Earth from outer space.

Conclusive evidence came on April 26, 1803, when fragments of a large meteorite pelted the French town of L'Aigle. The austere French Academy, whose members were among the last holdouts, sent the noted

Figure 16-10 [left]   A stony meteorite
Ninety-three percent of all meteorites that fall on the Earth are stones. Many freshly discovered specimens like the one shown here, are coated with dark fusion crusts. This particular stone fell near Plainview, Texas. (From the collection of Ronald A. Oriti)

Figure 16-11 [right]   A stone (cut and polished)
When cut and polished, some stony meteorites are found to contain tiny specks of iron mixed in the rock. This specimen was discovered near Neenach, California. (From the collection of Ronald A. Oriti)

Figure 16-12   A stony-iron meteorite
Stony-irons account for slightly less than 2 percent of all meteorites that fall on the Earth. This particular specimen is a variety of stony-iron called a pallasite. It fell near Antofagasta, Chile. (From the collection of Ronald A. Oriti)

physicist J. B. Biot to investigate the matter. His exhaustive report finally convinced the scientific community of the extraterrestrial nature of meteorites.

Meteorites are classified into three broad categories: stones, irons, and stony-irons. As their name suggests, **stony meteorites,** or **stones,** look like ordinary rocks at first glance, but they are sometimes covered with a **fusion crust** (see Figure 16-10). This crust is produced by the momentary melting of the meteorite's outer layers during its fiery descent through the atmosphere. When a stony meteorite is cut in two and polished, tiny flecks of iron are sometimes found in the rock (see Figure 16-11).

Although stony meteorites account for nearly 93 percent of all meteoritic material that falls on the Earth, stones are the most difficult specimens to find. If undiscovered and exposed to the weather for a few years, they become almost indistinguishable from common terrestrial rocks. Meteorites with a high iron content are much easier to find because they can be located with a metal detector. Consequently, iron and stony-iron meteorites dominate most museum collections.

As their name suggests, **stony-iron meteorites** consist of roughly equal amounts of rock and iron. Olivine commonly is the mineral suspended in the matrix of iron, as in the case of the **pallasite** shown in Figure 16-12.

Iron meteorites (see Figure 16-13), or irons, account for nearly 6 percent of the material that falls on the Earth. Iron meteorites have no stone inclusions, but many contain from 10 to 20 percent nickel.

In 1808, Count Alois von Widmanstätten discovered a conclusive test for the authenticity of the most common type of iron meteorite. About 75 percent of all iron meteorites are of a type called **octahedrites.** When an octahedrite is cut, polished, and briefly dipped into a dilute solution of acid, its unique crystalline structure is revealed. These crystalline designs are appropriately called **Widmanstätten patterns** (see Figure 16-14).

Widmanstätten patterns constitute conclusive proof of a meteorite's authenticity because nickel–iron crystals can grow to lengths of several centimeters only if the molten metal cools very slowly over many million years. Octahedrites cool at rates of 1 to 10 K per million years during the time that the crystals are forming. Therefore Widmanstätten patterns are never found in counterfeit meteorites, or "meteorwrongs" as they are humorously called.

Meteorites provide significant information about the formation of the solar system

The existence of Widmanstätten patterns strongly suggests that some asteroids were partly molten for a substantial period after their formation. Furthermore, the size of an octahedrite's parent asteroid can be estimated by calculating how much rock must have insulated the molten iron-nickel interior to produce its long-term cooling rate. Such calculations imply that typical meteorites are fragments of parent asteroids 200 to 400 km in diameter.

The three main types of meteorites may have come from different parts of a parent asteroid in the following manner. As soon as the asteroid had accreted from planetesimals $4\frac{1}{2}$ billion years ago, the rapid decay of short-lived radioactive isotopes heated the asteroid's interior to temperatures above the melting point of rock. Over the next few million years, chemical differentiation occurred: iron and associated siderophile elements sank toward the asteroid's center, thereby displacing the lighter silica-associated elements (called **lithophile elements**) upward toward the asteroid's surface. After the asteroid cooled and its core solidified, inter-asteroid collisions fragmented the parent body into meteoroids. Iron meteorites are specimens from the asteroid's core, whereas stones are samples of its crust. Stony-irons presumably come from intermediate regions between the asteroid's core and crust.

This theory is supported by analyses of the spectra of meteorites and asteroids. Different kinds of rock reflect or absorb characteristic fractions of incident light over various wavelength ranges. By comparing the spectrum of an asteroid with those of different kinds of meteorites, it is often possible to deduce the surface composition of the asteroid. Two examples are shown in Figure 16-15. The reflectance spectrum of Vesta strongly resembles the spectrum of a particular kind of stony meteorite called a calcium-rich achondrite. The spectrum of Amantis resembles that of a stony-iron meteorite, suggesting that the asteroid's rocky crust may have been torn away, revealing material closer to its core.

It is clear that meteorites derived from the fragmentation of large asteroids were subjected to substantial processing during the first billion years of solar-system formation. These meteoritic specimens are therefore not representative of the primordial material from which the solar system was originally created. In order to find primordial meteorites, we must search for specimens that show no evidence of having been subjected to the metamorphic processes that occurred inside the asteroids.

Figure 16-13 [left]   An iron meteorite
Irons are composed almost entirely of nickel–iron minerals. The surface of a typical iron is covered with thumbprintlike depressions caused by ablation during its highspeed descent through the atmosphere. This specimen was found near Henbury, Australia. (From the collection of Ronald A. Oriti)

Figure 16-14 [right]   Widmanstätten patterns
When cut, polished, and etched with a weak acid solution, most iron meteorites exhibit interlocking crystals in designs called Widmanstätten patterns. These patterns appear only in the type of iron meteorite called an octahedrite. This octahedrite was found near Henbury, Australia. (From the collection of Ronald A. Oriti)

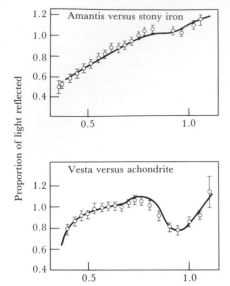

Proportion of light reflected

Amantis versus stony iron

Vesta versus achondrite

Wavelength (μ)

Figure 16-15 Asteroid versus meteorite spectra
These graphs show a comparison of the reflectance spectra of two kinds of meteorites (solid lines) with the spectra of two asteroids (circles with probable-error bars). From such comparisons, we can determine the composition of an asteroid's surface. (Adapted from C. R. Chapman)

Such specimens do exist; they are called **carbonaceous chondrites.** About 6 percent of all the stones that fall on the Earth are carbonaceous chondrites. Their primordial nature is inferred from their high content of volatile compounds, sometimes including as much as 20 percent water. Furthermore, carbonaceous chondrites are rich in complex organic compounds. The water and volatiles would have been driven out and the large organic molecules would have been broken down if these meteorites had been subjected to any significant heating.

Shortly after midnight on February 8, 1969, the night sky around Chihuahua, Mexico, was illuminated by a brilliant blue-white light moving across the heavens. The dazzling display was witnessed by hundreds of people, many of whom thought that the world was coming to an end. As the light moved across the sky, it exploded in a spectacular and noisy detonation that dropped thousands of rocks and pebbles over the terrified onlookers. Within hours, teams of scientists were on their way to collect specimens, collectively called the Allende meteorite after the locality of the finds.

Perhaps the most significant discovery to come from the Allende meteorite was made by Gerald J. Wasserburg and his colleagues at the California Institute of Technology. They found unmistakable evidence of the former presence of a radioactive isotope of aluminum, ^{26}Al. This particular isotope has a half-life of only 720,000 years and rapidly decays into a stable isotope of magnesium, ^{26}Mg.

Detectable amounts of short-lived ^{26}Al could have been included in the meteorite only if the radioactive aluminum was created very shortly before the formation of the solar system. If only a few hundred million years had elapsed, virtually all of the radioactive aluminum would have changed into magnesium before the meteorite formed. Astronomers were therefore faced with evidence of energetic nuclear processes that occurred in our vicinity roughly $4\frac{1}{2}$ billion years ago, about the time that the Sun was born.

One of nature's most violent and spectacular phenomena, called a **supernova explosion,** occurs during the death of a massive star. As we shall see in greater detail in Chapter 22, the doomed star blows itself apart in a cataclysm that hurls matter outward at tremendous speeds. During this detonation, violent collisions between nuclei produce a host of radioactive isotopes, including ^{26}Al. It therefore seems inescapable that a supernova occurred very near the Sun's location $4\frac{1}{2}$ billion years ago. In addition to contaminating the interstellar medium with radioactive ^{26}Al, the supernova's shock wave would have compressed the interstellar gas and dust, thereby triggering the birth of the solar system.

Besides telling us about the creation of the solar system, the study of meteorites may shed light on the origin of life on Earth. **Amino acids,**

Figure 16-16 Meteoritic swarms
Rock fragments and dust from "burned-out" comets continue to circle the Sun. (a) If the comet is only recently extinct, the particles are still tightly concentrated in a compact swarm. The most spectacular meteor showers occur when the Earth happens to pass through such a swarm. (b) Over the ages, the particles gradually spread out along the elliptical orbit. This configuration produces the most predictable meteor showers because the Earth must pass through the evenly distributed swarm on each trip around the Sun.

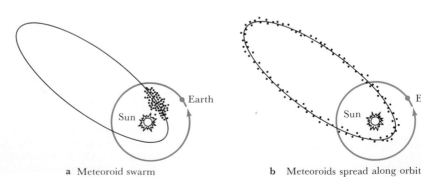

a Meteoroid swarm   b Meteoroids spread along orbit

the building blocks of proteins on which terrestrial life is based, are among the organic compounds found inside carbonaceous chondrites. Perhaps interstellar organic material falling on our planet played a role in the appearance of simple organisms on our planet nearly 4 billion years ago.

Of course, prebiological material (and perhaps genuine extraterrestrial organisms) continue to rain down on us to this day. Accordingly, British astrophysicist Fred Hoyle speculates that germs from outer space may have been responsible for the plagues and devastating diseases that suddenly appeared in Europe during the Middle Ages.

A comet is a dusty chunk of ice that is partly vaporized as it passes near the Sun

Although the vast majority of meteorites come from asteroids, the interplanetary particles that produce **meteor showers** are probably related to **comets.** Astronomers who think so note that meteor-shower-producing particles follow orbits around the Sun that are quite similar to those of comets.

As mentioned earlier in this chapter, asteroids travel around the Sun in roughly circular orbits that are largely confined to the asteroid belt and to the plane of the ecliptic. In sharp contrast, **comets** travel around the Sun along highly elliptical orbits inclined at random angles to the ecliptic plane. The swarms of particles that produce meteor showers also follow highly elliptical orbits (see Figure 16-16). In fact, some meteor showers are directly associated with "burned-out" comets.

Nearly a dozen meteor showers can be seen each year, the most prominent of which are listed in Box 16-2. Note that the **radiants** for these showers (that is, the places among the stars from which the meteors appear to come) are not confined to the constellations of the zodiac. Meteor showers come from various parts of the sky, just as comets are sighted at various locations without any correlation to the plane of the ecliptic.

Box 16-2   Meteor showers

At least ten notable meteor showers occur each year. They are listed with relevant information in the table. The date of maximum is the best time to observe a particular shower, although good displays are often seen a day or two before and after the maximum. The radiant is the location among the stars from which meteors seem to be coming. The hourly rate is that for a single observer under excellent conditions. The velocity is the average speed of meteoritic material striking the atmosphere.

| Shower name | Date of maximum | Radiant | | Hourly rate | Velocity (km/sec) |
		Right ascension	Declination		
Quadrantids	Jan. 3	$15^h\ 28^m$	$+50°$	40	40
Lyrids	Apr. 22	$18^h\ 16^m$	$+34°$	15	50
η Aquarids	May 4	$22^h\ 24^m$	$0°$	20	64
δ Aquarids	July 30	$22^h\ 36^m$	$-17°$	20	40
Perseids	Aug. 12	$3^h\ \ 4^m$	$+58°$	50	60

(continued)

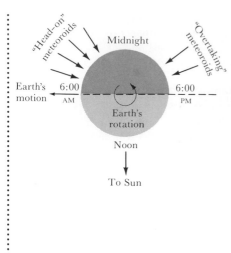

(Box 16-2, continued)

Orionids	Oct. 21	$6^h\ 20^m$	$+15°$	20	66
Taurids	Nov. 4	$3^h\ 32^m$	$+14°$	15	30
Leonids	Nov. 16	$10^h\ \ 8^m$	$+22°$	15	70
Geminids	Dec. 13	$7^h\ 32^m$	$+32°$	50	35
Ursids	Dec. 22	$14^h\ 28^m$	$+76°$	15	35

In order to see a fine meteor display, you need a clear, moonless sky. The Moon's presence above the horizon can significantly detract from the number of faint meteors you will be able to see. In addition, the early morning hours (between roughly 2 AM and dawn) are the best time to make your observations. As sketched in the diagram, you are on the leading side of the Earth during the early morning hours, and all meteor-producing particles in the Earth's path are swept into the atmosphere above you. On the other hand, you are on the trailing side of the Earth during the evening hours, where only high-speed particles manage to catch up with our planet to produce meteors.

Most astronomers agree that the solid part of a comet, called the **nucleus,** is essentially a chunk of ices, typically measuring a few kilometers across. Harvard astronomer Fred L. Whipple, a pioneer in comet research, coined the description "dirty iceberg" to reflect the fact that bits and pieces of dust and rocky material are mixed in with the ices. Frozen ammonia, methane, and water are the primary components of cometary ices, as indicated by their spectral lines.

As a comet approaches the Sun, solar heat begins to vaporize the ices. The liberated gases surrounding the icy nucleus soon begin to glow, pro-

Figure 16-17 ***The structure of a comet***
The solid part of a comet (the nucleus) is roughly 10 kilometers in diameter. The coma can be as large as 100,000 km to 1 million km across. The hydrogen envelope is typically 10 million kilometers in diameter. The comet's tail can be as long as 1 AU.

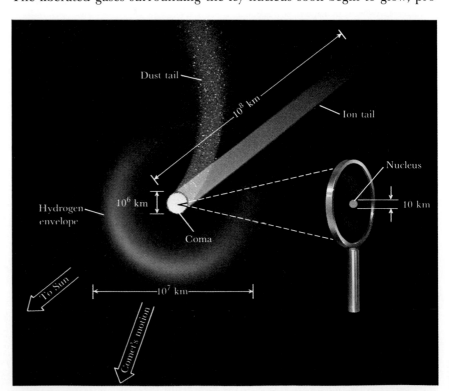

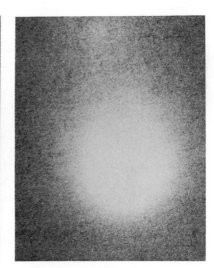

Figure 16-18 Comet Kohoutek and its hydrogen envelope
These two photographs of Comet Kohoutek are reproduced to the same scale. (left) The comet in visible light. (right) This view of the comet in ultraviolet wavelengths reveals a huge hydrogen cloud surrounding the comet's head. (John Hopkins University; Naval Research Laboratory)

Figure 16-19 [left] The head of Comet Brooks
A comet is always named after the person who first sights it. This comet, named Comet Brooks after its discoverer, had an exceptionally large, bright coma. It dominated the night skies during October 1911. (Lick Observatory)

Figure 16-20 [right] Comet Ikeya–Seki
This comet, named after its two Japanese codiscoverers, dominated the predawn sky during late October 1965. Although its coma was tiny, its tail was 1 AU long. (Lick Observatory)

ducing a fuzzy luminous ball called the **coma** that can eventually expand to a million kilometers in diameter. Continued action by the solar wind and radiation pressure blows these luminous gases outward into a long, flowing **tail.** The result is one of the most awesome sights ever visible in the nighttime sky.

Not visible to the human eye is the **hydrogen envelope,** a tenuous sphere of gas surrounding the comet's nucleus and measuring as much as 10 million kilometers in diameter. The overall structure of a comet is diagrammed in Figure 16-17. Figure 16-18 shows two views of Comet Kohoutek: as it appeared to Earth-based observers in 1973, and as it was photographed by an ultraviolet camera from a rocket. From this ultraviolet view, astronomers first discovered the enormous extent of the hydrogen envelope.

Comets come in a wide range of shapes and sizes. For example, the comet shown in Figure 16-19 had a large, bright coma but a short, stubby tail. In contrast, the comet seen in Figure 16-20 had an

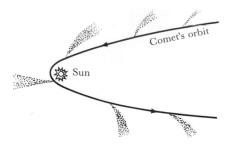

Figure 16-21   The orbit and tail of a comet
The solar wind and radiation pressure from sunlight blow the comet's dust particles and ionized atoms away from the Sun. Consequently, the comet's tail always points away from the Sun.

Figure 16-22 [right]   The two tails of Comet Mrkos
Comet Mrkos dominated the evening sky during August 1957. These three views, taken at two-day intervals, show dramatic changes in the comet's ion (Type I) tail. In contrast, the slightly-curved dust (Type II) tail remained fuzzy and featureless. (Palomar Observatory)

Figure 16-23 [below]   The antitail of Comet Arend–Roland
Comet Arend–Roland seen in April 1957, exhibited an "antitail." Actually, this antitail was merely the end of the dust tail. The Earth and the comet were oriented in such a way that the end of the arched dust tail seen past the comet's head looked like a spike sticking out of the comet's head. (Lick Observatory)

inconspicuous coma, but its tail had an astonishing length of 1 AU, long enough to reach all the way from the Earth to the Sun.

It has long been known that comet tails always point away from the Sun (see Figure 16-21), regardless of the direction of the comet's motion. The implication that something from the Sun was "blowing" the comet's gases radially outward led Ludwig Biermann to predict the existence of the solar wind a full decade before it was actually discovered in 1962 by Mariner 2.

In fact, the Sun usually produces two comet tails: an **ion tail,** and a **dust tail.** Ionized atoms (that is, atoms missing one or more electrons) are swept directly away from the Sun by radiation pressure. Micron-sized dust particles are blown away from the comet's coma by the solar wind. The relatively straight ion tail (sometimes called a **Type I tail**) can exhibit some dramatic structure that changes from night to night (see Figure 16-22). The more amorphous dust tail (sometimes called a **Type II tail**) is typically arched. On rare occasions, the geometry of the Earth,

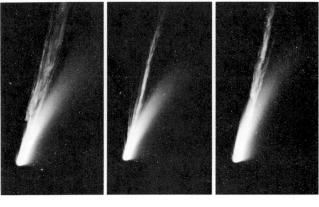

August 22   August 24   August 26

the comet, and its arched dust tail is such that the dust tail appears to be sticking out of the front of the comet (see Figure 16-23).

Astronomers discover at least a dozen comets in a typical year. Some are **short-period comets,** which circle the Sun in less than 200 years. Like the famous Halley's Comet (see Figure 4-13), they appear again and again at predictable intervals. The majority of comets discovered each year, however, are **long-period comets** that take 100,000 years to 1 million years to complete one orbit of the Sun. These comets travel along extremely elongated orbits and consequently spend most of their time at distances of 40,000 to 50,000 AU from the Sun—about one-fifth of the way to the nearest star.

Because astronomers discover long-period comets at a rate of roughly one per month, it is reasonable to suppose that there is an enormous population of comets out there at 50,000 AU from the Sun. This reservoir of cometary nuclei surrounding the Sun is called the **Oort cloud** after the Dutch astronomer Jan Oort, who first proposed its existence in the 1950s. Estimates of the numbers of "dirty icebergs" in the Oort cloud range from 1 million to more than 100,000 million. Only with such a large reserve of cometary nuclei can we understand why we see so many long-period comets even though each one takes up to 1 million years to travel once around its orbit.

Comets cannot survive very many perihelion passages. Eventually, a

| March 8 | March 12 | March 14 | March 18 | March 24 |

Figure 16-24 The fragmentation of Comet West
Shortly after passing near the Sun in 1976, the nucleus of Comet West broke into four pieces. This series of five photographs clearly shows the disintegration of the comet's nucleus. (New Mexico State University Observatory)

comet's ices are completely vaporized, and only a swarm of meteoritic dust and pebbles remain. A comet's nucleus will disintegrate more quickly if it happens to pass very near the Sun as what is called a **Sungrazing comet.** A comet's nucleus is sometimes observed to fragment (see Figure 16-24).

After a comet dies, its remaining dust and rock fragments spread out to form a **meteoritic swarm,** a loose collection of debris that continues to circle the Sun along the comet's orbit. If the Earth's orbit happens to pass near or through the swarm, a meteor shower is seen on the Earth.

As incredible as it may seem, a total of 300 tons of extraterrestrial rock and dust is estimated to fall on the Earth each day. The fluffy, low-density material from comets burns up in the atmosphere, and only denser specimens related to asteroids typically reach the ground. Nevertheless, there is evidence that a comet struck the Earth in the recent past.

On June 30, 1908, a spectacular explosion occurred over the Tunguska region of Siberia. Hundreds of square kilometers of forest were devastated (see Figure 16-25), and the blast was audible 1000 km away. The explosion was equivalent to the detonation of a tactical nuclear warhead with the destructive power of several hundred kilotons of TNT.

The most likely explanation of this event is that a small comet (perhaps a 100-m fragment of the short-period Comet Encke) collided with the Earth. No impact crater was formed, and the trees at "ground zero" were left standing upright, although they were completely stripped of branches and leaves. This is what would be expected from a loosely consolidated ball of cometary ices that vaporized with explosive force before

Figure 16-25 Aftermath of the Tunguska event
In 1908, a piece of a comet's nucleus struck the Earth's atmosphere over the Tunguska region of Siberia. Trees were blown down for many kilometers in all directions around the impact site. (Courtesy of Sovfoto)

striking the ground.

Large-sized objects occasionally strike the Earth with a destructive force that could easily be mistaken for the detonation of a nuclear weapon. A comet or small asteroid approaching our planet from the general direction of the Sun in the daytime sky would probably not be discovered by astronomers, so its impact would occur without any warning whatsoever. In spite of safeguards and diplomacy, Armageddon could be triggered by an entirely natural phenomenon.

Summary

. Bode's law is a numerical sequence that gives the distances of most of the planets (Mercury through Uranus) from the Sun in AU. This "law" inspired nineteenth-century astronomers to search for a planet in the gap between the orbits of Mars and Jupiter.

. Thousands of belt asteroids with diameters ranging from a few kilometers up to 1000 km circle the Sun between the orbits of Mars and Jupiter.

   Gravitational perturbations by Jupiter deplete certain orbits within the asteroid belt. The resulting gaps, called Kirkwood gaps, occur at simple fractions of Jupiter's orbital period.

   Jupiter's gravity also captures asteroids in two locations (called Lagrange points) along Jupiter's orbit.

   Some asteroids (called Amor and Apollo asteroids) move in elliptical orbits that cross the orbits of Mars and Earth.

. Small rocks in space are called meteoroids; if a meteoroid enters the Earth's atmosphere, it produces a fiery trail called a meteor; if part of the object survives the fall, the fragment that reaches the Earth's surface is called a meteorite.

   Meteorites are grouped in three major classes according to composition: iron, stony-iron, or stony meteorites.

   Rare stony meteorites called carbonaceous chondrites may be relatively unmodified material from the solar nebula; these meteorites often contain organic material and may have played a role in the origin of life on Earth.

   When a large meteoroid strikes the Earth, it forms an impact crater whose diameter depends on both the mass and speed of the impinging object.

   An asteroid may have struck the Earth 63 billion years ago causing the extinction of the dinosaurs and many other species.

   An analysis of isotopes in certain meteorites suggest that a nearby supernova explosion triggered the formation of the solar system $4\frac{1}{2}$ billion years ago.

. A comet is a chunk of ices and rock fragments that generally moves in a highly elliptical orbit about the Sun at a large inclination to the plane of the ecliptic.

   As a comet approaches the Sun, its icy nucleus develops a luminous coma surrounded by a vast hydrogen envelope; an ion tail and a dust tail extend from the comet, pushed away from the Sun by the solar wind and radiation pressure.

   Fragments of "burned out" comets produce meteor swarms; millions of cometary nuclei probably exist in the Oort cloud some 50,000 AU from the Sun.

Review questions

*1 Suppose that a planet were indeed 77.2 AU from the Sun as predicted by Bode's law. How long would it take such an object to orbit the Sun? At what rate (in degrees per year) would this object appear to move relative to the background stars? Would this motion be detectable with our telescopes?

2 Can you think of another place in the solar system in which occurs a phenomenon similar to Kirkwood's gaps in the asteroid belt? Explain.

3 Why do you suppose there aren't any asteroids at the L_1, L_2, or L_3 Lagrangian points in the Sun-Jupiter system?

4 Are there any examples in the solar system of objects being trapped at the L_4 and L_5 Lagrangian points other than in the Sun-Jupiter system?

5 Where on Earth might you find large numbers of stony meteorites which are not significantly weathered?

6 Suppose you found a rock that you suspect might be a meteorite. Describe some of the things you could do to see if it were a meteorite or a "meteorwrong."

7 List some of the difficulties of discovering a meteor shower that occurs in December and whose radiant is in the constellation of Scorpius.

8 Explain why comets are generally brighter after perihelion passage.

Advanced questions

*9 Suppose that a double asteroid is observed in which one member is 16 times brighter than the other. Suppose that both members have the same albedo and that the larger of the two is 200 kilometers in diameter. What is the diameter of the other member?

*10 Find the orbital periods of Sun-grazing comets whose aphelion distances arc: (a) 100 AU, (b) 1000 AU, (c) 10,000 AU, and (d) 100,000 AU. Assuming that these comets can survive only a thousand perihelion passages, calculate their lifetimes.

Discussion questions

11 Suppose that it were discovered that the asteroid Hermes had been perturbed in such a way as to put it on a collision course with Earth. Describe what you would do to counter such a catastrophe within the framework of present technology.

12 From the abundance of craters on the Moon and Mercury, we know that numerous asteroids and meteoroids struck the inner planets early in the history of the solary system. Is it reasonable to suppose that numerous comets also pelted the planets 3 to 4 billion years ago? Speculate about the effects of such a cometary bombardment, especially with regard to evolution of the primordial atmospheres of the terrestrial planets.

For further reading

Chapman, C. "The Nature of Asteroids." *Scientific American,* Jan. 1975.
Falk, S., and Schramm, D. "Did the Solar System Start with a Bang?" *Sky & Telescope,* July 1979, p. 18.
Hutchinson, R. *The Search for Our Beginnings.* Oxford U. Press, 1983. *A book on meteorites and the solar system.*
Morrison, D. "Asteroids." *Astronomy,* June 1976, p. 6.
Seargent, D. *Comets: Vagabonds of Space.* Doubleday, 1982.
Whipple, F. "The Nature of Comets." *Scientific American,* Feb. 1974.

17 The laws of light

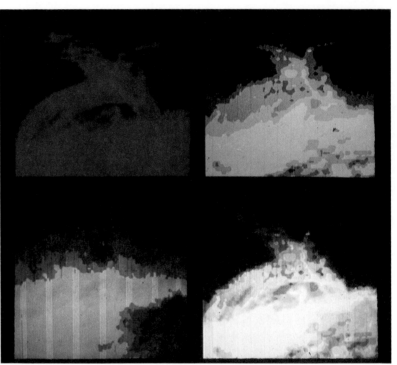

An eruption on the Sun
The light emitted by hot gas contains detailed information about physical conditions in the gas, such as temperature and chemical composition. These views of an eruption on the Sun's limb were taken by Skylab astronauts in 1973. The red image, taken at the wavelength of Lyman alpha at 1215 Å, shows hydrogen at a temperature of about 15,000 K. The green image was taken at 1032 Å and shows five-times ionized oxygen which corresponds to a temperature of about 300,000 K. The blue image was taken at 625 Å and shows nine-times ionized magnesium, characteristic of the 1,500,000 K temperature in the Sun's upper atmosphere. The whitish image at the lower-right was made by combining the other three images. (NASA)

We are ready to turn to the study of the universe beyond our own solar system. Our knowledge about distant stars and galaxies comes from analysis of the electromagnetic radiation they emit. This chapter provides more information about the properties of light—information that will be needed as background for the chapters that follow. We find that light has both wavelike and particlelike properties. We learn that the electromagnetic radiation emitted by an object is affected by the temperature of the object. Finally, we discuss spectral lines in more detail and explore their relationship to the structures of atoms.

Our knowledge about the universe outside our solar system comes from the study of light arriving at the Earth from distant stars. When we speak of "light" from stars and galaxies, we mean the full **electromagnetic spectrum** (see Figure 17-1). The human eye detects only the narrow band of wavelengths that we call visible light. Astronomers also view the universe with gamma rays, X rays, ultraviolet radiation, infrared radiation, microwaves, and radio waves. You will find it useful at this point to review Chapter 5; pay particular attention to Box 5-1, which describes the units used in measuring wavelengths.

Light has wavelike properties and travels through empty space at a speed of about 300,000 km/sec

The wavelike nature of light was convincingly demonstrated around 1801 by Thomas Young, an English physicist, physician, and general scholar. Young passed a bright beam of light through two thin, parallel slits in an opaque screen (see Figure 17-2*a*). On a white surface some distance beyond the slits, the light formed a pattern of alternating bright and dark bands. If a beam of light were a stream of particles (as argued earlier by Isaac Newton), the two beams of light from the slits should simply form bright images of the slits on the white surface. The observed pattern of bright and dark bands is just what would be expected, however, if light has wavelike properties.

Consider ocean waves pounding against a reef or breakwater that has two openings (see Figure 17-2*b*). A pattern of ripples is formed inside the barrier by the interference between waves coming through the two openings. At certain points along the shore, crests arrive simultaneously from the two openings, producing high waves. At intermediate points along the shore, a crest arrives from one opening at the same moment as a trough from the other opening; the waves from the two openings cancel each other, producing bands of still water. Similar interference between the light waves from the two slits can explain the pattern of bright and dark bands that Young observed on the white surface behind his slits.

Our modern understanding of the wavelike properties of light is based upon the work of the Scottish mathematician and physicist James Clerk Maxwell. In 1873, Maxwell showed that electrical and magnetic effects are two aspects of the same phenomenon, which we now call

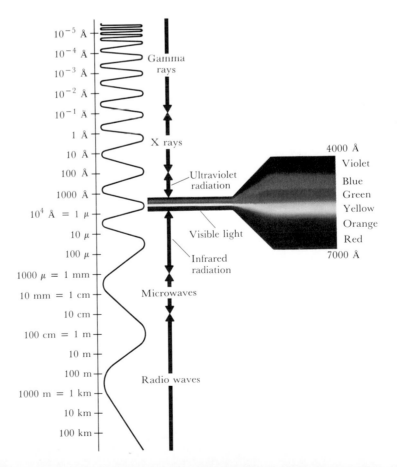

Figure 17-1 The electromagnetic spectrum
The full array of all types of electromagnetic radiation is called the electromagnetic spectrum. It extends from the shortest-wavelength gamma rays to the longest-wavelength radio waves. Visible light is only a tiny portion of the full electromagnetic spectrum.

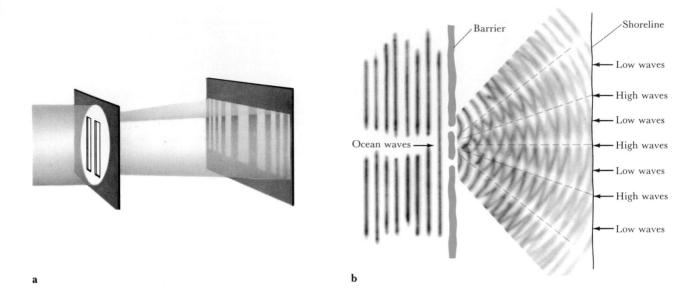

a   b

Figure 17-2   Young's double-slit experiment
(a) *In the modern laboratory, Thomas Young's classic double-slit experiment is most easily repeated by shining light from a laser onto two closely spaced, parallel slits. Alternating dark and light bands appear on a screen beyond the slits.* (b) *The intensity of light on the screen in the double-slit experiment is analogous to the height of water waves that strike a shore after passing through a barrier with two openings. In certain locations, ripples from both openings reinforce each other to produce high waves. At intermediate locations along the shoreline, the ripples cancel each other to produce still water.*

electromagnetism. From his equations, Maxwell showed that an electrically charged particle moving back and forth should emit electromagnetic waves. These waves can be pictured as oscillating electric and magnetic fields, with the electric field at right angles to the magnetic field (see Figure 17-3).

Maxwell showed mathematically that any **electromagnetic radiation** (regardless of its wavelength) should move through empty space at a very specific speed usually denoted by the symbol c. Maxwell calculated the value of c to be about 3×10^{10} cm/sec. He pointed out that this value of c was roughly equal to the best measurements then available for the speed of light (see Box 17-1). Therefore, Maxwell suggested that visible light is simply electromagnetic radiation. Furthermore, he suggested that infrared and ultraviolet radiation are forms of light with wavelengths such that they are not detected by our eyes. He went on to predict the existence of electromagnetic radiation with other wavelengths. This prediction was dramatically confirmed in 1888 when the German physicist Heinrich Hertz demonstrated the existence of radio waves.

Figure 17-3   Electromagnetic radiation
All forms of light consist of oscillating electric and magnetic fields that move through space at a speed of 3×10^{10} cm/sec. The distance between two successive crests is called the wavelength of the light and is usually designated by the Greek letter λ.

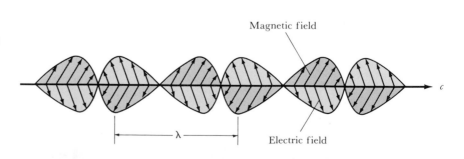

Box 17-1   Measuring the speed of light

The **speed of light** in empty space is one of the most important numbers in modern physical science. This value appears as a constant in Maxwell's equations describing electromagnetic phenomena. Furthermore, according to Albert Einstein's special theory of relativity, nothing in the universe can move at a speed greater than the speed of light.

The first accurate measurement of the speed of light was made in 1675 by the Danish astronomer Olaus Roemer. Roemer was studying the orbits of the Galilean satellites of Jupiter by timing carefully the moments when the satellites passed into and out of the planet's shadow. To his surprise, the orbital motions of the satellites seemed to depend upon the relative positions of Jupiter and the Earth. As the Earth moved closer to Jupiter (during the months just before an opposition), the Galilean satellites seemed to speed up slightly in their orbits about the planet. Then as the Earth moved away from Jupiter, they seemed to slow down slightly.

Roemer realized that this puzzling effect could be explained if light takes a measurable time to travel from Jupiter to the Earth. When the Earth is closer to Jupiter, the image of a Galilean satellite disappearing into the Jovian shadow arrives at our telescopes a little sooner than it does when Jupiter and the Earth are far apart, as shown in the first diagram. Roemer's measurements indicated that it takes about $16\frac{1}{2}$ minutes for light to travel across the diameter of the Earth's orbit (a distance of 2 AU). The size of the Earth's orbit was not accurately known in Roemer's day. Using the modern value of about 150 million kilometers for the astronomical unit, however, Roemer's method yields a value of about 300,000 km/sec for the speed of light.

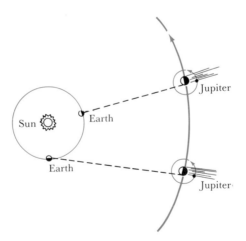

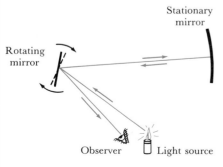

The first successful laboratory experiments to measure the speed of light were conducted in the mid-1800s in France. For example, in 1850, Hippolyte Fizeau and Jean Foucault built an apparatus consisting of a light source, a rotating mirror, and a stationary mirror as sketched schematically in the second diagram. Light from the light source is reflected from the rotating mirror toward the stationary mirror 20 m away. While the light is making the round trip, the rotating mirror moves slightly. Thus the returning light ray is deflected away from the initial beam by a small angle. By measuring this angle, and by knowing the dimensions of their apparatus, Fizeau and Foucault could deduce the speed of light. Once again, the answer was very nearly 300,000 km/sec.

(continued)

(Box 17-1, continued)

The most recent measurements of the velocity of light give a speed of

$c = 299{,}792.46$ km/sec

In most calculations, you can use $c = 3 \times 10^{10}$ cm/sec. Note, however, that this value is the *speed of light in a vacuum.* Light traveling through air, water, glass, or any transparent substance always moves more slowly than it does in a vacuum.

Recall from Box 5-1 that the **wavelength** (λ) of any wave is simply the distance between two successive wave crests. The **frequency** (ν) is the number of wave crests passing a stationary point each second. Because any electromagnetic wave in a vacuum travels with speed c, its wavelength is related to its frequency by the simple equation $\lambda\nu = c$.

Any object emits electromagnetic radiation with intensity and wavelengths related to the temperature of the object

The total amount of energy radiated by an object depends upon its temperature. The hotter the object, the more energy it emits in the form of electromagnetic radiation. The dominant wavelength of the emitted radiation also depends upon the temperature. A cool object emits most of the energy at long wavelengths, whereas a hotter object emits most of the energy at shorter wavelengths.

These basic phenomena are familiar to anyone who has watched a welder or blacksmith heat a bar of iron. As it becomes hot, the bar begins to glow with a deep red color. At the temperature rises, the bar begins to give off a brighter, reddish-orange light. At still higher temperatures, it shines with a brilliant, yellowish-white light. If the bar could be prevented from melting and vaporizing, it would emit a dazzling, blue-white light at very high temperatures.

In 1879, the Austrian physicist Josef Stefan summarized the results of his experiments on this phenomenon by stating that an object emits energy at a rate that is proportional to the fourth power of the object's temperature (measured in kelvins). If you double the temperature of an object (for example, from 500 K to 1000 K), then the energy emitted from the object's surface each second increases by a factor of $2^4 = 16$. If you triple the temperature (for example, from 500 K to 1500 K), the rate of energy emission increases by a factor of $3^4 = 81$.

The energy emitted from 1 cm² of an object's surface each second is called the **energy flux** (E) and Stefan's law can be written as

$$E = \sigma T^4$$

where σ ("sigma") is a constant, the temperature T is measured in kelvins, and the energy flux E is measured in ergs per square centimeter per second (usually written erg cm^{-2} sec^{-1}). Five years after Stefan announced this law, another Austrian physicist, Ludwig Boltzmann, showed how it could be derived mathematically from basic assumptions about atoms and molecules. Therefore, the law today is commonly known as the **Stefan–Boltzmann law.** The Stefan–Boltzmann constant σ is known from laboratory experiments to have the value

$$\sigma = 5.67 \times 10^{-5} \text{ erg cm}^{-2} \text{ K}^{-4} \text{ sec}^{-1}$$

Boltzmann showed that this law is obeyed only by an object that is a perfect absorber of energy. Such an object would be absolutely black at all wavelengths and is therefore called a **blackbody.** Since a blackbody does not reflect any light, its energy flux depends only on temperature according to the Stefan–Boltzmann law. Ordinary objects are not perfect absorbers (we can see them because they reflect some light) and thus the energy flux they emit is slightly different from the amount calculated from the Stefan–Boltzmann law.

A star efficiently absorbs all radiation falling on it from the outside. Thus a star behaves like a blackbody and astronomers can use the Stefan–Boltzmann law to relate its energy flux to its surface temperature. For example, consider the Sun. Astronomers have measured the average flux of solar energy arriving at the Earth; this value (called the solar constant) is 1.35×10^6 erg cm^{-2} sec^{-1}. Imagine a huge sphere of radius 1 AU with the Sun at its center. Each square centimeter of that sphere receives 1.35 million ergs of solar energy per second, so we can compute the total energy output of the Sun (multiply the solar constant by the sphere's area). The result, called the **luminosity** (L) of the Sun, turns out to be

$$L_\odot = 3.90 \times 10^{33} \text{ erg/sec}$$

(Astronomers use the symbol $\odot$ to represent the Sun.)

We know the size of the Sun, so we can now compute the energy flux emitted from its surface. The radius of the Sun is $R_\odot = 6.96 \times 10^{10}$ cm, and its surface area is $4\pi R_\odot^2$. Therefore, its energy flux is

$$E_\odot = \frac{L_\odot}{4\pi R_\odot^2} = 6.41 \times 10^{10} \text{ erg } cm^{-2} \text{ sec}^{-1}$$

Now we can use the Stefan–Boltzmann law to find the Sun's surface temperature:

$$T_\odot^4 = \frac{E_\odot}{\sigma} = 1.13 \times 10^{15} \text{ K}^4$$

Taking the fourth root (the square root of the square root) of this value, we find the surface temperature of the Sun to be

$$T_\odot = 5800 \text{ K}$$

Any object emits radiation over a wide range of wavelengths, but there is always a particular wavelength (λ_{max}) at which the emission of energy is strongest. This dominant wavelength gives a glowing hot object its characteristic color.

In 1893, a generation after the work of Stefan and Boltzmann, the German physicist Wilhelm Wien discovered a simple relationship between the temperature of a blackbody and the dominant wavelength (λ_{max}) of its energy emission:

$$\lambda_{max} = \frac{0.29}{T}$$

where λ_{max} is measured in centimeters, and T is measured in kelvins. This relation is today called **Wien's law.**

Wien's law states that, for any blackbody, λ_{max} (in cm) times T (in K) equals the constant value 0.29. If the temperature of the blackbody increases, then the dominant wavelength of its radiation must decrease so that the product remains constant. In other words, the hotter an object, the shorter the dominant wavelength of the electromagnetic radiation it emits.

From the Stefan–Boltzmann law, we see that any object with a temperature above absolute zero (0 K) emits some electromagnetic radiation. From Wien's law, we find that a very cold object with a temperature of a few kelvins emits primarily microwaves. An object at "room temperature" (about 300 K) emits primarily infrared radiation. An object with a temperature of a few thousand kelvins emits mostly visible light. An object with a temperature of a few million kelvins emits most of its radiation in the X-ray wavelengths.

Again consider the Sun's emission of energy. The Sun emits energy over a wide range of wavelengths, but the maximum intensity of sunlight is at a wavelength of roughly 5000 Å = 5×10^{-5} cm. From Wien's law, we find the Sun's surface temperature to be

$$T_\odot = \frac{0.29}{5 \times 10^{-5}} = 5800 \text{ K}$$

This result agrees with the value we computed using the Stefan–Boltzmann law. Wien's law is very useful for computations of the surface temperatures of stars because it does not require knowledge of the star's size or luminosity; all we need to know is the dominant wavelength of the star's electromagnetic radiation.

Explanations of the details of blackbody radiation require the assumption that light has particlelike properties

The Stefan–Boltzmann law and Wien's law describe only two basic properties of **blackbody radiation,** the electromagnetic radiation emitted by a hypothetical blackbody. A more complete picture is given by **blackbody curves** such as those in Figure 17-4. These curves show the relationship between the wavelength and intensity of light emitted by a blackbody at a given temperature.

The total area under a blackbody curve is proportional to the energy flux E; the wavelength corresponding to the peak of the curve is the dominant wavelength λ_{max}. Note that the blackbody curves clearly illustrate both of the laws we have discussed. A cool object has a low curve that peaks at a long wavelength; a hotter object has a much higher curve that peaks at a shorter wavelength.

Figure 17-5 shows how the intensity of sunlight varies with wavelength. This curve was obtained by measuring the intensity of sunlight at various wavelengths above the Earth's atmosphere. Note that the peak of the curve, as mentioned earlier, is at a wavelength of about 0.5 μ = 5000 Å. The blackbody curve for a temperature of 5800 K is also plotted in Figure 17-5. Note how closely the observed intensity curve for the Sun matches the blackbody curve. This close correlation between the observed intensity curves for most stars and the idealized blackbody curves is the reason that astronomers are interested in the physics of blackbody radiation.

By the end of the nineteenth century, physicists realized that they had reached an impasse. All attempts to explain the characteristic shape of blackbody curves in terms of then-known science had failed.

In 1900, the German physicist Max Planck discovered that he could derive a mathematical formula for the blackbody curves provided he assumed that electromagnetic energy is emitted in discrete packets of energy. This important assumption was verified in 1905 by Albert Einstein who used the idea of the particlelike nature of light to explain the **photoelectric effect.**

Figure 17-4 Blackbody curves
Three representative blackbody curves are shown here. Each curve shows the intensity of light at every wavelength emitted by a blackbody at a particular temperature. The range of wavelengths of visible light is indicated.

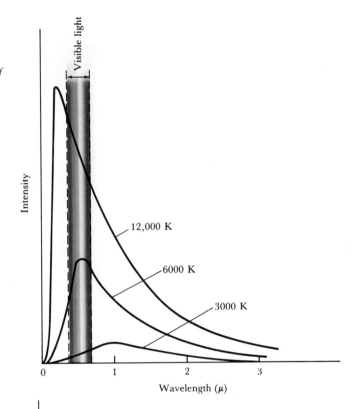

Figure 17-5 The Sun as a blackbody
This graph compares the intensity of sunlight over a wide range of wavelengths with the intensity of radiation from a blackbody at a temperature of 5800 K. Measurements of the Sun's intensity were made above the Earth's atmosphere. The Sun mimicks a blackbody remarkably well.

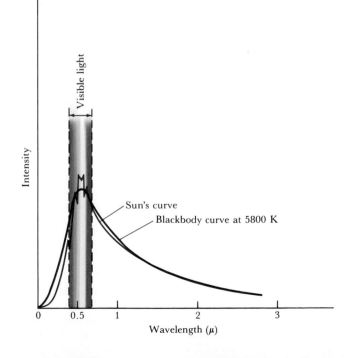

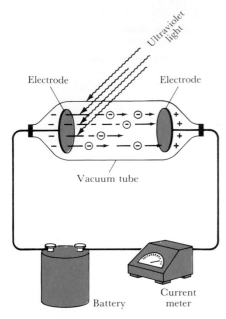

Figure 17-6 The photoelectric effect
The photoelectric effect can be demonstrated with a vacuum tube containing two electrodes attached to a battery, as sketched in this diagram. When no light falls on the electrode at the left, no electrons are liberated from the metal, and no current flows. When long-wavelength red light is shined on the electrode, still no electrons are released, and no current flows. When short-wavelength ultraviolet light falls on the electrode, however, many electrons are liberated from the metal. These negatively charged electrons are attracted to the positively charged electrode at the opposite end of the vacuum tube, and thus an electric current flows.

As shown in Figure 17-6, apparatus to demonstrate the photoelectric effect consists of two electrodes enclosed in a vacuum tube and connected to a battery. In the dark or when illuminated with visible light, no current flows through the circuit. If a beam of ultraviolet light is focused on the negatively charged electrode, however, a current does flow through the circuit. The ultraviolet light somehow liberates electrons from the electrode, allowing them to move across the vacuum to the positively charged electrode. It is not surprising that energy from light can knock electrons out of the metal. What is surprising is that a very bright red light has no effect on the apparatus, whereas a dim ultraviolet light causes a current to flow through the circuit.

Einstein explained the photoelectric effect by assuming that light exists in particlelike packets, or **photons,** of energy. The amount of energy in one **photon** of light is inversely proportional to the wavelength of the light. The longer the wavelength, the less energy in a photon. A single photon of red light, with its relatively long wavelength, does not have enough energy to knock an electron out of the metal. No matter how many photons of red light strike the metal, no collision of a photon with an atom can liberate an electron. A single photon of ultraviolet radiation, however, with its much shorter wavelength, does have enough energy to knock loose an electron. Thus, even a weak beam of ultraviolet light can liberate electrons and cause a current to flow. The photoelectric effect today plays an important part in some devices commonly used by astronomers (see Box 17-2).

The relationship between the energy E of each photon and the wavelength λ of the light can be expressed in a simple equation:

$$E = \frac{hc}{\lambda}$$

where c is the speed of light, and h is a constant value now called Planck's constant. Because $c/\lambda = \nu$, we can express this relationship in another form that is often called **Planck's law:**

$$E = h\nu$$

where ν is the frequency of the light. Note that both of these equations express a relationship between a particlelike property of light (the energy E of a photon) and a wavelike property (the wavelength λ or the frequency ν).

Laboratory experiments show that Planck's constant has the value

$$h = 6.625 \times 10^{-27} \text{ erg sec}$$

This is a very tiny value; by the standards of our everyday experience, a single photon carries a very tiny amount of energy.

Each chemical element produces its own unique set of spectral lines

In 1814, the German master optician Joseph Fraunhofer repeated the classic experiment of shining a beam of sunlight through a prism (recall Figure 5-2) and subjected the resulting rainbow-colored spectrum to intense magnification. To his surprise, Fraunhofer discovered that the solar spectrum contains hundreds of fine dark lines, called **spectral lines.** Fraunhofer counted over 600 such lines, and today we know more

Box 17-2   Astronomers' applications of the photoelectric effect

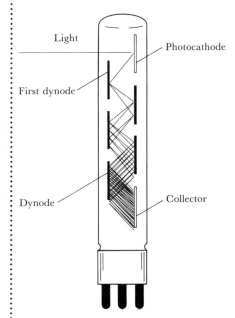

Light

Photocathode

First dynode

Dynode

Collector

Einstein's work with the photoelectric effect showed how photons liberate electrons from atoms. Pursuing this research, scientists soon learned how to turn a beam of light into an electric current, which is basically what a television camera does.

Astronomers have many uses for televisionlike devices based on the photoelectric effect. Astronomers are often interested in accurately measuring extremely low levels of light. To do this, they make use of devices called **photoelectric detectors.**

One of the most common types of photoelectric detectors is the **photomultiplier** shown in the diagram at left. The astronomer aims a telescope at a distant star or galaxy and focuses the feeble starlight onto the photocathode inside the photomultiplier. An incoming photon knocks an electron from the negatively charged photocathode. This electron is attracted to the first dynode, which has a positive charge. Upon hitting the first dynode, the electron knocks off a few more electrons, which are attracted to the second dynode. When these electrons hit the second dynode, still more electrons are liberated.

Each successive dynode carries a slightly greater positive charge, thereby ensuring that the electrons continue to cascade through the photomultiplier and finally arrive at the collector. In this way, a single photon can be converted into a current of a million electrons. The astronomer simply measures the resulting electric current, which is directly related to the brightness of the star or galaxy toward which the telescope is aimed. Special coatings on the photocathode make the photomultiplier sensitive to visible light.

In recent years, the quality of photomultipliers has improved dramatically, and astronomers can now measure the brightnesses of faint objects with unprecedented precision. With care, it is possible to measure a star's brightness with an accuracy of one-thousandth of a magnitude.

The **image intensifier** shown in the diagram below to the left is another important use of television-like technology. Incoming photons from a faint source strike the photocathode at one end of a vacuum tube and liberate numerous electrons. These electrons are accelerated toward a phosphorescent screen at the opposite end of the tube by charged electrodes that line the walls. The entire apparatus is surrounded by a magnetic field that focuses the electrons onto the phosphorescent screen. As the electrons strike the screen, a glowing image is produced in much the same way that a glowing image appears on the television set in your home. Because the electrons are considerably accelerated as they travel down the vacuum tube, the output image can be 10 to 100 times brighter then the original image from the telescope.

A wide range of superb photon detectors and image intensifiers are now commercially available. These instruments have greatly expanded the usefulness of telescopes and have dramatically enhanced our ability to probe the dim and distant reaches of the universe.

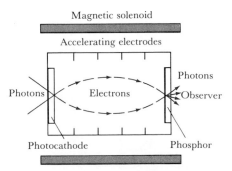

Magnetic solenoid

Accelerating electrodes

Photons

Electrons

Photons

Observer

Photocathode

Phosphor

than 20,000. A small portion of the Sun's spectrum is shown in Figure 17-7. Hundreds of spectral lines are visible.

Half a century later, chemists discovered that they could produce spectral lines in laboratory experiments. Chemists had long identified certain substances by the distinctive colors emitted when the substances are sprinkled into a flame. About 1857, the German chemist Robert Bunsen invented a special gas burner that produces a clean, colorless flame. This Bunsen burner was very useful for analyses of substances because its flame had no color of its own to be confused with the color produced when a substance is sprinkled in it.

Figure 17-7   A portion of the Sun's spectrum
The violet and deep-blue portion of the solar spectrum (from 3900 Å to 4700 Å) is shown here. The G, H, and K Fraunhofer lines are most prominent, but many other faint spectral lines are also visible. (Mount Wilson and Las Campanas Observatories)

A younger colleague, Prussian-born physicist Gustav Kirchhoff suggested that light from the colored flames could best be studied by passing it through a prism (see Figure 17-8). Bunsen and Kirchhoff collaborated in the design and construction of a spectroscope, a device consisting of a prism and several lenses, by which the spectra of flames might be magnified and examined (review Box 6-2). They promptly discovered that the spectrum from a flame consists of a pattern of thin bright spectral lines against a dark background. They soon found that *each chemical element produces its own characteristic pattern of spectral lines.* Thus was born in 1859 the technique of **spectral analysis,** the identification of chemical substances by their unique patterns of spectral lines.

New elements were discovered through spectral analysis. After Bunsen and Kirchhoff had recorded the prominent spectral lines of all the known elements, they soon began to discover other spectral lines in the spectra of mineral samples. In 1860, they found a new line in the blue portion of the spectrum of a sample of mineral water. After chemically isolating the previously unknown element responsible for the line, they

Figure 17-8   The Kirchhoff–Bunsen experiment
In the mid-1850s, Kirchhoff and Bunsen discovered that, when a chemical substance is heated and vaporized, the resulting spectrum exhibits a series of bright spectral lines. In addition, they found that each chemical element produces its own characteristic pattern of spectral lines.

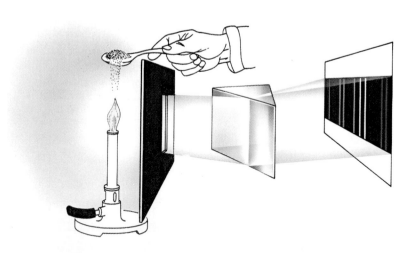

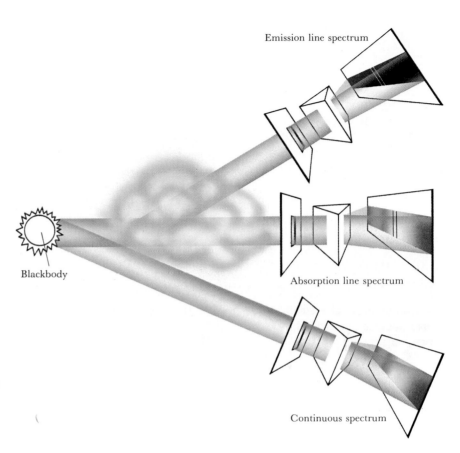

Emission line spectrum

Absorption line spectrum

Blackbody

Continuous spectrum

Figure 17-9 Kirchhoff's laws of spectral analysis
A hot, glowing object emits a continuous spectrum. If this white light is passed through some cool gas, then dark absorption lines appear in the resulting spectrum. If the same gas is examined at an oblique angle, bright emission lines are visible against a dark background.

named it cesium (from Latin *caesium,* "grey-blue"). The next year, a new line in the red portion of the spectrum of a mineral sample led them to the discovery of the element rubidium (from Latin *rubidium,* "red").

During the solar eclipse of 1868, astronomers found a new spectral line in light coming from the upper surface of the Sun while the main body of the Sun was hidden by the Moon. This line was attributed to a new element that was named helium (from Greek *helios,* "Sun"). Helium was not discovered on the Earth until 1895, when it was located in gases obtained from a uranium mineral.

By the early 1860s, Kirchhoff's experiments had progressed sufficiently for him to formulate three important statements about spectra that today are called **Kirchhoff's laws** of spectral analysis.

. **Law 1** A hot, glowing object emits a **continuous spectrum**—a complete rainbow of colors without any dark spectral lines.

. **Law 2** When a continuous spectrum is viewed through some cool gas, *dark* spectral lines (called **absorption lines**) appear in the continuous spectrum.

. **Law 3** If the gas is viewed at an angle away from the source of the continuous spectrum, a pattern of *bright* spectral lines (called **emission lines**) is seen against an otherwise black background.

Figure 17-9 illustrates Kirchhoff's laws.

The bright lines in the emission spectrum of a particular gas occur at exactly the same positions (wavelengths) as the dark lines in the absorption spectrum of the same gas. The only difference between the dark and bright patterns of spectral lines is the angle at which the gas is observed with respect to the light source.

Somehow, atoms in a gas extract light of very specific wavelengths from white light that passes through the gas or vapor. Thus dark absorption lines are created in the continuous spectrum of the white-light source. The atoms then radiate light of precisely these same wavelengths in all directions, so that an observer at an oblique angle (without the white-light source in the background) detects bright emission lines.

Why does an atom absorb only light of particular wavelengths? Why does it then emit only light of these same wavelengths? Classical theories of electromagnetism could not answer these questions. The answers came early in the twentieth century from the development of quantum mechanics and nuclear physics.

An atom consists of a small, dense nucleus surrounded by electrons

The first important clue about the internal structure of atoms came from an experiment conducted in 1910 by Ernest Rutherford, a gifted chemist and physicist from New Zealand. Rutherford and his colleagues at the University of Manchester in England were investigating the recently discovered phenomenon of radioactivity. Certain radioactive elements such as uranium and radium were known to emit particles. One type of particle, called an alpha (α) particle, is quite massive (about as massive as four hydrogen atoms) and is emitted from a radioactive substance with considerable speed.

In one series of experiments, Rutherford and his colleagues were using alpha particles as projectiles to probe the structure of solid matter. They directed a beam of alpha particles against a thin sheet of metal (see Figure 17-10). Almost all the alpha particles passed through the metal sheet with little or no deflection from the straight-line path. To the surprise of the experimenters, however, an occasional alpha particle bounced back from the metal sheet as though it had struck something very dense. Rutherford later remarked, "It was almost as incredible as if you fired a 15-inch shell at a piece of tissue paper and it came back and hit you."

Rutherford was able to explain the results of this experiment in only one way: most of the mass of an atom is concentrated in a very compact, very massive lump of matter that occupies only a small part of the atoms' volume. Most the alpha particles pass freely through the nearly empty space that makes up most of the atom, but a very few alpha parti-

Figure 17-10 Rutherford's experiment
Alpha particles from a radioactive source are directed against a thin metal foil. Most alpha particles pass through the foil with very little deflection. Occasionally, however, an alpha particle recoils dramatically, indicating that is has collided with the massive nucleus of an atom. This experiment provided the first evidence atoms have nuclei.

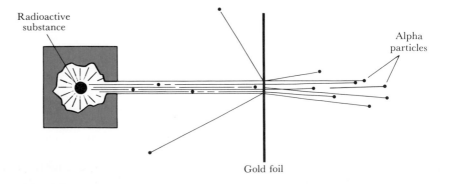

cles happen to strike the dense mass at the center of an atom and bounce back.

Rutherford proposed a new model of the structure of an atom. A very massive, positively charged **nucleus** at the center of the atom is orbited by tiny, negatively charged electrons (see Figure 17-11). Rutherford concluded that at least 99.98 percent of the mass of an atom is concentrated in a nucleus whose diameter is only about one ten-thousandth the diameter of the atom.

As discussed in Box 6-3, we know today that the nucleus contains two types of particles: protons and neutrons. A proton has almost the same mass as a neutron; each has about 2000 times as much mass as an electron. A proton has a positive electric charge, whereas a neutron has no electric charge. The attractive electrical force between the positively charged protons and the negatively charged electrons holds an atom together.

We know today that the alpha particle is identical to the nucleus of a helium atom; it consists of two protons and two neutrons, which happens to be a very stable combination. Another kind of particle commonly emitted by radioactive elements, the beta particle, is now known to be a single electron.

Figure 17-11 Rutherford's model of the atom
Electrons orbit the atom's nucleus, which contains most of the atom's mass. The nucleus contains two types of particles: protons and neutrons.

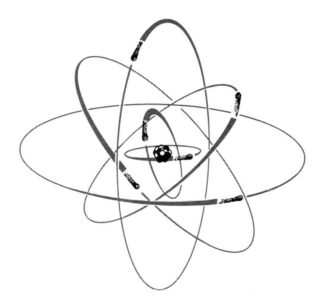

Spectral lines are produced when an electron jumps from one energy level to another within an atom

The task of reconciling Rutherford's atomic model with the observations of spectral analysis was undertaken by the young Danish physicist Niels Bohr, who joined Rutherford's group at Manchester in 1911.

Bohr began by trying to understand the structure of hydrogen, the simplest and lightest of the elements. A hydrogen atom consists of a single electron and a single proton. Hydrogen has a simple spectrum consisting of a pattern of lines that begins at 6563.1 Å and ends at 3645.6 Å. The first spectral line is called H_α, the second spectral line is called H_β, the third is H_γ, and so forth, ending with H_∞ at 3645.6 Å. The closer you get to 3645.6 Å, the more spectral lines you see.

The regularity in this spectral pattern was described mathematically in 1885 by Johann Jacob Balmer, an elderly Swiss school teacher. Balmer used trial and error to discover a formula from which the wavelengths

(λ) of the hydrogen lines can be calculated. Balmer's formula is usually written

$$\frac{1}{\lambda} = R\left(\frac{1}{4} - \frac{1}{n^2}\right)$$

where n is an integer greater than 2 and R is a number (R = 109,677 cm^{-1}) called the **Rydberg constant** in honor of the Swedish spectroscopist J. R. Rydberg. To get the wavelength of H$_\alpha$, you put n = 3 into Balmer's formula. To get H$_\beta$, use n = 4. To get H$_\gamma$, use n = 5. To get the convergence limit at 3645.6 Å, use n = ∞.

Since this formula is successful in giving the right answers, the spectral lines of hydrogen at visible wavelengths are today called the **Balmer lines** and the entire pattern from H$_\alpha$ to H$_\infty$ is called the **Balmer series.** The spectrum of the star HD 193182 shown in Figure 17-12 exhibits more than two dozen Balmer lines from H$_{13}$ through H$_{40}$.

Figure 17-12   Balmer lines in the spectrum of HD 193182
This portion of the spectrum of the star HD 193182 shows nearly two dozen Balmer lines. The series converges at 3645.6 Å, just to the left of H$_{40}$. This star's spectrum also contains the first twelve Balmer lines (H$_\alpha$ through H$_{12}$), but they are not visible in this particular spectrogram. (Mount Wilson and Las Campanas Observatories)

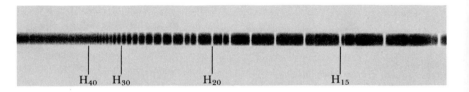

H$_{40}$ H$_{30}$ H$_{20}$ H$_{15}$

Bohr realized that, if he were to succeed in understanding the structure of the hydrogen atom, he should be able to derive Balmer's formula directly from his theory and calculations. He began by assuming that the electron in a hydrogen atom orbits the nucleus only in certain, specific orbits. As shown in Figure 17-13, it is customary to label these orbits n = 1, n = 2, n = 3 . . . etc. They are called the Bohr orbits.

In order for the electron to jump from one Bohr orbit to another, the hydrogen atom must gain or lose a very specific amount of energy. To go from an inner orbit to an outer orbit, the atom must absorb energy. Conversely, if the electron falls from a high orbit to one of the inner orbits, energy is released by the atom.

The energy gained or released by the atom when the electron jumps from the nth orbit to the mth orbit is the difference in energy between these two orbits. According to Planck and Einstein, the packet of energy

Figure 17-13   The Bohr model of the hydrogen atom
An electron circles the nucleus in allowed orbits n = 1, 2, 3, . . . A photon is absorbed by the atom as the electron jumps from an inner orbit to an outer orbit. A photon is emitted by the atom as the electron falls down to a low orbit. These absorptions and emissions of photons occur only at very specific wavelengths, thereby producing characteristic patterns of lines in the hydrogen spectrum.

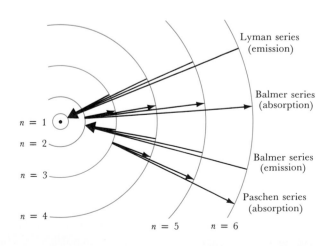

Lyman series (emission)

Balmer series (absorption)

Balmer series (emission)

Paschen series (absorption)

$n = 1$
$n = 2$
$n = 3$
$n = 4$
$n = 5$ $n = 6$

gained or released is a photon whose energy is given by $E = hc/\lambda$. Combining these ideas, Bohr mathematically proved that the wavelength (λ) of the photon emitted or absorbed as the electron jumps from level m to level n is:

$$\frac{1}{\lambda} = \left(\frac{2\pi^2 M e^4}{c h^3}\right)\left(\frac{1}{m^2} - \frac{1}{n_2}\right)$$

where M is the mass of the electron, e is the charge on the electron, h is Planck's constant and c is the speed of light.

Bohr's discovery elucidated the meaning of Balmer's formula: all the Balmer lines are produced by electron transitions between the second Bohr orbit and higher orbits ($n = 3, 4, 5, \ldots$ etc.).

In addition to giving the wavelengths of the Balmer series (when $m = 2$), Bohr's formula correctly predicts the wavelengths of other series of spectral lines that occur at non-visible wavelengths. For example, $m = 1$ gives the **Lyman series** which is entirely in the ultraviolet. Since $m = 1$, all of the spectral lines in this series involve electron transitions between the lowest Bohr orbit and all higher orbits ($n = 2, 3, 4, \ldots$ etc.). This pattern of spectral lines begins with L_α ("Lyman alpha") at 1215 Å and converges on L_∞ at 912 Å. At infrared wavelengths is the **Paschen series,** for which $m = 3$. It begins with P_α ("Paschen alpha") at 18,751 Å and converges on P_∞ at 8206 Å. Additional series exist at still longer wavelengths.

Bohr's ideas give us an explanation of Kirchhoff's laws. Each spectral line corresponds to one particular transition between the orbits of the atoms of a particular element. An absorption line occurs when an electron jumps from an inner orbit to an outer orbit, extracting the required photon from an outside source of energy such as the continuous spectrum of a hot, glowing object. An emission line is produced when the electron falls back down to a lower orbit and gives up a photon. Since these photons are the result of specific electron transitions, spectral lines can be used to determine the structure of the atoms.

Physicists today retain many features of the Bohr model of the atom, although they no longer picture electrons moving in specific orbits about the nucleus. Instead, electrons are said to occupy certain allowed **energy levels** in the atom. An extremely useful way of displaying the structure of a atom is with an **energy level diagram**, such as that shown in Figure 17-14 for hydrogen. The lowest energy level is called the ground state and corresponds to the n = 1 Bohr orbit. An electron can jump from the

Figure 17-14   The energy-level diagram of hydrogen
The structure of the hydrogen atom is conveniently displayed in a diagram showing the allowed energy levels above the ground state. A variety of electron jumps, or transitions, are shown—including those that produce some of the most prominent lines in the hydrogen spectrum.

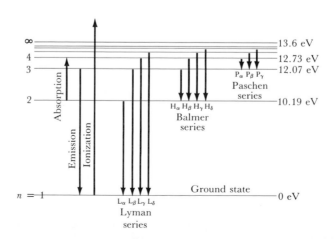

ground state up to the $n = 2$ level only if the atom absorbs a Lyman-alpha photon of wavelength 1216 Å. The energy of a photon, as determined by the relationship $E = h\nu = hc/\lambda$, is usually expressed in electron-volts (eV). (The common units of wavelength, frequency, and energy are discussed in Box 5-1.) The Lyman-alpha photon has an energy of 10.19 eV, so the energy level $n = 2$ is shown on the diagram as having an energy 10.19 eV above the energy of the ground state (conventionally assigned a value of 0 eV). Similarly, the $n = 3$ level is 12.07 eV above the ground state, and so forth up to the $n = \infty$ level at 13.6 eV. If the atom absorbs a photon of any energy greater than 13.6 eV, an electron from the ground state will be knocked completely out of the atom. This process, whereby high-energy photons knock electrons out of atoms, is called **ionization.**

The atoms of heavier elements have more complex energy-level diagrams. For example, Figure 17-15 shows the energy-level diagram of sodium. Astronomers find it useful to refer to such diagrams to identify the spectral lines they observe in the spectra of stars and nebulae.

With the work of people like Planck, Einstein, Rutherford and Bohr, the interchange between astronomy and physics had come full circle. Modern physics was born when Newton set out to understand the motions of the planets. Two and a half centuries later, physicists in their laboratories had probed the properties of light and the structures of atoms. The fruits of their labors would have immediate and direct astronomical applications.

Figure 17-15   The energy-level diagram of sodium

Complex atoms have complicated energy-level diagrams. The energy-level diagram of sodium is displayed here, along with the wavelengths of photons absorbed or emitted in some of the major electron transitions. At visible wavelengths, the sodium spectrum is dominated by two very strong lines (called the sodium D lines) at 5889.9 Å and 5895.9 Å. These two lines are strong because they correspond to two transitions that are the primary avenue through which electrons cascade from high orbits down to the ground state.

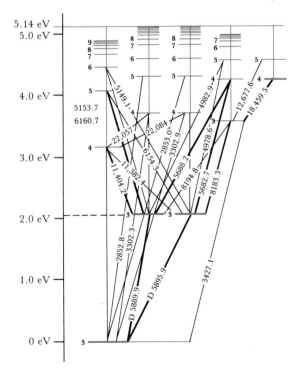

Summary

· Electromagnetic radiation (including visible light) has wavelike properties and travels through empty space at the constant speed $c = 3 \times 10^{10}$ cm/sec.

· The Stefan–Boltzman law states that any object radiates electromagnetic waves with a total energy flux (E) directly proportional to the fourth power of the absolute temperature (T) of the object: $E = \sigma T^4$.

. Wien's law states that the dominant wavelength at which an object emits electromagnetic radiation is inversely proportional to the absolute temperature of the object: λ_{max}(in cm) $= 0.29/T$.

. A blackbody is a hypothetical object that is a perfect absorber of electromagnetic radiation at all wavelengths; stars closely approximate the behavior of blackbodies.

The intensities of radiation emitted at various wavelengths by a blackbody at a given temperature are shown by a blackbody curve.

The shape of blackbody curves can be explained only by the assumption that light has particlelike properties; the particles of light are called photons.

. Planck's law relates the energy (E) of a photon to its frequency (ν) or wavelength (λ): $E = h\nu = hc/\lambda$, where h is Planck's constant.

. Kirchhoff's three laws of spectral analysis state that (1) a hot, glowing object emits a continuous spectrum containing all wavelengths; (2) when the continuous spectrum is viewed through a cool gas, dark spectral lines (absorption lines) characteristic of the elements present in the gas are observed; (3) the gas emits light that consists of a pattern of bright spectral lines (emission lines) corresponding exactly to the pattern of absorption lines.

. An atom consists of a small, dense nucleus (composed of protons and neutrons), surrounded by electrons that occupy only certain allowed energy levels.

When an electron jumps from one energy level to another, it emits or absorbs a photon of appropriate energy (and hence of a specific wavelength).

The spectral lines of a particular element correspond to the various electron transitions between allowed energy levels in atoms of that element.

The prominent spectral lines of hydrogen's Balmer series are produced by jumps between the second energy level of the atom and higher levels; the Lyman series in the far-ultraviolet is produced by jumps between the ground state (lowest level) and higher levels; the Paschen series in the infrared is produced by jumps between the third energy level and higher levels.

The process in which an atom absorbs an energetic photon and loses an electron entirely is called ionization.

Review questions

1 Describe an experiment in which light behaves like a wave.

2 Describe an experiment in which light behaves like a particle.

*3 Approximately how many times around the world could a beam of light travel in one second?

4 Using Wien's law and Stefan's law, explain the color changes that are observed as the temperature of a hot, glowing object increases.

5 Describe the experimental evidence that supports our current views of the structure of an atom.

*6 Calculate the wavelength of H_δ. Draw a schematic diagram of the hydrogen atom and indicate the electron transition that gives rise to this spectral line.

Discussion questions

*7 Imagine a star the same size as the Sun, but with a surface temperature twice that of the Sun. At what wavelength does that star emit most of its radiation? How many times brighter than the Sun is that star?

*8 Imagine a star whose diameter is ten times larger than the Sun and whose a surface temperature is 2900 K. Suppose both the Sun and this star were located at the same distance from you. Which would be brighter? By what factor?

9 Describe how the spectrum of a helium atom might appear if one of its two electrons were stripped off.

10 Why do different isotopes of the same element (such as Carbon 12 and Carbon 13) exhibit slightly different spectral line patterns?

Advanced questions

11 Compare chemical identification based on spectral line patterns with the identification of people by line patterns in their fingerprints.

12 Suppose you look up at the night sky and observe some of the brightest stars with your naked eye. Is there any way of telling which stars are hot and which are cool? Explain.

13 Galileo attempted to measure the speed of light and concluded that it travels instantaneously from one place to another. Why do you suppose he obtained such a result? What would you say to him now?

For further reading

Cline, B. *Men Who Made a New Physics*. Signet, 1965. *A history of the discoveries that led to our modern understanding of radiation and the atom.*

Gingerich, O. "Unlocking the Chemical Secrets of the Cosmos." *Sky & Telescope*, July 1981, p. 13.

van Heel, A., and Velzel, C. *What Is Light?* McGraw-Hill, 1968.

Weymann, R. "Extending the Visible Frontier: New Tools of the Optical Astronomer." *Mercury*, Sept./Oct. 1975, p. 2.

Wolf, F. *Taking the Quantum Leap*. Harper & Row, 1981. *A good introduction to quantum mechanics, the science of the atom, for the layperson.*

18 The nature of the stars

A cluster of stars
By analyzing starlight, an astronomer can determine details about a star such as its surface temperature, chemical composition, and luminosity. This photograph clearly shows color differences in a star cluster called NGC 3293. Reddish stars are comparatively cool and have surface temperatures around 3000 K. These stars are also very luminous and have very large diameters, typically 100 times larger than the Sun. Bluish and blue-white stars have much higher surface temperatures (15,000 to 30,000 K) but are roughly the same size as the Sun. (Anglo-Australian Observatory)

Although they appear only as brilliant pinpoints of light, the stars are known to be huge, massive spheres of glowing gas, much like our Sun. Astronomers have measured the distances to many of the nearer stars. In this chapter we also learn about ways in which astronomers determine the luminosities, surface temperatures, masses, and other properties of stars. We make our first acquaintance with the Hertzsprung–Russell diagram, which reveals the various fundamental types of stars. Finally, we turn to the topic of binary stars, those surprisingly common systems in which two stars orbit about each other. Binary stars are a very important source of information for the astronomer seeking data about the stars.

The night sky is spangled with thousands of stars, each appearing as a bright pinpoint of light. A telescope reveals many thousands of other stars too faint to be seen with the naked eye, but every star appears only as a bright point of light in even the most powerful telescope. A star is a huge, massive ball of hot gas like our Sun, held together by its own gravity. We now know that some stars are larger than our Sun, and some are smaller. Some stars are brighter than the Sun, and some are dimmer. Some stars are hotter than the Sun, and others are cooler.

The quest for information about the masses, luminosities, temperatures, and chemical compositions of the stars has been a major occupation of twentieth-century astronomers. In recent years, a remarkably complete picture has emerged. By understanding the stars, we gain insight into our relationship to the universe and our place in the cosmic scope of space and time.

Distances to nearby stars are determined by parallax

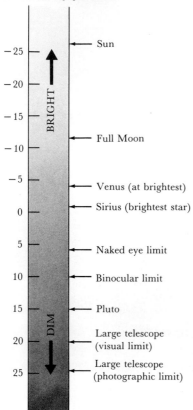

Figure 18-1   The apparent magnitude scale

Astronomers denote the brightnesses of objects in the sky by their apparent magnitudes. Most stars visible to the naked eye have magnitudes in the range +1 to +6. Photography through large telescopes reveals stars as faint as magnitude +24.

The system of magnitudes that astronomers use to denote the brightnesses of stars was invented in ancient Greece by the astronomer Hipparchus. The brightest stars Hipparchus saw in the sky he called first-magnitude stars. Those about one-half as bright he called second-magnitude stars, and so forth to the sixth-magnitude stars, which were the dimmest stars he could see. When telescopes came into use, astronomers extended Hipparchus's magnitude scale to larger magnitudes to describe the dimmer stars visible through their instruments.

In the nineteenth century, techniques were developed for measuring more exactly the amount of light energy arriving from a star. The branch of science dealing with such measurements is called **photometry.** Astronomers then set out to define the magnitude scale more precisely. Their measurement showed that a first-magnitude star is about 100 times as bright as a sixth-magnitude star. In other words, it would take 100 stars of magnitude +6 to provide as much light energy as we receive from a single star of magnitude +1. Therefore, the magnitude scale was redefined so that a magnitude difference of 5 corresponds exactly to a factor of 100 in the amount of light energy received. A magnitude difference of 1 therefore corresponds to a factor of 2.512 in light energy, because

$$2.512 \times 2.512 \times 2.512 \times 2.512 \times 2.512 = 100$$

Thus, for example, it takes about $2\frac{1}{2}$ third-magnitude stars to provide as much light as we receive from a single second-magnitude star.

Figure 18-1 illustrates the modern magnitude scale. Note, for example, that the dimmest stars visible through a pair of binoculars have magnitude +10, and the dimmest stars that can be photographed with a large telescope have magnitude +24. These magnitudes are properly called **apparent magnitudes** because they describe how bright an object *appears* to an Earth-based observer. Apparent magnitude is a measure of the energy arriving at the Earth.

Apparent magnitudes do not tell you about the actual brightnesses of the stars. A star that looks dim in the sky might actually be a very brilliant star that happens to be extremely far away. In order to determine the actual brightness of a star, we must first know how far away the star is.

The most straightforward way of measuring stellar distances involves

Figure 18-2   Parallax

Imagine looking at some nearby object (a tree) seen against a distant background (mountains). If you move from one location to another, the nearby object will appear to shift its location with respect to the distant background scenery. This familiar phenomenon is called parallax.

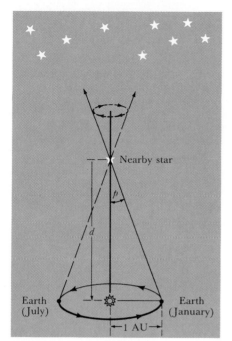

Figure 18-3 Stellar parallax
As the Earth orbits the Sun, a nearby star appears to shift its position against the background of distant stars. The parallax (p) of the star is equal to the angular radius of the Earth's orbit as seen from the star.

parallax, which is the apparent displacement of an observed object due to a change in the point of view. You experience parallax when nearby objects appear to shift their positions against a distant background as you move from one place to another (see Figure 18-2). Stars exhibit the same phenomenon. As the Earth orbits the Sun, nearby stars appear to move back and forth against the background of more distant stars.

The distance to a star can be determined by measuring the star's parallax. The **parallax** (*p*) of a star is one-half of the angle through which the star's apparent position shifts as the Earth moves from one side of its orbit to the other (see Figure 18-3). If the angle *p* is measured in arc sec, then the distance (*d*) to the star in parsecs is given by the equation

$$d = \frac{1}{p}$$

For example, a star whose parallax is 0.5 arc sec is 2 parsecs from Earth. This simple relationship between parallax and distance in parsecs is one of the main reasons why astronomers often measure cosmic distances in parsecs rather than light years. (As we saw in Box 1-3, 1 parsec equals 3.26 light years, and 1 light year is nearly 10^{13} km.)

The parallax method of determining stellar distances works only for nearby stars. The largest measured stellar parallax is less than 1 arc sec, a very tiny angle. There are slightly more than 1000 stars within 20 parsecs of the Earth whose parallaxes have been measured with a high degree of precision. Most of these nearby stars are invisible to the unaided eye. The majority of the familiar, bright stars in the nighttime sky are too far away to exhibit measurable parallax as the Earth orbits the Sun.

A star's luminosity can be determined from its apparent magnitude and its distance

The **absolute magnitude** of a star is the apparent magnitude that it would have if it were located at a distance of exactly 10 parsecs from the Earth. For example, if the Sun were moved to a distance of 10 parsecs from the Earth, it would have an apparent magnitude of +4.8. Therefore the absolute magnitude of the Sun is +4.8. The absolute magnitudes of other stars range from roughly −10 for the brightest to +15 for the dimmest. The Sun's absolute magnitude is about in the middle of this range, giving us our first hint that the Sun is an average star.

How do astronomers determine absolute magnitudes of stars? The farther away a source of light is, the dimmer it appears. This decrease of brightness with increasing distance is accurately described by the inverse square law (recall Box 8-2), which tells us that the apparent brightness of a light source is inversely proportional to the square of the distance between the source and the observer. Using this law, astronomers have derived a mathematical equation relating three quantities: a star's apparent magnitude (*m*), its absolute magnitude (*M*), and its distance (*d*) from the Earth. If you know two of these quantities (such as apparent magnitude and distance), then you can calculate the third quantity (such as absolute magnitude). Thus astronomers measure the apparent magnitude of a nearby star, find its distance by measuring its parallax, and then calculate its absolute magnitude. (For the reader familiar with logarithms, Box 18-1 shows the derivation of the equation relating *m*, *M*, and *d*.)

Absolute magnitude is directly related to **luminosity,** the amount of energy escaping from a star's surface each second (usually given in ergs

Box 18-1   Magnitude and brightness

The concept of a star's magnitude dates back to the days of ancient Greek astronomy. Modern astronomers often prefer to talk about a star's **brightness,** which is the energy flux arriving at the Earth. Brightness is usually expressed in ergs per square centimeter per second (erg cm^{-2} sec^{-1}). As explained in this chapter, each step in magnitude corresponds to a factor of 2.512 in brightness. In other words, your eyes receive 2.512 times more energy per square centimeter per second from a third-magnitude star than they do from a fourth-magnitude star.

Consider two stars with magnitudes m_1 and m_2 and brightnesses b_1 and b_2. The ratio of their brightnesses (b_1/b_2) corresponds to a difference in their magnitudes ($m_2 - m_1$). Because each step in magnitude corresponds to a factor of 2.512 in brightness, we can construct the table shown here.

Magnitude difference $(m_2 - m_1)$	Ratio of brightness (b_1/b_2)
1	2.512
2	$(2.512)^2 = 6.31$
3	$(2.512)^3 = 15.85$
4	$(2.512)^4 = 39.8$
5	$(2.512)^5 = 100$
10	$(2.512)^{10} = 10^4$
15	$(2.512)^{15} = 10^6$
20	$(2.512)^{20} = 10^8$

In general, the relationship between ($m_2 - m_1$) and (b_1/b_2) can be written as

$$\frac{b_1}{b_2} = 100^{(m_2 - m_1)/5}$$

Taking the logarithm of both sides of this equation and rearranging terms, we obtain

$$m_2 - m_1 = 2.5 \log \left(\frac{b_1}{b_2} \right)$$

The usefulness of these relationships and equations is best illustrated by several examples.

.   **Example 1**   At greatest brilliancy, Venus has a magnitude of -4. Compare the brightness of Venus with that of the dimmest stars visible to the naked eye.

The dimmest stars visible to the naked eye have a magnitude of $+6$. The magnitude difference is therefore $+6 - (-4) = 10$. This corresponds to a brightness ratio of $(2.512)^{10} = 10^4$. It would therefore take 10,000 sixth-magnitude stars to shine as brilliantly as Venus.

.   **Example 2**   In August 1975, a nova appeared in the constellation of Cygnus (the swan). A nova is a violent outburst involving a certain kind of star. In two days, its magnitude changed from $+15$ to $+2$. By what factor did its brightness increase?

The change in magnitude $(m_2 - m_1)$ is 13. Thus the ratio of brightness is

$$\frac{b_1}{b_2} = 100^{13/5} = 158{,}500$$

Thus, in two days, the nova's brightness increased by a factor of nearly 160,000.

Example 3 The variable star RR Lyrae periodically doubles its light output. How much does its magnitude change?

The brightness of the star varies by a factor of 2, so $b_1/b_2 = 2$. Thus

$$m_2 - m_1 = 2.5 \log (2) = 0.7$$

RR Lyrae periodically varies by seven-tenths of a magnitude.

These relationships can be extended to give a very useful relationship between a star's apparent magnitude, its absolute magnitude, and its distance from Earth. Let m and b be the apparent magnitude and brightness, respectively, of a star at a distance d from Earth. Let M and B be the apparent magnitude and brightness, respectively, of the star if it were 10 parsecs from Earth. By definition, M is the star's absolute magnitude. From our general relationship between brightness and apparent magnitude,

$$m - M = 2.5 \log \left(\frac{B}{b}\right)$$

According to the inverse square law (recall Box 8-2), the brightness of a light is inversely proportional to the square of the distance between the light and the observer. Thus

$$\frac{B}{b} = \left(\frac{d}{10}\right)^2$$

Combining these two equations and rearranging terms, we obtain

$$M = m - 5 \log \left(\frac{d}{10}\right)$$

where d is measured in parsecs.

As an example, consider the bright, nearby star Capella. Capella's apparent magnitude is $+0.05$, and its distance is 14 parsecs. Thus its absolute magnitude is

$$M = 0.05 - 5 \log (1.4) = -0.7$$

Comparing this value to the Sun's absolute magnitude $(+4.8)$, we see that Capella is actually $5\frac{1}{2}$ magnitudes brighter than the Sun. From

$$m_2 - m_1 = 5\frac{1}{2} = 2.5 \log \left(\frac{b_1}{b_2}\right)$$

we find $b_1/b_2 = 158$. Thus Capella emits about 160 times as much light energy as the Sun does.

per second). Many scientists prefer to speak of a star's luminosity rather than its absolute magnitude because luminosity is a measure of the star's energy output. There is a simple, direct relationship between absolute magnitude and luminosity (see Box 18-2), and astronomers can convert from one to the other as they see fit. For convenience, stellar luminosities are expressed in multiples of the Sun's luminosity ($L_\odot$), which equals 3.90×10^{33} erg/sec. The brightest stars in the sky (absolute magnitude = -10) have luminosities of $10^6 L_\odot$, which means that each of these stars has the energy output of a million Suns. The dimmest stars (absolute magnitude = $+15$) have luminosities of $10^{-4} L_\odot$.

Box 18-2 Bolometric magnitude and luminosity

In determining a star's absolute magnitude, astronomers must make allowances for nonvisible light and for the Earth's atmosphere. The apparent magnitude of a star we see in the sky could be misleading if the star happens to emit a significant fraction of its radiation at nonvisible wavelengths. For example, a very luminous and hot star with a surface temperature of 35,000 K appears deceptively dim to our eyes simply because most of the star's light is emitted at ultraviolet wavelengths. Furthermore, the Earth's atmosphere is opaque to many nonvisible wavelengths, and thus a sizeable fraction of the light from the hottest stars and the coolest stars simply does not penetrate the air to get to our eyes or telescopes.

To cope with this difficulty, astronomers have defined the bolometric magnitude. The **bolometric magnitude** of a star is the star's apparent magnitude as measured *above* the Earth's atmosphere and over *all* wavelengths.

In recent years, astronomical satellites have allowed us to determine the bolometric magnitudes of many stars. The absolute magnitude deduced from the bolometric magnitude is called the **absolute bolometric magnitude** (M_{bol}) of a star and is always brighter than the star's absolute visual magnitude (M) deduced from ground-based observations at visible wavelengths alone. By comparing satellite and ground-based data, astronomers have figured out how much they must subtract from a star's absolute visual magnitude to get its absolute bolometric magnitude. This correction factor is called the **bolometric correction** (BC) and we have the simple equation

$$M_{\text{bol}} = M - \text{BC}$$

The bolometric correction is very large for both the hottest stars and the coolest stars. For example, for stars hotter than 20,000 K or cooler than 3000 K, the bolometric correction is 3 magnitudes or greater. For stars such as the Sun that emit an overwhelming percentage of their radiation at visible wavelengths, the bolometric correction is almost zero. The Sun's absolute bolometric magnitude is $+4.75$, whereas its absolute visual magnitude is $+4.83$.

A star's absolute bolometric magnitude is directly related to the star's luminosity (L). From the equations in Box 18-1, it is possible to show that

$$M_{\text{bol}} = 4.72 - 2.5 \log (L/L_\odot)$$

Knowing a star's M_{bol} and using this equation, astronomers can calculate exactly how much energy is being released from the star's surface each second.

As an example, consider Sirius, which has an apparent magnitude of -1.46, making it the brightest star in the nighttime sky. The distance to Sirius is 2.7 parsecs, and Sirius's surface temperature is about 10,000 K, which corresponds to a bolometric correction of 0.6 magnitude. Suppose that you want to know the luminosity of Sirius.

First, you must calculate the absolute magnitude of Sirius as follows:

$$M = m - 5 \log (d/10)$$
$$= -1.46 - 5 \log 0.27$$
$$= -1.46 - 5(9.4314 - 10)$$
$$= 1.4$$

The absolute magnitude of Sirius is 1.4 and the bolometric correction is 0.6, so we find that the absolute bolometric magnitude is

$$M_{bol} = M - \text{BC} = 1.4 - 0.6 = 0.8$$

We know that

$$M_{bol} = 4.72 - 2.5 \log (L/L_\odot)$$

which is the same as

$$\log (L/L_\odot) = \frac{4.72 - M_{bol}}{2.5}$$
$$= \frac{4.72 - 0.8}{2.5}$$
$$= 1.6$$

Finally, we obtain

$$L/L_\odot = 10^{1.6} = 40$$

Thus Sirius is 40 times more luminous than the Sun.

A star's color reveals its surface temperature

Figure 18-4   Temperature and color
This schematic diagram shows the relationship between the color of a star and its surface temperature. The intensity of light emitted by three hypothetical stars is plotted against wavelengths (compare Figure 17-4). The range of visible wavelengths is indicated. The way in which a star's intensity curve is skewed determines the dominant color of its visible light.

One of the first things you notice when comparing stars in the nighttime sky is their differences in apparent magnitude. More careful examination, even with the naked eye, reveals that stars also have different colors. For example, in the constellation of Orion, you can easily note the difference between reddish Betelgeuse and bluish Rigel (see Figure 2-1).

As discussed in Chapter 17, a star's color is directly associated with its surface temperature through relationships such as Wien's law. The intensity of light from a cool star peaks at long wavelengths, and thus the star looks red (see Figure 18-4a). A hot star's intensity curve is skewed toward short wavelengths, and thus the star looks blue (see Figure 18-4b). The maximum intensity of a star of intermediate temperature (such as the Sun) occurs near the middle of the visible spectrum, giving the star a yellow-white color (see Figure 18-4c).

a   This star looks red

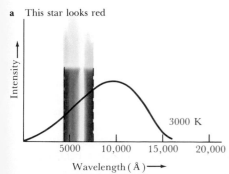

b   This star looks yellow–white

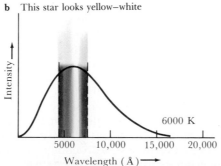

c   This star looks blue

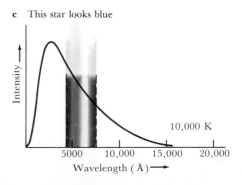

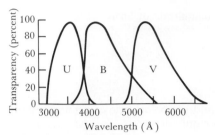

Figure 18-5 Light transmission through the UBV filters
This graph shows the wavelength ranges over which the standardized U, B, and V filters are transparent light. The U filter is transparent to light from 3000 to 4000 Å, a range called the near-ultraviolet because it lies just beyond the violet end of the visible spectrum. The B filter is transparent from about 3800 to 5500 Å, and the V filter is transparent from about 5000 to 6500 Å.

To measure accurately the colors of the stars, astronomers have developed a technique called **photoelectric photometry.** This technique uses a light-sensitive device (such as the photomultiplier discussed in Box 17-2) at the focus of a telescope, with a standardized set of colored filters. By far, the most commonly used filters are the UBV filters. Each of the three **UBV filters** is transparent in one of three broad wavelength bands: the ultraviolet (U), the blue (B), and the central region (V) of the visible spectrum (see Figure 18-5). (The transparency of the V filter fairly accurately mimics the sensitivity of the human eye.)

To do photometry, the astronomer aims a telescope at a star and measures the intensity of starlight that passes through each of the filters. This procedure gives three apparent magnitudes for the star, usually designated by the capital letters U, B, and V. The astronomer then compares the intensity of starlight in neighboring wavelength bands by subtracting one magnitude from another to form the combinations (B − V) and (U − B), which are called the star's **color indices.** The UBV magnitudes and color indices for several representative stars are given in Table 18-1.

TABLE 18-1 The UBV magnitudes and color indices of selected stars

Star name	V	B	U	(B − V)	(U − B)
Regulus (α Leo)	1.36	1.25	0.89	−0.11	−0.36
Altair (α Aql)	0.77	0.99	1.07	+0.22	+0.08
Bellatrix (γ Ori)	1.64	1.41	0.54	−0.23	−0.87
Alhena (γ Gem)	1.93	1.93	1.96	0.00	+0.03
Megrez (δ UMa)	3.31	3.39	3.46	+0.08	+0.07
Elnath (β Tau)	1.65	1.52	1.03	−0.13	−0.49

A color index tells you how much brighter (or dimmer) a star is in one wavelength band than in another. For example, the (B − V) color index tells you how much brighter (or dimmer) a star appears through the B filter than it does through the V filter.

A color index is important because it tells you about the star's surface temperature. If a star is very hot, its radiation is skewed toward the short-wavelength ultraviolet, which makes the star bright through the U filter, dimmer through the B filter, and dimmest through the V filter. The star Regulus (see Table 18-1) is an example of this case. Alternatively, if the star is cool, its radiation is peaked at long wavelengths, making the star brightest through the V filter, dimmer through the B filter, and dimmest through the U filter. The star Altair is an example.

The graph of Figure 18-6 gives the relationship between the (B − V) color index and temperature. If you know a star's (B − V) color index, you can use this graph to find the star's surface temperature. For example, the Sun's (B − V) index is +0.63, which corresponds to a surface temperature of 5800 K.

A word of caution is in order. The curve of Figure 18-6 was derived mathematically from the laws of blackbody radiation, but stars are not perfect blackbodies. Thus, the temperature deduced from a star's color index will be slightly different from temperatures deduced by other techniques. The details of these distinctions need not concern us in this book. Nevertheless, astronomers are careful to cite the method used to determine any stellar temperature.

A star's spectrum also reveals the star's surface temperature

The field of stellar spectroscopy was born in the 1860s when the Italian astronomer Angelo Secchi attached a spectroscope to his telescope and pointed it toward the stars. Secchi observed stellar spectral lines and made the important discovery that stars can be classified into various **spectral types** according to the appearances of their spectra.

In those days, the nature and cause of spectral lines were not well understood, and astronomers classified each star by assigning a letter from A through P, depending on the strength of the hydrogen Balmer lines in the star's spectrum. The A stars have the strongest Balmer lines, and the P stars have the weakest.

When Niels Bohr explained the structure of the hydrogen atom in the early 1900s (recall Figure 17-13), astronomers realized that the strength of the lines in a star's spectrum is directly related to the temperature of the gases in the star's outer layers. Hydrogen is by far the most abundant element in the universe, accounting for about three-quarters of the mass of a typical star. Hydrogen lines, however, do not necessarily show up in a star's spectrum. If the star is much hotter than 10,000 K, high-energy photons pouring out of the star's interior easily knock electrons out of the hydrogen atoms in the star's outer layers, ionizing the gas. The hydrogen ions have no electrons in their lower energy levels to absorb photons and produce Balmer lines. Conversely, if the star is much cooler than 10,000 K, the majority of photons escaping from the star do not possess enough energy to boost many electrons up from the ground state of the hydrogen atoms. These unexcited atoms also fail to produce Balmer lines. In summary, the star must be hot enough to excite the electrons out of the ground state, but not so hot that the atoms are ionized. A stellar surface temperature of 10,000 K results in the strongest Balmer lines.

A prominent set of Balmer lines is a clear indication that a star's surface temperature is about 10,000 K. At other temperatures, the spectral lines of other elements dominate a star's spectrum. For example, around 25,000 K, the spectral lines of helium are strong; at this temperature, photons have enough energy to excite helium atoms without tearing away the electrons altogether.

When a hydrogen atom is ionized, its only electron is torn away and no absorption lines can be produced. An atom of a heavier element, however, has two or more electrons. When one electron is knocked away, the remaining electrons can take over and produce a new and distinctive set of spectral lines. For example, in stars hotter than about 30,000 K, one of the two electrons in a helium atom is torn away. The remaining electron produces a set of spectral lines that is recognizably different from the lines produced by un-ionized helium. When the spectral lines of singly-ionized helium appear in a star's spectrum, we know that the star has a surface temperature greater than 30,000 K.

Astronomers designate an un-ionized atom with a Roman numeral I; thus H I is neutral hydrogen. A Roman numeral II is used to identify an atom with one electron missing; thus He II is singly ionized helium (He^+). Similarly, Si III is doubly ionized silicon (Si^{2+}), whose atoms are each missing two electrons.

In the early 1900s, E. C. Pickering, Annie Cannon, and their colleagues at Harvard Observatory set up the spectral classification scheme we use today. Many of Secchi's A-through-P categories were dropped, and the remaining spectral classes were reordered into a temperature sequence: **OBAFGKM.** This sequence has traditionally been memorized as "**O**h, **B**e **A F**ine **G**irl, **K**iss **M**e!" The O stars are the hottest; their surface temperatures are in excess of 35,000 K, and their spectra show

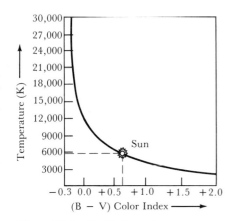

Figure 18-6 Blackbody temperature versus color index

The (B − V) color index is the difference between the B and V magnitudes of a star. If the star is hotter than about 10,000 K, it is a very bluish star, and its (B − V) index is less than zero. If a star is cooler than about 10,000 K, its (B − V) index is greater than zero. The Sun's (B − V) index is about 0.62, which corresponds to a temperature of 5800 K. After measuring a star's B and V magnitudes, an astronomer can estimate the star's surface temperature from a graph such as this one.

He II and Si IV. The M stars are the coolest; they have surface temperatures around 3000 K, which is so cool that atoms can stick together in molecules such as titanium oxide, whose spectral lines are prominent. Figure 18-7 shows representative examples of each spectral type.

Astronomers have found it useful to further subdivide the original OBAFGKM sequence into finer steps. These steps are indicated by the addition of an integer from 0 through 9. Thus, for example, we have . . . , F8, F9, G0, G1, G2, . . . , G9, K0, K1, K2, ... The Sun, whose spectrum is dominated by singly ionized metals (especially Fe II and Ca II) is a G2 star.

Figure 18-7 Principal types of stellar spectra
A star's spectrum is seen in the middle of each of these seven strips. The hydrogen lines are strongest in A stars, which have surface temperatures of about 10,000 K. The spectra of G and K stars exhibit numerous lines due to metals, indicating temperatures in the range of 4000 to 6000 K. The broad, dark bands in the spectrum of an M star are caused by titanium oxide, which can exist only if the temperature is cooler than about 3000 K. (Palomar Observatory)

The strength of a particular spectral line depends on both the ionization and the excitation of the atom responsible for that line. Ionization and excitation of atoms in a gas depend on the temperature of the gas as shown in Figure 18-8, allowing us to deduce a star's surface temperature from its spectral type. For example, a star exhibiting strong Ca II and Fe I lines in its spectrum is a K5 star with a surface temperature around 4500 K.

Figure 18-8 Spectral type and temperature
The strengths of the absorption lines of various elements are directly related to the temperature of the star's outer layers. For example, the Sun's spectrum has strong lines of singly ionized iron and calcium (Fe II and Ca II), corresponding to a spectral class of G2 and a surface temperature of about 5800 K. Note that hydrogen lines are strongest in A stars, whereas stars cooler than about 3500 K show absorption due to titanium oxide.

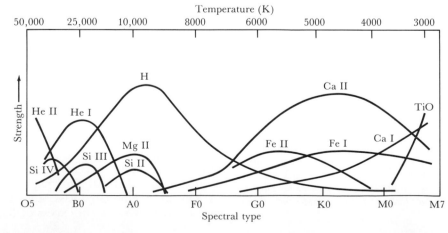

The Hertzsprung–Russell diagram demonstrates that there are different kinds of stars

The first accurate measurements of stellar parallax were made in the mid-nineteenth century, at about the time when astronomers began observing stellar spectra. During the next half-century, observing techniques improved, and the spectral types and absolute magnitudes of many stars became known.

Around 1905, the Danish astronomer Ejnar Hertzsprung pointed out that a regular pattern appears when the absolute magnitudes of stars are plotted versus their color indices on a graph. Almost a decade later, the American astronomer Henry Norris Russell independently discovered this regularity in a graph using spectral types instead of color indices. Plots of this kind are now known as **Hertzsprung–Russell diagrams,** or **H–R diagrams.**

Figure 18-9 is a typical Hertzsprung–Russell diagram. Each dot represents a star whose absolute magnitude and spectral type have been determined. Bright stars are near the top of the diagram; dim stars are near the bottom. Hot stars (O and B stars) are toward the left side of the graph; cool stars (M stars) are toward the right side.

The striking feature of the H–R diagram is that the data points are not scattered randomly all over the graph but are grouped in several distinct regions. The band stretching diagonally across the H–R diagram represents the majority of stars we see in the nighttime sky. This band is called the **main sequence** and extends from the hot, bright, bluish stars in the upper-left corner of the diagram down to the cool, dim, reddish stars in the lower-right corner. A star whose properties place it in this region of the H–R diagram is called a **main-sequence star.** For example, the Sun (spectral type G2, absolute magnitude +5) is such a star.

Toward the upper-right side of the H–R diagram, there is a second major grouping of data points. Stars represented by these points are both bright and cool. From Stefan's law, we know that a cool object radiates much less light per unit of surface area than a hot object does. Thus, in order to be so bright, these stars must be huge. As explained in Box 18-3, these stars are around 10 to 100 times as large as than the Sun. Most of these stars are around 100 times more luminous than the Sun and have surface temperatures around 3000 to 4000 K. They are called giants, or **red giants,** because they appear reddish in the nighttime sky. Aldebaran in the constellation Taurus and Arcturus in Boötes are good examples of red giants that you can easily see with the naked eye.

A few rare stars are considerably bigger and brighter than typical red giants. These superluminous stars are appropriately called **supergiants.**

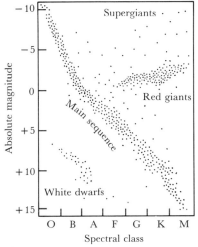

Figure 18-9 A Hertzsprung–Russell diagram
An H–R diagram is a graph on which the absolute magnitudes of stars are plotted versus their spectral types. Each dot on this diagram represents a star in the sky whose absolute magnitude and spectral class have been determined. Note that the data points are grouped in specific regions on the graph. This pattern reveals the existence of different types of stars in the sky: main-sequence stars, red giants, supergiants, and white dwarfs.

Box 18-3   Stellar radii

In Chapter 17, we saw that the energy flux from a blackbody is proportional to the fourth power of the blackbody's temperature. Specifically, the Stefan–Boltzmann law states that the emitted flux E (usually measured in erg cm^{-2} sec^{-1}) is related to the temperature T (in kelvins) by

$$E = \sigma T^4$$

where σ is a number called the Stefan–Boltzmann constant. Stars behave almost exactly like blackbodies, so (if you know a star's surface temperature) this equation tells how much energy is being radiated from each square centimeter of its surface each second.

(continued)

(Box 18-3, continued)

From elementary geometry, we know that a sphere's surface area is $4\pi R^2$, where R is the sphere's radius. A star is spherical, so we can use this expression for the surface area of a star. Multiplying the emitted flux E by the star's surface area, we get the total energy output of the star per second, which is the star's luminosity L:

$$L = 4\pi R^2 \sigma T^4$$

Rearranging this equation, we can find the radius of a star in terms of its luminosity and surface temperature:

$$R = \frac{1}{T^2}\sqrt{\frac{L}{4\pi\sigma}}$$

For example, consider the bright, reddish star Betelgeuse in the constellation Orion. Betelgeuse has a luminosity of $10,000L_\odot$ and a surface temperature of 3000 K. Substituting these values (and the value of the Sun's luminosity $L_\odot$) in the equation just derived, we find that the star's radius is 2.6×10^{13} cm, or nearly 2 AU. In other words, if Betelgeuse were located at the center of our solar system, the star would extend beyond the orbit of Mars.

In many calculations, it is convenient to relate everything to the Sun, which is a typical star. Specifically, for the Sun we have

$$L_\odot = 4\pi\,R_\odot^2\sigma T_\odot^4$$

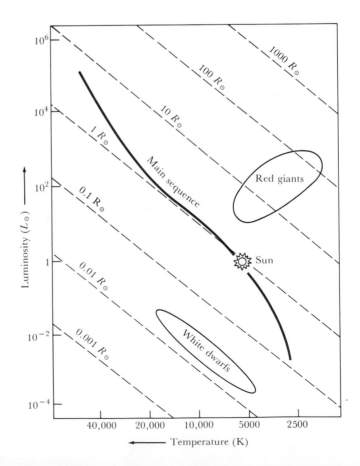

where $L_\odot$ is the Sun's luminosity, $R_\odot$ is the Sun's radius, and $T_\odot$ is the Sun's surface temperature (5800 K). Dividing the general equation for L by this specific equation for the Sun, we obtain

$$\frac{L}{L_\odot} = \left(\frac{R}{R_\odot}\right)^2 \left(\frac{T}{T_\odot}\right)^4$$

This is a very useful expression because all the constants such as π and σ have cancelled out, making the arithmetic much easier. Also, we can rearrange terms to arrive at an alternative useful equation:

$$\frac{R}{R_\odot} = \left(\frac{T_\odot}{T}\right)^2 \sqrt{\frac{L}{L_\odot}}$$

Again consider the example of Betelgeuse, for which $L = 10^4 L_\odot$ and $T = 3000$ K. Substituting this data into the previous equation, we get

$$R/R_\odot = \left(\frac{5800}{3000}\right)^2 \sqrt{10^4} = 370$$

In other words, Betelgeuse's radius is 370 times larger than the Sun's radius.

The Hertzsprung–Russell diagram is a graph of luminosity versus temperature, so we can use the general equation relating L, T, and R to draw lines on the H–R diagram representing various stellar radii, as shown here.

Note that most of the stars on the main sequence are around the same size as the Sun. Red giants have radii between $10R_\odot$ and $100R_\odot$, whereas the radii of supergiants range up to $1000R_\odot$. White dwarfs are roughly the same size as the Earth.

The radii of some stars have been measured with other techniques, some of which are discussed elsewhere in this text. The results of these other measurements are consistent with the radii calculated by the method described in this box.

Betelgeuse in Orion and Antares in Scorpius are examples of supergiants that you can find in the nighttime sky.

Finally, there is a third distinct grouping of data points toward the lower-left corner of the Hertzsprung–Russell diagram. These stars are both hot and dim, and so they must be small. They are appropriately called **white dwarfs.** These stars are roughly the same size as the Earth and can be seen only with the aid of a telescope.

There is another useful way to exhibit an H–R diagram. Instead of absolute magnitude, luminosity is plotted on the vertical axis of the graph; instead of spectral class, surface temperature is plotted on the horizontal axis. The resulting graph is still an H–R diagram, but observational quantities (such as spectral type) have been replaced by calculated quantities (such as temperature in kelvins).

Figure 18-10 shows this type of H–R diagram. Note that the temperature scale on the horizontal axis of the graph increases toward the left. This is because Hertzsprung and Russell drew their original diagrams with O stars on the left and M stars on the right. (This choice was made because the O stars have the simplest spectra.) Having hot stars toward the left and cool stars toward the right is a tradition that no one has seriously tried to change.

Figure 18-10 *An H–R diagram of the nearest and brightest stars*
It is often informative to draw an H–R diagram by plotting the luminosities of stars against their surface temperatures. The data for nearly 200 of the nearest and brightest stars are plotted on this diagram.

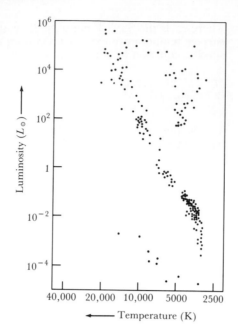

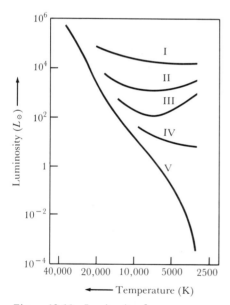

Figure 18-11 *Luminosity classes*
It is convenient to divide the H–R diagram into regions corresponding to the luminosity classes. This permits finer distinctions between giants and supergiants. Luminosity class V encompasses the main-sequence stars, including dim red stars called red dwarfs toward the lower-right side of the H–R diagram.

The H–R diagram in Figure 18-10 shows data for nearly 200 of the nearest and brightest stars in the sky. Even though this diagram contains far fewer data points than Figure 18-9 does, the main-sequence, white-dwarf, and red-giant regions are fairly clearly delineated.

The stars are classified into spectral types on the basis of the most prominent lines in their spectra. However, there are minor differences among the spectral patterns of stars having the same spectral type. In 1905, Hertzsprung pointed out that these minor spectral differences are related to the luminosities of the stars. In the 1930s, a system of **luminosity classes** was developed, based upon the minor differences in patterns of spectral lines. When the luminosity classes are plotted on the H–R diagram (see Figure 18-11), they provide a useful subdivision of the star types in the upper-right half of the diagram. Luminosity class I includes all of the supergiants, and luminosity class V includes the main-sequence stars. The intermediate classes distinguish giants of various luminosities as indicated in Table 18-2.

Astronomers commonly describe a star by combining its spectral type and its luminosity class in a shorthand description; for example, the Sun is said to be a G2V star. This notation supplies a great deal of information about the star. The spectral type is correlated with the star's surface temperature, and the luminosity class is correlated with its luminosity.

TABLE 18-2 *Stellar luminosity classes*

Luminosity class	Types of stars
I	Supergiants
II	Bright giants
III	Giants
IV	Subgiants
V	Main sequence

Thus an astronomer knows immediately that the G2V star is a main-sequence star with a luminosity of about $L_\odot$ and a surface temperature of nearly 6000 K. Similarly, the description of Aldebaran as a K5III star tells an astronomer that it is a red giant with a luminosity of around $500L_\odot$ and a surface temperature of about 4000 K.

The existence of fundamentally different types of stars is the first important lesson to come from the H–R diagram. As we shall see in following chapters, these different kinds of stars represent different stages of stellar evolution. We shall learn that a true appreciation of the H–R diagram involves an understanding of the life cycles of stars: how they are born, what happens as they mature, and where they go when they die.

Binary stars provide information about stellar masses

We now know something about the sizes, temperatures, and luminosities of stars. To complete our picture of the physical properties of the stars, we need only to know their masses. There is, however, no practical and direct way to measure the mass of an isolated star observed in the sky.

Fortunately for astronomers, about one-half of the visible stars in the nighttime sky are not isolated individuals. Instead, they are revealed by the telescope to be multiple-star systems in which two or more stars orbit about each other. By observing exactly how the stars orbit about each other, astronomers can glean important information about the stellar masses.

A pair of stars located very near each other in the sky is called a **double star.** William Herschel made the first organized search for such pairs, and between 1782 and 1821 published three catalogues listing more than 800 double stars. Later in the nineteenth century, his son John Herschel discovered 10,000 more doubles. Many of these double stars are true **binary stars,** or **binaries,** pairs in which the two stars are actually orbiting about each other.

In cases where astronomers actually see the two stars orbiting about each other, a binary is called a **visual binary** (see Figure 18-12). After

Figure 18-12 The binary star Kruger 60 *About one-half of the stars visible in the nighttime sky are revealed by the telescope to be double stars. This series of photographs shows the binary star called Kruger 60 in the constellation Cepheus. The orbital motion of the two stars about each other is apparent. This binary system has a period of $44\frac{1}{2}$ years. The maximum angular separation of the stars is about $2\frac{1}{2}$ arc sec, and their apparent magnitudes are +9.8 and +11.4. (Yerkes Observatory)*

1908

1915

1920

many years of patient observation, astronomers can plot the orbits of the stars in the binary pair (see Figure 18-13).

A binary-star system is held together by gravity. The gravitational force between the two stars keeps them in orbit about each other, and thus the details of the orbital motions are described by Newtonian mechanics. Specifically, the orbits obey Kepler's third law (see Box 4-3). For the binary-star system, Kepler's third law can be written as

$$M_1 + M_2 = \frac{a^3}{P^2}$$

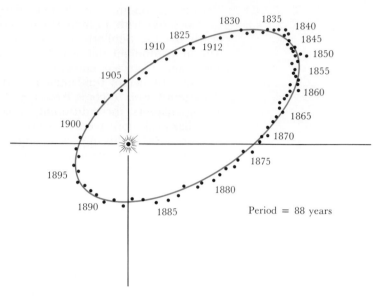

Figure 18-13   The orbit of 70 Ophiuchi
After plotting the observations of a binary star over the years, astronomers can draw the orbit of one star with respect to the other. Once the orbit is known, Kepler's third law can be used to deduce information about the masses of the stars. This illustration shows the orbit of a faint double star in the constellation Ophiuchus. In plotting the orbit, either star may be regarded as stationary—the shape and size of the orbit will be the same in either case.

where M_1 and M_2 are the masses of the two stars expressed in solar masses, P is the orbital period in years, and a is the semimajor axis (measured in AU) of the elliptical orbit of one star about the other plotted as in Figure 18-13.

In principle, it is easy to determine the period of a visual binary. All that is necessary is to keep observing until the two stars have returned to their original relative positions. As Figure 18-13 suggests, however, more than one lifetime of observations may be required.

Determining the semimajor axis of the orbit is somewhat more difficult. The angular separation between the stars can be determined by observation, but conversion of this angle into a linear distance (in AU) requires knowledge of the distance between the binary star and the Earth. Such information may in some cases be obtained through parallax measurements. Once the distance is known, the small-angle formula (Box 1-1) can be used to convert the angular separation to a distance in AU. Some care may be needed, however, to correct for the angle at which the orbit is viewed from the Earth.

Once both P and a are determined, Kepler's third law can be used to calculate $M_1 + M_2$, the *sum* of the masses of the two stars in the binary system. Note that this analysis provides no information about the individual masses of the two stars. To obtain the individual masses, more data about the binary orbit are needed.

Each of the two stars in the binary system actually moves in an elliptical orbit about the **center of mass** of the system (see Figure 18-14). The concept of the center of mass is analogous to two children on a seesaw. In order for the seesaw to balance properly, the center of mass of the two-child system must be located just above the support point (or fulcrum). As you know from experience, this center of mass is offset from the midpoint between the two children toward the heavier child.

Similarly, the center of mass of a binary system could (in imagination) be determined by placing the two stars at either end of a huge seesaw and determining where the fulcrum must be placed to balance the seesaw. The center of mass is always offset toward the more massive star by an amount proportional to the ratio of the two masses.

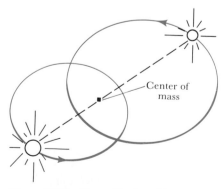

Figure 18-14   Star orbits in a binary
Each of the two stars in a binary system follows an elliptical orbit about their common center of mass. The center of mass is always nearer the more massive of the two stars.

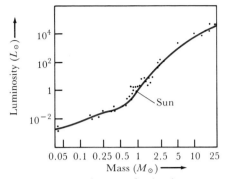

Figure 18-15 The mass–luminosity relation
For main-sequence stars, there is a direct correlation between mass and luminosity. The more massive a star, the more luminous it is.

In practice, the center of mass of a visual binary is determined with the aid of background stars. Using the background stars as reference points, the separate orbits of the two stars are plotted as in Figure 18-14. The center of mass is located by finding the common focus of the two elliptical orbits, and its offset from the midpoint between the stars is determined. This information yields the ratio M_1/M_2. The sum $M_1 + M_2$ is already known, so the individual masses can now be determined.

Years of careful, patient observations of many binaries have slowly but surely yielded the masses of many stars. As the data accumulated, an important trend began to emerge. For main-sequence stars, there is a direct correlation between mass and luminosity. The more massive the star, the more luminous it is. This **mass–luminosity relation** is conveniently displayed as a graph (see Figure 18-15). Note that the range of stellar masses extends from about one-twentieth of a solar mass to about 50 solar masses. The Sun's mass lies in the middle of this range, so we see again that our star is ordinary, or typical.

The mass–luminosity relation demonstrates that the main sequence on the H–R diagram is a progression in mass as well as in luminosity and surface temperature. The hot, bright, bluish stars in the upper-left corner of the H–R diagram are the most massive main-sequence stars in the sky. The dim, cool, reddish stars in the lower-left corner of the H–R diagram are the least massive. Main-sequence stars of intermediate temperature and luminosity have intermediate mass. This relationship of mass to the main sequence will play an important role in our discussion of stellar evolution.

Binary systems that cannot be resolved visually can also be detected and analyzed

Many binary stars are scattered throughout our galaxy, but only those that are nearby or that have a large separation between the two stars can be distinguished as visual binaries. The images of the two stars in a remote binary commonly are blended together to produce a visual image that looks like a single star. Spectroscopy provides the evidence that such binaries do not exist beyond the limits of visual resolution.

Spectral analysis of some stars yields incongruous spectral lines. For example, the spectrum of what appears to be a single star may include both strong hydrogen lines (characteristic of a type-A star) and strong absorption bands of titanium oxide (characteristic of a type-M star). A single star could not have the differing physical properties of these two spectral types, so the conclusion is that this star is actually a binary system too far away for resolution of its individual stars. A binary star that is detected in this way is called a **spectrum binary.**

If the orbital speeds of the two stars in a remote, unresolved binary are more than a few kilometers per second, then important additional information can be obtained from spectral analysis. As we saw in Box 7-3, the wavelength of a spectral line is affected by the relative speed between the source and the observer. If a light source is coming toward you, all of its spectral lines are displaced toward the short-wavelength (blue) end of the spectrum. Conversely, the spectral lines of a receding source are shifted toward the long-wavelength (red) end of the spectrum. This displacement of spectral lines is called the **Doppler effect.** If λ_0 is the unshifted wavelength of a spectral line and λ is the wavelength of that line in a star's spectrum, then

$$\frac{\lambda - \lambda_0}{\lambda_0} = \frac{v}{c}$$

where v is the star's velocity toward or away from us (along our line of sight), and c is the speed of light. With this equation, we can convert a measurement of wavelength displacement into information about a star's motion. It is important to emphasize that the Doppler effect applies only to motion along the line of sight. Motion perpendicular to the line of sight does not affect the wavelengths of spectral lines.

Some apparently single stars yield a spectrum in which two complete sets of spectra lines shift back and forth. Such stars are called **spectroscopic binaries.** The regular, periodic shifting of the spectral lines is due simply to the orbital motions of the stars as they revolve about their center of mass.

Figure 18-16 shows two spectra of a spectroscopic binary taken a few days apart. In Figure 18-16*a*, two sets of spectral lines are visible, slightly offset in opposite directions from the normal positions of these lines. One star is moving toward the Earth, and its lines are blueshifted; the other star is moving away from the Earth, and its lines are redshifted. A few days later, the stars have progressed along their orbits so that one star is moving toward the left and the other star toward the right. Neither star has any motion toward or away from the Earth, so there is no Doppler shifting, and both stars yield spectral lines at the same positions. Thus only one set of spectral lines appears in Figure 18-16*b*.

Significant information about the orbital velocities of the stars in a spectroscopic binary can be deduced from measurements of the shifts in spectral lines. This information is best displayed as a **radial velocity curve** that graphs radial velocity versus time for the binary system (see

Figure 18-16   A spectroscopic binary
A spectroscopic binary exhibits spectral lines that shift back and forth as the two stars revolve about each other. These two spectra show the behavior of the spectroscopic binary κ Arietis. (a) The stars are moving parallel to the line of sight (one star approaching Earth, the other star receding), producing two sets of shifted spectral lines. (b) Both stars are moving perpendicular to our line of sight. (Lick Observatory)

Figure 18-17). Radial velocity is the portion of a star's motion that is directed parallel to the line of sight between the Earth and the star.

In Figure 18-17, note that the wavy pattern repeats with a period of about 15 days, which is the orbital period of the binary. Also note that the entire wavy pattern is displaced upward from the zero-velocity line by about 12 km/sec, which is the overall motion of the binary system away from the Earth. Superimposed on this overall recessional motion (an overall redshift of the spectra lines) are the periodic approaches and recessions of the two stars as they orbit about the center of mass.

In many spectroscopic binaries, one of the stars is so dim that its spectral lines cannot be detected. The fact that the star is a binary, however, is obvious because its spectrum shows a single set of spectral lines that shift regularly back and forth. Such a **single-line spectroscopic binary** yields less information about its two stars than does a **double-line spectroscopic binary** such as that shown in Figure 18-16.

The orbital speeds of the two stars in a binary are related to the masses of the stars by Kepler's laws and Newtonian mechanics. From a radial velocity curve, one obtains a quantity involving the masses of both stars. The individual masses of the two stars can be determined only if the tilt of the binary orbits is known. The tilt of the orbits determines

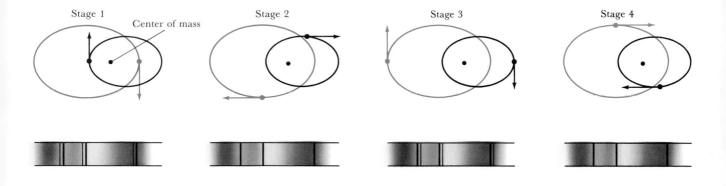

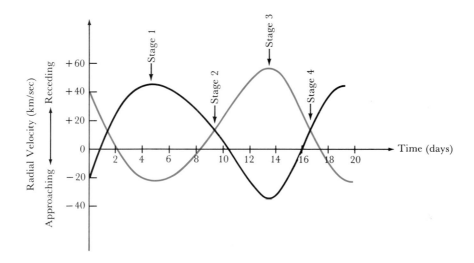

Figure 18-17 A radial velocity curve
The graph displays the radial velocity curve of the binary HD 171978. The drawings indicate the positions of the stars and their spectra at four selected moments during an orbital period.

how much of the true orbital speeds of the stars appears as radial velocity measured from the Earth.

If the two stars are observed to eclipse each other, then the orbit must be nearly edge-on as viewed from the Earth. As we shall see next, individual stellar masses can be determined if a spectroscopic binary also happens to be an **eclipsing binary.**

Light curves of eclipsing binaries provide detailed information about the stars

A small fraction of all binary systems are oriented so that the two stars periodically eclipse each other as seen from Earth. Such eclipsing binaries can be detected even when the two stars cannot be resolved visually as two distinct images in the telescope. The apparent brightness of the image of the binary dims momentarily each time that one star blocks out the other.

Using a photoelectric detector at the focus of a telescope, an astronomer can measure the light intensity very accurately. The data for an eclipsing binary are most usefully displayed in the form of **light curves** such as those shown in Figure 18-18. The overall shape of the light

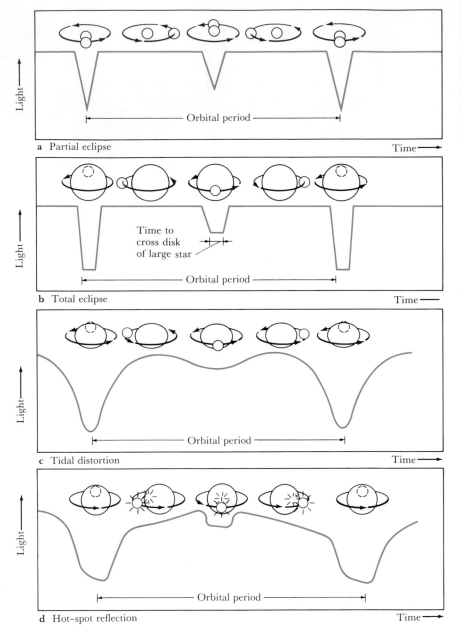

a Partial eclipse

b Total eclipse

Time to
cross disk
of large star

c Tidal distortion

d Hot–spot reflection

**Figure 18-18 Representative light curves
of eclipsing binaries**
*The shape of its light curve usually reveals
many details about an eclipsing binary. The
cases of* **(a)** *partial eclipse,* **(b)** *total eclipse,* **(c)**
tidal distortion, and **(d)** *hot-spot reflection are
illustrated here.*

curve for an eclipsing binary reveals at a glance such information as
whether the eclipse is total or partial (compare Figures 18-18*a* and 18-18*b*).

The light curve of the eclipsing binary provides detailed information
about the motion of the two stars. For example, the time between successive eclipses of the same star is the orbital period of the binary, which is
related by Kepler's third law to the separation of the two stars. Astronomers can use Newtonian mechanics to analyze such data obtained from
the light curve and thus calculate the combined masses of the two stars.
If spectroscopic data also are available (so that the individual orbital
speeds of the stars can be calculated from the Doppler effect), then astronomers have enough information to calculate the individual masses of
the two stars.

The light curve of a typical eclipsing binary can be analyzed to yield a surprising amount of detailed information. For example, the duration of an eclipse depends upon the size of the eclipsing star and the speed with which it moves. If the orbital speed is known, then the diameter of the star can be computed from the light curve. Even such details as tidal distortion of each star by the other can be detected in the shape of the light curve (see Figure 18-18c).

If one of the stars in an eclipsing binary is very hot, its radiation typically creates a "hot spot" on its cooler companion star. Every time this hot spot is exposed to our Earth-based view, we see receive a little extra light energy, which produces a characteristic "bump" on the binary's light curve (see Figure 18-18d).

Information about stellar atmospheres can be derived from light curves. Suppose that one star of a binary is a white dwarf, and the other star is a bloated red giant. By observing exactly how the light from the bright white dwarf is gradually cut off as it moves behind the edge of the red giant during the beginning of an eclipse, astronomers can infer many facts about the upper atmosphere of the red giant (for example, its pressure and density).

Binary stars are fascinating objects. A single star leads a straightforward, birth-to-death existence, but strange and exotic things can happen to stars in binary systems. One star in a binary might evolve very rapidly, become a bloated red giant, and have its outer layers stripped away by the gravitational pull of its companion. The result is an aging star with its interior expose to our view.

As we shall see in later chapters, many curious variations are possible as the two stars in a binary affect each other's evolutions. First, however, we shall take a close-up look at an isolated main-sequence star that happens to be only 150 million kilometers from the Earth.

Summary

. Distances to the nearer stars can be determined by parallax—the apparent shift of the star against the background stars as the Earth moves around its orbit.

. The absolute magnitude of a star is the apparent magnitude it would have when viewed from a distance of 10 parsecs; absolute magnitudes are calculated from the star's apparent magnitude and distance.

 Luminosity is the amount of energy escaping from the total surface area of a star each second; energy flux is the amount of energy emitted from each square centimeter of the star's surface each second; brightness is the energy flux (erg cm^{-2} sec^{-1}) arriving at the Earth from a star.

. Photoelectric photometry measures brightness through standard filters such as the UBV filters; color indices of a star are determined as differences between the brightness values obtained with different filters.

. The surface temperature of a star can be determined from its color indices or from its spectral type (O, B, A, F, G, K, or M), which is based upon the major patterns of spectral lines in its spectrum.

. The Hertzsprung–Russell (H–R) diagram is a graph plotting absolute magnitudes of stars versus their spectral types (or, equivalently, luminosities against surface temperatures); it reveals the existence of various types of stars such as main-sequence stars, red giants, supergiants, and white dwarfs.

. Binary stars are surprisingly common in the universe; those that can be resolved as two distinct star images in an Earth-based telescope are called visual binaries.

> The masses of the two stars in a binary system can be computed from measurements of the orbital period and orbital dimensions of the system.

> Each of the two stars in a binary moves in an elliptical orbit about the center of mass of the system.

. The mass–luminosity relation expresses a direct correlation between mass and luminosity for main-sequence stars.

. Some binaries can be detected and analyzed, even though the system is so distant or faint that two star images are not resolvable in Earth-based telescopes.

> A spectrum binary is one detected by the presence of spectral lines for two distinctly different spectral types in the spectrum of what appears to be a single star.

> A spectroscopic binary is one detected by the periodic shift of spectral lines due to the Doppler effect as the orbits of the stars carry them alternatively toward and away from the Earth.

> An eclipsing binary is one whose orbits are viewed nearly edge-on from the Earth, so that one star periodically eclipses the other.

> Surprisingly detailed information about the stars in an eclipsing binary can be obtained from a study of the binary's radial velocity curve and light curve.

....

Review questions

*1 Suppose that you can just barely see a 12th magnitude star through an amateur's 6-inch telescope. What is the magnitude of the dimmest star that you could see through a 60-inch telescope?

2 Which is the hottest star listed in Table 18-1? Which is the coolest? List the stars in Table 18-1 in order of decreasing surface temperature.

3 Explain why bolometric corrections are positive for both very hot and very cool stars.

4 Suppose that you want to determine the mass, temperature, diameter, and luminosity of a star. Which of these physical quantities require that you know the distance to the star? Explain.

5 Sketch the radial velocity curve of a binary whose stars are moving in a circular orbit that is **(a)** perpendicular and **(b)** parallel to our line of sight.

6 Sketch the light curve of an eclipsing binary having high orbital eccentricity in which **(a)** the major axes are pointed toward the Earth and **(b)** the major axes are perpendicular to our line of sight.

7 Estimate the mass of a main-sequence star that is 10,000 times as luminous as the Sun. What is the luminosity of a main-sequence star whose mass is one-tenth that of the Sun?

....

Advanced questions

*8 Suppose two stars have the same apparent magnitude, but one star is ten times farther away than the other. What is the difference in their absolute magnitudes?

*9 Suppose a star experiences an outburst in which its surface temperature doubles but its density decreases by a factor of eight. Find the new radius and luminosity of the star.

*10 The visual binary 70 Ophiuchi (see Figure 18-13) has a period of 87.7 years. The length of the semimajor axis is 4.5 arc seconds and the parallax of the system is 0.2 arc seconds. What is the sum of the masses of the two stars?

11 The bright star Capella (α Aur) has a spectral type of G8 and an absolute magnitude of -0.7. What are the star's surface temperature, luminosity, and diameter?

Discussion questions

12 Why do you suppose that stars of the same spectral type but different luminosity class exhibit slight differences in their spectra?

13 Earth-based astronomers can only measure stellar parallaxes larger than about 0.05 arc seconds with acceptable accuracy (10 percent or better). Discuss the advantages or disadvantages of parallax measurements from a space telescope in a large solar orbit, at the distance of Jupiter from the Sun. Assuming that this space telescope can also measure parallactic angles of 0.05 arc seconds, what is the distance of the most remote stars that can be accurately determined? How much bigger a volume of space would be covered compared to Earth-based observations? How many more stars would you expect to be contained in that volume?

For further reading

Ashbrook, J. "Visual Double Stars for the Amateur." *Sky & Telescope*, Nov. 1980, p. 379.

Evans, D., et al. "Measuring Diameters of Stars." *Sky & Telescope*, Aug. 1979, p. 130.

Gingerich, O. "A Search for Russell's Original Diagram." *Sky & Telescope*, Jul. 1982, p. 36.

Nielsen, A. "E. Hertzsprung—Measurer of Stars." *Sky & Telescope*, Jan. 1968, p. 4.

Page, T., and L. *Starlight—What It Tells About the Stars*. Macmillan, 1967.

Phillip, A., and Green, L. "Henry N. Russell and the H–R Diagram." *Sky & Telescope*, Apr. 1978, p. 306; May 1978, p. 395.

Upgren, A. "New Parallaxes for Old: Coming Improvements in the Distance Scale of the Universe." *Mercury*, Nov./Dec. 1980, p. 143.

The solar corona
This striking view of the Sun's outer atmosphere was prepared from data supplied by the Solar Maximum Mission satellite. Colors represent density in the corona and go from purple (densest) to yellow (least dense). The densest coronal regions (dark blue; purple) are located over sunspots. From one of these dense regions, a prominent coronal spike extends nearly 2 million kilometers westward from the Sun. Several other spikes can also be seen. Shortly after this picture was taken, a solar flare occurred in this field of view and the shape of the corona changed dramatically in a matter of minutes. (NASA)

The Sun is a typical main-sequence star. As we learn about the Sun in this chapter, we also begin our general study of stars. We first discuss the thermonuclear reactions that occur in the core of the Sun and the ways that energy moves from the core to the surface, where it is radiated into space. This discussion introduces a theoretical model of the Sun's interior structure. Next we turn to a variety of phenomena observed on the solar surface. We find that the 11-year sunspot cycle is only one aspect of a more-general 22-year solar cycle that affects many properties of the Sun. Finally, we mention some of the new tools that astronomers are using to study the Sun and thereby learn more about the general nature of stars.

The Sun is an average star. Its mass, size, surface temperature, and chemical composition lie roughly midway between the extremes exhibited by other stars. Unlike other stars, however, the Sun is available for detailed, close-up examination. Studying the Sun therefore offers excellent insights into the nature of similar main-sequence stars.

Although the Sun is a commonplace star, it is nevertheless the scene of fascinating and complicated processes. Beautiful and bewildering phenomena occur as columns of hot gases gush up to the solar surface, interact with the Sun's magnetic field, and dissipate energy into the Sun's outer atmosphere. The Sun is a dramatic arena where we observe the interaction of matter and energy on a colossal scale. The source of all this energy lies buried at the Sun's center.

Average Earth–Sun distance: 149,598,000 km (=1 AU)
Maximum Earth–Sun distance: 152,000,000 km
Minimum Earth–Sun distance: 147,100,000 km
Average angular diameter: 32 arc min
Radius: 696,000 km (=109 Earth radii)
Mass: 1.99×10^{30} kg (=3.3×10^5 Earth masses)
Average density: 1.41 g/cm^3
Luminosity: 3.90×10^{33} erg/sec (=3.90×10^{26} watts)
Surface temperature: 5800 K
Central temperature: 15.5×10^6 K
Spectral type: G2
Apparent magnitude: -26.8
Absolute magnitude: $+4.8$
Composition (by weight): 75 percent hydrogen
24 percent helium
1 percent all other elements
Distance to center of galaxy: 10 kpc (=30,000 light years)
Orbital period around center of galaxy: 225,000,000 years
Orbital velocity around center of galaxy: 250 km/sec
Light-travel time from Sun to Earth: 8.3 min

The Sun's energy is produced by thermonuclear reactions in the core of the Sun

During the nineteenth century, geologists and biologists found convincing evidence that the Earth must have existed in more-or-less its present form for hundreds of millions of years. This fact posed severe problems for physicists because it was impossible to explain how the Sun has been shining for so long, radiating immense amounts of energy into space. If the Sun were made of coal, for example, producing its heat and light by burning, it could last for only 3000 years.

The most satisfactory explanation of solar energy was proposed in the mid-1800s by Lord Kelvin (after whom the temperature scale is named) and Hermann von Helmholtz. They argued that the tremendous weight of the Sun's outer layers pressing inward from all sides should cause the Sun gradually to contract. As gravity causes the Sun to contract, its interior gases become compressed. Whenever a gas is compressed, its temperature rises. (Diesel engines operate on this principle; you can demonstrate it experimentally with a bicycle pump.) Thus gravitational contraction causes the Sun's gases to become hot enough to radiate energy into space.

This process, called **Kelvin–Helmholtz contraction,** does exist and it does convert gravitational energy (the energy associated with gravitational forces) into thermal energy (the energy associated with heat). Thermal energy in turn can be converted into electromagnetic energy. However, such contraction cannot be the major source of the Sun's energy. Helmholtz's calculations showed that the Sun must contract so rapidly to produce the energy it emits that it would have extended beyond the Earth's orbit only 25 million years ago. Again, physicists were unable to explain how the Earth could have existed in its present form for the hundreds of millions of years needed to produce its present surface features and life forms.

The first key to an explanation came in 1895 with the discovery of radioactivity. A radioactive element emits particles and energy, and the radioactivity of substances in the Earth provides the energy that has kept the Earth from cooling completely over its long history. The Sun's energy output, however, is so huge that it could be explained only by assuming that the Sun is composed almost entirely of radioactive elements such as uranium.

The second key was provided in 1905 by Albert Einstein's special theory of relativity. One of the implications of Einstein's theory is that matter and energy are interconvertible according to the simple equation

$$E = mc^2$$

In other words, a mass *(m)* can be converted into an amount of energy *(E)* equivalent to mc^2, where c is the speed of light. Because c is a large number and c^2 is huge, a small amount of matter can be converted into an awesome amount of energy.

Astronomers began to wonder if the Sun's energy output comes from conversion of matter into energy. In the 1920s the British astronomer Arthur Eddington showed that temperatures near the center of the Sun must be much greater than had previously been thought. Another British astronomer Robert Atkinson suggested that, under these conditions near the Sun's center, hydrogen nuclei might fuse together to produce helium nuclei in a reaction that would transform a tiny amount of mass into a very large amount of energy.

Recall that the nucleus of a hydrogen (H) atom consists of a single proton. The nucleus of a helium atom (He) consists of two protons and two neutrons. Neutrons and protons are very similar particles (see Box 6-3 for details) and can be interconverted in nuclear reactions. In the nuclear process

$$4\,H \rightarrow He$$

two of the four protons from hydrogen are changed into neutrons to produce a single helium nucleus.

This conversion of hydrogen into helium is a particularly interesting reaction because the ingredients (four hydrogen nuclei) weigh very slightly more than the product (one helium nucleus). Specifically, 0.7 percent of the mass of the hydrogen is lost during the reaction and does not show up in the mass of the helium. This lost mass is, of course, converted into energy in the amount predicted by the equation $E = mc^2$. For example, consider the conversion of 1 kg of hydrogen into helium. Although 1000 g of hydrogen go into the reaction, only 993 g of helium come out. Using Einstein's equation, we find that the missing 7 g of matter has been transformed into 6.3×10^{21} erg of energy. This is the same amount of energy released by the burning of 200 metric tons of coal.

To produce the observed luminosity of the Sun (3.9×10^{33} erg/sec), about 6×10^{11} kg of hydrogen must be converted into helium within the Sun each second. This is a prodigious rate, but the Sun contains a vast mass of hydrogen. In fact, the Sun contains enough hydrogen to continue the present rate of energy output for another 5 billion years.

This process of nuclear fusion, the conversion of hydrogen into helium at the Sun's center, is called **hydrogen burning** although nothing is actually burned in the conventional sense of that word. Box 19-2 describes the details of the nuclear reactions involved in hydrogen burning. These nuclear reactions occur at a star's core where it is very hot (15 million kelvins in the case of the Sun). Normally, the positive electric charge on protons is very effective in keeping protons far apart, because like charges repel each other. But in the extreme heat of the Sun's center, the protons are moving so fast that they can penetrate each other's electric repulsion and stick together. Hydrogen burning is therefore called a **thermonuclear reaction,** or **thermonuclear fusion,** because it can occur only at high temperature. In later chapters, we will see that other thermonuclear reactions (such as helium burning, carbon burning, and oxygen burning) occur late in the lives of many stars.

Box 19-2 The proton–proton chain and the CNO cycle

There are two principal reaction sequences by which hydrogen burning proceeds inside a star. In each case, four protons combine to form one helium nucleus, with a slight loss of mass that is converted into energy.

Temperature at the star's core determines which of the two sequences occurs. For stars with masses not greater than the Sun's mass, the central temperatures do not exceed 16 million kelvins, and hydrogen burning proceeds via the **proton–proton chain.** For stars more massive than the Sun, central temperatures are above 16 million kelvins, and hydrogen burning occurs through a series of reactions called the **CNO cycle.**

The proton–proton chain is essentially the direct fusion of hydrogen into helium. It occurs in three steps. First, two protons (each denoted by ^{1}H) combine to form an isotope of hydrogen called **deuterium** (^{2}H), which consists of a proton and a neutron bound together. In this step, one of the two protons turns into a neutron, with the release of an antielectron (e^+) and a neutrino (ν). Thus the first step is

$$^1H + {}^1H \rightarrow {}^2H + e^+ + \nu$$

An antielectron (e^+), or positron, is just like an ordinary electron (e^-) except that it has positive rather than negative electric charge. A neutrino is a particle to which most matter is virtually transparent. Consequently, neutrinos are very difficult to detect, and their properties are currently a topic of research and debate. Neutrinos have no charge but may have a very small mass. A neutrino is produced every time a proton is converted into a neutron and, conversely, an antineutrino ($\bar{\nu}$) is liberated when a neutron turns into a proton.

In the second step of the proton–proton chain, a third proton combines with the deuterium nucleus to produce a light-weight isotope of helium (^{3}He) whose nucleus contains two protons and one neutron. This reaction releases energy (γ stands for a gamma ray):

$$^1H + {}^2H \rightarrow {}^3He + \gamma$$

In the final step, two ^{3}He nuclei combine to produce an ordinary nucleus of helium (^{4}He = 2 protons + 2 neutrons) with the release of two protons:

$$^3He + {}^3He \rightarrow {}^4He + {}^1H + {}^1H$$

In massive stars, where the central temperatures exceed 16 million kelvins, carbon serves as a catalyst by which hydrogen burning proceeds. Cornell University physicist Hans Bethe was instrumental in discovering the steps whereby a carbon nucleus absorbs protons and finally spits out a helium nucleus. Along the way, the carbon is transformed into nitrogen and oxygen, and hence this process is called the CNO cycle. The CNO cycle occurs in six steps:

$$^{12}C + {}^1H \rightarrow {}^{13}N + \gamma$$
$$^{13}N \rightarrow {}^{13}C + e^+ + \nu$$
$$^{13}C + {}^1H \rightarrow {}^{14}N + \gamma$$
$$^{14}N + {}^1H \rightarrow {}^{15}O + \gamma$$
$$^{15}O \rightarrow {}^{15}N + e^+ + \nu$$
$$^{15}N + {}^1H \rightarrow {}^{12}C + {}^4He$$

(continued)

(Box 19-2, continued)

Like the proton–proton chain, the CNO cycle has the overall effect of converting four hydrogen nuclei (four protons) into one helium nucleus, two positrons, two neutrinos, and some high-energy, short-wavelength gamma radiation. The ^{12}C nucleus must be present to permit the CNO cycle to function, but it is restored at the end of the cycle, so carbon is not used up by the process.

The Sun's central temperature is about 16 million kelvins, and astronomers believe that the proton–proton chain is the dominant reaction from which the Sun draws its energy. To test these ideas about hydrogen burning, scientists have tried to detect the neutrinos coming from the nuclear reactions at the core of the Sun.

The most thorough neutrino experiment was performed in the 1970s by Raymond Davis of the Brookhaven National Laboratory. Davis used 100,000 gallons of cleaning fluid (C_2Cl_4) in a huge tank buried deep in a mine in South Dakota. Because matter is virtually transparent to neutrinos, most of the neutrinos from the Sun passed right through Davis's tank with no effect whatsoever. However, on rare occasions, a neutrino should be absorbed by one of the chlorine atoms in the cleaning fluid and be converted into an atom of argon. Davis measured the amount of argon to determine the number of neutrinos coming from the Sun.

To everyone's surprise, Davis's experiment detected only one-third the expected number of neutrinos. Various explanations have been proposed for this unexpected result, but further research is needed to decide among them. Further progress awaits funding for building other kinds of neutrino detectors.

A theoretical model of the Sun tells us how energy gets from the Sun's center to its surface

Although the Sun's interior is hidden from our view, we can use the laws of physics to calculate what is going on below the solar surface (see Figure 19-1). The results of such computations constitute a **model** of the Sun and tell us many of the Sun's internal characteristics such as its pressure, temperature, and density.

Figure 19-1   The Sun's surface
The Sun is the only star whose surface details can be examined through Earth-based telescopes. Astronomers always take great care (with extremely dark filters or by projecting the Sun's image on a screen) to avoid severe damage to their eyes. Never look directly at the Sun, with or without a telescope. (Celestron International)

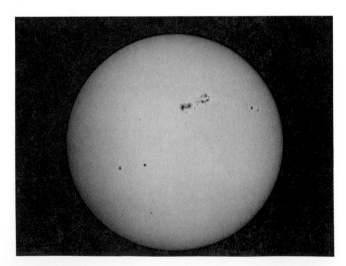

To develop a model of the Sun (or any stable star), we first note that the Sun is not undergoing any dramatic changes. The Sun is not exploding or collapsing; neither is it significantly heating or cooling. The Sun therefore is in balance in two ways, mechanically and thermally.

Mechanical balance, often called **hydrostatic equilibrium,** means simply that a star is supporting its own weight. Because of gravity, the tremendous weight of the Sun's outer layers pressing inward from all sides tries to make the star contract. As gravity compresses the star, however, gas pressure increases inside the star. The greater the compression, the higher the internal pressures. Hydrostatic equilibrium is achieved when the pressure at every depth in the star is exactly sufficient to support the weight of the overlaying layers.

Thermal balance, often called **thermal equilibrium,** means simply that a star keeps shining. Vast amounts of energy escape from the Sun's surface each second and, in a state of thermal equilibrium, this energy is constantly resupplied from the Sun's interior at a steady and persistent rate. But exactly how is energy transported from the Sun's center to its surface?

Experience teaches us that energy always flows from hot regions to cooler regions. For example, if you heat one end of a metal bar with a blowtorch, eventually the other end of the bar becomes warm. This method of energy transport is called **conduction.** Conduction varies significantly from one substance to another (copper is a good heat conductor, but wood is not), depending on the arrangement and interaction of the atoms. Calculations demonstrate that conditions inside stars such as the Sun are not favorable for conduction, and hence this is not an efficient means of energy transport. Nevertheless, in a later chapter, we shall see that conduction is important in very compact stars such as white dwarfs.

Inside main-sequence stars, energy moves from center to surface by two other means of energy transport: convection, and radiative diffusion. **Convection** involves the circulation of gases between hot and cool regions. Just as a hot-air balloon drifts skyward, so hot gases rise toward a star's surface while cool gases sink back down toward the star's center. The net effect of this physical movement of gases is to transfer heat energy from the center toward the surface.

In **radiative diffusion,** photons created in the thermonuclear inferno at a star's center leak, or diffuse, outward toward the star's surface. The paths of individual photons are quite random as they are knocked about between atoms and electrons inside the star. The overall photon migration, however, is outward from the very hot core where photons are constantly created toward the cooler surface where photons escape into space. In all, it takes nearly a million years for energy created at the Sun's center finally to reach the solar surface and escape as sunlight.

The concepts of hydrostatic equilibrium, thermal equilibrium, and energy transport can be expressed in the form of a set of mathematical equations, collectively called the **equations of stellar structure.** These equations can be solved to yield the conditions of pressure, temperature, and density that must exist inside the star if the equilibrium conditions are to be satisfied. Astrophysicists use high-speed computers to solve the equations of stellar structure, thus developing a detailed theoretical model of the structure of a star. The astrophysicist begins with astronomical data about the star's surface (for example, the Sun's surface temperature is 5800 K, its luminosity is 3.9×10^{33} erg sec, and the gas pressure and density are almost zero). Using the equations of stellar structure, the astrophysicist then calculates conditions layer by layer toward the star's center. In this way, the astrophysicist discovers how temperature, pressure, and density increase with increasing depth below the star's surface. In this way we have learned that, at the Sun's center, the temperature is 15.5 million kelvins, the pressure is 3.4×10^{11} atmospheres, and the density is 160 g/cm^3.

As a result of solving the equations of stellar structure, we know the behavior of various physical quantities (such as pressure, density, and temperature) throughout a star. This description *is* the **stellar model;** it can be presented in tables or graphs. Table 19-1 and Figure 19-2 present a theoretical model of the Sun.

Figure 19-2   A theoretical model of the Sun
The Sun's internal structure is displayed with graphs that show how the density, temperature, mass, and luminosity vary with distance from the Sun's center.

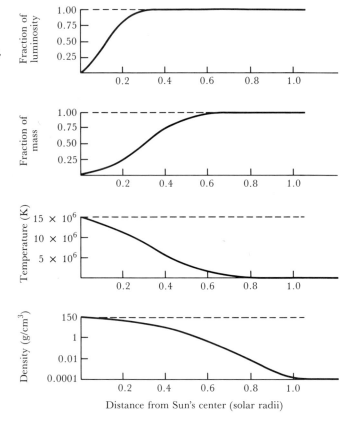

Examining Table 19-1, note that the fraction of luminosity rises to 100 percent at about one-quarter of the way from the Sun's center to its surface. This tells us that the Sun's energy production occurs within a volume extending out to ¼ solar radius.

Also note that the fraction of mass rises to nearly 100 percent at 0.8 solar radius, which is a distance of only 560,000 km from the Sun's center (recall that 1 solar radius is nearly 700,000 km). Consequently, the outermost 140,000 km of the Sun contain very little matter. That is why the pressure and the density are so low from 0.8 to 1.0 solar radius.

In this tenuous outer layer, convection dominates the energy flow. We therefore say that the Sun has a **convective zone,** or **convective envelope.** From the center out to 0.8 solar radius, energy is transported by radiactive diffusion; this inner region is therefore called the **radiative zone.** These aspects of the Sun's internal structure are sketched in Figure 19-3.

TABLE 19-1   A theoretical model of the Sun

Fraction of the radius	Temperature (× 10^6 K)	Fraction of central pressure	Density (g/cm^3)	Fraction of mass	Fraction of luminosity
0.0	15.5	1.00	160	0.00	0.00
0.1	13.0	0.46	90	0.07	0.42
0.2	9.5	0.15	40	0.35	0.94
0.3	6.7	0.004	13	0.64	1.00
0.4	4.8	0.007	4	0.85	1.00
0.5	3.4	0.001	1	0.94	1.00
0.6	2.2	0.003	0.4	0.98	1.00
0.7	1.2	4×10^{-5}	0.08	0.99	1.00
0.8	0.7	5×10^{-6}	0.02	1.00	1.00
0.9	0.3	3×10^{-7}	0.002	1.00	1.00
1.0	0.006	4×10^{-13}	3×10^{-7}	1.00	1.00

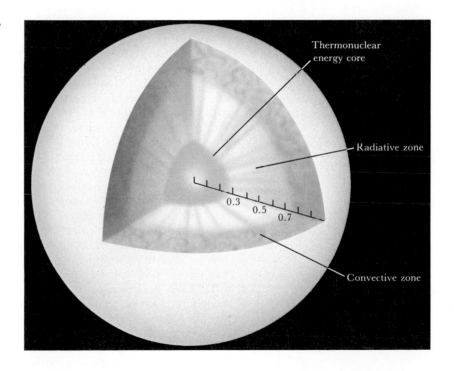

Figure 19-3   The Sun's internal structure
Thermonuclear reactions occur in the Sun's core, which extends out to a distance of 0.25 solar radius from the center. Energy is transported outward via radiative diffusion to a distance of 0.83 solar radius. Convection is responsible for energy transport in the Sun's outer layers.

The photosphere is the lowest of three main layers in the Sun's atmosphere

Although astronomers often speak of the solar surface, the Sun really does not have a surface at all. As you move in toward the Sun, you encounter ever more dense gases but no sharp boundary like the surface of the Earth or Moon. The Sun appears to have a surface (for example, see Figure 19-1) because there is a specific layer in the Sun's atmosphere from which most of the visible light comes. This 100-km-thick layer is appropriately called the **photosphere** ("sphere of light").

Convection in the Sun's outer layers affects the appearance of the photosphere. Under good observing conditions with a telescope (and special dark filters to protect your eyes), you can often see a blotchy

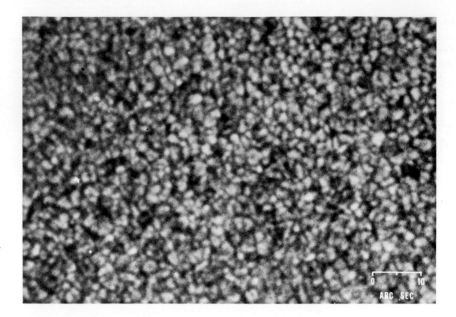

Figure 19-4   Solar granulation
High-resolution photographs of the Sun's sur-
face reveal a blotchy pattern called granula-
tion. Granules, each measuring about 1000
km across, are convection cells in the Sun's
outer layers. (San Fernando Observatory)

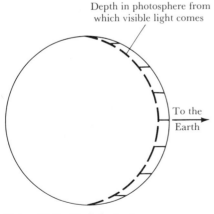

Depth in photosphere from
which visible light comes

To the
Earth

Figure 19-5   Limb darkening
Light reaching the Earth from the central re-
gions of the solar disk originates in deep,
warm layers of the photosphere. But light com-
ing to us from the Sun's limb originates in
higher, cooler layers of the photosphere. Conse-
quently, the center of the Sun's disk appears
brighter than the limb.

pattern called **granulation** (see Figure 19-4). Each light-colored **granule** measures about 1000 km across and is surrounded by a darkish boundary. High-resolution spectroscopy reveals a blueshifting of spectral lines in the central, bright regions of a granule and a redshifting along the dark, intergranule boundaries. These Doppler shifts show that hot gases rise upward in the granules, cool off, spill over the edges of the granules, and plunge back down into the Sun along the intergranule boundaries. The brightness difference between the center and edge of a granule corresponds to a temperature drop of 300 K.

Granules are individual convection cells in the Sun's outer layers. Time-lapse photography shows that granules form, disappear, and then reform in cycles lasting several minutes. At any one time, roughly 4 million granules cover the solar surface. Each granule occupies an area equal to Texas and Oklahoma combined (about 10^6 km^2).

The photosphere appears darker around the edge, or **limb,** of the Sun than it does toward the center of the solar disk. This phenomenon, called **limb darkening,** is caused by photon absorption in the solar atmosphere. Photons emitted in warm layers fairly deep in the solar atmosphere are able to escape into space only along the relatively short paths directed away from the Sun's center. The photons reaching the Earth from the edges of the Sun's visible disk come from cooler layers nearer to the top of the solar atmosphere, because the photons from the deeper layers are absorbed before escaping along the much longer paths they must travel toward the Earth. Thus, at the limbs of the Sun's disk, Earth-based photographs show less intense light from cooler, more shallow layers of the Sun (see Figure 19-5).

The photosphere shines with a continuous, nearly perfect blackbody spectrum corresponding to its average temperature of about 5800 K. Immediately above the photosphere is a layer of cooler gas about 500 km thick, where the temperature declines to about 4000 K. All of the absorption lines in the Sun's spectrum (recall Figure 17-7) are produced in this cool layer as atoms extract energy from the sunlight at various wavelengths.

The chromosphere is located between the photosphere and the Sun's outermost atmosphere

The cool region immediately above the photosphere is the lowest layer of the **chromosphere** ("sphere of color"), which is the second of the three major levels in the Sun's atmosphere. The chromosphere is visible during a total eclipse of the Sun, at the moment when the Moon blocks out the photosphere, as a pinkish strip around the edge of the dark Moon. This 2000-km-thick pinkish layer is the chromosphere.

The spectrum of the chromosphere is dominated by emission lines, many of which have exactly the same wavelengths as the absorption lines in the spectrum of the photosphere. (Recall Kirchhoff's laws of spectral analysis, summarized in Figure 17-9). When viewed against the glowing photosphere, the solar spectrum exhibits absorption lines. When seen with the dark sky as a background, however, the spectrum of the cooler chromosphere exhibits emission lines as the atoms in the cooler layer above the photosphere give up the energy they had extracted from the radiation pouring out of the Sun.

The characteristic reddish-pink color of the chromosphere is due to the Balmer line H_α at 6563 Å, one of the brightest emission lines in the chromosphere's spectrum. Other bright chromospheric emission lines include the so-called H and K lines of singly ionized calcium at 3968 Å and 3933 Å, both of which are very dark absorption lines in the photosphere's spectrum (see Figure 17-7). The chromospheric spectrum also exhibits emission lines of ionized helium and ionized metals not seen in the photospheric spectrum. These lines must originate in some regions of the chromosphere that are hotter than the photosphere.

The photosphere emits almost no light at the wavelengths of H_α and the calcium H and K lines; the photospheric spectrum has broad, dark absorption lines at these wavelengths. But the chromosphere is especially bright at these wavelengths. Thus, astronomers can study details of the chromosphere by viewing the Sun through special filters that are transparent to light only at the wavelengths of H_α or the calcium lines.

Figure 19-6 is a photograph of the solar surface taken through an H_α filter. Such photographs, usually called **filtergrams,** show us what is going on in the chromosphere. Note the numerous, dark, brushlike spikes that protrude upward. These spikes, called **spicules,** are jets of gas surging up out of the Sun. A typical spicule rises at the rate of 20 km/sec, reaches a height of about 7000 km, and then collapses and

Figure 19-6 Spicules and the chromosphere
Spicules are jets of cool gas that rise up into warmer regions of the Sun's outer atmosphere. Numerous spicules are visible in this H_α filtergram, which also shows many details of the chromosphere. Spicules are located along the irregularly shaped boundaries between supergranules. (Mount Wilson and Las Campanas Observatories)

fades away after a few minutes. At any one time, roughly 300,000 spicules exist, covering a few percent of the Sun's surface.

The location of spicules is directly associated with gas motions in the Sun's outer layers. Detailed observations of the solar surface in the 1960s, revealed the existence of large organized cells called **supergranules.** A typical supergranule is about 30,000 km in diameter and contains many hundreds of ordinary granules. Gases rise upward in the middle of a supergranule and move horizontally outward toward the edge of the supergranule, where they descend back into the Sun. Spicules outline the boundaries between supergranules.

The corona is the outermost layer of the Sun's atmosphere

The outermost region of the Sun's atmosphere is called the **corona.** It extends from the top of the chromosphere out to a distance of several million kilometers, where it begins to be called the solar wind. (As noted in Chapter 6, the solar wind consists of high-speed protons and electrons constantly escaping from the Sun). The **solar atmosphere** therefore has a total of three distinct layers (photosphere, chromosphere, and corona) as sketched schematically in Figure 19-7. Everything below the photosphere is called the **solar interior.**

The total amount of visible light emitted by the solar corona is comparable to the brightness of the Moon at full moon—only about one-millionth as bright as the photosphere. Consequently, the corona can be viewed only when the photosphere is blocked out during a total eclipse or in a specially designed telescope called a **coronagraph.** Figure 19-8 is an exceptionally detailed photograph of the corona. Numerous **coronal streamers** are visible, extending outward to a distance of 6 solar radii.

Around 1940, astronomers realized that the spectrum of the Sun's corona contains emission lines of highly ionized elements. For example, there is a prominent green line at 5303 Å due to Fe XIV (iron atoms each stripped of 13 electrons). Extremely high temperatures are required to strip that many electrons from atoms, so the corona must be very hot.

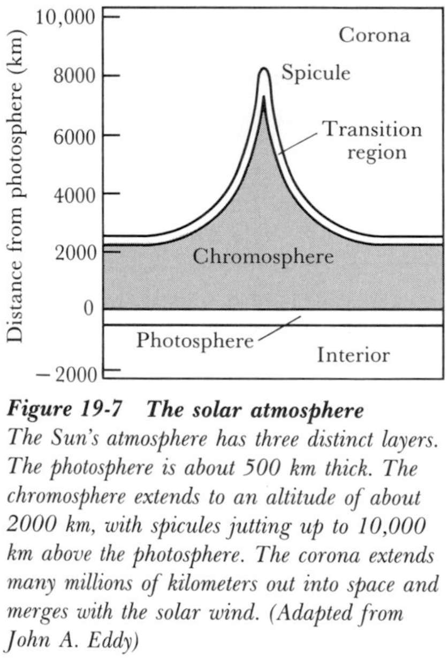

Figure 19-7   The solar atmosphere
The Sun's atmosphere has three distinct layers. The photosphere is about 500 km thick. The chromosphere extends to an altitude of about 2000 km, with spicules jutting up to 10,000 km above the photosphere. The corona extends many millions of kilometers out into space and merges with the solar wind. (Adapted from John A. Eddy)

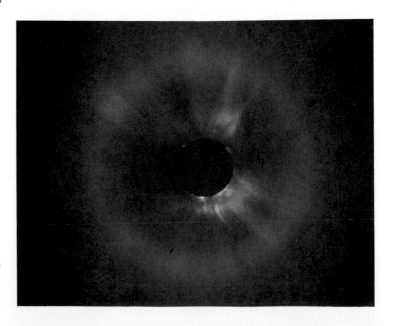

Figure 19-8   The solar corona
This extraordinary photograph was taken from a jet airplane flying 40,000 feet above Montana during the total solar eclipse of February 26, 1979. Numerous streamers are visible, extending outward to distances of 4 million kilometers above the solar surface. (Los Alamos Scientific Laboratory)

Figure 19-9 shows how temperature varies with distance above the photosphere. Coronal temperatures are in the range of 1 to 2 million kelvins.

The corona, however, is nearly a vacuum. The typical density of coronal gases is only about 10^5 particles per cubic centimeter. (The density of the photosphere is about 10^{17} particles per cubic centimeter; the density of the air we breathe is about 10^{19} particles per cubic centimeter.) Despite its very high temperature, the corona emits relatively dim light because of its very low density.

Recent observations from space show that the corona exhibits complicated structure and activity. For example, Figure 19-10 shows a huge, bubblelike disturbance called a **solar transient.** These short-lived protuberances erupt suddenly and expand rapidly outward through the corona. Solar transients probably occur as often as once a day, but they were unknown until the Sun was examined with a coronagraph carried aloft on board Skylab. They were later identified as the source of bursts of radio noise long known on the Earth.

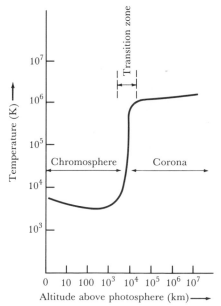

Figure 19-9   Temperature in the Sun's upper atmosphere
In a very narrow region about 10,000 km above the photosphere, the temperature rises abruptly to about 1 million kelvins. Pressure waves propagating outward from the Sun's convective zone provide the energy that heats the Sun's tenuous outer layers to extraordinarily high temperatures.

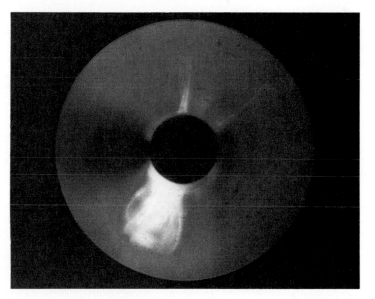

Figure 19-10   A solar transient
During the early 1970s, astronauts on Skylab discovered huge bubbles erupting outward through the corona. This solar transient was photographed on June 10, 1973. The leading edge of the bubble moved outward from the Sun at a speed of 500 km/sec. (NASA and the High Altitude Observatory)

X-ray photographs of the corona also were obtained during the Skylab missions. The corona has temperatures in the million-kelvin range, so Wien's law tells us that the corona should be shining brightly at X-ray wavelengths of roughly 30 Å. The Skylab pictures reveal a very blotchy, nonuniform inner corona (see Figure 19-11). Note the large dark area, which is called a **coronal hole** because it is devoid of the hot, glowing coronal gases. Many astronomers suspect that coronal holes are the main corridors through which particles of the solar wind escape from the Sun.

In addition to dark coronal holes, X-ray photographs also reveal numerous bright spots that are hotter than the surrounding corona. Temperatures in these bright points occasionally reach 4 million kelvins. Many of the bright coronal hot spots seen in X-ray pictures such as Figure 19-11 are located over sunspots.

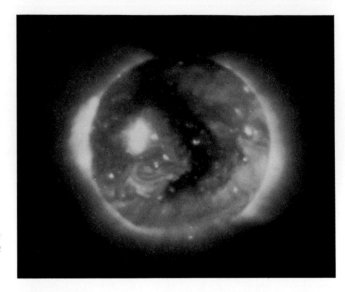

Figure 19-11   A coronal hole in X rays
*This X-ray picture of the Sun was taken by
Skylab astronauts on August 21, 1973. A
huge, dark, boot-shaped coronal hole dominates
this view of the inner corona. Numerous bright
points also are visible. (NASA and Harvard
College Observatory)*

Sunspots are one of many phenomena associated with the 22-year solar cycle

Superimposed on the basic structure of the solar atmosphere are a host
of phenomena that vary with a 22-year period. Irregularly shaped dark
regions called **sunspots** are the most easily recognized of these phenomena because they occur in the photosphere (see Figure 19-12). The very
dark central core of a sunspot is called the **umbra;** it usually is surrounded by a less-dark border called the **penumbra.** Although they vary
greatly in size, typical sunspots measure a few tens of thousands of kilometers across.

On very rare occasions, a sunspot group is so large that it can be seen
with the naked eye (always be sure to use special dark filters or other
means to protect your eyes when observing the Sun). Ancient Chinese

Figure 19-12   A sunspot group
*This photograph was taken by a balloon-borne
telescope at an altitude of roughly 50,000 ft.
Note the featherlike appearance of the
penumbrae. Also notice the granulation in the
surrounding, undisturbed photosphere. (Project
Stratoscope; Princeton University)*

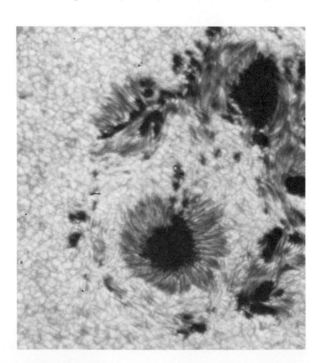

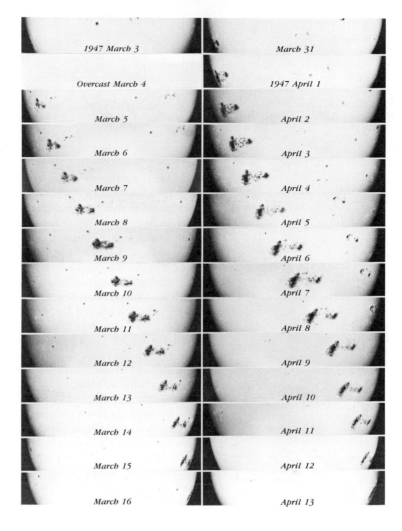

Figure 19-13   The Sun's rotation
By observing the same group of sunspots from one day to the next, Galileo found that the Sun rotates once in about four weeks. The equatorial regions of the Sun actually rotate somewhat faster than the polar regions. This series of photographs shows the same sunspot group over 1½ solar rotations. (Mount Wilson and Las Campanas Observatories)

astronomers recorded such sightings 2000 years ago. Of course, a telescope gives a much better view, and thus Galileo was the first person to examine sunspots in detail. In fact, Galileo discovered that he could determine the Sun's rotation rate by following sunspots as they move across the solar disk (see Figure 19-13). He found that the Sun rotates once in about four weeks. A typical sunspot group lasts about two months, so it can be followed for two solar rotations.

More careful observations by the British astronomer Richard Carrington in 1859 demonstrated that the Sun does not rotate as a rigid body. Carrington found that the equatorial regions rotate more rapidly than the polar regions do. A sunspot near the solar equator takes only 25 days to go once around the Sun. At 30° north or south of the equator, a sunspot takes 27½ days to complete a rotation. The rotation period at 75° north or south of the equator is about 33 days, and at the poles it may be as long as 35 days. This phenomenon is called **differential rotation,** because different parts of the Sun rotate at slightly different rates.

Careful observations over many years reveal that the numbers of sunspots change in a periodic fashion. In some years, there are many sunspots; in other years, there are almost none. This phenomenon, first noticed by the German astronomer Heinrich Schwabe in 1843, is called the

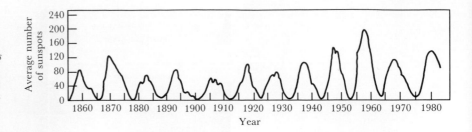

Figure 19-14   The Sunspot cycle
The number of sunspots on the Sun varies in a periodic fashion. Large numbers of sunspots were seen in 1959, 1970, and 1981. Exceptionally few sunspots were observed in 1954, 1965, and 1976. The next sunspot minimum is due in 1987; the next sunspot maximum should occur in 1992.

sunspot cycle. As shown in Figure 19-14, the average number of sunspots varies with a period of about 11 years. A time of exceptionally many sunspots is called a **sunspot maximum;** such maxima occurred in 1959, 1970, and late 1980. The Sun was almost devoid of sunspots (times of **sunspot minimum**) in 1954, 1965, and 1976.

The locations of the sunspots also vary in a periodic fashion over the sunspot cycle. At the beginning of a cycle, just after sunspot minimum, sunspots start to appear at latitudes 30° north and south of the solar equator. Over succeeding years, the sunspots occur closer and closer to the equator; at the end of the cycle, they are virtually on the solar equator. Observations showing this equatorial migration of sunspots are displayed in Figure 19-15. The data on this graph cover areas shaped somewhat like butterflies, and the graph is called a **Maunder butterfly diagram** after E. Walter Maunder, who discovered this phenomenon of sunspot migration in 1904.

A few years later (in 1908), the American astronomer George Ellery Hale made the important discovery that sunspots are associated with very intense magnetic fields on the Sun. When Hale focused a spectroscope on sunlight coming from a sunspot, he found that each spectral line in the normal solar spectrum is flanked by additional, closely spaced spectral lines not usually observed (see Figure 19-16). This "splitting" of a single spectral line into two or more lines is called the **Zeeman effect** after the Dutch physicist Pieter Zeeman, who first observed it in his labo-

Figure 19-15   The Maunder butterfly diagram
The location of sunspots varies in a regular fashion over the sunspot cycle. The first sunspots of a cycle appear at large distances from the solar equator, whereas the last spots of a cycle, are formed very near the equator. At sunspot maximum in the middle of the cycle, most sunspots occur at latitudes of 10° to 15° north and south of the equator.

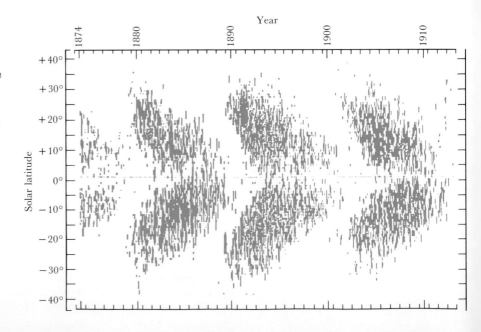

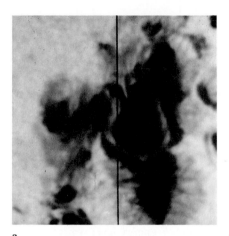

Figure 19-16 Zeeman splitting by a sunspot's magnetic field
(a) *The black line drawn across the sunspot indicates the location toward which the slit of a spectroscope was aimed.* **(b)** *In resulting spectrogram, one line in the middle of the normal solar spectrum is split into three components. The separation between the three lines corresponds to a magnetic field strength in excess of 4000 G. (Kitt Peak National Observatory)*

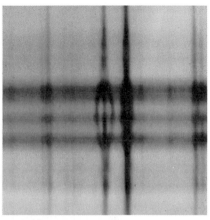

a   b

ratory in 1896. Zeeman showed that the effect is produced when the light source is inside an intense magnetic field.

The more intense the magnetic field, the wider is the separation of the split lines in the Zeeman effect (see Box 19-3). The strength of a magnetic field usually is expressed in terms of a unit called the **gauss** (G). For example, the strength of the Earth's magnetic field at the north and south magnetic poles is about 0.7 G. The splitting of the Fe I spectral line in Figure 19-16 into three lines corresponds to a very intense magnetic field of 4130 G.

Box 19-3 The Zeeman effect

In the 1890s, the Dutch physicist Pieter Zeeman placed some hot gases between the poles of a powerful magnet and examined the spectrum of the glowing gases through a spectroscope. He found that each spectral line was split into two, three, or more closely spaced lines, depending on the particular gases used in the experiment.

Electrons circle the nuclei of atoms at specific energy levels dictated by the laws of quantum mechanics. As we saw in Chapter 17, only certain specific energy levels are allowed. Spectral lines are produced when electrons jump from one energy level to another.

An electron moving around a nucleus is actually a tiny electric current, which produces a tiny magnetic field. When an atom is placed in a magnetic field, the tiny magnetic fields due to the atom's moving electrons try to line up with the overall field. According to quantum mechanics, this alignment occurs only at specific angles of inclination. Each angle of inclination corresponds to a slightly different energy level for the electron. The number of allowed inclinations increases for higher energy levels.

The diagram shows an example of how the Zeeman effect is produced. When the magnetic field is not present, electrons jumping from a lower energy level up to a higher energy level absorb photons of a specific wavelength, producing a single absorption line in the spectrum. When the magnetic field is turned on around the light source, the lower energy level is

(continued)

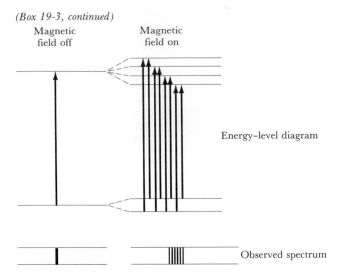

(Box 19-3, continued)

Magnetic field off   Magnetic field on

Energy–level diagram

Observed spectrum

replaced by two slightly different energy levels, corresponding to different alignments of the moving electron in the outside magnetic field. Similarly, the higher energy level is replaced by four slightly different energy levels. Eight different electron jumps are now possible, corresponding to absorption of photons of eight slightly different wavelengths. In this case, some of the jumps have such similar energy differences that only six distinct spectral lines can be resolved in the spectrum with the magnetic field turned on.

The stronger the magnetic field, the greater the difference in the energy levels. Thus by measuring the separations of the lines, astronomers can deduce the strength of the magnetic field surrounding the atoms or ions that emit the light.

Hale's discovery demonstrates that sunspots are places where a powerful, concentrated magnetic field protrudes through the hot gases of the photosphere. Because of the temperature, many of the atoms in the photosphere are ionized, so that the photosphere is a mixture of electric charges (ions and electrons) technically called a **plasma.** A plasma is an extremely good conductor of electricity, and it interacts vigorously with magnetic fields. Specifically, a magnetic field restricts and constrains the motions of a plasma. In a sunspot, the intense magnetic field greatly inhibits the natural convective motions. Energy cannot flow freely upward from the Sun's convective zone, and the gases in this region of the photosphere cool off. Indeed, temperatures in a sunspot are typically 4000 to 4500 K, or more than 1000 K cooler than the surrounding, undisturbed photosphere. Because of this lowered temperature, sunspots look dark in contrast to their brighter surroundings.

A host of exotic phenomena occur around and above sunspots as a direct result of their intense magnetic fields. Huge, arching columns of gas called **prominences** often appear above sunspot regions (see Figure 19-17). Some prominences, called **quiescent prominences,** hang suspended for days above the solar surface. In contrast, **eruptive prominences** blast material outward from the Sun at speeds of roughly 1000 km/sec.

The most violent, eruptive events on the Sun, called **solar flares,** occur in complex sunspot groups. In only a few minutes, temperatures in a compact region soar to 5 million kelvins. Vast quantities of particles and radiation are blasted out into space. The flare is usually over within 20 minutes. Nevertheless, when these ejected particles arrive at the Earth a day or so later, they interfere with radio communication and produce beautiful, shimmering aurorae. Most astronomers suspect that prominences and flares involve concentrated portions of the Sun's magnetic field.

Astronomers can construct artificial pictures called **magnetograms** that display the magnetic fields in the solar atmosphere by combining two photographs taken at wavelengths on either side of a magnetically split spectral line. The magnetogram in Figure 19-18 shows a large sunspot group. Dark blue indicates the area of the photosphere covered by one magnetic polarity (north); yellow indicates the area covered by the opposite (south) magnetic polarity.

Many sunspot groups are said to be **bipolar,** meaning that they have roughly comparable areas covered by north and by south magnetic polarities. The sunspots on the side of the group toward which the Sun is rotating are called the **preceding** members of the group. The remaining spots, which follow behind, are called the **following** members.

After years of studying the solar magnetic field, Hale was able to piece together a remarkable magnetic description of the solar cycle. First of all, Hale discovered that the preceding spots of all sunspot groups in one solar hemisphere have the same magnetic polarity. We now know that this polarity is the same as that of the hemisphere in which the group is located. In other words, in the hemisphere of the Sun's north magnetic pole, the preceding members of all sunspot groups have north magnetic polarity. In the south magnetic hemisphere, all the preceding members have south magnetic polarity.

In addition, Hale found that the polarity pattern completely reverses itself every 11 years. In other words, the hemisphere that has a north magnetic polarity at one solar maximum has a south magnetic polarity at the next solar maximum. For this reason astronomers prefer to speak of a **solar cycle** with a period of 22 years, rather than the sunspot cycle with its period of 11 years.

In 1960, the American astronomer Horace Babcock proposed a description that seems to account for many aspects of the 22-year solar cycle. Babcock's scenario, called a **magnetic-dynamo** model, makes use of

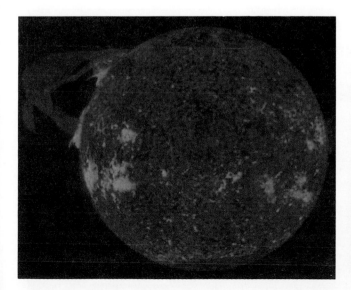

Figure 19-17 [left]   A prominence
A huge prominence arches above the solar surface in this Skylab photograph taken on December 19, 1973. The radiation that exposed this picture is from singly ionized helium (He II) at a wavelength of 304 Å and corresponds to a temperature of about 50,000 K. (Naval Research Laboratory)

Figure 19-18 [right]   A magnetogram of a sunspot group
This artificially colored picture displays the intensity and polarity of the magnetic field associated with a large sunspot group. One side of a typical sunspot group has one magnetic polarity, whereas the other side has the opposite magnetic polarity. (Kitt Peak National Observatory)

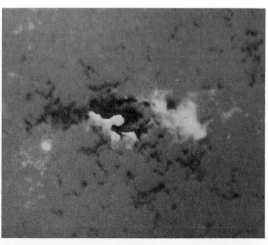

two basic properties of the Sun: its differential rotation, and its convective envelope. Differential rotation causes the magnetic field in the photosphere to become wrapped around the Sun (see Figure 19-19). The magnetic field becomes concentrated at latitudes on either side of the solar equator. Convection in the photosphere causes the concentrated magnetic field to become tangled, and kinks erupt through the solar surface. Sunspots appear at locations where the magnetic field protrudes through the photosphere. Careful examination of the orientation of the magnetic field in Figure 19-19 reveals that the preceding members of a sunspot group should indeed have the same magnetic polarity as that of the hemisphere in which they are located.

There are many things that the magnetic-dynamo model fails to explain. Most embarrassing is a basic lack of understanding of the physics of sunspots. Astronomers still do not know what holds a sunspot together for week after week. Our best calculations predict that a sunspot should break up and disperse as soon as it forms.

Our understanding of sunspots is further confounded by irregularities of the solar cycle. For example, the overall reversal of the Sun's magnetic field is often piecemeal and haphazard. At the time of reversal, one pole may change long before the other. Thus, for several weeks, the Sun may have two north poles and no south pole at all. To make matters worse, there is strong historical evidence that all traces of sunspots and the sunspot cycle have vanished for many years. For example, virtually no sunspots were seen from 1645 through 1715. This period is called the **Maunder minimum,** and apparently similar sunspot-free periods occurred at irregular intervals in earlier times.

The Maunder minimum offers the best evidence for a connection between the Sun and the Earth's weather. During this sunspot-free period, Europe experienced years of record low temperatures often called the Little Ice Age, while the western United States was subjected to severe drought. Some scientists speculate that solar activity may have an ongoing, although subtle, effect on terrestrial climates.

Not so subtle would be the effects of a varying solar diameter. Jack Eddy of the High Altitude Observatory in Colorado and other astrono-

Figure 19-19 Babcock's magnetic dynamo
A possible partial explanation for the sunspot cycle involves the wrapping of a magnetic field around the Sun by differential rotation. Sunspots appear where the concentrated magnetic field has broken through the solar surface.

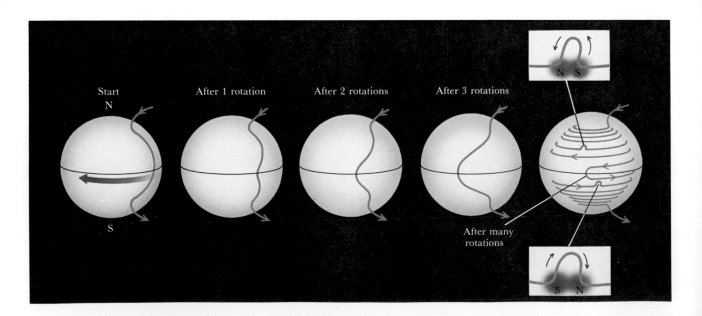

Start
N

S

After 1 rotation

After 2 rotations

After 3 rotations

N S

After many
rotations

S N

mers have recently reported controversial observations suggesting that the Sun's diameter may be shrinking at the remarkably fast rate of 0.1 percent per century. This topic is currently a matter of heated debate because astronomers are at a complete loss to explain such changes in the Sun's size. Nevertheless, some scientists suspect that the Sun may undergo very slow, long-term variations in its physical properties.

Solar seismology and new satellites are among the latest tools for solar research

Puzzles and enigmas in solar research have prompted astronomers to question many long-standing ideas about the Sun. For example, the magnetic field that produces sunspots might be bits and pieces of a **primordial field** that was trapped deep inside the Sun as it formed $4\frac{1}{2}$ billion years ago. To search for such a magnetic field or to study other deeply buried phenomena, astronomers need a way to probe the solar interior.

Geologists are able to determine the Earth's interior structure by using a seismograph to record vibrations during an earthquake. Although there are no true sunquakes, the Sun does vibrate at a variety of frequencies, somewhat like a ringing bell. These vibrations were first noticed in 1960 by Robert Leighton at Caltech; Leighton's sensitive Doppler-shift measurements revealed that portions of the Sun's surface move up and down by about 10 km every 5 minutes. In the mid-1970s, Henry Hill at the University of Arizona discovered slower vibrations with periods ranging from 20 minutes to nearly an hour. More recently, a variety of long-period oscillations have been found, and the possibility of detecting extremely slow pulsations has inspired astronomers who call themselves helioseismologists to set up telescopes at the South Pole, where the Sun can be observed continuously for many days. This new field of solar research is called **solar seismology.**

No one knows why the Sun vibrates. In 1970, Roger Ulrich at UCLA pointed out that the 5-minute oscillations are similar to sound waves but are more complicated. Various segments of the Sun's surface rise and fall as the waves move back and forth in the Sun's convective zone. Slower oscillations, such as the 30-minute "g mode" are believed to originate at very deep levels in the Sun. Many helioseismologists are hopeful that observations of these long-period oscillations will yield detailed information about solar interior.

While some solar astronomers are probing the Sun's core, others have concentrated on observations of the solar atmosphere from Skylab and from Earth-orbiting satellites. During the 1970s, a series of satellites called the Orbiting Solar Observatories provided a wealth of data about the Sun at ultraviolet, X-ray, and gamma-ray wavelengths. In 1980, a satellite called the Solar Maximum Mission spacecraft (see Figure 19-20) was launched to study solar flares. Unfortunately, the attitude-control system that keeps the satellite pointed toward the Sun failed after only 10 months of operation. In 1984, NASA scientists used the Space Shuttle to repair SMM so that its instruments can continue to send back spectra and pictures of the most violent phenomena occurring on the surface of our star.

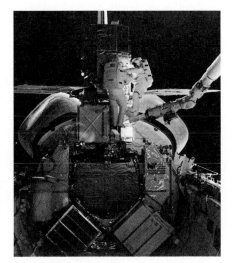

Figure 19-20 The Solar Maximum Mission spacecraft
The SMM spacecraft was specifically designed to study solar flares during the most recent sunspot maximum. It was launched in February 1980 and recorded data for 10 months. Two astronauts are seen repairing the satellite above the open cargo bay doors of the Space Shuttle. (NASA)

Summary

. The Sun's energy is produced by the thermonuclear process called hydrogen burning, in which four hydrogen nuclei combine to produce a single helium nucleus, with the release of energy.

The energy released in a nuclear reaction corresponds to a slight reduction of mass, according to Einstein's equation $E = mc^2$.

A thermonuclear reaction is a nuclear reaction that occurs only at very high temperatures; the hydrogen-burning reactions occur only at temperatures of more than about 8 million kelvins.

. A stellar model is a theoretical description of a star's interior derived from calculations based upon the laws of physics.

The solar model suggests that hydrogen burning occurs in a core that extends from the Sun's center to about 0.25 solar radius.

The core is surrounded by a radiative zone extending to about 0.8 solar radius, in which energy travels outward through the process of radiative diffusion.

The radiative zone is surrounded by a convective zone of gases at relatively low temperatures and pressures, in which energy travels outward primarily through the process of convection.

The visible surface of the Sun is a 100-km-thick layer called the photosphere at the bottom of the solar atmosphere; the gases in this layer shine with almost-perfect blackbody radiation corresponding to a temperature of 5800 K; convection produces features called granules in the photosphere.

Above the photosphere is a layer of cooler and less-dense gases called the chromosphere; the gases of this layer produce the absorption lines in the solar spectrum; spicules extend upward from the photosphere into the chromosphere along the boundaries of supergranules.

The outermost layer of thin gases in the solar atmosphere is called the corona; the corona blends into the solar wind at great distances from the Sun; the gases of the corona are very hot but at very low density; solar transients, coronal streamers, and coronal holes are other features associated with the corona.

. The solar surface features vary periodically in a 22-year solar cycle.

Sunspots are cool regions produced by local concentrations of the Sun's magnetic field; the average number of sunspots increases and decreases in a regular 11-year cycle.

A solar flare is a brief eruption of very hot, ionized gases from a sunspot group.

. The magnetic-dynamo model suggests that many features of the solar cycle are caused by the effects of differential rotation and convection on the Sun's magnetic field. Space missions to observe the Sun from outside the Earth's atmosphere and studies of solar seismology are expected to yield further understanding of the Sun (and hence of stellar processes in general) in the next few decades.

Review questions

1 Using the data given in Box 19-1, calculate the average density of the Sun. Compare your answer with the average densities of the Jovian planets.

2 Describe the dangers in attempting to observe the Sun. How have astronomers circumvented these observational problems?

3 Give an everyday example of hydrostatic equilibrium. Give an everyday example of thermal equilibrium.

4 Give some everyday examples of conduction, convection, and radiative diffusion.

5 Suppose that you want to determine the Sun's rotation rate by observing sunspots. Is it necessary to take the Earth's orbital motion into account? Why or why not?

6 When do you think the next sunspot maximum and minimum will occur? Explain.

7 It was once thought that sunspots were giant holes into the Sun's cooler interior. Why don't astronomers believe this anymore?

8 Why do you suppose that the Sun's energy generation occurs only around the Sun's center?

Advanced questions

9 What would happen if the Sun were not in a state of hydrostatic or thermal equilibrium?

*10** Assuming that the current rate of hydrogen burning in the Sun remains constant, what fraction of the Sun's mass will be converted into helium over the next five billion years? How will this affect the chemical composition of the Sun?

Discussion questions

11 Discuss the extent to which cultures around the world have worshipped the Sun as a diety throughout history. Why do you suppose that such widespread worship has occurred?

12 Discuss some of the difficulties of correlating solar activity with changes in the terrestrial climate.

13 Describe some of the advantages and disadvantages of observing the Sun **(a)** from space, and **(b)** from the South Pole. What kinds of phenomena and issues do solar astronomers want to explore from Earth-orbiting and antarctic observatories?

For further reading

Eddy, J. *A New Sun*. NASA SP-402, 1979. *A superb, lavishly illustrated introduction to the Sun.*
Eddy, J. "The Case of the Missing Sunspots." *Scientific American*, May 1977.
Frazier, K. *Our Turbulent Sun*. Prentice-Hall, 1983.
Levine, R. "The New Sun." In J. Cornell and P. Gorenstein, eds., *Astronomy from Space*. MIT Press, 1983.
Mitton, S. *Daytime Star*. Scribners, 1981.
Noyes, R. *The Sun, Our Star*. Harvard U. Press, 1982.
Robinson, L. "The Disquieting Sun: How Big, How Steady?" *Sky & Telescope*, Apr. 1982, p. 354.
Wolfson, R. "The Active Solar Corona." *Scientific American*, Feb. 1983.

20

The birth of stars

Reflection and emission nebulae
The two main blue objects, NGC 6589 (top) and NGC 6590 (below), are reflection nebulae surrounding main-sequence stars of spectral type B5 and B6. Interstellar dust around these two stars efficiently reflects their bluish light. Several smaller reflection nebulae are scattered around the large reddish patch of ionized hydrogen gas, IC 1283-4. Dust mixed with the gas dilutes the intense red emission with a soft blue haze. (Anglo-Australian Observatory)

Observations of stars along with calculations of stellar models have led astronomers to formulate a theory of stellar evolution. In this chapter we learn how protostars form in cold clouds of interstellar dust and gas and then evolve into main-sequence stars. We find that giant gas clouds scattered about the galaxy are the sites of star formation. Finally, we learn that the spiral arms of our galaxy and the explosions of supernovae are important factors in triggering the birth of stars.

Consider the legend of the Ephemera, a race of remarkable insects who inhabited a great forest. These noble creatures were blessed with great intelligence, yet cursed with tragically short life spans of less than a day.

To the Ephemera, the forest seemed eternal and unchanging. Many generations lived out their brief lives without ever noticing any alteration in the surrounding foliage. Nevertheless, careful observation and reasoning led some Ephemera to postulate that the forest is not static. They began to suspect that small green shoots grow to become huge trees and that mature trees eventually die, topple over, and litter the forest with rotting logs, enriching the soil for future trees. Although unable to witness this transformation personally, the Ephemera became aware of life cycles stretching over grand and awesome periods of many years.

An analogous situation faces astronomers. Gazing across the galaxy, we see a vast array of stars and nebulae. At first, the heavens seem eternal and unchanging; the views that greet us are virtually indistinguishable from those our ancestors saw. This permanence, however is an illusion. Stars emit vast amounts of radiation, and such expenditures must produce changes that cause the stars to evolve. With careful observation and calculation, astronomers have assembled a theory of **stellar evolution.** Stars are born in great interstellar clouds of gas and dust. They mature, grow old, and eventually blow themselves apart in death throes that enrich the interstellar gas for future stellar generations. The stars seem unchanging only because major stages in their lives last for millions or billions of years.

Protostars form in cold, dark nebulae

For many years, astronomers have suspected that stars are born in cold, dark clouds of interstellar gas. As we saw in Box 7-2, the temperature of a gas is directly related to the average speed of its atoms and molecules. If an interstellar cloud is warm, its atoms are moving about so rapidly that there is no chance for a protostar to condense from the agitated gases. If the temperature is low, however, then the atoms are moving slowly enough to allow denser portions of the cloud to contract gravitationally into clumps that collapse to form new stars.

Many of these cold clouds are scattered across the Milky Way. In some cases, they appear as dark regions silhouetted against glowing background nebulosity, such as the famous Horsehead Nebula seen in Figure 20-1. In other cases, they appear as dark blobs that obscure background stars (see Figure 20-2). These **dark nebulae** are called **Barnard objects** (after the American astronomer E. E. Barnard, who first discovered them around 1900). Some 200 relatively small and round dark nebulae are called **Bok globules** (after the Dutch-American astronomer Bart Bok, who called attention to them in the 1940s).

Figure 20-1 [left]   The Horsehead Nebula
Dust grains block light from background nebulosity whose glowing gases are excited by ultraviolet radiation from young, massive stars. The nebula is located in Orion at a distance of roughly 490 parsecs from Earth. The bright star to the left of center is Alnitak (ζ Ori), the easternmost star in the "belt" or Orion. (Royal Observatory, Edinburgh)

Figure 20-2 [right]   A dark nebula
This dark nebula, called Barnard 86, is located in Sagittarius. It is visible in this photograph simply because it blocks out light from the stars beyond it. The cluster of bluish stars to the left of the globule is called NGC 6520. (Anglo-Australian Observatory)

A typical Bok globule measures 1 to 2 light years across and has a mass between 20 and 200 solar masses. The chemical composition of these dark clouds is the standard "cosmic abundance" of about 75 percent (by mass) hydrogen, 23 percent helium, and 2 percent heavier elements (recall Table 6-4). Infrared observations demonstrate that the internal temperatures of globules are in the range of 5 to 15 K. This is so cold that dense regions inside the globule contract gravitationally, and the globule gradually coalesces into lumps called **protostars.**

Protostars evolve into young main-sequence stars

As early as the 1950s, astrophysicists such as L. Henyey in the United States and C. Hayashi in Japan began calculating the structure and evolutionary history of protostars. At first, a protostar is a cool blob of gas several times larger than the solar system. Low pressures inside the protostar are incapable of supporting all this cool gas, so the protostar begins to contract. As the protostar contracts, gravitational energy is converted into thermal energy (recall the discussion of Kelvin–Helmholtz contraction in Chapter 19), and the gases heat up and start glowing. Hayashi's calculations indicate that convection efficiently transports energy from the interior of the warming protostar to its surface. After only a few thousand years of gravitational contraction, the surface temperature has reached 2000 to 3000 K. At this point, the protostar is still quite large, and thus its glowing gases produce substantial luminosity. After only 1000 years of contraction, a protostar of 1 solar mass would be 20 times larger in diameter and 100 times brighter than the Sun.

Astrophysicists use high-speed computers and the equations of stellar structure described in Chapter 19 to calculate conditions inside a contracting protostar. The results tell how the protostar's luminosity and surface temperature change at various stages during contraction. With this information, we can graph the **evolutionary track** of the protostar on a Hertzsprung–Russell diagram.

When a protostar begins to shine at visible wavelengths, it is both luminous and cool. Thus, evolutionary tracks of protostars begin near the upper-right corner of the H–R diagram (see Figure 20-3). Continued gravitational contraction causes the protostar to shift rapidly away from this region of the diagram. A protostar more massive than five Suns becomes hotter without much change in overall luminosity, because the effect of decreasing surface area (which alone would decrease luminosity) is counterbalanced by an increase in surface temperature (which alone would increase luminosity). Thus, the evolutionary tracks of massive protostars traverse the H–R diagram horizontally from right to left. In less massive contracting protostars, the increase in surface temperature is less rapid and does not fully compensate for the shrinking surface area, thus causing luminosity to decrease.

A protostar continues to shrink until the temperature at its center reaches a few million degrees, when hydrogen burning begins. As we saw in Chapter 19, this thermonuclear process releases enormous amounts of energy (recall Box 19-2). The outpouring of energy creates conditions inside the protostar that finally halt the gravitational contraction. Hydrostatic and thermal equilibrium are eventually established, and a stable star is born. At this stage, the wandering evolutionary track ends on the main sequence, as shown in Figure 20-3.

We now know what the main sequence represents. Main-sequence stars are relatively young stars, inside of which hydrogen burning has only "recently" begun. This is a very stable situation for most stars. For example, the Sun will remain on or very near the main sequence, quietly burning hydrogen at its core, for a total of 10 billion years.

Note that the evolutionary tracks in Figure 20-3 end at locations along the main sequence that agree with the mass–luminosity relation (recall Figure 18-15): the most-massive stars are the most luminous, and the least-massive stars are the least luminous. Protostars less massive than about 0.08 solar masses never manage to develop the necessary pressures and temperatures to start hydrogen burning at their cores. Instead, these small protostars contract to become planetlike objects. Protostars with masses greater than 100 solar masses ($100M_\odot$) rapidly develop such extremely high temperatures that radiation pressure becomes the dominant

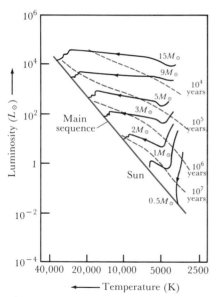

Figure 20-3 Pre–main-sequence evolutionary tracks
The evolutionary tracks of seven stars of different masses are shown on this H–R diagram. The dashed lines indicate the stage reached after the indicated numbers of years of evolution. Note that all tracks terminate on the main sequence at locations in agreement with the mass–luminosity relation (recall Figure 18-15). (Based on stellar model calculations by I. Iben)

force supporting the star against gravitational collapse. Such stars are very unstable and very uncommon.

The evolutionary tracks of protostars begin in the red-giant region of the H–R diagram, but these protostars are not red giants. An H–R diagram such as Figure 18-9 shows where stars spend *most* of their lives. Protostars spend only a tiny fraction of their existence in the red-giant region. A $15M_\odot$ protostar takes only 10,000 years to become a main sequence star; a $1M_\odot$ protostar takes a few million years to ignite hydrogen burning at its core. By astronomical standards, these are such brief intervals that it is very unusual to discover a protostar in its earliest stages of formation.

We also are unlikely to observe the birth of a star at visible wavelengths because the surrounding globule or interstellar cloud shields the protostar from our view. The vast amount of visible light emitted by the protostar is absorbed by interstellar dust in the surrounding **cocoon nebula,** which becomes heated to a few hundred kelvins. The warmed dust reradiates the energy at infrared wavelengths. Consequently, infrared observations reveal what is going on inside a "stellar nursery."

Figure 20-4 shows views of a stellar nursery taken at infrared and visible wavelengths. The dark cloud near the center of Figure 20-4b contains roughly 1000 solar masses of hydrogen and is a site of active star formation. This same region appears very bright in the infrared view of Figure 20-4a. Although interstellar dust in the cloud absorbs nearly all the visible light emitted by the newborn stars, infrared radiation passes through the obscuring material. Recent infrared observations by American astronomers Charles J. Lada and Bruce A. Wilking reveal a cluster of 20 new stars embedded deep within the dust cloud. Astronomers are very excited by the ease with which infrared observations probe these clouds. Better pictures and data will come from the Shuttle Infrared Telescope Facility (SIRTF) to be launched in the late-1980s.

Figure 20-4   Formation of new stars near ρ Ophiuchi
(a) *A wide-angle infrared view from IRAS of the region near the star ρ Ophiuchi, covering nearly 13° × 13° at a wavelength of 100 μ. The white rectangle indicates the smaller region covered by the photograph in visible wavelengths.* (b) *The white circle indicates the location of at least 20 newborn stars that probably are less than 1 million years old. The bluish star near the top of this view at visible wavelengths is ρ Ophiuchi; Antares (α Scorpius) appears at the lower left. Many stars in this region have a fuzzy appearance because of reflection nebulae produced by interstellar dust. (National Aerospace Laboratory of the Netherlands; Royal Observatory, Edinburgh)*

a

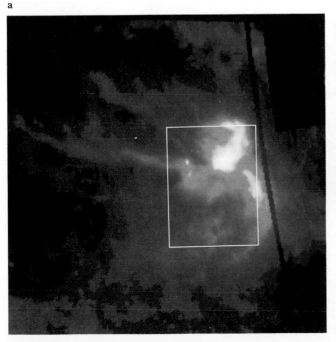

b

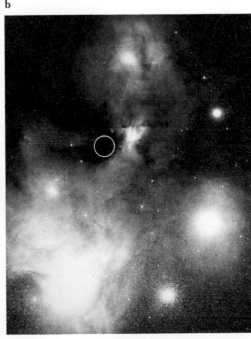

Young star clusters are found in H II regions

From the evolutionary tracks of protostars in Figure 20-3, we see that the most-massive stars are the first to form. The more massive the protostar, the sooner it develops the necessary central pressures and temperatures to ignite hydrogen burning. These massive main-sequence stars (spectral types O and B) are also the hottest and most luminous stars in the sky. Their surface temperatures are typically 15,000 to 35,000 K, and thus they emit vast quantities of ultraviolet radiation.

The energetic ultraviolet photons from newborn massive stars easily ionize the surrounding hydrogen gas. For example, radiation from an O5 star can knock electrons from hydrogen atoms in a volume roughly 1000 light years across. This has a dramatic effect on the globule or dark nebula in which a cluster of stars is forming. While some hydrogen atoms are being knocked apart by ultraviolet photons, other hydrogen atoms are being reassembled as some of the free protons and electrons manage to get back together. During this **recombination** of hydrogen atoms, the captured electrons cascade downward through the atom's energy levels toward the ground state. These downward quantum jumps release numerous photons, many visible wavelengths. The nebula begins to glow. Particularly prominent is the transition from $n = 3$ to $n = 2$, which produces H_{α} photons at 6563 Å in the red portion of the visible spectrum. Thus the nebulosity around a newborn star cluster shines with a distinctive reddish hue.

Figure 20-5   An H II region
Because of its shape, this emission nebula is called the Eagle nebula. It surrounds the star cluster called M16 or NGC 6611 in the constellation of Serpens at a distance of 2000 parsecs from Earth. Several bright, hot O and B stars are responsible for the ionizing radiation that causes the gases to glow. (Anglo-Australian Observatory)

Figure 20-5 shows one of these **emission nebulae.** Because these nebulae are predominantly ionized hydrogen, they are also called **H II regions** (several famous examples are listed in Box 20-1). The collections of a few hot, bright O and B stars near the core of the nebula that produces the ionizing ultraviolet radiation is called an **OB association.**

Box 20-1 Famous H II regions

H II regions are some of the most beautiful nebulae in the sky. Some of the most famous H II regions are listed in the table with information about their sizes, masses, distances, and densities.

Name of nebula	Distance from Earth (light years)	Diameter (light years)	Mass ($M_\odot$)	Density (atoms/cm³)
Lagoon (M8)	4000	30	1000	80
Eagle (M16)	6000	20	500	90
Omega (M17)	5000	30	1500	120
Trifid (M20)	3000	12	150	100
Orion (M42)	1600	16	300	600

Both the Lagoon and Trifid nebulae are shown in the photograph, which covers an area $2° \times 2\frac{1}{2}°$ in Sagittarius. The Lagoon Nebula is on the right, and the smaller Trifid Nebula is on the left.

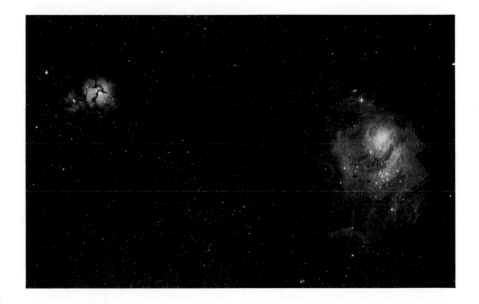

Observations of individual stars in a young cluster yield further information about stars in their infancy. Figure 20-6 shows a beautiful emission nebula surrounding the cluster called NGC 2264. By observing each star in the cluster and measuring its magnitude and color index, astronomers can deduce the star's luminosity and surface temperature. The data for all the stars in the cluster can then be plotted on an H–R diagram as shown in Figure 20-7. Note that the hottest stars (surface temperatures around 20,000 K) are on the main sequence. These are the rapidly evolving, massive, hot stars whose radiation is causing the surrounding gases to glow. The stars cooler than about 10,000 K, however, have not yet quite arrived at the main sequence. These less-massive stars are in the final stages of pre–main-sequence contraction and are just now

Figure 20-6   A young star cluster
*This H II region in the constellation of Mono-
ceros contains a young star cluster called NGC
2264. It is located about 800 parsecs from
Earth and contains numerous T Tauri stars
that are about to begin hydrogen burning in
their cores. (Anglo-Australian Observatory)*

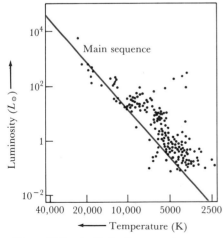

**Figure 20-7   An H–R diagram of
NGC 2264**
*Each dot plotted on this H–R diagram repre-
sents a star in NGC 2264 whose luminosity
and surface temperature have been determined.
Note that most of the cool, low-mass stars have
not yet arrived at the main sequence. The star
cluster probably started forming only 2 million
years ago. (Based on observations by Merle
Walker)*

beginning thermonuclear reactions at their centers. The locations of data points on Figure 20-7 suggest that the cluster has an age roughly 2 million years.

Spectroscopic observations of the cooler stars in NGC 2264 show that many are vigorously ejecting gas, a very common phenomenon in most stars just before they reach the main sequence. Such gas-ejecting stars are called **T Tauri stars,** named after the first example discovered in the constellation of Taurus. Some astronomers suggest that the onset of hydrogen burning is preceded by vigorous chromospheric activity, with enormous spicules and flares that propel the star's outermost layers back out into space. In fact, an infant star going through its T Tauri stage can lose as much as 0.4 solar masses before it settles down on the main sequence.

During these final spasmodic stages in the birth of a star, material in the surrounding cocoon nebula may become excited and start to glow. Such activity is probably responsible for the Herbig–Haro objects, named after astronomers George Herbig and Guillermo Haro who began discovering them in the 1940s. As shown in Figure 20-8, a **Herbig–Haro object** consists of several bright knots. From year to year, these clumps of glowing gases change slightly in size, shape, and brightness. Presumably, these alterations are due to fluctuations in nearby infant stars undergoing their final stages of pre–main-sequence contraction. Indeed, Herbig–Haro objects are often found in the vicinity of T Tauri stars.

Sporadic activity and mass loss can continue after a star has arrived on the main sequence. For example, Figure 20-9 shows a young star cluster called the Pleiades that is easily visible to the unaided eye in the constellation of Taurus. All of the stars in this picture are on the main sequence. The cluster's age is presumably about 100 million years, because that is how long it takes for the least massive stars finally to begin hydrogen burning in their cores. One of the brighter stars in the cluster, Pleione, is still experiencing minor outbursts. Spectroscopic observations reveal that Pleione blew off clouds of gas in 1888, 1938, and 1972. Fig-

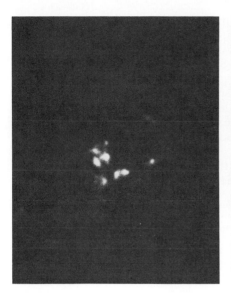

Figure 20-8 [left]   Herbig–Haro objects
Herbig–Haro objects consist of bright knots of glowing gas that vary slightly over the years. They probably are sites of active star formation and are typically found in the vicinity of T Tauri stars. (Lick Observatory)

Figure 20-9 [right]   The Pleiades
This open cluster is 126 parsecs from Earth and can easily be seen with the naked eye. The cluster has a diameter of about 1½ parsecs and is about 100 million years old. (U.S. Naval Observatory)

ure 20-10 is a dramatic photograph of a young star experiencing significant mass loss.

Note the distinctly bluish color of the nebulosity around the Pleiades. This glow is called a **reflection nebula** because it is caused by starlight reflected from dust grains in the interstellar material. The nebulosity is blue for the same reason that the daytime sky is blue: particles (in a nebula or in the air) scatter short wavelength light much more efficiently than longer wavelength radiation. Thus blue light is bounced around and reflected back toward us much more readily than is light of other colors.

In many respects, the Pleiades cluster is a mature version of the H II regions seen in Figures 20-5 and 20-6. Many years of T Tauri activity and vigorous stellar winds have widely dispersed the surrounding gases so that they no longer glow as an emission nebula. Eventually these gases

Figure 20-10   The mass-loss star HD 148937
This unusual star in the constellation of Norma is the brightest member of a triple system of stars in orbit about one another. This extremely hot, massive star is losing its outer layers continuously. More vigorous outbursts in the past gave rise to the symmetric shells (called NGC 6164 and 6165) on either side of the star. (Anglo-Australian Observatory)

will become so widely scattered that the reflection nebulosity also will fade from view.

A loose collection of stars such as the Pleiades or NGC 2264 is called an **open cluster,** or **galactic cluster.** Such clusters possess barely enough mass to hold themselves together by gravitational forces. Occasionally, a star moving faster than the average speed will escape, or "evaporate," from a cluster. By the time the stars are a few billion years old, they may be so widely separated that a cluster no longer truly exists. If the group of stars is gravitationally unbound from the beginning—that is, if the stars are moving away from one another so rapidly that gravitational forces cannot pull them back—then the apparent cluster that initially exists is called simply a **stellar association.**

Star birth begins in giant molecular clouds

Astronomers agree that H II regions are stellar nurseries, but where do the H II regions come from? This question was finally answered in the 1970s when radio astronomers began discovering enormous clouds of gas scattered about our galaxy.

We know that hydrogen is by far the most abundant element in the universe. In the cold depths of interstellar space, hydrogen atoms combine to form hydrogen molecules (H_2). As described in Box 20-2, a molecule vibrates and rotates at specific frequencies dictated by the laws of quantum mechanics. As a molecule goes from one vibrational or rotational state to another, it emits or absorbs a photon. The situation is entirely analogous to an atom that emits or absorbs a photon as an electron jumps from one energy level to another. Many interstellar molecules emit photons with wavelengths of a few millimeters. Consequently, observations with radio telescopes tuned to wavelengths in this range have greatly increased our knowledge of the **interstellar medium,** or interstellar matter, in recent years.

Although hydrogen molecules are scattered abundantly across space, they are difficult to detect. The hydrogen molecule is symmetric (two atoms of equal mass joined together), and quantum mechanics explains that such molecules do not efficiently emit photons at radio frequencies. Instead, radio astronomers more easily detect asymmetric molecules such as carbon monoxide (CO), which consists of two atoms of unequal mass joined together. In particular, carbon monoxide emits photons at a wavelength of 2.6 mm, which corresponds to a transition between two rates of rotation of the molecule.

Carbon monoxide is especially useful in probing the interstellar medium. Astronomers have reason to believe that the ratio of carbon monoxide to hydrogen is reasonably constant in space: for every CO molecule, there are about 10,000 H_2 molecules. Consequently, carbon monoxide is an excellent "tracer" for hydrogen gas. Wherever astronomers detect strong emission from CO, they know that hydrogen gas must be abundant.

The first systematic surveys of our galaxy looking for 2.6-mm CO radiation were undertaken in 1974 by Philip Solomon at the State University of New York. Mapping the locations of CO emission, astronomers soon realized that vast amounts of hydrogen are concentrated in huge regions called **giant molecular clouds.** These clouds have masses in the range of 10^5 to 2×10^6 solar masses and diameters that range from 20 to 80 parsecs. Inside one of these clouds, the density is about 200 hydrogen molecules per cubic centimeter. Astronomers estimate that our galaxy contains about 5000 of these enormous clouds.

Box 20-2   Interstellar molecules

A molecule is a combination of two or more atoms. A molecule vibrates and rotates at specific frequencies allowed by the laws of quantum mechanics. A molecule can speed up or slow down its rate of vibration or rotation by either absorbing or emitting a photon. Astronomers therefore discover interstellar molecules by detecting the radiation from these vibrational or rotational transitions.

Although the first discovery of an interstellar molecule was made at visual wavelengths, most molecules are strong emitters of radiation with wavelengths of around 1 to 10 mm. Consequently, observations with radio telescopes tuned to "millimeter wavelengths" have greatly increased the rate of discovery of interstellar molecules in recent years. Nearly 100 different kinds of interstellar molecules have been discovered so far, and the list is constantly growing. A partial listing is given here.

Two-atom molecules:

- CH
- CN
- CO
- CS
- H_2
- OH
- SiO
- SO

Three-atom molecules:

- H_2O (water)
- H_2S
- SO_2
- HCN
- HCO
- OCS

Four-atom molecules:

- NH_3 (ammonia)
- H_2CO (formaldehyde)
- $HNCO$
- H_2CS
- HC_2H (acetylene)

Five-atom molecules:

- H_2CHN
- H_2NCN
- $NCOOH$ (formic acid)
- HC_3N
- H_2C_2O

Six-atom molecules:

- CH_3OH (methyl alcohol)
- CH_3CH (methyl cyanide)
- $HCONH_2$

Seven-atom molecules:

- CH_3NH_2
- CH_3C_2H
- $HCOCH_3$
- H_2CCHCN (vinyl cyanide)

Note that the vast majority of these substances contain carbon. Such molecules are called organic molecules. Apparently, organic chemistry is the chemistry of interstellar space as well as the chemistry of life on Earth. Astronomers have had such success in recent years in detecting interstellar molecules that they often boast of someday discovering every conceivable chemical somewhere in the universe. In fact, several organic molecules have been discovered in space that do not exist here on Earth.

It is intriguing to speculate that interstellar molecules may have assembled somewhere to form living organisms. After all, it took no more than a billion years for the first primitive lifeforms to develop on Earth from the organic molecules in the primordial oceans. Interstellar clouds in our Galaxy have had 15 billion years to accomplish the same thing. Is it possible that clouds drifting between the stars contain extraterrestrial life?

Many galaxies, including our own and the one shown in Figure 20-11, have **spiral arms,** which are huge, arching lanes of glowing gas and stars. In Chapter 25, when we discuss the details of our galaxy, we shall learn that these spiral arms are caused by compressional waves that squeeze the interstellar gases through which they pass. When one of these waves passes through a giant molecular cloud, it compresses the cloud, and vigorous star formation begins in the densest regions. Massive

 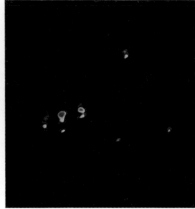

stars, which are the first to form, emit ultraviolet light that soon ionizes the surrounding hydrogen, and an H II region is born.

A newborn H II region buried deep inside a giant molecular cloud is not at first visible at optical wavelengths. But the ionized hydrogen emits radio waves as the electrons are decelerated in the electric fields of the protons, and high-resolution radio observations have revealed H II regions inside giant molecular clouds. The radio picture from the VLA in Figure 20-12 shows H II regions so small and compact that they probably formed within the past 100 years. Figure 20-12 is therefore a remarkable view of the earliest stages of star formation.

With the continued development of new stars, an H II region grows and eventually becomes visible at optical wavelengths. The famous Orion nebula (see Figure 20-13), like many H II regions, is a small, bright "hot spot" in an enormous molecular cloud. Four hot, massive O and B stars at the heart of the Orion nebula are responsible for the ionizing radiation that causes the surrounding gases to glow.

The OB association at the core of the H II region affects the rest of the giant molecular cloud. Vigorous stellar winds, along with ionizing ultraviolet radiation from the O and B stars, carve out a cavity in the cloud. Much of this outflow is supersonic, and thus a **shock wave** forms where the outer edge of the expanding H II region impinges on the rest of the giant molecular cloud. The shock wave compresses the hydrogen

Figure 20-11 [left]   The spiral galaxy NGC 2997
As compressional waves pass through the molecular clouds, they trigger vigorous star birth. This process produces numerous H II regions and clusters of O and B stars that outline the spiral arms. Each pinkish speck seen along the spiral arms of this galaxy is an H II region. (Anglo-Australian Observatory)

Figure 20-12 [right]   Newborn, compact H II regions
This radio picture from the VLA shows several hot, bright H II regions deep within a giant molecular cloud in the constellation of Aquila. The cloud is about 15,000 parsecs from Earth, and the H II regions are only about 0.04 parsecs in diameter. Because of their very small size, these H II regions are thought to be very young. (Observations by William J. Welch and John W. Dreher)

Figure 20-13   A mature H II region
This famous H II region, the Orion Nebula, can be seen with the naked eye. It is located 460 parsecs from Earth and has a diameter of roughly 5 parsecs. The mass of this nebulosity is estimated at 300 solar masses. Four bright, massive stars at the center of the nebula are producing the ultraviolet light that causes the gases to glow. These four stars, called the Trapezium, are separated from each other by only 0.04 parsec. (Anglo-Australian Observatory)

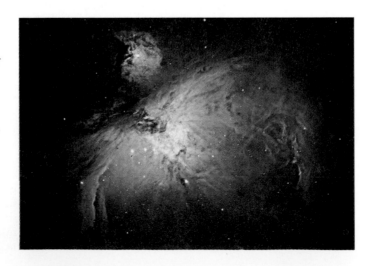

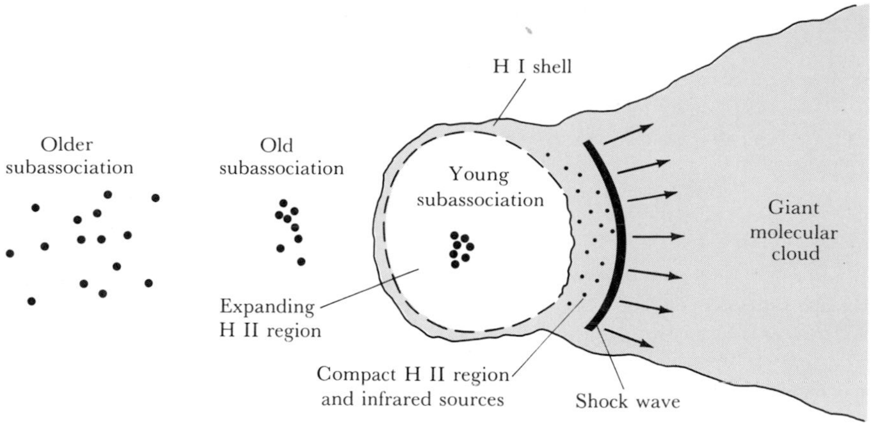

Figure 20-14   The evolution of an OB association
Ultraviolet radiation from young O and B stars produces a shock wave that compresses gas further into the molecular cloud and stimulates new star formation deeper into the cloud. Meanwhile, older stars are left behind. (Adapted from C. Lada, L. Blitz, and B. Elmegreen)

gas through which it passes, thereby stimulating a new round of star birth. Indeed, compact H II regions and infrared sources are often found in the swept-up layer immediately behind the shock wave. These compact H II regions soon form their own O and B stars, which start carving out a new cavity deeper into the giant molecular cloud. Meanwhile, the older O and B stars are left behind and begin to disperse (see Figure 20-14). In this way, an OB association "eats into" a giant molecular cloud, "spitting out" stars in its wake.

Examples of this evolutionary process are visible in many H II regions: a compact, bright, young OB subassociation is on one side of the nebula, and an older, more widely scattered OB subassociation is on the other side of the nebular. In the Trifid nebula (see Figure 20-15), we find emission and reflection nebulosity side by side. Presumably, the reddish emission nebula surrounds young stars, whereas the bluish reflection nebulosity surrounds slightly more mature stars.

Recent radio observations of giant molecular clouds by Patrick Thaddeus and his colleagues at Columbia University suggest that it takes only 10^7 years for one of our galaxy's spiral arms to pass through a giant molecular cloud. By that time, the bright, massive O and B stars have rushed through their brief lives and faded from view. With no more

Figure 20-15   The Trifid Nebula
This nebulosity in Sagittarius is 1000 parsecs from Earth and measures about 4 parsecs across. The youngest stars are enveloped in reddish emission nebulosity. Bluish reflection nebulosity surrounds slightly more mature stars. (Kitt Peak National Observatory)

ultraviolet light to ionize hydrogen atoms, the beautiful emission nebulae also fade and disappear. In other words, by the time that a spiral arm leaves a giant molecular cloud, all of the dramatic activity associated with star birth has come to an end.

The compression wave associated with a spiral arm, however, continues to revolve about the galaxy and soon encounters new giant molecular clouds, where a new round of star birth is triggered. This is why we can see spiral arms: they are outlined by glowing H II regions that surround newborn O and B stars. In many galaxies like our own, star birth is an ongoing process intimately related to the majestic, pinwheellike rotation of the spiral arms.

Supernova explosions also compress the interstellar medium and thereby trigger star birth

Presumably, any mechanism that compresses interstellar clouds can trigger the birth of stars. As we shall see in Chapter 22, massive stars sometimes end their lives with a violent detonation called a **supernova explosion.** In a matter of seconds, the core of the doomed star collapses, releasing vast quantities of particles and energy that blow the star apart. The star's outer layers are blasted outward into space at speeds of several thousand kilometers per second.

Astronomers find many nebulosities across the sky that are the shredded funeral shrouds of these dead stars. Such nebulae, like the Cygnus

Figure 20-16   A supernova remnant
This remarkable nebula called the Cygnus Loop is the remnant of a supernova explosion that occurred about 20,000 years ago. The expanding spherical shell of gas now has a diameter of about 120 light years. (Palomar Observatory)

Loop shown in Figure 20-16, are called **supernova remnants.** Many supernova remnants have a distinctly arched appearance, as would be expected for an expanding shell of gas. This wall of gas is typically still moving away from the dead star at supersonic speeds, and passage through surrounding interstellar medium excites the atoms, causing the gases to glow.

Supersonic motion is always accompanied by a shock wave that abruptly compresses the medium through which it passes. If the expanding shell of a supernova remnant encounters an interstellar cloud, it can squeeze the cloud, thereby stimulating star birth. This kind of star birth is observed in the stellar association seen in Figure 20-17. This stellar nursery is located along a luminous arc of gas about 30 parsecs long that is presumably the remnant of an ancient supernova explosion. In fact, this arc is part of an almost-complete ring of glowing gas with a diameter of about 60 parsecs. Spectroscopic observations of the stars along the arc reveal substantial T Tauri and chromospheric activity indicative of newborn stars experiencing mass loss in their final stages of pre–main-sequence contraction.

As we learned in Chapter 16, there is strong evidence that the Sun was once a member of one of these stellar associations created by a supernova explosion. These loose associations do not remain intact for long. Individual stellar motions soon carry the stars in various directions away from their birthplace. Nearly 5 billion years have passed since the birth of our star, and thus the Sun's brothers and sisters are now widely scattered across the galaxy.

Our understanding of star birth has improved dramatically in recent years, primarily due to infrared and millimeter-wavelength observations. Nevertheless, many puzzles and mysteries remain. For example, astronomers generally assumed that there must be a lot of interstellar dust shielding a stellar nursery to protect it from the disruptive effects of external sources of ultraviolet light—that's what we seem to find in our own galaxy. In a neighboring galaxy called the Large Magellanic Cloud, however, there are young OB associations with virtually no dust. Does the process of star birth differ slightly from one galaxy to another?

Another problem involves the fact that different methods of star birth tend to produce different percentages of different kinds of stars. Specifically, the passage of a spiral arm through a giant molecular cloud tends to produce an abundance of massive O and B stars. In contrast, the shock wave from a supernova seems to produce fewer O and B stars, but many more of the less massive A, F, G, and K stars. We do not know why this is so.

There may be additional methods of star birth that have yet to be explored or discovered. For example, Robert Loren at the University of Texas has pointed out that a simple collision between two interstellar clouds should create new stars. When two such clouds collide, compression must occur at the interface, and vigorous star formation is sure to follow. Another intriguing possibility studied by Bruce Elmegreen at Columbia University and others is that light from a star or group of stars may exert strong-enough radiation pressure on interstellar clouds to cause compression followed by star formation.

In spite of unanswered questions, it is now clear that star birth involves mechanisms on a colossal scale—from the deaths of massive stars to the rotation of the entire galaxy. In many respects, we have just begun to appreciate these cosmic processes. The study of cold, dark stellar nurseries will certainly be an active and exciting area of astronomical research for many years to come.

Figure 20-17   The Canis Major R1 association
This luminous arc of gas is about 100 light years long and is studded with numerous young stars. This stellar association illustrates how the shock wave from a supernova explosion can trigger star formation in the interstellar clouds through which it passes. (Palomar Observatory)

....

Summary
- Enormous cold clouds of gas, called giant molecular clouds, are scattered about our galaxy.

  A cloud that is visible as a dark blot against distant stars is called a dark nebula, or Barnard object; a small, round dark nebula is called a Bok globule.

- Star formation can begin when gravitational attraction causes a protostar to coalesce within a giant molecular cloud.

  As a protostar grows by gravitational accretion of gases, the process of Kelvin–Helmholtz contraction causes it to begin glowing; its low temperature and high luminosity place it near the upper-right corner of the H–R diagram.

  As evolution of the protostar continues, it moves toward the main sequence on the H–R diagram; when core temperatures become high enough to begin hydrogen burning, the protostar becomes a main-sequence star.

  The most massive protostars are the first to become main-sequence stars (O and B stars); they emit strong ultraviolet radiation that ionizes hydrogen in the surrounding cloud, creating reddish emission nebulae called H II regions.

  Ultraviolet radiation and stellar winds from the OB association at the core of an H II region create shock waves moving outward through the gas cloud, compressing the gas to trigger formation of more protostars.

  Shock waves associated with the spiral arms of our galaxy and with supernova explosions also compress gas clouds to trigger star formation.

- In the final stages of pre–main-sequence contraction, as thermonuclear reactions are about to begin in the core of a protostar, the star may undergo vigorous chomospheric activity that ejects large amounts of matter into space; such gas-ejecting stars are called T Tauri stars.

  Material in the cocoon nebula surrounding T Tauri stars may begin to glow, producing the Herbig-Haro objects that consist of bright "knots," or clumps of glowing gases.

  A bluish reflection nebula is typically observed around a cluster of T Tauri stars, produced by reflection of starlight from the dust grains in the interstellar material.

- A collection of newborn stars may form an open cluster, or galactic cluster, in which stars are held together by gravitational attractions; occasionally, a star moving more rapidly than the average speed will escape, or "evaporate," from such a cluster.

  A stellar association is a group of newborn stars that are moving apart so rapidly that their gravitational attractions to one another cannot pull them into orbit about one another.

....

Review questions

1 Why are low temperatures necessary in order for protostars to form inside dark nebulae?

2 Describe the energy source that causes a protostar to shine. How does this differ from the energy source inside a true star?

3 Explain why thermonuclear reactions occur only at the center of a main sequence star and never on its surface.

4 What is an "evolutionary track" and in what way can evolutionary tracks help us interpret the H–R diagram?

5 Why are infrared and millimeter observations so much more useful in exploring interstellar clouds than are observations at visible wavelengths?

6 If you took a spectrum of a reflection nebula, would you see absorption lines, emission lines, or no lines? Explain your answer and explain how the spectrum demonstrates that the light is reflected from nearby stars.

7 Briefly describe four mechanisms that compress the interstellar medium and trigger star formation.

8 Speculate on why a shock wave from a supernova seems to produce relatively few high-mass O and B stars compared to the lower-mass A, F, G, and K stars.

Advanced questions

*9 Find the density (in atoms per cubic centimeter) in a globule having a radius of 1 light year and a mass of 100 solar masses. How does your result compare with the densities listed in Box 20-1? (Assume that the globule is made of pure hydrogen.)

10 How would you distinguish a newly formed protostar from a red giant, in view of their identical location on the H-R diagram?

*11 The density of ethyl alcohol (atomic weight = 46) in a typical molecular cloud is about 1 molecule per 10^{14} cm^3. What volume of the cloud would contain enough alcohol to make a martini (about 10 gm of alcohol)?

Discussion questions

12 What do you think would happen if the solar system passed through a giant molecular cloud? Do you think that the Earth has ever passed through such clouds?

13 Speculate about the possibility of life forms and biological processes occurring in giant molecular clouds. In what ways might conditions in giant molecular clouds favor or hinder biological evolution?

For further reading

Blitz, L. "Giant Molecular Cloud Complexes in the Galaxy." *Scientific American*, Apr. 1982.

Bok, B. "Early Phases of Star Formation." *Sky & Telescope*, Apr. 1981, p. 284.

Cohen, M. "Stellar Formation." *Astronomy*, Sept. 1979, p. 66.

Herbst, W. "Canis Major Rl: A Stellar Nursery." *Mercury*, July/Aug. 1979, p. 86.

Lada, C. "Energetic Outflows from Young Stars." *Scientific American*, July 1982.

Loren, R., and Vrba, F. "Starmaking with Colliding Molecular Clouds." *Sky & Telescope*, June 1979, p. 521.

Reipurth, B. "Bok Globules." *Mercury*, Mar./Apr. 1984, p. 50.

Robinson, L. "Orion's Stellar Nursery." *Sky & Telescope*, Nov. 1982, p. 430.

Rodriquez, L. "Searching for the Energy Source of the Herbig–Haro Objects." *Mercury*, Mar./Apr. 1981, p. 34.

Scoville, N., and Young, J. "Molecular Clouds, Star Formation, and Galactic Structure." *Scientific American*, Apr. 1984.

Zeilik, M. "The Birth of Massive Stars." *Scientific American*, Apr. 1978.

21

Stellar maturity and old age

A mass-loss star
Old stars become giants and supergiants whose bloated outer atmospheres shed matter into space. This star, HD 65750, is losing matter at a high rate and is surrounded by a reflection nebula (IC 2220) caused by starlight reflected from dust grains. These dust grains may have condensed from material shed by the star. A typical red giant can lose 10^{-7} solar masses per year and many are surrounded by circumstellar shells of matter they have ejected. (Anglo-Australian Observatory)

We turn now to the evolutionary history of stars after they leave the main sequence. First, we learn how stars become red giants after their core hydrogen is exhausted. Then we discuss the thermonuclear process of helium burning that occurs in the cores of aging stars. We find that helium burning ignites gradually in high-mass stars but explosively in low-mass stars. Next we see how an H–R plot of the stars in a cluster can be used to estimate the age of the cluster. Finally, we learn that some stars become unstable and pulsate in old age.

A main-sequence star is a young star whose radiated energy comes from the thermonuclear process of hydrogen burning in its core. The main-sequence star is in thermal equilibrium: energy liberated in the core is balanced by energy radiated from the surface. Eventually, however, all of the hydrogen in the core of the star is used up. **Core hydrogen burning** then must cease, with dramatic effects upon the star's equilibrium, structure, and evolution.

When core hydrogen burning ceases, a main-sequence star becomes a red giant

As we have seen in Chapters 19 and 20, a very massive protostar quickly builds up the necessary core temperature and pressure for hydrogen burning to commence, so that the most massive protostars are the first to become main-sequence (O and B) stars. Furthermore, the more massive main-sequence stars are the most luminous stars; this rapid emission of energy must correspond to a rapid depletion of hydrogen in the core of the massive stars. Thus, even though a massive O or B star contains much more hydrogen fuel than a less-massive main-sequence star, it consumes its hydrogen far more rapidly. The main-sequence lifetime of a massive star thus is considerably shorter than that of a less-massive star.

Hydrogen burning has continued in the Sun's core for the past 5 billion years. Initially, the Sun's chemical composition was roughly 75 percent hydrogen and 25 percent helium (plus a smattering of heavy elements). The continued fusion of hydrogen into helium in the Sun's core, however, has dramatically altered the core composition. As shown in Figure 21-1, there is now more helium than hydrogen at the Sun's center.

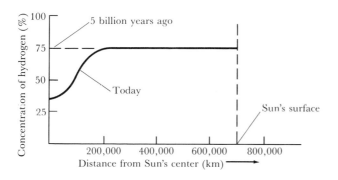
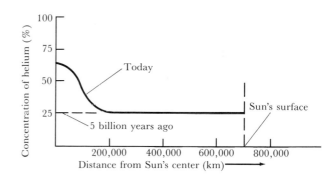

Figure 21-1 **The Sun's chemical composition**
Like all stars, the Sun began with a composition of nearly 75 percent hydrogen and 25 percent helium. However, 5 billion years of thermonuclear reactions at the Sun's center have depleted the concentration of hydrogen and increased that of helium.

As we saw in Chapter 19, enough hydrogen remains in the Sun's core for another 5 billion years of core hydrogen burning. That brings the Sun's total lifetime on the main sequence to 10 billion years. Table 21-1 shows how long other stars take to exhaust the supplies of hydrogen in their cores (see Box 21-1 for a more detailed discussion). Note that high-mass stars gobble up their hydrogen fuel in only a few million years, whereas low-mass stars take hundreds of billions of years to accomplish the same thing.

As the supply of hydrogen at a star's center dwindles, the star begins to have difficulty supporting the weight of its outer layers. This

TABLE 21-1 *Main-sequence lifetimes*

Mass ($M_\odot$)	Surface temperature (K)	Luminosity ($L_\odot$)	Time on main sequence (10^6 years)
25	35,000	80,000	3
15	30,000	10,000	15
3	11,000	60	500
1.5	7,000	5	3,000
1.0	6,000	1	10,000
0.75	5,000	0.5	15,000
0.50	4,000	0.03	200,000

Box 21-1   Main-sequence lifetimes

During hydrogen burning, a portion of a star's mass is being converted into energy. We can use Einstein's famous equation relating mass and energy to calculate how long a star remains on the main sequence.

Suppose that M is the mass of a star, and f is the fraction of the star's mass that is converted into energy as a result of hydrogen burning. The total energy E supplied by hydrogen burning is

$$E = fMc^2$$

where c is the speed of light.

This energy is released gradually over many years. Specifically, suppose that L is the star's luminosity, and T is the total time over which hydrogen burning occurs. Then

$$E = LT$$

From these two equations, we see that

$$T = fc^2M/L$$

In other words, a star's lifetime on the main sequence is proportional to its mass divided by its luminosity:

$$T \propto M/L$$

We can carry this analysis further by recalling that main sequence stars obey the mass–luminosity relation (see Figure 18-15). The mass–luminosity relation tells us that a star's luminosity is roughly proportional to the fourth power of its mass:

$$L \propto M^4$$

Substituting this relationship into the previous proportionality, we find that

$$T \propto \frac{1}{M^3}$$

This approximate relationship can be used to obtain rough estimates of how long a star remains on the main sequence. It is often convenient to relate these estimates to the Sun (a typical $1M_\odot$ star), which will spend 10^{10} years on the main sequence. For example, consider a star whose mass is $2M_\odot$. Because $2^3 = 8$, this star will be on the main sequence for only about one-eighth of the time that the Sun is on the main sequence. Thus, a $2M_\odot$ star will burn hydrogen in its core for about 1¼ billion years.

enormous weight pressing inward from all sides compresses the star's core slightly. The compressed gases become warmer, allowing hydrogen burning to move outward from the core. In other words, during its final years on the main sequence, a star becomes involved in a final attempt to maintain thermal equilibrium by enlarging its hydrogen-burning region. There is still plenty of fresh hydrogen surrounding the star's center. By tapping this supply, the star manages to eke out a few million more years on the main sequence.

Finally, all the hydrogen in the core of an aging main-sequence star is used up, and hydrogen burning therefore ceases in the star's core. However, hydrogen burning does continue in a thin spherical shell surround-

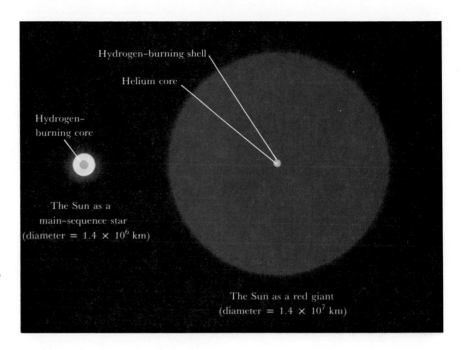

Hydrogen–burning shell

Helium core

Hydrogen–
burning core

The Sun as a
main–sequence star
(diameter = 1.4×10^6 km)

The Sun as a red giant
(diameter = 1.4×10^7 km)

Figure 21-2   The Sun today and as a red giant
Today, the Sun's energy is produced in a hydrogen-burning core whose diameter is about 300,000 km. When the Sun becomes a red giant, in some 5 billion years, it will draw its energy from a hydrogen-burning shell surrounding a compact helium-rich core. The helium core will have a diameter of only 10,000 km.

ing the core. This **shell hydrogen burning** initially occurs only in the hot region just outside the core, where hydrogen fuel has not yet been exhausted.

Because no thermonuclear reactions are now producing energy in the star's core, the heat flowing out of the hot core is no longer replaced. The core therefore gradually contracts, converting gravitational energy into thermal energy to maintain thermal equilibrium. As the hydrogen-burning shell slowly works its way outward from the original core, more helium is added to the core, causing further core contraction and heating. Over the course of a few million years, the core of a $1M_\odot$ star is compressed to about one-fiftieth of its original radius, and the central temperature increases from around 15 million kelvins to about 100 million kelvins.

As the core shrinks and becomes hotter, its increased energy output causes the star's outer layers to expand, so that the overall size of the star increases. The diameter of the star may increase by a factor of ten. As the star's outer atmosphere expands farther and farther into space, its gases cool. Soon the temperature of the star's bloated surface has fallen to about 3500 K and, in accordance with Wien's law, the gases glow with a reddish hue. Such stars are appropriately called **red giants**.

In the Sun's case, it will take about 5 billion years more to finish converting hydrogen into helium at the Sun's core. As the Sun's core contracts, its atmosphere will expand to encompass Mercury, then Venus, and finally our own planet. The red-giant Sun will swell to a diameter of 2 AU, and its surface temperature will decline to about 3500 K. Although the surface temperature will be some 2300 K cooler than it is today, the Sun will be so huge that its luminosity will be much greater than it is today. As a full-fledged red giant (see Figure 21-2), our star will shine with the brightness of a hundred Suns. While the inner planets are vaporized, the thick atmospheres of the outer planets will boil away to reveal tiny, rocky cores. Thus in its later years, the aging Sun will destroy the planets that have accompanied it since its birth.

Helium burning begins at the center of a red giant

Helium is the "ash" of hydrogen burning. Thus, when a star first becomes a red giant, its hydrogen-burning shell surrounds a small, compact core of almost pure helium. In a moderately low-mass red giant (like the Sun 5 billion years from now), the dense helium core is about the same size as the Earth, and the star's bloated surface has roughly the same diameter as the Earth's orbit.

At first, no thermonuclear reactions occur in the helium core of a red giant. The hydrogen-burning shell continues to move outward in the star, adding mass to the helium core. The core slowly contracts, thereby forcing the star's central temperature to climb.

Finally, when the central temperature reaches 100 million kelvins, **helium burning** is ignited at the star's center. This new thermonuclear reaction occurs in two steps. First, two helium nuclei combine to form an isotope of beryllium:

$$^4He + {}^4He \rightarrow {}^8Be$$

This particular isotope of beryllium is very unstable and breaks down into two helium nuclei almost as soon as it forms. However, in the star's dense core, a third helium nucleus can strike the ^8Be nucleus before it has a chance to fall apart. This collision creates a stable, common isotope of carbon:

$$^8Be + {}^4He \rightarrow {}^{12}C + \gamma$$

with the release of a gamma-ray photon (γ).

During the pioneering days of nuclear-physics research, helium nuclei were called **alpha particles;** hence, the fusion of three helium nuclei to form a carbon nucleus is still called the **triple alpha process.** Some of the carbon created in this process can fuse with an additional helium nucleus to produce oxygen:

$$^{12}C + {}^4He \rightarrow {}^{16}O + \gamma$$

Thus, both carbon and oxygen make up the "ash" of helium burning.

The second step in the triple alpha process and the process of oxygen formation release energy (γ rays). Thus, for the first time since leaving the main sequence, the aging star again has a central energy source. This source, properly called **core helium burning** because of its central location, establishes thermal equilibrium, thereby preventing any further gravitational contraction of the star's core. A mature red giant burns helium in its core for about 5 to 20 percent as long as the time that it spent burning hydrogen as a main-sequence star. For example, in the distant future, the Sun will consume helium in its core for about 1 billion years.

The way in which helium burning begins at a red giant's center depends on the mass of the star. For high-mass stars (those with masses greater than $2M_\odot$), the ignition of helium burning begins gradually as temperatures in the star's core finally reach 100 million degrees. In low-mass stars (those with masses less than $2M_\odot$), however, helium burning begins explosively and suddenly, an event called the **helium flash.**

The helium flash occurs because of unusual conditions that develop in the core of a low-mass star on its way to becoming a red giant. To appreciate these conditions we must first understand how an ordinary gas behaves, and then we must explore how the densely packed electrons at the star's center alter this behavior.

Usually, when a gas is compressed, it becomes denser and warmer. For convenience, scientists have invented the concept of a **perfect gas** that has a simple relationship between pressure, temperature, and density. Specifically, the pressure exerted by a perfect gas is directly proportional to both the density and temperature of the gas. Many real gases behave like a perfect gas over a wide range of temperatures and densities.

In most circumstances, the gases inside a star act like a perfect gas: if the gas is compressed, it heats up; if the gas expands, it cools down. This behavior serves as a safety valve, ensuring that the star does not explode. For example, if energy production overheats the star's core, then the core expands, thereby cooling the gases and slowing the rate of thermonuclear reactions. Conversely, if too little energy is being created to support the star's overlying layers, the core becomes compressed, and the increased temperatures then speed up the thermonuclear reactions and increase the energy output.

In a low-mass red giant, the core must undergo considerable gravitational compression in order to drive the temperatures high enough to begin helium burning. At the extreme pressures and temperatures deep inside the star, the atoms are completely ionized, and thus the gases of the stellar core consist of dissociated nuclei and electrons. In this highly compressed core, the electrons are so closely crowded together that a law of quantum mechanics called the **Pauli exclusion principle** comes into play. This principle, formulated in 1925 by the Austrian physicist Wolfgang Pauli, explains that two identical particles cannot simultaneously occupy the same "quantum state." A quantum state is a particular set of circumstances (involving location and speed) available to a particle. In the submicroscopic world of atoms and particles, the Pauli exclusion principle is analogous to the larger-scale observation that you can't have two things in the same place at the same time.

Just before the onset of helium burning, the electrons in the core of a low-mass star are so closely crowded together that any further compression would violate the Pauli exclusion principle. Because they cannot be squeezed any closer together, the electrons produce a powerful pressure that resists further core contraction.

This phenomenon, whereby closely packed particles resist compression because of the Pauli exclusion principle, is called **degeneracy.** Astronomers therefore say that the helium-rich core of a low-mass red giant is **degenerate** and is supported by **degenerate-electron pressure**.

When the temperature in the core of a low-mass red giant reaches the high level required for the triple alpha reaction, gamma-ray photons begin to be released by this reaction. At first, these photons interact primarily with the helium nuclei, but not with the degenerate electrons. The helium nuclei therefore become heated, which causes the triple alpha process to proceed more rapidly. The degenerate electrons, which are responsible for the pressure in the star's core, have not been significantly heated, and thus the pressure does not change. Without the "safety valve" of increasing pressure rising temperature causes the helium to burn at an ever increasing rate, producing an event called the **helium flash.** Eventually, however, the temperature becomes so high that electron degeneracy is removed. The electrons then behave like a perfect gas and the star's core expands, thereby terminating the helium flash. These events occur extremely rapidly and the helium flash is over in only a few seconds.

Evolutionary tracks on the H–R diagram reveal the ages of star clusters

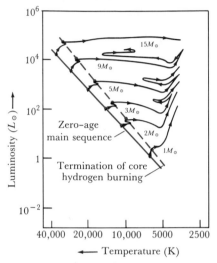

Figure 21-3   *Post–main-sequence evolution*

The evolutionary tracks of six stars are shown on this H–R diagram. In the high-mass stars, core helium burning ignites where the evolutionary tracks make a sharp turn in the red-giant region of the diagram. The evolutionary tracks for low-mass stars ($1M_\odot$ and $2M_\odot$) are shown only up to the points where the helium flash occurs at their centers. (Adapted from I. Iben)

Figure 21-4   *A globular cluster*

A globular cluster is a spherical cluster that typically contains a few hundred thousand stars. This particular cluster, call M13, is located in the constellation of Hercules roughly 25,000 light years from Earth. (U.S. Naval Observatory)

It is very enlightening to follow the post–main-sequence evolution of mature stars by plotting their evolutionary tracks on a Hertzsprung–Russell diagram (see Figure 21-3). The **zero-age main sequence** (or ZAMS) is the location on the H–R diagram where stars first begin core hydrogen burning. In subsequent years, the evolutionary tracks slowly inch away from the ZAMS as the hydrogen-burning core grows in search of fresh fuel. The dashed line on Figure 21-3 shows the locations of the stellar models when all the core hydrogen has been consumed.

After the cessation of core hydrogen burning, the points representing high-mass stars move rapidly from left to right across the H–R diagram. During this transition, the star's core contracts, while its outer layers expand in response to increased energy output from the star's hydrogen-burning shell. Although the star's surface temperature is decreasing, its surface area is increasing in such a way that the star's overall luminosity remains roughly constant.

Just before the ignition of core helium burning, the evolutionary tracks of high-mass stars turn upward in the red-giant region of the H–R diagram. After core helium burning begins, the evolutionary tracks back away from these temporary peak luminosities. The tracks then wander back and forth in the red-giant region as the stars readjust to their new energy sources.

The evolutionary tracks of two low-mass stars also appear in Figure 21-3. However, these two tracks are shown only to the point where the helium flash occurs. Although the helium flash releases a sudden flood of energy, the effects of this outburst take thousands of years to reach the star's surface. By that time, additional internal changes have become more important in determining the star's external appearance.

With the removal of degeneracy immediately after the helium flash, the star's superheated core expands because it is again behaving like a perfect gas. Temperatures around the expanding core fall, thereby cooling the hydrogen-burning shell and reducing its energy output. As the energy output declines, the star's outer layers contract and heat up. Consequently, a post–helium-flash star should be smaller and be hotter at the surface than is a red giant.

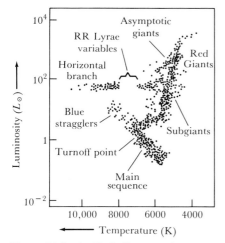

Figure 21-5 An H–R diagram of the globular cluster M3
Each dot of this graph represents a star in the globular cluster M3 whose luminosity and surface temperature have been determined. Note that the upper half of the main sequence is missing. The horizontal branch stars are believed to be low-mass stars that recently experienced the helium flash in their cores and now exhibit core helium burning and shell hydrogen burning. (Adapted from H. L. Johnson and A. R. Sandage)

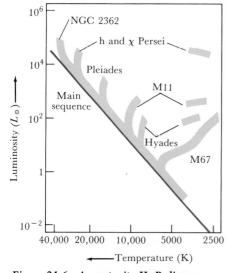

Figure 21-6 A composite H–R diagram
The shaded bands indicate where data from various star clusters fall on H–R diagram. The age of a cluster can be estimated from the location of the turnoff point where the cluster's most massive stars are just leaving the main sequence for the red-giant region.

Examples of these post–helium-flash stars are found in old star clusters, called **globular clusters** because of their spherical shape. A typical globular cluster, like the one shown in Figure 21-4, contains up to 1 million stars in a volume less than 100 parsecs across. Astronomers know that such clusters are old because they contain no high-mass main-sequence stars. If you measure the luminosity and surface temperature of many stars in a globular cluster and plot the data on an H–R diagram as shown in Figure 21-5, you find that the upper half of the main sequence is missing. All of the high-mass main-sequence stars have long ago evolved into red giants, leaving behind only low-mass, slowly evolving stars that still have core hydrogen burning.

The H–R diagram of a globular cluster typically shows a horizontal grouping of stars in the upper-left to upper-center portion of the diagram. As shown in Figure 21-5, these stars form a horizontal row at luminosities in the range of $50L_\odot$ to $100L_\odot$. They are called **horizontal-branch stars** and are believed to be post–helium-flash low-mass stars. In years to come, these stars will move back toward the red-giant region as both core helium burning and shell hydrogen burning continue to devour fuel.

An H–R diagram of a cluster can be used to determine the age of the cluster. In the diagram for a very young cluster (such as those discussed in Chapter 20), the entire main sequence is intact. But as a cluster gets older, stars begin to leave the main sequence. The high-mass, high-luminosity stars are first to become red giants, and thus the main sequence starts to burn down like a candle. As the years pass, the main sequence gets shorter and shorter. The top of the surviving portion of the main sequence is called the **turnoff point.** Stars at the turnoff point are just now exhausting the hydrogen in their cores, and their main-sequence lifetime (recall Table 21-1) is equal to the age of the cluster. For example, in the case of the cluster M3 (see Figure 21-5), $0.8M_\odot$ stars have just left the main sequence, and thus the cluster's age is roughly 15 billion years.

On H–R diagrams of globular clusters such as Figure 21-5, there are typically a few stars on the main sequence just above the turnoff point. They are called **blue stragglers** because they are hotter and thus bluer than the stars at the turnoff point and because they should have left the main sequence long ago. The nature of the blue stragglers is a subject of current research. Some astronomers suspect that these are stars in which hydrogen has somehow been mixed from outer layers into the core, so that core hydrogen burning can continue long after other stars of similar mass have become red giants.

Data for several star clusters are plotted on Figure 21-6. All the young clusters (those with their main sequences still intact) are open clusters in the plane of our galaxy, where star formation is an ongoing process. Stars in these young clusters are said to be **metal-rich** because their spectra contain many prominent spectral lines of heavy elements. This material came from dead stars that long ago exploded, enriching the interstellar gases with the heavy elements formed in their cores. In contrast, most of the oldest clusters are globular clusters. Globular clusters are generally located outside the plane of our galaxy, and their spectra show only weak lines of heavy elements. These ancient stars are therefore said to be **metal-poor.** They were created long ago from interstellar gases that had not yet been substantially enriched with heavy elements.

The young, metal-rich stars are commonly called **population I** stars. The Sun is such a star. In contrast, the old, metal-poor stars are called

population II stars. As we shall see in later chapters, hydrogen and helium were the only two elements to emerge from the birth of the universe. Thus, the most ancient stars should have no spectral lines of any heavy elements whatsoever. Astronomers are searching for these very old, metal-free stars. When found, they probably will be called **population III** stars.

Supergiants and red giants typically show mass loss

Both supergiant and red-giant stars are so enormous that their bloated outer layers constantly leak gases into space. At times, this **mass loss** is quite significant.

Mass loss can be detected spectroscopically. Escaping gases coming toward us exhibit narrow absorption lines that are slightly blueshifted. According to the Doppler effect (recall Box 7-3), this small shift toward shorter wavelengths corresponds to a speed of 10 km/sec. This value is typical of the expansion velocities with which gases leave the tenuous outer layers of red giants. A typical mass-loss rate for a red giant is roughly 10^{-7} solar masses per year.

Betelgeuse in the constellation of Orion is a good example of a red giant experiencing mass loss. Betelgeuse is 470 light years away and has a diameter roughly equal to the diameter of Mars's orbit. Recent spectroscopic observations show that this star is losing mass at the rate of 1.7×10^{-7} solar masses per year and is surrounded by a huge **circumstellar shell** that is expanding with a speed of 10 km/sec. These escaping gases have been detected out to distances of 10,000 AU from the star. Consequently, the expanding circumstellar shell has an overall diameter of $\frac{1}{3}$ light year.

Supergiant stars (that is, stars brighter than 10^5 Suns) are involved in mass loss throughout most of their existence with mass-loss rates comparable to those of red giants. Figure 21-7 shows a supergiant star losing mass. This particular star is a member of a class called **Wolf–Rayet stars,** named after two nineteenth-century astronomers who first drew attention to bright emission lines in their spectra. Wolf–Rayet stars have masses in the range of $30 M_\odot$ to $50 M_\odot$ and lie near the main sequence on the H–R diagram. Thus, they must be fairly young stars. A very large

Figure 21-7   The Wolf–Rayet star HD 56925
This beautiful nebulosity surrounds a supergiant star that is experiencing significant mass loss. The ejected material collides and interacts with the surrounding interstellar gas and dust, thereby producing the cosmic bubble seen here. The nebulosity, called NGC 2359, is located in the constellation of Canis Major. (Anglo-Australian Observatory)

percentage of these rare and beautiful stars have been confirmed to be members of close binary systems. In fact, many Wolf–Rayet stars have nearby companions whose gravity plays an important role in detaching gases from the supergiants' outer atmospheres.

In the next chapter, we will see that dying stars eject vast quantities of material into space. Nevertheless, mass loss from supergiants and red giants accounts for roughly one-fifth of all the matter returned by stars to the interstellar medium.

Many mature stars pulsate

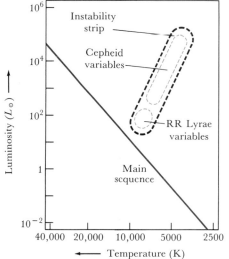

Figure 21-8 The instability strip
The instability strip occupies a region between the main sequence and the red-giant branch on the H–R diagram. A star passing through this region along its evolutionary track becomes unstable and pulsates.

After core helium burning begins, mature stars wander across the middle of the H–R diagram. Figure 21-3 shows the evolutionary tracks of high-mass stars crisscrossing the H–R diagram. Post–helium-flash low-mass stars on the horizontal branch also cross the middle of the H–R diagram as they return to the red-giant region.

During these transitions across the H–R diagram, a star can become unstable and pulsate. In fact, there is a region on the H–R diagram between the main sequence and the red-giant branch that is called the **instability strip** (see Figure 21-8). When a star passes through this region as it moves along its evolutionary track, the star pulsates. As it pulsates, its brightness varies periodically.

The first pulsating variable was discovered in 1784 by John Goodricke, a 19-year-old English amateur astronomer. This star, δ Cephei, regularly varies in apparent magnitude from 4.4 to 3.7, with a period of 5.4 days. It is the prototype of a very important class of pulsating stars called **Cepheid variables,** or simply Cepheids.

A Cepheid variable is recognized by the characteristic way in which its light output varies: rapid brightening followed by gradual dimming. This behavior is most easily displayed in the form of a **light curve,** which is a graph of the star's brightness plotted against time (recall Figure 18-18). The light curve of δ Cephei is shown in Figure 21-9a.

A Cepheid variable brightens and fades because of a cyclic expansion and contraction of the whole star. This behavior is deduced from spectroscopic observations. In 1894, the Russian astronomer A. A. Belopolsky noticed that spectral lines in the spectrum of δ Cephei shift back and forth with the same 5.4-day period as that of the magnitude variations.

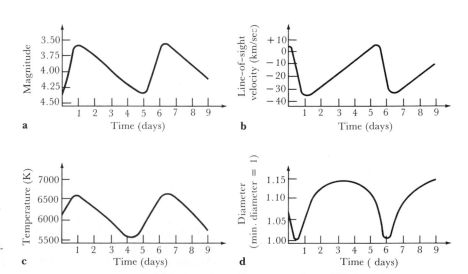

Figure 21-9 The details of δ Cephei
These four graphs display details of the pulsations of δ Cephei: **(a)** *the star's light curve;* **(b)** *the velocity curve;* **(c)** *periodic variations in the star's surface temperature; and* **(d)** *periodic variations in the star's diameter.*

Box 21-2   The period–density relation for Cepheids

We can use Kepler's third law to derive an approximate relationship between the period of a Cepheid variable and its average density. Consider the star when it is at maximum size. When the gas pressure can no longer hold up the star's outer layers, they fall inward under the influence of the star's gravity. At least temporarily, the atoms in the star's outer atmosphere are freely falling, which means that the atoms are moving along orbits that obey Kepler's laws. In Box 4-3, we saw that Kepler's third law can be written as

$$P^2 = \left(\frac{4\pi^2}{GM}\right)a^3$$

where G is the universal constant of gravitation. In the case of the atoms at the surface of a Cepheid variable, P is the period of pulsation, M is the star's mass, and a (the semimajor axis of the atom's orbit) is the star's radius, R. Thus, the square of period is proportional to the cube of the star's radius divided by its mass:

$$P^2 \propto \frac{R^3}{M}$$

By definition, the average density ρ of a star equals its mass M divided by its volume $4\pi R^3/3$, so that

$$M = \frac{4}{3}\pi R^3 \rho$$

Substituting this expression into the proportionality for P^2, we find that

$$P^2 \propto \frac{1}{\rho}$$

or

$$P\sqrt{\rho} = \text{constant}$$

Of course, the particles in the Cepheid's outer layers move along Keplerian orbits for only a fraction of the star's period. Thus, the actual period of an individual Cepheid is different from that calculated by this procedure. Nevertheless, to a reasonable approximation, the *ratio* of the pulsation periods of *two* Cepheids is inversely proportional to the ratio of the square roots of their average densities because the numbers contained in the constant cancel out. In other words, if P_1 and ρ_1 are the period and average density of one Cepheid, and P_2 and ρ_2 are the period and average density of another Cepheid, then

$$\frac{P_1}{P_2} = \sqrt{\frac{\rho_2}{\rho_1}}$$

From the Doppler effect (recall Box 7-3), we can translate these wavelength shifts into speeds and draw a **velocity curve** (see Figure 21-9b). Negative speeds mean that the star's surface is expanding toward us; positive speeds mean that the star's surface is receding from us. Note that rapid expansion of the star's surface occurs at the same time as the rapid increase in the star's brightness. Similarly, a gradual contraction of the star accompanies the gradual decline in the star's brightness.

When a Cepheid variable pulsates, the star's surface oscillates up and down like a spring. During these cyclical expansions and contractions, the star's gases alternately heat up and cool down. The resulting temperature changes of the star's surface are displayed in Figure 21-9c. Finally, the periodic changes in the star's diameter are shown in Figure 21-9d.

Just as a bouncing ball eventually comes to rest, a pulsating star would soon stop pulsating without some sort of mechanism to keep the oscillations going. In 1941, the British astronomer Sir Arthur Eddington explained that a Cepheid variable feeds energy into pulsation by a valvelike action involving periodic ionization and deionization of hydrogen and helium in its outer layers. When the star is at its largest diameter, its surface temperature is at a minimum, and ions combine with electrons to produce neutral atoms. These electrically neutral gases freely radiate energy into space, causing gas pressure to decline and the star's outer layers to start to fall inward. As the gases of the shrinking star become more compressed, they absorb energy from the star's interior and become reionized. This absorption of energy heats the gases and raises the pressure, thereby pushing the star's surface back out to start the cycle again. As explained in Box 21-2, the period of pulsation is related to the star's average density.

Cepheid variables are very important to astronomers because there is a direct relationship between a Cepheid's period and its average luminosity. Dim Cepheid variables pulsate rapidly; they have periods of 1 to 2 days and average brightness of a few hundred Suns. The most luminous Cepheids are the slowest variables; they have periods of 100 days and average brightnesses equal to 10,000 Suns. This connection between period and brightness is called the **period–luminosity relation.** As we shall see in Chapter 26, this relationship played an important role in determining the overall size and structure of the universe.

Details of a Cepheid's pulsation depend on the abundance of heavy elements in its atmosphere. The average luminosity of metal-rich Cepheids is roughly four times greater than the average luminosity of metal-poor Cepheids of the same period. Thus there are two classes: **Type I Cepheids,** which are the brighter, population I, metal-rich stars; and **Type II Cepheids,** which are the dimmer, population II, metal-poor stars. The period–luminosity relation for both types of variables is shown in Figure 21-10.

The evolutionary tracks of mature, high-mass stars pass back and forth through the upper end of the instability strip on the H–R diagram. These stars become Cepheids when the ionization layers occur at just the right depth to drive the pulsations. For stars on the high-temperature side of the instability strip, the ionization zone is too close to the surface and contains an insignificant fraction of the star's mass. For stars on the cool side of the instability strip, convection becomes important in the star's outer layers and prevents the storage of energy needed to drive the pulsations. Thus Cepheids exist only in a narrow range of effective temperatures on the H–R diagram.

Low-mass, post–helium-flash stars pass through the lower end of the instability strip as they move in the horizontal branch along their evolutionary tracks. These stars become **RR Lyrae variables,** named after the prototype in the constellation of Lyra. RR Lyrae variables all have periods shorter than one day, and they all have roughly the same average brightness as stars on the horizontal branch. In fact, as shown in Figure 21-5, the RR Lyrae region of the instability strip is a segment of the horizontal branch.

Stellar pulsations can in rare cases be quite substantial. In some cases, the expansion velocity exceeds the star's escape velocity, and the star's

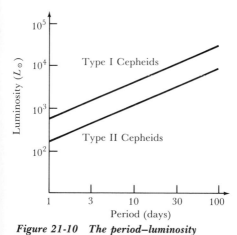

Figure 21-10 The period–luminosity relation

The period of a Cepheid variable is directly related to its average luminosity. Metal-rich (Type I) Cepheids are brighter than the metal-poor (Type II) Cepheids.

outer layers are ejected completely. As we shall see in the next chapter, significant mass ejection accompanies the death of stars in a sometimes violent process that renews and enriches the interstellar medium for future generations of stars.

Summary

. The more massive a star is, the shorter is its main-sequence lifetime; the Sun has been a main-sequence star for about 5 billion years and should remain so for about another 5 billion years.

. Core hydrogen burning ceases when hydrogen is exhausted in the core of a main-sequence star, leaving a core of nearly pure helium surrounded by a shell where hydrogen burning works its way outward in the star; the core shrinks and becomes hotter, while the star expands and becomes a red giant.

> When the central temperature of a red giant reaches about 100 million kelvins, the thermonuclear process of helium burning begins there; this process (also called the triple alpha process) converts helium to carbon and oxygen.

> In a more-massive red giant, helium burning begins gradually; in a less-massive red giant, it begins suddenly in a process called a helium flash.

. As a star moves through the red-giant phase of its evolution, its evolutionary track moves rapidly to the right from the zero-age main sequence (ZAMS) on the H–R diagram at fairly constant luminosity.

. In a high-mass red giant, luminosity increases sharply just before core helium burning begins and then decreases sharply again just after core helium burning begins; the star's evolutionary track then wanders back and forth across the red-giant region of the H–R diagram. After the helium flash in a low-mass red giant, the star moves to the region called the horizontal branch on the H–R diagram.

. The age of a stellar cluster can be estimated by plotting its stars on an H–R diagram; the upper portion of the main sequence will be missing because more-massive main-sequence stars have become red giants; the age of the cluster is equal to the age of main-sequence stars at the turnoff point (the remaining upper end of the main-sequence).

. Relatively young population I stars are metal-rich; ancient population II stars are metal-poor.

. Supergiants and red giants undergo extensive mass loss, sometimes producing circumstellar shells of ejected material around the stars.

. When a star's evolutionary track carries it through a region called the instability strip in the H–R diagram, the star becomes unstable and begins to pulsate.

> Pulsating variables are only one kind of variable star.

> Cepheid variables are high-mass pulsating variables having a regular relationship between period of pulsation (or period of variation in brightness) and luminosity.

> RR Lyrae variables are low-mass pulsating variables with short periods.

Review questions

1 Why do you suppose that the vast majority of the stars we see in the sky are main sequence stars?

2 On what grounds are astronomers able to say that the Sun has about 5 billion years remaining in its main sequence stage?

3 What does it mean when an astronomer says that a star "moves" from one place to another on an H–R diagram?

*4 Estimate the main-sequence lifetime of **(a)** 100-solar-mass star, and **(b)** a 0.1-solar-mass star.

*5 Assuming that the Sun's luminosity remains roughly constant during its 10 billion years on the main sequence, what fraction of the Sun's hydrogen is converted into helium?

*6 When the Sun becomes a red giant, its luminosity will be 100 times greater than it is today. Assuming that this luminosity is due *only* to burning the Sun's remaining hydrogen, calculate how long our star will be a red giant.

7 Why do astronomers attribute the observed Doppler shifts of a Cepheid variable to pulsation, rather than to some other cause such as orbital motion?

8 Suppose that an oxygen nucleus were fused with a helium nucleus. What element would be formed? Look up the relative abundance of this element and comment on whether such a process is likely.

Advanced questions

9 What observations would you make of a star to determine whether its primary source of energy were hydrogen or helium burning?

10 Some astronomers have suggested that blue stragglers are members of close binary systems in which the companion star has recently dumped matter onto a previously low-mass star. How would you test this idea?

Discussion questions

11 The half-life of the ^{8}Be nucleus is 2.6×10^{-16} seconds, which is the average time that elapses before this unstable nucleus breaks into two alpha particles. How would the universe be different if the ^{8}Be half-life were zero? How would the universe be different if the ^{8}Be nucleus were stable?

12 What observational consequences would we find in H–R diagrams for star clusters if the universe had a finite age? Could we use these consequences to establish some constraints on the possible age of the universe? Explain.

For further reading

Johnson, B. "Red Giant Stars." *Astronomy,* Dec. 1976, p. 26.
Kaufmann, W. *Stars and Nebulas.* W. H. Freeman, 1979.
Kippenhahn, R. *100 Billion Suns: The Birth, Life and Death of Stars.* Basic Books, 1983.
Percy, J. "Observing Variable Stars for Fun and Profit." *Mercury,* May/June 1979, p. 45.
Weymann, R. "Stellar Winds. *Scientific American,* Aug. 1978.
Wyckoff, S. "Red Giants: The Inside Scoop." *Mercury,* Jan./Feb. 1979, p. 7.

22 Stellar death

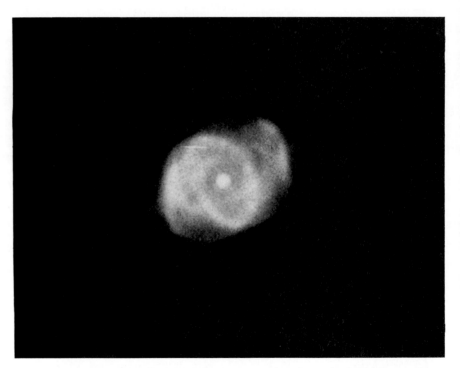

A planetary nebula
Dying stars often eject their outer layers. A low-mass star can lose half its mass in a comparatively gentle process that produces a planetary nebula. The exposed stellar core typically has a surface temperature of about 100,000 K and is roughly one-tenth the size of the Sun. Ultraviolet radiation from the hot stellar core causes the surrounding gases to glow. The greenish color of this planetary nebula, NGC 6543, is due to doubly ionized oxygen. The central star in this nebular has exhausted all its nuclear fuel and is gravitationally contracting to become a white dwarf. (Lick Observatory)

After its old age as a red giant, a star approaches the end of its life. In this chapter, we learn that stars eject significant amounts of matter into space as they die. We find that low-mass stars eject their outer layers relatively gently, producing planetary nebulae, whereas high-mass stars explode violently as supernovae. We discuss the many different thermonuclear reactions that occur during the final stages of stellar life, producing a wide variety of heavy elements. Finally, we talk briefly about some exotic techniques now being developed in order to learn more about the detailed stages of stellar death.

From infancy through adulthood, a star leads a fairly placid life with hydrogen burning in its core. As old age approaches, however, the star takes on a schizophrenic character with a compressed core and a bloated atmosphere, and it becomes a red giant. Erratic and fitful behavior becomes even more pronounced as the star devours its remaining nuclear fuels and begins to die.

Low-mass stars die by gently ejecting their outer layers, thereby creating planetary nebulae

Carbon and oxygen are the "ashes" of helium burning. After the helium flash in a low-mass red giant, substantial amounts of these two elements begin to accumulate at the star's center as a result of core helium burning. Eventually, all the helium at the center of a low-mass star is used up, and core helium burning ceases. By this time, thermonuclear reactions have also ceased in the hydrogen-burning shell. The star's core therefore begins to contract. This compression heats the helium-rich gases surrounding the carbon–oxygen core, and helium burning soon begins in a thin shell around the core. This process is called **shell helium burning.** The star's internal structure now consists of a helium-burning shell inside the former hydrogen-burning shell, all within a volume roughly the size of the Earth (see Figure 22-1).

When shell hydrogen burning first began, the outpouring of energy caused the star to expand and become a red giant. Describing the evolutionary track of the star on the H–R diagram, astronomers say that the star ascended the red-giant branch for the first time. Then came the helium flash, and the star shifted over to the horizontal branch. But now, with shell helium burning, a renewed outpouring of energy causes the star to expand again. The star ascends the red-giant branch for a second and final time (along the asymptotic-giant branch shown in Figure 21-5) to become a **red supergiant.** Such stars typically have diameters as big as the orbit of Mars and shine with the brightness of 10,000 Suns. A star experiencing this furious rate of energy loss cannot live much longer.

The star's impending death is signaled by instabilities that develop in its helium-burning shell. As with the helium flash discussed in the previous chapter, there is again a thermal runaway, but the details are quite different. The helium flash occurred because the star's core was degenerate. The helium-burning shell is not compressed to high-enough density to be degenerate. Instead, this thermal runaway occurs because the shell is thin. A slight increase in the energy output from a thin shell does little to relieve the pressure from the star's overlaying layers. Instead, the temperature in the shell goes up a little. This further increases the energy output, which continues to drive up the temperature. This vicious circle produces a thermal runaway that ends only after the helium-burning shell becomes thick and relieves the pressure of the star's outer layers.

These **helium-shell flashes** were investigated by M. Schwarzschild and R. Härm at Princeton University in the 1960s. The results of their

Figure 22-1 The structure of an old low-mass star
Near the end of its life, a low-mass star becomes a red supergiant, whose diameter is almost as large as the diameter of the orbit of Mars. The star's dormant hydrogen-burning shell and active helium-burning shell are contained within a volume roughly the size of the Earth.

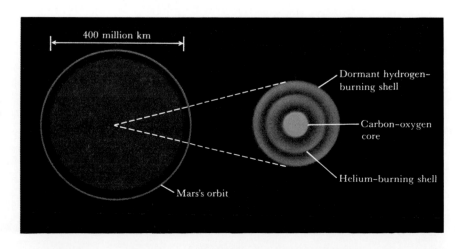

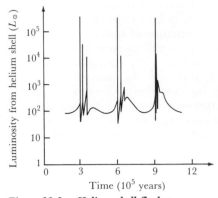

Figure 22-2   Helium-shell flashes
This graph shows how the energy output of the helium-burning shell varies in a red supergiant. Brief periods of runaway helium burning produce thermal pulses that can eventually cause the star to eject its outer layers completely. (Adapted from M. Schwarzschild and R. Härm)

calculations are displayed in Figure 22-2. During the flashes, the energy output from the helium-burning shell jumps from 100 Suns to roughly 100,000 Suns in a rapid series of brief bursts called **thermal pulses.** As shown in Figure 22-2, these bursts are separated by relatively quiet intervals lasting about 300,000 years. During this period of thermal pulses, thermonuclear reactions begin again in the hydrogen-burning shell.

During one of these periods of thermal oscillation, the dying star's outer layers can separate completely from the carbon–oxygen core. As the ejected material expands into space, dust grains condense out of the cooling gases. Radiation pressure from the star's hot, burned-out core acting on the specks of dust continues to propel them outward, and the star sheds its outer layers altogether. A star can lose more than one-half of its mass in this fashion.

As a dying star ejects its outer layers, the star's hot core is exposed and is emitting ultraviolet radiation intense enough to ionize the expanding shell of ejected gases. The gases therefore glow, producing a so-called **planetary nebula,** or "planetary." Planetary nebulae have nothing to do with planets; this unfortunate term was introduced in the eighteenth century when these glowing objects, viewed through small telescopes, were thought to look like distant planets.

Many planetary nebulae, such as the one shown in Figure 22-3, have a distinctly spherical appearance due to the symmetrical way in which the gases were ejected. In other cases, as shown in Figure 22-4, the rate

Figure 22-3   The planetary nebula NGC 7293
This beautiful object, often called the Helix nebula, covers an area of the sky equal to half of the full moon. The star that ejected these gases is seen at the center of the glowing shell. The greenish color is due to ionized oxygen; the pink and red are caused by ionized nitrogen and hydrogen. The small radial blobs in the red shell (thought to be about 150 AU across) give the object its alternate name, the Sunflower Nebula. It is in the constellation of Aquarius 400 light years from Earth. (Anglo-Australian Observatory)

Figure 22-4 [right]   The planetary nebula NGC 6302
At the end of their lives, low-mass stars shed much of their mass, leaving behind a tiny, hot, burned-out stellar core. Although most cases involve a fairly symmetrical ejection of gases, there are many examples in which the gas has expanded unevenly. An irregular planetary nebula such as NGC 6302 in Scorpius (shown here) is the result. Spectroscopic observations of NGC 6302 indicate that the gases are moving toward us at 400 km/sec, indicating a particularly violent initial ejection. (Anglo-Australian Observatory)

of expansion is not the same in all directions, and the resulting nebula takes on an hourglass or dumbbell appearance.

Planetary nebulae are quite common. Several well-known examples are listed in Box 22-1. Astronomers estimate that there are 20,000 to 50,000 planetaries in our galaxy alone. Spectroscopic observations of these nebulae show bright emission lines of ionized hydrogen, oxygen, and nitrogen. From the Doppler shifts of these lines, astronomers conclude that the expanding shell of gas is moving outward from the dying star with speeds of 10 to 30 km/sec. A typical planetary nebula has a

---

Box 22-1   Famous planetary nebulae

Many planetary nebulae are scattered across the sky. Some of the most famous planetaries are listed in the table with information about their distances and diameters.

Name of nebula	Distance from Earth (light years)	Apparent magnitude of central star	Diameter (arc min)
Dumbbell (M27)	3,500	13	8
Ring (M57)	4,000	15	1
"Little Dumbbell" (M76)	15,000	17	1
Owl (M97)	12,000	13	3
Saturn (NGC 7009)	3,000	12	1
Helix (NGC 7293)	400	13	15
Clown (NGC 2392)	10,000	10	$\frac{1}{2}$

Planetary nebulae are favorite observational objects of amateur astronomers. In many cases (such as the Ring Nebula shown in the photograph), a small telescope reveals the nebulosity but not the central star. The central star, destined to become a white dwarf, supplies the ultraviolet radiation that causes the surrounding gases to glow.

diameter of roughly 1 light year, and thus it must have begun expanding about 10,000 years ago.

By astronomical standards, a planetary nebula is a very short-lived object. After about 50,000 years, the nebula has spread over distances so far from the cooling central star that the nebulosity simply fades from view. The gases then mingle and mix with the surrounding interstellar medium. Astronomers estimate that all the planetary nebulae in the galaxy return a total of $5M_\odot$ to the interstellar medium each year. This is about 15 percent of all the matter expelled by all sorts of stars each year. Because of this significant contribution, planetary nebulae must play an important role in the evolution of the galaxy as a whole.

The burned-out core of a low-mass star cools and contracts to become a white dwarf

Stars less massive than about $3M_\odot$ never develop the necessary central pressures or temperatures to permit thermonuclear reactions using carbon or oxygen as fuel. Instead, mass ejection strips most outer layers away from the carbon–oxygen core, which simply cools off.

The evolutionary tracks of three burned-out stellar cores are shown in Figure 22-5. The initial red supergiants had masses between $0.8M_\odot$ and $3.0M_\odot$. During final spasms, these dying stars eject between 25 and 60 percent of their matter. During the ejection phase, the outward appearance of these stars changes very rapidly, and they race along their evolutionary tracks across the H–R diagram, sometimes executing loops that correspond to stellar pulses. Finally, as the ejected nebulae fade and the stellar cores cool, their evolutionary tracks take a sharp turn downward toward the white-dwarf region of the H–R diagram.

There is no possibility of igniting any additional nuclear fuels inside one of these dead stars, so the crushing weight of gases pressing inward

Figure 22-5 Evolution from red supergiants to white dwarfs
The evolutionary tracks of three low-mass red supergiants are shown as they eject planetary nebulae. The table gives the extent of mass loss in each of the three cases. The dots on this graph represent the central stars of planetary nebulae whose surface temperatures and luminosities have been determined. The crosses are white dwarfs for which similar data exist. (Adapted from B. Paczynski)

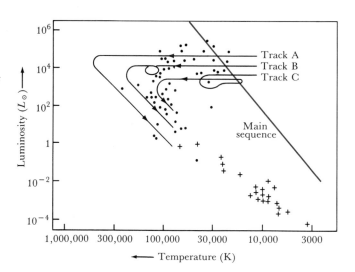

Evolutionary track on Figure 22-5	Supergiant mass $(M_\odot)$	Mass of ejected nebula $(M_\odot)$	White dwarf mass $(M_\odot)$
Track A	3.0	1.8	1.2
Track B	1.5	0.7	0.8
Track C	0.8	0.2	0.6

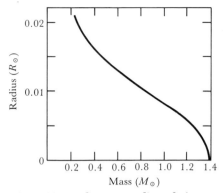

Figure 22-6 The mass–radius relation for white dwarfs
The more massive a white dwarf is, the smaller it is. This unusual relationship between mass and radius is a result of the degenerate electron pressure that supports the star. The maximum mass of a white dwarf, called the Chandrasekhar limit, is $1.4M_\odot$ Incidentally, $0.01R_\odot = 6960$ km $= 1.09R_\oplus$, where $R_\oplus$ is the radius of the Earth.

from all sides severely compresses the stellar corpse. Density inside the dead star skyrockets until the electrons are so closely packed that they become degenerate throughout most of the star. Degenerate-electron pressure then produces the necessary conditions to support the star, and the gravitational contraction is halted. This happens when the star has been squeezed down to a sphere roughly the same size as the Earth. Such a star is called a **white dwarf.**

The density of matter in one of these Earth-sized stellar corpses is typically 10^6 g/cm^3. In other words, a teaspoonful of white-dwarf matter brought to Earth would weigh as much as a truck. In addition, as we saw in Chapter 21, degenerate gases are governed by unusual relationships between pressure, density, and temperature. Consequently, these stars have some unusual properties. For example, the radius of a white dwarf is inversely proportional to the cube root of its mass.

$$R \propto \frac{1}{M^{1/3}}$$

Thus, the more massive a white dwarf is, the smaller it is.

Figure 22-6 displays the **mass–radius relation** for white dwarfs. Note that, the more degenerate matter you pile onto a white dwarf, the smaller it becomes. Also note that there is an upper limit to the mass that a white dwarf can have. This maximum mass is called the **Chandrasekhar limit** (named after Subrahmanya Chandrasekhar at the University of Chicago who pioneered theoretical studies of white dwarfs) and is equal to $1.4M_\odot$. This is the maximum mass that can be supported by degenerate electron pressure, and thus all white dwarfs must have masses less than $1.4M_\odot$.

Several hundred white dwarfs are scattered across the sky. All are too faint to be seen with the naked eye. One of the first white dwarfs to be discovered is a companion to the bright star Sirius. The binary nature of Sirius was first deduced in 1844 by the German astronomer Friedrich Bessel, who noticed that the star was moving back and forth slightly, as if orbited by an unseen object. This companion, called Sirius B, was first glimpsed in 1863 and is shown in Figure 22-7. Recent satellite observations at ultraviolet wavelengths (where white dwarfs emit most of their light) demonstrate that the surface temperature of Sirius B is about 30,000 K.

The material inside a white dwarf consists of highly ionized atoms floating in a sea of degenerate electrons. As the dead star cools off, the particles slow down, and electric forces between the ions begin to dominate over random thermal motions. The ions no longer move freely, and (according to calculations by Hugh Van Horn and Malcolm Savedoff at the University of Rochester) the ions then arrange themselves in orderly rows to form a crystal lattice. The situation is roughly analogous to a cooling solution in which crystals form as the ions in the solution arrange themselves in orderly patterns, as dictated by the electric forces between ions. From this time on, you could say that the star is "solid." Furthermore, the degenerate electrons move freely around this crystalline lattice, just as they move freely through an electrically conducting metal here on Earth. Thus, the material inside an old white dwarf has many properties similar to copper or silver.

As a white dwarf cools off, both its luminosity and its surface temperature decline. Its energy now comes only from cooling at constant size. Consequently, the evolutionary tracks of aging white dwarfs point toward

Figure 22-7 Sirius and its white-dwarf companion
Sirius, the brightest-appearing star in the sky, is actually a double star. The secondary star is a white dwarf, seen in this photograph at the "five o'clock" position, buried in the glare of Sirius. The spikes and rays around Sirius are created by optical effects within the telescope. (Courtesy of R. B. Minton)

the lower-right corner of the H–R diagram (see Figure 22-8). As billions of years pass, white dwarfs get dimmer and dimmer as their surface temperatures drop toward absolute zero. This will be the final fate of our Sun—a cold, dark, dense sphere of degenerate gases rich in oxygen and carbon, about the size of the Earth.

High-mass stars die violently by blowing themselves apart in supernova explosions

High-mass stars end their lives very differently than low-mass stars do. First, a high-mass star is capable of igniting a host of additional thermonuclear reactions in its core. After core helium burning ceases, gravitational compression drives the star's central temperature up to 600 million kelvins, and **carbon burning** begins. This thermonuclear process produces neon, magnesium, oxygen, and helium:

$$^{12}C + ^{12}C \rightarrow ^{20}Ne + ^{4}He$$
$$^{12}C + ^{12}C \rightarrow ^{24}Mg + \gamma$$
$$^{12}C + ^{12}C \rightarrow ^{16}O + 2\ ^{4}He$$

The internal temperature needed for carbon burning develops in stars whose main-sequence mass was at least $4M_\odot$ (before mass ejection began).

The more massive a star is, the greater the temperatures that can be achieved in its core by gravitational compression. High temperatures are required for nuclear reactions involving heavy nuclei because the large electric charges of these nuclei exert strong forces tending to keep the nuclei apart. Only at the very great speeds associated with very high temperatures are the nuclei traveling so fast that they can penetrate each other's repulsive electric fields and fuse together.

If the star had a main-sequence mass of at least $9M_\odot$ (before mass ejection), its central temperature will rise to 1 billion kelvins, at which temperature the process of **neon burning** begins:

$$^{20}Ne + \gamma \rightarrow ^{16}O + ^{4}He$$
$$^{20}Ne + ^{4}He \rightarrow ^{24}Mg + \gamma$$

This process uses up the neon accumulated from carbon burning, further increasing the concentrations of oxygen and magnesium in the star's core.

If the central temperature of the star reaches about 1.5 billion kelvins, **oxygen burning** begins. The principal product of oxygen burning is sulfur:

$$^{16}O + ^{16}O \rightarrow ^{32}S + \gamma$$

As the star consumes increasingly heavier nuclei, however, the thermonuclear reactions produce a wider variety of products. Oxygen burning also produces two isotopes of silicon, another isotope of sulfur, phosphorus, and more magnesium:

$$^{16}O + ^{16}O \rightarrow \begin{cases} ^{28}Si + ^{4}He + \gamma \\ ^{31}P + ^{1}H + \gamma \\ ^{31}S + n + \gamma \\ ^{24}Mg + 2\ ^{4}He \\ ^{30}Si + 2\ ^{1}H + \gamma \end{cases}$$

This is an example of the way that thermonuclear reactions in the cores of massive stars create many different elements.

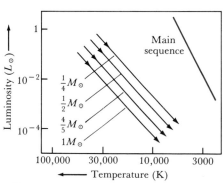

Figure 22-8 White-dwarf "cooling curves"
The evolutionary tracks of four white dwarfs of different masses are shown. As these dead stars lose heat at constant size, they become dimmer and cooler, moving toward the lower right on the H–R diagram.

Note that one of the oxygen-burning reactions produces neutrons (n). Many other reactions in the star's core also produce neutrons.

It is important to remember that a neutron is very much like a proton except that it does not have an electric charge. Therefore, a neutron is not affected by the electric fields that surround nuclei. Consequently, neutrons can easily collide and combine with nuclei, thereby creating new isotopes. As described in Box 22-2, this absorption of neutrons can happen in a rapid **r process** (producing **r-process isotopes**) or in a slow **s process** (producing **s-process isotopes**). In this way, **neutron capture** creates many elements and isotopes that are not produced directly in fuel-burning reactions.

Box 22-2   The s and r processes

A nucleus is composed of protons and neutrons. Generally, there are a few more neutrons in a nucleus than protons. This happens because neutrons have no electric charge, and thus they can be packed together more easily than protons. In fact, neutrons act as buffers between the protons with their mutually repulsive electric fields. If a nucleus has either too many protons or too many neutrons, it is *unstable* and will eject surplus particles until a stable balance of neutrons and protons has been restored.

There are many reactions in stellar cores that produce free neutrons. These neutrons can be absorbed by nuclei, thereby transmuting one isotope into another and creating new elements. For example, successive neutron captures by ^{110}Cd can produce four stable isotopes of cadmium:

$$^{110}\text{Cd} + {}^{1}\text{n} \rightarrow {}^{111}\text{Cd}$$
$$^{111}\text{Cd} + {}^{1}\text{n} \rightarrow {}^{112}\text{Cd}$$
$$^{112}\text{Cd} + {}^{1}\text{n} \rightarrow {}^{113}\text{Cd}$$
$$^{113}\text{Cd} + {}^{1}\text{n} \rightarrow {}^{114}\text{Cd}$$

The specific isotopes produced by a nucleus absorbing neutrons depend on how many neutrons are bombarding the nucleus. If the rate of neutron bombardment is low, then unstable nuclei have a chance to decay radioactively between neutron absorptions. For example, absorption of a neutron by ^{114}Cd, forms the unstable nucleus ^{115}Cd, which decays into an isotope of indium:

$$^{114}\text{Cd} + {}^{1}\text{n} \rightarrow {}^{115}\text{Cd}$$
$$^{115}\text{Cd} \rightarrow {}^{115}\text{In} + e^{-} + \bar{\nu}$$

Then, the ^{115}In nucleus can absorb another neutron creating an unstable isotope (^{116}In), which decays into a stable isotope of tin (^{116}Sn):

$$^{115}\text{In} + {}^{1}\text{n} \rightarrow {}^{116}\text{In}$$
$$^{116}\text{In} \rightarrow {}^{116}\text{Sn} + e^{-} + \bar{\nu}$$

This production of ^{115}In and ^{116}Sn occurs only if the rate of neutron absorption is low. This *slow* reaction is called an **s process,** and these two isotopes are called **s-process isotopes.**

Occasionally, a flood of neutrons is released in a star's core, so that a very rapid absorption of neutrons can occur. In fact, a nucleus can soak up a large number of neutrons without having the chance to decay radioactively between absorptions. Such a *rapid* reaction is called an **r process,** and it produces **r-process isotopes** that are often different from those created by the corresponding s process.

(continued)

(Box 22-2, continued)

For example, again consider the stable isotope of cadmium, ^{114}Cd. A deluge of neutrons could produce a highly unstable isotope,

$$^{114}\text{Cd} + 8\ ^1\text{n} \rightarrow\ ^{122}\text{Cd}$$

which decays into an unstable isotope of indium:

$$^{122}\text{Cd} \rightarrow\ ^{122}\text{In} + e^- + \bar{\nu}$$

The indium then decays into a stable isotope of tin:

$$^{122}\text{In} \rightarrow\ ^{122}\text{Sn} + e^- + \bar{\nu}$$

This particular isotope of tin (^{122}Sn) can be produced only by the r process, just as ^{116}Sn can be produced only by the s process. Both isotopes are stable and are found on Earth. By measuring the relative abundances of various isotopes in the world around us, astronomers can deduce conditions that existed inside the stars that manufactured the elements of which our planet is composed.

When a nuclear fuel is exhausted in the core of a massive star, gravitational contraction to ever higher densities drives up the star's central temperatures, thereby igniting the "ash" of the previous burning stage and/or igniting the outlying shell of unburned fuel. Each successive thermonuclear reaction occurs with increasing rapidity. For example, detailed calculations for a $25M_\odot$ star by Stanford E. Woosley and Thomas A. Weaver at Lawrence Livermore Laboratory demonstrate that carbon burning occurs for 600 years, neon burning occurs for 1 year, and oxygen burning occurs for only 6 months.

After $\frac{1}{2}$ year of core oxygen burning in a $25M_\odot$ star, a final gravitational compression forces the central temperature up toward 3 billion kelvins, and **silicon burning** begins. This thermonuclear process proceeds so furiously that the entire core supply of silicon in a $25M_\odot$ star is used up in 1 day.

Silicon burning involves many hundreds of nuclear reactions. The major final product, or "ash," of this process is a stable isotope of iron, ^{56}Fe. Because ^{56}Fe does not act as a fuel for any thermonuclear reactions, this is where the sequence of burning stages ends. In order for an element to be a thermonuclear fuel, energy must be given off when its nuclei collide and fuse together. This energy comes from packing the neutrons and protons together more tightly in the ash nuclei than in the fuel nuclei. The 56 protons and neutrons inside the iron nuclei are so tightly bound together that no further energy can be extracted by fusing still more nuclei with iron.

The build-up of an inert, iron-rich core signals the impending violent death of a massive star. Surrounding this iron core, successive layers of shell burning consume the star's remaining reserves of fuel (see Figure 22-9). The entire energy-producing region of the star is contained in a volume as big as the Earth, whereas the star's enormously bloated atmosphere is nearly as big as the orbit of Jupiter.

Because iron does not "burn," the electrons in the core now must support the star's outer layers by the brute strength of degeneracy pressure alone. Soon, however, continued deposition of fresh iron from the silicon-burning shell causes the core's mass to exceed the Chandrasekhar limit.

Electron degeneracy suddenly becomes unable to support the star's enormous weight, and the star's core begins to collapse.

Any star with mass greater than $10M_\odot$ is capable of developing an iron core that at some stage will exceed the Chandrasekhar limit, triggering a very rapid series of cataclysms that tear the star apart in a few seconds. For purposes of illustration, we shall focus our attention on the death of a $25M_\odot$ star.

In a $25M_\odot$ star, degenerate-electron pressure fails when density inside the iron core reaches 1 billion grams per cubic centimeter. The core immediately begins to collapse. Central temperatures promptly soar to almost inconceivable heights. In roughly a tenth of a second, the temperature exceeds 5 billion kelvins. Photons associated with this intense heat have so much energy that they begin to break the iron nuclei into alpha particles, a process called **photodisintegration**.

Within another tenth of a second, as densities continue to climb, the electrons are forced to combine with protons to produce neutrons in a process called **neutronization**, which also releases a flood of neutrinos (ν):

$$e^- + p \rightarrow n + \nu$$

At about 0.25 seconds after the collapse begins, the density in the core reaches 4×10^{14} g/cm^2. This is **nuclear density,** the density with which neutrons and protons are packed together inside nuclei.

Matter at nuclear density is virtually incompressible. Thus, when the neutron-rich material of the core reaches nuclear density, it suddenly becomes very stiff, and the core collapse comes to an equally sudden halt.

At this critical stage, the star's unsupported inner regions are plunging inward at speeds of 10 to 15 percent of the speed of light. As this material crashes onto the now-rigid core, enormous temperatures and pressures develop, causing the falling material to bounce. In a fraction of a second, a wave of matter begins to move back out toward the star's

**Figure 22-9   The structure of
an old high-mass star**
Near the end of its life, a high-mass star becomes a red supergiant almost as big as the orbit of Jupiter. The star's energy comes from six concentric burning shells, all contained within a volume roughly the same size as the Earth.

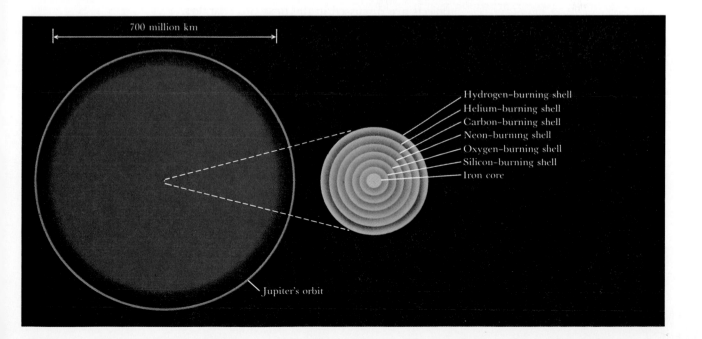

700 million km

Hydrogen-burning shell
Helium-burning shell
Carbon-burning shell
Neon-burning shell
Oxygen-burning shell
Silicon-burning shell
Iron core

Jupiter's orbit

surface. This wave rapidly accelerates as it encounters less and less resistance, and it soon becomes a shock wave. After a few days, the shock wave reaches the star's surface, by which time the star's outer layers have begun to lift away from the core. In this way, the star becomes a supernova.

This particular description of the death of a $25M_\odot$ star is based on detailed computer calculations of an evolving stellar model. Key stages in the star's evolution are summarized in Table 22-1.

TABLE 22-1 *Evolutionary stages of a 25$M_\odot$ star*

Stage	Temperature (K)	Density (g/cm^3)	Duration of stage
Hydrogen burning	4×10^7	5	7×10^6 yrs
Helium burning	2×10^8	700	5×10^5 yrs
Carbon burning	6×10^8	2×10^5	600 yrs
Neon burning	1.2×10^9	4×10^6	1 yr
Oxygen burning	1.5×10^9	10^7	6 mos
Silicon burning	2.7×10^9	3×10^7	1 day
Core collapse	5.4×10^9	3×10^9	$\frac{1}{4}$ sec
Core bounce	2.3×10^{10}	4×10^{14}	milliseconds
Explosive	about 10^9	varies	10 sec

The final stages in the evolution of other massive stars probably follow similar scenarios, although many details may be different. This particular $25M_\odot$ star ejects $24M_\odot$, leaving behind a $1M_\odot$ corpse called a **neutron star**. Under slightly different conditions, a massive star might blow itself completely apart, leaving no corpse at all. For example, the bounce and subsequent shock wave might develop at the center of the star rather than at the surface of the neutronized core. Alternatively, with a different set of initial conditions inside the star, a very massive burned-out core might gravitationally collapse to form a **black hole**. These exotic objects are discussed in the next two chapters.

Observations of supernovae and supernova remnants reveal details of the deaths of massive stars

As the outer layers of a massive dying star are blasted into space, the star's luminosity suddenly increases by a factor of 10^8. This is equivalent to a jump of 20 magnitudes in brightness. An outburst of this enormity is called a **supernova**. For a few days following the explosion, the star can shine as brightly as an entire galaxy.

Astronomers often find supernovae in distant galaxies (see Figure 22-10). Indeed, most of our understanding of supernovae comes from observing outbursts in remote galaxies. Such observations reveal that supernovae fall into two categories, designated Type I and Type II. In both cases, the eruption begins with a sudden rise in brightness that occurs in less than a day (see Figure 22-11). A **Type I supernova** typically reaches an absolute magnitude of −19 at peak brightness, whereas a **Type II supernova** is about 2 magnitudes fainter. Type I supernovae then settle into a gradual decline that lasts for more than a year, whereas the Type II light curve has a steplike appearance due to alternating periods of steep and gradual declines in brightness.

Although the original definition of Type I and Type II supernovae was based on the shape of light curves, astronomers have recently real-

a　　　　　　　　　　b

Figure 22-10　A supernova
Sometime during 1940, a supernova exploded in the galaxy NGC 4725 in the constellation of Coma Berenices. **(a)** *The galaxy before the outburst.* **(b)** *By the time this photograph was taken in 1941, the supernova (indicated by arrow) had faded from its maximum brightness. (Mount Wilson and Las Campanas Observatories)*

ized that there are major spectroscopic differences between these two kinds of dying stars. Apparently, Type I supernovae are created by metal-poor (population II) stars, whereas Type II supernovae result from the deaths of metal-rich (population I) stars. This distinction is supported by calculation of supernova light curves predicted by detailed evolutionary models that describe events occurring as the star is torn apart. In metal-rich supernovae, there is an explosive production of the unstable isotope ^{56}Ni. Subsequent decay of this radioactive nickel releases energy that temporarily slows the decline in the supernova's brightness, thereby producing the steps in a Type II light curve.

Astronomers find many **supernova remnants** scattered across the skies. A beautiful example is the Veil Nebula seen in Figure 22-12. The doomed star's outer layers were blasted into space with such violence that they are still traveling at supersonic speeds through the interstellar medium. As this expanding shell of gas plows through the interstellar medium, collision with interstellar atoms excites the gas, causing it to glow.

Many supernova remnants are quite large and cover sizable fractions of the sky. The largest is the Gum Nebula, named after Colin Gum who first noticed its faint glowing wisps on photographs of the southern sky (see Figure 22-13). It is also called the Vela supernova remnant; it has a diameter of 60° centered around the constellation of Vela.

The nebula looks so big because it is nearby. Its near side is only about 100 pc from Earth, and the center of the nebulosity is about 460 parsecs away. Studies of the nebula's expansion rate suggest that the supernova exploded around 9000 BC. It could have been witnessed by people then living in places such as Egypt and India. At maximum brilliancy, the exploding star reached an apparent magnitude of −10, equal to the brightness of the Moon at first-quarter.

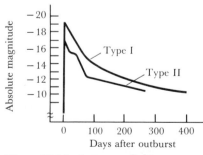

Figure 22-11　Supernova light curves
Type I supernovae exhibit a gradual decline in brightness and are thought to be caused by the death of metal-poor population II stars. Type II supernovae have alternating intervals of steep and gradual decline. These Type II supernovae are probably caused by the death of metal-rich population I stars.

Figure 22-12 [left]   The Veil Nebula
This nebulosity is a portion of the Cygnus Loop (see Figure 20-16), which is the remnant of a supernova that exploded about 20,000 years ago. The distance to the nebula is about 500 pc, and the overall diameter of the loop is about 22 pc. (Palomar Observatory)

Figure 22-13 [right]   The Gum Nebula
The Gum Nebula is the largest known supernova remnant and spans 60° of the sky. Only the central regions of the nebula are shown here. The nearest portions of this expanding nebula are only 100 pc from the Earth. The supernova explosion occurred about 11,000 years ago, and the supernova remnant now has a diameter of about 700 pc. (Royal Observatory, Edinburgh)

Many supernova remnants are virtually invisible at optical wavelengths. Nevertheless, as the expanding gases collide with the interstellar medium, they do radiate energy at a wide range of wavelengths from X rays through radio waves. For example, Figure 22-14 shows both X-ray and radio images of the supernova remnant Cassiopeia A. Optical photographs of this part of the sky reveal only a few, small, very faint wisps. In fact, radio searches for supernova remnants are more fruitful than optical searches. Whereas only two dozen supernova remnants have been found on photographic plates, more than 100 remnants have been discovered by radio astronomers.

From the expansion rate of the nebulosity in Cassiopeia A, astronomers conclude that the supernova explosion occurred about 300 years ago. Although numerous telescopes were in wide use during the late 1600s, no one saw the outburst. In fact, the last supernova seen in our galaxy occurred in 1604 and was observed by Johannes Kepler. A few years earlier, in 1572, Tycho Brahe also recorded the sudden appearance of an exceptionally bright star in the sky. To find any other accounts of supernova explosions, we must delve into ancient astronomical records that are almost 1000 years old.

At first glance, this apparent lack of nearby supernovas may seem puzzling. Astronomers have seen more than 600 supernovae in distant galaxies. From this frequency, it is reasonable to suppose that a galaxy such as our own should have about five supernovae per century. Where are they?

Vigorous stellar evolution is occurring primarily in the disk of our galaxy where, for example, the spiral arms sweep through giant molecular clouds. Our galaxy's disk is therefore the place where massive stars are born and where supernovae explode. This region of our galaxy is so filled with interstellar gas and dust, however, that we simply cannot see very far into space in those directions occupied by the Milky Way. In other words, supernovae probably do erupt every few years in remote parts of our galaxy, but their detonations are hidden from our view by intervening interstellar debris.

In recent years, astronomers have devised an ingenious way of hunting for these unseen supernovae. We have seen that a supernova explosion is triggered by the collapse of a massive star's core. This core collapse produces a flood of neutrinos that are created as electrons combine with protons to produce neutrons. Neutrinos are very elusive particles that hardly interact with matter at all. In fact, under most conditions, matter is transparent to neutrinos. Consequently, when a supernova explodes, vast quantities of neutrinos pour out into space and stream across the galaxy, easily passing through any gas, dust, or nebulae they happen to encounter. At some point, this deluge of neutrinos rains down on and through the Earth. By detecting the arrival of these particles, astronomers would know about the supernova.

Detecting neutrinos is a difficult and tricky business. Current attempts to build a **neutrino telescope** involve large quantities of water, either the ocean or in large drums or tanks. Water (H_2O) contains lots of protons (the nuclei of hydrogen atoms), and astronomers hope to detect neutrinos by observing the recoil of protons struck by the neutrinos.

The neutrinos from a supernova explosion are travelling very near the speed of light and carry a lot of energy. Most of them pass right through the Earth as though it were not here. On rare occasions, however, a neutrino hits a proton in the water-filled drum. The proton recoils with such a high speed that it emits a brief flash of light called **Cerenkov radiation**. This type of radiation, first observed in 1934 by the Russian physicist Pavel A. Cerenkov, occurs whenever a particle moves through a medium (such as water) at a speed faster than the speed of light in that medium. Such motion does not violate the tenet that the speed of light *in a vacuum* (3×10^{10} cm/sec) is the ultimate speed limit in the universe. The speed of light in a substance such as water or glass is much less than that in a vacuum. Recoiling protons in the water-filled drums should easily exceed this reduced light speed, thereby producing the flashes that reveal the existence of an erupting supernova somewhere in our galaxy.

Ordinary telescopic observations of a supernova explosion show us only the expanding outer layers of the dying star. Even X-ray or radio observations fail to see through the hot gases being blasted into space. Thus, ordinary techniques do not allow us to observe the extraordinary events occurring in and around the doomed star's core. Neutrinos, however, easily penetrate the star's outer layers. These particles carry information about the conditions under which they were created. By measuring the energy and momentum carried by the escaping neutrinos, we should be able to learn many facts about the star's collapsing core.

Figure 22-14 Cassiopeia A
*Supernova remnants, such as Cassiopeia A, are typically strong sources of X rays and radio waves. **(a)** An X-ray picture of "Cas A" taken by the Einstein Observatory. **(b)** A corresponding radio image produced by the VLA. The supernova explosion that produced this nebula occurred 300 years ago 3000 pc from Earth. (Smithsonian Institution and the Very Large Array)*

a

b

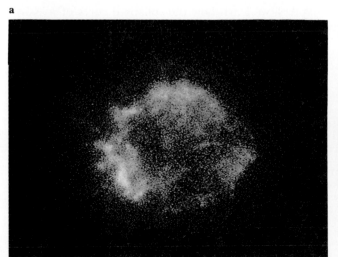
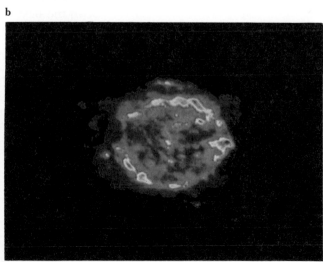

Several teams of astronomers around the world are working hard to perfect neutrino telescopes. With a little luck, they might detect a supernova explosion before the end of the century.

Another exotic technique being developed to probe a supernova explosion involves **gravitational radiation**. During core collapse and subsequent core bounce, vast amounts of matter crushed to nuclear density are hurled about at enormous speeds. This dense, massive material is surrounded by a strong gravitational field. Because this matter moves very rapidly during the explosion, the gravitational field around the star changes rapidly.

As we saw in Chapter 4, Einstein's general theory of relativity is the most accurate description of gravity we have. According to general relativity, gravity actually curves the fabric of space. During a supernova explosion, the sudden and vigorous changes in gravity around the star create ripples in the geometry of space. These ripples move outward from the supernova at the speed of light and are called **gravitational waves,** or gravitational radiation.

Like neutrinos, gravitational waves should bear the imprints of conditions in the star's collapsing core. Unfortunately, gravitational radiation carries very little energy, and thus the waves are very difficult to detect. Nevertheless, important techniques have been developed by Joseph Weber at the University of Maryland. Because gravitational waves are ripples in the geometry of space, objects wiggle slightly as a gravitational wave passes through. Weber has constructed a gravitational-wave antenna by gluing sensitive crystals onto a large aluminum cylinder. If the cylinder oscillates, the crystals produce an elecric voltage that can be amplified and recorded. In this way, the passage of a gravitational wave may be detected.

Teams of physicists around the world are working hard to perfect techniques of detecting gravitational waves. Many scientists, such as Kip Thorne at Caltech, are optimistic that highly sensitive antennas will be operational during the 1990s. These antennas should be able to detect bursts of gravitational waves from the collapsing cores of supernovas as far away as 30 million parsecs (about 100 million light years). This volume of space is so huge and contains so many galaxies that astronomers might be able to detect supernova explosions as frequently as once a month. By analyzing a burst of gravitational waves from a supernova, astronomers should be able to figure out many details of exactly how the star's core collapsed.

In the next two chapters, we learn how the collapsed cores of massive dead stars can become bizarre stellar corpses called neutron stars and black holes.

Summary

. A low-mass star becomes a red giant when shell hydrogen burning begins; it becomes a horizontal-branch star when core helium burning begins; and it becomes a red supergiant when the helium in the core is exhausted and shell helium burning begins.

Thermal runaways in the helium-burning shell produce thermal pulses at intervals of about 300,000 years; more than one-half of the star's mass may be ejected into space.

Ultraviolet radiation from the hot carbon—oxygen core ionizes and excites the ejected gases, producing a planetary nebula.

The burned-out core of a low-mass star becomes a degenerate, very dense sphere about the size of the Earth called a white dwarf, which glows because of thermal radiation; as the sphere cools, it becomes dimmer and eventually becomes cold.

. A high-mass star undergoes a sequence of different thermonuclear reactions in its core and shells; these are carbon burning, neon burning, oxygen burning, and silicon burning.

In the last stages of its life, the high-mass star has an iron-rich core surrounded by concentric shells of the various thermonuclear reactions.

The high-mass star dies in a violent cataclysm that ejects most of the star's matter into space at very high speeds; the luminosity of the star increases suddenly by a factor of around 10^8 during this explosion, producing a supernova.

The ejected matter, moving at supersonic speeds through interstellar gases and dust, glows as a nebula called a supernova remnant.

If the core of the star survives the supernova explosion, it may become a neutron star or a black hole (discussed in later chapters).

. Most supernovae occurring in our galaxy are hidden from our view by interstellar dust and gases.

Study of neutrinos and gravitational waves may enable us to detect supernovae in our galaxy and to learn more about the details of the last stages in the lives of high-mass stars.

Review questions

1 Why is the temperature in a star's core so important in determining which nuclear reactions can occur?

2 On an H-R diagram, sketch the evolutionary track that the Sun will follow as it leaves the main sequence and becomes a white dwarf. Approximately how much mass will the Sun have when it becomes a white dwarf? Where will the rest of the mass have gone?

3 Why do you suppose that all of the white dwarfs known to astronomers are relatively close to the Sun?

*4 Find the average density of a 1 $M_\odot$ white dwarf having the same size as the Earth.

*5 Suppose that the central star in a newly formed planetary nebula has a luminosity of 1000 $L_\odot$ and a surface temperature of 100,000 K. How big is the star?

6 Suppose that the brightness of a star becoming a supernova increases by 20 magnitudes. Prove that this corresponds to an increase of 10^8 in luminosity.

7 Why do astronomers suggest that a supernova visible to the naked eye is long overdue?

8 Why do you suppose that radio search for supernovae remnants have been more fruitful than optical searches?

Advanced questions

9 Suppose you wanted to determine the age of a planetary nebula. What observations would you make and how would you use the resulting data?

10 What reasons can you think of to explain why the rate of expansion of the gas shells in some planetary nebulae is non-uniform?

11 What kinds of stars would you monitor if you wished to observe a supernova explosion from its very beginning? Look up tabulated lists of the brightest and nearest stars. Which, if any, of these stars are possible supernova candidates? Explain.

Discussion questions

12 Suppose that you discover a small glowing disk of light while searching the sky with a telescope. How would you observationally decide if this object is a planetary nebula? Could your object be something else? Explain.

13 Why are astronomers interested in building "neutrino telescopes" and instruments to detect gravitational waves?

For further reading

Clark, D. *Superstars.* McGraw-Hill, 1984. *A book on supernovae.*
Irwin, J. "The Case of the Degenerate Dwarf." *Mercury,* Nov./Dec. 1978, p. 125.
Kaler, J. "Planetary Nebulae and Stellar Evolution." *Mercury,* July/Aug. 1981, p. 114.
Kaufmann, W. "Listening for the Whisper of Gravity Waves." *Science 80,* May/June 1980, p. 64.
Maran, S. "The Gum Nebula." *Scientific American,* Dec. 1971.
Reddy, F. "Supernovae: Still a Challenge." *Sky & Telescope,* Dec. 1983, p. 485.
Tierney, J. "Quest for Order: Profile of S. Chandrasekhar." *Science 82,* Sept. 1982, p. 68.
Wallerstein, G., and Wolff, S. "The Next Supernova?" *Mercury,* Mar./Apr. 1981, p. 44.
Weiler, K. "A New Look at Supernova Remnants." *Sky & Telescope,* Nov. 1979, p. 414.

23 Neutron stars

Four views of SS433

SS433 is believed to be a neutron star in a binary star system. Gas from the companion star is captured into an accretion disk around the neutron star. Matter is ejected from the accretion disk along two oppositely directed, high-speed jets. These four radio views, taken in early 1981, show blobs of gas in the jets extending out to one-sixth of a light year on either side of SS433. Three-quarters of the radio emission comes from SS433 itself (the red central blob) which is located at the center of a supernova remnant 18,000 light years from Earth. (Very Large Array)

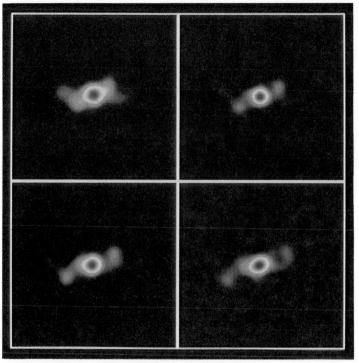

When a supernova explosion leaves behind a stellar corpse of intermediate mass (between $1.4M_\odot$ and about $3M_\odot$), that corpse becomes a neutron star— a dense sphere of degenerate neutrons. In this chapter, we learn about the different forms in which neutron stars are observed: pulsars, pulsating X-ray sources, and bursters. We see how current theories explain these different phenomena and what these models tell us about the properties of neutron stars. We discover that both kinds of stellar corpses—white dwarfs and neutron stars— can undergo explosive thermonuclear reactions on their surfaces under certain circumstances. Finally, we learn that there is an upper limit to the mass of a neutron star, just as the Chandrasekhar limit is the maximum mass of a white dwarf.

The brightening eastern sky was aglow with pink and red as Yang Wei-T'e waited patiently for the sunrise. The date would be officially recorded as "the day of Ch'ih Ch'iu in the fifth moon of the first year of the Shih-huo period"; we call it July 4, 1054. As imperial astronomer to the Chinese court during the Sung dynasty, Yang was thoroughly familiar with the constellations. He therefore immediately realized that something quite extraordinary had happened during that night. Preceding the Sun by only a few minutes, a dazzling object ascended above the eastern horizon. It was far brighter than Venus and considerably more resplendent than any star he had ever seen in the heavens. Yang dutifully recorded his observations and thoughts on this unusual celestial event:

> I make my kowtow. I observed the phenomenon of a guest star. Its color was slightly iridescent. Following an order of the Emperor, I respectfully make the prediction that the guest star does not disturb Aldebaran. This indicates that . . . the Empire will gain great power. I beg to store this prediction in the Department of Historiography.

Yang's "guest star" was so brilliant that it could easily be seen during broad daylight for the rest of July. After a year, however, "it faded and became invisible." Actually the supernova had occurred some 5000 years earlier, its light having taken this long to reach the Earth. Today we realize that Yang Wei-T'e had witnessed the creation of a neutron star.

Pulsars are rapidly rotating neutron stars with intense magnetic fields

The neutron was discovered during laboratory experiments in 1932. Within a year, two astronomers predicted the existence of neutron stars. Inspired by the realization that white dwarfs are supported by degenerate electron pressure, Fritz Zwicky at the California Institute of Technology and his colleague Walter Baade at Mount Wilson Observatory proposed that a highly compact ball of neutrons could similarly produce a powerful pressure. This **degenerate neutron pressure** also could support a stellar corpse, perhaps even more massive than the Chandrasekhar limit. "With all reserve," Zwicky and Baade theorized, "we advance the view that supernovae represent the transition from ordinary stars into **neutron stars,** which in their final stages consist of extremely closely packed neutrons." In other words, there could be at least two types of stellar corpses: white dwarfs, and neutron stars.

This prophetic proposal was politely ignored by most scientists for many years. After all, a neutron star must be a rather weird object. As we saw in Chapter 22, in order to transform protons and electrons into neutrons, the density in the star must be equal to nuclear density, 10^{14} g/cm^3. Thus a thimbleful of neutron-star matter brought back to Earth would weigh 100 million tons. Furthermore, an object compacted to nuclear density would be very small. A $1M_\odot$ neutron star would have a diameter of only 30 km, or about the same size as a large city such as San Francisco or Manhattan. The surface gravity on one of these neutron stars would be so strong that the escape velocity would equal one-half the speed of light. All of these conditions seemed so outrageous that few astronomers paid any serious attention to the subject of neutron stars—at least, not until 1968.

As a young graduate student at Cambridge University, Jocelyn Bell had spent many months assisting in the construction of an array of radio antennas covering $4\frac{1}{2}$ acres of the English countryside. By the fall of 1967, the instrument was completed, and Bell and her colleagues began detecting radio emissions from various celestial sources. In November, while scrutinizing data from the new telescope, Bell noticed that the antenna had detected regular "beeps" from one particular location in the sky. Careful repetition of the observations demonstrated that the radio pulses were arriving with a very regular period of 1.33733011 sec (see Figure 23-1).

The regularity of this pulsating radio source was so striking that the Cambridge team suspected that they might be detecting signals from an

Figure 23-1   A recording of the first pulsar

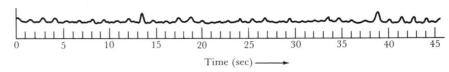

Time (sec) ⟶

Figure 23-1   A recording of the first pulsar
This chart recording shows the intensity of radio emission from the first pulsar, CP1919 (this name means: Cambridge pulsar at a right ascension of 19^h 19^m). Note that some pulses are weak and others are strong. Nevertheless, the spacing between the pulses is exactly 1.3373011 sec. (Adapted from Antony Hewish)

advanced alien civilization. This possibility was soon discarded as several more of these pulsating radio sources, or **pulsars,** were discovered across the sky. In all cases, the periods were extremely regular and ranged from about $\frac{1}{4}$ sec for the fastest to about $1\frac{1}{2}$ sec for the slowest.

The discovery of pulsars was officially announced in early 1968, and many astronomers around the world began proposing all sorts of possible explanations for the regular radio pulsations. Many of these theories were quite bizarre, and arguments raged among astronomers for several months. However, by late 1968, all of this controversy was laid to rest with the discovery of a pulsar in the middle of the Crab Nebula.

Yang Wei-T'e and his colleagues were quite precise about the location of the supernova of 1054: "the guest star appeared several inches southeast of T'ieng Kuang." When we turn a telescope toward this location in the constellation of Taurus, we find the Crab Nebula shown in Figure 23-2. This object looks like an exploded star and is a supernova remnant. The pulsar at the center of the Crab Nebula is called the Crab pulsar.

The Crab pulsar is one of the fastest pulsars ever discovered. Its period is 0.033 sec, which means that it beeps 30 times each second. The fact that a pulsar is located in a supernova remnant tells us that pulsars probably are associated with dead stars.

Before the discovery of pulsars, most astronomers believed that all stellar corpses are white dwarfs. There seemed to be a sufficient number of white dwarfs in the sky to account for all the stars that have died since our galaxy was formed, and it was generally assumed that all dying stars—even the most massive ones that produce supernovae—somehow manage to eject enough matter so that their corpses do not exceed the Chandrasekhar limit.

Figure 23-2   The Crab Nebula
This beautiful nebula, named for the armlike appearance of its filamentary structure, is the remnant of the supernova of 1054 AD. The distance to the nebula is about 2000 pc, and thus its present angular size (4 by 6 arc min) corresponds to linear dimensions of about 7 by 10 light years. See also Figure 1-6. (Palomar Observatory)

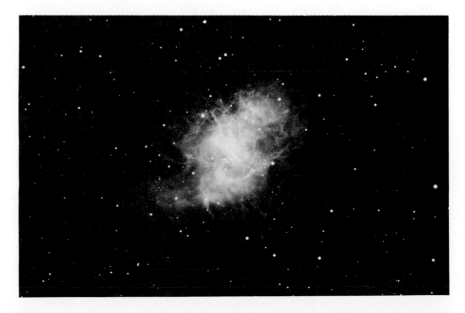

Box 23-1 The conservation of angular momentum

Angular momentum is a measure of the momentum carried by an object because of its rotation about an axis. To appreciate the meaning of angular momentum, consider a mass m revolving about an axis, as shown in the diagram. If r is the distance from the mass to the axis of rotation and v is the linear speed of the mass, then the angular momentum L is

$$L = mvr$$

Furthermore, when speaking about rotation, it is often convenient to define the **angular velocity** ω as

$$\omega = \frac{v}{r}$$

Thus we can write

$$L = mr^2\omega$$

The conservation of angular momentum requires that both the size and direction of the angular momentum never change. For a mass revolving about an axis, this means that the value of L is constant and that the axis of rotation remains fixed in space.

For a rotating object of some size (rather than an idealized point mass), the object's angular momentum is given by

$$L = I\omega$$

where I is the object's moment of inertia, an expression of the way that matter is distributed throughout the object. For example, consider a sphere of uniform density with mass M and radius R. Its moment of inertia is given by

$$I_{sphere} = \frac{2}{5}MR^2$$

Thus the angular momentum of a uniform sphere rotating with an angular velocity ω can be written as

$$L = \frac{2}{5}MR^2\omega$$

Suppose that a star collapses from solar dimensions ($R = R_\odot = 7 \times 10^5$ km) down to neutron-star dimensions ($R = 16$ km). Assuming no mass loss (so that M is constant), the conservation of angular momentum tells us that the original angular velocity (ω_0) is related to the final angular velocity (ω_f) by

$$\omega_0(7 \times 10^5)^2 = \omega_f(16)^2$$

or

$$\frac{\omega_f}{\omega_0} = \left(\frac{7 \times 10^5}{16}\right)^2 = 2 \times 10^9$$

In other words, the star is rotating 2 billion times faster after the collapse than it was before. For example, suppose the star (like the Sun) rotates about once a month. Because 1 month = 3×10^6 sec, the star is rotating nearly 1000 times per second after becoming a neutron star. In the early 1980s, astronomers discovered two pulsars spinning at approximately this rate. The first to be discovered (PSR 1937 + 214) has a period of 0.0015577 sec, or about 1.6 milliseconds. The other millisecond pulsar has a period of 0.0061337 sec, or about 6.1 milliseconds.

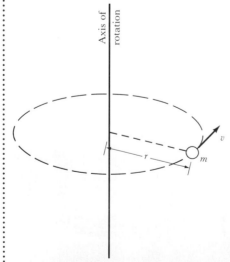

These conservative opinions were dashed with the discovery of the Crab pulsar. It was immediately apparent to astrophysicists that white dwarfs are too big and bulky to produce 30 signals per second. Calculations demonstrated that a white dwarf can neither rotate that fast nor vibrate that fast. The Crab pulsar clearly indicates that the stellar corpse at the center of the Crab nebula is much smaller and more compact than a white dwarf. As Thomas Gold of Cornell University emphasized, astronomers would have to face the prospect of neutron stars seriously.

Exactly what would a neutron star be like? As we have seen, it would be small and dense. It should also be rotating rapidly. All stars rotate, most of them quite leisurely. For example, our Sun takes nearly 1 month to rotate once about its axis. Just as an ice skater doing a pirouette speeds up when she pulls in her arms, however, a collapsing star also speeds up as its size shrinks. This is a direct consequence of a law of physics called the **conservation of angular momentum** (see Box 23-1 for details). In fact, an ordinary star rotating once a month would be spinning faster than once a second by the time it is compressed to the size of a neutron star.

In addition to rapid rotation, we expect that a neutron star could have an intense magnetic field. It is probably safe to say that every star has a magnetic field. As in the case of the Sun, the strength of these stellar magnetic fields is typically quite low. A major reason for this weakness is that a star's magnetic field is spread out over millions upon millions of square kilometers of the star's surface. However, if a star of solar dimensions collapses down to a neutron star, its surface area (which is proportional to the square of its radius) shrinks by a factor of 10^9. Because the magnetic field becomes concentrated onto one-billionth of the original area, the strength of the magnetic field increases by 1 billion.

Magnetic field strength is usually measured in the unit called the gauss (G) named after the famous German mathematician and astronomer Karl Fredrich Gauss. For example, Zeeman splitting of lines in the solar spectrum reveals that the Sun's overall magnetic field has a strength in the range of 1 to 2 G. Similar measurements of Zeeman splitting in the case of white dwarfs typically reveal field strengths of 1 million gauss or more—thousands of times stronger than anything ever produced in the laboratory. By our usual standards, the billion-gauss fields surrounding neutron stars are extremely powerful.

Finally, we would expect the axis of rotation of a typical neutron star to be inclined at some angle to the magnetic axis connecting the north and south magnetic poles (see Figure 23-3). In 1969, Peter Goldreich at the California Institute of Technology pointed out that, like a giant electric generator, the combination of a powerful magnetic field and rapid rotation should create extremely intense electric fields near the star's surface. At the star's surface, there are plenty of protons and electrons because the pressures are too low to combine them into neutrons. The powerful electric fields acting on these charged particles cause them to flow out from the neutron star's polar regions along the curved magnetic field, as sketched in Figure 23-3. As the particles stream along the curved field, they are accelerated and emit energy. The final result is two very thin beams of radiation pouring out of the neutron star's north and south magnetic polar regions.

A rotating, magnetized neutron star is somewhat like a lighthouse beacon. As the star rotates, the beams of radiation sweep around the sky. If the Earth happens to be located in the right direction, a brief flash is observed each time the beam sweeps past our line of sight.

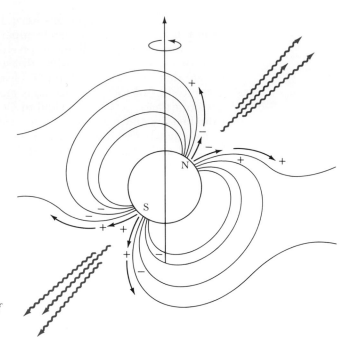

Figure 23-3 A rotating, magnetized neutron star
It is reasonable to suppose that a neutron star is rotating rapidly and possesses a powerful magnetic field. Charged particles accelerated near the star's magnetic poles produce two oppositely directed beams of radiation. As the star rotates, the beams sweep around the sky. If the Earth happens to lie in the path of the beams, we see a pulsar.

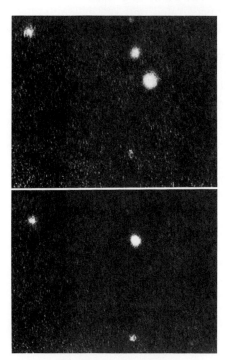

Figure 23-4 The Crab pulsar
These two photographs show the Crab pulsar in its "on" and "off" states. Like the radio pulses, these visual flashes have a period of 0.033 sec. These pictures of the center of the Crab Nebula can be used to identify the neutron star in Figure 23-2. (Lick Observatory)

When this scenario was first proposed in the late 1960s, a team of astronomers at the University of Arizona began wondering if pulsars emit flashes at wavelengths other than radio frequencies. After aiming a telescope at the center of the Crab Nebula, they used a spinning disk with a slit to "chop," or interrupt, the incoming beam of light at rapid intervals. To their surprise, they found that one of the stars at the center of the nebula is actually flashing on and off 30 times each second (see Figure 23-4).

Since these pioneering days, the Crab pulsar has been observed over a wide range of wavelengths. For example, Figure 23-5 is an X-ray picture from the Einstein Observatory. Figure 23-6 shows the light curves of the Crab pulsar at optical, X-ray, and radio wavelengths. Note that, in addition to the main pulse, there is a secondary pulse about halfway through the pulsar's period. Many pulsars exhibit both main and secondary pulses simply because neutron stars have both north and south magnetic poles. In general, one of the magnetic poles is more directly aligned toward the Earth than is the other. Hence we detect one strong pulse and one weak pulse during each rotation of the neutron star.

During the 1970s, radio astronomers discovered about 300 pulsars scattered across the sky. Presumably, each one is the neutron-star corpse of a massive extinct star. In the early 1980s, astronomers discovered two pulsars with periods of only a few milliseconds. One of them (called PSR 1937 + 214, a name that gives its position in the sky) is blinking on and off at visible wavelengths, like the Crab pulsar. Also visibly flashing is the Vela pulsar located at the core of the Gum Nebular (see Figure 22-13). With a period of 0.089 sec, the Vela pulsar is the slowest pulsar ever detected at visible wavelengths.

The Crab pulsar is one of the youngest pulsars, its creation having been observed some 900 years ago. Of course, a few supernovae have been seen since then (such as those noted by Tycho Brahe and Johannes

Figure 23-5   The Crab Nebula and pulsar in X Rays
This view of the Crab Nebula was obtained with an X-ray telescope on the Earth-orbiting Einstein Observatory. The bright spot is the pulsar. (Harvard-Smithsonian Center for Astrophysics)

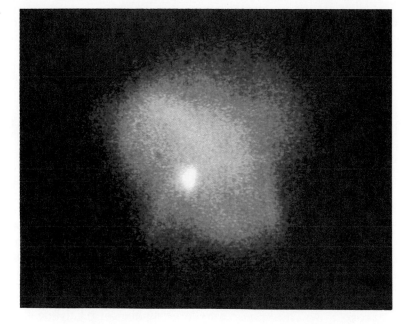

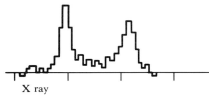

X ray

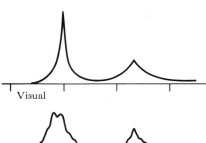

Visual

Radio

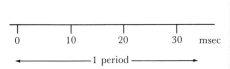

0   10   20   30   msec

◄──────── 1 period ────────►

Figure 23-6 [above]   Light curves of the Crab pulsar
These three graphs show the intensity of radiation emitted by the Crab pulsar at X-ray, optical, and radio wavelengths. Incidentally, 1 msec = 1 millisecond = 10^{-3} sec. Thus, the pulsar's period is 33 msec.

Figure 23-7 [right]   The Vela pulsar in X rays
The Vela pulsar is located at the middle of the Gum Nebula shown in Figure 22-13. Like the Crab pulsar, the Vela pulsar is detected at X-ray, optical, and radio wavelengths. This view was obtained from one of the X-ray telescopes on the Einstein Observatory. (Harvard-Smithsonian Center for Astrophysics)

Kepler), but no pulsars are found at these locations. Perhaps the stellar corpses are not neutron stars, or perhaps the beams simply do not sweep past the Earth. Indeed, there must be many pulsars that we will never discover because they are oriented at unfavorable angles.

The Vela pulsar (see Figure 23-7) also is quite young, its creation having occurred roughly 11,000 years ago. We conclude that pulsars slow down as they get older and that only the very youngest pulsars are energetic enough to emit optical flashes along with their radio pulses.

Superfluidity and superconductivity are among the strange properties of neutron stars

Many of the details of how pulsars emit radiation are still poorly understood. Nevertheless, the concept of a rapidly rotating, magnetized neutron star has proved very successful in explaining many phenomena that were once quite puzzling. For example, without a spinning neutron star, it is virtually impossible to explain why the Crab Nebula shines the way it does.

The diffuse part of the Crab Nebula (not the reddish filaments) shines with an eerie light identified by the Russian astronomer Iosif Shklovskii as synchrotron radiation. Synchrotron radiation was first observed in 1947 in a particle accelerator built at the General Electric Company in Schenectady, New York. This machine, called a synchrotron, was designed to accelerate electrons to speeds very close to the speed of light for experiments in high-energy nuclear physics. The electrons were whirled about a circular path, held in orbit by powerful magnets. Through a small window in the side of the doughnut-shaped machine, scientists noticed a strange light being emitted by the circulating beam of electrons. This light, called **synchrotron radiation,** is emitted whenever high-speed electrons move along curved paths through a magnetic field. Because their velocities are near the speed of light, these electrons are said to be "relativistic." Consequently, the Crab Nebula must contain a very large number of relativistic electrons gyrating wildly in an extensive magnetic field.

The total energy output of the Crab Nebula in synchrotron radiation is 3×10^{38} erg/sec. The Sun emits 4×10^{33} erg/sec; thus, the Crab Nebula is 75,000 times more luminous than the Sun. In 1966 (two years before the discovery of pulsars), John A. Wheeler at Princeton University and Franco Pacini from Italy speculated that the ultimate source of this prodigious energy output might be a spinning neutron star. Their prophetic ideas were confirmed with the discovery that the Crab pulsar is slowing down.

Although pulsars were first noted for their regular periods, careful timing measurements at radio telescopes soon revealed that all pulsars are gradually slowing down. The period of a typical pulsar increases by a few ten-billionths of a second each day. The Crab pulsar, which has one of the shortest periods, is slowing more quickly than most other pulsars: its period increases by 3×10^{-8} seconds each day. This may sound like a trivial detail, but a rapidly spinning neutron star possesses so much rotational energy that even the slightest slowdown corresponds to a tremendous loss of energy. In fact, the observed slowdown rate for the Crab pulsar corresponds to an energy loss equal to the luminosity of the Crab Nebula. In other words, although the details are poorly understood, the pulsar's rotational energy is constantly being converted into radiant energy, and that is why the Crab Nebula shines.

While making timing observations, astronomers noticed that pulsars sometimes exhibit a sudden and unexpected speedup called a **glitch.** For example, Figure 23-8 shows accurate period measurements of the Vela pulsar during 1975 and 1976. The pulsar's slowdown is obvious on this graph. However, in September 1975, there was an abrupt speedup, after which the pulsar continued to slow down at its usual rate. This was the third glitch observed for the Vela pulsar; similar glitches have been seen with the Crab pulsar.

According to the conservation of angular momentum, a rotating object speeds up when it contracts. A neutron star is so dense that a typical glitch (which involves a sudden speedup causing the period to decrease by 10^{-7} sec) corresponds to a settling of the star's crust by as little as 1 mm. Because the speedup was sudden, the movement of the star's

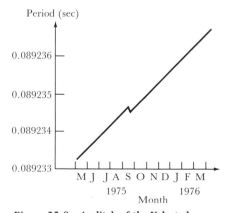

Figure 23-8 A glitch of the Vela pulsar
This graph shows how the period of the Vela pulsar increased during the mid-1970s. The sudden speedup during September 1975 is called a glitch and is probably due to a star-quake, during which the star's crust cracks and settles.

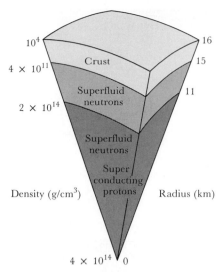

Figure 23-9   A model of a neutron star
This possible model for a 1.3M$_\odot$ neutron star has a superfluid, superconducting core 22 km in diameter. The core is surrounded by a superfluid mantle of neutrons 4 km thick. The star's crust is probably composed of heavy nuclei (such as iron) and free electrons. (Adapted from Pandharipande, Pines, and Smith)

crust must have been equally abrupt, suggesting that the crust of a neutron star is very brittle. A glitch occurs when the crust cracks and settles. This phenomenon is appropriately called a **starquake.** The most powerful earthquakes have a Richter magnitude of 8½. The glitches that occur with the Vela and Crab pulsars correspond to starquakes with magnitudes in the range of 23 to 25 on the Richter scale.

Starquakes tell us about the crust of a pulsar, but astrophysicists use elaborate calculations to surmise conditions inside a neutron star. As we have seen, the interior of a neutron star is dominated by degenerate neutron pressure caused by closely packed neutrons obeying the Pauli exclusion principle. Detailed neutron-star models strongly suggest that this sea of closely packed neutrons can flow without any friction whatsoever. This phenomenon, called **superfluidity,** is observed in laboratory experiments with liquid helium cooled to temperatures very near absolute zero. Because it is frictionless, a superfluid exhibits such strange properties as creeping up the walls of a container in apparent defiance of gravity. Motions of the superfluid neutrons inside a neutron star may be responsible for some glitches that arise from starquakes below the crust.

Although a neutron star is predominantly made of neutrons, there are some protons and electrons scattered throughout the star's interior. Indeed, a pulsar's magnetic field is anchored to the neutron star by these charged particles. Of course, a neutron is electrically neutral, so neutrons cannot hold onto the star's magnetic field. Without the protons and electrons in its interior, a neutron star would rapidly lose its magnetic field.

Different teams of astrophysicists have calculated slightly different details for the internal structure of a neutron star. Nevertheless, many of these models strongly suggest that the protons in the star's core can move around without experiencing any electrical resistance whatsoever. This phenomenon, called **superconductivity,** is observed in the laboratory with certain metals cooled to near absolute zero.

As shown in Figure 23-9, the structure of a neutron star consists of a superfluid and superconducting core surrounded by a superfluid mantle, which in turn is surrounded by a brittle crust only 1 km thick.

Pulsating X-ray sources are neutron stars in close binary systems

During the 1960s, astronomers obtained tantalizing views of the X-ray sky during brief rocket and balloon flights that momentarily lifted X-ray detectors above the Earth's atmosphere. A number of strong X-ray sources were discovered; each was named after the constellation in which it is located. For example, Scorpius X-1 is the first X-ray source found in the constellation of Scorpius.

Astronomers were so intrigued by these preliminary discoveries that they began designing an Earth-orbiting, X-ray-detecting satellite that could make observations 24 hours a day. Their hopes and dreams were realized with the launch of Explorer 42 on December 12, 1970 (see Figure 23-10). Because it was to be placed in an equatorial orbit, the satellite was launched from Kenya in Africa. In recognition of the hospitality of the Kenyan people, Explorer 42 was christened Uhuru which means "freedom" in Swahili.

Uhuru gave us our first comprehensive look at the X-ray sky. As the satellite slowly rotated, its X-ray detectors swept across the heavens. Each time an X-ray source came into view, signals were transmitted to receiving stations on the ground. Before its battery and transmitter failed in early 1973, Uhuru had succeeded in locating 339 X-ray sources.

The discovery of pulsars was still fresh in everyone's mind, so the

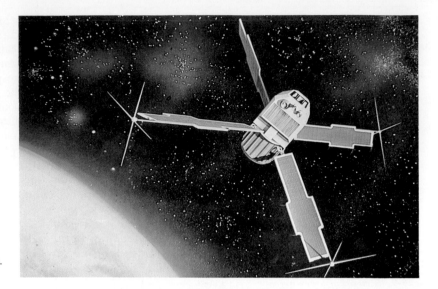

Figure 23-10   Uhuru
Uhuru was a small satellite designed to detect astronomical sources of X rays. During three years of flawless operation, it observed more than 300 different X-ray objects across the sky. (NASA)

Uhuru team headed by Riccardo Giacconi was very excited by the detection of X-ray pulses from Centaurus X-3 in early 1971. Figure 23-11 shows data from one sweep of Uhuru's detectors across Centaurus X-3. The pulses have a regular period of 4.84 sec. A few months later, similar pulses were discovered coming from a source called Hercules X-1 that has a period of 1.24 sec. Because the periods of these two X-ray sources are so short, astronomers began to suspect that they had found some rapidly rotating neutron stars.

It soon became clear, however, that systems such as Centaurus X-3 and Hercules X-1 are not ordinary pulsars like the Crab or Vela pulsars. Centaurus X-3 turns on and off periodically. Every 2.087 days, Centaurus X-3 turns off for almost 12 hours. This fact suggests that Centaurus X-3 is an eclipsing binary and that it takes nearly 12 hours for the X-ray source to pass behind its companion star.

The binary nature of Hercules X-1 is even more compelling. In addition to an "off" state corresponding to a 6-hour eclipse every 1.7 days, careful timing of the X-ray pulses shows a periodic Doppler shifting every 1.7 days. This is direct evidence of orbital motion about a companion star: when the X-ray source is approaching us, its pulses are separated by slightly less than 1.24 sec; when the source is receding from us, slightly more than 1.24 sec elapses between the pulses.

Careful optical searches of the location of Hercules X-1 soon revealed a dim star called HZ Herculis. The magnitude of this star varies between 13 and 15, with a period of 1.7 days. Because this period is exactly the same as the orbital period of the X-ray source, astronomers conclude

Figure 23-11   X-Ray pulses from Centaurus X-3
This graph shows the intensity of X rays detected by Uhuru as Centaurus X-3 moved across the satellite's field of view. Successive pulses are separated by 4.84 sec. The gradual variation in the height of the pulses from left to right is due to the changing orientation of Uhuru's X-ray detectors toward the source as the satellite rotates. (Adapted from R. Giacconi and colleagues)

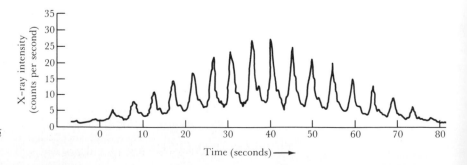

that HZ Herculis is the companion star around which Hercules X-1 is orbiting.

Putting all the pieces together, astronomers now realize that systems such as Centaurus X-3 and Hercules X-1 are examples of double stars in which one of the stars is a neutron star. All of these binaries have very short orbital periods. Consequently, the distance between the ordinary star and the neutron star in each of these systems must be very small. This proximity puts the neutron star in a position to capture gases escaping from the ordinary companion star.

In the mid-1800s, the French mathematician Edward Roche pointed out that you can draw a figure-eight curve around two stars in a binary to portray the gravitational domain of each star. If gases should happen to leak over the figure-eight-shaped surface surrounding one of the stars, this material is no longer bound to the star and is free to escape into space. This figure-eight curve (which is one of the equipotential contours described in Box 16-1) is often called the **critical surface,** and each half of the curve is called a **Roche lobe.**

To explain pulsating X-ray sources such as Centaurus X-3 or Hercules X-1, astronomers assume that the ordinary star either fills or nearly fills its Roche lobe. Either way, matter escapes from the star. This mass loss occurs by direct "Roche-lobe overflow" if the star fills its lobe, as in the case of Hercules X-1, or it occurs by a stellar wind if the star's surface lies just inside its lobe, as in the case of Centaurus X-3 (see Figure 23-12). A typical rate of mass loss from the ordinary star is roughly 10^{-9} solar masses per year.

As in the case of an ordinary pulsar, the neutron star in a pulsating X-ray source is rapidly rotating and has a powerful magnetic field inclined to the axis of rotation (recall Figure 23-3). Because of its strong gravity, the neutron star easily captures much of the gas escaping from the companion star. As the gas falls toward the neutron star, the

Figure 23-12 The model of a pulsating X-ray source
Gas escaping from the ordinary star is captured by the neutron star. The infalling gas is funneled down onto the neutron star's magnetic poles and strikes the star with enough energy to create two X-ray–emitting hot spots. As the neutron star spins, beams of X rays from the hot spots sweep around the sky.

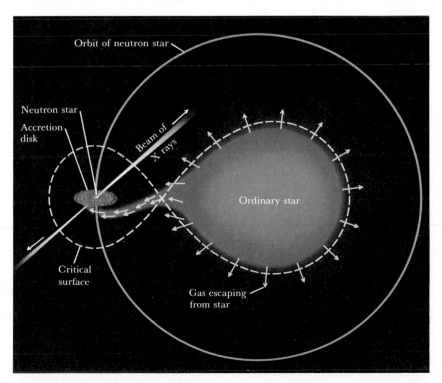

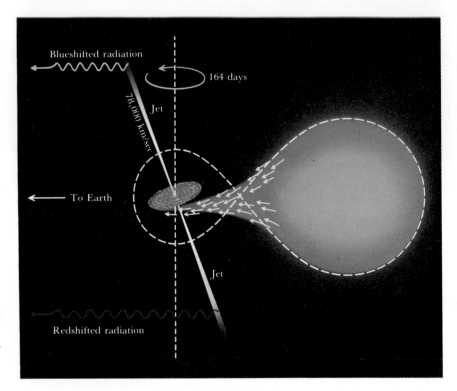

Figure 23-13   A model of SS433
Gas from a normal star is captured into an accretion disk about a neutron star. Two high-speed, oppositely directed jets of gas are ejected from the faces of the disk. Because the disk is tilted, the gravitational pull of the normal star causes the jets to precess with a period of 164 days.

magnetic field funnels the incoming matter down onto the star's north and south magnetic polar regions. The star's gravity is so strong that the gas is traveling at nearly half the speed of light by the time it crashes onto the star's surface. This violent impact creates hot spots at both poles with temperatures of about 10^8 K; these hot spots therefore emit abundant X rays. Indeed the X-ray luminosity is roughly 10^{38} erg/sec, which is nearly 100,000 times brighter than the Sun. As the neutron star rotates, the beams of X rays from the polar caps sweep around the sky. If the Earth happens to be in the path of one of the two beams, we observe a pulsating X-ray source. The pulse period is thus equal to the neutron star's rotation period. For example, the neutron star in Hercules X-1 is spinning at the rate of once every 1.24 sec.

There are a few interesting details to this basic scenario. First, X rays from the neutron star can heat the exposed hemisphere of the companion star. Thus one side of the companion star is hot and bright, while the other side is cooler and dimmer. This is why the brightness of HZ Herculis varies periodically; as the system rotates, the two hemispheres are alternately exposed to our view. Second, gases captured by the neutron star's gravity may go into orbit about the neutron star. The result is a rotating disk of material called an **accretion disk,** as shown in Figure 23-12.

Accretion disks have been detected in many close binary systems where mass transfer is occurring between the two stars. Under certain circumstances, some bizarre things can happen. With ordinary pulsating X-ray sources such as Hercules X-1, the rate at which gas falls onto the neutron star is fairly low. This material falls onto the neutron star from the inner edge of the accretion disc at a rate low enough to allow the resulting X rays to escape.

If the companion star is dumping vast amounts of material onto the neutron star, however, then the resulting energy cannot easily escape. Instead, tremendous pressures build up in the gases crowding down onto the neutron star. These pressures meet strong resistance in the plane of the accretion disk, where newly arrived gases are constantly spiraling in toward the neutron star. The "path of least resistance" is perpendicular to the plane of the accretion disk. Consequently, pressure around the neutron star is relieved by squirting material out along this perpendicular direction. The final result can be two powerful beams of high-velocity hot gas, as shown in Figure 23-13. This is apparently the explanation for the weird star called SS433. (SS433 is so named because it is the 433rd star on a list of similar objects published a year earlier by C. Bruce Stephenson and Nicholas Sanduleak of Case Western Reserve.)

In the autumn of 1978, Bruce Margon and his colleagues at UCLA began taking a series of spectrograms of the star SS433, which has been noted for its strong emission lines. To everyone's surprise, the spectrum of SS433 contains several complete sets of spectral lines. One set of lines was very redshifted from its usual wavelengths, and another set was comparably blueshifted. Somehow, SS433 is "coming and going at the same time." To make matters even more puzzling, the wavelengths of these redshifted and blueshifted lines change dramatically from one night to the next.

Astronomers had never seen anything like this, and soon many were observing SS433. By mid-1979, it was clear that the system's redshifted and blueshifted lines are actually moving back and forth across the spectrum of SS433 with a period of 164 days. British astrophysicists Andrew Fabian and Martin Rees of Cambridge were quick to point out that the two sets of spectral lines could be caused by two oppositely directed jets of gas, one pointing toward us and the other pointing away from us. Furthermore, the 164-day variation could be explained by a precession of the two jets. As the two jets circle about the sky every 164 days, we see a periodic variation in the Doppler shift.

All of these features come together in the model sketched in Figure 23-13. To explain the large redshifts and blueshifts discovered by Margon, gas in the two oppositely directed jets must have a speed of 78,000 km/sec—roughly one-quarter of the speed of light. In addition, the accretion disk must be tilted with respect to the orbital plane of the two stars of the binary system. Just as the tilt of the Earth's axis with respect to the plane of the ecliptic causes the Earth to precess, the tilt of the accretion disk results in the 164-day precession of the two jets.

Figure 23-14 is a high-resolution radio view of SS433. Note the two

Figure 23-14 SS433
This view of SS433 was made with the VLA. It shows two oppositely directed features emerging from a central object. SS433 is located roughly 4000 pc from Earth in the constellation of Aquila. At visual wavelengths, SS433 has a magnitude of +14. (Observations by R. M. Hjellming and K. J. Johnston)

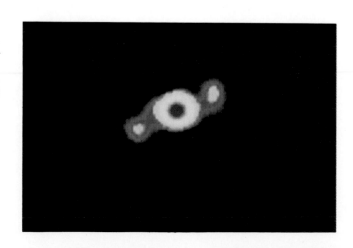

oppositely directed appendages emerging from the central source. As we shall see in Chapter 28, many quasars and peculiar galaxies have a very similar radio structure, although on a much larger scale. Quasars are incredibly far away, and thus they are difficult to study. The real significance of SS433 may be that it gives us a miniature quasarlike object right in our own celestial backyard.

Explosive thermonuclear processes on white dwarfs and neutron stars produce novae and bursters

With all of the bizarre and fascinating phenomena associated with neutron stars, you might be wondering about white dwarfs. Low-mass stars are far more common than high-mass stars, so white dwarfs are far more common than neutron stars. Do all these white dwarfs do anything more dramatic than simply cool off?

The answer definitely is yes. Occasionally, some star in the sky suddenly brightens by a factor of 10^6. This phenomenon is called a **nova** (not to be confused with a supernova, which involves a much greater increase in brightness). Novae are fairly common. The abrupt rise in brightness is followed by a gradual decline that may stretch on for several months or more (see Figures 23-15 and 23-16).

Painstaking observations of numerous novae by Robert Kraft, Merle Walker and colleagues at Lick Observatory strongly suggest that all novae are members of close binary systems containing a white dwarf. Gradual mass transfer from the ordinary companion star (which presumably fills its Roche lobe) deposits fresh hydrogen onto the white dwarf. Because of the strong gravity, this hydrogen is compacted into a dense layer covering the hot surface of the white dwarf. As more gas is deposited, the temperature in the hydrogen layer increases. Finally, when the temperature reaches about 10^7 K, hydrogen burning ignites throughout the layer, embroiling the white dwarf's surface in a thermonuclear holocaust that we see as a nova.

Figure 23-15   Nova Herculis 1934
These two pictures show a nova (a) shortly after peak brightness as a magnitude +3 star, and (b) two months later, when it had faded to magnitude +12. Novae are named after the constellation and year in which they appeared. (Lick Observatory)

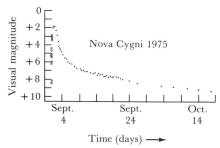

Figure 23-16   The light curve of Nova Cygni 1975
This graph shows the history of a nova that blazed forth in the constellation of Cygnus in September 1975. The rapid rise followed by a gradual decline in magnitude is characteristic of all novae.

Figure 23-17   X rays from a burster
A burster emits a constant low intensity of X rays interspersed with occasional powerful bursts of X rays. This particular burst is typical. It was recorded on September 28, 1975, by an X-ray telescope on board the Astronomical Netherlands Satellite while the telescope was pointed toward the globular cluster NGC 6624. About one-third of all known bursters are located in globular clusters. (Adapted from Walter Lewin)

A very similar phenomenon occurs with neutron stars. Beginning in late 1975, astronomers analyzing data from X-ray satellites realized that their instruments had detected sudden, powerful bursts of X rays from objects in the sky. The record of a typical burst is shown in Figure 23-17. The source emits X rays at a constant low level; suddenly, without warning, there is an abrupt increase in X rays, followed by a more gradual decline. Typically, an entire burst lasts for only 20 sec, and sources that behave in this fashion are called **bursters.** Several dozen bursters have been discovered, most of them located toward the center of our galaxy.

Bursters, like novae, are believed to involve close binaries experiencing mass transfer. With a burster, however, the stellar corpse is a neu-

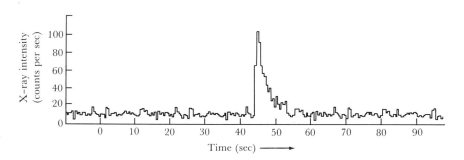

tron star rather than a white dwarf. Gases escaping from the ordinary companion star fall onto the neutron star. The energy released as this gas crashes down onto the neutron star's surface produces the low-level X rays that are continuously emitted by the burster.

Most of the gas falling onto the neutron star is hydrogen, which becomes compressed against the hot surface of the star by the star's powerful surface gravity. In fact, temperatures and pressures in this accreting layer are so high that the arriving hydrogen promptly is converted into helium by the hydrogen-burning process. This constant hydrogen burning soon produces a layer of helium that covers the entire neutron star.

Finally, when the helium layer is about 1 m thick, helium burning ignites explosively, and we observe a sudden burst of X rays. In other words, whereas explosive hydrogen burning on a white dwarf produces a nova, explosive helium burning on a neutron star produces a burster. In both cases, the burning is explosive because the fuel is so strongly compressed against the star's surface that it is degenerate, like the star itself. As we saw with the helium flash inside red giants, the ignition of a degenerate thermonuclear fuel always involves a sudden thermal run away because the usual safety valve between temperature and pressure is not operating.

Just as there is an upper limit to the mass of a white dwarf, there is also an upper limit to the mass of a neutron star. Above this limit, degenerate neutron pressure cannot support the overpowering weight of the star's matter pressing inward from all sides. Whereas the Chandrasekhar limit for a white dwarf is $1.4 M_\odot$, the corresponding upper limit for a neutron star is probably 2.5–$3 M_\odot$.

Before the discovery of pulsars, most astronomers believed that all dead stars are white dwarfs. Dying stars were thought somehow to eject enough material so that their corpses would be below the Chandrasekhar limit. Obviously, this idea proved incorrect. Inspired by this lesson, astronomers soon began wondering what might happen if a dying massive star failed to eject enough matter to get below the upper limit for a neutron star. For example, what might a $5M_\odot$ stellar corpse be like?

The gravity associated with a neutron star is so strong that the escape velocity is roughly one-half the speed of light. With a stellar corpse greater than $3M_\odot$, there is so much matter crushed into such a small volume that the escape velocity exceeds the speed of light. Because nothing can travel faster than light, nothing—not even light—can leave the dead star. The star therefore disappears from the universe, its powerful gravity leaving a hole in the fabric of space and time. In this way, the discovery of neutron stars inspired astrophysicists seriously to examine one of the most bizarre and fantastic objects ever predicted by modern science, the black hole.

Summary

· A neutron star is a very dense stellar corpse consisting of closely packed, degenerate neutrons.

  A neutron star typically has a diameter of about 30 km, a mass less than $3M_\odot$, a magnetic field a billion times stronger than that of the Sun, and a rotation period of roughly 1 sec.

  The neutron star consists of a superfluid and superconducting core surrounded by a superfluid mantle and a thin brittle crust.

  Energy pours out of the north and south polar regions of a neutron star in intense beams produced by streams of charged particles moving in the star's intense magnetic field.

· A pulsar is a source of radio radiation exhibiting periodic pulses of increased intensity; it is thought to be a neutron star, with the pulses caused as a beam of radio waves from the star's polar region sweeps past the Earth.

  The steady slowing of a pulsar's pulse rate presumably reflects the gradual slowing of the rotation rate of the neutron star; glitches (sudden speedups of the pulse rate) correspond to starquakes.

· Some X-ray sources exhibit regular pulses as well as the effects of orbital motion; these objects are thought to be neutron stars in very close binary systems with ordinary stars.

  Gases from the ordinary star in such a system fall onto the dense neutron star at very great speeds, creating hot spots near its poles; these hot spots radiate intense beams of X rays.

  Gases falling onto the neutron star can form an accretion disk around the neutron star; in some systems, beams of very-high-velocity gases may shoot out from the faces of the accretion disk, producing intensely blueshifted or redshifted radiation.

· Material from the ordinary star in a binary pair can fall onto the surface of the companion white dwarf or neutron star to produce a surface layer in which thermonuclear reactions can occur; because this layer is degenerate at the high pressure produced by the intense gravitational field of the stellar corpse, the thermonuclear process will be explosive.

Explosive hydrogen burning may occur in the surface layer of a companion white dwarf, producing the sudden increase in luminosity that we call a nova (the brightness increase is far less than that observed in the case of a supernova).

Explosive helium burning may occur in the surface layer of a companion neutron star, producing the sudden increase in X-ray radiation that we call a burster.

Review questions

1 The distance to the Crab Nebula is about 2000 parsecs. When did the star actually explode?

2 What is the difference between a nova and a supernova? On the basis of our understanding of stellar evolution, would novae or supernovae be more common? Explain.

3 Why do you suppose that most of the angular momentum contained in the solar system resides with the planets rather than with the far more massive Sun?

4 During the weeks immediately following the discovery of the first pulsar, one suggested explanation was that the pulses are signals from an extraterrestrial civilization. Why do you suppose that astronomers discarded this idea?

5 How do we know that the Crab pulsar is really embedded in the Crab Nebula and is not simply located at a different distance along the same line of sight?

6 How do you think that astronomers have deduced that the Vela pulsar is about 11,000 years old?

7 If the model for Hercules X-1 discussed in the text is correct, at what orientation of the binary system do we see the maximum optical brightness? Explain your answer.

Advanced questions

*8 To determine accurately the period of a pulsar, astronomers must take the Earth's orbital motion about the Sun into account. Explain why. Knowing that the Earth's orbital velocity is 30 km/sec, calculate the maximum correction to a pulsar's period due to the Earth's motion. Explain why the size of the correction is greatest for pulsars located near the ecliptic.

9 The mass of a neutron is about 1.7×10^{-24} g and its radius is about 10^{-13} cm. Compare the density of matter in a neutron with the density of a neutron star.

10 Propose an explanation for the fact that X-ray pulsars are speeding up while ordinary (radio) pulsars are slowing down.

*11 From the data given in the caption to Figure 23-2, calculate the rate of expansion of the Crab Nebula. Assuming that your telescope can distinguish features as small as 1 arc second, how long would you have to wait to see a change in the size of the Crab Nebula?

Discussion questions

12 Compare novae and bursters. What do they have in common? In what ways are they different?

13 How might astronomers be able to detect the presence of an accretion disk in a close binary system?

For further reading

Anderson, L. "X-Rays from Degenerate Stars." *Mercury*, Sept./Oct. 1976, p. 6; Nov./Dec. 1976, p. 2.

Greenstein, G. *Frozen Stars: Of Pulsars, Black Holes, and the Fate of Stars.* Freundlich Books, 1984.

Grindlay, J. "New Bursts in Astronomy." *Mercury*, Sept./Oct. 1977, p. 6. *Article on bursters.*

Helfand, D. "Pulsars." *Mercury*, May/June 1977, p. 2.

Margon, B. "The Bizarre Spectrum of SS 433." *Scientific American*, Oct. 1980.

Ruderman, M. "Solid Stars." *Scientific American*, Feb. 1971.

Trimble, V. "How to Survive Cataclysmic Binaries." *Mercury*, Jan./Feb. 1980, p. 8

24 Black holes

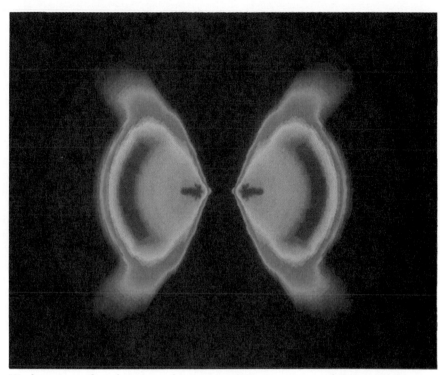

An accretion disk around a black hole
This false color image shows (in cross section) how matter distributes itself around a black hole. A Cray 1 "supercomputer" was used to solve the complicated relativistic equations that govern the infall of matter toward the black hole which is at the center of this diagram, where the two broad angles point. Many astrophysicists believe that supercomputers present a new and revolutionary way of doing research. With a supercomputer, you can simulate a wide range of physical situations that are not easily observable. Furthermore, what you see in nature is usually a complicated combination of many physical effects. With a supercomputer simulation, you can study the behavior of individual physical quantities (such as pressure, density, temperature, or velocity) in an otherwise complicated phenomenon. (Courtesy of L. Smarr and J. Hawley)

We have seen that the mass of a neutron star cannot exceed $3M_\odot$. In this chapter we learn that more-massive stellar corpses become compacted to a single point in space of infinite density, surrounded by gravity so intense that nothing—not even light—can escape. We discuss the theory of relativity that predicts the existence of these black holes, and we learn how astronomers search for evidence of them. We find that a black hole has only three physical properties, and that the stellar corpse in the center of a black hole has disappeared forever from our direct observation.

Suppose that the mass of a dying star's burned-out core exceeds 3 solar masses ($3M_\odot$). This mass is well above the Chandrasekhar limit, so degenerate electron pressure could not possibly support the resulting stellar corpse. Similarly, because $3M_\odot$ is also above the mass limit for neutron stars, degenerate neutron pressure also is incapable of supporting the crushing weight of the burned-out matter pressing inexorably inward toward the dead star's center. If it can be neither a white dwarf nor a neutron star, what might this massive stellar corpse become?

As we saw in the previous chapter, a typical neutron star consists of roughly $1M_\odot$ of matter compressed by its own gravity to nuclear density in a sphere roughly 30 km in diameter. This gravity is so strong that the escape velocity from the neutron star's surface is one-half the speed of light.

A massive stellar corpse, whose weight overpowers degenerate neutron pressure, easily compresses its matter to densities greater than nuclear density. It does not take very much further compression to cause the escape velocity to exceed the speed of light. For example, if $3M_\odot$ of matter is squeezed inside a sphere 18 km in diameter, the escape velocity from the object is greater than the speed of light. Nothing can travel faster than the speed of light, and thus nothing—not even light—can manage to escape from the dead star. The star has disappeared from the observable universe, although some of its effects can be detected.

The general theory of relativity describes gravity in terms of the geometry of space and time

To appreciate fully the nature of a massive stellar corpse, we must use the best theory of gravity at our disposal. The gravitational field around one of these massive dead stars is so strong that the theory of Isaac Newton gives the wrong answers. Instead, we must turn to Albert Einstein's general theory of relativity.

According to the classical physics of Newton, space is perfectly uniform and fills the universe like a rigid framework. Similarly, time passes at a monotonous, unchangeable rate. It is always possible to know how fast you are moving through this rigid fabric of space and time, and the results of your observations depend on your state of motion. Significant differences between the observations of a stationary person and a moving person arise at speeds near the speed of light.

Albert Einstein began a revolution in physics with his **special theory of relativity** in 1905. His goal was to reformulate electromagnetic theory so that it does not depend on the motion of an observer. In other words, using the special theory of relativity, both you on Earth and your friend in a rocketship traveling near the speed of light would have the same logical and complete description of electricity and magnetism, devoid of any pitfalls or paradoxes due to your relative motions.

Neither our location in space and time nor our motion through space and time shall prejudice our description of physical reality—this was the lofty principle that guided Einstein. To achieve this goal, Einstein found that he had to abandon the old-fashioned, rigid notions of space and time. For example, imagine your friend whizzing across the solar system in a rocketship while you remain here at rest on the Earth. In order for both of you to agree on the same coherent description of reality, Einstein proved that you must say that your friend's clocks are ticking more slowly than your own and that her rulers (when held parallel to the direction of motion) have become shorter than yours. In short, an observer always finds that moving clocks seem to be slowed and moving rulers seem to be shortened in the direction of motion. Some of the details of the special theory of relativity are discussed in Box 24-1.

After developing the special theory of relativity, Einstein turned his attention to gravity. He began by demonstrating that it is not necessary to think of gravity as a force. According to Newton's theory, an apple falls to the ground because the force of gravity pulls the apple down. Einstein pointed out that the apple would behave exactly in the same way in free space far from any gravity if the floor were accelerating upward. In other words, the floor comes up to meet the apple, as sketched in Figure 24-1.

This is an example of Einstein's **principle of equivalence** which explains that, in a small volume of space, the downward pull of gravity can be accurately and completely duplicated by an upward acceleration of the observer. Indeed, the two gentlemen in their closed compartments Figure 24-1 have no way of telling who is at rest on the Earth and who is in the elevator moving upward at a constantly increasing speed.

This approach allowed Einstein to focus entirely on motion (rather than force) in discussing gravity. From his special theory of relativity, he knew exactly how rulers and clocks are affected by motion, and he could therefore describe gravity entirely in terms of the effects on space and time. Far from a source of gravity, the acceleration is small, and thus the effect on clocks and rulers is small; nearer a source of gravity, the acceleration is larger, and thus the distortion of clocks and rulers is larger. In this way, Einstein "generalized" his special theory and arrived at the **general theory of relativity.**

The general theory of relativity describes gravity entirely in terms of the geometry of space and time. Far from a source of gravity, space is "flat," and clocks tick at their normal rate. As you move closer to a source of gravity (a mass), however, clocks slow down and space becomes increasingly curved, as seen by someone outside the region affected by the mass.

Figure 24-1 The equivalence principle
The equivalence principle asserts that you cannot distinguish between being at rest in a gravitational field and being accelerated upward in a gravity-free environment. This idea was an important step in Einstein's quest to develop the general theory of relativity.

The Earth

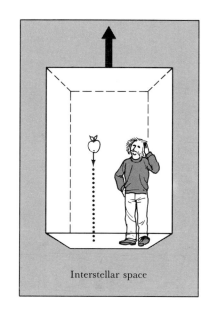

Interstellar space

In a weak gravitational field, Einstein's general relativity gives the same results as the classical theory of Newton. But in stronger gravity, such as that near the Sun's surface, the two theories give different predictions. Examples include the precession of Mercury's perihelion and the deflection of a light ray grazing the Sun, which we discussed in Chapter 4 (recall Figures 4-15 and 4-17). In these and other situations, general relativity has withstood numerous tests. It is by far the most elegant and accurate description of gravity ever devised.

Box 24-1   Some comments on special relativity

The special theory of relativity describes the way in which measurements of time, distance, and mass are affected by motion. Einstein proved that these measurements must depend on the speed of the observer in order for all people (both moving and stationary) to agree on certain basic physical phenomena, especially those involving the behavior of light.
(continued)

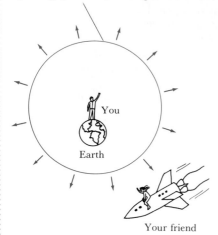

Expanding spherical shell of light

You

Earth

Your friend

(*Box 24-1, continued*)

Imagine that you are sitting at rest on the Earth while your friend is traveling across the solar system at a very high speed, as shown in the sketch. You set off a flash bulb that emits a sudden bright flash of light. The radiation moves away from you equally in all directions, and thus you "see" an expanding spherical shell of light.

What does your high-speed friend see? Einstein argued that your friend also sees an expanding spherical shell of light. For example, she does not see an expanding cube or an expanding football-shaped shell.

By requiring that *both* people observe a *spherical* shell, Einstein derived a series of equations that related specific measurements of time and distance between two people. These equations are called the **Lorentz transformations,** named after the famous Dutch physicst Hendrik Antoon Lorentz, a contemporary of Albert Einstein, who developed these equations independently from other considerations. These equations tell us exactly how a moving person's clocks slow down and rulers shrink.

To appreciate the Lorentz transformations, again imagine that you are on the Earth while your friend is moving at a speed v with respect to you. Suppose that you both observe the same phenomenon on the Earth, which appears to occur over an interval of time. According to your (stationary) clock, the phenomenon lasts for T_0 seconds; according to your friend's moving clock, the same phenomenon lasts for T seconds. The Lorentz transformation for time tells us that these two time intervals are related by

$$T = \frac{T_0}{\sqrt{1 - (v^2/c^2)}}$$

where c is the speed of light. For example, suppose that your friend is moving at 98 percent of the speed of light. Then

$$\frac{v}{c} = 0.98$$

so that

$$T = \frac{T_0}{\sqrt{1 - (0.98)^2}} = 5T_0$$

Thus, a phenomenon that lasts for 1 sec on a stationary clock is stretched out to 5 sec on a clock moving at 98 percent of the speed of light. This phenomenon is often called the **dilation of time.**

The Lorentz transformation for time is plotted in the graph. This graph shows how 1 sec on a stationary clock is stretched out as measured by a moving clock. Note that significant differences between the stationary and moving clocks occur only at speeds near the speed of light. For speeds less than about one-half the speed of light, the mathematical factor $\sqrt{1 - (v^2/c^2)}$ is almost exactly equal to 1, and thus stationary and slowly moving clocks tick at almost exactly the same rate.

In the language of relativity, we say that a clock at rest measures **proper time** (T_0) and a ruler at rest measures **proper distance** (L_0). According to the Lorentz transformations, distances perpendicular to the direction of motion are unaffected. However, a ruler of proper length L_0 held parallel to the direction of motion shrinks to a length L given by

$$L = L_0\sqrt{1 - (v^2/c^2)}$$

For example, if your friend is traveling at 98 percent of the speed of light relative to you, you conclude that her clocks are ticking only one-fifth as fast as yours and that her 1-ft ruler is only about $2\frac{1}{2}$ in. long when held parallel to the direction of motion. This shrinkage of length is often called **Fitzgerald contraction.**

Albert Einstein also demonstrated that measurements of mass are affected

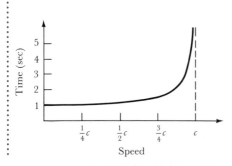

by the velocity of the observer. Specifically, suppose that an object has a **proper mass** M_0 when at rest. If this same object is moving with a velocity v, then it appears to have a mass M given by

$$M = \frac{M_0}{\sqrt{1 - (v^2/c^2)}}$$

Thus a 1-kg brick traveling at 98 percent of the speed of light appears to a stationary observer to behave as though it has a mass of 5 kg.

The mathematical expression $\sqrt{1 - (v^2/c^2)}$ becomes significantly different from 1 only when v is nearly as large as c. Thus the effects of relativity on time, distance, and mass become noticeable only at extremely high speeds. Particle accelerators (such as cyclotrons and synchrotrons) that propel electrons and protons to velocities near the speed of light have tested these predictions of special relativity to a very high degree of accuracy.

Finally, we can see why it is impossible to travel faster than the speed of light. Suppose that your friend climbs aboard a rocketship containing an infinite supply of fuel. With rocket engines blazing, your friend attempts to "break the light barrier." You keep in touch with her by radio and, as she gets closer and closer to the speed of light, you begin to notice the effects of the dilation of time. For example, her speech becomes drawn out to a leisurely drawl as her clock slows down relative to yours. In the same way, the rocket engines appear to you to shut down. The rate at which fuel pours into the rocket engines (in gallons per minute) is directly associated with the passage of time and is therefore subject to time dilation. As the velocity approaches the speed of light, the factor $\sqrt{1 - (v^2/c^2)}$ approaches zero. Your friend therefore never gets the chance to burn that last drop of fuel that would put her past the speed of light.

A black hole is a very simple object that has only a "center" and a "surface"

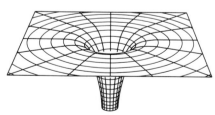

Figure 24-2 The geometry of a black hole
This diagram shows how the shape of space is distorted by the gravitational field of a black hole. Far from the hole, gravity is weak, and space is "flat". Near the hole, gravity is strong, and space is highly curved. This diagram uses a two-dimensional surface as an analogy for the three-dimensional fabric of space.

Imagine a dying star too massive to become either a white dwarf or a neutron star. The overpowering weight of the star's burned-out matter pressing inward from all sides causes the star to contract rapidly. The strength of gravity around the collapsing star increases dramatically as the star's matter is compressed to enormous densities inside the rapidly shrinking sphere. According to the general theory of relativity, distortions of space and time become increasingly pronounced around the dying star. Finally, the escape velocity from the star's surface equals the speed of light, and the star seems to disappear from the universe. At this stage, space has become so severely curved that a hole is punched in the fabric of the universe. The dying star disappears into this hole in space, leaving behind only a **black hole.**

The geometry of space around a black hole is sketched in Figure 24-2. Note that space is flat far from the hole because gravity is weak there. Near the hole, however, gravity is strong, and the curvature of space is severe.

The location in space where the escape velocity from the black hole equals the speed of light is called the **event horizon.** This sphere is also sometimes called the surface of the black hole. Once a massive dying star collapses inside its event horizon, it permanently disappears from the universe. Indeed, the term "event horizon" is very appropriate; this surface is literally a horizon in the geometry of space beyond which we cannot see any events.

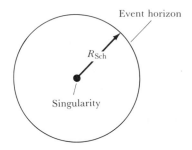

Figure 24-3   The structure of a black hole
A black hole has only two parts: a singularity surrounded by an event horizon. Inside the event horizon, the escape velocity exceeds the speed of light, and thus the event horizon is a one-way surface. Things can fall in, but nothing can get out.

In addition to curving space, gravity causes time to slow down. If you stood at a safe distance, and watched your friend fall toward a black hole, you would note that her clocks tick more and more slowly. In fact, at the event horizon, you would conclude that her clocks have stopped entirely.

Once a dying star has contracted inside its event horizon, no forces in the universe can prevent the complete collapse of the star down to a single point at the center of the black hole. The star's entire mass is crushed to infinite density at this point, called the **singularity.** The singularity is often called the center of the black hole.

The structure of a black hole is therefore very simple. As sketched in Figure 24-3, it has only two parts: a singularity (the center) surrounded by an event horizon (the surface). The distance between the singularity and the event horizon is the **Schwarzschild radius,** named after the German astronomer Karl Schwarzschild who in 1916 was the first person to completely solve Einstein's equations of general relativity. The Schwarzschild radius (R_{Sch}) is related to the mass M of the black hole by

$$R_{Sch} = \frac{2GM}{c^2}$$

where c is the speed of light and G is the universal constant of gravitation. For example, for a $10M_\odot$ black hole, the Schwarzschild radius is 30 km.

To understand why the complete collapse of the doomed star is inevitable, think about your own life here on Earth, far from any black holes. You have the freedom to move as you wish through the three dimensions of space. You can move up and down, left and right, or forward and back. But you do not have freedom to move at will through the dimension of time. Whether we like it or not, we are all carried inexorably from the cradle to the grave.

Inside a black hole, powerful gravity distorts the shape of space and time so severely that the directions of space and time become interchanged. In a limited sense, inside a black hole you can have freedom to move through time. It does you no good, however, because you lose a corresponding amount of freedom to move through space. Whether you like it or not, you are dragged inexorably from the event horizon to the singularity. Just as no force in the universe can prevent the forward march of time (past to future) outside a black hole, no force in the universe can prevent the inward march of space (event horizon to singularity) inside a black hole.

At the singularity, the strength of gravity is infinite, and thus the curvature of space and time is infinite. This means that space and time at the singularity are all jumbled up. Space and time do not exist as separate, identifiable entities.

This confusion of space and time has profound implications for what goes on inside a black hole. All of the laws of physics require a clear and distinct background of space and time. Indeed, without this identifiable background, we could not even speak rationally about the arrangement of objects in space or the ordering of events in time. Because space and time are all jumbled up at the center of a black hole, the singularity does not obey the laws of physics. The singularity behaves in a random and capricious fashion, totally devoid of any rhyme or reason.

Fortunately, we are shielded from the singularity by the event horizon. In other words, although irrational things happen at the singularity, none of the effects manage to escape to the outside universe. Consequently the outside universe remains understandable and predictable.

The irrational, random behavior of the singularity is so disturbing to physicists that, in 1969, the British mathematician Roger Penrose and colleagues proposed a **law of cosmic censorship:** "Thou shalt not have naked singularities." In other words, every singularity must be completely surrounded by an event horizon; otherwise, these singularities would affect the universe in an unpredictable and random way.

The structure of a black hole is completely described with only three numbers

In addition to shielding us from singularities, the event horizon prevents us from ever knowing many details about the material that fell into a black hole. For example, there is no way we could ever discover the chemical composition of the massive star whose collapse produced a black hole. Even if someone went into a black hole and made a measurement or chemical test, there is no way that observer could get any of this information back out to the outside world. Indeed, a black hole is an "information sink," because infalling matter carries many properties (chemical composition, texture, color, shape, size) that are forever removed from the universe.

Because a black hole removes information from the universe, there is no way that this information can affect the structure or properties of the hole. For example, consider two black holes—one made from the gravitational collapse of $10M_\odot$ of iron, and the other made from the gravitational collapse of $10M_\odot$ of peanut butter. Obviously, very different substances went into the creation of the two holes. Once the event horizons formed, however, both the iron and the peanut butter permanently disappeared from the universe. As seen from the outside, the two holes are absolutely identical, and we cannot tell which ate the peanut butter and which ate the iron. In this way, a black hole is unaffected by the information that it destroys.

Because a black hole is an information sink, you might wonder whether we can know or measure anything about a black hole. In other words, what properties characterize a black hole?

First, we can measure the total **mass** of a black hole. We could do this, for example, by placing a satellite in orbit about the hole. Kepler's third law (recall Box 4-3) relates a satellite's orbital period and semimajor axis to the mass around which the satellite is moving. Thus, after measuring the size and period of the satellite's orbit, we can use Kepler's third law to determine the mass of the hole. This mass is equal to the total mass of all the material that went into the hole.

Incidentally, science fiction abounds with nasty rumors that black holes are evil things that go around gobbling everything in the universe. Not so! The bizarre effects of highly warped space and time are limited to a few million kilometers around the hole. Farther from the hole, gravity is sufficiently weak that Newtonian physics adequately describes everything. For example, at a distance of only a few astronomical units from a $10M_\odot$ black hole, the behavior of gravity is identical to that of any ordinary $10M_\odot$ star.

In addition to the total mass, we can also measure the total **electric charge** possessed by a black hole. Like the gravitational force, electric force is a long-range interaction, and thus its effects are felt in the space around the hole. Appropriate equipment on a space probe passing near the hole could measure the intensity of the electric field around the hole, and the electric charge can be determined from these measurements.

Box 24-2 Gravitational radiation

A gravitational wave is a ripple in the overall geometry of space and time. As an example of how these ripples are produced, think of a man whose mass is 80 kg. All matter has gravity, and general relativity tells us that gravity curves space and time. Thus, the 80-kg man is surrounded by a slight warping of space and time commensurate with his mass. Now suppose that the man begins waving his arms. Although his total mass does not change, the details of how his mass is distributed do change. The geometry of space and time must respond to these changes because the gravitational field of the man with his hands over his head is slightly different than that of the man with his hands by his side. These minor readjustments appear as tiny ripples in the overall geometry of space and time surrounding the man. In the same way, a bouncing ball, the Moon going around the Earth, or two stars in a binary all produce gravitational waves. From the equations of general relativity, it is possible to prove that gravitational radiation moves outward from its source at the speed of light.

Gravitational waves are very difficult to detect because they carry very little energy. To appreciate how weak gravitational waves are, imagine two electrons separated by a small distance. Because they possess both mass and charge, these electrons exert both gravitational and electric forces on each other. The gravitational force is about 10^{42} times weaker than the electric force. If these two electrons are caused to wiggle back and forth, they will radiate both gravitational and electromagnetic waves. Because gravity is so much weaker than electromagnetism, the resulting gravitational waves are subdued by a factor of 10^{-42} compared to the electromagnetic waves.

Joseph Weber at the University of Maryland has pioneered the design and construction of gravitational wave antennas. Because gravitational waves are ripples in the geometry of space, they distort the shapes of objects through which they pass. Weber's antennas consist of large aluminum cylinders covered with pressure-sensitive crystals that produce a small voltage in response to any changes in the exact shape of the cylinders. In this way, the passage of a gravitational wave is converted into an electric signal that can be amplified and recorded.

Processes involving dramatic changes in intense gravitational fields produce the strongest bursts of gravitational radiation. For example, the collapse of a massive star's core during a supernova explosion emits substantial gravitational radiation. Of course, using ordinary telescopes we cannot observe the actual core collapse because the outer layers of the supernova emit an overpowering amount of light. However, gravitational waves from the collapsing core carry detailed information about how this dense matter is being rearranged. Thus, with a gravitational wave antenna, we should be able to observe directly the creation of a neutron star or black hole. Many physicists are hopeful that Weber-type antennas can be improved to the extent that we will be able to detect these awesome processes before the end of the century.

Although a burst of gravitational waves has not yet been conclusively detected, many astronomers feel that the effects of gravitational radiation have been observed. In 1974, Joseph Taylor and his colleagues at the University of Massachusetts discovered a pulsar in a binary system. In fact, the system apparently consists of two neutron stars separated by only 2.8 solar radii. One of the neutron stars emits radio pulses every 0.059 sec, and the orbital period of the two stars about each other is only 7.75 hours. As a result, the average orbital velocity is about 0.1 percent of the speed of light. Because these two stars have strong gravitational fields and because they are moving so rapidly, the entire system should be a substantial source of gravitational waves. As gravitational radiation carries energy away from the system, the two stars spiral in closer and closer to each other, and thus the orbital period of the two stars should decrease. Because one of the stars is a pulsar, radio astronomers have been able to measure the orbital period with extreme accuracy. These observations suggest that the two stars are indeed spiraling in toward each other due to the emission of gravitational waves.

In reality, we would not expect a black hole to possess any appreciable electric charge. For example, if a hole did start off with a sizable positive charge, it would vigorously attract vast numbers of negatively charged electrons from the interstellar medium, and the hole's charge would soon be neutralized. For this reason, astronomers neglect electric charge when discussing real black holes.

Although a black hole might have a tiny electric charge, it cannot have any magnetic field whatsoever. It can be mathematically proved that Einstein's equations do not permit a north-pole/south-pole asymmetry in the geometry of space around a black hole. In the creation of a realistic black hole, we would expect that the collapsing star might possess an appreciable magnetic field. This magnetic field must be radiated away in the form of electromagnetic and **gravitational waves** before the dead star can settle down inside its event horizon. Gravitational waves are ripples in the overall geometry of space, and some physicists are exploring the possibility of observing the creation of black holes by detecting bursts of **gravitational radiation** emitted by collapsing massive stars. A discussion of gravitational radiation appears in Box 24-2.

In addition to mass and electric charge, we can also measure a black hole's total angular momentum. Because of the conservation of angular momentum (recall Box 23-1), we expect a black hole to be spinning very rapidly. Einstein's theory makes the startling prediction that this rotation causes space and time to be dragged around the hole. A spinning black hole is therefore surrounded by space that rotates with the hole. Indeed, around the event horizon of every rotating black hole is a region where this dragging of space and time is so severe that it is impossible to stay fixed in the same place. No matter what you do, you get pulled around the hole along with the rotating geometry of space and time. This region, where it is impossible to be at rest, is called the **ergosphere** (see Figure 24-4).

To measure a black hole's angular momentum, we could place two satellites in orbit about the hole. Suppose that one satellite circles the hole in the same direction as the hole rotates, while the other satellite circles in the opposite direction. One satellite is therefore carried along with the rotating geometry of space and time while the other is constantly fighting its way "upstream." Thus, the two satellites will have different orbital periods. From a comparison of the periods, the total angular momentum of the hole can be deduced.

And that is all. A black hole possesses no qualities other than mass, charge, and angular momentum. This is the essence of the famous **no-hair theorem** first formulated in the early 1970s: "Black holes have no hair." To put it more formally, a hole's structure is completely specified by only three numbers: its mass, its charge, and its angular momentum. Any and all additional properties carried by the matter that fell into the hole have disappeared from the universe and thus can have no effect on the structure of the hole.

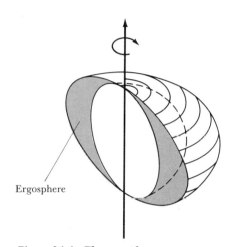

Figure 24-4   The ergosphere
A rotating black hole is surrounded by a region where the dragging of space and time around the hole is so severe that it is impossible for anything to remain at a fixed location. Because the ergosphere (the shaded region in this cross-section diagram) is outside the event horizon, this bizzare region is accessible to us and can be traversed by astronauts or asteroids without disappearing into the black hole. Detailed calculations demonstrate that objects grazing the ergosphere can be catapulted back out into space at tremendous speeds.

Ergosphere

Black holes have been detected in binary-star systems

Finding black holes in the sky is a difficult and often frustrating business. Obviously, you cannot directly observe a black hole in the same sense that you can observe a star or a planet. The best you can hope for is to detect the effects of the hole's powerful gravity.

One option is that a black hole should distort the appearance of background objects. For example, suppose that the Earth, a black hole, and a background star are in nearly perfect alignment, as sketched in Figure

24-5. Because of the warped space around the hole, there are two paths along which light rays can travel from the star to us here on Earth. Thus we should see two images of the background star.

This distortion of background stars is called a **gravitational lens.** It is virtually the only way we can hope to find an isolated black hole in our galaxy. Unfortunately, in order for this effect to be noticeable, the alignment between the Earth, the black hole, and the remote star must be almost perfectly straight. Without a nearly perfect alignment, the secondary image of the background star is too faint to be noticed. Although there are many stars in our galaxy, the probability of finding a significant number of alignments is virtually zero. Thus, although isolated black holes almost certainly exist in interstellar space, they are very difficult to detect.

Figure 24-5   A gravitational lens
A black hole should deflect light rays from a distant star so that an observer sees two images of the background star. Several gravitational lenses have been discovered in which light from a remote quasar is deflected by an intervening galaxy. No gravitational lenses caused by black holes, however, have yet been discovered.

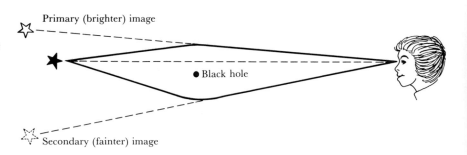

We have a much better chance of finding a black hole in a double-star system. For example, a black hole might capture gas from its companion star, and the fate of this material might reveal the existence of the hole.

Shortly after the launch of Uhuru in the early 1970s, astronomers became intrigued with an X-ray source called Cygnus X-1. The source is highly variable and irregular. Its X-ray emission flickers on time scales as short as one-hundredth of a second. One of the fundamental concepts in physics is that nothing can travel faster than the speed of light (recall Box 24-1). This means that an object cannot vary its brightness or flicker faster than travel time of light across the object. Because light travels 3000 km in 0.01 sec, Cygnus X-1 must be smaller than the Earth.

The X-ray detectors on the Uhuru satellite have poor resolution, and thus astronomers could not pinpoint the location of Cygnus X-1 from X-ray data alone. However, in the spring of 1971, Cygnus X-1 produced major X-ray fluctuations that were accompanied by the appearance of a weak radio source in the same part of the sky. Suspecting that the X-ray fluctuations and the radio source were caused by the same object, astronomers turned to large radio telescopes capable of high resolution to locate Cygnus X-1. The 140-ft dish at the National Radio Astronomy Observatory was used to identify the star HDE 226868 (see Figure 24-6) at the site of the radio source.

Spectroscopic observations promptly revealed that HDE 226868 is a B0 supergiant. Such stars do not emit significant amounts of X rays, and thus HDE 226868 alone cannot be Cygnus X-1. Double stars are very common, however, and—as far as anyone could tell, both HDE 226868 and Cygnus X-1 are at the same location. Thus astronomers began to suspect that the visible star and the X-ray source are in orbit about each other.

Figure 24-6   HDE 226868
This star is the optical companion of the X-ray source Cygnus X-1. The star is a B0 supergiant and is located 8000 light years from Earth. Many astronomers agree that Cygnus X-1 is probably a black hole. This photograph was taken with the 200-in. telescope at Palomar. (Courtesy of J. Kristian, Mt. Wilson and Las Campanas Observatories)

Further spectroscopic observations soon showed that the spectral lines in the spectrum of HDE 226868 shift back and forth with a period of 5.6 days. This is characteristic of a single-line spectroscopic binary; the companion of HDE 226868 is too dim to produce its own set of spectral lines. The clear implication is that HDE 226868 and Cygnus X-1 are the two components of a double-star system.

From the mass–luminosity relation, HDE 226868 is estimated to have a mass of roughly $30M_\odot$ This information implies that Cygnus X-1 also must be fairly massive; otherwise it would not exert enough gravitational pull to make the B0 star wobble by the amount deduced from the periodic Doppler shifting of its spectral lines. Specifically, Cygnus X-1 must have a mass greater than $6M_\odot$. This is too massive to be either a white dwarf or a neutron star, so the only remaining possibility is a black hole.

Of course, the X rays from Cygnus X-1 do not come from the black hole itself. Gas captured from HDE 226868 goes into orbit about the hole, forming an accretion disk about 4 million kilometers in diameter (see Figure 24-7). As material in the disk spirals in toward the hole, friction heats the gas to temperatures approaching 2 million kelvins. In the final 200 km above the hole, these extremely hot gases emit the X rays that we detect with our satellites. Presumably, the X-ray flickering is due to small "hot spots" on the rapidly rotating inner edge of the accretion disk. In this way, the black hole's existence is announced by doomed gases just before they plunge to oblivion.

Details of the inner edge of the accretion disk, as well as the development of hot spots, were elucidated in the 1984 by Larry L. Smarr and John F. Hawley of the University of Illinois. Using a Cray-1 supercomputer in West Germany, Smarr and Hawley solved the complicated equations that describe how accreting gas is distributed around a black

Figure 24-7 The Cygnus X-1 system
A stellar wind from HDE 226868 pours matter onto an accretion disk surrounding a black hole. The infalling gases are heated to high temperatures as they spiral in toward the hole. At the inner edge of the disk, just above the black hole, the gases are so hot that they emit vast quantities of X rays.

hole. Their calculations demonstrate that a fat accretion disk (resembling a doughnut) is more stable than a thin disk (resembling Saturn's rings). The inner edge of the accretion disk occurs where the centrifugal force due to the angular momentum of the gas just balances the inward pull of gravity. Along this inner edge, the flow of gas is unstable and hot bubbles can develop (see Figure 24-8).

In the early 1980s, a binary system similar to Cygnus X-1 was identified in a nearby galaxy called the Large Magellanic Cloud. The X-ray source, called LMC X-3, exhibits rapid fluctuations just like those of Cygnus X-1. LMC X-3 circles a B3 main-sequence star every 1.7 days. From the orbital data, astronomers conclude that the mass of the compact X-ray source is probably about $9M_\odot$. Once again, a black hole is the only reasonable candidate for such a massive, compact object.

There are several other rapidly flickering X-ray sources in binary systems. All are excellent black-hole candidates. They include Circinus X-1, as well as GX 339-4 in Scorpius. With these and similar objects, a great deal of observational effort is necessary to rule out all non–black-hole explanations of the data. Only then can we feel confident that additional black holes have been discovered.

Figure 24-8 The development of an accretion disk
Three stages in the evolution of an accretion disk are shown in false color by computer graphics. The black hole is the dot in the center of each view. In (b) bubbles and fingers of hot gas develop because of instabilities along the inner edge of the disk. In (c) the bubbles have extended into the disk. (Courtesy of Larry L. Smarr and John F. Hawley)

a   b   c

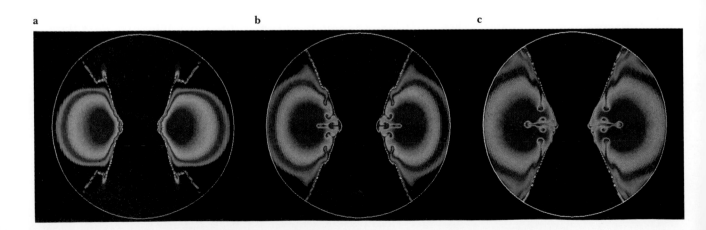

Although finding black holes is a tedious and tricky business, it is becoming clear that black holes should be moderately common. Of course, a black hole is created by the gravitational collapse of a burned-out star whose mass exceeds $3M_\odot$, but that is not the only way a black hole can form. A white dwarf or neutron star in a binary system can be transformed into a black hole by accreting enough matter from its companion star. This can happen when the companion star becomes a red giant and dumps a significant fraction of its mass over its Roche lobe. Another possibility is that two dead stars could coalesce to form a black hole. For example, imagine a binary system consisting of two neutron stars, such as the binary pulsar discussed in Box 24-2. Because of the emission of gravitational radiation, the two stars gradually spiral in toward each other. Eventually, they will merge. If their total mass exceeds $3M_\odot$, the entire system will become a black hole.

A black hole is one of the most bizarre and fantastic objects ever to emerge from modern physical science. Although the idea of black holes was initially met with skepticism, it is now clear that many of the stars we see in the sky are doomed someday to disappear from the universe, leaving only black holes behind. Even more astounding is the idea that enormous black holes—black holes containing millions or even billions of solar masses—are located at the centers of many galaxies and quasars. Indeed, as we shall see in the next chapter, one of these monstrosities may be lurking at the center of the Milky Way, only 10 kiloparsecs from the Earth.

Summary

.   The special theory of relativity asserts that an observer will note a slowing of clocks and a shortening of rulers that are moving with respect to the observer; this effect becomes significant only if the clock or ruler is moving with a speed near the speed of light.

.   The general theory of relativity asserts that gravity causes space to become curved and time to slow down; these effects are significant only in the vicinity of large masses or very compact objects.

.   If a stellar corpse has a mass greater than about $3M_\odot$, gravitational compression will make the object so dense that the escape velocity exceeds the speed of light; the corpse then contracts rapidly to a single point called a singularity.

>   The singularity is surrounded by a surface called the event horizon (where the escape velocity equals the speed of light); nothing—not even light—can escape from the region inside the event horizon.

>   A black hole (a singularity surrounded by an event horizon) has only three physical properties: mass, electric charge, and angular momentum.

>   In a region called the ergosphere around the outside of the event horizon, space and time themselves are dragged along with the rotation of the black hole.

.   The general theory of relativity predicts the existence of gravitational radiation; gravitational waves are ripples in the overall geometry of space and time that are produced by moving masses.

>   It may be possible to study the collapse of stars by studying the gravitational waves emitted during these events; effective antennas for gravitational radiation are under development.

. An isolated black hole might be detected by its effect as a gravitational lens, distorting the image of a star behind it.

. Some are thought to contain binary-star systems a black hole; in such a system, gases captured from the companion star by the black hole emit detectable X rays.

Review questions

1 Under what circumstances are degenerate electron pressure and degenerate neutron pressure incapable of preventing the complete gravitational collapse of a dead star?

2 In what way is a black hole blacker than black ink or a black piece of paper?

3 If the Sun suddenly became a black hole, how would the Earth's orbit be affected?

4 According to general relativity, why can't some sort of yet-undiscovered degenerate pressure prevent the matter inside a black hole from collapsing all the way down to the singularity?

5 Why do you suppose that all of the black hole candidates mentioned in the text are members of very short period binary systems?

6 As a binary system loses energy by emitting gravitational waves, why do its members *speed up* and why does the period become *shorter*?

Advanced questions

*7 To what density must the matter of a dead $10M_\odot$ star be compressed in order for the star to disappear inside its event horizon?

*8 Prove that the density of matter needed to produce a black hole is inversely proportional to the square of the mass of the hole. If you wanted to make a black hole from matter compressed to a density of only $1g/cm^3$, how much mass would you need?

*9 Find the total mass of the neutron star binary system described in the text (see Box 24-2) for which the orbital period is 7.7 hours and the average distance between the neutron stars is 2.8 solar radii. Is your result reasonable for a pair of neutron stars? Explain.

Discussion questions

10 Discuss why the fact that the speed of light is the same for all observers regardless of their motion requires us to abandon the Newtonian view of space and time.

11 Describe the kinds of observations you might make in order to locate and identify black holes.

12 Speculate on the effects you might encounter on a trip to the center of a black hole.

For further reading

Goldsmith, D. "When Time Slows Down." *Mercury,* May/June 1975, p. 2.
            On special relativity.
Kaufmann, W. *Black Holes and Warped Spacetime.* W. H. Freeman, 1979.
Kaufmann, W. *Cosmic Frontiers of General Relativity.* Little Brown, 1977.
Stokes, G., and Michalsky, J. "Cygnus X-1." *Mercury,* May/June. 1979,
            p. 60.
Thorne, K. "The Search for Black Holes." *Scientific Amercian,* Dec. 1974.
Weisberg, J., et al. "Gravitational Waves from an Orbiting Pulsar." *Scientific American,* Oct. 1981.

25 Our galaxy

The galaxy NGC 253
A spiral galaxy, such as our own or NGC 253, contains a significant amount of interstellar dust, primarily concentrated in the galaxy's disk. In our own galaxy, this dust blocks light from remote stars and severely restricts our visual observations in the plane of the Milky Way. In other galaxies, the dust appears as dark blotches and mottling, as shown here. The pinkish spots scattered around this galaxy are H II regions that outline the spiral arms. NGC 253 belongs to a cluster of galaxies called the Sculptor group which is nearly 10 million light years from Earth. (Anglo-Australian Observatory)

The hazy band across the night sky called the Milky Way is our edge-on view of the disk of our own galaxy. In this chapter we follow the studies in which astronomers have learned about the size, shape, and rotation of our galaxy. We learn about the processes that produce the spiral arms of our galaxy, and we probe the mysterious source of energy at the galactic center. Our studies of our own galaxy in this chapter prepare us to explore other galaxies in the next chapter.

On a clear, moonless night far from city lights, you can often see a hazy, luminous band stretching across the sky. Ancient peoples devised fanciful myths to account for this "Milky Way" among the constellations. Today we realize that this hazy band is actually our inside view of a vast disk-shaped assemblage of several hundred billion stars that includes the Sun.

In studying the Milky Way, we explore the universe on a large scale and ask comprehensive questions. Instead of examining individual stars, we look at an entire system of stars. Instead of focusing on the location and life of an isolated star, we look at the overall arrangement and history of a huge stellar community, of which the Sun is a member.

The Sun is located in the disk of the galaxy, about 10,000 parsecs from the galactic center

Galileo was the first person to look at the Milky Way through a telescope. He immediately discovered that it is composed of countless dim stars. The Milky Way stretches all the way around the sky in a continuous band that is almost perpendicular to the plane of the ecliptic. Figure 25-1 is a mosaic of several wide-angle photographs showing roughly one-third of the Milky Way.

Because the Milky Way completely encircles us, astronomers in the eighteenth century began to suspect that the Sun and all the stars in the sky are part of an enormous disk-shaped assemblage called **the Milky Way galaxy.** In the 1780s, William Herschel attempted to deduce the Sun's location in the galaxy by counting the number of stars in 683 regions of the sky. Herschel reasoned that he should see the greatest density of stars toward the galaxy's center, whereas a lesser density of stars should be seen toward the edge of the galaxy.

Figure 25-1   The Milky Way
This mosaic of five wide-angle photographs shows a portion of the Milky Way extending from Sagittarius to Cassiopeia. Note the dark lanes and blotches; this mottling is caused by interstellar gas and dust that obscure the light from background stars. (Palomar Observatory)

Herschel found roughly the same density of stars all along the Milky Way. He therefore concluded that we are at the center of our galaxy (see Figure 25-2). In the 1920s, the Dutch astronomer J. C. Kapteyn analyzed the available observations and obtained essentially the same result.

Both Herschel and Kapteyn were wrong about the Sun being at the center of our galaxy. The reason for their mistake was discovered in the 1930s by R. J. Trumpler. While studying star clusters, Trumpler discov-

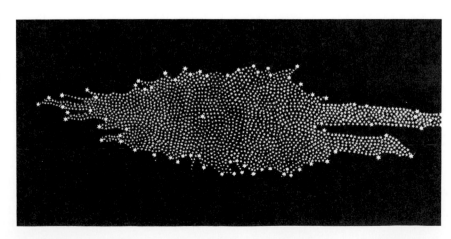

Figure 25-2   Herschel's map of our galaxy
William Herschel attempted to map the Milky Way galaxy by counting the numbers of stars in various parts of the sky. Because of interstellar extinction, he erroneously concluded that the Sun is at the center of the galaxy. (Yerkes Observatory)

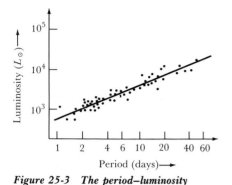

Figure 25-3 The period–luminosity relation for classical Cepheids
This graph shows the relationship between the periods and luminosities of classical (Type I) Cepheid variables. Each dot represents a Cepheid in the Small Magellanic Cloud whose brightness and period have been measured. The line is the "best fit" to the data. (Adapted from H. C. Arp)

ered that remote clusters appear unusually dim—dimmer than would be expected from their distance alone. Trumpler therefore concluded that interstellar space is not a perfect vacuum: it contains dust that absorbs light from distant stars. Like the stars, this obscuring material is concentrated in the plane of the galaxy. Great patches of this interstellar dust are clearly visible in wide-angle photographs such as Figure 25-1.

The dimming of light by the interstellar medium is called **interstellar extinction.** In the plane of our galaxy, the average interstellar extinction is about 1 magnitude per kiloparsec. For example, a star in the Milky Way 5 kpc from Earth appears 5 magnitudes dimmer than it should from its distance alone. In regions of dense interstellar clouds, such as those toward the galactic center, the extinction is even greater. Indeed, at optical wavelengths, the center of the galaxy is totally obscured from our view. Interstellar extinction misled Herschel and Kapteyn. Because they were actually seeing only the nearest stars in the galaxy, they had no true idea either of the enormous size of the galaxy or of the vast number of stars concentrated around the galactic center.

Because interstellar dust is concentrated in the plane of our galaxy, interstellar extinction is strongest in those parts of the sky covered by the Milky Way. However, our view is relatively unobscured to either side of the Milky Way. Knowledge of our position in the galaxy was to come from observations of globular clusters in these unobscured portions of the sky. Before we turn to those observations, we must review some background information on variable stars.

In 1912, the American astronomer Henrietta Leavitt reported her important discovery of the period–luminosity relation for classical (Type I) Cepheid variables. As we saw in Chapter 21 (recall Figure 21-9), Cepheid variables are pulsating stars that vary their brightness in a characteristic way. Leavitt studied numerous Cepheids in the Small Magellanic Cloud (a small galaxy very near the Milky Way) and found that their periods are directly related to their average luminosities (see Figure 25-3). Today, astronomers realize that there are two kinds of Cepheid variables: the metal-rich Type I Cepheids that Leavitt studied, and the metal-poor Type II Cepheids. As shown in Figure 21-10, the Type II Cepheids are slightly dimmer than the Type I.

The period–luminosity law is a very important tool in astronomy because it can be used to determine distances. For example, suppose you find a Cepheid variable in the sky. By measuring its period and using a graph such as Figure 25-3, you promptly discover the star's average luminosity. This is a measure of the star's true brightness and can easily be expressed as an absolute magnitude. Meanwhile, you can observe the star's apparent magnitude. Since you now know both the apparent and absolute magnitudes, you can calculate the star's distance from equations such as those in Box 18-1.

Shortly after Leavitt's discovery, Harlow Shapley, a young astronomer at the Mount Wilson Observatory in southern California, concluded that Cepheid variables are pulsating stars. Actually, Shapley was very interested in a closely related family of pulsating stars called **RR Lyrae variables.** The light curve of an RR Lyrae variable (see Figure 25-4) is very similar to that of a Cepheid, and Shapley believed RR Lyrae variables to be simply short-period classical Cepheids like those Leavitt had studied. RR Lyrae variables are commonly found in globular clusters (see Figure 25-5).

By 1915, Shapley had noticed a peculiar property of globular clusters. Ordinary stars and open star clusters are rather uniformly spread along the Milky Way. The majority of the 93 globular clusters that Shapley

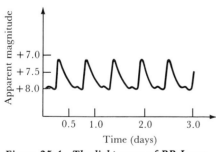

Figure 25-4 The light curve of RR Lyrae
An RR Lyrae variable is recognized by the characteristic way in which it varies its brightness. An RR Lyrae light curve is similar to that of a Cepheid variable (compare Figure 21-9). All RR Lyrae variables have periods of roughly ½ day and vary their brightnesses by about 1 magnitude. The average luminosity of an RR Lyrae variable is approximately 100 Suns. The example shown here is the curve of RR Lyrae, the prototype of this class of stars.

Figure 25-5 [left]   The globular cluster M55

Three RR Lyrae variables are identified in this globular cluster located in the constellation of Sagittarius. From the average apparent brightness (as seen in this photograph) and the average true brightness (known to be roughly 100 suns), astronomers deduce that the distance to this cluster is 20,000 light years. (Harvard Observatory)

Figure 25-6 [right]   A view toward the galactic center

More than 1 million stars are visible in this photograph. This view looks toward a relatively clear "window" just 4° south of the galactic nucleus in Sagittarius. There is surprisingly little obscuring matter in this tiny section of the sky. The two globular clusters are NGC 6522 and NGC 6528. (Kitt Peak National Observatory)

studied, however, were preferentially located in one half of the sky, widely scattered around the portion of the Milky Way in Sagittarius. Figure 25-6 shows two globular clusters in this part of the sky.

Shapley used the period–luminosity relation to determine the distances to the then-known 93 globular clusters in the sky. From their directions and distances, he mapped out the three-dimensional distribution of these clusters in space. By 1917, Shapley had discovered that the globular clusters form a huge spherical system that is not centered on the Earth. Instead, the globular clusters are centered about a point in the Milky Way toward the constellation of Sagittarius. Shapley then made the bold conjecture (subsequently confirmed) that the globular clusters outline the true size and extent of the galaxy.

Today we know that the **disk** of our galaxy is about 100,000 light years in diameter and about 2000 light years thick (see Figure 25-7). The **galactic nucleus** is about 30,000 light years from Earth and is surrounded by a spherical distribution of stars called the **central bulge.** The spherical distribution of globular clusters outlines the **halo** of the galaxy. If we could view our galaxy edge-on from a very great distance, it would probably look very much like NGC 4565, shown in Figure 25-8.

Figure 25-7   The structure of our galaxy (edge-on view)

There are three major components of our galaxy: a thin disk, a central bulge, and a halo. The disk contains gas and dust along with metal-rich (population I) stars. The halo is composed almost exclusively of old, metal-poor (population II) stars. The central bulge is a mixture of population I and population II stars.

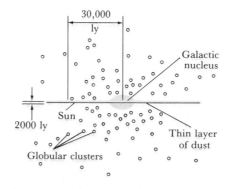

Figure 25-8   The edge-on view of galaxy NGC 4565
If we could view our galaxy edge-on from a very great distance, it would probably look like this galaxy in the constellation of Coma Berenices. The thin layer of dust and gas is clearly visible in the plane of the galaxy. Also note the reddish color of the bulge that surrounds the galaxy's nucleus. (U.S. Naval Observatory)

The spiral structure of our galaxy is determined from radio observations

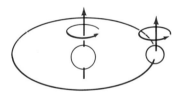

Parallel spins

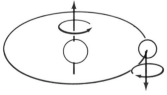

Antiparallel spins

Figure 25-9   The hyperfine structure of the hydrogen atom
In the lowest orbit of the hydrogen atom, the electron and the proton can be spinning either in the same direction or in opposite directions. When the electron flips over, the atom either gains or loses a tiny amount of energy. This energy is either absorbed or emitted as photons with wavelengths of 21 cm.

Because interstellar dust effectively obscures our visual views in the plane of our galaxy, a detailed understanding of the structure of the galactic disk had to wait for the development of radio astronomy. Because of their long wavelengths, radio waves easily penetrate the interstellar medium without being scattered or absorbed. As we shall see in this section, radio observation reveal that our galaxy has **spiral arms,** spiral-shaped concentrations of gas and dust unwinding from the center in a shape reminiscent of a pinwheel.

We have seen that hydrogen is by far the most abundant element in the universe. Hence, by looking for concentrations of hydrogen gas, we should detect important clues about the structure of the disk of our galaxy. Unfortunately, the major electron transitions in the hydrogen atom (recall Figure 17-13) produce photons at ultraviolet and visible wavelengths that do not penetrate the interstellar medium. What hope do we have of detecting all this hydrogen?

In addition to mass and charge, particles such as protons and electrons possess a tiny amount of angular momentum commonly called **spin.** Indeed, an electron or a proton can be crudely visualized as a tiny spinning sphere. According to the laws of quantum mechanics, the electron and proton in a hydrogen atom can be spinning either in the same direction or in opposite directions (see Figure 25-9), but they can have no other relative spin orientations (the spin direction is quantized). If the electron flips over from one configuration to the other, the hydrogen atom must gain or lose a tiny amount of energy. For example, in going from parallel to antiparallel spins, the atom emits a low-energy photon whose wavelength is 21 cm.

In 1944, the Dutch astronomer H. C. van de Hulst predicted that 21-cm radiation from this **spin-flip transition** in interstellar hydrogen could be detected as soon as appropriate radio telescopes were constructed. In 1951, a team of astronomers at Harvard succeeded in detecting the faint hiss of 21-cm radio static from interstellar hydrogen.

The detection of 21-cm radio radiation was a major breakthrough that permitted astronomers to map the structure of the galactic disk. To see how this is done, suppose that you aim your radio telescope along a particular line of sight across the galaxy as sketched in Figure 25-10.

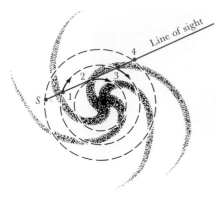

Figure 25-10   A technique for mapping our galaxy
Hydrogen clouds at different locations along our line of sight are moving at slightly different speeds. Radio waves from various gas clouds are therefore subjected to slightly different Doppler shifts, permitting radio astronomers to sort out the gas clouds and map the galaxy.

Your radio receiver picks up 21-cm emission from hydrogen clouds at points 1, 2, 3, and 4; point *S* is the location of the Sun. However, the radio waves from these various clouds are Doppler-shifted by slightly different amounts because they are moving away from you at slightly different speeds as they travel with the rotating galaxy.

It is important to remember that the Doppler shift reveals only motion parallel to the line of sight (recall Box 7-3). Note that cloud 2 has the highest line-of-sight speed because it is moving directly away from us. The radio waves from cloud 2 therefore exhibit a larger Doppler shift than those from any other clouds along the line of sight. Clouds 1 and 3 are at the same distance from the galactic center and thus have the same orbital velocity. The fraction of their velocity parallel to the line of sight is also the same, and thus their radio waves exhibit the same Doppler shift (which is less than the Doppler shift of cloud 2). Cloud 4 is at the same distance from the galactic center as the Sun. It is therefore orbiting the galaxy at the same speed as the Sun, and thus there is no net motion along the line of sight. Radio waves from cloud 4 (as well as radio waves from hydrogen gas near the Sun) are not Doppler-shifted at all.

The final result is that the 21-cm radiation is smeared out over a range of wavelengths. Because radio waves from gas in different parts of the galaxy arrive at the telescope with slightly different wavelengths, it is possible to sort out the various gas clouds and thereby map the galaxy. Figure 25-11 is a map of the galaxy produced in this fashion.

The arrangement of long, curved lanes of hydrogen gas in the map clearly suggests that our galaxy has a spiral structure. Further clues about these spiral arms are obtained from studying the locations of OB associations and H II regions in the sky. Unfortunately, interstellar ab-

Figure 25-11   A map of our galaxy
This map was based on radio-telescope surveys of 21-cm radiation. The distribution of hydrogen gas across our galaxy clearly reveals a spiral structure. The Sun's location is indicated with a cross (+). Details in the large, blank, wedge-shaped region (toward the bottom of the map) are unknown because gas in this part of the sky is moving perpendicular to our line of sight and thus does not exhibit a detectable Doppler shift. (Courtesy of G. Westerhout)

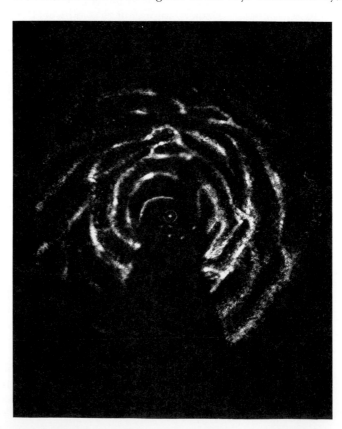

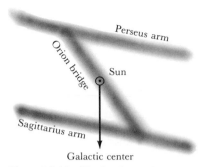

Figure 25-12   The local spiral arms
The Sun is located between two major spiral arms of our galaxy: the Sagittarius arm, and the Perseus arm (named for constellations through which they pass). Some astronomers believe that the Sun may be part of a bridge of stars (called the Orion bridge, or the Orion spur) that connects the two spiral arms.

sorption limits visual observations in the plane of the galaxy to a range of less than 3000 pc from the Earth. Nevertheless, from looking at photographs of other galaxies, we know that these very bright stars and nebulae outline spiral arms. From mapping the locations of nearby O and B stars and H II regions, astronomers have discovered that the Sun is located between two major spiral arms (see Figure 25-12). The Sagittarius arm is nearer to the galactic center than the Sun is. This is the arm you see during the summer months when you look at the portion of the Milky Way stretching across Scorpius and Sagittarius. During the winter months, when our nighttime view is directed away from the galactic center, the Milky Way we see is part of the Perseus arm. If we could view our galaxy from afar, it would probably look somewhat like the spiral galaxy shown in Figure 25-13.

Figure 25-13   The spiral galaxy NGC 6946
If we could view our galaxy face-on from a very great distance, it would look somewhat like this spiral galaxy in the constellation of Cygnus. The spiral arms are outlined by numerous H II regions and OB associations. (U.S. Naval Observatory)

Moving at half a million miles per hour, the Sun takes 230 million years to complete one orbit of our galaxy

Spiral arms suggest that galaxies rotate. Our galaxy must be rotating; otherwise, all the stars would fall into the galactic center. However, detecting and measuring the rotation of our galaxy has been a difficult business.

Radio observations of 21-cm radiation from hydrogen gas give important clues about our galaxy's rotation. By measuring Doppler shifts, astronomers can at least determine speeds parallel to our line of sight across the galaxy. These observations clearly indicate that our galaxy does not rotate like a rigid body but rather exhibits **differential rotation:** stars at different distances from the galactic center travel at different orbital speeds about the galaxy.

Further clues to this rotation come from examining the motions of stars in the sky. Because of differential rotation, the Sun is like a car on a circular freeway with the fast lane on one side and the slow lane on the other side. As sketched in Figure 25-14, stars in the fast lane are passing the Sun and thus appear to be moving in one direction, while stars in the slow lane are being overtaken by the Sun and therefore appear to be moving in the opposite direction.

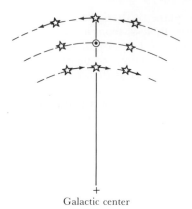

Figure 25-14   Differential rotation of the galaxy
Different parts of the galaxy are rotating at different speeds. Thus, stars traveling more quickly than the Sun appear to be moving in one direction, while stars traveling more slowly than the Sun appear to be moving in the opposite direction.

This analysis of stellar motions was pioneered by the Dutch astronomer Jan H. Oort. Unfortunately, like the 21-cm observations, this study reveals only how fast things are moving relative to the Sun. Of course, the Sun itself is moving. To get a complete picture of our galaxy's rotation, we must find out how fast the Sun is traveling.

A method of doing this was proposed by the Swedish astronomer Bertil Lindblad. Not all of the stars in the sky move in the orderly pattern sketched in Figure 25-14. Distant galaxies and some components of our galaxy (such as globular clusters and RR Lyrae variables) do not seem to participate in the general rotation of the galaxy; instead they have more-or-less random motions. Lindblad took the average of these random motions as a background, using Doppler shifts of spectral lines to measure the Sun's speed with respect to the background in various directions. From these measurements, he concluded that the Sun moves along its orbit about the galactic center at a speed of 250 km/sec, or about $\frac{1}{2}$ million miles per hour.

We know that we are 30,000 light years from the galactic center, so we can use the Sun's speed v to calculate the Sun's orbital period T, the time required for one trip about the Sun's orbit:

$$T = \frac{2\pi r}{v} = \frac{2\pi \times 30,000 \text{ ly}}{250 \text{ km/sec}} \times \frac{9.46 \times 10^{12} \text{ km}}{1 \text{ ly}}$$

$$= 7.1 \times 10^{15} \text{ sec} = 2.3 \times 10^8 \text{ years}$$

Traveling at $\frac{1}{2}$ million miles per hour, it takes us 230 million years to complete one trip around the galaxy. These results demonstrate how vast our galaxy is.

Because we now know the details of our orbit around the galaxy, we can use Kepler's third law (see final equation in Box 4-3) to estimate the mass of the galaxy. The relevant version of Kepler's third law is

$$M = rv^2/G$$

where M is the mass of the galaxy interior to the Sun's orbit, r is the radius of the Sun's orbit, v is the Sun's orbital speed, and G is the gravitational constant. Putting in the numbers, we obtain a mass of $1.3 \times 10^{11} M_\odot$.

This estimate must be too low because Kepler's law gives us only the mass inside the Sun's orbit. The matter exterior to the Sun's orbit does not affect the Sun's motion and thus does not enter into Kepler's law. But obviously there is matter out there. In recent years, astronomers have been astonished to discover how much matter actually lies beyond the orbit of the Sun.

Because we know the true speed of the Sun, we can convert the Doppler shifts measured by radio astronomers into actual speeds of the spiral arms. This gives us a **rotation curve,** a graph of the velocity of galactic rotation measured outward from the galactic center (see Figure 25-15).

This graph is surprising because the rotation curve keeps rising, even out to a distance of 60,000 light years from the galactic nucleus. This means that we still have not detected the actual edge of our galaxy. According to Kepler's third law, the orbital speeds of stars or gas clouds beyond the confines of most of the galaxy's mass should decrease with distance away from the galaxy's center, just as the orbital speeds of the planets decrease with distance away from the Sun (remember that most of the solar system's mass is in the Sun). Instead, galactic orbital speeds continue to climb well beyond the visible edge of the galactic disk. This

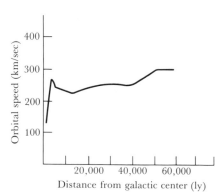

Figure 25-15   The galaxy's rotation curve
The orbital velocity of stars and gas is observed to increase out to a distance of at least 60,000 light years from the galactic center. This means that vast quantities of subluminous matter must surround our Galaxy.

indicates that a surprising amount of matter must be scattered around the edges of our galaxy. In fact, our galaxy's mass could easily be at least $6 \times 10^{11} M_\odot$. To make matters even more mysterious, this outlying matter is dark; it does not show up on photographs. Many astronomers suspect that this hidden matter is spherically distributed all around the galaxy along with the globular clusters. Thus our galaxy's halo is more massive than previously expected. The nature of this hidden mass (black holes? gas? dim stars?) is a complete mystery.

Spiral arms are caused by density waves that sweep across the galaxy

That spiral arms should exist at all was another mystery that confounded astronomers for many years. Many galaxies exhibit these beautiful arching arms outlined by brilliant H II regions and OB associations (recall Figure 25-13). As we think through the effects of a galaxy's rotation, a dilemma arises. All spiral galaxies have rotation curves similar to our own. As Figure 25-15 demonstrates, the velocity of stars and gas is fairly constant over a large portion of a galaxy's disk. However, the farther stars are from a galaxy's center, the farther they must travel to complete one orbit of the galaxy. Thus stars and gas in the outskirts of the galaxy take much longer to complete an orbit than does material near the galaxy's center. Consequently, the spiral arms should eventually wind up. After a few galactic rotations, the spiral structure should disappear altogether.

The enigma of spiral structure intrigued several brilliant and ingenious scientists. Among the first was Bertil Lindblad, who in the 1920s argued that the spiral arms of a galaxy are merely a *pattern* that moves among the actual stars. For example, think about waves on the ocean. As the waves move across the surface of the water, the individual water molecules simply bob up and down in little circles. A cork in the water bobs up and down as the waves ripple by. The waves are simply a pattern that moves across the water; no water actually travels along with the wave pattern. Indeed, Lindblad spoke of **density waves** in discussing the possible cause of spiral structure.

This density-wave theory was greatly elaborated and mathematically embellished by the American astronomers C. C. Lin and Frank Shu in the mid-1960s. Lin and Shu argued that density waves passing through the disk of a galaxy cause material to "pile up" temporarily. A spiral arm therefore is simply a temporary enhancement or compression of the material in a galaxy.

The situation is somewhat analogous to a traffic jam. Imagine some workers painting a line down a busy freeway. The cars normally cruise down the freeway at 55 mph. Because of the crew of painters, however, there is a temporary bottleneck. The cars must slow down temporarily to avoid hitting anyone. As seen from the air, there is a noticeable congestion of cars around the painters. An individual car spends only a few moments in the traffic jam before resuming the usual 55-mph speed. The traffic jam, however, lasts all day long, inching its way along the road. The traffic jam—which would be seen so clearly from an airplane—is simply a temporary enhancement of the number of cars in a particular location.

To better understand how a density wave operates in a galaxy think once again about the ocean. If the water molecules were left completely undisturbed, the surface of the ocean would be perfectly smooth. In reality, however, these molecules are constantly buffeted by small disturbances, or **perturbations,** such as the wind. Because of a perturbation, one

a Water wave

b Kinematic wave

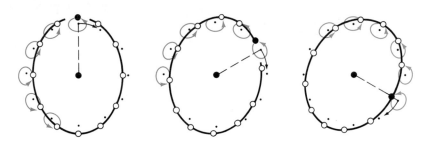

Figure 25-16   Water waves and kinematic waves
(a) *In a water wave, each molecule rotates about a point on the undisturbed water level in a tiny ellipse.* **(b)** *Similarly, a small disturbance in the orbit of a star can cause the star to oscillate in tiny ellipses about its original orbit. The actual path of the star is a precessing ellipse. (Adapted from A. Toomre)*

molecule pushes the next, which pushes the next, and so on. The result is a water wave. Individual molecules on the surface of the ocean move in tiny elliptical paths as the wave pattern moves across the water (see Figure 25-16a).

In a galaxy, the stars are separated by vast distances. Collisions between stars virtually never happen. Nevertheless, stars do interact because they are affected by each other's gravity. In water waves or sound waves, molecular forces are responsible for orchestrating the motions of molecules. In a galaxy, the force of gravity controls the interactions between stars.

Seen from above, the undisturbed orbit of a star about the center of a galaxy would be nearly a perfect circle. However, the motions of other matter in the galaxy produce small gravitational perturbations that cause the star to deviate from its undisturbed orbit. Just as a water molecule bobs up and down on the surface of the ocean, the star oscillates back and forth about its undisturbed orbit. Lindblad demonstrated that these oscillations can be described by thinking of the star as attached to a tiny epicycle. As sketched in Figure 25-16b, the star rotates counterclockwise around the epicycle while the epicycle itself moves clockwise along the undisturbed path. The final path of the star is a **precessing ellipse,** an ellipse that rotates. Of course the gravity of this star affects the motions of its neighbors, and thus a wave disturbance, called a **kinematic wave,** propagates from one stellar orbit to the next.

In 1973, Agris J. Kalnajs in Australia bridged the gap between Lindblad's orbits and the density-wave theory of Lin and Shu. Kalnajs argued that the precessing elliptical orbits of stars are not randomly oriented, as sketched in Figure 25-17a. Instead, there is a strict correlation between orbits: each precessing elliptical orbit is tilted with respect to its neighbor through a specific angle. The result is shown in Figure 25-17b; a beautiful spiral pattern emerges.

The spiral pattern arises in those locations where the ellipses are bunched closest together. Of course, stars in a galaxy are scattered randomly along their orbits. With correlated orbits, however, some of the stars happen to get close together along the huge arching spiral arms. After all, these spiral arms are where the stars' orbits are closest together for the longest stretches.

The temporary enhancement of stars has a profound effect on the interstellar gas and dust. Because of the presence of extra stars, there is

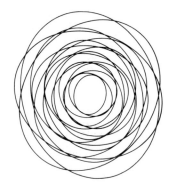

a

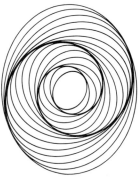

b

Figure 25-17   The origin of spiral density waves
Both drawings have exactly the same number of ellipses, each ellipse representing the orbit of a star. **(a)** *Randomly oriented ellipses.* **(b)** *Ellipses with a correlation between the orientations of adjacent ellipses.*

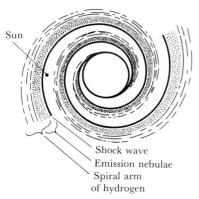

Sun

Shock wave
Emission nebulae
Spiral arm
of hydrogen

Figure 25-18 **Shock fronts and density waves in our galaxy**
A shock wave accompanies the spiral density wave as it sweeps around the galaxy. The resulting spiral arms trail behind the galactic rotation (clockwise in this drawing).

an increased gravitational attraction all along the spiral. This has almost no effect on the massive stars, which simply continue to lumber along their orbits. The lightweight atoms and molecules in the interstellar medium, however, are readily sucked into the gravitational well along the spiral, forming the crest of the density wave.

As Kalnajs's spiral patterns precess, the density waves move through the material of a galaxy at a speed of roughly 30 km/sec (more slowly than the stars are moving). On its own, however, the interstellar gas can transport a disturbance (such as a slight compression) at a speed of only 10 km/sec, which is the speed of sound in the interstellar medium. Thus, the density wave is **supersonic,** because its speed through the interstellar gas is greater than the speed of sound in that gas. As in the case of a supersonic airplane traveling through the air, a **shock wave** is created along the leading edge of the density wave. Shock waves are characterized by a sudden and abrupt compression of the medium through which they move (you hear a sonic boom from a supersonic airplane). The interstellar medium is violently compressed by the cosmic sonic boom of the density wave.

This density-wave theory seems to explain many of the properties of spiral structure. As the spiral density waves sweep through the plane of a galaxy, they recycle the interstellar medium. Old gas and dust left behind from ancient, dead stars are compressed into new nebulae in which new stars are formed. As first noted by M. Fujimoto and W. W. Roberts in the late 1960s, the sprawling dust lanes alongside the string of emission nebulae that outlines a spiral arm attest to the recent passage of a compressional shock wave. Because the material left over from the deaths of ancient stars is enriched in heavy elements, new generations of stars are more metal-rich than were their ancestors. The overall structure of spiral shocks and density waves in our own galaxy is shown schematically in Figure 25-18.

You would be sorely misled to think that we have solved all the questions about spiral structure in galaxies. Many astronomers believe that the density-wave theory is on the right track, but many gaps and loopholes remain in our understanding. For example, what keeps the density waves going? Why don't the density waves fade away? Density waves expend an enormous amount of energy to compress the interstellar gas and dust. In order to keep the density waves going, there must be a constant replenishing of that energy. We really do not understand where this energy comes from, but the nuclei of galaxies seem to be the place to look.

The galactic nucleus may contain a supermassive black hole

The nucleus of our galaxy is a very active, crowded place. The stars in Figure 25-6 give some hint of the stellar congestion. If you lived on a planet near the galactic center, you would see a million stars as bright as Sirius, the brightest single star in our own nighttime sky. The total intensity of starlight from all those nearby stars would be equivalent to 200 of our full moons. Night would never really fall on a planet near the center of our galaxy.

Because of severe interstellar absorption at visual wavelengths, some of our most important information about the galactic nucleus comes from radio and infrared observations rather than from visible light. The pioneering radio observations were made by Jan H. Oort and G. W. Rougoor in 1960. By observing Doppler shifts of 21-cm radiation, Oort and Rougoor discovered two enormous expanding arms of hydrogen gas. One arm is located between us and the galactic center and is ap-

proaching us at a speed of 53 km/sec. The other arm is on the other side of the galactic nucleus and is receding from us with a speed of 135 km/sec. The total amount of hydrogen in these expanding arms is at least several million solar masses. Something quite extraordinary must have happened about 10 million years ago in order to expel such an enormous amount of gas from the center of our galaxy.

In addition to the 21-cm radiation from the expanding arms, radio astronomers also detect a vast amount of continuous-spectrum radio noise coming directly from the galactic nucleus. The radio noise does not come from hydrogen, however. Instead it is produced by high-speed electrons spiraling around a magnetic field. This kind of radio emission is called **synchrotron radiation,** and the powerful source at the galactic center is named **Sagittarius A.** Sagittarius A is one of the brightest radio sources in the entire sky. Recent observations suggest that Sagittarius A is very small. In spite of its enormous energy output, Sagittarius A is only 40 light years in diameter.

Figure 25-19 shows three different views of the center of our galaxy. Figure 25-19*a* is a wide-angle photograph of the Milky Way around Sagittarius at visible wavelengths. Figure 25-19*b* is an infrared view of the field indicated by the white rectangle in Figure 25-19*a*. The prominent band across this view from IRAS is the infrared image of a thin layer of dust in the plane of the galaxy (recall Figure 25-7). Figure 25-19*c* is a detailed IRAS view of the galactic center, covering the area outlined by the white rectangle in Figure 25-19*b*. Note the numerous streamers of dust (in blue) surrounding the galactic nucleus. The strongest infrared emission (in white) comes from Sagittarius A, which is also a very powerful source of radio waves. The galactic center is believed to coincide with the western portion of this source and is called **Sagittarius A West.**

The center of our galaxy is a very active and dynamic place. Numerous infrared sources scattered along the galactic plane and around the galactic center are believed to be giant clouds of gas and dust heated by groups of young O and B stars. In addition, much of the gas around the

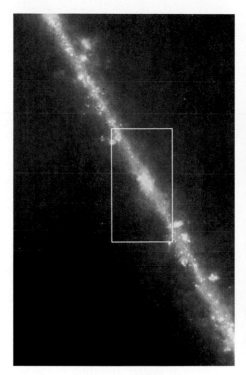

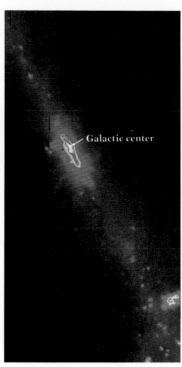

Galactic center

Figure 25-19(b) [left]   The galactic center at infrared wavelengths
This "false-color" infrared view from IRAS covers a field about 48° × 33°. Black represents the dimmest regions of infrared emission, with blue the next dimmest, followed by yellow and red, with white for the strongest emission. The numerous knots and blobs along the plane of the galaxy are interstellar clouds of gas and dust heated by nearby stars. (NASA)

Figure 25-19(c) [right]   The galactic nucleus at infrared wavelengths
Streamers of dust around the galactic center are visible in blue in this "false-color" view from IRAS that covers the region indicated by the white rectangle in Figure 25-19b. The bright object near the lower right is a large H II region (NGC 6357) about 7° southwest of the galactic center. The prominent band across this and the previous IRAS view is a layer of dust in the plane of the galaxy. (NASA)

galactic nucleus is extremely rich in molecules. Indeed, the giant molecular cloud called Sagittarius B2 is a favorite hunting ground for radio astronomers searching for new interstellar chemicals (recall Box 20-2).

The motions of gas clouds within a few light years of the galactic center were studied in the late 1970s by Charles H. Townes and colleagues from the University of California. They found that a spectral line of singly ionized neon, which normally has a wavelength of about 12.80 μ in the infrared, is extremely broad. The shape of this Ne II emission line is shown in Figure 25-20.

Townes and his colleagues interpret this broadening of the Ne II line as a result of the speed at which ionized gas orbits the galactic nucleus. Radiation from gas coming toward us is blueshifted, while radiation from

Figure 25-20   An emission line in the infrared
This graph shows the intensity of an emission line of singly ionized neon in the central parsec of the galaxy. This line is broadened by motions of gas clouds in which neon is a minor constituent. Motion towards us is indicated by negative velocities, while motion away from us is designated with positive velocities. (Adapted from T. R. Geballe)

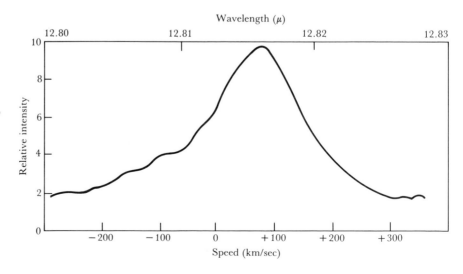

receding gas is redshifted. The final result is to smear out a spectral line over a range of wavelengths corresponding to a range of line-of-sight velocities. As indicated in Figure 25-20, the broadening of the Ne II line reveals a velocity spread of about 400 km/sec. On one side of the galactic nucleus, gas is coming toward us at speeds up to 200 km/sec; on the other side of the galactic nucleus, it is rushing away from us at speeds up to 200 km/sec.

Something must be holding this high-speed gas in orbit about the galactic center. Using Kepler's third law, it is estimated that a $10^6 M_\odot$ of material is needed to prevent this gas from flying off into interstellar space. Townes's observations therefore imply that an object with the mass of a million Suns is concentrated at Sagittarius A West. This object must be extremely compact—much smaller than the few light years over which Townes's observations extended. Many astronomers, such as Martin Ryle of Cambridge University, vigorously argue that an object this massive and this compact could only be a black hole. Because of its enormous mass, it is called a **supermassive black hole.**

Intriguing evidence supporting the idea of a supermassive black hole at the galactic center came in 1983 with high-resolution radio observations at the Very Large Array. A view of the galactic center is shown in Figure 25-21. The bright S-shaped feature strongly suggests oppositely directed jets of high-speed particles squirting out of a central object.

If there is a supermassive black hole at the center of our galaxy, we would expect it to be surrounded by an accretion disk of gravitationally captured matter spiraling in toward the hole. If the rate of infall is high, pressures in the inner regions of the accretion disk expel material in a perpendicular direction to the disk. The situation is entirely analogous to the model of SS433 discussed in Chapter 23 (recall Figure 23-13). Two oppositely directed jets emerge from the accretion disk simply because that is the path of least resistance offered by the material crowding in toward the hole.

Recent measurements of the Doppler shifts of the Ne II line in the

Figure 25-21 **The galactic nucleus**
This color-coded radio map shows the appearance of the center of our galaxy at a wavelength of 6 cm. The strongest radio emission is shown in red; weaker emission is colored green through blue. The bright red object at the middle of this view is a compact radio source precisely at the center of our galaxy. The area covered in this view measures about 10 light years across. (Courtesy of K. Y. Lo and M. J. Claussen)

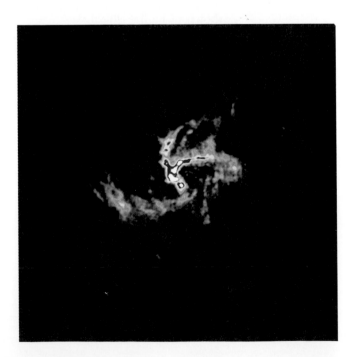

vicinity of the galactic center indicate that the axis of rotation of gas orbiting the galactic nucleus is inclined at a large angle to the plane of the galaxy. If the accretion disk around the supermassive black hole is similarly inclined, then the gravitational field of the rest of the galaxy tends to shift the axis of the accretion disk. The situation is again analogous to that for SS433; the accretion disk precesses. As the accretion disk slowly precesses, the two jets also gradually sweep around the sky. This precession may explain the S-shaped stream seen in Figure 25-21.

Whether our galaxy's nucleus actually harbors a supermassive black hole will be decided only after many years of observation, analysis, and debate. All non–black–hole explanations of the data must be ruled out, or at least rendered ridiculously improbable. Nevertheless, as we shall see in Chapter 27, astronomers find violent activity occurring at the nuclei of other galaxies, also seeming to indicate the presence of supermassive black holes at those galactic centers.

Summary

. Our galaxy has a disk about 100,000 light years in diameter and about 2000 light years thick; there is a high concentration of interstellar dust and gas in the galactic disk.

The galactic nucleus is surrounded by a spherical distribution of stars called the central bulge; the entire galaxy is surrounded by a spherical distribution of globular clusters called the halo of the galaxy.

OB associations and H II regions in the galactic disk outline huge spiral arms.

From studies of the rotation of the galaxy, astronomers estimate that the total mass of the galaxy is probably at least $6 \times 10^{11} M_\odot$, with much of this mass in some nonvisible, unknown form spread outside the edge of the luminous material of the galaxy.

. The Sun is located about 30,000 light years from the galactic nucleus, between two spiral arms.

The Sun moves in its orbit at a speed of about $\frac{1}{2}$ million miles per hour and takes about 230 million years to complete one orbit about the galaxy.

. Interstellar dust obscures our view at visual wavelengths along lines of sight that lie in the plane of the galactic disk.

Hydrogen clouds are detected despite intervening interstellar dust by the 21-cm radio waves emitted in the spin-flip transition.

The galactic nucleus has been effectively studied through its infrared emissions (which also pass readily through intervening interstellar dust).

. Spiral arms are caused by density waves that sweep around the galaxy.

Each star moves in a precessing ellipse about the galactic nucleus; because the orbits are correlated, a spiral pattern is created.

The gravitational attraction of this spiral pattern compresses interstellar clouds through which it passes; this compression triggers the formation of OB associations and H II regions that illuminate the spiral arms.

. A supermassive black hole with a mass of about $10^6 M_\odot$ possibly exists at the galactic center.

This supermassive black hole is surrounded by a huge accretion disk of inrushing gases, producing oppositely directed jets of outrushing material from the faces of the disk.

Review questions

1 Why do you suppose that the Milky Way is far more prominent in July than in December?

2 How would the Milky Way appear if the Sun were located at the edge of the galaxy?

3 Why don't astronomers detect 21-cm radiation from the hydrogen in giant molecular clouds?

4 Describe the Doppler shifts of the 21-cm line that you would observe if the galaxy were rotating like a rigid body.

5 Explain why globular clusters spend most of their time in the galactic halo, even though their eccentric orbits take them very close to the galactic center.

*6 Approximately how many times has the solar system orbited the center of the galaxy since the Sun and planets were formed?

7 How would you estimate the total number of stars in the galaxy?

8 What can you surmise about galactic evolution from the fact that the galactic halo is dominated by population II stars while population I stars are predominately found in the galactic disk?

Advanced questions

*9 The mass of the galaxy interior to the Sun's orbit is calculated from the radius of the Sun's orbit and its orbital speed. By how much would this mass be in error if the calculated distance to the galactic center is off by 10 percent? By how much would this mass be in error if the calculated orbital velocity is off by 10 percent?

*10 The galaxy is about 30,000 parsecs in diameter and 600 parsecs thick. If supernovae occur randomly in the galaxy at the rate of about five each century, how often on the average would we expect to see a supernova within 300 parsecs (1000 light years) of the Sun?

11 Speculate on the reasons for the rapid rise in the galaxy's rotation curve (see Figure 25-15) at distances close to the galactic center.

Discussion questions

12 From what you know about stellar evolution, the interstellar medium, and the density wave theory, explain the appearance and structure of spiral arms in spiral galaxies.

13 What observations would you propose in order to determine the nature of the hidden mass in our galaxy's halo?

For further reading

Bok, B., and P. *The Milky Way*, 5th ed. Harvard U. Press, 1981.
Chaisson, E. "Journey to the Center of the Galaxy." *Astronomy*, Aug. 1980, p. 6.
Weaver, H. "Steps Toward Understanding the Large-Scale Structure of the Milky Way." *Mercury*, Sept./Oct. 1975, p. 18; Nov./Dec. 1975, p. 18; Jan./Feb. 1976, p. 19.

26 Galaxies

The spiral galaxy M66
This galaxy is one member of a cluster of galaxies in Leo and is at a distance of about 9.3 Mpc (30 million light years) from Earth. The galaxy's distorted appearance is due to gravitational interactions with its neighbors, which include the spiral galaxies M65 and NGC 3628. Spiral galaxies are usually found in sprawling clusters with few members. Very rich clusters containing thousands of members are usually dominated by elliptical galaxies. (U.S. Naval Observatory)

Turning our attention to the universe beyond our galaxy, we find that galaxies exist in many shapes and sizes. We learn that galaxies are grouped in clusters rather than being scattered randomly through space, and we discuss some of the spectacular phenomena associated with collisions of galaxies. We find that there is a simple relationship between the distance to a galaxy and the speed with which it moves away from us (as measured by its redshift). In our discussion of this relationship, we learn something about the measurements of intergalactic distances.

William Parsons was the third Earl of Rosse. He was rich, he liked machines, and he was fascinated with astronomy. Accordingly, he set about the business of building gigantic telescopes. In February 1845, his *pièce de resistance* was finished. The telescope's massive mirror measured 6 ft in diameter and was mounted at one end of a 60-ft tube that was controlled by cables, straps, pulleys, and cranes. For a brief period of time, Lord Rosse's contraption enjoyed the dubious reputation of being the largest and most dangerous telescope in the world.

During this reign, Lord Rosse examined many of the nebulae that had been discovered and catalogued by William Herschel. Lord Rosse observed that some of these nebulae have a distinct spiral structure. Perhaps the best example is M51 (also called NGC 5194, the 5194th object in the "New General Catalogue," which is a list of all the nebulae and star clusters observed by William Herschel, his son John Herschel, and others).

Because Lord Rosse did not have any photographic equipment, he had to make drawings of what he saw. His drawing of M51 is shown in Figure 26-1, and a modern photograph appears in Figure 26-2. Views such as this inspired Lord Rosse to echo the famous German philosopher Immanuel Kant, who in 1755 suggested that these objects might be "island universes"—vast collections of stars far beyond the confines of the Milky Way.

Many astronomers did not subscribe to this notion of island universes. Many of the objects listed in the NGC were in fact nebulae and star clusters scattered about the Milky Way and it seemed just as likely that these intriguing spiral nebulae could also be members of our galaxy.

The astronomical community became increasingly divided over the nature of the spiral nebulae. A debate on the topic was scheduled in April 1920 at the National Academy of Sciences in Washington, D.C. On one side was Harlow Shapley, a young, brilliant astronomer renowned for his recent determination of the size of the Milky Way galaxy. Shapley believed that the spiral nebulae are relatively small, nearby objects scattered around our galaxy as are the globular clusters he had studied. Opposing Shapley was Heber D. Curtis of the Lick Observatory near San Jose, California. Curtis championed the island-universe theory and argued that each of these spiral nebulae is a rotating system of stars much like our own galaxy.

The Shapley–Curtis debate generated much heat but little light. Nothing was decided because no one could present any firm arguments or observations to demonstrate exactly how far away the spiral nebulae are. Astronomy desperately needed a definitive method of measuring the distances to the spiral nebulae with which no one could quarrel. This was the first great achievement of a young lawyer who abandoned his law practice in Kentucky and moved to Chicago to study astronomy. His name was Edwin Hubble.

Figure 26-1 [left]   Lord Rosse's sketch of M51
Using a large telescope of his own design, Lord Rosse of England was able to distinguish spiral structure in this "spiral nebula." (Courtesy of Lund Humphries)

Figure 26-2 [right]   The spiral galaxy M51 (also called NGC 5194)
This spiral galaxy in the constellation of Canes Vantici is often called the Whirlpool galaxy because of its distinctive appearance. Its distance from Earth is about 15 million light years. The "blob" at the end of one of the spiral arms is a companion galaxy called NGC 5195. (U.S. Naval Observatory)

Hubble discovered the distances to galaxies and devised a system for their classification

Edwin Hubble joined the staff of the Mount Wilson Observatory in Pasadena, California, and on October 6, 1923, took an historic photograph of the "Andromeda nebula." A modern photograph of this object appears in Figure 26-3; it was one of the spiral nebulae around which so much controversy raged. Careful examination of the photographic plate revealed what was at first thought to be a nova. Reference to previous

Figure 26-3 The Andromeda galaxy (called M31 or NGC 224)
This nearby galaxy covers an area of the sky roughly five times as large as the full moon. Under good observing conditions, the galaxy's bright central bulge can be glimpsed with the naked eye in the constellation of Andromeda. The distance to the galaxy is $2\frac{1}{4}$ million light years. The white rectangle outlines the area shown in Figure 26-4. (Palomar Observatory)

plates of the region soon showed that the object actually is a Cepheid variable. Scrutiny of additional plates over the next several months revealed many additional Cepheids, two of which are identified in Figure 26-4.

As we saw in the previous chapter, Cepheid variables help astronomers determine distances. From the period–luminosity relation (recall Figure 21-10), astronomers can determine the average absolute magnitude of a Cepheid variable. From both the absolute magnitude and the apparent magnitude seen in the sky, a star's distance can be deduced.

Cepheid variables are intrinsically very bright. They typically have luminosities of a few thousand Suns. Hubble realized that, in order for these luminous stars to appear as dim as they do on his photographs of the "Andromeda nebula," they must be extremely far away. Straightforward calculations (recall Box 18-1) using modern data about Cepheids demonstrate that M31 is $2\frac{1}{4}$ million light years away, proving it is not a traditional nebula, but an enormous stellar system located far beyond the confines of the Milky Way.

Hubble's results were presented at a meeting of the American Astronomical Society on December 30, 1924, settling the Shapley–Curtis debate once and for all. The universe was recognized to be far larger and populated with far bigger objects than anyone had seriously imagined. Hubble had discovered the realm of the galaxies.

There are millions of galaxies all across the sky. Galaxies are seen in every unobscured direction. A typical spiral galaxy contains 100 billion stars and measures 100,000 light years in diameter. Galaxies are the biggest individual objects in the universe.

Hubble found that galaxies can be classified into four broad categories. These categories form the basis for the **Hubble classification scheme.** They are ellipticals, spirals, barred spirals, and irregulars.

Both M51 and M31 seen in Figures 26-2 and 26-3 are examples of **spiral (S) galaxies.** While studying spiral galaxies, Hubble noted that this class can be further subdivided according to the size of the central bulge and the winding of the spiral arms. Spirals with tightly wound spiral arms and a prominent, fat central bulge are called **Sa galaxies.** Those with moderately wound spiral arms and a moderate-sized central bulge

Figure 26-4 Cepheid variables in the Andromeda galaxy
Two Cepheid variables are identified in this view of the outskirts of the Andromeda galaxy. Because these stars appear so faint (although intrinsically they are very luminous), Hubble successfully demonstrated that the "Andromeda nebula" is extremely far away. (Mount Wilson and Las Campanas Observatories)

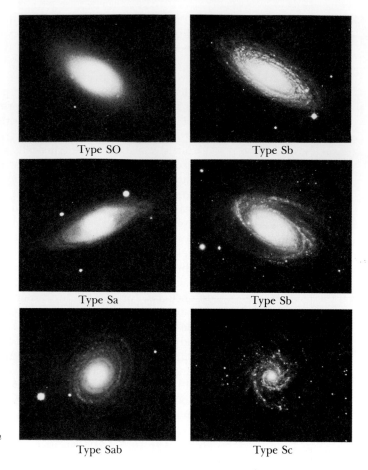

Type SO

Type Sb

Type Sa

Type Sb

Figure 26-5 Various spiral galaxies
Hubble classified spiral galaxies according to the winding of the spiral arms and the size of the central bulge. The designation "Type Sab" means the galaxy looks as though it is halfway between an Sa and an Sb galaxy. (Mt. Wilson and Las Campanas Observatories)

Type Sab

Type Sc

are called **Sb galaxies.** And finally, loosely wound spirals with a tiny central bulge are **Sc galaxies.** Both M31 and M51 are Sb spirals. Several examples of other types of spirals are displayed in Figure 26-5.

Fortunately, the size of the central bulge and the degree of winding of the spiral arms go hand in hand, so we can classify even those spiral galaxies that are viewed nearly edge-on. Thus, for example, M104 (see Figure 26-6) must be an Sa galaxy because of its huge central bulge. In contrast, the edge-on galaxy NGC 4565, which appears in Figure 25-8, must be an Sb galaxy because of its smaller central bulge. The tiny central bulge of an Sc would hardly be noticeable at all in an edge-on view.

In **barred spiral (SB) galaxies** the spiral arms originate at the ends of a bar running through the galaxy's nucleus rather than from the nucleus itself (see Figure 26-7). As in the case of ordinary spirals, Hubble subdivided barred spirals according to the size of the central bulge and the winding of the spiral arms. An **SBa galaxy** has a large central bulge and tightly wound spiral arms. A barred spiral with a moderate central bulge and moderately wound spiral arms is an **SBb galaxy,** and an **SBc galaxy** has loosely wound spiral arms and a tiny central bulge.

As we saw in the previous chapter, spiral arms are caused by density waves sweeping through great clouds of interstellar gas and dust. The passage of a density wave triggers star formation, which produces glowing H II regions and OB associations that outline the spiral arms. The development of a bar across a galaxy's nucleus is apparently related to details of how stars in the galaxy's disk are moving. Computer simula-

Figure 26-6 The Sombrero galaxy (called M104 or NGC 4594)
Because of the large size of its central bulge, this galaxy is an Sa. If we could see it face-on, we would find that the spiral arms are tightly wound around the voluminous bulge. This galaxy is in the constellation of Virgo and is tilted by only 6° from our line of sight. (Kitt Peak National Observatory)

tions, such as that by Frank Hohl of NASA shown in Figure 26-8, suggest that gravitational interactions between stars in a huge rotating disk can cause these stars occasionally to "pile up" along a barlike structure.

The tightness of spiral windings is related to the dust and gas content of the galaxy. Sa and SBa galaxies tend to have lower abundances of gas and dust than do Sc and SBc galaxies. Galaxies gradually exhaust their gas and dust content as stars form. Although supernovae and planetary

Figure 26-7 Various barred spiral galaxies
As with spiral galaxies, Hubble classified barred spirals according to the winding of the spiral arms and the size of the central bulge. An SB0 galaxy has a bar but no spiral arms. (Mt. Wilson and Las Campanas Observatories)

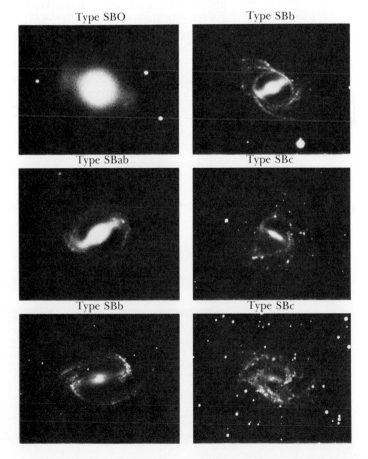

Type SB0 Type SBb

Type SBab Type SBc

Type SBb Type SBc

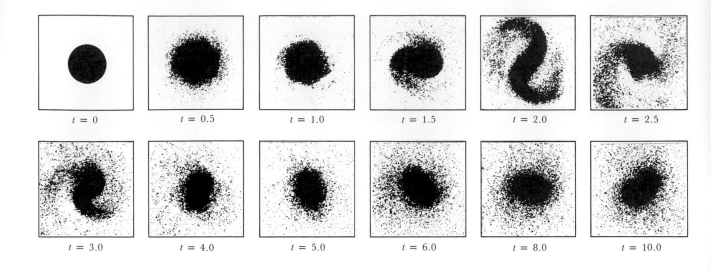

| $t = 0$ | $t = 0.5$ | $t = 1.0$ | $t = 1.5$ | $t = 2.0$ | $t = 2.5$ |

| $t = 3.0$ | $t = 4.0$ | $t = 5.0$ | $t = 6.0$ | $t = 8.0$ | $t = 10.0$ |

Figure 26-8 A computer simulation of a bar instability

A computer was used to follow the motions of 100,000 points, each point representing a star in the disk of a galaxy. Because of their mutual gravitational interactions, the points tend to "pile up" along spiral arms. Note that a bar occasionally appears briefly, as in the view t = 2.0, producing a pattern that looks like an SBb galaxy. These fleeting barlike structures are often called bar instabilities. (Courtesy of F. Hohl; NASA)

nebulae return some of this material to the interstellar medium, as a galaxy ages, more and more of its gas and dust become locked up inside stars. To some astronomers, these circumstances suggest an evolutionary trend. The so-called S0 and SB0 galaxies seen in Figures 26-5 and 26-7 may be former spirals and barred spirals that have finally converted all their interstellar material into stars.

Elliptical (E) galaxies—so named because of their distinctly elliptical shapes—have no spiral arms. Hubble subdivided elliptical galaxies according to how round or flattened they look. The roundest elliptical galaxies are called **E0 galaxies,** and the flattest elliptical galaxies are called **E7 galaxies.** Elliptical galaxies with intermediate amounts of flattening receive intermediate designations (see Figure 26-9).

Of course, an E1 or E2 galaxy might actually be a very flattened disk of stars that just happens to be viewed face-on, or a cigar-shaped E7 galaxy might look spherical when viewed end-on. The Hubble scheme clas-

Figure 26-9 Various elliptical galaxies

Hubble classified elliptical galaxies according to how round or flattened they look. An E0 galaxy is round, and the flattest elliptical galaxies are E7s. (Yerkes Observatory)

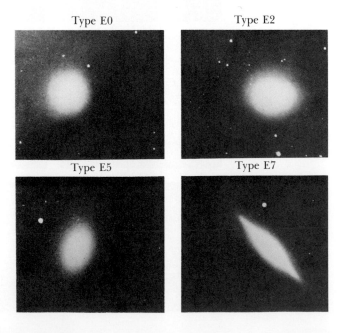

Type E0   Type E2

Type E5   Type E7

sifies galaxies entirely by their appearance in our Earth-bound view. Statistical studies have shown, however, that elliptical galaxies of various actual degrees of ellipticity do exist.

Elliptical galaxies look far less dramatic than their spiral and barred spiral cousins because they are virtually devoid of interstellar gas and dust. In addition, there is no evidence of young stars in most elliptical galaxies; there is no material from which stars could have recently formed. Star formation in elliptical galaxies ended a long time ago.

Elliptical galaxies exist in enormous ranges of size and mass. Both the biggest and the smallest galaxies in the universe are ellipticals. M87 (see Figure 26-10) is an example of a **giant elliptical galaxy.** This huge galaxy is located near the middle of a large cluster of galaxies in the constellation of Virgo. We shall have more to say about the fascinating case of M87 in the next chapter.

Whereas giant ellipticals are rather rare, **dwarf elliptical (dE) galaxies** are extremely common. Dwarf ellipticals are only a fraction the size of their normal counterparts, and each may contain only a few million stars. Some nearby dwarf ellipticals contain so few stars that these galaxies are completely transparent. You can actually see straight through the galaxy's nucleus and out the other side. Figure 26-11 shows a typical dwarf elliptical.

Several years ago, many astronomers suspected that the flattened appearance of some elliptical galaxies (for example, E5s, E6s, and E7s) might be the result of rotation. Recent spectroscopic observations demonstrate that this idea is wrong. Because of the Doppler effect, the average motions of stars in a galaxy can be deduced from a detailed examination of spectral lines in the galaxy's spectrum. These studies demonstrate that the motions of stars in elliptical galaxies are quite random. In the roundest elliptical galaxies, this randomness is **isotropic,** which means "equal in all directions." Because the stars are whizzing around equally in all directions, the result is a galaxy that is genuinely spherical. In flattened elliptical galaxies, the randomness of the stellar motions is **anisotropic,** which means that the range of star speeds is different in different directions. That's why E5s, E6s, and E7s are fatter in one direction than another.

You might suspect that the central bulges of spiral galaxies are

Figure 26-10 [left]   A giant elliptical galaxy
This huge elliptical galaxy (called M87 or NGC 4486) sits near the center of a rich cluster of galaxies in the constellation of Virgo. Although this galaxy is classified by some astronomers an an E0, it has unusual properties. For example, it is a strong source of X rays and radio waves. (Palomar Observatory)

Figure 26-11 [right]   A dwarf elliptical galaxy
This nearby elliptical (called NGC 147) is in the constellation of Cassiopeia and is classified as a dE5 galaxy. Millions of individual stars are visible in this remarkably clear photograph taken through the Palomar 200-in. telescope. Many dwarf galaxies are even smaller and contain far fewer stars than does NGC 147. (Palomar Observatory)

somehow related to elliptical galaxies. Such a connection—if it exists—is not simple. The central bulges of spirals and barred spirals are flattened by an amount consistent with rotation. For example, consider the central bulge of the Andromeda galaxy, shown in Figure 26-12. Its elliptical shape is exactly what would be expected from the rotation rate observed in the rest of the galaxy.

Possible relationships between spirals, barred spirals, and ellipticals are topics of current research and debate among astronomers today. Nevertheless, Hubble connected these three types of galaxies in his now-famous tuning-fork diagram (see Figure 26-13). Note that the S0 and SB0 galaxies are transition types between ellipticals and the two kinds of spirals according to this scheme.

Finally, Hubble found that galaxies that do not fit into his scheme of spirals, barred spirals, and ellipticals are usually "oddballs." He called these **irregular galaxies.** Examples include the Large Magellanic Cloud (LMC) and the Small Magellanic Cloud (SMC), both of which can be seen with the naked eye from southern latitudes. Both of these galaxies are nearby companions of our Milky Way galaxy.

Figure 26-12 [above]   The central bulge of the Andromeda galaxy
Although the central bulges of spiral galaxies resemble elliptical galaxies, there are some basic differences. The flattened shape of a central bulge is due to the galaxy's rotation, whereas the flattened shape of some elliptical galaxies is caused by anistropy of random stellar motions. A satellite trail is visible in this photograph of M31. (Kitt Peak National Observatory)

Figure 26-13 [right]   Hubble's tuning-fork diagram
Hubble summarized his classification scheme for regular galaxies with this tuning-fork diagram. The S0 and SB0 galaxies are transition types between ellipticals and spirals.

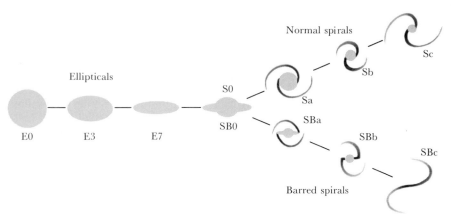

Telescopic views of the Magellanic clouds easily distinguish individual stars (see Figures 26-14 and 26-15). Note that the SMC does not exhibit any geometrical symmetry characteristic of spirals or ellipticals; it is therefore a true irregular. However, the LMC does apparently have a barlike structure, and it is therefore classified as a transition type between a barred spiral and an irregular. The Magellanic clouds are the nearest members of a cluster of galaxies to which the Milky Way and Andromeda galaxies belong.

Figure 26-14   The Large Magellanic Cloud
At a distance of only 150,000 light years, this galaxy is the nearest companion of our Milky Way galaxy. Note the huge H II region (called the Tarantula Nebula or 30 Doradus) just above the middle of the photograph. With a diameter of 800 light years and a mass of 300,000 Suns, it is the largest known H II region. (Courtesy of R. J. Talbot, R. J. Dufour, and E. B. Jensen)

Figure 26-15   The Small Magellanic Cloud
The SMC is only slightly farther away from us than the LMC is. Because of its sprawling, unsymmetrical shape, the SMC is classified as an irregular galaxy. Note that the SMC is rich in young blue stars. (Courtesy of R. J. Talbot, R. J. Dufour, and E. B. Jensen)

Galaxies are grouped in clusters

Galaxies are not scattered randomly across the universe but are grouped in **clusters.** Clusters are said to be **poor** or **rich,** depending on how many galaxies they have. For example, the Milky Way, the Andromeda galaxy, and the Large and Small Magellanic clouds belong to a poor cluster affectionately called the **local group.** As listed in Box 26-1, the local group contains nearly two dozen galaxies, most of which are dwarf ellipticals.

Box 26-1   The local group

The Andromeda galaxy is the biggest and most massive galaxy in the local group. The Milky Way galaxy takes second place. Scattered around these two primary galaxies are 20 smaller galaxies (see the table on the next page). The nine galaxies closest to us are satellites of the Milky Way galaxy. Similarly, the Andromeda galaxy has eight satellites. The map shows the galaxies of the local group. Additional dwarf galaxies are constantly being discovered.

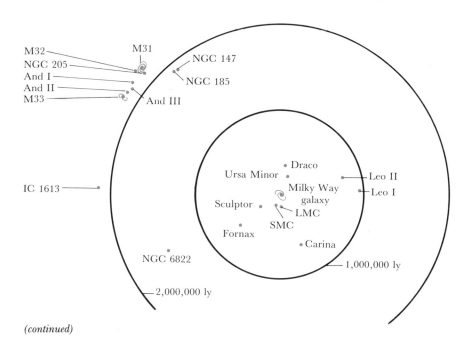

(continued)

(Box 26-1, continued)

Galaxy	Type	Distance (10^3 ly)	Diameter (10^3 ly)	Luminosity ($10^6 L_\odot$)	Mass ($10^6 M_\odot$)
Milky Way	Sb	——	100	20,000	200,000
LMC	Irr	170	30	3,000	25,000
SMC	Irr	190	25	600	6,000
Draco system	dE2	220	3	1	0.1
Ursa Minor system	dE4	220	3	0.4	0.1
Sculptor system	dE3	270	8	2	3
Carina system	dE3	500	4	?	?
Fornax system	dE3	650	20	23	20
Leo I system	dE4	700	6	3	4
Leo II system	dE0	700	4	1	1
NGC 6822	Irr	1,630	10	100	1,000
NGC 147	E6	2,000	10	70	350
NGC 185	E2	2,000	8	90	450
IC 1613	Irr	2,100	15	65	250
NGC 205	E5	2,250	16	330	3,000
NGC 221 (M32)	E3	2,250	8	250	2,000
Andromeda Galaxy (M31 = NGC 224)	Sb	2,250	130	25,000	300,000
Andromeda I	dE0	2,250	2	2	2
Andromeda II	dE0	2,250	2	2	2
Andromeda III	dE0	2,250	2	2	2
NGC 598 (M33)	Sc	2,350	60	4,000	40,000
Maffei I	S0	3,300	100	10,000	200,000

The nearest fairly rich cluster is the Virgo cluster. It is a sprawling collection of over 1000 galaxies covering an area of the sky $10° \times 12°$. The distance to the Virgo cluster is about 50 million light years—too far away for Cepheid variables to be seen from our galaxy. Instead, the distance to the Virgo cluster has been determined by the apparent faintness of O and B supergiant stars, by the brightnesses of globular clusters surrounding some of the galaxies, and by the angular sizes of H II regions in some of the cluster's spiral galaxies. The overall diameter of the Virgo cluster is about 7 million light years.

The central region of the Virgo cluster is dominated by three giant elliptical galaxies. Two of them (M84 and M86) appear in Figure 26-16; the third (M87) appears in Figure 26-10. These enormous galaxies may be as large as 2 million light years in diameter—20 times as large as an ordinary elliptical or spiral. In other words, one giant elliptical is roughly the same size as the entire local group.

In addition to the rich-versus-poor classification, astronomers often classify clusters of galaxies as **regular** or **irregular,** depending on the overall shape of the cluster. The Virgo cluster is called irregular because

Figure 26-16 [left]   The center of the Virgo cluster

This fairly rich cluster lies about 50 million light years from us. Only the center of this huge cluster appears in this view. Note the two giant elliptical galaxies (M84 and M86). The giant elliptical M87 (see Figure 26-10) also is a member of this cluster. (Kitt Peak National Observatory)

Figure 26-17 [right]   The Coma cluster

This is the nearest rich regular cluster in the sky. Hundreds of galaxies are visible in this wide-angle view. Regular clusters are predominately composed of elliptical and S0 galaxies and are commonly sources of X rays. (Palomar Observatory)

of the unsymmetrical way its galaxies are scattered about a sprawling region of the sky. In contrast, a regular cluster has a distinctly spherical appearance, with a marked concentration of galaxies at its center.

The nearest example of a regular cluster is the Coma cluster, located 350 million light years from us toward the constellation of Coma Berenices. It also is a very rich cluster. In spite of the great distance to this cluster, more than 1000 bright galaxies are easily visible on photographic plates (see Figure 26-17). Certainly, many thousands of dwarf ellipticals are too faint to be detected from our distance. The total membership of the Coma cluster may therefore be as many as 10,000 galaxies.

The regular versus-irregular classification is directly correlated with the dominant types of galaxies in a cluster. Rich regular clusters, such as the Coma cluster, contain mostly elliptical and S0 galaxies. Only 15 percent of the Coma cluster's galaxies are spirals and irregulars. Irregular clusters, such as the Virgo cluster, have a more even mixture of galaxy types. Of the 200 brightest galaxies in the Virgo cluster, 68 percent are spirals, 19 percent are ellipticals, and the rest are irregulars.

A cluster of galaxies must be a gravitationally bound system. In other words, there must be enough matter in the cluster to produce enough gravity to prevent the galaxies from wandering away. Nevertheless, careful examination of a rich cluster, such as the Coma cluster, typically reveals that the mass of the visually luminous matter is not sufficient to bind the cluster gravitationally. The observed line-of-sight speeds of the cluster galaxies (measured by Doppler shifts) are so large that more mass than that observed is needed to keep them bound in orbits about the center of the cluster. This dilemma is called the **missing-mass problem.** A lot of nonluminous matter in some form must be scattered about each of the clusters; otherwise, the galaxies would long ago have wandered away in random directions, and the cluster would not exist today. Analyses demonstrate that the total mass needed to bind a typical rich cluster is ten times greater than the mass of visible material that shows up on visual photographs such as Figure 26-17.

Some of this mystery has been solved recently by X-ray astronomers. Satellite observations of rich clusters reveal that X rays pour from the space between galaxies in rich clusters (see Figure 26-18). This is evidence of substantial amounts of hot intergalactic gas at temperatures between 10 and 100 million degrees. Analyses demonstrate that the mass

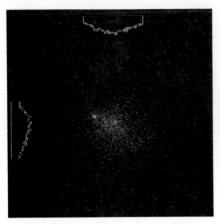

Figure 26-18   An X-ray view of a cluster of galaxies
This view from the Einstein Observatory shows the rich cluster A1367 (that is, the 1367th cluster in a list of rich clusters catalogued by UCLA astronomer George O. Abell). The X-ray emission comes from hot gas between the galaxies. (Harvard-Smithsonian Center for Astrophysics)

of this hot intergalactic gas is typically as great as the combined mass of the visible galaxies in the cluster.

Unfortunately, the discovery of hot intergalactic gas in rich clusters solves only part of the missing-mass problem. Most astronomers agree that a great deal of matter remains to be discovered in rich clusters. One popular speculation is that these clusters may contain a lot of undetected dim stars. These faint stars could be located in extended halos surrounding individual galaxies, and they could also be scattered throughout the space between galaxies of a cluster.

Evidence for extended halos comes from the rotation curves of galaxies. As mentioned in Chapter 25, many galaxies have rotation curves similar to that of our Milky Way galaxy (recall Figure 25-15). These rotation curves remain remarkably flat to surprisingly large distances from the galaxy's center. For example, Figure 26-19 shows the rotation curve for the Sc spiral galaxy UGC 2885. Note that the orbital speed is fairly constant out to 60 kpc (60,000 pc) from the galactic center. Beyond this distance, the galaxy's stars and H II regions are so dim and sparsely scattered that reliable measurements are not possible. Nevertheless, we still have not detected the true edge of this and many similar galaxies. In the outer portions of a galaxy, we should see a decline in orbital speed, in accordance with Kepler's third law. Because this decline is not observed, astronomers conclude that there must be a considerable amount of dark material extending well beyond the visible portion of a galaxy's disk. Although many proposals have been made (including cold gas, black holes, and massive particles), an extended halo of dim stars seems to be one reasonable candidate for the nature of this outlying subluminous matter.

Galaxies sometimes collide and merge

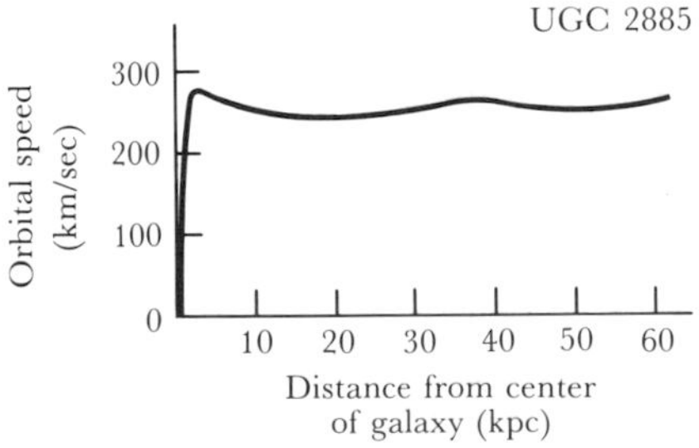

Figure 26-19   The rotation curve of a large spiral galaxy
This graph shows the orbital speed of material in the disk of the galaxy UGC 2885 (that is, the 2885th galaxy listed in the Uppsala General Catalogue). Many galaxies have flat rotation curves, indicating the presence of extended halos of low-luminosity material. (Adapted from V. Rubin, K. Ford, and N. Thonnard)

A loosely dispersed sea of stars scattered across intergalactic space also seems to be a reasonable consequence of galactic collisions within a cluster. All of the galaxies in a cluster are in orbit about their common center of mass, and occasionally two galaxies might pass close enough to each other to collide.

When two galaxies collide, their stars pass by each other. There is so much space between the stars that the probability of two stars crashing into each other is extremely small. However, huge clouds of interstellar gas and dust cannot interpenetrate. Thus, when two galaxies collide, their interstellar clouds slam into each other, producing strong shock waves. Whereas the stars keep right on going, the colliding interstellar clouds are stopped in their tracks. In this way, two colliding spiral galaxies are stripped of their interstellar gas and dust and thereby are converted into S0 galaxies. Because of the violence of the collision between interstellar clouds, the gas stripped from these galaxies is heated to extremely high temperatures. This may be a major source of the hot intergalactic gas often observed in rich regular clusters (recall Figure 26-18).

Although stars pass by each other during the collision of two galaxies, their gravitational interactions can produce some spectacular results. Gravitational forces can hurl thousands of stars out into intergalactic space along huge arching streams. This is dramatically illustrated in computer simulations by Alar Toomre at MIT and Juri Toomre at the Joint Institute for Laboratory Astrophysics in Colorado. One of these simulations is shown in Figure 26-20; note the striking similarity with the photograph of Figure 26-21.

As you might expect, the most dramatic results occur when two galax-

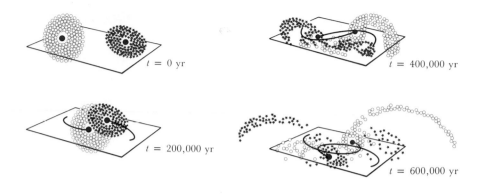

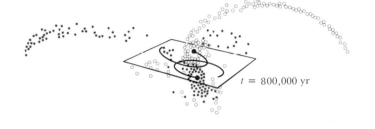

Figure 26-20 A simulated collision between two galaxies
Two galaxies are each simulated with 350 mass points, each point representing a star. As the two galaxies pass through each other, the gravitational forces between these masses spew out the stars along long streamers. The drawings represent the structure at intervals of 200 million years. (Adapted from Alar and Juri Toomre)

ies in a cluster suffer a head-on collision. However, in a rich cluster there must be many near misses. If, as we suspect, galaxies are surrounded by extended halos of dim stars, then these near misses should efficiently strip the galaxies of their outlying stars. In this way, a loosely dispersed sea of dim stars might come to populate the void between galaxies in a cluster. Searching for these dim stars in extended halos and in intergalactic space is one of the main projects for the Space Telescope in the late 1980s.

During a collision between galaxies, some of the stars are flung far and wide, scattering material into intergalactic space. However, other stars suffer a loss of energy and momentum. Because these stars are slowed down, the galaxies may merge. Several dramatic examples of **galaxy mergers** have recently been discovered by François Schweizer at the Cerro Tololo Inter-American Observatory in Chile (see Figure 26-22).

Figure 26-21 A colliding pair of galaxies with "antennae"
Many pairs of colliding galaxies exhibit long "antennae" of stars ejected by the collision. This particular system, called NGC 2623, is located in the constellation of Cancer. Many of these colliding pairs are strong sources of radio radiation. (Palomar Observatory)

Figure 26-22   The giant radio galaxy NGC 1316
This huge galaxy is located in the constellation of Fornax at a distance of 100 million light years from Earth. This system seems to be an excellent example of two galaxies merging. Note the loops and swirls of material churned up by the collision. This system is a strong source of radio radiation. (Courtesy of François Schweizer)

When two galaxies merge, the result is a bigger galaxy. If this galaxy is located in a rich cluster, it may capture and devour additional galaxies and thereby grow to enormous dimensions. This phenomenon is called **galactic cannibalism.** Cannibalism differs from mergers in that the dining galaxy is bigger than its dinner, whereas merging galaxies are about the same size.

Many astronomers suspect that galactic cannibalism is the reason why giant ellipticals are so huge. As we have seen, giant ellipticals typically occupy the centers of rich clusters. Two giant ellipticals (M84 and M86) near the center of the Virgo cluster are visible in Figure 26-16, and careful examination of Figure 26-17 reveals two giant elliptical galaxies (NGC 4874 and NGC 4889) at the center of the Coma cluster. In many cases, smaller galaxies are located around the giant ellipticals. As they pass through the extended halo of a giant elliptical, these smaller galaxies slow down and eventually are devoured by the larger galaxy.

The Hubble law is a simple relationship between the redshifts of galaxies and their distances

Whenever an astronomer finds an object in the sky that can be seen or photographed, the natural inclination is to attach a spectrograph to the telescope and take a spectrum. As long ago as 1914, V. M. Slipher working at the Lowell Observatory in Arizona began taking spectra of the "spiral nebulae." He was surprised to discover that, of the 15 spiral nebulae he studied, 11 had their spectral lines shifted toward the red end of the spectrum, indicating substantial recessional velocities. Indeed, this marked dominance of redshifts was presented by Curtis at the Shapely–Curtis debate as evidence that these spiral nebulae could not be ordinary nebulae in our Milky Way Galaxy.

During the 1920s, Edwin Hubble and Milton Humason photographed the spectra of many galaxies with the 100-in. telescope on Mount Wilson. Five representative elliptical galaxies and their spectra are shown in Figure 26-23. As indicated in this illustration, there seemed to be a direct

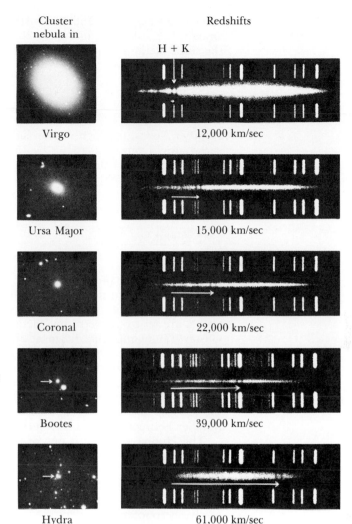

Cluster nebula in   Redshifts

H + K

Virgo   12,000 km/sec

Ursa Major   15,000 km/sec

Coronal   22,000 km/sec

Bootes   39,000 km/sec

Hydra   61,000 km/sec

Figure 26-23 Five galaxies and their spectra
The photographs of these five elliptical galaxies all have the same magnification. The spectrum of each galaxy is the hazy band between the comparison spectra. In all five cases, the so-called H and K lines of singly ionized calcium are seen. The recessional velocity (calculated from the Doppler shifts of the H and K lines) is given below each spectrum. Note that, the more distant a galaxy is, the greater is its redshift. (Mt. Wilson and Las Campanas Observatories)

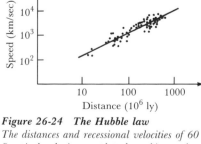

Figure 26-24 The Hubble law
The distances and recessional velocities of 60 Sc spiral galaxies are plotted on this graph. The straight line is the "best fit" for the data. This linear relationship between distance and speed is called the Hubble law. (Adapted from Sandage and Tammann)

correlation between the distance to a galaxy and the size of its redshift. In other words, nearby galaxies are moving away from us slowly, whereas more distant galaxies are rushing away from us much more rapidly. This universal recessional movement is sometimes called **Hubble flow.**

Using various techniques, Hubble estimated the distances to a number of galaxies. Using the Doppler formula (recall Box 7-3 on the Doppler shift), Hubble calculated the speed with which each galaxy is receding from us. When he plotted the data on a graph of distance versus speed, he found that the points lie near a straight line. Figure 26-24 is a version of Hubble's graph based on modern data.

This relationship between the distances to galaxies and their redshifts is one of the most important astronomical discoveries of the twentieth century. As we shall see in Chapter 28, this relationship tells us that we are living in an expanding universe. Hubble published his discovery in 1929, and it is now known as the **Hubble law.**

The Hubble law is most easily stated as a formula:

$$v = H_0 r$$

where v is the recessional velocity, r is the distance, and H_0 is a constant commonly called the **Hubble constant.** This formula is equivalent to the straight line displayed in Figure 26-25. The Hubble constant H_0 tells us the slope of the line in Figure 26-25. From the data plotted on this graph we find that

$$H_0 = \frac{15 \text{ km/sec}}{10^6 \text{ ly}}$$

In other words, for each million light years to a galaxy, the galaxy's speed away from us increases by 15 km/sec. For example, a galaxy located 100 million light years from Earth should be rushing away from us with a speed of 1500 km/sec.

Incidentally, many astronomers prefer to speak of megaparsecs (Mpc) rather than millions of light years. Using that unit,

$$H_0 = \frac{50 \text{ km/sec}}{1 \text{ Mpc}}$$

It should be emphasized that the exact value of the Hubble constant is a topic of heated debate and controversy among astronomers today. The data plotted in Figure 26-24, as well as the values of H_0 given above, are all from the recent meticulous work of Allan Sandage and Gustav Tammann, who did most of their observations with the 200-in. Palomar telescope. Other prominent astronomers, however, such as Sidney van den Bergh in Canada and Gerard de Vaucouleurs of the University of Texas, strongly feel that the Sandage–Tammann value for H_0 is too low. They argue that the true value for H_0 is about $(30 \text{ km/sec})/10^6 \text{ ly} = (100 \text{ km/sec})/\text{Mpc}$. Many astronomers simply use a number between the two extremes, such as $(75 \text{ km/sec})/\text{Mpc}$.

The Hubble constant is one of the most important numbers in all physical science. As we shall see in Chapter 28, it expresses the rate at which the universe is expanding and tells us the age of the universe. As described in Box 26-2, the Hubble law and the Hubble constant are used to determine distances to extremely remote objects such as quasars. Naturally, astronomers are very interested in an accurate determination of H_0.

Unfortunately, determining the value of H_0 is an extremely difficult and tricky business. To appreciate the difficulties, think about what you would have to do to measure H_0. You would first have to observe many galaxies. For each galaxy, you would have to determine the galaxy's recessional velocity from the redshifts of its spectral lines. By some means, you would also have to determine the distance to each galaxy. Finally, you would plot all the data on a graph like Figure 26-24. The slope of the line that best fits through the data gives you H_0. Unfortunately, you would encounter numerous pitfalls along the way.

There are pitfalls with the redshifts. Measuring the recessional velocity of a galaxy sounds easy enough. You take a spectrum, see how far the spectral lines are shifted, and use the Doppler formula to get the speed. An example is worked out in Box 26-2. Galaxies come in clusters, however, and all of the galaxies in a cluster are in orbit about their common center of mass. Because of these orbital motions, some galaxies are coming toward you, while others are moving away. Thus, tacked onto the true recessional velocity that you are trying to determine is an additional velocity (either toward you or away from you) because of the galaxy's individual motion around its cluster.

One way of coping with this problem is to turn your attention to extremely distant galaxies. For very remote galaxies, the recessional velocity

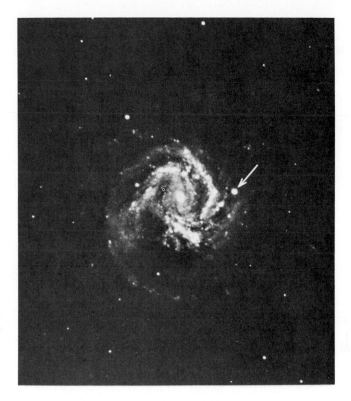

Figure 26-25   A supernova in NGC 4303
In 1961, a supernova erupted in the spiral galaxy NGC 4303 (also called M61), which is a member of the Virgo cluster. Supernovae can be seen in extremely remote galaxies and are important "standard candles" used to determine the distances to these faraway galaxies. (Lick Observatory)

due to the expansion of the universe far outweighs the orbital motions in clusters. Thus, your redshift measurements of remote galaxies more accurately reflect the Hubble flow. Indeed, astronomers trying to determine H_0 commonly restrict their observations to galaxies farther than 100 million light years from Earth. Unfortunately, new pitfalls arise: you are no longer exactly sure about the distances to these remote galaxies.

The distances to nearby galaxies are determined by fairly reliable methods. For example, Cepheid variables can be seen out to 20 million light years from Earth. The distances to galaxies in this nearby volume of space can therefore be determined from the period–luminosity law.

Beyond 20 million light years, even the brightest Cepheid variables (which have absolute magnitudes of about −6) fade from view. Astronomers then turn to more luminous stars. The brightest red supergiants have absolute magnitudes of −8, and the brightest blue supergiants have absolute magnitudes of −9. These two types of stars can be seen out to distances of 50 million and 80 million light years, respectively. Out to these limits, you can determine the distances to galaxies from the apparent magnitudes of these luminous supergiants.

Beyond 80 million light years, individual stars are no longer discernible. Astronomers therefore turn to entire clusters and nebulae. The brightest globular clusters have a total absolute magnitude of about −10 and can be seen out to 130 million light years from Earth. The brightest H II regions have absolute magnitudes of −12 and can be detected out to 300 million light years. From the faintness of these clusters and nebulae, distances to remote galaxies can be estimated.

Finally, to get beyond 300 million light years, astronomers must wait for supernova explosions. The brightest supernovae reach an absolute magnitude of −19 at the peak of their outbursts (see Figure 26-25).

Box 26-2 The Hubble law as a distance indicator

Suppose you aim your telescope at an extremely distant galaxy. You take a spectrum of the galaxy and find that the spectral lines are shifted toward the red end of the spectrum. For example, a spectral line that normally has a wavelength λ_0 is observed at a longer wavelength λ. Thus, the spectral line has been shifted by an amount $\Delta\lambda = \lambda - \lambda_0$. The **redshift** of the galaxy (usually denoted by z) is

$$z = \frac{\Delta\lambda}{\lambda_0} = \frac{\lambda - \lambda_0}{\lambda_0}$$

According to the Doppler effect (recall Box 7-3), this wavelength shift corresponds to a speed v, where

$$z = \frac{v}{c}$$

and c is the speed of light.

According to the Hubble law, the recessional velocity v of a galaxy is related to its distance r from Earth by

$$v = H_0 r$$

where H_0 is the Hubble constant. Thus, the distance to a galaxy is related to its redshift by

$$r = \frac{zc}{H_0}$$

As a specific example, suppose that you observe the giant elliptical galaxy NGC 4889 in the Coma cluster (recall Figure 26-17). The so-called K line of singly ionized calcium normally has a wavelength of 3933 Å. In the spectrum of NGC 4889, you find this spectral line at 4018 Å. Thus, the redshift of the galaxy is

$$z = \frac{4018 - 3933}{3933} = 0.0216$$

The galaxy is therefore moving away from us with a speed of

$$v = zc = (0.0216)(3 \times 10^5 \text{ km/sec}) = 6500 \text{ km/sec}$$

Using $H_0 = (15 \text{ km/sec})/10^6$ ly, the Hubble law gives you the distance to the galaxy:

$$r = \frac{zc}{H_0} = \frac{6500}{15} = 430 \text{ million light years}$$

Note that, if you had used a Hubble constant of $(30 \text{ km/sec})/10^6$ ly, you would have concluded that the distance is

$$r = \frac{zc}{H_0} = \frac{6500}{30} = 220 \text{ million light years}$$

Unfortunately, this is the kind of uncertainty that astronomers must deal with. We look forward to an accurate determination of the Hubble constant on which everyone can agree. Most astronomers suspect that the true value of H_0 lies between the extremes of Sandage [$(15 \text{ km/sec})/10^6$ ly] and van den Bergh [$(30 \text{ km/sec})10^6$ ly]. Thus, the true distance to NGC 4889 is probably somewhere between 220 million and 430 million light years. A compromise distance of 350 million light years is quoted in the text.

These brilliant outbursts can be seen out to distances of 8 billion light years from Earth.

These various objects—Cepheid variables and the most luminous supergiants, globular clusters, H II regions, and supernovae—are commonly called **standard candles.** As you might suspect, astronomers go to great lengths to check the reliability of their standard candles. After all, a tiny mistake in the absolute magnitude of a supergiant star or globular cluster can lead to an error of many millions of light years in the distance to a remote galaxy.

The major obstacle in determining the Hubble constant is this: the farther we look out into space, the fewer standard candles we have. For example, the distance to a nearby galaxy can be cross-checked in many ways. The distance computed from the period–luminosity relation can be compared to the distance determined from the magnitudes of the most luminous supergiants. These results can be further compared with the magnitudes of the galaxy's globular clusters and the angular sizes of its H II regions. After all this work, the results from these various methods can be averaged to obtain a distance to the galaxy in which astronomers may have confidence.

As we turn to more distant galaxies, however, we can see fewer and fewer standard candles. For example, the nearest rich regular cluster is the Coma cluster seen in Figure 26-17. These galaxies are so far away that it is impossible to see individual stars. With fewer standard candles, fewer cross-checks are possible. The distance to these remote galaxies therefore becomes less certain. Unfortunately, remote galaxies are precisely the objects whose distances we must determine in order to find the value of the Hubble constant. Uncertainty in distance determinations is the cause of our uncertainty in the exact value of H_0.

Astronomers hope that the Space Telescope will solve many of these problems. The Space Telescope (recently renamed the Hubble Space Telescope) will produce views of galaxies far sharper than anything that is possible from the ground and give us important information about the fundamental nature of our universe.

Summary

· Galaxies can be grouped into four major categories: ellipticals, spirals, barred spirals, and irregulars (the Hubble classification).

  Spiral galaxies like our own are sites of active star formation; barred spiral galaxies appear to be temporary stages in the normal development of spiral galaxies.

  Elliptical galaxies are virtually devoid of interstellar gas and dust; no star formation is occurring in these galaxies.

· Galaxies are grouped into clusters rather than being scattered randomly through the universe.

  A rich cluster contains hundreds or thousands of galaxies; a poor cluster may contain only a few dozen galaxies.

  A regular cluster has a nearly spherical shape with a central concentration of galaxies; an irregular cluster has an unsymmetrical distribution of its galaxies.

  Our galaxy is a member of a poor irregular cluster called the local group.

Rich regular clusters contain mostly elliptical and S0 galaxies; irregular clusters contain far more spirals and irregulars.

Giant elliptical galaxies are often found near the centers of rich clusters.

. The observed mass of a cluster of galaxies is not large enough to account for the observed motions of the galaxies; a large amount of unobserved mass must be present between the galaxies ("the missing-mass problem").

Hot intergalactic gases emit X rays in rich clusters; extended halos of dim stars probably surround all galaxies.

. When two galaxies collide, their stars pass each other, but their interstellar media collide violently, stripping the gas and dust from the galaxies and heating the intergalactic gas.

Gravitational effects during a galactic collision can throw stars out of the galaxies into intergalactic space.

Galactic mergers may occur; a large galaxy in a rich cluster may tend to grow steadily by galactic cannibalism, perhaps producing a giant elliptical galaxy.

. There is a simple linear relationship between the distance from the Earth to a galaxy and the redshift of that galaxy (which is a measure of the speed with which it is receding from us); this relationship is the Hubble law, $v = H_0 r$.

Because of difficulties in measuring distances to galaxies, the value of the Hubble constant H_0 is not known with certainty; this leads to uncertainties in our knowledge about the rate at which the universe is expanding and about the age of the universe.

Standard candles such as Cepheid variables and the most luminous supergiants, globular clusters, H II regions, and supernovae in a galaxy are used in estimating intergalactic distances.

Review questions

1 What types of galaxies are most likely to have new stars forming? Describe the observational evidence that supports your answer.

2 How is it possible that galaxies in our local group remain to be discovered? In what part of the sky would these galaxies be located? What sorts of observations might reveal these galaxies?

3 Are there any galaxies besides our own that can be seen with the naked eye? If so, which one(s)?

4 How would you distinguish star images from unresolved images of remote galaxies on a photographic plate?

5 Explain why the "missing mass" in galaxy clusters could not be neutral hydrogen.

6 Why do some galaxies in the local group exhibit blueshifted spectral lines? Is this a violation of the Hubble law? Explain.

7 Why do you suppose there are such discordant determinations of H_0?

8 What kinds of stars would you expect to populate space between galaxies in a cluster?

Advanced questions

*9 In 1937, a Type I supernova was observed in the galaxy IC 4182. At maximum brilliance, the supernova reached an apparent magnitude of $+8$. Assuming that the absolute magnitude at maximum light was -19, calculate the distance to the galaxy.

*10 Suppose that you mistook a Type I Cepheid for a Type II Cepheid in a distant galaxy. How great an error would you make in determining the distance to that galaxy?

11 How might you determine what part of a galaxy's redshift is due to the galaxy's orbital motion about the center of mass of its cluster?

Discussion questions

12 Discuss the advantages and disadvantages of using the various distance indicators to obtain extragalactic distances.

13 Discuss the idea that the various Hubble types of galaxies represent some sort of evolutionary sequence.

For further reading

Berendzen, R., et al. *Man Discovers the Galaxies.* Neale Watson, 1976.

DeVaucouleurs, G. "The Distance Scale of the Universe." *Sky & Telescope,* Dec. 1983, p. 511.

Field, G. "The Hidden Mass in Galaxies." *Mercury,* May/June 1982, p. 74.

Gorenstein, P., and Tucker, W. "Rich Clusters of Galaxies." *Scientific American,* Nov. 1978.

Hausman, M. "Galactic Cannibalism." *Mercury,* Nov./Dec. 1979, p. 119.

Kaufmann, W. *Galaxies and Quasars,* W. H. Freeman, 1979.

Rubin, V. "Dark Matter in Spiral Galaxies." *Scientific American,* June 1983.

Smith, R. *The Expanding Universe: Astronomy's Great Debate.* Cambridge U. Press, 1982.

Tenn, J. "Cosmic Distances and QSO's." *Mercury,* Jul./Aug. 1979, p. 67.

Tully, R. "Unscrambling the Local Supercluster." *Sky & Telescope,* June 1982, p. 550.

Owen Gingerich

The Shapley–Curtis debate

Owen Gingerich is an astrophysicist at the Smithsonian Astrophysical Observatory in Cambridge and a Professor of Astronomy and the History of Science at Harvard University. His research interests have included modeling the solar atmosphere, interpretation of stellar spectra, and the recomputation of ancient Babylonian mathematical tables. In ancient astronomy, he is a leading authority on the works of Johannes Kepler and Nicholas Copernicus. He is also actively involved with work on recent astronomy, having coedited *A Source Book in Astronomy and Astrophysics,* 1900-1975, and serving currently as editor of the 20th century part of the International Astrophysical Union's *General History of Astronomy.* He is an active participant in such scientific organizations as the American Astronomical Society and the International Astronomical Union and has several hundred general interest, technical, and historical publications to his credit. The most accessible of these are his *Scientific American* articles and his bimonthly column in *Sky and Telescope.*

On April 26, 1920, two astronomers debated the scale of the universe before the National Academy of Sciences. Although their views were very different, both argued for a universe far larger than had generally been accepted, and astronomy has never been the same since.

Heber D. Curtis was an authority on spiral nebulae. Working at the Lick Observatory he had photographed hundreds of these faint objects. He (and others) had recently discovered 25 "new stars" or novae in the spirals, most of them in the large Andromeda nebula, which was presumably the closest of all these cosmic pinwheels. By comparing the faintness of these novae with about 30 specimens from within the Milky Way, Curtis concluded that the spirals lay between 500,000 and 10,000,000 light years away. These were enormous numbers to scientists accustomed to think at most in sizes of 20,000 light years.

Harlow Shapley had spoken first about his studies of the Milky Way itself. He, too, had applied new distance ranging techniques. At the Mount Wilson Observatory he had explored the pulsating cepheid variable stars in the globular star clusters. By assuming that a cepheid which varied its brightness with a period of, say, 6 days had the same intrinsic luminosity no matter where it was found in the universe, he could plumb the distance to these clusters. Not only did he find them to be quite far away—many as distant as 50,000 and some as far as 220,000 light years—but he also noticed that these globular clusters were heavily concentrated in a specific direction in the sky, toward the constellation Sagittarius.

Shapley made the bold speculative leap that the globular clusters were concentrated about a distant nucleus of the Milky Way system. Lying perhaps 50,000 light years in the direction of Sagittarius, the nucleus was too far away to be seen directly. But because the Milky Way seemed to be such an enormous unit all by itself, Shapley was exceedingly reluctant to accept the spirals as comparable systems. He pointed out that measurements of the angular rotations of the spirals would lead to absurdly high spin velocities if the spirals were really some millions of light years away.

Curtis, on the other hand, simply did not believe Shapley's measurements. (In retrospect he was quite right, though to this day no one is sure what went wrong with those seemingly careful determinations!) Curtis tended to agree that if the Milky Way were as big as Shapley claimed, then it would be strange to have the rest of space filled with distant but very much smaller spiral systems. Therefore, he reasoned,

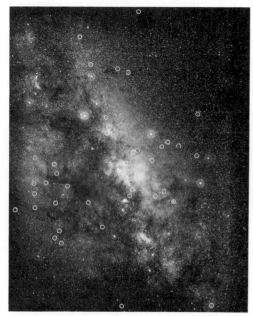

The circled objects in this plate are globular clusters. More than one-third of those in our galaxy are found within 5 percent of the sky as shown here. This concentrated distribution led Shapley to conclude that the Sun was off-center in the Milky Way system. (Brookhaven National Laboratory)

Shapley must be wrong about his distance scale for the Milky Way, and in the debate he tried to undermine the period–luminosity relation for the cepheid variables on which Shapley's distance determinations depended.

Curtis not only hoped to discredit Shapley's distance-measuring technique, but also to scotch Shapley's claim that the sun lay far from the center of our own Milky Way star system. He proposed that at most the Milky Way would be 30,000 light years across, compared to the 300,000 light years advocated by his opponent, and he believed the sun was near the center of our galactic star system.

It is interesting that both speakers agreed that the composition of the universe should be uniform—the same mix of stars should be found nearby the sun as in distant realms. It was in trying to establish this coherency that their ideas clashed. Both were using the concept that "faintness means farness" to fathom vastly greater distances than astronomers had been able to find by geometrical procedures, and both helped to teach us the photometric techniques needed to explore the great reaches of the Milky Way as well as the depths of extragalactic space.

Why did these West Coast rivals reach such differing conclusions? Neither astronomer appreciated the insidious, light-obscuring effects of interstellar material. As a consequence, Shapley placed the globular clusters too far away, because he did not realize that part of their faintness came from absorption rather than farness. Thus he overestimated the size of the Milky Way by a factor of three. Nevertheless, he was right in general about the off-center position of the Sun in a large Milky Way.

Meanwhile, Curtis underestimated the luminosities of the novae in the Milky Way, not realizing that part of their seeming faintness was caused by absorption. Thus, in order to compensate for his underestimation of the intrinsic brightnesses of the novae, he placed the Andromeda nebula and other spirals at about a quarter of their actual distances from us. Nevertheless, he was right in general about the great distances of the spirals.

The Milky Way system in cross section showing our Sun's position far from the center, from J. S. Plaskett's Halley Lecture June 5, 1935 as published by Oxford at Clarendon Press that year.

Not until 1952 did astronomers sort out the proper distance scale to the Andromeda galaxy. Until then our own Milky Way seemed to be by far the largest galaxy known, an inconsistency that puzzled astronomers, but not enough for them to appreciate instantly that something was wrong. The Shapley–Curtis debate teaches us that bold and speculative insights, which may appear simple in retrospect, take a good deal of intellectual effort, and even require courage to ignore apparently conflicting observational evidence.

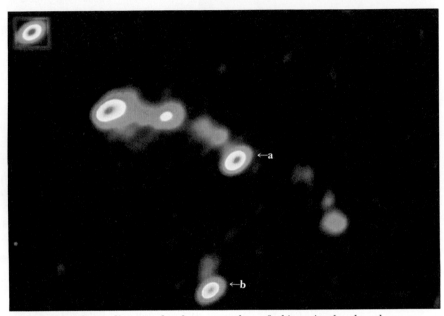

A gravitational lens

*Two images (**a** and **b**) of the same quasar are seen in this radio view from the Very Large Array. The light of a distant quasar is deflected to either side of a massive galaxy located between us and the quasar (recall Figure 24-5) thereby producing two images. The jet-like feature protruding from the upper image does not appear alongside the lower image because the jet is too far away from the required quasar–galaxy–Earth alignment. Both quasar images have the same spectral lines and redshift. The angular separation between the two quasar images is 6 arc sec. (VLA; NRAO)*

Astronomers have discovered a large number of objects in the sky whose extreme redshifts indicate very large recessional velocities and therefore, according to the Hubble law, very great distances from the Earth. To be observable at all at such huge distances, these quasi-stellar objects (quasars) and active galaxies must be extremely luminous. In this chapter, we learn about evidence indicating that a typical quasar emits the energy output of 100 galaxies from a volume roughly the size of the solar system. We discuss the puzzling features of these objects and the controversy about whether the Hubble law can be trusted in calculating their distances. Finally, we learn about recent theories that supermassive black holes are involved in the intense energy production of these luminous objects.

The development of radio astronomy in the late 1940s ranks among the most important scientific accomplishments of the twentieth century. Prior to that time, everything known about the distant universe was gleaned from visual observations. Radio telescopes provided a view of the universe in a wavelength range far removed from visible light. Many unexpected and surprising discoveries emerged from this new ability to examine the previously invisible universe.

The first radio telescope was built in 1937 by an amateur astronomer, Grote Reber, in his backyard in Illinois. By 1944, Reber had detected strong radio emission from Sagittarius, Cassiopeia, and Cygnus. Two of these sources (nicknamed Sgr A and Cas A) happen to be in our own galaxy; they are the galactic nucleus and a supernova remnant. The precise location of the third source, called Cygnus A (Cyg A), was established in 1951 using a newly constructed radio interferometer. Armed with precise coordinates, Walter Baade and Rudolph Minkowski used the 200-in. optical telescope on Palomar Mountain to discover a strange-looking galaxy at that position. Figure 27-1 is a photograph of the optical counterpart of Cyg A.

The peculiar galaxy associated with Cygnus A is very dim. Nevertheless, Baade and Minkowski managed to photograph its spectrum, which shows a number of bright spectral lines, all shifted by 5.7 percent toward the red end of the spectrum. A redshift of $z = 0.057$ corresponds to a

Figure 27-1   Cygnus A (also called 3C 405) *This strange-looking galaxy was discovered at the location of the radio source Cygnus A. This galaxy has a substantial redshift ($z = 0.057$), which means that it must be far away (about 1 billion light years from Earth). Because Cyg A is one of the brightest radio sources in the sky, the energy output of this remote galaxy must be enormous. (Palomar Observatory)*

speed of 17,000 km/sec. According to the Hubble law, this speed corresponds to a distance of 1 billion light years!

The enormous distance to Cygnus A astounded astronomers because Cyg A is one of the brightest radio sources in the sky. Although Cyg A is barely visible through the 200-in. telescope at Palomar, its radio waves can be picked up by amateur astronomers with backyard equipment. The energy output in radio waves of Cyg A must therefore be enormous. Indeed, Cyg A shines with a radio luminosity that is 10^7 times as bright as that of an ordinary galaxy such as M31 in Andromeda. Obviously, the object corresponding to Cyg A must be something quite extraordinary.

Quasars look like stars but have huge redshifts

During the late 1950s and early 1960s, radio astronomers were busy making long lists of all the radio sources they were finding across the sky. One of the most famous lists is called the *Third Cambridge Catalogue* (the first two catalogs produced by the British team were filled with inaccuracies) and was published in 1959. It lists 471 radio sources. Even today, astronomers often refer to these sources by their "3C numbers." With the discovery of the extraordinary luminosity of Cyg A (also called 3C 405, because it is the 405th source on the Cambridge list), astronomers were eager to learn whether any other sources in the 3C catalog have similarly extraordinary properties.

One interesting case is 3C 48. In 1960, Allan Sandage used the 200-in. telescope to discover a "star" at the location of this radio source (see Figure 27-2). Since ordinary stars are not strong sources of radio emission, 3C 48 must be something unusual. Indeed, its spectrum shows a series of bright spectral lines that no one could identify. Although 3C 48 was clearly an oddball, many astronomers thought it was just another strange star in our galaxy.

But another such "star" was discovered in 1962. In a continuing effort to identify sources in the *Third Cambridge Catalogue,* astronomers at the Australian National Radio Observatory observed several lunar occultations of 3C 273. (A lunar occultation occurs when the Moon eclipses a background object.) Australian astronomers managed to determine the exact location of 3C 273 by noting when the radio waves were blocked by the Moon.

3C 273 was found to have several peculiar characteristics. As shown in Figure 27-3, a luminous "jet" protrudes from one side of 3C 273. And, as was the case with 3C 48, the "star" contains a series of bright spectral lines that no one could identify.

Astronomers had difficulty identifying the emission lines in the spectra of 3C 48 and 3C 273 because they assumed these starlike objects were peculiar stars nearby in our galaxy. After all, they certainly look like stars.

In 1963, Maarten Schmidt at the California Institute of Technology was examining the spectrum of 3C 273 and realized that four of its brightest spectral lines are positioned relative to one other exactly as four very familiar spectral lines of hydrogen are arranged. However, the

Figure 27-2   The quasar 3C 48
For several years, astronomers believed erroneously that this object is simply a nearby peculiar star that happens to emit radio waves. Actually, the redshift of this starlike object is so great ($z = 0.367$) that, according to the Hubble law, it must be roughly 5 billion light years away. (Palomar Observatory)

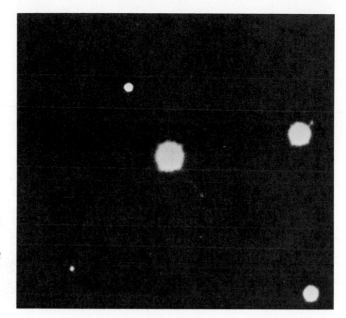

Figure 27-3   The quasar 3C 273
This greatly enlarged view shows the starlike object associated with the radio source 3C 273. Note the luminous jet to one side of the "star." By 1963, astronomers discovered that the redshift of this "star" is so great ($z = 0.158$) that its distance, according to the Hubble law, is nearly 3 billion light years from Earth. (Kitt Peak National Observatory)

emission lines of 3C 273 are at much longer wavelengths than the usual positions of the Balmer lines. In other words, 3C 273 has a substantial redshift.

Stellar spectra exhibit comparatively small Doppler shifts, because a star in our galaxy cannot be moving extremely fast relative to the Sun— if it were, the star would soon escape from the galaxy. Schmidt therefore conjectured that perhaps 3C 273 is *not* a nearby star. Pursuing this hunch, he promptly identified all four spectral lines as hydrogen lines that have suffered enormous redshift of almost 16 percent, corresponding to a speed of 45,000 km/sec (that is, 15 percent of the speed of light). According to the Hubble law, this huge redshift implies the incredible distance to 3C 273 of roughly 3 billion light years.

Because of their starlike appearance, 3C 48 and 3C 273 were dubbed **quasistellar radio sources.** This term was soon shortened to **quasars.**

Figure 27-4 shows the spectrum of 3C 273. Instead of using photography to record a spectrum, many observatories use a charge-coupled device at the focus of the spectrograph. As discussed in Box 6-2, the output of the machine is a graph of intensity versus wavelength, on which emission lines appear as peaks. The graph in Figure 27-4 was obtained in this way; the four hydrogen lines are identified.

These spectral lines of 3C 273 are brighter than the intensity of the background radiation at other wavelengths. The background is called the **continuum,** and the bright lines are emission lines caused by excited atoms that are emitting radiation at specific wavelengths. Spectra of ordinary galaxies (recall Figure 26-23) are dominated by dark absorption lines. Most quasars and many peculiar galaxies exhibit strong emission lines in their spectra—a sign that something unusual is going on.

Once the four hydrogen lines in the spectrum of 3C 273 were identified, Schmidt recognized the remaining spectral lines as those of carbon and oxygen. Two other Caltech astronomers, Jesse Greenstein and T. A. Matthews, then identified the spectral lines of 3C 48 as having suffered a redshift of $z = 0.367$. That corresponds to a velocity of nearly one-third the speed of light, which (according to the Hubble law) places

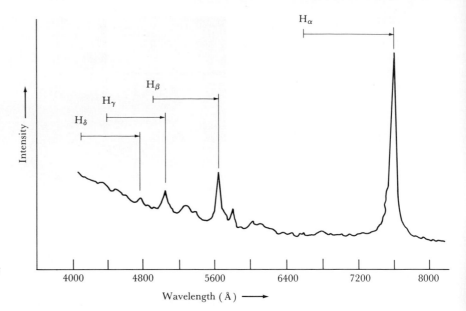

Figure 27-4   The spectrum of 3C 273
Four bright emission lines due to hydrogen dominate the spectrum of 3C 273. The arrows indicate how far these spectral lines are redshifted from their usual wavelengths.

3C 48 twice as far away as 3C 273, or approximately 6 billion light years from Earth.

Incidentally, these distances to quasars assume that the Hubble constant (H_0) is (15 km/sec)/10^6 ly. As mentioned in the previous chapter, some astronomers believe that this value is too low; they would prefer to use (30 km/sec)/10^6 ly for H_0. Doubling the Hubble constant would halve these quasar distances. Also, in dealing with the enormous recessional velocities of quasars, the equations relating redshift and speed must be modified in accordance with the special theory of relativity, as described in Box 27-1.

Hundreds of quasars have been discovered since those pioneering days of the early 1960s. All quasars look like stars, and all have enor-

Box 27-1   The relativistic redshift

By definition, the redshift (z) is

$$z = \frac{\lambda - \lambda_0}{\lambda_0}$$

where λ is the observed wavelength of a spectral line in the spectrum of a star, galaxy, or quasar, and λ_0 is the unshifted wavelength of that same spectral line, as deduced from laboratory experiments.

For example, consider the quasar called PKS 2000–330. Its spectrum, obtained in 1982 by a team of Australian astronomers, is shown in the diagram below. The strongest spectral line is the Lyman-alpha line of hydrogen, observed at a wavelength of 5825 Å. From laboratory measurements, we know that this spectral line is normally seen in the ultraviolet at a wavelength of 1216 Å. Thus, this quasar has a redshift of

$$z = \frac{5825 - 1216}{1216} = 3.79$$

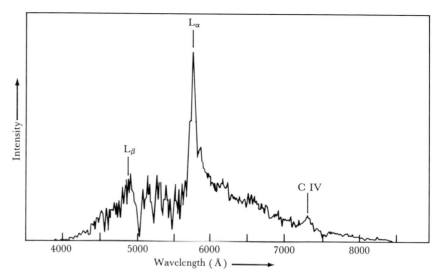

The average redshift for this quasar, determined from measurements of a number of spectral lines, is $z = 3.78$. As this book goes to press, this quasar holds the record for the largest redshift ever observed.

The Hubble law states that the redshift z of an object is directly related to its recessional speed v. In dealing with the enormous recessional velocities of quasars, however, we must modify our ideas and equations in accordance with the special theory of relativity (recall Box 24-1). For low velocities, we used the nonrelativistic equation:

$$z = \frac{v}{c}$$

where c is the speed of light. Thus, for example, a 5 percent shift in wavelength ($z = 0.05$) corresponds to a velocity of 5 percent of the speed of light ($v = 0.05c$). For high velocities, however, we must use the relativistic equation

$$z = \sqrt{\frac{c + v}{c - v}} - 1$$

An equivalent and useful form of this relationship is

$$\frac{v}{c} = \frac{(z + 1)^2 - 1}{(z + 1)^2 + 1}$$

This relativistic relationship between z and v is displayed in the graph at the left. Note that z approaches infinity as v approaches the speed of light. For very low velocities, these complicated relativistic equations both reduce to the simpler, nonrelativistic equation $v = cz$.

As an example of the full, relativistic equation, again consider the quasar PKS 2000-330. Using $z = 3.78$, we obtain

$$\frac{v}{c} = \frac{(4.78)^2 - 1}{(4.78)^2 + 1} = \frac{21.85}{23.85} = 0.92$$

In other words, this quasar is receding from us with a velocity of 92 percent of the speed of light.

To estimate the distance to this quasar, we can use the Hubble law:

$$r = \frac{v}{H_0} = \frac{0.92(3 \times 10^5 \text{ km/sec})}{(15 \text{ km/sec})/10^6 \text{ ly}} = 18.4 \text{ billion light years}$$

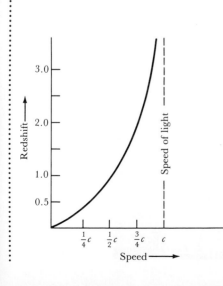

(continued)

Incidentally, the highest possible redshift ($z = \infty$) corresponds to $v = c$, which in turn corresponds to a distance of

$$r = \frac{c}{H_0} = \frac{3 \times 10^5 \text{ km/sec}}{(15 \text{ km/sec})/10^6 \text{ ly}} = 20 \text{ billion light years}$$

It is therefore impossible to see anything farther than 20 billion light years from Earth.

As we shall see in Chapter 28, the age of the universe can be estimated as

$$\text{age of universe} = \frac{1}{H_0} = 20 \text{ billion years}$$

Thus a redshift of $z = \infty$ corresponds to the distance back to the creation of the universe. That is, light now reaching us from an object 20 billion light years away must have been emitted at the time the universe began.

mous redshifts. There are even a few quasars with redshifts greater than $z = 3$. For example, the quasar OH 471 shown in Figure 27-5 has a redshift of $z = 3.4$. According to the special theory of relativity (refer to Box 27-1), this redshift corresponds to a speed slightly greater than 90 percent of the speed of light. From the Hubble law, it follows that the distance to OH 471 is 18 billion light years. Its light has taken 18 billion years to reach us. When we look at OH 471, we are seeing an object as it existed when the universe was very young.

Figure 27-5   The quasar OH 471
This quasar has one of the largest redshifts ever discovered (z = 3.4). This redshift corresponds to a speed slightly greater than 90 percent of the speed of light. According to the Hubble law, OH 471 must be 18 billion light years away. (Palomar Observatory)

A quasar emits a huge amount of energy from a very small volume

Galaxies are big and bright. A typical large galaxy like our own or M31 contains several hundred billion stars and shines with a luminosity of 10 billion Suns. The most gigantic and most luminous galaxies (such as the giant elliptical M87) are only ten times brighter, shining with the brilliance of 100 billion Suns. A redshift of 60 percent corresponds to a distance of nearly 9 billion light years, beyond which even the brightest galaxies are to faint to be detected. That is why so few galaxies have been discovered with redshifts greater than about $z = 0.6$. Most ordinary galaxies are too dim to be detected at one-half that distance.

Although it is difficult to find high-redshift galaxies, high-redshift quasars are quite common. Therefore quasars must be incredibly luminous—far more luminous than galaxies. Indeed, a typical quasar is 100 times brighter than the brightest galaxies we have ever seen. The visual luminosity of some quasars is as much as 10^{46} erg/sec, more than 10^{12} times the output of the Sun. Recent X-ray observations from the Einstein Observatory show that the X-ray luminosity of quasars is even more impressive, perhaps as great as 10^{47} erg/sec.

In the mid-1960s, several astronomers discovered that some of the newly identified quasars had been photographed inadvertently in the past. For example, 3C 273 was found on numerous photographs, including one taken in 1887. By carefully examining the images of quasars on these old photographs, astronomers found that quasars fluctuate in brightness; they occasionally flare up. For example, data from old photographs of 3C 279 are plotted in Figure 27-6. Note the prominent outbursts that occurred around 1937 and 1943. During these outbursts, the luminosity of 3C 279 increased by a factor of at least 25. Because of the enormous distance to 3C 279, this quasar must have been shining with a brilliance 10,000 times as great as that of the entire Andromeda galaxy at the peak of each of these outbursts.

Fluctuations in brightness allow astronomers to place strict limits on the maximum sizes of quasars, because an object cannot vary in brightness faster than light's travel time across that object. For example, an object that is 1 ly in diameter cannot vary in brightness with a period of less than 1 year.

The brightness of many quasars may vary over periods of only a few weeks or months. Some quasars fluctuate from night to night. Recent X-ray data from the Einstein Observatory reveal large variations in as little as three hours.

This rapid flickering means that quasars are small. The energy-emitting region of a typical quasar—the "powerhouse" that blazes with

Figure 27-6 The brightness of 3C 279
This graph shows the magnitude variations of the quasar 3C 279. The data were obtained from careful examination of old photographic plates in the files of the Harvard College Observatory. Note the large outburst in 1937. (Adapted from L. Eachus and W. Liller)

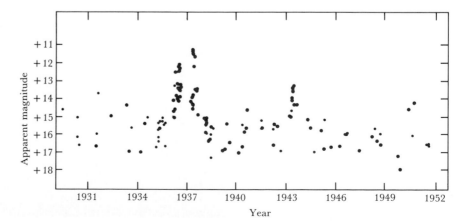

the luminosity of 100 galaxies—is less than 1 light-day in diameter. If quasars are at the huge distances indicated by their redshifts, then a quasar must be producing the luminosity of 100 galaxies in a volume not much farther across than the diameter of our solar system.

Does the Hubble law apply to quasars?

That something so small as a quasar could produce so much energy is an enigma so troublesome that some astronomers prefer to speculate that quasars are *not* located at the vast distances inferred from their redshifts. If the Hubble law does not apply to quasars, then they could be much nearer to us than their redshifts lead us to believe. If quasars are nearby, then they need not be extremely luminous—thereby relieving astronomers of the difficult task of accounting for enormous energy output.

A new problem then arises, however: If quasars are nearby, what causes their large redshifts? The known laws of physics cannot explain how a nearby quasar could have such a large redshift. From the traditional viewpoint, the only thing that makes sense is the Hubble-law interpretation: a big redshift means a big distance. Consequently, maverick astronomers who profess the anti-Hubble-law interpretation of quasar redshifts often argue that we are at the brink of discovering *new* laws of physics. This "new physics" would then explain how quasars can be nearby and still have large redshifts.

The radical approach to the quasar enigma is not popular. Most astronomers feel that the Hubble law is a valid indicator of distance. Conventional astronomers believe that quasars are indeed located at the enormous distances indicated by their huge redshifts. These astronomers accept the challenge of trying to explain how the luminosity of 100 galaxies can be generated in a volume not much larger than our solar system.

Figure 27-7   NGC 4319 and Markarian 205
The galaxy NGC 4319 has a small redshift that corresponds to the relatively nearby distance of 120 million light years. The quasarlike object Mk 205 indicated by the arrow has a redshift 11 times as large as that of the galaxy, but appears to be attached to the galaxy. If these two objects are really connected, they would be evidence of a major violation of the Hubble law. (Courtesy of H. C. Arp)

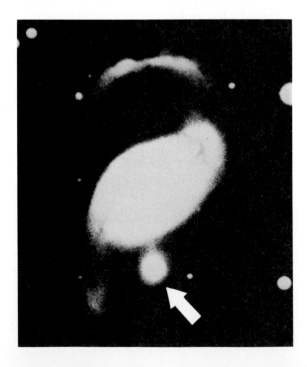

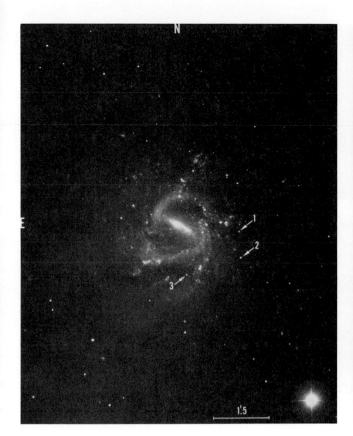

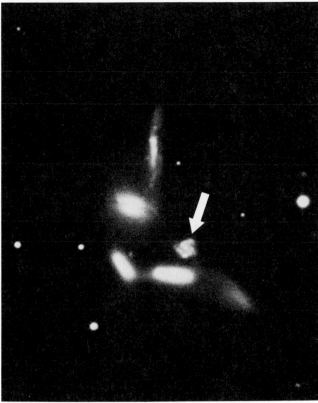

***Figure 27-8 [left]   NGC 1073 and
three quasars***

*This photograph, taken with the 200-in. tele-
scope at Palomar, shows the beautiful barred
spiral galaxy NGC 1073 in the constellation
of Cetus. Three quasars (identified by arrows)
appear to be nestled in the galaxy's spiral
arms. Most astronomers believe that this is just
a "projection effect" caused by three very remote quasars that happen to be in the same
part of the sky (as seen from Earth) as the galaxy. (Courtesy of H. C. Arp)*

Figure 27-9 [right]   Seyfert's Sextet

*Five of the galaxies in this cluster have nearly
the same redshift (z = 0.015), but the remaining galaxy (indicated by an arrow) has a
much higher redshift (z = 0.067). (Palomar
Observatory)*

The maverick astronomers, however, have produced some interesting
observations that might indicate something is wrong with the traditional
approach. For example, in 1972, Halton C. Arp of the Mount Wilson
and Las Campanas Observatories found a quasarlike object that seems to
be attached to a nearby galaxy by a luminous bridge of gas (see Figure
27-7). The galaxy (NGC 4319) has a small redshift (z = 0.006), corresponding to a speed of only 1800 km/sec. According to the Hubble law,
NGC 4319 is only 120 million light years away. The quasarlike object
(called Markarian 205), however, has a much larger redshift (z = 0.07),
corresponding to a speed of 21,000 km/sec. If Markarian 205 is really
connected to NGC 4319, then the quasarlike object must be much closer
to us than the distance implied by its redshift and the Hubble law.

Conventional astronomers argue that this is a case of "chance alignment." The galaxy NGC 4319 is where it belongs: at the distance indicated by its redshift. The quasar Markarian 205, which has a redshift 11
times as large as that of the galaxy, is where it belongs: 11 times as far
away. They only *look* connected because they happen to lie in nearly the
same direction in the sky.

Undaunted, Arp and his colleagues continue to find intriguing cases
across the sky. For example, Figure 27-8 shows the barred spiral galaxy
NCG 1073. Three quasars are nestled in its spiral arms. Arp argues that
the probability of finding three quasars this bright in this small a region
of the sky is exceedingly low and therefore that the quasars must be
somehow associated with the galaxy.

Arp and his colleagues have also discovered several cases where galaxies with widely differing redshifts appear to be located together in space.
Figure 27-9 shows Seyfert's Sextet, a grouping of six galaxies first seen

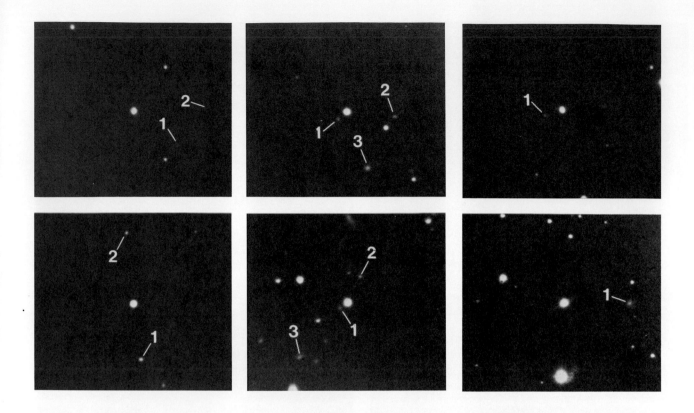

Figure 27-10   Quasars in remote clusters of galaxies
A quasar is located at the center of each of these photographs. Very distant galaxies (identified by numbers) are faintly visible near each of the quasars. In each case, the redshift of the quasar is virtually the same as the redshifts of the galaxies that surround it. (Courtesy of Alan Stockton)

by the American astronomer Carl Seyfert in 1954. Although they seem to be clustered together, one galaxy has a redshift more than four times as large as the redshifts of the others. According to the Hubble law, the high-redshift galaxy must be a background object, four times as far away as the other five galaxies. The apparent tight grouping seen in Figure 27-9, however, makes some people wonder whether something may be wrong with the Hubble law.

Powerful evidence supporting the conventional approach came in 1978 when Alan Stockton of the University of Hawaii published his observations of the regions surrounding 27 low-redshift quasars. Because of their low redshifts, these quasars should be comparatively nearby—perhaps near enough to detect ordinary galaxies in the vicinity of the quasars. In eight cases, Stockton found galaxies huddled around the quasars (see Figure 27-10). In each of these situations, the redshifts of the galaxies are virtually the same as the redshift of the quasar around which they are clustered. Apparently, quasars sometimes exist in poor clusters of galaxies, just as giant elliptical galaxies sometimes exist at the centers of rich clusters of galaxies. Because the Hubble law works for ordinary galaxies, and because each of Stockton's eight quasars has nearly the same redshift as the galaxies that surround it, the redshifts of the quasars themselves almost certainly obey the Hubble law.

These kinds of observations may be the deciding factor in the long-standing redshift debate between the conventionalists and the mavericks. Hubble discovered Cepheid variables scattered around several enigmatic "spiral nebulae" and thereby settled the Shapley–Curtis debate. Many astronomers expect that the Space Telescope will find ordinary galaxies with high redshifts grouped around distant quasars, thereby settling the redshift debate.

Active galaxies bridge the gap between ordinary galaxies and quasars

In the 1960s, the gap in energy output between ordinary galaxies and quasars seemed so huge that some astronomers preferred to question the Hubble law rather than accept the existence of such highly luminous objects. In recent years, however, astronomers have discovered various kinds of peculiar galaxies whose luminosities fall between those of ordinary galaxies and quasars. Some of these strange galaxies have unusually bright, starlike nuclei. Some have strong emission lines in their spectra. Some are highly variable. Some have jets or beams of radiation eminating from their cores. Most of these objects are more luminous than ordinary galaxies, and all are called **active galaxies.**

The first active galaxies were discovered in 1943, during a survey of spiral galaxies, by Carl Seyfert at the Mount Wilson Observatory. Called **Seyfert galaxies,** these luminous objects have bright, starlike nuclei and strong emission lines in their spectra. For example, NGC 4151 (see Figure 27-11) has an extremely rich emission spectrum; 28 percent of the galaxy's light is concentrated in emission lines. These emission lines include Fe X and Fe XIV (iron atoms with 9 or 13 electrons stripped away), indicating that NGC 4151 contains some extremely hot gas. Seyfert galaxies also commonly exhibit variability in brightness; the magnitude of NGC 4151 changes every few months.

Many more Seyfert galaxies have been discovered in recent years, largely through the efforts of Wallace Sargent at Caltech. Approximately 10 percent of the most luminous galaxies in the sky are Seyfert galaxies. Some of the brightest Seyfert galaxies shine as brightly as faint quasars, and thus many astronomers suspect that remote Seyfert galaxies could easily be interpreted as quasars.

Seyfert galaxies are divided into two categories based on their spectra. In the spectrum of a **Type 1 Seyfert galaxy,** the hydrogen lines are

Figure 27-11   The Type 1 Seyfert galaxy NGC 4151

This is one of the best-studied Seyfert galaxies. Because of its bright, starlike nucleus and emission-line spectrum, NGC 4151 might be mistaken for a quasar if it were very far away. The redshift of this galaxy is 0.0033, suggesting a distance (according to the Hubble law) of 70 million light years from Earth. (Palomar Observatory)

Figure 27-12   The Type 2 Seyfert galaxy NGC 1068 (also called M77 or 3C 71)
This Seyfert galaxy is renowned for its extraordinary infrared brightness. Like most Seyfert galaxies, NGC 1068 varies its brightness every few months. The redshift of this galaxy is 0.0036, which means that the galaxy is probably the same distance from Earth as NGC 4151 is (see Figure 27-11). (Lick Observatory)

much broader than the other emission lines, probably because of turbulence at the galaxy's center where the hydrogen emission lines are produced. In the spectrum of a **Type 2 Seyfert galaxy,** the hydrogen lines have roughly the same widths as the other emission lines. NGC 4151 (see Figure 27-11) is an example of a Type 1 Seyfert, whereas NGC 1068 (see Figure 27-12) is an example of a Type 2 Seyfert.

An interesting feature of NGC 1068 is its extraordinary brightness at infrared wavelengths. The total infrared luminosity of this galaxy equals 10^{11} Suns. Frank Low of the University of Arizona has detected varia-

Figure 27-13   The exploding galaxy NGC 1275 (also called 3C 84)
This Seyfert galaxy, located in the Perseus cluster, is a strong source of X rays and radio radiation. The galaxy's redshift is 0.018, suggesting a distance (according to the Hubble law) of nearly 400 million light years from Earth. Note the streamers of gas. Spectroscopic observations confirm significant mass ejection from the galaxy's center. (Kitt Peak National Observatory)

Figure 27-14   An X-ray image of NGC 1275
This picture from the Einstein Observatory shows the X-ray appearance of the Seyfert galaxy NGC 1275. Most of the X-ray emission comes from a point source at the galaxy's nucleus. (Harvard-Smithsonian Center for Astrophysics)

Figure 27-15 [left]   The peculiar galaxy NGC 5128 (also called Centaurus A)
This extraordinary galaxy is located in the constellation of Centaurus, roughly 13 million light years from Earth. Note the broad dust lane across the face of the galaxy. Vast quantities of radio radiation pour from extended regions of the sky on either side of the dust lane. (Cerro Tololo Inter-American Observatory)

Figure 27-16 [right]   An X-ray image of NGC 5128
This picture from the Einstein Observatory shows that NGC 5128 has a bright X-ray nucleus. An X-ray jet protrudes from the galaxy's nucleus along a direction perpendicular to the galaxy's dust lane. Diffuse X-ray emission comes from regions surrounding the galaxy's center. (Harvard-Smithsonian Center for Astrophysics)

tions in infrared brightness of 7×10^9 Suns over only a few weeks. In other words, the infrared power output of the nucleus of NGC 1068 rises and falls by an amount equal to the total luminosity of our entire galaxy.

Many Seyfert galaxies exhibit the vestiges of violent, explosive phenomena in their nuclei. Note the twisted, tormented galaxy NGC 1275 in Figure 27-13. Filaments of gas tens to thousands of light years long protrude from the galaxy in all directions. Spectroscopic studies indicate that the gas is being blasted away from the galaxy's nucleus at 3000 km/sec. The nucleus of this galaxy is a strong source of X rays and radio waves; Figure 27-14 is an X-ray picture.

Other kinds of peculiar galaxies than Seyferts also seem to be ejecting enormous amounts of matter and energy. Note the extraordinary dust lane across the face of the disrupted galaxy NGC 5128 in Figure 27-15. Recent X-ray pictures reveal an X-ray jet (see Figure 27-16) sticking out of the galaxy's nucleus. The jet is perpendicular to the galaxy's dust lane.

NGC 5128 also is one of the brightest sources of radio waves in the sky. It was one of the first sources discovered when radio telescopes were built in Australia; as a radio source, it is called Centaurus A,

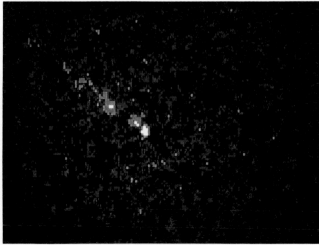

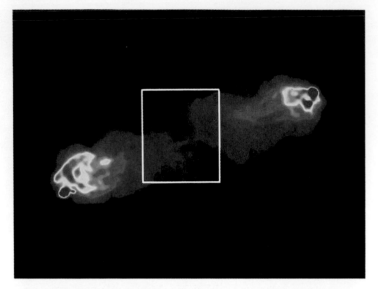

Figure 27-17 A radio image of Cygnus A
This color-coded radio picture was produced at the Very Large Array. Most of the radio emission from Cygnus A comes from the radio lobes located on either side of the visible peculiar galaxy. These two radio lobes are each about 160,000 light years from the optical galaxy, and each contains a brilliant, condensed region of radio emission. The small rectangle indicates the area visible in Figure 27-1. (National Radio Astronomy Observatory)

because of its location among the constellations. In part, the brightness of Centaurus A at radio wavelengths is due to its proximity to Earth—only 13 million light years away. Radio waves pour from two very large regions located on either side of the galaxy's dust lane. These regions cover many square degrees of the sky, corresponding to a volume spanning 2 million light years.

By 1970, radio astronomers realized that Centaurus A and Cygnus A (discussed at the beginning of this chapter; recall Figure 27-1) belong to a class of objects called **double radio sources,** in which radio emission is concentrated in two "radio clouds," or radio lobes. A peculiar galaxy, typically resembling a giant elliptical galaxy, is located between the two radio lobes in many cases (see Figure 27-17). Cygnus A puts out 1000 times more energy than Centaurus A does, and the peculiar galaxy seen in Figure 27-1 is located between the two radio lobes. Further evidence for a high level of activity in the visible galaxy comes from its spectrum, which exhibits many emission lines and resembles an over-excited Seyfert galaxy.

Some double radio sources are enormous. The scale drawing in Figure 27-18 shows that the largest (such as 3C 236) are as big as an entire cluster of galaxies.

Every double radio source seems to have some sort of central "engine" that squirts particles (probably electrons) and magnetic field outward along two oppositely directed jets at speeds very near the speed of light. After traveling many thousands or even millions of light years, this ejected material has slowed down, and interactions between the electrons and the magnetic field produce the radio radiation that we detect. As we saw in Chapter 23 (recall the discussion of the Crab Nebula), a specific type of radio emission called synchrotron radiation occurs whenever electrons spiral within a magnetic field. The radio waves that come from the lobes of a double radio source have all the characteristics of synchrotron radiation.

The idea that a double radio source involves powerful jets of relativistic particles (recall that "relativistic" means "traveling near the speed of light") is supported by the existence of **head–tail sources,** named because each such source appears to have a region ("head") of concentrated radio emission followed by a weaker tail of emission. A good

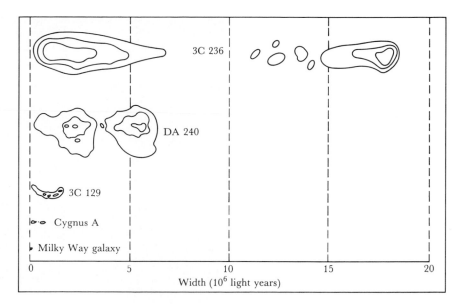

3C 236

DA 240

3C 129

Cygnus A

Milky Way galaxy

0   5   10   15   20

Width (10^6 light years)

Figure 27-18 Giant double radio sources
Some double radio sources are enormous. This scale drawing shows our Galaxy and Cygnus A, along with several very large sources whose dimensions are comparable to the size of an entire cluster of galaxies. (Adapted from R. G. Strom, G. K. Miley, and J. Oort)

example is the active elliptical galaxy NGC 1265 in the Perseus cluster of galaxies. NGC 1265 is known to be moving at a high speed (2500 km/sec) relative to the cluster as a whole. Figure 27-19 is a radio map of NGC 1265. Note that the radio emission has a distinctly windswept appearance. Just as smoke pouring from an old-fashioned locomotive trails a rapidly moving train, particles ejected along two jets from the galaxy are deflected by the galaxy's passage through the sparse intergalactic medium.

At radio wavelengths, the double radio sources are among the brightest objects in the universe. Using some basic physics, astronomer

Figure 27-19 The head–tail source NGC 1265
The active elliptical galaxy NGC 1265 would probably be an ordinary double radio source except that the galaxy is moving at a high speed through the intergalactic medium. Because of this motion, the two jets trail the galaxy, giving this radio source a distinctly windswept appearance. (National Radio Astronomy Observatory)

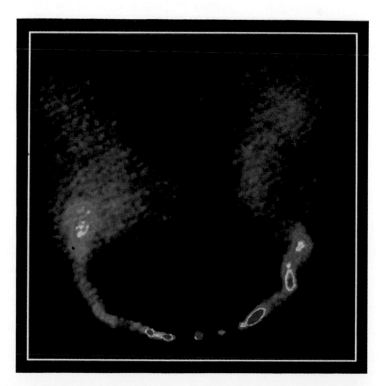

Figure 27-20   BL Lacertae
*This superb photograph shows "fuzz" around
BL Lacertae. BL Lacertae objects appear to be
giant elliptical galaxies with bright, starlike
nuclei—much as Seyfert galaxies are spiral
galaxies with quasarlike nuclei. BL Lacertae
objects contain much less gas and dust than do
Seyfert galaxies. (Courtesy of T. D. Kinman;
Kitt Peak National Observatory)*

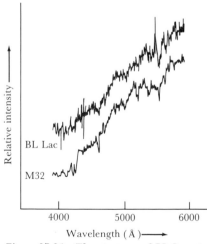

Figure 27-21   The spectrum of BL Lacertae
*The spectrum of the "fuzz" surrounding BL
Lac is shown, along with the spectrum of
M32, a small elliptical galaxy in the local
group. The slight differences between these two
spectra can be explained by assuming that BL
Lac is a giant elliptical galaxy. The spectrum
of BL Lac is redshifted by an amount corresponding to a distance of 1 billion light years
from Earth. (Adapted from J. S. Miller, H. B.
French, and S. A. Hawley)*

Geoffrey Burbidge at Kitt Peak National Observatory has demonstrated
that the energy contained in the radio lobes of a typical source must exceed 10^{61} ergs—equal to the energy released by 10 billion supernova
explosions.

A final class of active galaxies is the **N galaxies,** so-named because
they have bright nuclei. Extreme examples of such galaxies are the **BL
Lacertae objects,** named after their prototype, BL Lac, in the constellation of Lacerta ("the Lizard").

BL Lacertae (see Figure 27-20) was first discovered in 1929, when it
was mistaken for a variable star; its brightness varies by a factor of 15 in
only a few months. BL Lac's most intriguing characteristic is a totally
featureless spectrum that exhibits neither absorption nor emission lines.

Careful examination of BL Lac revealed some "fuzz" around its
bright, starlike core. In the early 1970s, Joseph Miller at Lick Observatory blocked out the light from the bright center of BL Lac and managed to obtain a spectrum of the "fuzz." The spectrum contains many
spectral lines and strongly resembles the spectrum of an elliptical galaxy
(see Figure 27-21). In other words, a BL Lacertae object is an elliptical
galaxy with a very bright starlike center, much as a Seyfert galaxy is a
spiral galaxy with a quasarlike center.

These discoveries have prompted many astronomers to suspect that
quasars are the superluminous centers of very-distant, very-active galaxies. Painstaking observations by Susan Wyckoff, Peter Wehinger, and others have in fact revealed faint galaxylike "fuzz" around several quasars.
Of course, quasars are extremely far away and thus are very difficult to
observe. Therefore, clues about the "engine" that powers a quasar might
come from studies of such nearby active galaxies such as the giant elliptical M87 in the Virgo cluster.

Supermassive black holes may power quasars and active galaxies

How do quasars and active galaxies produce such enormous amounts of energy from such small volumes? As long ago as 1968, the British astronomer Donald Lynden-Bell, working at the California Institute of Technology pointed out that an extremely massive black hole could be the "engine" that powers a quasar or active galaxy. After all, nothing could possibly be more compact than a black hole and, as explained in Box 27-2, extremely massive black holes might be very common objects in the universe. At the center of a quasar or active galaxy, nature may be drawing upon the tremendous amounts of energy tied up in the black hole's powerful gravitational field.

Various realistic scenarios about how to tap the gravitational energy of an extremely massive black hole have been worked out by Richard Lovelace at Cornell University and Roger Blandford at Caltech. In essence, these schemes are simply scaled-up versions of the explanation of Cygnus X-1 discussed in Chapter 24. A large black hole is needed to ensure a large energy output.

Imagine a $10^9 M_\odot$ black hole sitting at the center of a galaxy. The center of a galaxy is a very congested place, and thus we expect this **supermassive black hole** to be surrounded by a huge accretion disk of matter captured by the hole's gravity. According to Kepler's third law, the inner regions of the accretion disk would orbit the hole more rapidly than would the outer regions. Thus the rapidly spinning inner regions are constantly rubbing against the more slowly moving gases in the outer regions. The resulting friction heats the gases, which spiral toward the hole as sketched in Figure 27-22.

Because of the constant inward crowding of hot gases, the gas pressure and radiation pressure surrounding the hole are tremendous. Of course, some of the inflowing material is swallowed by the hole, but the pressures around the center of the accretion disk are so great that most of the hot gas never really gets near the hole. Instead, in an attempt to relieve the pressure, matter is violently ejected along the perpendicular to the accretion disk, the direction along which the hot gases experience the least resistance. The result is two oppositely directed beams of relativistic particles.

How is this ejected matter confined to narrow jets? Recent studies in the physics of fluids demonstrate that there is a natural "self-focusing" effect whenever a high-speed jet penetrates a medium rather than squirting into empty space. For example, consider the action of an ordinary garden hose to which a nozzle is attached. As water squirts out of the nozzle, the unconfined stream broadens, and the spray fans out through a wide angle. If the nozzle is placed in a swimming pool, however, the stream of water does not fan out as much in the water as it does in air. Similarly, as the two jets of hot gas leave the vicinity of the black hole, they must blast their way through some gas still crowding inward. Passage through this material causes the jets to become extremely narrow, concentrated beams.

Perhaps this self-focusing effect explains double radio sources and some of the jets and beams we see protruding from active galaxies and from some quasars. To account for the luminosity of a quasar (10^{47} erg/sec), a $10^9 M_\odot$ black hole would have to accrete matter at the rate of $10 M_\odot$ per year.

As we saw in Chapter 24, finding black holes is a tricky and difficult business. At best, we can see only the effects of the hole's gravity and try to rule out all non–black-hole explanations of the data. This is the general approach some astronomers have taken with the giant elliptical galaxy M87.

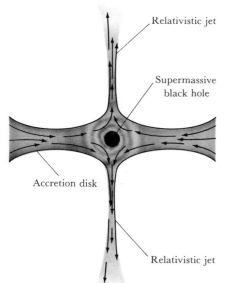

Figure 27-22   A supermassive black hole
The energy output of active galaxies and quasars may involve extremely massive black holes that accrete matter from their surroundings. In the scenario depicted here, the inflow of material through an accretion disk produces two powerful, relativistic jets.

Relativistic jet

Supermassive black hole

Accretion disk

Relativistic jet

Box 27-2 Comments about the plausibility of extremely massive black holes

Despite their exotic properties, the creation of massive black holes does not necessarily require exotic circumstances. To see why this is so, we shall examine the density of material needed to produce a very massive black hole. However, first a word of caution: space around a black hole is highly curved. Therefore, simple geometrical equations (such as a volume $V = (4/3)\pi r^3$ for a sphere of radius r) are not exactly correct. Nevertheless, they are accurate enough to ensure that the following arguments are valid.

We want to know how tightly we must compress a mass M inside a sphere of radius R in order to create a black hole. Just before the creation of the hole, the average density (ρ) of the compressed matter is the mass M divided by the volume $(4/3)\pi R^3$. Thus

$$\rho = \frac{3M}{4\pi R^3}$$

As explained in Chapter 24, however, the radius of a black hole is related to its mass by Schwarzschild's equation:

$$R = \frac{2GM}{c^2}$$

Substituting this relation into the previous equation, we obtain

$$\rho = \frac{3c^6}{32\pi G^3 M^2}$$

The important point is that the required density is inversely proportional to the square of the mass of the hole. In other words, as the mass increases, the density needed to make a black hole drops dramatically.

For example, to make a $1M_\odot$ black hole, the equation tells us that we must compress that one solar mass to a density of roughly 10^{16} g/cm^3—slightly greater than the typical density inside a neutron star.

If we want to make a $10^9 M_\odot$ black hole, however, the required density is lowered by a factor of 10^{18}. Thus we need squeeze the matter to a density of only 0.01 g/cm^3—only one-hundredth the density of water.

Incidentally, it is interesting to think about the biggest possible black hole. As mentioned in Box 27-1, the age of the universe can be estimated as $1/H_0$. Thus, the "distance to the creation" is

$$\frac{c}{H_0} = 20 \text{ billion light years}$$

Using this distance as the radius R of a black hole, we obtain

$$\rho = \frac{3M}{4\pi R^3} = \frac{3c^2}{8\pi G R^2} = \frac{3H_0^2}{8\pi G}$$

Using a Hubble constant of (20 km/sec)/10^6 ly, we find

$$\rho = \frac{3H_0^2}{8\pi G} = 8 \times 10^{-30} \text{ g/cm}^3$$

which is equivalent to about five hydrogen atoms per cubic meter of space. As we shall see in the final chapter, that is roughly the average density of matter across the universe. In other words, we just might be living *inside* the biggest black hole.

M87 is an active galaxy that sits near the center of the Virgo cluster only 50 million light years from Earth. Figure 26-10 shows the full extent of this galaxy. In 1918, H. D. Curtis made a brief photographic exposure of M87 that revealed a bright, starlike nucleus from which a jet protrudes. Two views of this jet appear in Figure 27-23.

In the mid-1970s, several Caltech astronomers headed by Wallace Sargent made detailed spectroscopic observations of M87, carefully examining the widths of spectral lines across its face. As sketched in Figure 27-24, a spectral line is broadened because of the motions of the individual stars along the line of sight. Indeed, the spread in stellar velocities (called the **velocity dispersion**) is directly related to the overall width of a spectral line.

Sargent and his colleagues found that spectral lines in the center of M87 are exceptionally broad, indicating a very large range of stellar velocities. Some stars near the center of M87 are coming toward us and others rushing away from us at speeds far greater than those we would expect in an ordinary galaxy. This observation suggests that these stars are rapidly circling the center of M87 along very compact, high-speed orbits, which is possible only if some sort of extremely massive object is

Figure 27-23 The core and jet of M87
A short time exposure reveals a luminous jet surging out of the bright, starlike nucleus of M87. (a) A single, ordinary photograph made with the 200-in. Palomar telescope. (b) A computer-enhanced exposure made from several photographic plates. The jet is about 6500 light years long. (Courtesy of H. C. Arp and J. J. Lorre)

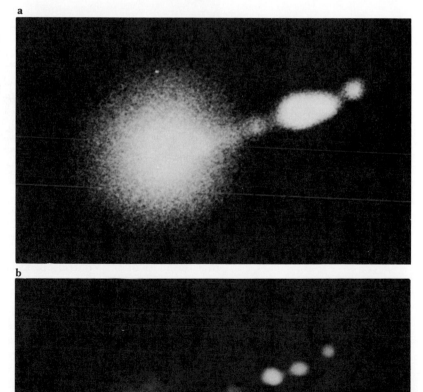

Figure 27-24 *The broadening of galaxy absorption lines*
The spectral lines of individual stars are Doppler-shifted according to how fast the stars are moving parallel to our line of sight. From Earth, however, we see the combined light of all the stars along our line of sight. The spectral line that we see is therefore broadened by an amount directly related to the spread in stellar velocities.

located at the galaxy's center. After all, without the gravity of such an object to keep the stars in their high-speed orbits, these stars would long ago have escaped from the galaxy's core. Sargent's observations imply that the mass of this central object is about 5 billion solar masses. That much matter confined to such a small volume strongly suggests the existence of a supermassive black hole.

While Sargent made spectroscopic observations, a second Caltech team headed by Peter Young carefully measured the brightness across the face of M87. Using Newtonian mechanics, it is possible to calculate how stars should be distributed on the average in an elliptical galaxy. Calculations of this type were done in 1966 by Ivan King at the University of California at Berkeley. King's theoretical models tell us what the brightness should be across the face of an ordinary elliptical galaxy.

Young found that his observations do not agree with King's calculations. As graphed in Figure 27-25, M87's core is much brighter than would be expected from a simple distribution of billions of stars alone. To account for the bright core, Young had to add a powerful, centrally located source of gravity to King's calculations. With this source of gravity, stars naturally crowd around the center of M87 thereby producing a bright nucleus. Once again, the data suggest the presence of a $5 \times 10^9 M_\odot$ black hole.

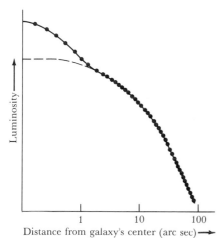

Figure 27-25 The luminosity profile of M87
This graph shows the brightness across M87, measured outward from the galaxy's center. Each dot represents an individual measurement. The dashed curve is the standard model calculated by King. The solid curve (which fits the data!) is the standard model revised to include a supermassive black hole at the galaxy's center. (Adapted from P. Young and colleagues)

Properties of jets, such as that protruding from M87, were elucidated in 1984 by Larry L. Smarr of the University of Illinois and his colleagues in West Germany. Using a Cray-1 supercomputer at the Max Planck Institute near Munich, they calculated details of supersonic flow. Figure 27-26 shows jets flowing at Mach 3 and Mach 6 (3 and 6 times the speed of sound in the surrounding medium). Because of instabilities in the flow, the jets are lumpy; red spots indicate knots of high pressure. The general appearance of these jets is strikingly similar to the knotty structure of the jet in M87 (compare with Figure 27-23).

In the early 1980s, Matthew Malkan at Caltech demonstrated that combined infrared, visual, and ultraviolet observations of quasars strongly suggest the presence of massive accretion disks orbiting supermassive black holes. As an example of Malkan's analysis, Figure 27-27 shows the spectrum of the quasar PKS 0405−123 from 10 μ in the infrared to 1000 Å in the ultraviolet. This spectrum is typical of most quasars and Type 1 Seyfert galaxies. Note the steep decline in the infrared region. This decline slows at visible wavelengths, only to resume in the ultraviolet region. The resulting broad plateau at visible wavelengths (on which Balmer emission lines are superimposed), often called the "3000-Å bump," is a common feature in the spectra of many quasars and active galaxies.

a

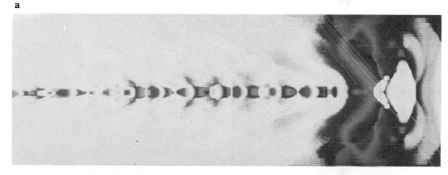

b

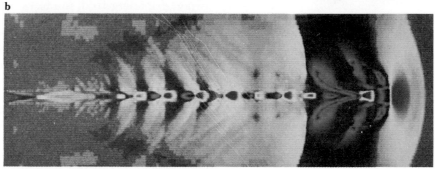

Figure 27-26 Details of supersonic jets
Computer graphics are used to display the pressure in two supersonic jets. The flow is from left to right at speeds of (a) Mach 3 and (b) Mach 6. Red indicates high pressure; blue and green are at lower pressures. Note the knotty structure of the jets and compare with Figure 27-23. (Courtesy of Larry Smarr, Michael Norman, and Karl-Heinz Winkler)

The steep decline in the infrared region is characteristic of radiation from high-speed electrons interacting with their environment and with magnetic fields. These processes produce **nonthermal radiation,** such as synchrotron radiation, whose intensity S declines with frequency ν according to the **power law:**

$$S = \kappa \nu^{-\alpha}$$

where κ and α are constants. This nonthermal contribution to the

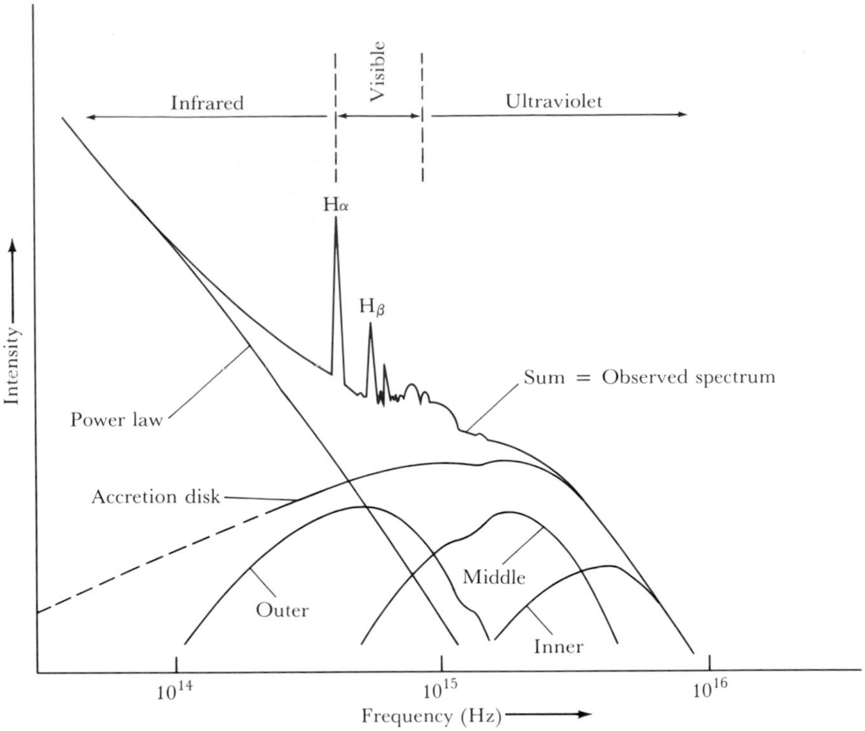

Figure 27-27   *The spectrum of the quasar PKS 0405-123*

Like the spectra of most quasars and Type 1 Seyfert galaxies, the spectrum of PKS 0405-123 (color line) shows an overall decline with increasing frequency. The prominent Balmer emission lines H_α and H_β are identified. The quasar's spectrum can be explained as the sum of a nonthermal spectrum ("power law") and the spectrum of hot gas in an accretion disk. In this model, the accretion disk has a diameter of about 1 light year. (Adapted from M. Malkan)

spectrum of PKS 0405-123 is shown in Figure 27-27 as the line labeled "power law."

To explain the 3000-Å bump, Malkan used theoretical models of accretion disks developed in the 1970s by Caltech physicist Kip Thorne. For ease of calculation, a disk is divided into three regions: a hot inner region extending from the black hole out to 10 Schwarzschild radii, a warm middle region from 10 to 50 Schwarzschild radii, and a cooler outer region from 50 to 500 Schwarzschild radii. (Recall from Chapter 24 that the Schwarzschild radius is the distance from the singularity to the event horizon of a nonrotating black hole). In accordance with its temperature, each of these three regions makes a specific thermal contribution to the quasar's light. The three contributions are shown on Figure 27-27, along with their sum (which is labeled "accretion disk").

When the nonthermal radiation is added to the thermal radiation from the accretion disk, the entire spectrum of PKS 0405–123 is reproduced with remarkable accuracy. In this case, the mass of the black hole is calculated to be between 200 million and 500 million solar masses, depending on how fast the black hole is rotating. An accretion rate of between $10 M_\odot$ to $20 M_\odot$ per year explains the power output of this quasar. Other quasar spectra examined by Malkan imply similar black-hole masses and accretion rates. In each case, the observed spectrum is accurately reproduced by a power-law spectrum plus an accretion-disk spectrum, thereby giving strong support to a black-hole explanation of quasars.

The possible existence of supermassive black holes in quasars and active galaxies is one of the most exciting topics in modern astronomy. Wallace Sargent points out that the Space Telescope will be used to probe the nuclei of galaxies with much higher resolution than is possible from the ground. These observations will undoubtedly reveal many details about the powerful "engines" at the cores of active galaxies and quasars.

Summary

- A quasar (or quasistellar object) is an object in the sky that looks like a star but has a huge redshift corresponding (by the Hubble law) to a very great distance from the Earth.

  To be seen from Earth, a quasar must be very luminous—typically about 100 times brighter than an ordinary bright galaxy.

  Relatively rapid fluctuations in the brightnesses of quasars indicate that they cannot be much larger in size than the diameter of our solar system.

- An active galaxy is an extremely luminous galaxy that has one or more unusual features: an unusually bright, starlike nucleus; strong emission lines in its spectrum; extreme variations in luminosity; or jets or beams of radiation emanating from its core.

  An active galaxy with a bright, starlike nucleus and strong emission lines in its spectrum is called a Seyfert galaxy.

  Most double radio sources seem to have an active galaxy located between the two radio lobes that distinguish this type of radio source.

  A head–tail radio source seems to show evidence of relativistic particle jets emerging from an active galaxy.

  N galaxies and BL Lacertae objects are active galaxies with very bright nuclei whose cores show relatively rapid variations in luminosity.

  A BL Lacertae object is probably an elliptical galaxy with a quasar-like center; a Seyfert galaxy is probably a spiral galaxy with a quasarlike center; quasars may be simply very distant objects much like active galaxies.

- The strong energy emission from quasars, active galaxies, and double radio sources may be produced as matter falls toward a supermassive black hole at the center of the object.

Review questions

1 Suppose you saw an object in the sky that you suspected might be a quasar. What sort of observations might you perform to find out if it were indeed a quasar?

2 Explain why astronomers do not use any of the standard candles described in Chapter 26 to determine the distances to quasars.

3 How would you distinguish between thermal and non-thermal radiation.

4 In the 1960s, it was suggested that quasars might be compact objects ejected at high speeds from the centers of nearby ordinary galaxies. Why does the absence of blueshifted quasars disprove this hypothesis?

*5 The quasar 3C 175 has a redshift of 0.768. How fast is this object receding from us? How far away is this object? Be sure to state the value of the Hubble constant you use.

*6 Calculate the Schwarzschild radius of a $10^9 M_\odot$ black hole. How does your answer compare with the size of the solar system?

7 Why do you suppose there are no quasars relatively near our galaxy?

8 Compare and contrast SS433 with a typical double radio source.

Advanced questions

*9 The quasar 3C 245 has a redshift of $z = 1.03$ and an apparent magnitude of $+17$. Find the quasar's **(a)** radial velocity, **(b)** distance from Earth, **(c)** absolute magnitude, and **(d)** energy output.

10 Verify by direct calculation the statement in the text that for distances beyond about 9 billion light years, even the brightest galaxies are too faint to be detected.

Discussion questions

11 Some quasars show several sets of absorption lines whose redshifts are less than the redshift of the quasars' emission lines. For example, the quasar PKS 0237−23 has five sets of absorption lines with redshifts in the range 1.364 to 2.202 while the quasar's emission lines have a redshift of 2.223. Propose an explanation for these sets of absorption lines.

12 Speculate on the possibility that quasars, double radio sources, giant elliptical galaxies, etc. form some sort of evolutionary sequence.

For further reading

Arp, H. "Related Galaxies with Different Redshifts." *Sky & Telescope*, Apr. 1983, p. 307.

Balick, B. "Quasars with Fuzz." *Mercury*, May/June 1983, p. 81.

Blandford, R., et al. "Cosmic Jets." *Scientific American*, May 1982.

Ferris, T. "The Spectral Messenger: The Redshift Controversy." *Science 81*, Oct. 1981, p. 66.

Kaufmann, W. "Exploding Galaxies and Supermassive Black Holes." *Mercury*, Sep./Oct. 1978, p. 97.

Tananbaum, H., and Lightman, A. "Cosmic Powerhouses: Quasars and Black Holes." In Cronell, J., and Lightman, A., eds. *Revealing the Universe*. MIT Press, 1982.

Wyckoff, S., and Wehinger, P. "Are Quasars Luminous Nuclei of Galaxies?" *Sky & Telescope*, Mar. 1981, p. 200.

28 Cosmology and the creation of the universe

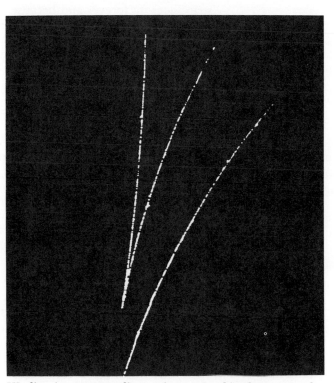

Pair production
A gamma ray (from below) enters a bubble chamber—a device filled with liquid hydrogen and designed to make the path of a charged particle visible as a long row of tiny bubbles. Near the bottom of the photograph, the energy carried by the gamma ray is converted into an electron and an antielectron. Because of a magnetic field surrounding the bubble chamber, the electron is deflected to the right while the antielectron veers toward the left. The path of a stray electron is also seen. The path of the gamma ray is not visible because photons are electrically neutral. (Courtesy of Lawrence Berkeley Laboratory)

We live in an expanding universe. In this chapter, we learn that the expansion of our universe began with an explosion of space at the beginning of time. We find that the universe is filled with microwave photons left over from the early moments of the universe. We learn about the possibility that all of the matter and energy in the universe emerged from a pure vacuum within a second after the universe began. In fact, we find that the matter existing today is just a tiny fraction of that formed originally; the microwave background is the faint remnant of the abundant matter and antimatter of the early universe. Finally, we discuss some of the intriguing speculations about the origins and evolution of the universe, and we find out why the question of the mass of the neutrino is a very important one for those who study the evolution of the universe.

As foolish as it may seem, one of the most profound questions you can ask is, "Why is the sky dark at night?" This question apparently haunted Kepler as long ago as 1610 and was popularized in the early 1800s by the German amateur astronomer Heinrich Olbers.

To appreciate the problem, you must begin by assuming that the universe is infinite and that stars are scattered more-or-less randomly across this infinite expanse of space. Isaac Newton suggested that no other assumption makes sense. If the universe were not infinite or if stars were grouped in only one part of the universe, then the gravitational forces between the stars would soon cause all this matter to fall together into a very compact blob. Obviously, this has not happened. Thus, according to classical Newtonian mechanics, we must be living in a universe that is

519

infinite and static. Only then does each star feel a uniform gravitational pull from every part of the sky, due to all the other stars in the universe. According to this model, the universe can exist forever without major changes in its structure.

Imagine looking out into space in this static, infinite universe. Because space goes on forever and because stars are scattered throughout space, your line of sight eventually must hit a star. No matter where you look in the sky, you should see a star. The entire sky should therefore be as bright as an average star. Even at night, the entire sky should be blazing like the surface of the Sun. Obviously, this is not so. This dilemma is called **Olbers's paradox.**

We live in an expanding universe that began with an explosion called the Big Bang

Olbers's paradox tells us there is something very wrong with the classical Newtonian idea of an infinite, static universe. Edwin Hubble's discovery of the recession of the galaxies provides a resolution of this dilemma. As we saw in Chapter 26 (recall Figure 26-24), Hubble discovered that remote galaxies are rushing away from us with speeds proportional to their distances from the Earth. Specifically, the recessional velocity v of a galaxy is related to its distance r from Earth by the equation

$$v = H_0 r$$

where H_0 is the Hubble constant.

To illustrate the meaning of the Hubble law, astronomers often use the raisin-cake analogy sketched in Figure 28-1. Suppose that a baker mixes dough, yeast, and raisins to make a raisin cake as shown in Figure 28-1a. Unknown to the baker, however, an intelligent bug is sitting on one of the raisins. This bug measures the distances between her raisin and several other raisins, as sketched in the drawing. Now some time passes while the yeast causes the dough to rise.

Eventually, the raisin cake has doubled in size, as shown in Figure 28-1b. The intelligent bug now repeats her measurements and discovers that all the distances between her raisin and the other raisins have doubled. For example, the raisin that was 2 cm away is now 4 cm away, and the raisin that was 10 cm away is now 20 cm away.

After a little more thought, the intelligent bug realizes that the nearby raisins moved away from her slowly, whereas the more remote raisins rushed away more rapidly. For example, the raisin that started 2 cm away moved only 2 cm in the same time interval that the raisin initially 10 cm away moved an additional 10 cm. The bug's conclusion is just like the Hubble law, which tells us that nearby galaxies are moving away from us slowly, whereas more remote galaxies are rushing away from us more rapidly. Note that a bug on any raisin would reach the same conclusion.

This story reveals the significance of the Hubble law. A simple, linear relationship between distance and speed (such as $v = H_0 r$) is exactly what we mean by the word "expansion." The Hubble law therefore tells us that we live in an expanding universe.

It is interesting to think back into the distant past. The universe has been expanding for billions of years, so there must have been a time in the ancient past when all the matter in the universe was concentrated in a state of infinite density. Presumably, some sort of colossal explosion must have occurred to start the expansion of the universe. This explosion, commonly called the **Big Bang,** marks the creation of the universe.

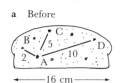

a Before

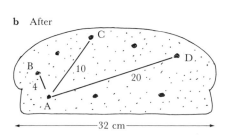

b After

Figure 28-1   The raisin-cake analogy
(a) *A bug on raisin A measures the distances to raisins B, C, and D.* (b) *Because of yeast in the dough, the raisin cake has now expanded to twice its original size. The bug repeats her measurements and discovers a Hubble-law—like relationship between the distances to the raisins and the speed with which they have receded from her.*

To calculate the time elapsed since the Big Bang, think of any two galaxies separated by a distance r and receding from each other with a velocity v. These galaxies have been moving apart since the moment of the Big Bang, which occurred T years ago. Thus we have the simple equation

$$r = vT$$

or, alternatively,

$$v = \frac{r}{T}$$

This second equation *is* the Hubble law ($v = H_0 r$), provided that

$$T = \frac{1}{H_0} = \frac{1}{(15 \text{ km/sec})/10^6 \text{ ly}} = 20 \text{ billion years}$$

Actually, the age of the universe must be slightly less than 20 billion years. Because of their mutual gravitational attraction, galaxies have not been flying away from each other with a constant velocity v. Instead, because of gravity, the speed of separation between galaxies has been gradually decreasing since the Big Bang. Thus, the expansion rate of the universe has been decreasing. An age of 20 billion years was calculated by assuming no deceleration, so this value would be valid only for an empty universe that contains no matter and thus no gravity to slow the expansion. The true age of the universe is probably between 15 and 20 billion years.

The finite age of the universe offers a resolution of Olbers's paradox. Because the universe is less than 20 billion years old, we cannot see any stars that are more than 20 billion light years away. The universe may indeed be infinite, with galaxies scattered throughout its limitless expanse. However, the light from stars more than 20 billion light years away has simply not had enough time to get here.

You can think of the Earth as being at the center of an enormous sphere with a radius of roughly 20 billion light years (see Figure 28-2). The surface of this sphere is called the **cosmic particle horizon.** The entire **observable universe** is located inside this sphere. We cannot see anything beyond the cosmic particle horizon because the travel time for light from these incredibly remote objects is greater than the age of the universe. Throughout the observable universe, the distribution of galaxies is sufficiently sparse that most of our lines of sight do *not* hit any stars. Thus, the night sky is dark.

The finite age of the universe is an important way out of Olbers's paradox. However, there is a second effect that contributes significantly to the darkness of the night sky: the redshift. The Hubble law tells us that the redshift of a galaxy is directly related to its distance from Earth. The greater the distance, the greater the redshift. Recall from Chapter 17, however, that the energy E carried by a photon is related to its wavelength λ by the expression

$$E = \frac{hc}{\lambda}$$

where h is Planck's constant and c is the speed of light.

When a photon is redshifted, its wavelength is lengthened, and thus its energy is decreased. The greater the redshift, the less the energy.

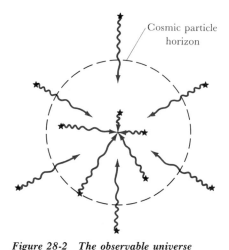

Figure 28-2   The observable universe
The radius of the cosmic particle horizon is equal to the distance that light has traveled since the Big Bang. The Big Bang occurred about 20 billion years ago, so the cosmic particle horizon is about 20 billion light years away. We cannot see stars beyond the cosmic particle horizon because their light has not had enough time to reach us.

Consequently, although there are many galaxies far from the Earth, they have large redshifts, and thus their light does not carry much energy. Indeed, a galaxy at the cosmic particle horizon has an infinite redshift, and thus its light does not carry any energy at all. This decrease in photon energy due to the expansion of the universe decreases the brilliance of remote galaxies, thereby contributing to the darkness of the night sky.

The Big Bang was an explosion of space at the beginning of time

The idea of a Big-Bang origin of the universe is both the simplest concept and one of the most complex concepts in modern astronomy. It is a simple idea because it is the most straightforward, logical consequence of an expanding universe. If you just imagine far enough into the past, you arrive at time nearly 20 billion years ago when the density of matter throughout the universe was infinite. Indeed, the entire universe was like the center of a black hole. For this reason, a better name for the Big Bang is the **cosmic singularity.**

There are, however, many misconceptions about the Big Bang. For example, it was *not* like an exploding bomb. When a bomb explodes, pieces of debris fly off *into space* from a central location. If you could trace all the pieces back to their origin, you could find out exactly where the bomb had been located. This is *not* possible with the universe because the universe itself consists of *all space*—it always has, and it always will. There is genuinely nothing—not even space—beyond the "edge" of the universe. For this reason, the universe logically cannot have an "edge." In this chapter and the next, you will learn to appreciate why the universe has neither an edge nor a center. You will understand that questions such as "What's beyond the edge of the universe?" or "What is the universe expanding into?" are meaningless gibberish.

To understand the Big Bang, we must first appreciate the real meaning of the redshifts of the galaxies. The redshift caused by the expansion of the universe is properly called the **cosmological redshift** to distinguish it from ordinary Doppler shifts. This important distinction is closely related to the classical, Newtonian pitfalls that led us into the dilemma of Olbers's paradox.

According to the classical, Newtonian picture of reality, space is laid out in any direction like a great, inflexible, flat sheet of rectangular graph paper. This rigid, flat space stretches on and on, totally independent of stars or galaxies or anything else. Similarly, a Newtonian clock ticks steadily and monotonously forever, never slowing down or speeding up. Furthermore Newtonian space and Newtonian time are unrelated to each other; measurements made with rulers are independent of measurements made with clocks.

Albert Einstein demonstrated that this view of space and time is wrong. In his special theory of relativity (recall Box 24-1), he proved that measurements with clocks and rulers are interrelated and that they depend on the motions of the observer. As explained in Chapter 24, the general theory of relativity tells us that the shape of space is profoundly influenced by the masses occupying that space. John A. Wheeler at the University of Texas puts it this way: Matter tells space how to curve, and curved space tells matter how to move.

It is more accurate to take a relativistic view of the universe than it is to take a Newtonian view. For example, think about viewing the stars at night. If you observe a star that is 100 light years away, you are actually seeing how that star looked 100 years ago because its light took that long to reach your eyes. This statement reveals the intimate connection that exists between time and space: as you gaze out into space, you are also

Figure 28-3   The expanding-balloon analogy
The expanding universe can be compared to the expanding surface of an inflating balloon. All spots on the balloon recede from one another as the balloon expands, just as all the galaxies recede from one another as the universe expands.

looking back in time. The relativistic view properly describes the intimate relationships between space, time, and matter.

Shortly after formulating his general theory of relativity in 1915, Albert Einstein applied his ideas to the structure of the universe. The prevailing (Newtonian) view was that the universe is infinite and static. To Einstein's dismay, his calculations could not produce a truly static universe; they indicated that the universe must be either expanding or contracting. The prevailing opinions were so strong that Einstein doubted the validity of his equations. In desperation, he added a term to his equations (called the cosmological constant) to ensure that his calculations would predict a static universe. "It was the greatest blunder of my life," he later lamented. Because he doubted his original equations, Einstein missed the opportunity to postulate that we live in an expanding universe. He could have beat Hubble to the punch by at least ten years.

According to relativity, space and time are not rigid and unchanging. In a relativistic model of our universe, as time elapses, *space itself expands*. Indeed, the expansion of the universe *is* the expansion of space. A good analogy is a person blowing up a balloon, as sketched in Figure 28-3. As the balloon expands, the amount of space between dots on the balloon gets larger and larger.

Imagine sitting on one of the dots on the expanding balloon of Figure 28-3. As you look around, you see all the other dots moving away from you. Specifically, you observe that nearby dots move away from you slowly, whereas more distant dots move away from you more rapidly. This is true no matter which dot you call home. And of course, your home dot is certainly not at the center of the balloon. Indeed, the *surface* of the balloon does not have a center. You could move around the surface of the balloon forever without finding any center. Similarly, the surface of the balloon has no edge. You could explore every inch of the balloon's surface and never find an edge.

The expanding surface of the balloon is a two-dimensional example of the expanding three-dimensional space that fills our universe. Just as the surface of the balloon has neither a center nor an edge, our three-dimensional universe also has no center or edge. Even though we see galaxies rushing away from us in all parts of the sky, we are not at the center of the universe. No matter what galaxy you call home, all the other galaxies recede from you.

The realization that the expansion of the universe is the expansion of space gives us a new way of understanding cosmological redshifts. Imagine a photon coming toward us from a remote galaxy. As the photon travels through space, space itself is expanding, so the photon's wavelength becomes stretched. When the photon reaches our eyes, we see a longer wavelength than usual—that is the redshift. The longer the photon's journey, the more its wavelength has been stretched. Thus photons from distant galaxies have larger redshifts than those of photons from nearby galaxies—that is the Hubble law.

This is a vast improvement over the old-fashioned Newtonian approach in which we try to imagine galaxies moving away from each other against a rigid, static background of space. In that model, the only interpretation of the galaxies' redshifts is the Doppler effect, and we fall into the trap of thinking of the Big Bang as a bomblike explosion.

Comparisons of the Big Bang to the center of a black hole help us appreciate many aspects of the creation of the universe. As we saw in Chapter 24, matter is crushed to infinite density at the center of a black hole. This location, called the singularity, is characterized by infinite curvature of space and time. Because of this infinite curvature, space and time are all tangled up, and thus it is impossible to use the laws of physics to describe or predict the behavior of the singularity. Quite simply, you cannot figure out *when* and *where* something is going to happen if you do not have a clear background of time and space.

At the moment of the Big Bang, a state of infinite density filled the universe. Thus space and time throughout the universe are completely jumbled up in a state of infinite curvature, like that at the center of a black hole. We therefore cannot use the laws of physics to tell us exactly what happened at the moment of the Big Bang. We certainly cannot use science to tell us what existed *before* the Big Bang. These things are fundamentally unknowable. Indeed, the terms "*before* the Big Bang" or "at the *moment* of the Big Bang" are virtually meaningless because time itself did not really exist.

A very short time after the Big Bang, space and time did exist in the way we think of them today. This interval, called the **Planck time** (t_P), is given by the expression

$$t_P = \sqrt{\frac{Gh}{c^5}} = 1.35 \times 10^{-43} \text{ sec}$$

where G is the constant of gravitation, h is Planck's constant and c is the speed of light.

From the moment of the Big Bang (at time $t = 0$) to the Planck time 1.35×10^{-43} sec later, the density of matter throughout the universe was so great and the resulting gravitational interactions so vigorous that general relativity fails us. Most physicists believe that a quantum theory of gravity is needed to describe properly the nature of gravity under these extreme circumstances. No such theory exists, and thus no one knows how to describe the universe up to the Planck time.

Nevertheless, some physicists such as John A. Wheeler, speculate that space and time may have had a foamlike consistency during this incredibly brief interval. Space and time as we know them today burst forth from this seething mishmash. During these earliest moments, the expansion of the universe had not yet begun to slow, and thus the expansion of space began with incredible violence and vigor. Indeed, you can think of the Big Bang as an explosion of space at the beginning of time.

The 3-K microwave background is evidence for a hot Big Bang

One of the major successes of modern astronomy involves the origin of the heavy elements. We know today that all the heavy elements are created in the fiery infernos at the centers of stars (recall Box 22-1 for some typical details). As astronomers in the 1960s began understanding details of thermonuclear synthesis, however, a new problem arose: there is too much helium around. For example, consider the Sun. By weight, the Sun consists of about 73 percent hydrogen, 24 percent helium, and 3 percent all remaining heavier elements combined. This 3 percent can be understood as material produced at the centers of ancient stars that long ago cast these heavy elements into space. Certainly, some freshly made helium accompanied these heavy elements, but not nearly enough helium to account for one-quarter of the Sun's mass.

Shortly after World War II, George Gamow at George Washington University proposed that, immediately following the Big Bang, the universe was so incredibly hot that thermonuclear reactions could occur everywhere throughout space. Following up this idea in 1960, Princeton physicists Robert Dicke and P. J. E. Peebles discovered that they could indeed account for today's high abundance of helium by assuming that

Figure 28-4   The Bell Labs horn antenna
This horn antenna at Holmdel, New Jersey, was used by Arno Penzias and Robert Wilson to detect the cosmic microwave background. (Bell Labs)

the early universe was at least as hot as the Sun's center (where helium is currently being produced). This means that the early universe must have been filled with many high-energy, short-wavelength photons, which formed a radiation field whose temperature is given by Planck's blackbody law (recall Figure 17-4).

The universe has expanded considerably since those ancient times, and all those short-wavelength photons have become so stretched that they are now low-energy, long-wavelength photons. Thus, the blackbody temperature of this cosmic radiation field should now be quite low, perhaps only a few kelvins above absolute zero. According to Wien's law, blackbody radiation at this low temperature should have its peak intensity at microwave wavelengths of a few centimeters. Dicke, Peebles, and their colleagues promptly began building an antenna to detect this radiation.

Meanwhile, a few miles away from Princeton University, Arno Penzias and Robert Wilson of Bell Telephone Laboratories were working on a new microwave horn antenna (see Figure 28-4). This antenna was designed to relay telephone calls to Earth-orbiting communication satellites. Penzias and Wilson were deeply puzzled because no matter where they pointed their antenna in the sky, they detected a faint background noise. Thanks to an intermediary, Penzias and Wilson soon learned of the work of Dicke and Peebles. Penzias and Wilson had detected the cooled-down cosmic background radiation left over from the hot Big Bang.

Since those pioneering days back in 1965, several teams of scientists have made many measurements of this background radiation. A few important details are now known. First, by measuring the intensity of the radiation at a variety of wavelengths, we now know that this radiation field has a blackbody spectrum with a temperature of nearly 3 K. As indicated in Figure 28-5, there are some slight departures from the blackbody curve, but it is not clear whether these tiny deviations are significant. Because of its various properties, this radiation field that fills all space is commonly called the **3-K cosmic microwave background.**

A second important feature of the microwave background is that its intensity is almost perfectly isotropic (recall that "isotropic" means "the

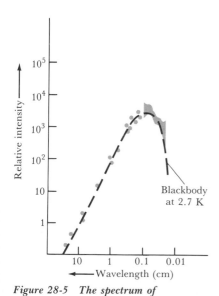

Figure 28-5   The spectrum of the cosmic microwave background
Measurements of intensity at various wavelengths (indicated by dots and the shaded area) demonstrate that the cosmic microwave background is blackbody radiation with a temperature of nearly 3 K.

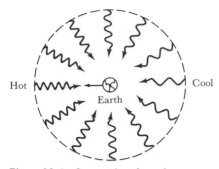

Figure 28-6   Our motion through the microwave background
Because of the Doppler effect, the microwave background is slightly warmer in that part of the sky toward which we are moving. Recent measurements indicate that our galaxy along with the rest of the local group is moving in the general direction of the Virgo cluster.

same in all directions"). In other words, we detect nearly the same intensity from all parts of the sky. However, extremely accurate measurements made in the late 1970s by Richard Muller and George Smoot in a high-flying U-2 airplane revealed a very slight anisotropy. The microwave background is about 0.0035 K warmer than average in the direction of the constellation of Leo. Exactly 180° away, in the opposite direction in the sky (toward Aquarius), the microwave background is about 0.0035 K cooler than average. As we scan from the warm spot in Leo to the cool spot in Aquarius, we find that the background temperature across the sky declines in a smooth fashion.

This temperature variation can be explained as a result of the Earth's overall motion through the cosmos. Of course, we are going around the Sun, the Sun is orbiting around the center of our galaxy, and the galaxy is moving in some fashion with respect to neighboring galaxies. If we were at rest with respect to the microwave background, the radiation would be truly isotropic. Because we are moving through this radiation field, however, we see a Doppler shift. As sketched in Figure 28-6, we see shorter-than-average wavelengths in the direction toward which we are moving. A decrease in wavelength corresponds to an increase in photon energy and thus to an increase in temperature. Specifically, a temperature excess of 0.0035 K corresponds to a speed of 390 km/sec.

Conversely, we see longer-than-average wavelengths in the part of the sky from which we are receding. An increase in wavelength corresponds to a decline in photon energy and thus to a decline in temperature. Consequently, we are traveling from Aquarius toward Leo at a speed of 390 km/sec. Subtracting the known velocity of the Sun around the center of our galaxy, we find that the entire Milky Way galaxy is moving at 520 km/sec in the general direction of the Virgo cluster.

Our motion toward the Virgo cluster is probably due to the gravitational attraction of this massive, rich cluster. However, the local group is not moving directly toward the Virgo cluster; apparently, the gravitational pulls of other galaxy clusters also are influencing our motion.

Figure 28-7   The Cosmic Background Explorer (COBE)
This satellite, due to be launched in 1989, will study the spectrum and angular distribution of the cosmic background radiation over a wavelength range of 1 μ to 1 cm. COBE (pronounced co-bee) will look for deviations from a perfect backbody spectrum and from perfect isotropy. (Courtesy of John Mather; NASA)

To examine this motion and many other properties of the microwave background, such as possible deviations from a perfect blackbody spectrum, astronomers anxiously look forward to the launch of the Cosmic Background Explorer satellite (see Figure 28-7) in the late 1980s. Unhampered by the Earth's atmosphere, COBE's instruments will enable astronomers to explore many aspects of this extremely ancient radiation.

The cosmic background radiation dominated the universe during the first million years

It is very enlightening to examine what exists in the universe today and then to extrapolate back in time toward the Big Bang. Everything in the universe falls into one of two categories: matter or energy. The matter is contained in objects such as stars, planets, and galaxies, which are ultimately composed of particles such as electrons, protons, and neutrons. If we take a large volume V of space and add up the total mass M of all the stars and galaxies we can see in it, we can calculate the **average density ρ_m of matter** in the universe by simply dividing the volume into the mass: $\rho_m = M/V$. Such measurements yield the value

$$\rho_m = 3 \times 10^{-31} \text{ g/cm}^3$$

In other words, if we took all the *luminous* matter we can find and smeared it out uniformly throughout space we would have 3×10^{-31} in every cubic centimeter. The mass of a hydrogen atom is 1.7×10^{-24}g, so ρ_m is equivalent to about one hydrogen atom in every 5 m^3 of space.

This density refers only to *luminous* matter that we can actually observe. Many astronomers strongly suspect that there is a lot of nonluminous matter in space. For example, as discussed in Chapter 26, the famous missing-mass problem applies to clusters of galaxies: the matter we can actually detect is not sufficient to hold the clusters together gravitationally. The true matter density of the universe (including luminous and nonluminous forms) may be as high as 10^{-29} g/cm^3. That value is equivalent to about six hydrogen atoms per cubic meter of space. As we shall see in the final chapter, a density this great has profound implications for the ultimate fate of the universe.

The radiation energy in the universe consists of photons. Of course there are many starlight photons traveling across space, but the vast majority of photons in the universe are members of the 3-K microwave background.

To compare the matter in the universe with the radiation in the universe, we recall Einstein's famous equation $E = mc^2$, which tells us that we can think of the energy in the universe as equal to a mass multiplied by the square of the speed of light. This connection between matter and energy allows us to speak of the **mass density of radiation, ρ_{rad}.** Einstein's equation ($E = mc^2$) can be combined with Stefan's law (recall Chapter 17) to give

$$\rho_{rad} = \frac{aT^4}{c^2}$$

where a is a known value called the radiation constant. For $T = 3$ K, this equations yields

$$\rho_{rad} = 6 \times 10^{-34} \text{ g/cm}^3$$

Note that ρ_m is at least 500 times as large as ρ_{rad}. In other words, the matter density (which we have probably underestimated) is clearly much greater than the mass density of radiation. For this reason we say we are living in a **matter-dominated universe.**

Matter dominates over radiation today only because the energy carried by the microwave photons is very small. Nevertheless, the number of photons in the microwave background is astounding. From the physics of blackbody radiation, it can be demonstrated that there are today 550 million photons in every cubic meter of space. In other words, photons outnumber atoms by roughly a billion to one. Thus, in terms of numbers of particles, the universe consists almost entirely of microwave photons. This radiation field no longer has much "clout" because its photons have been redshifted to long wavelengths and low energies after nearly 20 billion years of stretching by the expansion of the universe.

Although matter dominates the universe today, this was not always the case. To see why, think back in time toward the Big Bang. As we go back in time, we find the universe more compressed, so that the matter density was greater. The photons in the background radiation also were crowded more closely together in earlier times, but an additional effect must be considered. In earlier times, these photons were less redshifted and therefore had shorter wavelengths and higher energy than they do today. Because of this added energy, the mass density of radiation (ρ_{rad}) increases more quickly than the density of matter (ρ_m) as we go back in time toward the Big Bang. In fact, as shown in Figure 28-8, there is a time in the ancient past when $\rho_{rad} = \rho_m$. Before this time, $\rho_{rad} > \rho_m$, and the universe was a **radiation-dominated universe.**

This transition from a radiation-dominated universe to a matter-dominated universe occurred about 1 million years after the Big Bang, a time that corresponds to a redshift of $z = 1000$. In other words, the wavelengths of photons from that time have been stretched by a factor of 1000. Today these microwave photons typically have wavelengths of about 1 mm, but they had wavelengths of about 0.001 mm back when the universe was only 1 million years old.

We can use Wien's law to calculate the temperature of the cosmic background radiation at the time of this transition from a radiation-dominated universe to a matter-dominated universe. Just as a peak wavelength λ_{max} of 1 mm corresponds to a blackbody temperature of 3 K, a peak wavelength of 0.001 mm corresponds to 3000 K. In other words,

Figure 28-8 [left]   *The evolution of density*
During the first million years, the mass density of radiation (ρ_{rad}) exceeded the matter density (ρ_m), and the universe was radiation dominated. At later times, however, continued expansion of the universe caused ρ_{rad} to become less than ρ_m and the universe became matter-dominated.

Figure 28-9 [right]   *The evolution of temperature*
As the universe expands, the photons of the radiation background become increasingly redshifted, and thus the temperature of the radiation falls. Roughly 1 million years after the Big Bang, when the temperature fell below 3000 K, hydrogen atoms formed, and the radiation field "decoupled" from the matter in the universe.

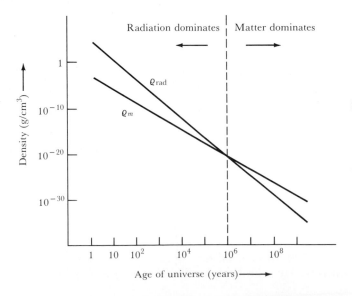

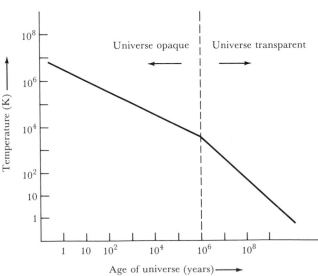

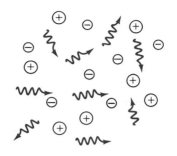

a Before recombination

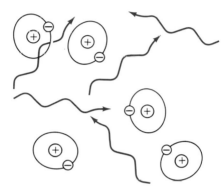

b After recombination

Figure 28-10 The era of recombination
(a) Before recombination, photons in the cosmic background prevented protons and electrons from forming hydrogen atoms. (b) As soon as hydrogen atoms could survive, the universe became transparent. This transition from an opaque to a transparent universe occurred roughly 1 million years after the Big Bang.

the temperature of the radiation background is proportional to the redshift and has been declining over the ages, as shown in Figure 28-9. It just so happens that $T = 3000$ K when $z = 1000$ and $\rho_m = \rho_{rad}$.

The nature of the universe changed in a fundamental way when $z = 1000$. To understand this change, recall that hydrogen is by far the most abundant element in the universe; hydrogen atoms outnumber helium atoms about 16 to 1. Of course, a hydrogen atom consists of a single proton orbited by a single electron. It does not take much energy to knock the proton and electron apart. In fact, a radiation field warmer than about 3000 K easily ionizes hydrogen. Thus, hydrogen atoms could not exist before $z = 1000$ (that is, at times earlier than about 1 million years after the Big Bang). As sketched in Figure 28-10a, the background photons prior to $t = 1$ million years, had energies great enough to prevent electrons and protons from getting together to form hydrogen atoms. Only after $t = 1$ million years have these photons been redshifted enough to permit hydrogen atoms to exist (see Figure 28-10b).

Prior to $t = 1$ million years, the universe was completely filled with a shimmering expanse of high-energy photons colliding vigorously with protons and electrons. This state of matter is called a **plasma,** and it is opaque—just as the glowing gases inside a discharge tube (such as a neon advertising sign) or a fluorescent light bulb are opaque. P. J. E. Peebles coined the phrase **primordial fireball** to describe the universe during this time.

After $t = 1$ million years, the photons no longer had enough energy to keep the protons and electrons apart. As soon as the temperature of the radiation field fell below 3000 K, protons and electrons everywhere began combining to form hydrogen atoms. Hydrogen is transparent, so the universe suddenly became transparent! All those photons, which a few minutes earlier had been vigorously colliding with charged particles, could now stream unimpeded across space. We see these same photons today in the microwave background.

This dramatic moment, when the universe went from being opaque to being transparent, is called the **era of recombination**. Because the universe was opaque prior to $t = 1$ million years, we cannot see any farther into the past than the era of recombination. The microwave background, whose photons have suffered a redshift of $z = 1000$, contains the most ancient photons we can ever possibly observe. This microwave background is today only a ghostly relic of its former dazzling splendor.

All the mass and energy in the universe may have burst forth from a vacuum during the Big Bang

Where did all the matter and radiation in the universe come from in the first place? Recent intriguing theoretical research by physicists such as Steven Weinberg of Harvard and Ya. B. Zel'dovich in Moscow suggest that the universe began as a perfect vacuum and that all the particles of the material world were created from the expansion of space. To see how this might have happened, we must first understand what quantum mechanics tells us about empty space.

Quantum mechanics is the branch of physics that deals with submicroscopic phenomena such as the behavior of individual particles and photons. Quantum mechanics tells us how to calculate the structure of atoms and the interactions between nuclei.

There are significant differences between the ordinary world around us and the submicroscopic world of quantum mechanics. In the ordinary, macroscopic world we have no trouble knowing where things are. You know where your house is; you know where your car is; you know where your mother is. When you peer into the subatomic world of electrons and nuclei, however, you can no longer speak with this same confidence and surety. A certain amount of fuzziness and uncertainty enter into the description of reality. At the incredibly small dimensions of the quantum world, things are no longer sharp and distinct.

To appreciate the reasons for this fuzziness and uncertainty, imagine trying to measure the position of a single electron. In order to find out where the electron is located, you must observe it. In order to observe it, you shine a light on it. The electron is so tiny and has such a small mass, however, that the photons in your beam of light possess enough energy to give the electron a mighty kick. As soon as a photon strikes the electron (so that you can observe the reflected photon to measure the electron's position), this same photon imparts a huge momentum to the electron, causing it to recoil in some unknown direction. Consequently, in trying to measure the location of an electron with great precision, you necessarily introduce a large amount of uncertainty into the speed or momentum of that electron.

These ideas are at the heart of the famous **Heisenberg uncertainty principle,** first formulated by Werner Heisenberg in 1927. This principle states that there is a reciprocal fuzziness between position and momentum. The more precisely you try to measure the position of a particle, the more unsure you are of how the particle is moving. Conversely, the more accurately you determine the speed of a particle, the less sure you are of the particle's location. This same fuzziness does not enter into our large-scale everyday world simply because macroscopic objects (such as your house, car, and mother) have enormous masses compared with electrons and therefore are not significantly affected by the light that shines on them.

There is also an analogous uncertainty, or fuzziness, involving energy and time. You cannot know the exact energy of a quantum mechanical system with infinite precision at every moment in time. Over very short time intervals, there can be large uncertainty in the amounts of energy in the subatomic world. Specifically, let ΔE be the uncertainty in energy over a short interval of time Δt. Then

$$\Delta E \times \Delta t \geq \frac{h}{2\pi}$$

where h is Planck's constant.

We can look upon the Heisenberg uncertainty principle as merely an unfortunate limitation on our ability to know everything with infinite precision. Alternatively, we can explore how the uncertainty principle might unlock new doors in our quest to understand the universe.

We have seen that one of the important conclusions of Einstein's special theory of relativity is the equivalence of mass and energy: $E = mc^2$. There is nothing uncertain about the speed of light (c). If there is an uncertainty in the energy of a physical system, therefore, Einstein's equation allows us to express this fuzziness as an uncertainty Δm in the mass.

Thus

$$\Delta E = c^2 \Delta m$$

Combining this expression with the previous equation, we obtain

$$\Delta m \times \Delta t \geq \frac{h}{2\pi c^2}$$

This astonishing result means that, in a very brief interval Δt of time, we cannot be sure how much matter there is in a particular location—even in "empty space." During any brief moment, particles and antiparticles can spontaneously appear and disappear.

There is nothing mysterious about antimatter. A particle and an antiparticle are identical in almost every respect, except that they carry opposite electric charges. For example, an antielectron (or positron, e^+) has a positive charge, whereas an ordinary electron (e^-) carries a negative charge. Particles and antiparticles are always created or destroyed in equal numbers, thereby ensuring that the total electric charge in the universe remains constant, regardless of how many pairs of particles and antiparticles come and go. Physicists often refer to this kind of balance between matter and antimatter as a **symmetry.**

The time interval over which this process occurs is incredibly brief. For example, consider an electron and an antielectron, each of which has a mass

$$m_e = 9.1 \times 10^{-28}\,\text{g}$$

To see how long an electron–antielectron pair might exist without violating the conservation of mass, we note that a mass $\Delta m = 2m_e$ must briefly appear. The time Δt over which this can happen then is

$$\Delta t = \frac{h}{4\pi m_e c^2} = 6.4 \times 10^{-21}\,\text{sec}$$

In other words, during a time interval shorter than 6.4×10^{-21} sec, an electron and an antielectron can spontaneously appear and then disappear without violating any laws of physics.

This process can happen absolutely *anywhere* at *any time.* It can also occur with more massive particles, but because of the "reciprocal" nature of the uncertainty principle, these massive particles can exist only for correspondingly shorter time intervals. For example, the proton is about 2000 times heavier than the electron. Therefore, pairs of protons and antiprotons can appear and disappear spontaneously if they exist for only 1/2000 as long as the pairs of electrons and antielectrons exist.

One of the fundamental doctrines of subatomic physics (usually attributed to Murray Gell-Mann of Caltech) is, "If it is not strictly forbidden, then it must occur." "It" refers to any quantum process. Therefore, pairs of every conceivable particle and antiparticle are constantly being created and destroyed at every location all across the universe. Of course, we have no way to observe these pairs of particles and antiparticles directly; that is forbidden by the uncertainty principle. The particle–antiparticle pairs exist for such short time intervals that direct observation is impossible in theory as well as in practice. For this reason, they are called **virtual pairs.** They don't "really" exist; they "virtually" exist.

Although virtual pairs of particles and antiparticles cannot be observed directly, their effects have been detected. Imagine an electron in orbit about the nucleus of an atom, such as a hydrogen atom. Ideally, the electron should follow its orbit in a smooth and unhampered fashion. Because of the constant brief appearance and disappearance of pairs of particles and antiparticles, however, tiny electric fields exist for extremely short intervals of time. These tiny, fleeting electric fields cause the electron to jiggle slightly in its orbit. This jiggling was first detected in 1947 by W. E. Lamb and R. C. Retherford, who noticed a tiny shift in the spectrum of light from the hydrogen atom. This phenomenon, called the **Lamb shift,** provides powerful support for the idea that every point

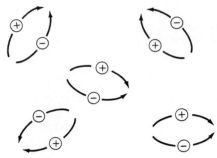

Figure 28-11   Virtual pairs
Pairs of particles and antiparticles can appear and then disappear anywhere in space, provided that each pair exists only for a very short time interval, as dictated by the uncertainty principle.

in space—all across the universe—is seething with virtual pairs of particles and antiparticles. The constant appearance and disappearance of particles and antiparticles is sketched schematically in Figure 28-11.

The concept of virtual pairs also provides a ready explanation for the phenomenon of pair production. It has been known for many years that highly energetic gamma rays can convert their energy into pairs of particles and antiparticles according to the equation $E = mc^2$. Quite simply, the gamma ray disappears (upon colliding with a second photon), and a particle and an antiparticle appear in its place. This process is commonly observed in high-energy nuclear experiments. Indeed, it is one way in which nuclear physicists can manufacture many exotic species of particles and antiparticles. This process of **pair production** and the inverse process of **annihilation** are sketched schematically in Figure 28-12.

Where do these particles and antiparticles come from? They spring from nature's ample supply of virtual pairs. The gamma rays provides a virtual pair with so much energy that the virtual particles can materialize and appear as *real* particles in the real world. The only requirement is that nature's balance sheet be satisfied. To create a particle and an antiparticle of total mass M, the incoming gamma-ray photons must possess an amount of energy E that satisfies the condition $E \geq Mc^2$. If the photons carry too little energy, pair production will not proceed. Conversely, the more energetic the photon, the more massive are the particles and antiparticles that can be manufactured.

Around 1980, physicists began applying these ideas to the creation of the universe. Think about the universe immediately after the Big Bang. Space is violently expanding with explosive vigor. Yet, as we have seen, all space is seething with virtual pairs of particles and antiparticles. Normally, a particle and antiparticle have no trouble getting back together in a time interval (Δt) short enough so that the conservation of mass is satisfied under the uncertainty principle. During the Big Bang, however, space was expanding so fast that particles were rapidly pulled away from their corresponding antiparticles. Deprived of the opportunity to recombine, these virtual particles had to become real particles in the real world. Where did the energy come from to achieve this materialization?

Recall that the Big Bang was like the center of a black hole. A vast supply of gravitational energy was therefore associated with the intense gravity of this cosmic singularity. This resource provided ample energy to completely fill the universe with all conceivable kinds of particles and antiparticles. Thus, immediately after the Planck time, the universe was flooded with particles and antiparticles created by the violent expansion of space.

Figure 28-12   Pair production and annihilation
Pairs of virtual particles can be converted into real particles by high-energy photons. Conversely, a particle and an antiparticle can annihilate each other by giving up energy in the form of gamma rays.

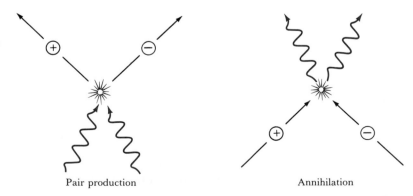

Pair production   Annihilation

Most of the matter and antimatter in the early universe annihilated each other as the primordial fireball cooled off

As soon as the flood of matter and antimatter appeared in the universe, collisions between particles and antiparticles produced numerous high-energy gamma rays. An equilibrium was soon achieved. For example, for every proton (p) and antiproton ($\overline{p}$) that annihilated each other to create gamma rays, two gamma rays collided somewhere else to produce a proton and an antiproton. Thus, reactions such as

$$p + \overline{p} \rightarrow \gamma + \gamma$$

and

$$\gamma + \gamma \rightarrow p + \overline{p}$$

proceed with equal vigor. This condition, wherein the number of particles and antiparticles is just high enough so that precisely as many are being created each second as are being destroyed, is called **thermal equilibrium.**

Now let us think ahead in time. The universe continued to expand; all the gamma-ray photons became increasingly redshifted, and thus the temperature of the radiation field fell. Eventually, the temperature became so low that the gamma rays no longer had enough energy to create particular kinds of particles and antiparticles. Collisions between particles and antiparticles added photons to the cosmic-radiation background, but collisions between the photons no longer replenished the supply of particles and antiparticles.

For example, again consider protons and antiprotons, each of which has a mass (or rest mass)

$$m_P = 1.7 \times 10^{-24} \, g$$

According to $E = mc^2$, the **rest energy** associated with the mass of each of these particles is

$$E = m_P c^2 = 1.5 \times 10^{-3} \text{erg} = 938 \text{ MeV}$$

where MeV stands for "million electron volts," a unit of energy commonly used by physicists (recall Box 5-1). This value tells us how much energy the colliding gamma-ray photons must have in order to create a proton and an antiproton. If the combined energy of the two photons is less than 2×938 MeV, the reaction $\gamma + \gamma \rightarrow p + \overline{p}$ will not occur.

A photon energy of 938 MeV corresponds to a particular temperature, according to the equation

$$E = kT$$

where k is the Boltzmann constant ($k = 1.38 \times 10^{-16}$ erg/K $= 8.6 \times 10^{-5}$ eV/K). For gamma-ray photons of energy 938 MeV, the temperature is 10.9 trillion kelvins. Thus, as soon as the temperature of the radiation field falls below 10.9×10^{12} K, proton–antiproton pairs are no longer created. This temperature is therefore called the **threshold temperature** for protons and antiprotons.

There are many different kinds of particles, and each kind of particle has its own threshold temperature. For example, electrons and antielectrons have masses 1800 times smaller than protons and antiprotons. Thus the threshold temperature for the creation of electron–antielectron pairs is 1/1800 of that for proton–antiproton pairs. Consequently, when the radiation temperature falls below 5.9 billion kelvins, electron–

antielectron pairs no longer can be created by the reaction $\gamma + \gamma \rightarrow e^+ + e^-$. We therefore say that the threshold temperature for electrons and antielectrons is 5.9×10^9 K. Several common types of particles are listed in Box 28-1, along with their masses and threshold temperatures.

When the universe was about 0.0001 sec old (that is, at time $t = 10^{-4}$ sec after the Big Bang), the temperature of the radiation fell below 10^{13} K. The universe was then cooler than the threshold temperatures of both protons and neutrons, so the annihilation of protons with antiprotons and of neutrons with antineutrons occurred vigorously everywhere throughout space. This wholesale annihilation dramatically lowered the matter content (particles and antiparticles) of the universe, while simultaneously increasing the radiation (photon) content.

Similarly, when the universe was about 1 sec old, the temperature fell below 6×10^9 K, which is the threshold temperature for electrons and antielectrons. Consequently, the annihilation of pairs of electrons and antielectrons began occurring everywhere, thereby further decreasing the matter content of the universe while raising the radiation content.

Now we have a dilemma. If the symmetry between particles and antiparticles were truly valid, then for every proton there would have been an antiproton. For every electron, there would have been an antielectron. Thus, by the time the universe was 1 sec old, every particle would have annihilated with an antiparticle, leaving no matter in the universe whatsoever.

Obviously, this did not happen. Physicists therefore say that a **symmetry breaking** occurred during the very earliest moments of the universe: the number of particles emerging from the Big Bang was not exactly equal to the number of antiparticles. Specifically, for every billion

---

Box 28-1   Properties of some elementary particles

Physicists classify all particles into two categories, leptons and hadrons, depending on details of their behavior. **Leptons** typically are low-mass particles related to electrons. **Hadrons** typically are more massive; this category includes protons and neutrons. Most physicists believe that all hadrons are composed of more-basic particles called **quarks.**

The **rest energy** of a particle equals its mass times the square of the speed of light. It is the energy that would be released if all the mass of the particle were converted into energy. The **threshold temperature** is the rest energy divided by Boltzmann's constant. It is the temperature above which a particle can be freely created out of thermal radiation.

The table lists several types of particles, along with their rest energies and threshold temperatures.

	Particle	Symbol	Rest energy (MeV)	Threshold temperature (10^9 K)
Leptons	Neutrino	$v, \bar{v}$	0.0001 (?)	0.001
	Electron	e^-, e^+	0.5110	5.930
	Muon	μ^-, μ^+	105.66	1,226.2
Hadrons	Pi meson	π°	134.96	1,556.2
		π^+, π^-	139.57	1,619.7
	Proton	$p, \bar{p}$	938.26	10,888
	Neutron	$n, \bar{n}$	939.55	10,903

antiprotons, a billion plus one ordinary protons were created. For every billion antielectrons, a billion plus one ordinary electrons were created. This particular slight excess of matter over antimatter is suggested by the fact that (as mentioned earlier in this chapter) there are roughly a billion photons today in the microwave background for each proton and neutron in the universe.

....

A neutrino background and most of the helium in the universe are relics of the primordial fireball

The early universe must have been populated with vast numbers of neutrinos (ν) and antineutrinos ($\bar{\nu}$). These particles are involved in nuclear reactions that transform neutrons into protons and vice versa. For example, under laboratory conditions, the neutron decays into a proton by emitting an electron and an antineutrino:

$$n \rightarrow p + e^- + \bar{\nu}$$

The half-life for this radioactive decay is about 12 minutes, which is why we do not find any free neutrons floating around in the universe today.

In early universe, however, collisions between particles kept the numbers of protons approximately equal to the numbers of neutrons according to the two-way reactions

$$p + e^- \rightleftarrows n + \nu$$

and

$$p + \bar{\nu} \rightleftarrows n + e^+$$

This balance was possible only as long as the density of matter in the universe was high. By the time the universe was about 2 sec old, matter was thinned out enough so that neutrinos and antineutrinos no longer collided with protons and neutrons in significant numbers. The natural tendency for neutrons to decay into protons then took over, and the number of neutrons began to decline.

The first step in creating helium involves combining a proton and a neutron to produce deuterium (^2H), sometimes called "heavy hydrogen," according to the reaction

$$p + n \rightarrow {}^2H$$

The proton and the neutron, however, do not stick together very well. In early universe, high-energy gamma rays easily broke down deuterons into independent protons and neutrons. Because deuterons could not survive, helium could not be created. This block to helium creation is called the **deuterium bottleneck.**

Finally, when the universe was about 3 min old, the photons in the background radiation became sufficiently redshifted that they could no longer break up the deuterium. By this time, the decay of neutrons into protons had shifted the neutron–proton balance to about 14 percent neutrons and 86 percent protons. Because deuterons could now survive, all of these remaining neutrons combined with protons and, through the usual series of reactions (recall Box 19-2), rapidly produced helium. The final result is 1 helium atom for every 16 hydrogen atoms that we find in the universe today.

What happened to all those primordial neutrinos and antineutrinos that interacted vigorously with protons and neutrons before the universe was 2 sec old? As mentioned in the discussion of the Sun (refer to Box 19-2), neutrinos and antineutrinos are very difficult to detect because they do not interact very strongly with matter. Indeed the Earth is virtually transparent to the neutrinos from the Sun.

When the universe was about 2 sec old, matter was sufficiently spread out that the universe became transparent to neutrinos and antineutrinos. From that time on, neutrinos and antineutrinos could travel unimpeded across the universe.

This "decoupling" of the neutrino–antineutrino background from the matter in the universe was very much like the decoupling of the photon background during the era of recombination (recall Figure 28-9). In fact, physicists estimate that these neutrinos and antineutrinos may be approximately as populous today as the photons in the microwave background (550 million per cubic meter). The neutrino–antineutrino background should be slightly cooler than the photon background, which received an extra amount of energy from electron–antielectron annihilations. Physicists estimate that the current temperature of the neutrino–antineutrino background is about 2 K, versus 3 K for the microwave background.

Until recently, no one paid much attention to this elusive neutrino–antineutrino background. It was generally believed that these particles, like photons, are massless and, because they interact so weakly with ordinary matter, they just did not seem very important. However, around 1980, experiments by E. T. Tretyakov and colleagues in the Soviet Union as well as Frederick Reines and colleagues in the United States suggested that neutrinos and antineutrinos do have mass. It is an extremely tiny mass, perhaps only 10^{-7} of the mass of the proton. Later experiments, however, have failed to confirm the existence of the neutrino rest mass.

There are so many neutrinos and antineutrinos spread throughout space that these particles would account for *most* of the matter in the universe if this tiny rest mass does exist. Indeed, the matter contained in neutrinos and antineutrinos may be 100 times greater than all the matter in stars, planets, and galaxies. This means that the true average density of matter (ρ_m) may be as great as 10^{-29} g/cm^3. As we shall see in the next chapter, a density this great is sufficient to someday stop the expansion of the universe. In that case, the gravity content of the universe is so great that we are actually living inside an enormous black hole.

....

Summary

. The universe began as an infinitely dense cosmic singularity that expanded explosively in the event called the Big Bang, which can be described as an explosion of space at the beginning of time.

   The Hubble law describes the continuing expansion of space; it is meaningless to speak of an edge or center to the universe.

   The observable universe extends about 20 billion light years in every direction from the Earth; we cannot see objects beyond the cosmic particle horizon at the distance of 20 billion years because light from these objects has not had enough time to reach us.

   Before the Planck time (about 10^{-43} sec after the projected time of the Big Bang), the universe was so dense that known laws of physics do not properly describe the behavior of space, time, and matter.

. The 3-K cosmic microwave background radiation, corresponding to radiation from a blackbody at a temperature of 3 K, is the greatly redshifted remnant of the very hot universe that existed about 1 million years after the Big Bang.

Slight differences in the temperature of the 3-K cosmic background in various directions suggest that our cluster of galaxies is moving in the general direction of the Virgo cluster.

During the first million years of the universe, matter and energy formed an opaque plasma (the primordial fireball); therefore, the 3-K microwave background radiation contains the oldest photons in the universe.

About 1 million years after the Big Bang, expansion and redshift caused the temperature of the universe to fall below 3000 K so that protons and electrons could combine to form hydrogen atoms; this time is called the era of recombination.

. The observable matter in the universe has an average density of about 3×10^{-31} g/cm^3, equivalent to about one hydrogen atom per 5 m^3 of space.

The radiation in the universe has a mass equivalent (computed from Einstein's relation $E = mc^2$) of about 6×10^{-34} g/cm^3, so the universe today is matter-dominated; during the first million years, before the background radiation was greatly redshifted, the universe was radiation-dominated.

To account for clusters of galaxies, astronomers must assume the presence of large amounts of nonluminous and undetected matter, so that the actual average density of matter in the universe today may be as great as 10^{-29} g/cm^3, equivalent to about six hydrogen atoms per cubic meter.

. Heisenberg's uncertainty principle states that the uncertainty in the speed of a particle increases as its position is known more precisely, and vice versa; a similar reciprocal relationship exists between mass and time.

Because of the uncertainty principle, particle–antiparticle pairs can form spontaneously and disappear again within a fraction of a second; these pairs can never be detected directly, so they are called virtual pairs.

A virtual pair can become a real particle–antiparticle pair when photons collide in the process of pair production; the photons disappear and their energy is replaced by the mass of the particle–antiparticle pair.

A corresponding process of annihilation involves the disappearance of a colliding particle–antiparticle pair and the appearance of photons.

Just after the Planck time, the universe was filled with particles and antiparticles formed by pair production and numerous high-energy photons formed by annihilation; a state of thermal equilibrium existed in this hot plasma.

As the universe expanded, the temperature decreased; as the temperature fell below the threshold temperature for production of each kind of particle, annihilation dominated production for that kind of particle.

The present domination of matter over antimatter results because particles and antiparticles were not created in exactly equal numbers just after the Planck time; this imbalance is called a symmetry breaking.

. Production of helium could not begin until the cosmological redshift eliminated most of the high-energy gamma rays that created the deuterium bottleneck by breaking down deuterons before they could combine to form helium.

· If the neutrino has a small rest mass, the density of the universe may be so great that we actually live inside a vast black hole.

Review questions

1 From the data given in Figure 28-1, draw a graph of distance versus speed as observed by the bug sitting on raisin A.

2 Repeat Exercise #1, but use some other raisin instead of raisin A as "home." (You may have to make some reasonable assumptions about the distances between raisins B, C, and D in the "before" and "after" views.) Is your new graph the same as the last one?

***3** Calculate the maximum age of the universe for a Hubble constant of (50 km/sec) /Mpc, (75 km/sec) /Mpc, and (100 km/sec) /Mpc. In view of your answers, explain how the ages of globular clusters could be used to place a limit on the maximum value of the Hubble constant.

4 How does modern cosmology preclude the possibility of a "center" or an "edge" of the universe?

5 In what way would it be possible to say that the universe was matter-dominated immediately after the Big Bang? How did this matter-dominated period differ from the matter-dominated universe in which we today live?

Advanced questions

6 Prior to the discovery of the 3-K cosmic microwave background, it seemed possible that we might be living in a "steady state universe" whose overall properties do *not* change with time. The steady state model—like the Big Bang model—assumes an expanding universe, but does not assume a "creation event". Instead, in the steady state theory matter is assumed to be created continuously everywhere in space to ensure that the average density of the universe remains constant. Explain why the cosmic microwave background is a major blow to the steady state theory.

***7** Does the fact that nothing can travel faster than the speed of light put an upper limit on the radius of the observable universe? Is so, what is the value of that radius?

8 How can astronomers be certain that the 3-K cosmic microwave background is not some sort of localized phenomenon? Explain.

Discussion questions

9 Suppose we were living in a radiation-dominated universe. Discuss how such a universe would be different from what we now observe.

10 Discuss the theological implications of the idea that we cannot use science to tell us what existed before the Big Bang.

For further reading

Barrow, J., and Silk, J. *The Left Hand of Creation: Origin and Evolution of the Universe.* Basic Books, 1983.
Harrison, E. "The Paradox of the Dark Night Sky." *Mercury,* July/Aug. 1980, p. 83.
Wagoner, R., and Goldsmith, D. *Cosmic Horizons: Understanding the Universe.* W. H. Freeman, 1983.
Webster, A. "The Cosmic Background Radiation." *Scientific American,* Aug. 1974.

J. Richard Gott, III

Experiences of a cosmologist

J. Richard Gott, III was born in Louisville, Kentucky, in 1947. He graduated summa cum laude in physics from Harvard in 1969 and received his Ph.D. in astrophysics from Princeton in 1972. He was a postdoctoral fellow at CalTech and a visiting fellow at Cambridge University before returning to Princeton where he is now an associate professor of astrophysics. Dr. Gott has been a Sloan Foundation Fellow and in 1975 received the R. J. Trumpler Award in Astronomy. He lives in Princeton Junction, New Jersey, with his wife and daughter. (Photo courtesy Princeton University)

High in the French Alps in the summer of 1979 astronomers and physicists from around the world gathered to swap theories about the early universe. I was there to give a series of lectures on the growth of structure in the universe. Near the end of the conference a Belgian physicist named Francois Englert gave a talk about work he and his colleagues had done on how there could be an early phase in the history of the universe where the density remained constant while the universe expanded exponentially by a large factor. This model was particularly exciting because it solved some long-standing problems with the standard big bang model by simply changing the dynamics of the universe's expansion at early times. During its exponentially expanding phase the model approximated a piece of de Sitter space, named after the Dutch astronomer who first investigated its mathematical properties in 1917.

Now, one of the interesting things about a complete de Sitter space is that it has event horizons. An event horizon occurs whenever there are events in curved spacetime that we can never see no matter how long we wait. In 1974 Stephen Hawking showed that the existence of event horizons causes the production of thermal radiation. Hawking's work was probably the single most important development in general relativity theory since Einstein. I was interested in this because I thought this might be a nice way to make the initial thermal radiation in the universe. Unfortunately, because the model Englert presented utilized only a part of de Sitter space, it did not, after all, have any event horizons. Something slightly different was needed.

Then in 1981 Alan Guth showed how an inflationary epoch in which the universe expanded exponentially would occur naturally in the then most promising Grand Unified Theory of particle physics. He also pointed out quite clearly all the advantages inflation would have for cosmology. Guth's paper got the particle physicists interested, becoming the most influential paper in cosmology in recent years. In Guth's model the inflation occurred when the universe became trapped in a symmetric vacuum state with a high constant density. Then there has to be a phase transition to the ordinary zero density (asymmetric) vacuum state in which we live today. Guth noted that this phase transition would naturally occur by forming bubbles of low density vacuum in the high density de Sitter space, just as bubbles of steam appear in boiling water. Guth pointed out that this was a real problem because our universe does not look like a disconnected froth of bubbles.

I went back to thinking about this problem by trying to find a solution whose geometry at late times looked like a standard big bang model and at early times like de Sitter space. I wanted the transition between the two geometries to be casually traceable to a single event and I wanted a model with an event horizon in it. I found that there was only one way to do this. It led to the formation of an open universe. An important part of this was drawing a space–time diagram of what the overall solution looked like (see the figure). I was thinking about this on a day when my wife and I went out for a snack and I remember drawing some of the first versions of this diagram on napkins at the Howard Johnson's restaurant. It was clear from the diagram that one could make many universes out of one initial de Sitter space. Our universe looked just like an expanding bubble in the high density de Sitter space. Our universe was *one* of the bubbles. Out there beyond our event horizon there can be other bubbles that we never see no matter how long we wait. My model required that our bubble form slowly so that there would be enough inflation within it as it formed to explain the current flatness of the universe. Shortly after my work on this, Linde in the U.S.S.R. and Albrect and Steinhart in the U.S.A. independently proposed detailed particle physics scenarios that led to slow-forming single bubble models. This solved the problem pointed out by Guth because from inside one of the bubbles everything looks uniform just as we observe. For me it was very exciting that both I and these particle physicists were led to the same type of model from completely different considerations. They were looking directly at the particle physics while as an astronomer I was trying to produce models with the right geometrical properties from the general relativity standpoint. This bubble universe picture is now one of several inflationary scenarios being investigated around the world.

Theoretical cosmology is a field of ideas. After you have an idea and publish it, it really has a life of its own. One paper I wrote on cosmology ended up in Gregory Benford's science fiction novel *Timescape*. Right there on page 271 in the year 1998 the hero reads *my* paper and this helps him to save the world from ecological disaster. You can never tell what will eventually happen to your ideas! Cosmology is at an exciting point now. The new inflationary ideas have opened up many new possibilities currently under investigation, perhaps allowing us to look one level further back toward the origin of the universe.

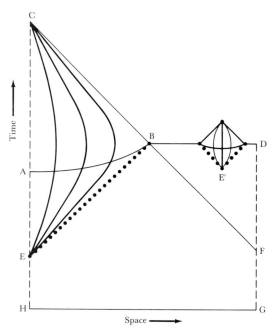

Gott's space–time diagram of the formation of bubble universes as it appeared on the cover of Physics News *in 1982 published by the American Institute of Physics. A single space coordinate runs horizontally while the time coordinate runs vertically. The map is constructed so that light signals travel upward to the left or right at a 45° angle to the vertical. Our universe is formed at the event E. The lines connecting E and C represent world-lines (paths) of observers in our universe. The convergence of these lines at point C is just an artifact of this special map projection; actually it represents a universe that expands forever. The line AB marks the completion of the phase transition and the region ABC evolves just like a standard big bang model. The left half of our diamond shaped universe has been suppressed for convenience. Another bubble universe is shown forming at E′ beyond the event horizon (BF) marking the limit of what observers in our universe will ever be able to see. GH represents the start of the de Sitter phase. It is hard to determine what preceded this because the de Sitter space quickly forgets the initial conditions that produced it.*

GUTs, inflation, and the fate of the universe

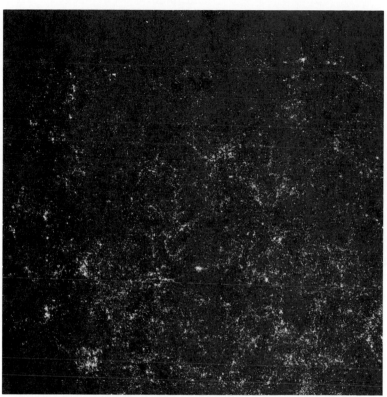

A million galaxies
This map was based on a survey made by C. Donald Shane and Carl A. Wirtanen of the Lick Observatory and includes nearly a million galaxies brighter than magnitude +19. The area shown here is centered on the north celestial pole and measures approximately 100° on each side. To produce this map, the sky was divided into squares a sixth of a degree on a side and the number of galaxies in each square is represented by a shade of gray. Although individual galaxies are not seen at this scale, superclusters are recognized as white blotches. Note that the superclusters are not smoothly distributed across the sky but rather form a lacy, filamentary structure (compare with Figure 29-11). (Courtesy of Edward J. Groth et al.)

As we complete our survey of astronomy with a look at the universe across the largest scales of time and distance, we find a surprising number of connections between the infinite and the infinitesimal. We learn in this chapter that the most fundamental properties of the universe and its ultimate fate were determined during the first fraction of a second after the Big Bang. Events just after the Planck time determined whether the universe will collapse to a Big Crunch or will simply expand forever. We find that the origin of clusters of galaxies depends upon the distribution and properties of the neutrino, the least-massive particle in the universe. We end our discussion with speculations about the extremely distant future; our knowledge about the ultimate fate of the universe remains uncertain because of our uncertainty about the average density of the universe and about the value of the deceleration parameter.

Newton's assumption of an isotropic and static universe led him directly to the conclusion that the universe must be infinite. Gravitational forces in a finite universe would cause all of the matter to fall into a huge clump at the center, so a finite universe could not be static according to the laws of Newtonian mechanics. Later, more detailed studies showed that even the infinite universe could not be static; matter would accrete into ever-larger objects, so that the structure of the universe would change over time. Newtonian mechanics is consistent with an infinite, static, and isotropic universe only if the universe contains no matter—obviously not the case with the real universe.

As we have seen, Newtonian mechanics is not valid for the very high speeds and very large masses that exist in the universe. Today, cosmologists (people who study the structure and evolution of the universe) use the general theory of relativity set forth by Einstein. As emphasized in Chapters 24 and 28, general relativity fully describes the intimate relationships between matter, energy, gravity, and the geometry of space and time.

Even with general relativity, however, cosmologists have been forced to abandon the assumption that the universe is static. As we have seen, the universe has been expanding since its origin in the Big Bang, and its structure has been evolving. The universe is infinite in the sense that it extends forever in all directions and thus has no center and no edge. However, the *observable* universe is finite because we can see only those stars and galaxies near enough for their light to have reached us since the Big Bang.

We are now ready to extend out theories into the future and ask how the universe will change. Finally, we can tackle the ultimate cosmological question: What will be the final fate of the universe?

The future of the universe is determined by the average density of matter in the universe

During the 1920s, general relativity was applied to cosmology by Alexandre Friedmann in Russia, Georges Lamaître in France, Willem de Sitter in the Netherlands, and, of course, Einstein himself. The resulting picture of the structure and evolution of the universe, called **relativistic cosmology,** is in surprisingly good agreement with our intuitive notion that gravity should be slowing the cosmological expansion.

A good analogy involves a rocket blasting off from the surface of the Earth. If the rocket's speed is less than the escape velocity (recall that the escape velocity from the Earth is 11 km/sec, as discussed in Box 7-2), then the spacecraft will fall back down to Earth. If the rocket's speed equals the escape velocity, the spacecraft just barely manages not to fall back to Earth. If the rocket's speed exceeds the escape velocity, then the spacecraft easily leaves the Earth and never falls back, despite the relentless pull of gravity.

A very similar set of circumstances surrounds the ultimate fate of the universe. Instead of talking about some sort of "mutual escape velocity" between galaxies, however, astronomers speak of the **average density** of matter in the universe.

If the average density of matter throughout space is small, then the gravity associated with this matter is weak, and the expansion of the universe will continue forever. Even infinitely far into the future, galaxies will continue to rush away from each other. In this case, we say that the universe is **unbounded.**

Conversely, if the density of matter across space is large, then the resulting gravity is sufficiently strong eventually to halt the expansion of the universe. The universe will reach a maximum size and then will begin contracting as gravity starts to pull the galaxies back toward each other. In this case, we say that the universe is **bounded.**

Separating these two scenarios is the case where the universe is **marginally bounded.** In this case, the density of matter across space exactly equals the **critical density** (ρ_c), and the galaxies just barely manage to keep moving away from each other. This case is analogous to a rocket leaving the Earth with a speed exactly equal to the escape velocity. As discussed in Box 27-2, the critical density is given by the expression

$$\rho_c = \frac{3H_0{}^2}{8\pi G}$$

Where H_0 is the Hubble constant and G is the gravitational constant. Using a Hubble constant of $(15 \text{ km/sec})/10^6 \text{ ly}$, we find

$$\rho_c = 4.5 \times 10^{-30} \text{ g/cm}^3$$

which is equivalent to about three hydrogen atoms per cubic meter of space.

As mentioned in Chapter 28, the average density of luminous matter we see in the sky seems to be about $3 \times 10^{-31} \text{ g/cm}^3$, clearly less than the critical density. However, both the missing-mass problem for clusters of galaxies and the possibility of a cosmic background of neutrinos possessing mass suggest that the true matter density across space may be greater than ρ_c. In short, present-day observations and estimates of the density are not accurate enough to tell us whether the universe is bounded or unbounded.

The deceleration of the universe can be determined from observations of extremely distant galaxies

Another way of determining the future of the universe is to measure the rate at which the cosmological expansion is slowing down. The deceleration of the universe shows up as a deviation from the straight-line relationship predicted by the Hubble law. To see why, examine the Hubble-law curve displayed in Figure 26-24. This graph shows the usual linear relationship between the distance to a galaxy and its recessional velocity from the Earth. Note, however that the graph is based on data extending only to 600 million light years from Earth. Over a distance less than 10^9 ly, the deceleration is virtually undetectable. All we see in Figure 26-24 is a straight line, which tells us that we live in an expanding universe.

However, suppose you measure the redshifts of galaxies several billion light years from Earth. Light from these galaxies has taken billions of years to get to your telescope, so your measurements reveal how fast the universe was expanding billions of years ago. Because the universe was expanding faster in the past than it is today, your data will deviate slightly from the straight-line Hubble law.

The graphs in Figure 29-1 display the relationship between the deceleration of the universe and the Hubble law extended to great distances.

Figure 29-1 Deceleration and the Hubble diagram
These two graphs compare the evolution of the universe and the appearance of the Hubble diagram. The case $q_0 = 0$ is an empty universe ($\rho = 0$). The case $q_0 = \frac{1}{2}$ is a marginally bounded universe ($\rho = \rho_c$). If $0 < q_0 < \frac{1}{2}$, then the universe is unbounded and will expand forever. If $q_0 > \frac{1}{2}$, the universe is bounded and will someday collapse.

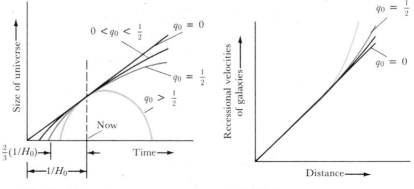

a The size of the universe

b Hubble–law curves

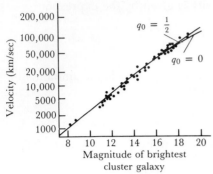

Figure 29-2   The Hubble diagram
This graph shows the Hubble diagram extended to include extremely remote galaxies. The magnitude of the brightest galaxy in a cluster is directly correlated with the distance to the cluster. If the data points fall between the curves marked $q_0 = 0$ and $q_0 = \frac{1}{2}$, then the universe is unbounded. If the data points fall above the curve marked $q_0 = \frac{1}{2}$, then the universe is bounded. (Adapted from A. Sandage)

Astronomers denote the amount of deceleration by the **deceleration parameter** q_0. Appropriately, $q_0 = 0$ corresponds to no deceleration at all. This is possible only if the universe is completely empty and thus has no gravity to slow down the expansion. As sketched in Figure 29-1a the $q_0 = 0$ universe expands forever at a constant rate. As explained in Chapter 28, the present age of this empty universe is exactly $1/H_0$, or about 20 billion years.

The case $q_0 = \frac{1}{2}$ corresponds to a marginally bounded universe. Such a universe just barely manages to expand forever, and it contains matter at the critical density ρ_c. In this case, the present age of the universe is $\frac{2}{3}(1/H_0)$, or about 13 billion years.

If q_0 is in the range between 0 and $\frac{1}{2}$, then the universe is unbounded and will continue to expand forever. Such a universe contains matter at less than the critical density, and it has a present age between 13 and 20 billion years.

If q_0 is greater than $\frac{1}{2}$, then the universe is bounded and is filled with matter of a density greater than ρ_c. The present age of such a universe is less than $\frac{2}{3}(1/H_0)$. This universe is doomed to collapse in upon itself in the extremely distant future.

In principle, it should be possible to determine q_0 by measuring the redshifts and distances of many remote galaxies and plotting the data on a Hubble diagram like the one in Figure 29-1b. If the data points fall above the $q_0 = \frac{1}{2}$ line, then the universe is bounded. If the data points fall between the $q_0 = 0$ and $q_0 = \frac{1}{2}$ lines, then the universe is unbounded.

Unfortunately, such observations are extremely difficult. Galaxies nearer than 10^9 ly are of no help in determining q_0. Beyond 10^9 ly, galaxies are so faint and indistinct that uncertainties cloud the distance and redshift determinations. For example, Figure 29-2 shows data obtained by Allan Sandage of the Mount Wilson and Las Campanas Observatories. Note how the data points are scattered about the $q_0 = \frac{1}{2}$ line. This scatter, due to observational uncertainties, prevents us from determining conclusively whether the universe is bounded or unbounded. Nevertheless, because the data lie close to the $q_0 = \frac{1}{2}$ line, we conclude that the universe is not far from marginally bounded. Many astronomers hope that the Space Telescope will enable them to do a better job of determining the value of q_0.

The shape of the universe is related to the deceleration parameter and to the average matter density in the universe

There is another way of tackling the issue of the fate of the universe. Our understanding of the universe is based on Einstein's general theory of relativity, which explains that gravity curves the fabric of space. The gravity of all matter scattered across space should give the universe some overall "shape." Gravity also determines the fate of the universe. Thus, by measuring the shape (or geometry) of space, we should be able to discover whether the universe is bounded or unbounded.

To see what astronomers mean by the geometry of the universe, imagine that we shine two powerful laser beams out into space. Furthermore, suppose that we align these two beams so that they are perfectly parallel as they leave the Earth. Finally, suppose that nothing gets in the way of these two beams, so that we can follow them for billions of light years across the universe—across the space whose curvature we wish to detect.

There are only three possibilities. First, we might find that our two beams of light remain perfectly parallel, even after traversing billions of light years. In this case, space is not curved: the universe has **zero curvature,** and space is **flat.**

Alternatively, we might find that our two beams of light gradually converge. The two beams gradually get closer and closer together as they move across the universe, so that they will eventually intersect at some enormous distance from Earth. In this case, space is not flat. Recall that lines of longitude on the Earth's surface are parallel at the equator but intersect at the poles. Thus, in this case, the three-dimensional geometry of the universe must be analogous to the two-dimensional geometry of a spherical surface. We therefore say that space is **spherical** and that the universe has **positive curvature.**

The third and final possibility is that the two parallel beams of light gradually diverge, becoming farther and farther apart as they move across the universe. In this case, also, the universe must be curved, but it must be curved in the opposite sense from that of the spherical case. We therefore say that the universe has **negative curvature.** In the same way that a sphere is a positively curved two-dimensional surface, a saddle is a good example of a negatively curved two-dimensional surface. Just as parallel lines drawn on a sphere always converge, parallel lines drawn on a saddle always diverge. Mathematicians say that saddle-shaped surfaces are hyperbolic. Thus, in a negatively curved universe, we say that space is **hyperbolic.**

These three cases are sketched in Figure 29-3. To illustrate these cases, three surfaces are drawn: a plane, a sphere, and a saddle. Of course, real space is three-dimensional, but it is much easier to visualize a two-dimensional surface. Thus, as you examine the drawings in Figure 29-3, remember that the real universe has one more dimension. For example, if the universe is hyperbolic, then the geometry of space is the three-dimensional analog of the two-dimensional surface of a saddle. Three-dimensional space curves through a fourth dimension perpendicular to each of the other three—something that few people can visualize except by analogy.

Each of the three possible geometries corresponds to a different behavior and fate of the universe. Flat space corresponds to the marginally bound case ($q_0 = \frac{1}{2}$) in which the galaxies just barely manage to keep receding from each other. This flat-space scenario divides the positive-curvature cases from the negative-curvature cases. If the density across space is greater than the critical density, then $q_0 > \frac{1}{2}$, and space is positively curved. Conversely, if the density across space is less than the critical density, then $0 < q_0 < \frac{1}{2}$, and space is negatively curved. These relationships are summarized in Table 29-1.

Note that both the flat and the hyperbolic universes are infinite. They

TABLE 29-1 *The geometry and fate of the universe*

Geometry of space	Curvature of space	Average density throughout space	Deceleration parameter (q_0)	Type of universe	Ultimate future of the universe
Spherical	Positive	Greater than the critical density	Greater than $\frac{1}{2}$	Closed	Eventual collapse
Flat	Zero	Exactly equal to the critical density	Exactly equal to $\frac{1}{2}$	Flat	Perpetual expansion (just barely)
Hyperbolic	Negative	Less than the critical density	Between 0 and $\frac{1}{2}$	Open	Perpetual expansion

Figure 29-3   The geometry of the universe
*The "shape" of space is determined by the mat-
ter contained in the universe. The curvature is
either positive, zero, or negative, depending on
whether the density is greater than, equal to,
or less than the critical density.*

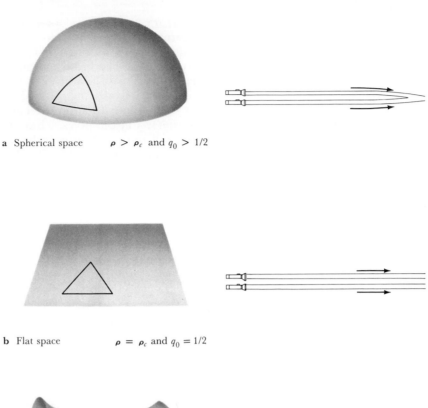

a Spherical space $\rho > \rho_c$ and $q_0 > 1/2$

b Flat space $\rho = \rho_c$ and $q_0 = 1/2$

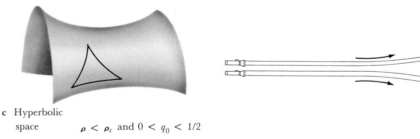

c Hyperbolic
space $\rho < \rho_c$ and $0 < q_0 < 1/2$

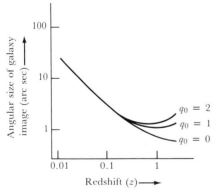

**Figure 29-4   The angular-diameter–
redshift relation**
*The curvature of the universe can magnify the
images of remote galaxies. The amount of
magnification depends on the distance to a
galaxy and the value of* q_0. *This effect is noticeable only for galaxies with redshifts greater
than about* z = 1.

extend forever in all directions, and thus they have neither an "edge"
nor a "center." In contrast, the spherical universe is finite. Nevertheless,
it also lacks a center or edge. A good analogy is the Earth. The Earth
has a finite surface area (511 million square kilometers), but you could
walk forever around our planet without ever finding a center or an
edge. Relativistic cosmology strictly rules out any possibility of a center
or edge to the universe.

There are no easy, practical ways of measuring the curvature of space
across the universe. In theory, you might imagine drawing an enormous
triangle whose sides are each 10^9 light years long. You could then measure the three angles of the triangle. If their sum equals 180°, then space
is flat. If the sum is greater than 180°, then space is spherical. If the
sum of the three angles is less than 180°, then space is hyperbolic. Unfortunately, this direct method of measuring curvature is not practical.

There is an alternative, more feasible method. Light rays from distant
galaxies are bent slightly because of the curvature of space. This deflection of light rays distorts the appearance of remote galaxies. Normally,
the more distant an object is, the smaller it appears. However, for extremely remote galaxies, the bending of light by the curvature of the
universe can actually magnify images. Thus, an extremely distant galaxy
can appear bigger than would normally be expected.

The graph is Figure 29-4 shows how the diameters of the images of galaxies depend on q_0. Unfortunately, we must examine galaxies with redshifts greater than $z = 1$ in order to decide between various possible values of q_0. These galaxies are so far away that they appear as faint, hazy blobs through Earth-based telescopes. Furthermore, galaxies evolve, and thus we are forced to compare galaxies at different stages of evolution. Astronomers hope that the Space Telescope will shed light on galaxy evolution and, along with sharp galaxy images, will allow us to use tests such as the diameter–redshift relation to determine the ultimate fate of the cosmos.

The flatness of the universe and the isotropy of the microwave background suggest that a period of vigorous inflation followed the Big Bang

Ever since Hubble discovered that the universe is expanding, astronomers have struggled to determine the deceleration parameter. During the 1960s and 1970s, various teams of astronomers determined various values for the deceleration parameter, some slightly larger than $q_0 = \frac{1}{2}$ and some smaller than $q_0 = \frac{1}{2}$. Because $q_0 = \frac{1}{2}$ is the special case of a flat universe that separates a bounded cosmological model from an unbounded model, the predicted fate of the universe swung back and forth. According to some data, the universe seems to be just barely open and infinite, whereas other data indicate that the universe is just barely closed and doomed to collapse.

Motivated by the fact that q_0 may be nearly equal to $\frac{1}{2}$, physicists began looking for special conditions associated with a flat universe in which the average density of matter is exactly equal to the critical density, ρ_c. Specifically, suppose that the average density of matter during the Big Bang were slightly larger (or smaller) than the critical density. How would this deviation grow (or decrease) as the universe evolves?

In the previous chapter we saw that the earliest understandable moment in the universe was the Planck time, about 10^{-43} sec after the Big Bang. Between $t = 0$ and $t = 10^{-43}$ sec, the universe was so dense and particles were interacting so violently that no known theory properly describes what happened. However, immediately after the Planck time, the fate of the universe was very sensitive to the density of matter. Calculations demonstrate that the slightest deviation from precisely the critical density would mushroom very rapidly, multiplying itself every 10^{-43} sec. If the density were slightly less than ρ_c, the universe would soon become wide open and virtually empty. If the density were slightly greater than ρ_c, the universe would soon become tightly closed and so packed with matter that the entire cosmos would rapidly collapse into a black hole. In other words, immediately after the Big Bang, the fate of the universe hung in the balance—like a pencil teetering on its point—so that the tiniest deviation from the precise equality $\rho = \rho_c$ would rapidly propel the universe away from the special case of $q_0 = \frac{1}{2}$.

Observations reveal that q_0 is today approximately $\frac{1}{2}$. Consequently, the density of the universe immediately after the Big Bang must have been equal to the critical density to an incredibly high degree of precision. Calculations demonstrate that, in order for q_0 to be roughly $\frac{1}{2}$ today, ρ must have been equal to ρ_c to more than 50 decimal places!

What could have happened immediately after the Planck time to ensure that $\rho = \rho_c$ to such an astounding degree of accuracy? Because $\rho = \rho_c$ means that space is flat, this enigma is called the **flatness problem.**

A second enigma closely related to the flatness problem is the isotropy of the 3-K cosmic microwave background. As we saw in Chapter 28, the microwave background is so incredibly uniform across the sky that sensitive temperature measurements reveal our motion through this radiation field (recall Figure 28-6). Subtracting the effects of our motion, we find that the temperature of the microwave background is the same in all parts of the sky to an accuracy of 1 part in 10,000.

To appreciate the dilemma, think about microwave radiation coming to us from two opposite parts of the sky. This radiation is left over from the primordial fireball and has been traveling toward us for nearly 20 billion years. The total distance between opposite sides of the observable universe is roughly 40 billion light years, and thus these widely separated regions have absolutely no connection with each other. So, why do these unrelated parts of the universe have the same temperature?

In the early 1980s, Alan Guth working at Stanford University offered a remarkable solution to the problems of the flatness of the universe and the isotropy of the microwave background. Guth analyzed the suggestion that the universe experienced a brief period of extremely rapid expansion shortly after the Planck time. During the **inflationary epoch,** as this period is called, the universe ballooned outward in all directions at speeds much greater than the speed of light to become many billions of times its original size. This **inflation** places much of the material that was originally near our location far *beyond* the edge of the *observable* universe today. Thus the observable universe is now expanding *into* space containing matter and radiation that was once in close contact with our location.

An inflationary epoch accounts for the isotropy of the microwave background. In looking at microwaves from opposite parts of the sky, we are seeing radiation from parts of the universe that were originally in intimate contact with each other. That is why they have the same temperature.

Inflation also accounts for the flatness of the universe. To see why, think about a small portion of the Earth's surface, such as your backyard. For all practical purposes, it is impossible to detect the Earth's curvature over such a small area; your backyard looks very flat. Similarly, the observable universe is such a tiny fraction of the inflated universe that any overall curvature is virtually undetectable. Like your backyard, our observable segment of space looks very flat.

Grand unified theories (GUTS) explain that all the forces had the same strength immediately after the Big Bang

All of the behavior and interactions of everything in the universe can be understood as the result of only *four* physical forces: gravity, electromagnetism, and the strong and weak nuclear forces. We are all intimately familiar with the force of gravity; it is a long-range force that dominates the universe over astronomical distances. The electromagnetic force also is a long-range force (in principle, its influence extends to infinity as gravity does), but it is much stronger than the gravitational force. Just as the force of gravity holds the Moon in orbit about the Earth, the electromagnetic force holds electrons in orbit about nuclei in atoms. We do not generally observe the long-distance effects of the electromagnetic force, however, because (in most circumstances) there is a negative electric charge for every positive charge and a south magnetic pole for every north magnetic pole. Thus, over large volumes of space, the net effects of electromagnetism effectively cancel. A similar canceling does not occur with gravity because there is no "negative mass."

Both the strong and weak nuclear forces are said to be "short-range" because their influence extends only over distances less than about 10^{-13} cm. The **strong nuclear force** holds protons and neutrons together inside the nuclei of atoms. Without the strong nuclear force, nuclei would disintegrate because of the electromagnetic repulsion of the positively charged protons. Thus, the strong nuclear force overpowers the electromagnetic force inside nuclei.

The weak nuclear force is so weak that it does not hold anything together. Instead, the **weak nuclear force** is at work in certain kinds of radioactive decay, such as the transformation of a neutron (n) into a proton (p), with the accompanying release of an electron (e⁻) and an antineutrino ($\bar{\nu}$):

$$n \rightarrow p + e^- + \bar{\nu}$$

Numerous experiments in nuclear physics strongly suggest that protons and neutrons are composed of more basic particles called **quarks,** the most common varieties being "up" quarks and "down" quarks. A proton is composed of two up quarks and one down quark, whereas a neutron is made of two down quarks and one up quark.

In the 1970s, the concept of quarks gave rise to a more fundamental description of the strong and weak nuclear forces. In its most basic form, the strong nuclear force is the force that holds quarks together. Similarly, the weak nuclear force is at work whenever a quark changes from one variety to another. For example, when a neutron decays into a proton, one of the neutron's down quarks (d) changes into an up quark (u). Thus, the weak nuclear force is responsible for transformations such as

$$d \rightarrow u + e^- + \bar{\nu}$$

In the 1940s, physicists Richard P. Feynman, Julian S. Schwinger, and Sinitiro Tomonaga succeeded in developing a very basic description of what we mean by "force." They focused their attention on the electromagnetic force and tried to describe exactly what happens when two charged particles interact. According to their theory, called **quantum electrodynamics,** charged particles interact by exchanging virtual photons. Virtual photons, like virtual particles, cannot be observed directly because they exist only for extremely short time intervals. The process of exchanging virtual photons is often described with the aid of a **Feynman diagram** (named after its inventor) as in Figure 29-5, which schematically shows two electrons interacting by exchanging a photon. After the exchange, the electrons are traveling in different directions than they were before the exchange.

Quantum electrodynamics has proven to be one of the most successful theories in modern physics; it accurately describes many details of the electromagnetic interaction between charged particles. Inspired by these successes, physicists have tried to develop similar theories for the other three forces. Specifically, the weak nuclear force occurs when particles exchange **weakons,** the gravitational force occurs when particles exchange **gravitons,** and quarks stick together by exchanging **gluons.** Thus we may summarize the four forces as indicated in Table 29-2.

In recent years, important progress has been made in understanding the weak nuclear force—primarily through the efforts of physicists Steven Weinberg, Sheldon Glashow, and Abdus Salam. Their theory predicted the existence of three types of weakons (W⁺, W⁻, and Z°), which are exchanged in various manifestations of the weak force (see Figure 29-6). All three types of weakons were discovered in high-energy nuclear-physics experiments in the early 1980s, thereby providing strong support for the theory.

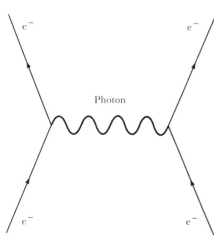

Figure 29-5 A Feynman diagram of the electromagnetic force
According to quantum electrodynamics, electric and magnetic forces occur as the result of virtual photons jumping back and forth between charged particles. This exchange occurs during time intervals so short that the photons cannot be directly detected or observed while in flight between charged particles.

TABLE 29-2   The four forces

Force	Relative strength	Particles exchanged	Particles acted upon	Range	Example
Strong	1	Gluons	Quarks	10^{-13} cm	Holds nuclei together
Electromagnetic	1/137	Photons	Charged particles	Infinite	Holds atoms together
Weak	1/10,000	Weakons	Quarks, electrons, neutrinos	$<10^{-14}$ cm	Radioactive decay
Gravity	6×10^{-39}	Gravitons	Everything	Infinite	Holds the solar system together

One of the startling predictions of the Weinberg–Glashow–Salam theory is that the weak force and the electromagnetic force should be identical to each other at energies greater than 100 GeV. In other words, if particles are slammed together with total energy greater than 100 billion electron volts, then electromagnetic interactions are indistinguishable from weak interactions. We therefore say that, "above 100 GeV," the electromagnetic force and the weak force are "unified."

This unification occurs because, above 100 GeV, the three types of weakons behave just like photons. Physicists describe this similarity by saying that "symmetry is restored" above 100 GeV. In the world around us, however, particles interact with much less energy than 100 GeV. Below 100 GeV, weakons behave like massive particles, whereas photons are always massless. Because the similarity does not exist at low energies, we say that the "symmetry is broken" below 100 GeV, which is why electromagnetic and weak forces behave so differently in the world around us.

In the 1970s, Sheldon Glashow and Howard Georgi proposed a **grand**

Figure 29-6   Feynman diagrams of the weak nuclear force
The weak nuclear force exists when particles emit or exchange weakons (W^+, W^-, Z°), which are also called intermediate-vector bosons. **(a)** *A neutron (n) decays into a proton (p) by emitting a W^- weakon, which rapidly turns into an electron (e^-) and an antineutrino ($\bar{v}$).* **(b)** *When a neutron collides with a neutrino (v), a W^- weakon is exchanged between the particles, transforming the neutron to a proton and the neutrino to an electron. However, in a Feynman diagram, an outgoing antineutrino is exactly equivalent to an incoming ordinary neutrino. Therefore, this reaction is exactly equivalent to the neutron decay shown in part a.* **(c)** *This diagram also is equivalent to those of parts a and b because a W^+ weakon going from right to left is equivalent to a W^- weakon going from left to right.* **(d)** *A neutron or proton interacts with a neutrino or antineutrino through exchange of a Z° weakon.*

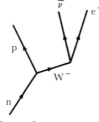

a Decay of neutron
$(n \longrightarrow p + e^- + \bar{v})$

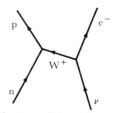

c Neutron collides with neutrino
$(n + v \longrightarrow p + e^-)$

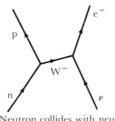

b Neutron collides with neutrino
$(n + v \longrightarrow p + e^-)$

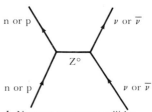

d Neutron or proton collides
with neutrino or antineutrino

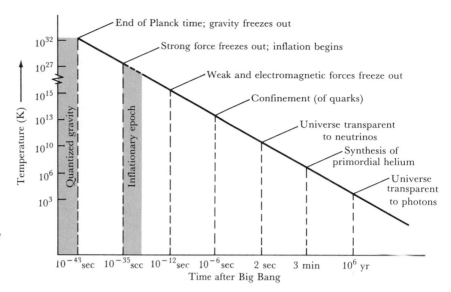

Figure 29-7 **The early history of the universe**
As the universe cooled, the four forces "froze out" of their unified state as a result of the spontaneous symmetry breaking described in the text. The inflationary epoch lasted from 10^{-35} sec to 10^{-24} sec after the Big Bang. (Adapted from David Schramm)

unified theory (or **GUT**), which predicts that the strong, weak, and electromagnetic forces are unified at energies above 10^{14} GeV. In other words, if particles were colliding at energies greater than 100 trillion GeV, then the strong, weak, and electromagnetic interactions would be indistinguishable from each other.

Many physicists suspect that all four forces are unified at energies greater than 10^{19} GeV. If particles were colliding at these colossal energies, there would be no difference between gravitational, electromagnetic, and nuclear forces. However, no one has yet succeeded in working out the details of such a **supergrand unified theory** (or **superGUT**).

Particle accelerators exist with energies great enough to test the unification of the weak and electromagnetic forces around 100 GeV. Physicists see no hope of ever constructing machines that could slam particles together with energies of trillions of GeV. Thus, it is impossible to test the grand and supergrand unified theories in a laboratory. However, immediately after the Big Bang, the universe was so hot and particles were moving with such high speeds that they did indeed collide with energies of trillions of GeV. Thus the earliest moments of the universe become the laboratory wherein scientists try to test some of the most elegant and most sophisticated theories in physics.

Many of the ideas connecting particle physics and cosmology are very new and still quite speculative. Nevertheless, it is possible to summarize our understanding with the aid of Figure 29-7. During the Planck time (from $t = 0$ to $t = 10^{-43}$ sec), particles collided with energies greater than 10^{19} GeV, and all four forces were unified. Because we do not have a superGUT that properly describes the behavior of gravitons and quantized gravity, we are ignorant of what was going on during the first 10^{-43} sec of the universe's existence. By the end of the Planck time, however, the energy of particles in the universe had fallen to 10^{19} GeV, below which gravity is no longer unified with the other three forces. We therefore say that, at $t = 10^{-43}$ sec, there was a **spontaneous symmetry breaking** in which gravity "froze out" of the otherwise unified hot soup that filled all space. As we saw in Chapter 28, energy (E) is related to temperature (T) by $E = kT$, where k is the Boltzmann constant (roughly 10^{-4} eV/K). Thus, the temperature of the universe was 10^{32} K when gravity emerged as a separate force.

At $t = 10^{-35}$ sec, the energy of particles in the universe had fallen to 10^{14} GeV (equal to a temperature of 10^{27} K), below which the strong nuclear force is no longer unified with the electromagnetic and weak nuclear forces. Thus, at $t = 10^{-35}$ sec, there was a second spontaneous symmetry breaking, and the strong nuclear force made its appearance, freezing out of an otherwise unified hot soup. Calculations suggest that the inflationary epoch lasted from $t = 10^{-35}$ sec to about $t = 10^{-24}$ sec, during which the universe increased its size by a factor of between 10^{20} and 10^{30} over what it would have normally been.

At $t = 10^{-12}$ sec, the temperature of the universe had dropped to 10^{15} K, the energy of particles had fallen to 100 GeV, and there was a final spontaneous symmetry breaking and "freeze-out" as the electromagnetic force separated from the weak nuclear force. From that moment on, all four forces interacted with particles essentially as they do today.

The next significant event occurred at $t = 10^{-6}$ sec, when the temperature was 10^{13} K and particles collided with energies of roughly 1 GeV. Prior to this moment, particles collided so violently that individual protons and neutrons could not exist because they were constantly being fragmented into quarks. After this moment, appropriately called **confinement,** quarks could finally stick together to form individual protons and neutrons.

The remaining significant events proceeded as outlined in Chapter 28. The universe became transparent to neutrinos at $t = 2$ sec, and all the primordial helium was produced by $t = 3$ min. Finally, the universe became transparent to photons at $t = 1$ million years. We next face the difficult problem of explaining why the matter in the universe then became concentrated in galaxies.

Galaxies formed from density fluctuations in the early universe

The distribution of matter in the universe today is quite "lumpy." Stars are grouped together in galaxies, galaxies are grouped together in clusters, and clusters of galaxies are grouped together in **superclusters.** A typical supercluster contains dozens of individual clusters spread over a volume up to 100 million light years across.

Although there is a lot of lumpiness in the universe today, the early universe must have been exceedingly smooth. To see why this is so, think back to the era of recombination that occurred roughly 1 million years after the Big Bang. Before recombination, high-energy photons were constantly and vigorously colliding with charged particles throughout all space. After recombination, the universe became transparent, and these photons stopped interacting with the matter in the universe. Astronomers therefore say that matter "decoupled" from radiation during the era of recombination. Because the 3-K microwave background is extremely isotropic, we conclude that the matter with which these photons once so-frequently collided must also have been spread smoothly across space.

Although the distribution of matter across space during the early universe must have been very smooth, it could not have been *perfectly* uniform. If it had been absolutely uniform billions of years ago, then it would still be absolutely uniform today. There would be neither stars nor galaxies—only a few atoms per cubic meter throughout all space. Consequently, a slight lumpiness–or **density fluctuations** in the distribution of matter—must have existed in the early universe. Through the action of gravity, these fluctuations eventually grew to become the galaxies and clusters of galaxies that we see throughout the universe today.

An understanding of how gravity can cause density fluctuations to

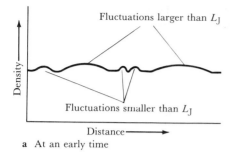

a At an early time

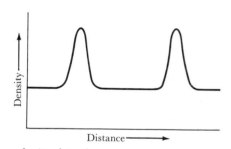

b At a later time

Figure 29-8 The growth of density fluctuations
(a) *Small density fluctuations in the distribution of matter shortly after the era of recombination.* **(b)** *If the size of a fluctuation is greater than the Jean's length (L$_J$), then it becomes gravitationally unstable and can grow in amplitude.*

Figure 29-9 A globular cluster
A typical globular cluster contains 10^5 to 10^6 stars. The diameters of these clusters range from about 20 to 400 ly. Because these parameters are comparable to the Jean's length (L$_J$) during the era of recombination, astronomers believe that globular clusters were among the first objects to form in the universe. (NASA)

grow dates back to 1902, when the British physicist James Jeans solved a problem first proposed by Isaac Newton. Suppose that you have a uniform, static distribution of matter of density ρ at temperature T. Also, suppose that you perturb this medium by introducing slight fluctuations in density as shown in Figure 29-8. These regions of enhanced density have an enhanced gravity that attracts nearby material. As this happens, however, gas pressure inside these regions increases, resisting further compression and growth. The question therefore is this: under what conditions does gravity overwhelm gas pressure so that a permanent object can condense out of the medium?

James Jeans proved that an object will gravitationally grow from a density perturbation provided that the overall size of the original disturbance is greater than the **Jeans' length** L_J given by

$$L_J = \sqrt{\frac{\pi k T}{m G \rho}}$$

where k is the Boltzmann constant, G is the gravitational constant, and m is the mass of a single particle in the medium.

Using the conditions during the era of recombination ($T = 3000$ K, $\rho = 10^{-21}$ g/cm^3) and m as the mass of the hydrogen atom ($m = 2 \times 10^{-24}$ g), we find that $L_J = 100$ ly, the diameter of a typical globular cluster. Furthermore, the mass contained in this volume is $\rho L_J^3 = 5 \times 10^5 M_\odot$, the mass of a typical globular cluster. For these reasons, Robert Dicke and P. J. E. Peebles of Princeton University have proposed that globular clusters (see Figure 29-9) were among the first objects to form after radiation decoupled from matter. Indeed, as we saw in Chapter 21, globular clusters are composed of the most ancient stars we can find in the sky. Nevertheless, these calculations do not shed light on how galaxies and clusters of galaxies formed.

An important clue about the formation of clusters and superclusters came in the early 1980s when astronomers began discovering enormous **voids** in space, regions where exceptionally few galaxies are found. The first of these voids was noticed by Robert Kirshner and colleagues working at the Kitt Peak National Observatory. They were involved in a program of measuring redshifts of galaxies in several regions of the sky. In one section of Boötes, they found only one galaxy in the redshift range of 12,000 to 18,000 km/sec. Using a Hubble constant of (15 km/sec) /10^6 ly,

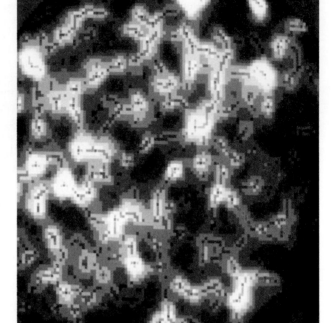

Figure 29-10   The distribution of 400,000 galaxies
This map shows the locations of nearly half a million galaxies across 100° of the sky. The north pole of our galaxy (in the constellation of Coma Berenices) is at the center of the map. Each small square (pixel) covers 10 × 10 arc min. Color indicates the number of galaxies found in each pixel. Light colors indicate a high density of galaxies, whereas darker regions indicate few galaxies. Green pixels emphasize filaments along which galaxies are concentrated, and single red pixels (near the centers of white regions) indicate the points where the galaxy counts reach a maximum. (Courtesy of E. L. Turner, J. E. Moody, and J. R. Gott)

this redshift range corresponds to a volume roughly $\frac{1}{3}$ billion light years across.

Numerous additional voids soon were discovered by carefully analyzing the locations of hundreds of thousands of galaxies in the sky. Figure 29-10 shows the distribution of 400,000 galaxies covering approximately one-quarter of the sky. On this map, the sky is divided into tiny squares (called pixels), each measuring about 10 arc min across; colors are used to indicate the number of galaxies found in each pixel. Light-colored regions indicate high densities of galaxies, whereas darker regions denote areas where exceptionally few galaxies are found. Thcse brown and black areas are the voids. Note that many of the white regions seem to be connected to each other by bridges, called **filaments,** which are emphasized by green squares.

Maps such as Figure 29-10 demonstrate that galaxies are concentrated in enormous sheets and filaments that are typically 100 million light years long, containing up to 1 million galaxies with a mass of roughly 10^{16} solar masses. Densely populated clumps seem to occur where two filaments of sheets intersect. Finally, interspersed between these huge structures are voids, virtually free of galaxies, that are between 100 million and 400 million light years across.

In the early 1980s, Yakov Zel'dovich at Moscow University, Joseph Silk at Berkeley, and their colleagues made significant progress in explaining why the matter in the universe is concentrated in sheets and filaments separated by voids. They began by assuming that density fluctuations of *all* sizes were initially present in the universe. However, maps such as Figure 29-10 tell us that only extremely large fluctuations (containing 10^{15} to 10^{16} solar masses and stretching for 100 million light years) managed to survive. They therefore had to explain what happened to erase all the smaller fluctuations.

Very small fluctuations were easily erased by collisions between matter and photons before the era of recombination. However, photons could not travel very far before recombination without hitting a charged particle, and thus photons were incapable of erasing the larger fluctuations. To explain the erasure of all but the very largest fluctuations, physicists had to turn to particles that can penetrate dense matter far more easily than photons can. Neutrinos were the answer.

If a neutrino has no mass, then it will always travel at the speed of light (just as a photon does). Freely streaming across space, neutrinos would eventually smooth any density fluctuations that initially existed, even the very largest. However, if a neutrino has a little mass, then it must slow down as the universe cools, and eventually it may be gravitationally captured in a growing density fluctuation. Thus, at some time, the smoothing effect of neutrinos ceased.

Recent controversial experiments, primarily conducted in the Soviet Union and involving the radioactive decay of tritium, suggest that the neutrino has a mass between 20 and 40 eV, or about one ten-thousandth of the mass of an electron. This result immediately attracted the attention of astronomers, because it implies that the cosmic neutrino background contains an enormous amount of matter. Today, on the average, there should be roughly 100 neutrinos per cubic centimeter. If each neutrino has a mass of about 30 eV, then the cosmic neutrino background alone has an average density very nearly equal to the critical density. This assumption would explain why q_0 is nearly equal to $\frac{1}{2}$, even though the density of matter we can see in the universe is much less than the critical density. Also, because neutrinos possessing mass would eventually be gravitationally captured in clusters of galaxies, these particles could be the solution to the missing-mass problem.

A tiny neutrino mass was precisely what Zel'dovich, Silk, and their colleagues needed because it permits calculation of the maximum distances over which neutrinos can freely stream before being gravitationally captured. That maximum distance corresponds to the smallest-sized fluctuations that are not erased. If the neutrino mass is 30 eV, then the smallest surviving fluctuations should today be 100 million light years across and should contain 10^{15} to 10^{16} solar masses—exactly what we observe in maps such as Figure 29-10.

Assuming that neutrinos have a small mass, many physicists have performed computer simulations showing how all but the very largest primordial neutrino fluctuations are smoothed out. A typical simulation is shown in Figure 29-11. Matter gravitationally follows the growing

Figure 29-11   The evolution of large-scale fluctuations
This computer simulation shows how large-scale fluctuations develop if neutrinos have nonzero mass. Empty regions grow as matter becomes concentrated in long filaments. The resulting patterns resemble the observed distribution of voids and superclusters. This simulation was done by George Efstathiou at the University of Cambridge.

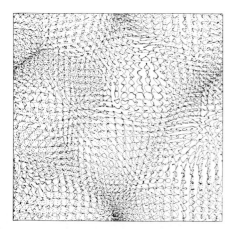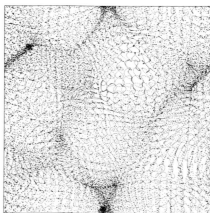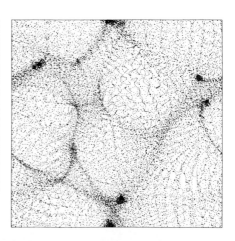

neutrino fluctuations and collapses into long filaments, while the remaining regions of space become increasingly empty. High concentrations of matter occur where filaments intersect, just as in the map of Figure 29-10. Thus, the least-massive particle apparently had a profound influence on the most-massive structures in the universe.

If the universe is bounded, we are living inside a huge black hole that will collapse in a Big Crunch

If the average density of matter across space is greater than the critical density, then the universe will someday stop expanding. Eventually, the mutual gravitational attraction between galaxies will cause them to start moving back toward each other. The redshifts we see today in galactic spectra will be replaced by blueshifts as the universe begins to contract.

Specific events leading up to the death of a contracting universe mimic those of the birth of the universe, but in reverse. For many billions of years, this contraction will be noticeable only to astronomers. At first, only the nearest clusters of galaxies will exhibit blueshifts in their spectra because we see these galaxies in the most recent past. Further into the future, however, increasing numbers of more remote galaxy clusters will have their redshifts replaced by blueshifts.

The contraction of the universe will also cause the temperature of the radiation background to increase. Just as the expansion of the universe is, in reality, the expansion of space, so the contraction of the universe is actually the contraction of space. Photons traveling across this contracting space are subjected to a shortening of their wavelengths. Their energies increase, and thus the temperature of the radiation field increases.

Just as the expanding universe began with a Big Bang, the contracting universe will end with a **Big Crunch.** About 70 million years before the Big Crunch, the temperature of the background will be up to 300 K, and galaxies will be so closely crowded together that the night sky will be as bright as day.

At 700,000 years before the Big Crunch, the radiation temperature will be up to 3000 K. The photons will then possess enough energy to begin dissociating atoms and molecules. With three weeks to go before the Big Crunch, the temperature will be 10 million kelvins, and stars and planets will dissolve. With three minutes to go, the temperature will climb above 10 billion kelvins, and the resulting extremely energetic photons will break up nuclei. Finally, all matter and radiation will be crushed out of existence, as a cosmic singularity will envelop all space and time.

Just as we cannot use physics to tell us what existed before the Big Bang, so science cannot reveal what might happen after the Big Crunch. The infinite warping of space and time prevents us from extrapolating any further into the future.

If the universe expands forever, then black-hole evaporations will occur in the extremely distant future

If the average density of matter across space is less than or equal to the critical density, then the universe will expand forever. Extrapolating into the extremely distant future is a very risky business. Nevertheless, it is reasonable to predict a few significant events that will occur in a universe that exists forever.

Although the universe today consists mostly of hydrogen and helium, these gases will eventually be used up, consumed in the thermonuclear fires of generation after generation of stars. Astronomers estimate that this will occur in about 10^{12} years. Star birth will then cease because

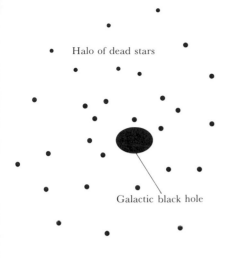

Halo of dead stars

Galactic black hole

Figure 29-12   A dead galaxy
*Close encounters can eject stars from a galaxy.
Over a span of 10^{27} years, a galaxy might
lose up to 99 percent of its stars in this manner. Meanwhile, the remaining stars coalesce
into an enormous black hole. This galactic
black hole might typically have a mass of about
10^{11} solar masses.*

there will be neither hydrogen nor helium in the interstellar medium.
Galaxies will grow dim as the final generation of stars dies off. Thus,
roughly a trillion years after the Big Bang, all the matter in the universe
will consist of either dead stars (white dwarfs, neutron stars, and black
holes) or cold lumps (meteorites, used spaceships, rocks and so forth).

Still further into the future, the effects of highly improbable events
begin to become important. For example, as noted in an earlier chapter,
the probability of two stars colliding somewhere in our galaxy is extremely small. Indeed, the probability of two stars even passing near
each other is almost infinitesimal. Over trillions upon trillions of years,
however, these events will occur.

A close encounter between two stars is significant because one of the
stars can easily transfer enough energy and momentum to kick the other
star out of the galaxy. The remaining star then falls into a lower-energy
orbit closer to the galaxy's nucleus. Eventually, because of the emission
of gravitational radiation (recall Box 24-2), these remaining stars will
gradually spiral into the galaxy's center, coalescing into one enormous
black hole. The mass of the resulting **galactic black hole** would be
roughly $10^{11}M_\odot$. The time needed for all this to occur is about 10^{27}
years. Thus, when the universe is a billion billion billion years old, galaxies will consist of very large black holes surrounded by halos of dead
stars (see Figure 29-12).

The orbital motions of galaxies in clusters also emit gravitational radiation. As a result of this energy loss, entire galaxies spiral toward each
other and eventually will coalesce into **supergalactic black holes.** It will
take roughly 10^{31} years for these colossal objects to form; each of them
might contain $10^{15}\,M_\odot$.

Recent research involving grand unified theories suggests that the
proton might not be a stable particle. Instead, the proton may decay into
an antielectron (e^+) and two gamma rays. The proton seems to be a
stable particle only because its half-life is extremely long—at least 10^{32}
years. Thus, 10^{32} years into the future, protons not swallowed by black
holes will decay in large numbers. The final result will be a low-density
mixture of electrons and antielectrons so widely separated that they
never manage to annihilate each other.

Black-hole evaporations will become important even further into the
future. As discussed in Box 29-1, a black hole emits particles. Although
this quantum-mechanical phenomenon is significant only for low-mass
black holes at usual time scales, even the most-massive black holes will
eventually evaporate completely if the universe exists forever. From the
equations given in Box 29-1, we find that ordinary black holes (such as
Cygnus X-1) evaporate after roughly 10^{67} years. Galactic black holes are
gone after 10^{97} years. Finally, supergalactic black holes evaporate after
10^{106} years. All of these evaporations end the same way—in a burst of
particles and radiation equivalent to the detonation of a billion
1-metagon hydrogen bombs (roughly 10^{22} ergs).

Astronomers usually think of a black hole as an **information sink:**
material falls in, and nothing gets out. Specifically, as explained in Chapter 24, all of the "information" carried by infalling matter (such as its
shape, color, and chemical composition) is removed from the universe.
This view is justified only for ordinary, massive black holes where the
quantum mechanical emission of particles is negligible. However, during
the final stages of evaporation, vast quantities of particles pour out of a
black hole. Thus the black hole behaves like a **white hole,** dumping new
material into the universe. This material carries information (color,
shape, size, and so on), and hence a white hole is an **information source.**

Box 29-1   The evaporation of black holes

During the early 1970s, Stephen W. Hawking of Cambridge University proposed that large density fluctuations during the Big Bang could have created numerous small black holes. Theoretically, these black holes could have had masses as small as 10^{-5} g. Because they were created during the Big Bang, they are called **primordial black holes.**

Most astronomers feel that primordial black holes probably do not exist. Only very special kinds of density fluctuations during the earliest moments of the universe could have led to the copious creation of very small black holes. We have no reason to suspect that the early universe obliged us with the necessary special conditions. Nevertheless, theoretical investigations of the properties of very small black holes produced some surprising and significant discoveries.

First, it is instructive to realize that low-mass black holes are indeed extremely tiny. As we saw in Chapter 24, the radius of a black hole is given by

$$r = \frac{2GM}{c^2}$$

where M is the mass of the hole, G is the gravitational constant, and c is the speed of light. For example, if the Earth (mass 6×10^{27} g) were crushed to become a black hole, it would be about the size of a pingpong ball. A typical large asteroid (mass 10^{20} g) crushed to become a black hole would be about the size of an atom. A black hole made from a billion tons of matter (roughly 10^{15} g) would be about the size of a proton.

Because low-mass black holes are submicroscopic, we must use quantum mechanics in order to understand their properties. As we saw in Chapter 28, the **Heisenberg uncertainty principle** is a basic tenet of submicroscopic physics. This principle explains that you cannot determine precisely both the position and the speed of a subatomic particle. Over extremely small distances or times, a certain amount of "fuzziness" is built into the nature of reality.

We saw also that the Heisenberg uncertainty principle leads logically to the concept of **virtual pairs.** At every point in space, particles and antiparticles are constantly being created and destroyed. This process occurs over such incredibly brief time intervals, however, that these virtual particles and antiparticles are not directly observable.

Think about a very tiny black hole. Furthermore, think about the momentary creation of a virtual proton–antiproton pair inside the hole's event horizon. Because the hole itself is roughly the size of a proton, it is possible that one of the virtual particles is briefly located *outside* the black hole and is therefore free to escape from the hole. Its partner is then deprived of a counterpart with which to annihilate. The virtual particles therefore must become *real* particles. To accomplish this conversion, some of the energy of the black hole's gravity must be converted into matter according to $E = mc^2$. Because one of the particles escapes from the hole, the mass of the black hole decreases. In this way, particles can quantum-mechanically "leak" out of a black hole, carrying some of the hole's mass with them.

The smaller a black hole is, the more easily particles can leak out through its event horizon. Stephen Hawking proved that you can speak of the **temperature** of a black hole as a way of describing the rate of particle leakage. This temperature is given by

$$T = \frac{hc^3}{16\pi^2 kGM}$$

where h is Planck's constant, k is the Boltzmann constant, and M is the mass of the hole. For example, a 1-billion-ton black hole emits particles and energy

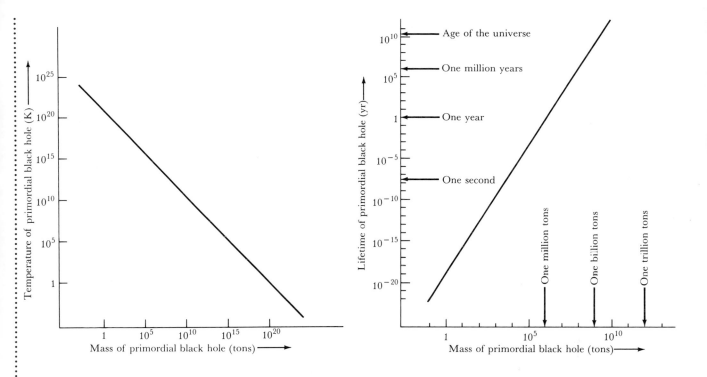

as if its temperature is nearly 10^{12} K. This relationship between the mass of a black hole and its temperature is graphed in the diagram on the left.

Incidentally, for ordinary black holes such as Cygnus X-1, this effect is negligible over time spans of billions of years. For example, the temperature of a 10 $M_\odot$ black hole is about 10^{-7} K above absolute zero. In other words, particles hardly ever manage to escape from ordinary black holes.

As particles escape from a small black hole, the mass of the black hole decreases and thus its temperature goes up. As the temperature goes up, still more particles escape, thereby further decreasing the hole's mass and forcing the temperature still higher. This is a runaway process that causes the black hole to **evaporate** completely. During the final seconds of evaporation, the hole gives up the last of its mass in a violent burst of energy equal to the detonation of a billion 1-megaton hydrogen bombs.

The smaller a black hole is, the hotter it is, and hence the faster it evaporates. Hawking proved that the lifetime t of a black hole is

$$t = 10240\pi^2\left(\frac{G^2M^3}{hc^4}\right)$$

This value tells us that very-low-mass primordial black holes evaporated fairly soon after the Big Bang. For example, a 1-million-ton black hole lasts for only 3000 years. The relationship between the mass of a black hole and its lifetime is graphed in the diagram on the right.

Because the universe is about 20 billion years old, it is possible to calculate the minimum mass needed for a primordial black hole to survive up to the present time. The answer is roughly 200 million tons. In other words, only those primordial black holes with masses greater than 200 million tons are cool enough and evaporating slowly enough to have survived to the present time.

Although few people think that such small black holes actually exist, the ideas and formulas developed by Stephen Hawking reveal a fundamental interrelation between thermodynamics, quantum mechanics, and general relativity.

Indeed, British physicist Stephen W. Hawking has proved that, during the final moments of evaporation, a black hole is a white hole that randomly pumps new matter into the universe in an unpredictable fashion.

To emphasize the totally unpredictable way in which evaporating black holes randomly dump new information into the universe, Hawking humorously proposed that (although it is very improbable), one of these holes "could emit a television set or the works of Proust in 10 leather volumes." Thus, as we stand at the frontiers of human knowledge, groping dimly into the remote future for an understanding that might not even exist, it is perhaps wise to recall the words of Mark Twain:

> There is something fascinating about science. One gets such a wholesale return of conjecture out of such a trifling investment of fact.

Life on the Mississippi (1874)

Summary

- If the average density of matter in the universe is greater than the critical density ρ_c, then space is spherical (with positive curvature), the deceleration parameter q_0 has a value greater than $\frac{1}{2}$, and the universe is closed (bounded) and ultimately will collapse in a Big Crunch.

- If the average density of matter in the universe is less than ρ_c, then space is hyperbolic (with negative curvature), q_0 has a value less than $\frac{1}{2}$, and the universe is open (unbounded) and will continue to expand forever.

- If the average density of matter in the universe is exactly equal to ρ_c, then space is flat (with zero curvature), q_0 is exactly equal to $\frac{1}{2}$, the universe is marginally bounded, and expansion will just barely continue forever (decreasing toward a zero rate of expansion at an infinite time in the future).

- Present techniques do not permit measurements of ρ_c or q_0 precise enough to determine which of the preceding cases exists; the best available evidence suggests that the average density is very near ρ_c and that q_0 is very near $\frac{1}{2}$.

- That the universe is nearly flat and that the 3-K microwave is almost perfectly isotropic may be explained as the result of a brief period of very rapid expansion (the inflationary epoch) occurring from 10^{-35} to 10^{-24} sec after the Big Bang; during this tiny fraction of a second, the universe expanded to a size some 10^{25} times larger than it would have reached through its normal expansion rate.

  During the inflationary period, much of the material originally near our location moved far beyond the limit of our observable universe; the observable universe thus is today expanding into space containing matter and radiation that was in close contact with our matter and radiation during the first instant after the Big Bang.

- Four basic forces explain all of the interactions observed in the universe: gravity, electromagnetism, the strong nuclear force, and the weak nuclear force.

  Grand unified theories (GUTs) are attempts to explain two or more forces in terms of a single consistent set of physical laws; a super-GUT would explain all four forces.

  GUTs suggest that all four forces were equivalent just after the Big Bang; because we have no satisfactory superGUT, we can say nothing about the nature of the universe during this period before the Planck time.

At the Planck time ($t = 10^{-43}$ sec after the Big Bang), gravity "froze out" to become a distinctive force in a spontaneous symmetry breaking; at $t = 10^{-35}$ sec, there was a second spontaneous symmetry breaking in which the strong nuclear force became a distinctive force.

At $t = 10^{-12}$ sec, the final spontaneous symmetry breaking separated the electromagnetic force from the weak nuclear force; from that moment on, the universe behaved according to the physical laws with which we are familiar today.

At $t = 10^{-6}$ sec, confinement of quarks occurred, permitting the quarks to combine into protons and neutrons.

At $t = 2$ sec, the universe became transparent to neutrinos; all of the original helium had been produced by $t = 3$ sec; the universe became transparent to photons at $t = 10^6$ yr.

. The formation of clusters and superclusters of galaxies can be explained as a result of tiny fluctuations in the neutrino density that existed at $t = 10^6$ yr; the universe today contains regions of high galaxy density, connected by filaments and sheets of high density, surrounding voids of very low galaxy density.

The explanations of the present distribution of galaxies and clusters in the universe are based upon the as-yet-unproven assumption that the neutrino has a very small mass.

. If the universe is closed (bounded), then the universe will collapse in a Big Crunch during which the events following the Big Bang will recur in reverse sequence.

. If the universe is open (unbounded) or marginally bounded (flat), then matter will eventually fall into galatic and supergalactic black holes and protons may decay, leaving behind a still-expanding universe containing only a very low density of widely separated electrons and positrons (antielectrons).

In the infinite time that an open or flat universe will exist, black holes will eventually evaporate, pouring matter and energy randomly back into the universe.

Review questions

1 Suppose that the universe will expand forever. What will eventually become of the microwave background radiation?

2 Describe an example of each of the four basic interactions in the physical universe. Do you think it is possible that a fifth force might someday be discovered? Explain your answer.

3 Why are neutrinos important in modern cosmology?

4 What is the observational evidence for **(a)** the Big Bang, **(b)** the inflationary epoch, **(c)** the era or recombination, **(d)** the confinement of quarks?

5 The curvature of the universe can magnify images (recall Figure 29-4). How might this affect our distance determinations to remote galaxies?

6 Under what circumstances does a black hole behave like a white hole?

7 Various GUTs predict that the proton is unstable. What would it be like to live at a time when protons were decaying in large numbers?

Advanced questions

8 With a diagram show how you would expect the observed number of galaxies to depend on distance from Earth for various values of q_0. Discuss some of the problems of using such a diagram to determine q_0.

*9 Suppose that the average density of neutrinos throughout space is 100 neutrinos per cubic centimeter and that these neutrinos are responsible for giving the universe a density equal to the critical density. What range of neutrino masses corresponds to the often-quoted range of the Hubble constant, namely (15–30 km/sec) /10^6 ly? Compare your answer with 20–40 eV range suggested by the 1980 data of V. Lubimov and E. Tretyakov. Assuming the Soviet data to be correct, do your calculations suggest any limitations on the Hubble constant?

Discussion questions

10 Suppose that numerous primordial black holes were produced during the Big Bang. What kinds of observations might you perform to search for these black holes? Is it absurd to suggest that one of these black holes might be relatively nearby (perhaps in our solar system!) and have escaped being discovered by astronomers? How might you use data from a gamma ray survey of the sky to place an upper limit on the number of primordial black holes created during the Big Bang?

11 Do you think that there can be "other universes," regions of space and time that are not connected to our universe? Should astronomers be concerned with such possibilities? Why or why not?

For further reading

Chicarini, G., and Rood, H. "The Cosmic Tapestry." *Sky & Telescope*, May 1980, p. 364.

Davies, P. "The Anthropic Principle and the Early Universe." *Mercury*, May/June 1981, p. 66.

Dicus, D., et al. "The Future of the Universe." *Scientific American*, Mar. 1983.

Gregory, S., and Thompson, L. "Superclusters and Voids in the Distribution of Galaxies." *Scientific American*, Mar. 1982.

Guth, A., and Steinhardt, P. "The Inflationary Universe." *Scientific American*, May 1984.

Hawking, S. "The Quantum Mechanics of Black Holes." *Scientific American*, Jan. 1977.

Islam, J. "The Ultimate Fate of the Universe." *Sky & Telescope*, Jan. 1979, p. 13.

Kaufmann, W. *Black Holes and Warped Spacetime*. W. H. Freeman, 1979.

Overbye, D. "The Universe According to Guth." *Discover*, June 1983. p. 92.

Schechter, B. "The Prodigal Particle: The Neutrino." *Discover*, Mar. 1981, p. 20.

Shu, F. "The Expanding Universe and the Large-Scale Geometry of Space-Time." *Mercury*, Nov./Dec. 1983, p. 162.

Silk, J., et al. "The Large-Scale Structure of the Universe." *Scientific American*, Oct. 1983.

Wagoner, R., and Goldsmith, D. "Quarks, Leptons, and Bosons." *Mercury*, July/Aug. 1983, p. 98.

S. W. Hawking

The edge of spacetime

Stephen Hawking is a theoretical physicist at the University of Cambridge. He was born in Oxford in 1942 and took his B.A. at Oxford in 1962 and went on to take his doctorate working on gravity and cosmology under Dr. Dennis W. Sciama at Cambridge. He continued his work on these subjects at Cambridge and, during 1974 and 1975, at California Institute of Technology. A Fellow of the Royal Society since 1974, he has received many honors and awards including the Eddington Medal of the Royal Astronomical Society, The Dannie Heinemann Prize of the American Institute of Physics and the American Physical Society, the Maxwell Medal and Prize of the Institute of Physics, and the Einstein Medal given in Berne, Switzerland in 1979. Also in 1979 he was elected to Newton's old chair, Lucasian Professor at Cambridge.

His contributions have been in the areas of general relativity, gravitation, and quantum theory as it relates to blackholes and their thermodynamics. Though afflicted by a progressive nervous disease since 1961, which has confined him to a wheelchair for a decade, he continues to be phenomenally productive as both a writer and a researcher.

From the dawn of civilization people have asked questions like: "Did the universe have a beginning in time? Will it have an end? Is the universe bounded or infinite in spatial extent?" In this essay I shall outline some of the answers to these questions which are suggested by modern developments in science. Most of what I describe is now fairly generally accepted though some of it was controversial. My final conclusion, however, is based on some very recent work on which there has not yet been time to reach a consensus.

In most of the early mythological or religious accounts the universe, or at least its human inhabitants, was created by a Divine Being at some date in the fairly recent past like 4004 B.C.. Indeed the necessity of a "First Cause" to account for the creation of the universe was used as an argument to prove the existence of God. The Greek philosophers like Plato and Aristotle, on the other hand, did not like the thought of such direct Divine intervention in the affairs of the world and so mostly preferred to believe that the universe had existed and would exist forever. Most people in the ancient world believed that the universe was spatially bounded. In the earliest cosmologies the world was a flat plate with the sky as a pudding basin overhead. The Greeks, however, realized that the world was round. They constructed an elaborate model in which the Earth was a sphere surrounded by a number of spheres that carried the Sun, Moon, and the planets. The outermost sphere carried the so-called fixed stars which maintain the same relative positions but which appear to rotate across the sky.

This model with the Earth at the center was adopted by the Christian Church. It had the great attraction that it left plenty of room outside the sphere of the stars for Heaven and Hell, though quite how these were situated was never clear. It remained in favor unitl the seventeenth century when the observations of Galileo showed that this model of the universe had to be replaced by the Copernican model in which the Earth and the other planets orbited around the Sun. Not only did this get rid of the spheres but it also showed that the "fixed stars" must be at a very great distance because they did not show any apparent movement as the Earth went round the Sun, apart from that caused by the rotation of the Earth about its own axis. Having realized this and having abandoned the belief that the Earth was at the center of the universe, it was fairly natural to postulate that the stars were other Suns like our own and that they were distributed roughly uniformly throughout an infinite universe. This, however, raised a problem: according to Newton's theory of gravity, published in 1687, each star would be attracted toward every other star in the universe. Why, then, did not the stars all fall together to a single point? Newton himself tried to argue that this would indeed happen for a bounded collection of stars but that if one had an infinite universe, the gravitational force on a star caused by the attraction of the stars on one side of it would be balanced by the force arising from the stars on the other side. The net force on any star would therefore be zero and so the stars could remain motionless. This argument is in fact an example of the fallacies one can fall into when one adds up an infinite number of quantities: by adding them up in different orders one can get different results. We now know that an infinite distribution of stars cannot remain motionless if they are all attracting each other; they will start to fall toward each other. The only way that one can have a static infinite universe is if the force of gravity becomes repulsive at large distances. Even then, the universe is unstable because if the stars get slightly nearer each other, the attraction wins out over the repulsion and the stars fall together. On the other hand, if they get slightly farther away from each other, the repulsion wins and they move away from each other.

Despite these and other difficulties, nearly everyone in the eighteenth and nineteenth centuries believed that the universe was essentially unchanging in time. For such a universe the question of whether it had a beginning was metaphysical: one could equally well believe that it had existed forever or that it had been created in its present form a finite time ago. The belief in a static universe still persisted in 1915 when Einstein formulated his general theory of relativity which modified Newton's theory of gravity to make it compatible with discoveries about the propagation of light. He therefore added a so-called cosmological constant, which produced a repulsive force between particles at a great distance. This repulsive force could balance the normal gravitational attraction and allow a static uniform solution for the universe. This solution was unstable but it had the interesting property that in it space was finite but unbounded, just as the surface of the Earth is finite in area but does not have any boundary or edge. Time in this solution, however, could be infinite.

Einstein's static model of the universe was one of the great missed opportunities of theoretical physics: if he had stuck to his original version of general relativity without the cosmological constant he could have predicted that the universe ought to be either expanding or collapsing. As it hap-

pened, however, it was not realized that the universe was changing with time until astronomers like Slipher and Hubble began to observe the light from other galaxies. Visible light is made up of waves, like radio waves only with a much shorter wavelength or distance between wave crests. If one passes the light through a triangualr shaped piece of glass called a prism it is decomposed into its constituent wavelengths—colors like a rainbow. Slipher and Hubble found the same characteristic patterns of wavelengths or colors as for the light from stars in our own galaxy but the patterns were all shifted toward the red or longer wavelength end of the rainbow or spectrum. The only reasonable explanation of this was that the galaxies were moving away from us. In this case the distance between the wave crests would be crowded up and the wavelength would be reduced. This effect, known as the Doppler shift, is used by the police to measure the speed of cars.

During the 1920s Hubble observed the remarkable fact that the red shift was greater the further the galaxy was from our own. This meant that other galaxies were moving away from us at rates that were roughly proportional to their distance from us. The universe was not static as had been previously thought but was expanding. The rate of expansion is very low: it will take something like twenty thousand million years for the separation of two galaxies to double but it completely changes the nature of the discussion about whether the universe has a beginning or an end. This is not just a metaphysical question as in the case of a static universe: as I shall describe, there may be a very real physical beginning or end to the universe.

The first model of an expanding universe that was consistent with Einstein's general theory of relativity and Hubble's observations of red shifts was proposed by the Russian physicist and mathematician, Alexander Friedmann, in 1922. However, it received very little attention until similar models were discovered by other people toward the end of the 1920s. The Friedmann model and its later generalizations assumed that the universe was the same at every point in space and in every direction. This is obviously not a good approximation in our immediate neighborhood; there are local irregularities like the Earth and the Sun and there are many more visible stars in the direction of the center of our galaxy than in other directions. However, if we look at distant galaxies, we find that they are distributed roughly uniformly throughout the universe, the same in every direction. Thus it does seem to be a good approximation on a large scale. Even better evidence comes from observations of the background of microwave radiation that was discovered in 1965 by two scientists at the Bell Telephone Laboratories. The universe is very transparent to radio waves of a few inches wavelength so this radiation must have traveled to us from very great distances. Any large-scale irregularities in the universe would cause the radiation reaching us from different directions to have different intensities. Yet the observed intensity is the same in every direction to a very high degree of accuracy.

There are three kinds of generalized Friedmann models. In one of them, the galaxies are moving apart sufficiently slowly that the gravitational attraction between them will eventually stop them moving apart and start them approaching each other. The universe will expand to a maximum size and then re-collapse. In the second model, the galaxies are moving apart so fast that gravity can never stop them and the universe expands forever. Finally, there is a third model in which the galaxies are moving apart at just the critical rate to avoid re-collapse. In principle we could determine which model corresponds to our universe by comparing the present rate of expansion with the present average mass density. The mass of the matter in the universe that we can observe directly is not enough to stop the expansion. However, we have indirect evidence that there is more mass that we cannot see. Whether this "invisible" mass could be enough to stop the expansion eventually remains an open question.

In the Friedmann model which re-collapses eventually, space is finite but unbounded, like in the Einstein static model. In the other two Friedmann models, which expand forever, space is infinite. Time, on the other hand, has a boundary or edge. In all the models the expansion starts from a state of infinite density called the Big Bang singularity. In the model that re-collapses there is another singularity called the Big Crunch at the end of the re-collapse. Singulariites are places where the curvature of spacetime is infinite and the concepts of space and time cease to have any meaning. Scientific theories are formulated on a spacetime background so they will all break down at a singularity. If there were events before the Big Bang, they would not enable one to predict the present state of the universe because predictability would break down at the Big Bang. Similarly, there is no way that one can determine what happened before the Big Bang from a knowledge of events after the Big Bang. This means that the existence or non-existence of events before the Big Bang is purely metaphysical; they have no consequences for the present state of the universe. One might as well apply the Principle of Economy, known as Occam's Razor, to cut them out of the theory and say that time began at the Big Bang. Similarly, there is no way that we can predict or influence any events after the Big Crunch, so one might as well regard it as the end of time. This beginning and possible end of time that are predicted by the Friedmann solutions are very different from earlier ideas. Prior to the Friedmann solutions, the beginning or end of time was something that had to be imposed from outside the universe; there was no necessity for a beginning or an end. In the Friedmann models, on the other hand, the beginning and end of time occur for dynamical reasons. One could still imagine the universe being created by an external agent in a state corresponding to some time after the Big Bang but it would not have any meaning to say that it was created *before* the Big Bang. From the present rate of expansion of the universe we can estimate that the Big Bang should have occurred between ten and twenty thousand million years ago.

Many people disliked the idea that time had a beginning or an end because it smacked of Divine Intervention. There were therefore a number of attempts to avoid this conclusion. One of these was the "steady state" model of the universe proposed in 1948 by Herman Bondi, Thomas Gold, and Fred Hoyle. In this model it was proposed that, as the galaxies moved further away from each other, new galaxies were formed in-between out of matter that was being "constantly created." The universe would therefore look more or less the same at all times and the density would be roughly constant. This model had the great virtue that it made definite predictions that could be tested by observations. Unfortunately, observations of radio sources by Martin Ryle and his collaborators at Cambridge in the 1950s and early 1960s showed that the number of radio sources must have been greater in the past, contradicting the steady state model. The final nail in the coffin of the steady state theory was the discovery of microwave background radiation in 1965. There was no way this radiation could be accounted for in the model.

Another attempt to avoid a beginning of time was the suggestion that maybe the singularity was simply a consequence of the high degree of symmetry of the Friedmann solutions. This restricted the relative motion of any two galaxies to be along the line joining them. It would therefore not be surprising if they all collided with each other at some time. However, in the real universe, the galaxies would also have some random velocities perpendicular to the line joining them. These transverse velocities might be expected to cause the galaxies to miss each other and to allow the universe to pass from a contracting phase to an expanding one without the density ever becoming infinite. Indeed, in 1963 two Russian scientists claimed that this would happen in nearly every solution of the equations of general relativity. They based this claim on the fact that all the solutions with a singularity that they constructed had to satisfy some constraint or symmetry. They later realized, however, that there was a more general class of solutions with singularities that did not have to obey any constraint or symmetry.

This showed that singularities *could* occur in general solutions of general relativity but it did not answer the question of whether they necessarily *would* occur. However, between 1965 and 1970 a number of theorems were proved which showed that any model of the universe which obeyed general relativity, satisfied one or two other reasonable assumptions and contained as much matter as we observe in the universe, must have a Big Bang singularity. The same theorems predict that there will be a singularity that will be an end of time if the whole universe re-collapses. Even if the universe is expanding too fast to collapse in its entirety, we nevertheless expect some localized regions, such as massive burnt-out stars, to collapse and form black holes. The theorems predict that the black holes will contain singularities which will be an end of time for anyone unfortunate or foolhardy enough to fall in.

Einstein's general theory of relativity is probably one of the two greatest intellectual achievements of the twentieth century. It is however incomplete because it is what is called a classical theory; that is, it does not incorporate the uncertainty principle of the other great discovery of this century, quantum mechanics. The uncertainty principle states that certain pairs of quantities, such as the position and velocity of a particle, cannot be predicted simultaneously with an arbitrary high degree of accuracy. The more accurately one predicts the position of the particle, the less accurately one will be able to predict its velocity and vice versa. Quantum Mechanics was developed in the early years of this century to describe the behavior of very small systems such as atoms or individual elementary particles. In particular, there was a problem with the structure of the atom, which was supposed to consist of a number of electrically charged particles called electrons orbit-

ing around a central nucleus, like the planets around the Sun. The previous classical theory predicted that the electron would radiate light waves because of their motion. The waves would carry away energy and so would cause the electrons to spiral inwards until they collided with the nucleus. However, such behavior is not allowed by quantum mechanics because it would violate the uncertainty principle: if an electron were to sit on the nucleus, it would have both a definite position and a definite velocity. Instead, quantum mechanics predicts that the electron does not have a definite position but that the probability of finding it is spread out over some region around the nucleus with the probability density remaining finite even at the nucleus.

The prediction of classical theory of an infinite probability density of finding the electron at the nucleus is rather similar to the prediction of classical general relativity that there should be a Big Bang singularity of infinite density. Thus one might hope that if one were able to combine general relativity and quantum mechanics into a theory of quantum gravity, one would find that the singularities of gravitational collapse or expansion were smeared out like in the case of the collapse of the atom. The first indication that this might be the case came with the discovery that black holes, formed by the collapse of localized regions such as stars, were not completely black if one took into account the uncertainty principle of quantum mechanics. Instead, a black hole would radiate particles and radiation like a hot body with a temperature higher than the smaller mass of the black hole. The radiation would carry away energy and so would reduce the mass of the black hole. This in turn would increase the rate of emissions. Eventually, it seems that the black hole will disappear completely in a tremendous burst of emission. All the matter that collapsed to form the black hole and any astronaut who was unlucky enough to fall into the black hole would disappear, at least from our region of the universe. However, the energy that corresponded to his mass by Einstein's famous equation $E = mc^2$ would be emitted by the black hole in the form of radiation. Thus the astronaut's mass-energy would be recycled to the universe. However, this would be rather a poor sort of immortality as the astronaut's subjective concept of time would almost certainly come to an end and the particles out of which he was composed would not in general be the same as the particles that were re-emitted by the black hole. Still, black hole evaporation did indicate that gravitational collapse might not lead to a complete end of time.

The real problem with spacetime having an edge or boundary at a singularity is that the laws of science do not determine the initial state of the universe at the singularity but only how it evolves thereafter. This problem would remain even if there were no singularity and time continued back indefinitely: the laws of science would not fix what the state of the universe was in the infinite past. In order to pick out one particular state for the universe from among the set of all possible states that are allowed by the laws, one has to supplement the laws by boundary conditions that say what the state of the universe was at an initial singularity or in the infinite past. Many scientists are embarrassed at talking about the boundary conditions of the universe because they feel that it verges on metaphysics or religion. After all, they might say, the universe could have started off in a completely arbitrary state. That may be so, but in that case it could also have evolved in a completely arbitrary manner. Yet all the evidence that we have suggests it evolves in a well-determined way according to certain laws. It is therefore not unreasonable to suppose that there may also be simple laws that govern the boundary conditions and determine the state of the universe.

In the classical general theory of relativity, which does not incorporate the uncertainty principle, the initial state of the universe is a point of infinite density. It is very difficult to define what the boundary conditions of the universe should be at such a singularity. However, when quantum mechanics is taken in account, there is the possibility that the singularity may be smeared out and that space and time together may form a closed four-dimensional surface without boundary or edge, like the surface of the Earth but with two extra dimensions. This would mean that the universe was completely self-contained and did not require boundary conditions. One would not have to specify the state in the infinite past and there would not be any singularities at which the laws of physics would break down. One could say that the boundary conditions of the universe are that it has no boundary.

It should be emphasized that this is simply a *proposal* for the boundary conditions of the universe. One cannot deduce them from some other principle but one can merely pick a reasonable set of boundary conditions, calculate what they predict for the present state of the universe, and see if they agree with observations. The calculations are very difficult and have been carried out so far only in simple with a high degree of symmetry. However, the results are very encouraging. They predict that the universe must have started out in a fairly smooth and uniform state. It would have undergone a period of what is called exponential or "inflationary" expansion during which its size would have increased by a very large factor but the density would have remained the same. The universe would then have become very hot and would have expanded to the state that we see it today, cooling as it expanded. It would be uniform and the same in every direction on very large scales but would contain local irregularities that would develop into stars and galaxies.

What happened at the beginning of the expansion of the universe? Did spacetime have an edge at the Big Bang? The answer is that if the boundary conditions of the universe are that it has no boundary, time ceases to be well-defined in the very early universe just as the direction "North" ceases to be well-defined at the North Pole of the Earth. Asking what happens before the Big Bang is like asking for a point one mile North of the North Pole. The quantity that we measure as time had a beginning but that does not mean spacetime has an edge, just as the surface of the Earth does not have an edge at the North Pole, or at least, so I am told: I have not been there myself.

If spacetime is indeed finite but without boundary or edge, this would have important philosophical implications. It would mean that we could describe the universe by a mathematical model which was determined completely by the laws of science alone; they would not have to be supplemented by boundary conditions. We do not yet know the precise form of the laws: at the moment we have a number of partial laws which govern the behavior of the universe under all but the most extreme conditions. However, it seems likely that these laws are all part of some unified theory that we have yet to discover. We are making progress and there is a reasonable chance that we will discover it by the end of the century. At first sight it might appear that this would enable us to predict everything in the universe. However our powers of prediction would be severely limited, first by the uncertainty principle that states that certain quantities cannot be exactly predicted but only their probability distribution, and secondly, and even more importantly, by the complexity of the equations which makes them impossible to solve in any but very simple situations. Thus we would still be a long way from Omniscience.

Appendixes

1 The planets: Orbital data

Planet	Semimajor axis		Sidereal period		Synodic period (days)	Mean orbital speed (km/sec)	Orbital eccentricity	Inclination of orbit to ecliptic
	(AU)	(10⁶ km)	(tropical years)	(days)				
Mercury	0.3871	57.9	0.2408	87.97	115.88	47.9	0.206	7°.00
Venus	0.7233	108.2	0.0615	224.70	583.96	35.0	0.007	3.39
Earth	1.0000	149.6	1.0000	365.26	—	29.8	0.017	0.0
Mars	1.5237	227.9	1.8809	686.98	779.87	24.1	0.093	1.85
(Ceres)	2.7671	414	4.603		466.6	17.9	0.077	10.6
Jupiter	5.2028	778	11.86		399	13.1	0.048	1.31
Saturn	9.588	1427	29.46		378	9.6	0.056	2.49
Uranus	19.191	2871	84.07		370	6.8	0.046	0.77
Neptune	30.061	4497	164.82		367	5.4	0.010	1.77
Pluto	39.529	5913	248.6		367	4.7	0.248	17.15

The terrestrial worlds
This montage of photographs taken by various spacecraft shows the terrestrial planets and six large moons at the same scale. (Prepared for NASA by Stephen P. Meszaros)

2 The planets: Physical data

Planet	Diameter (km)	Diameter (Earth = 1)	Mass (Earth = 1)	Mean density (g/cm³)	Rotation period (days)	Inclination of equator to orbit	Surface gravity (Earth = 1)	Albedo	Brightest visual magnitude	Escape velocity (km/sec)
Mercury	4878	0.38	0.055	5.43	58.6	0°.0	0.38	0.106	−1.9	4.3
Venus	12,104	0.95	0.82	5.24	−243.0	177.4	0.91	0.65	−4.4	10.4
Earth	12,756	1.00	1.00	5.52	0.997	23.4	1.00	0.37	—	11.2
Mars	6794	0.53	0.107	3.9	1.026	25.2	0.38	0.15	−2.0	5.0
Jupiter	142,796	11.2	317.8	1.3	0.41	3.1	2.53	0.52	−2.7	60
Saturn	120,000	9.41	94.3	0.7	0.43	26.7	1.07	0.47	+0.7	36
Uranus	50,800	3.98	14.6	1.3	−0.65	97.9	0.92	0.50	+5.5	21
Neptune	50,450	3.81	17.2	1.5	0.77	29	1.18	0.5	+7.8	24
Pluto	3400	0.27	0.0023	0.5 (?)	6.387	90	0.03	0.5	+15.1	1

3 Satellites of the planets

Planet	Satellite	Discovered by	Mean distance from planet (km)	Sidereal period (days)	Orbital eccentricity	Diameter of satellite* (km)	Approximate magnitude at opposition
Earth	Moon	—	384,404	27.322	0.055	3476	−12.5
Mars	Phobos	A. Hall (1877)	9,380	0.319	0.021	25	+12
	Diemos	A. Hall (1877)	23,500	1.262	0.003	13	13
Jupiter							
	Almalthea	Barnard (1892)	181,300	0.498	0.003	240	13
	Io	Galileo (1610)	421,600	1.769	0.000	3640	5
	Europa	Galileo (1610)	670,900	3.551	0.000	3130	6
	Ganymede	Galileo (1610)	1,070,000	7.155	0.002	5270	5
	Callisto	Galileo (1610)	1,880,000	16.689	0.008	4840	6
	Himalia	Perrine (1904)	11,470,000	250.57	0.158	(170)	14
	Elara	Perrine (1905)	11,800,000	259.65	0.207	(40)	18
	Lysithea	Nicholson (1938)	11,850,000	263.55	0.130	(10)	19
	Leda	Kowal (1974)	11,110,000	239.2	0.147	(8)	20
	Aranke	Nicholson (1951)	21,200,000	631.1	0.169	(10)	18
	Carme	Nicholson (1938)	22,600,000	692.5	0.207	(15)	19
	Pasiphae	Melotte (1908)	23,500,000	738.9	0.378	(25)	17
	Sinope	Nicholson (1914)	23,700,000	758	0.275	(15)	18
Saturn	Mimas	W. Herschel (1789)	185,500	0.942	0.020	390	13
	Enceladus	W. Herschel (1789)	237,900	1.370	0.004	500	12
	Tethys	Casini (1684)	294,700	1.888	0.000	1050	10
	Dione	Cassini (1684)	377,400	2.737	0.002	1120	10
	Rhea	Cassini (1672)	526,700	4.518	0.001	1530	10
	Titan	Huygens (1655)	1,222,000	15.945	0.029	5120	8
	Hyperion	Bond (1848)	1,481,000	21.277	0.104	310	14
	Iapetus	Cassini (1671)	3,560,000	79.331	0.028	1440	11
	Phoebe	W. Pickering (1898)	12,930,000	550.45	0.163	40	16
Uranus	Miranda	Kuiper (1948)	123,000	1.414	0	(200)	17
	Ariel	Lassell (1851)	191,700	2.520	0.003	(600)	14
	Umbriel	Lassell (1851)	267,000	4.144	0.004	(400)	15
	Titania	W. Herschel (1787)	438,000	8.706	0.002	(1000)	14
	Oberon	W. Herschel (1787)	585,960	13.463	0.001	(900)	14

(continued)

3 Satellites of the planets (continued)

Planet	Satellite	Discovered by	Mean distance from planet (km)	Sidereal period (days)	Orbital eccentricity	Diameter of satellite* (km)	Approximate magnitude at opposition
Neptune	Triton	Lassell (1846)	353,400	5.877	0.000	6000	13
	Nereid	Kuiper (1949)	5,560,000	359.881	0.749	(500)	19
Pluto	Charon	Christy (1978)	17,000	6.387	0	(1200)	17

NOTE: *This table does not include several very small satellites about Jupiter and Saturn that were discovered by Voyagers 1 and 2.*
A diameter of a satellite given in parentheses is estimated from the amount of sunlight it reflects.

4 The nearest stars

Name	Parallax (arc sec)	Distance (ly)	Spectral type	Radial velocity (km/sec)	Proper motion (arc sec/yr)	Apparent visual magnitude	Luminosity (Sun = 1.0)
Sun			G2 V			−26.7	1.0
α Cen A	0.750	4.3	G2 V	−22	3.68	−0.01	1.6
B			K0 V			1.3	0.45
C	0.772	4.2	M5e			11.0	0.00006
Barnard's star	0.552	5.9	M5 V	−108	10.30	9.5	0.00045
Wolf 359	0.431	7.6	M8e	+13	4.84	13.5	0.00002
Lalande 21185	0.402	8.1	M2 V	−84	4.78	7.5	0.0055
Luyten 726-8A	0.1387	8.4	M6e	+30	3.35	12.5	0.00006
B(UV Ceti)			M6e			13.0	0.00004
Sirius A	0.377	8.6	A1 V	−8	1.32	−1.5	23.5
B			wd			8.7	0.003
Ross 154	0.345	9.4	M5e	−4	0.74	10.6	0.00048
Ross 248	0.314	10.3	M6e	−81	1.82	12.3	0.00011
ε Eri	0.303	10.7	K2 V	+16	0.97	3.7	0.30
Luyten 789-6	0.302	10.8	M7e	−60	3.27	12.2	0.00014
Ross 128	0.301	10.8	M5	−13	1.40	11.1	0.00036
61 Cyg A	0.292	11.2	K5 V	−64	5.22	5.2	0.083
B			K7 V			6.0	0.040
ε Ind	0.291	11.2	K5 V	−40	4.67	4.7	0.13
Procyon A	0.287	11.4	F5 IV–V	−3	1.25	0.4	7.65
B			wd			10.7	0.00055
Σ 2398 A	0.284	11.5	M3.5 V		2.29	8.9	0.0028
B			M4 V			9.7	0.0013
Groombridge 34 A	0.282	11.6	M1 V		2.91	8.1	0.0058
B			M6 V			11.0	0.00040
Lacaille 9352	0.279	11.7	M2 V		6.87	7.4	0.013
τ Ceti	0.273	11.9	G8 V	−16	1.92	3.5	0.45
BD + 5° 1668	0.266	12.2	M5	+26	3.73	9.8	0.0015
L725-32 (YZ Ceti)	0.262	12.4	M5e		1.31	11.6	0.0002
Lacaille 8760	0.260	12.5	M1		3.46	6.7	0.028
Kapteyn's star	0.256	12.7	M0 V	+245	8.79	8.8	0.0040
Kruger 60 A	0.254	12.8	M4	−26	0.87	9.8	0.0017
B			M5e			11.3	0.00044

5 *The brightest stars*

Star	Name	Apparent visual magnitude	Spectral type	Absolute magnitude	Distance (ly)	Radial velocity (km/sec)	Proper motion (arc sec/yr)
α CMa A	Sirius	−1.46	A1 V	+1.42	8.7	−8	1.324
α Car	Canopus	−0.72	F0 I–II	−3.1	98	+21	0.025
α Boo	Arcturus	−0.06	K2 III	−0.3	36	−5	2.284
α Cen A	Rigil Kentaurus	0.01	G2 V	+4.39	4.3	−25	3.676
α Lyr	Vega	0.04	A0 V	+0.5	26.5	−14	0.345
α Aur	Capella	0.05	G8 III (?)	−0.6	45	+30	0.435
β Ori A	Rigel	0.14	B8 Ia	−7.1	900	+21	0.001
α CMi A	Procyon	0.37	F5 IV–V	+2.7	11.3	−3	1.250
α Ori	Betelgeuse	0.41	M2 Iab	−5.6	520	+21	0.028
α Eri	Achernar	0.51	B3 V	−2.3	118	+19	0.098
β Cen AB	Hadar	0.63	B1 III	−5.2	490	−12	0.035
α Aql	Altair	0.77	A7 IV–V	+2.2	16.5	−26	0.658
α Tau A	Aldebaran	0.86	K5 III	−0.7	68	+54	0.202
α Vir	Spica	0.91	B1 V	−3.3	220	+1	0.054
α Sco A	Antares	0.92	M1 Ib	−5.1	520	−3	0.029
α PsA	Fomalhaut	1.15	A3 V	+2.0	22.6	+7	0.367
β Gem	Pollux	1.16	K0 III	+1.0	35	+3	0.625
α Cyg	Deneb	1.26	A2 Ia	−7.1	1600	−5	0.003
β Cru	Beta Crucis	1.28	B0.5 III	−4.6	490	+20	0.049
α Leo A	Regulus	1.36	B7 V	−0.7	87	+4	0.248

Glossary

absolute magnitude   The apparent magnitude a star would have at a distance of 10 pc.

absolute zero   A temperature of −273°C (or 0 K) where all molecular motion stops; the lowest possible temperature.

absorption spectrum   Dark lines superimposed on a continuous spectrum.

acceleration   A change in velocity.

accretion   The gradual accumulation of matter in one location, typically due to the action of gravity.

active galactic nucleus   The center of a galaxy that is emitting exceptionally large amounts of energy; the center of a Seyfert galaxy or a quasar.

active Sun   The Sun during times of frequent solar activity such as sunspots, flares, and associated phenomena.

albedo   The fraction of sunlight that a planet, asteroid, or satellite reflects.

alpha decay   The decay of a radioactive isotope by the emission of alpha particles.

alpha particle   The nucleus of a helium atom, consisting of two protons and two neutrons.

angstrom (Å)   A unit of length equal to 10^{-8} cm.

angular diameter   The angle subtended by the diameter of an object.

angular momentum   A measure of the momentum associated with rotation.

annihilation   The process by which the masses of a particle and antiparticle are converted into energy.

annular eclipse   An eclipse of the Sun in which the Moon is too distant to completely cover the Sun, so that a ring of sunlight is seen around the Moon at mid-eclipse.

antielectron   A positron.

antimatter   Matter consisting of antiparticles such as antiprotons, antielectrons (positrons), and antineutrons.

aperture   The diameter of an opening; the diameter of the primary lens or mirror of a telescope.

aphelion   The point in its orbit where a planet is farthest from the Sun.

apogee   The point in its orbit where a satellite or the Moon is farthest from the Earth.

Apollo asteroid   An asteroid whose orbit brings it closer to the Sun than the distance of the Earth's orbit.

apparent magnitude   A measure of the brightness of light from a star or other object as measured at Earth.

apparent solar day   The interval between two successive transits of the Sun's center across the local meridian.

apparent solar time   Time reckoned by the position of the Sun in the sky.

ascending node   The point along an orbit where an object crosses a reference plane (usually the ecliptic or the celestial equator) from south to north.

association   A loose cluster of stars whose spectra, motions, or positions suggest that they had a common origin.

asteroid   One of tens of thousands of small, rocky, planetlike objects in orbit about the Sun.

asthenosphere   A warm, plastic layer of the mantle beneath the lithosphere of the Earth.

astigmatism   An optical defect of a lens or mirror whereby light rays in different planes do not focus at the same point.

astrometry   The branch of astronomy dealing with the precise determination of the positions and motions of celestial objects.

astronomical unit (AU)   The semimajor axis of the Earth's orbit; the average distance between the Earth and the Sun.

astronomy   The branch of science dealing with objects and phenomena that lie beyond the Earth's atmosphere.

astrophysics   That part of astronomy dealing with the physics of astronomical objects and phenomena.

atom   The smallest particle of an element that has the properties characterizing that element.

atomic mass unit (amu)   A unit of mass (1.67×10^{-24} g) nearly equal to the mass of a hydrogen atom.

atomic number   The number of protons in the atom of a particular element.

atomic weight   The mass of an atom in atomic mass units.

aurora   Light radiated by atoms and ions in the Earth's upper atmosphere, mostly in the polar regions.

autumnal equinox   The intersection of the ecliptic and the celestial equator where the Sun crosses the equator from north to south.

Balmer lines   Emission or absorption lines in the hydrogen spectrum involving electron transitions between the second and higher energy levels.

bar   A unit of pressure.

barred spiral galaxy   A spiral galaxy in which the spiral arms begin from the ends of a "bar" running through the nucleus rather than from the nucleus itself.

Big Bang   An explosion of all space roughly 20 billion years ago from which the universe emerged.

binary star   Two stars revolving about each other.

black body   A hypothetical perfect radiator that absorbs and reemits all radiation falling upon it.

black hole An object whose gravity is so strong that the escape velocity exceeds the speed of light.

Bode's law A numerical sequence that gives the approximate distances of the planets from the Sun in astronomical units.

Bohr atom A model of the atom, described by Niels Bohr, in which electrons revolve about the nucleus in certain allowed circular orbits.

bolometric correction The difference between the visual and bolometric magnitudes of a star.

bolometric magnitude A measure of the brightness of a star or other object as detected by a device above the Earth's atmosphere and sensitive to all wavelengths of radiation.

breccia A rock formed by the sudden amalgamation of various rock fragments under pressure.

burster A nonperiodic X-ray source that emits powerful bursts of X rays.

Callisto One of the four Galilean satellites of Jupiter.

carbon cycle A series of nuclear reactions, involving carbon as a catalyst, by which hydrogen is transformed into helium.

Cassegrain focus An optical arrangement in a reflecting telescope in which light rays are reflected by a secondary mirror to a focus behind the primary mirror.

CCD Charge-coupled device; a type of solid-state silicon wafer designed for the detection of photons.

celestial equator A great circle on the celestial sphere 90° from the celestial poles.

celestial mechanics The branch of astronomy dealing with the motions and gravitational interactions of objects in the solar system.

celestial poles Points about which the celestial sphere appears to rotate.

celestial sphere A sphere of very large radius centered on the observer; the apparent sphere of the sky.

center of mass That point in an isolated system that moves at a constant velocity in accordance with Newton's first law.

Cepheid variable One of two types (Type I and Type II) of yellow, supergiant, pulsating stars.

Ceres The largest asteroid and the first to be discovered.

Chandrasekhar limit The maximum mass of a white dwarf.

chromatic aberration An optical defect whereby different colors of light passing through a lens are focused at different locations.

chromosphere A layer in the solar atmosphere between the photosphere and the corona.

cluster of galaxies A collection of galaxies containing a few to several thousand member galaxies.

color index The difference between the magnitudes of a star measured in two different spectral regions.

color–magnitude diagram A plot of the magnitudes of stars in a cluster against their color indices.

coma (of a comet) The diffuse gaseous component of the head of a comet.

comet A small body of ice and dust in orbit about the Sun. While passing near the Sun, a comet's vaporized ices give rise to a coma and tail.

conduction The transfer of heat by directly passing energy from atom to atom.

conic section The curve of intersection between a circular cone and a plane; this curve can be a circle, ellipse, parabola, or hyperbola.

conservation of angular momentum A law of physics stating that the total amount of angular momentum in an isolated system remains constant.

constellation A configuration of stars, often named after an object, person, god, or animal.

continental drift The gradual movement of the continents over the surface of the Earth due to plate tectonics.

continuous spectrum A spectrum of light over a range of wavelengths without any spectral lines.

convection The transfer of energy by moving currents of fluid or gas containing that energy.

Coriolis effect The deflection of moving objects on a rotating surface.

corona The Sun's outer atmosphere.

cosmic rays Atomic nuclei (mostly protons) that strike the Earth with extremely high speeds.

cosmological model A specific theory about the structure and evolution of the universe.

cosmology The study of the structure and evolution of the universe.

coudé focus A reflecting telescope in which a series of mirrors direct light to a remote focus away from the moving parts of the telescope.

crater A circular depression on a planet or satellite, caused by the impact of a meteoroid.

crescent moon One of the phases of the Moon in which less than one-half of its illuminated surface is visible from the Earth.

critical density The average density throughout the universe at which space is flat and galaxies just barely continue receding from each other infinitely far into the future.

cyclonic motion Circular wind motion (counterclockwise in the Earth's northern hemisphere) in a planet's atmosphere due to the Coriolis effect.

dark nebula A cloud of interstellar gas and dust that obscures the light of more distant stars.

daughter isotope An isotope that results from the radioactive decay of another isotope.

daylight saving time A time one hour more advanced than standard time, usually adopted during the summer months.

decay series A sequence of isotopes that radioactively decay one into another.

deceleration parameter (q_0) A quantity that characterizes the rate at which the expansion of the universe is slowing.

declination Angular distance of a celestial object north or south of the celestial equator.

deferent   A stationary circle in the Ptolemaic system along which another circle (an epicycle) moves, carrying a planet, the Sun, or Moon.

degenerate gas   A gas in which all the allowed states for particles (electrons or neutrons) have been filled, thereby causing the gas to behave differently than ordinary gases do.

density   The ratio of the mass of an object to its volume.

density-wave theory   An explanation of spiral arms in galaxies proposed by C. C. Lin and colleagues.

descending node   A point along an orbit where an object crosses a reference plane (usually the ecliptic or celestial equator) from north to south.

deuterium   An isotope of hydrogen whose nuclei each contain one proton and one neutron; heavy hydrogen.

differential rotation   The rotation of a nonrigid object in which parts adjacent to each other do not always stay close together.

differentiation (geological)   The separation of different kinds of material in different layers inside a planet.

diffraction   The spreading-out of light passing the edge of an opaque object.

diffraction grating   A system of closely spaced slits or reflecting strips (usually on a piece of glass) used to produce a spectrum.

diffuse nebula   A reflection or emission nebula consisting of interstellar gas and dust.

distance modulus   The difference between the apparent and absolute magnitudes of an object.

diurnal   Daily.

diurnal motion   Motion in one day.

Doppler effect   The apparent change in wavelength of radiation due to relative motion between the source and the observer along the line of sight.

double radio source   An extragalactic radio source characterized by two large regions of radio emission, typically located on either side of an active galaxy.

dyne   A unit of force; the force needed to accelerate 1 g by 1 cm/sec^2.

eccentricity (of an ellipse)   The value $\sqrt{1 - b^2/a^2}$, where a is the semimajor axis of the ellipse and b is its semiminor axis.

eclipse   The cutting off of part or all the light from one celestial object by another.

eclipse path   The track of the tip of the Moon's shadow along the Earth's surface during a total or annular solar eclipse.

eclipse season   A period during the year when a solar or lunar eclipse is possible.

eclipsing binary star   A binary system in which, as seen from Earth, the stars periodically pass in front of each other.

ecliptic   The apparent annual path of the Sun on the celestial sphere.

electromagnetic radiation   Radiation consisting of oscillating electric and magnetic fields including gamma rays, X rays, visible light, ultraviolet and infrared radiation, radio waves, and microwaves.

electromagnetic spectrum   The entire array or family of electromagnetic radiation.

electron   A negatively charged subatomic particle usually found in orbits about the nuclei of atoms.

electron volt   The energy acquired by an electron accelerated through an electric potential of one volt.

element   A substance that cannot be decomposed by chemical means into simpler substances.

ellipse   A conic section obtained by cutting completely through a circular cone with a plane.

elliptical galaxy   A galaxy with an elliptical shape and no conspicuous interstellar material.

elongation   The angular distance between a planet and the Sun as viewed from Earth.

emission line   A bright spectral line.

emission nebula   A glowing gaseous nebula whose light comes from fluorescence caused by a nearby star.

emission spectrum   A spectrum that contains emission lines.

energy   The ability to do work.

energy level (in an atom)   A particular amount of energy possessed by an atom above the atom's least-energetic state.

epicycle   A moving circle in the Ptolemaic system about which a planet revolves.

equation of state   An equation relating the temperature, pressure, and density of a gas.

equation of time   The difference between apparent and mean solar time.

equinox   One of the intersections of the ecliptic and the celestial equator.

erg   A unit of energy; the work done by a force of 1 dyne moving through a distance of 1 cm.

ergosphere   The region of space immediately outside the event horizon of a rotating black hole where it is impossible to remain at rest.

escape velocity   The speed needed by one object to achieve a parabolic orbit away from a second object and thereby permanently move away from the second object.

Europa   One of the Galilean satellites of Jupiter.

event horizon   The location around a black hole where the escape velocity equals the speed of light; the surface of a black hole.

excitation   The process of imparting energy to an atom or ion.

eyepiece   A magnifying lens used to view the image produced at the focus of a telescope.

extinction   The attenuation of light due to absorption by material between the source and the observer.

extragalactic   Beyond our galaxy.

filtergram   A photograph of the Sun taken through special filters that are transparent only to certain wavelengths.

flare   A sudden, temporary outburst of light from an extended region of the solar surface.

focal length   The distance from a lens or mirror to the point where converging light rays meet.

focus   The point where light rays converged by a lens or mirror meet.

force   That which can change the momentum of an object.

frequency   The number of wave crests or troughs that cross a given point per unit time; the number of vibrations per unit time.

full moon   A phase of the Moon during which its full daylight hemisphere can be seen from Earth.

galactic cannibalism   A collision between two galaxies of unequal mass and size in which the smaller galaxy seems to be absorbed into the larger galaxy.

galactic cluster   A loose association of young stars in the disk of our galaxy.

galactic equator   The intersection of the principal plane of the Milky Way with the celestial sphere.

galaxy   A large assemblage of stars.

Galilean satellite   Any one of the four large moons of Jupiter.

gamma rays   The most energetic form of electromagnetic radiation.

Ganymede   One of the Galilean satellites of Jupiter.

general theory of relativity   A description of gravity formulated by Albert Einstein, which explains that gravity affects the geometry of space and the flow of time.

geomagnetic   Referring to the Earth's magnetic field.

gibbous moon   A phase of the Moon in which more than one-half, but not all, of the Moon's daylight hemisphere is visible from Earth.

globular cluster   A large spherical cluster of stars, typically found in the outlying regions of a galaxy.

globule   A small, dense, dark nebula.

gluon   A particle that is exchanged between quarks.

granulation   The "rice-grain"–like structure of the solar photosphere.

gravitation   The tendency of matter to attract matter.

gravitational waves   Oscillations of space produced by changes in the distribution of matter.

greatest elongation   The largest possible angle between the Sun and an inferior planet.

greenhouse effect   The trapping of infrared radiation near a planet's surface by the planet's atmosphere.

Greenwich meridian   The meridian of longitude that passes through the old Royal Greenwich Observatory near London; the longitude of 0°.

H I region   A region of neutral hydrogen in interstellar space.

H II region   A region of ionized hydrogen in interstellar space.

half-life   The time required for one-half of the radioactive nuclei of an isotope to disintegrate.

Heisenberg uncertainty principle   A principle of quantum mechanics that places limits on the precision of simultaneous measurements.

helio-   A prefix referring to the Sun.

heliocentric   Centered on the Sun.

helium flash   The nearly explosive beginning of helium burning in the dense core of a red-giant star.

Hertzsprung–Russell (H–R) diagram   A plot of the absolute magnitude (or luminosity) of stars against their surface temperatures.

horizontal branch   A group of stars on the Hertzsprung–Russell diagram of a typical globular cluster, near the main sequence and having roughly constant absolute magnitude.

Hubble constant (H_0)   The constant of proportionality in the relation between the velocities of remote galaxies and their distances.

Hubble law   The empirical relationship stating that the redshifts of remote galaxies are directly proportional to their distances from Earth.

hydrostatic equilibrium   A balance between the weight of a layer in a star and the pressure that supports it.

hyperbola   A comic section formed by cutting a circular cone with a plane at an angle steeper than the sides of the cone.

igneous rock   A rock that formed from the solidification of molten lava or magma.

image   The optical representation of an object produced through the focusing of light rays by lenses or mirrors.

inertia   The property of matter that requires a force to act on it to change its state of motion.

inferior conjunction   The configuration when an inferior planet is between the Sun and Earth.

inferior planet   A planet that is closer to the Sun than the Earth is.

inflationary epoch   A brief period shortly after the Big Bang during which the scale of the universe increased very rapidly.

infrared radiation   Electromagnetic radiation of wavelength longer than visible light, yet shorter than radio waves.

interplanetary medium   The sparse distribution of gas and dust particles in interplanetary space.

interstellar dust   Microscopic solid grains of various compounds in interstellar space.

interstellar gas   Sparse gas in interstellar space.

interstellar medium   Interstellar gas and dust.

Io   One of the Galilean satellites of Jupiter.

ion   An atom that has become electrically charged due to the addition or loss of one or more electrons.

ionization   The process by which an atom loses electrons.

ionization potential   The energy required to remove an electron from an atom.

ionosphere   A layer in the Earth's upper atmosphere in which many of the atoms are ionized.

ion tail (of a comet) The relatively straight tail of a comet produced by the solar wind acting on ions.

irregular galaxy An unsymmetrical galaxy having neither spiral arms nor an elliptical shape.

isotope Any of several forms for the same chemical element whose nuclei all have the same number of protons but different numbers of neutrons.

isotropic The same in all directions.

Jovian planet Any of the four largest planets: Jupiter, Saturn, Uranus, or Neptune.

Kepler's laws Three statements, discovered by Johannes Kepler, that describe the motions of the planets.

kiloparsec (kpc) One thousand parsecs; about 3260 ly.

Kirkwood's gaps Gaps in the spacing of asteroid orbits discovered by Daniel Kirkwood.

Lagrangian points Five points in the orbital plane of two bodies revolving about each other in circular orbits where a third object of negligible mass can remain in equilibrium.

leap year A calendar year with 366 days.

libration A slight rocking of the Moon in its orbit whereby an Earth-based observer can, over time, see slightly more than one-half of the Moon's surface.

light Electromagnetic radiation that is visible to the eye.

light curve A graph that displays variations in the brightness of a star or other astronomical object.

light year (ly) The distance light travels in a vacuum in 1 yr.

limb (of Sun or Moon) The apparent edge of the Sun or Moon as seen in the sky.

limb darkening The phenomenon whereby the Sun is darker near its limb than near the center of its disk.

limiting magnitude The faintest magnitude that can be observed with a certain telescope under certain conditions.

line of nodes A line connecting the nodes of an orbit.

lithosphere The solid, upper layer of the Earth; essentially the Earth's crust.

local group The cluster of galaxies of which our galaxy is a member.

LMC The Large Magellanic Cloud.

luminosity The rate at which electromagnetic radiation is emitted from a star or other object.

luminosity class A classification of a star of a given spectral type according to its luminosity.

lunar Referring to the Moon.

lunar eclipse An eclipse of the Moon by the Earth; a passage of the Moon through the Earth's shadow.

Lyman lines A series of spectral lines of hydrogen produced by electron transitions to and from the lowest energy state of the hydrogen atom.

Magellanic clouds Two nearby galaxies visible to the naked eye from southern latitudes.

magnetic field A region of space near a magnetized body within which significant magnetic forces can be detected.

magnetometer A device for measuring magnetic fields.

magnetosphere The region around a planet occupied by its magnetic field.

magnifying power The number of times larger in angular diameter an object appears through a telescope than when viewed with the naked eye.

magnitude A measure of the amount of light received from a star or other luminous object.

main sequence A grouping of stars on the Hertzsprung–Russell diagram extending diagonally across the graph from the hottest, brightest stars to the dimmest, coolest stars.

major axis (of an ellipse) The longest diameter of an ellipse.

mantle (of a planet) That portion of a terrestrial planet located between its crust and core.

mare Latin "sea;" a large, relatively crater-free plain on the Moon.

maria Plural of "mare."

mascon A localized concentration of dense material beneath the lunar surface.

mass A measure of the total amount of material in an object.

mass function A numerical relationship involving the masses of the stars in a binary system and the angle of inclination of their orbit in the sky.

mass–luminosity relation A relationship between the masses and luminosities of main-sequence stars.

mass–radius relation A relationship between the masses and radii of white-dwarf stars.

mean solar day The interval between successive meridian passages of the mean Sun; the average length of a solar day.

mean solar time Time reckoned by the location of the mean Sun.

mean Sun A fictitious object that moves eastward at a constant speed along the celestial equator, completing one circuit of the sky with respect to the vernal equinox in one tropical year.

mechanics The branch of physics dealing with the behavior and motions of objects acted upon by forces.

megaparsec (Mpc) One million parsecs.

meridian (local) The great circle on the celestial sphere that passes through an observer's zenith and the north and south celestial poles.

mesosphere A layer in a planet's atmosphere above the stratosphere.

metamorphic rock A rock whose properties and appearance have been transformed by the action of pressure and heat beneath the Earth's surface.

meteor The luminous phenomenon seen when a meteoroid enters the Earth's atmosphere; a "shooting star."

meteor shower Many meteors that seem to radiate from a common point in the sky.

meteorite A fragment of a meteoroid that has survived passage through the Earth's atmosphere.

meteoroid A small rock in interplanetary space.

micrometeorite A very small meteoroid; a grain of interplanetary dust.

microwave Short-wavelength radio waves.

Milky Way Our galaxy; the band of faint stars seen from the Earth in the plane of our galaxy's disk.

minor axis (of an ellipse) The smallest diameter of an ellipse.

minor planet An asteroid.

model atmosphere The results of a theoretical calculation that gives the values of temperature, pressure, density, and so forth throughout the outer layers of a star.

molecule A combination of two or more atoms.

momentum A measure of the inertia of an object; an object's mass multiplied by its velocity.

monochromatic Of one wavelength or color.

muon A subatomic particle that behaves like a heavy electron.

nadir The point on the local meridian 180° from the zenith.

nanosecond One-billionth (10^9) second.

neap tide An ocean tide that occurs when the Moon is near first-quarter or last-quarter phase.

nebula A cloud of interstellar gas and dust.

neutrino A subatomic particle with no electric charge and little or no mass, yet one that is important in many nuclear reactions.

neutron A subatomic particle with no electric charge and with a mass nearly equal to that of the proton.

neutron star A very compact, dense star composed almost entirely of neutrons.

New General Catalogue (NGC) A catalogue of star clusters, nebulae, and galaxies first published in 1888.

new moon A phase of the Moon when it is nearest the Sun in the sky, so that only its dark hemisphere is visible from Earth.

Newtonian focus An optical arrangement in a reflecting telescope in which a small mirror reflects converging light rays to a focus on one side of the telescope tube.

Newton's laws The laws of mechanics and gravitation formulated by Isaac Newton.

node The intersection of an orbit with a reference plane such as the plane of the celestial equator or the ecliptic.

nonthermal radiation Radiation emitted by charged particles moving through a magnetic field; synchrotron radiation.

nova A star that experiences a sudden outburst of radiant energy, temporarily increasing its luminosity roughly a thousand-fold.

nuclear Referring to the nucleus of an atom.

nuclear bulge The central region of our galaxy.

nucleus (of an atom) The massive part of an atom, composed of protons and neutrons, about which electrons revolve.

nucleus (of a comet) A collection of ices and dust that constitute the solid part of a comet.

nucleus (of a galaxy) The concentration of stars and dust at the center of a galaxy.

nutation A small periodic wobbling of the Earth's axis superimposed on precession.

OB association An association of very young, massive stars predominantly of spectral types O and B.

objective The principle lens or mirror of a telescope.

oblateness A measure of how much a flattened sphere (or spheroid) differs from a perfect sphere.

obliquity (of the ecliptic) The angle between the planes of the celestial equator and the ecliptic (about $23\frac{1}{2}°$).

obscuration (interstellar) The absorption of starlight by interstellar dust.

occultation The eclipsing of an astronomical object by the Moon or a planet.

opacity The ability of a material to impede the passage of light.

open cluster A loose association of young stars in the disk of our galaxy; a galactic cluster.

opposition The configuration of a planet when it is at an elongation of 180°, and thus appears opposite the Sun in the sky.

optics The branch of physics dealing with the behavior and properties of light.

orbit The path of an object that is moving about a second object or point.

outgassing Volcanic processes by which gases escape from a planet's crust into its atmosphere.

Pallas The second asteroid to be discovered.

Pangaea The name of a hypothetical continent that fragmented into several of the continents on Earth today.

parabola A conic section formed by cutting a circular cone at an angle parallel to one of the sides of the cone.

parallax The apparent displacement of an object due to the motion of the observer.

parent isotope An unstable isotope that radioactively decays into another isotope.

parsec (pc) A unit of distance; 3.26 ly.

partial eclipse A lunar or solar eclipse in which the eclipsed object does not appear completely covered.

Pauli exclusion principle A principle of quantum mechanics stating that two identical particles cannot have the same position and momentum.

penumbra The portion of a shadow in which only part of the light source is covered by an opaque body.

penumbral eclipse A lunar eclipse in which the Moon passes only through the Earth's penumbra.

perfect gas An idealized gas that obeys a very simple equation of state.

perigee The point in its orbit where a satellite or the Moon is nearest the Earth.

perihelion The point in its orbit where a planet is nearest the Sun.

period The interval of time between successive repetitions of a periodic phenomenon.

periodic table A listing of the chemical elements according to their properties, invented by Dmitri Mendeleev.

period–luminosity relation A relationship between the period and average density of a pulsating star.

perturbation A small disturbing effect.

phases of the Moon The appearances of the Moon at different times as it orbits the Earth.

photometry The measurement of light intensities.

photon A discrete unit of electromagnetic energy.

photosphere The region in the solar atmosphere from which most of the visible light escapes into space.

plage A bright region in the solar atmosphere as observed in the monochromatic light of a spectral line.

Planck's constant (h) The constant of proportionality between a photon's energy and its frequency.

Planck's radiation law A relationship between the intensity of radiation emitted by a blackbody, its wavelength, and the temperature of the blackbody.

planetary nebula A luminous shell of gas ejected from an old, low-mass star.

plasma A hot ionized gas.

plate tectonics The motions of large segments (plates) of the Earth's surface over the underlying mantle.

population I star A star whose spectrum exhibits spectral lines of many elements heavier than helium; a metal-rich star.

population II star A star whose spectrum exhibits comparatively few spectral lines of elements heavier than helium; a metal-poor star.

positron An electron with a positive rather than negative electric charge; an antielectron.

precession (of the Earth) A slow, conical motion of the Earth's axis of rotation caused by the gravitational pull of the Moon and Sun on the Earth's equatorial bulge.

precession (of the equinoxes) The slow westward motion of the equinoxes along the ecliptic due to precession of the Earth.

prime focus The point in a telescope where the objective focuses light.

primeval fireball The extremely hot gas that filled the universe immediately following the Big Bang.

principle of equivalence A principle of general relativity stating that, in a small volume, it is impossible to distinguish between the effects of gravitation and acceleration.

prism A wedge-shaped piece of glass that is used to disperse white light into a spectrum.

prominence Flamelike protrusions seen near the limb of the Sun and extending into the solar corona.

proper motion The annual change in the location of a star on the celestial sphere.

proton A heavy, positively charged subatomic particle that is one of two principle constituents of atomic nuclei.

proton–proton chain A sequence of thermonuclear reactions by which hydrogen nuclei are built up into helium nuclei.

proto- A prefix referring to the embryonic stage of a young astronomical object (planet, star, etc.) that is still in the process of formation.

pulsar A pulsating radio source believed to be associated with a rapidly rotating neutron star.

pulsating variable A star that pulsates in size and luminosity.

quantum mechanics The branch of physics dealing with the structure and behavior of atoms and their interaction with light.

quark One of several hypothetical particles presumed to be the internal constituents of certain heavy subatomic particles such as protons and neutrons.

quarter moon A phase of the Moon when it is located 90° from the Sun in the sky, so that one-half of its daylit hemisphere is visible from the Earth.

quasar A starlike object with a very large redshift; a quasistellar object or quasistellar source.

r process A process of nuclear transformation initiated when certain isotopes rapidly capture large numbers of neutrons.

RR Lyrae variable One class of pulsating stars with periods less than one day.

radar A technique of reflecting radio waves from a distant object.

radial velocity That portion of an object's velocity parallel to the line of sight.

radial-velocity curve A plot showing the variation of radial velocity with time for a binary star or variable star.

radiant (of a meteor shower) The point in the sky from which meteors of a particular shower seem to originate.

radiation Electromagnetic energy; photons.

radiation pressure The transfer of momentum carried by radiation to an object on which the radiation falls.

radio astronomy That branch of astronomy dealing with observations at radio wavelengths.

radio galaxy A galaxy that emits an unusually large amount of radio waves.

radio telescope A telescope designed to detect radio waves.

radioactivity The process whereby certain atomic nuclei naturally decompose by spontaneously emitting particles.

ray (lunar) Any one of a system of bright, elongated streaks on the lunar surface.

recurrent nova A nova that has erupted more than once.

red giant A large, cool star of high luminosity.

reddening (interstellar) The reddening of starlight as it passes through the interstellar medium.

redshift The shifting to longer wavelengths of the light from remote galaxies and quasars; the doppler shift of light from a receding source.

reflecting telescope   A telescope in which the principal optical component is a concave mirror.

reflection   The return of light rays by a surface.

reflection nebula   A comparatively dense cloud of dust in interstellar space that is illuminated by a star.

refracting telescope   A telescope in which the principal optical component is a lens.

refraction   The bending of light rays passing from one transparent medium to another.

regression of nodes   The slow motion of the line of nodes of an orbiting body due to gravitational perturbations.

relativistic particle   A particle moving at nearly the speed of light.

resolution   The degree to which fine details in an optical image can be distinguished.

resolving power   A measure of the ability of an optical system to distinguish, or resolve, fine details in the image it produces.

retrograde motion   The apparent westward motion of a planet with respect to background stars.

revolution   The motion of one body about another.

right ascension   A coordinate for measuring the east–west positions of objects on the celestial sphere.

rille   A trenchlike depression in the lunar surface.

Roche limit   The smallest distance from a planet or other object at which a second object can be held together by purely gravitational forces.

rotation   The turning of a body about an axis passing through the body.

s process   A process of nuclear transformation that occurs when certain isotopes capture neutrons at a slow rate.

saros   A particular cycle of similar eclipses that recur about every 18 years.

satellite   A body that revolves about a large one.

Schmidt telescope   A reflecting telescope invented by Bernard Schmidt that is used to photograph large areas of the sky.

Schwarzschild radius   The distance from the singularity to the event horizon in a nonrotating black hole.

sedimentary rock   A rock that is formed from material deposited by rain or winds, or on the ocean floor.

seismic waves   Vibrations traveling through a terrestrial planet usually associated with earthquakelike phenomena.

seismograph   A device used to record and measure seismic waves, such as those produced by earthquakes.

seismology   The study of earthquakes and related phenomena.

semimajor axis   One-half of the major axis of an ellipse.

Seyfert galaxy   A spiral galaxy with a bright nucleus whose spectrum exhibits emission lines.

shell star   A star, usually of spectral type A to F, that is surrounded by a shell of gas.

shock wave   An abrupt, localized region of compressed gas caused by an object traveling through the gas at a speed greater than the speed of sound.

sidereal day   The interval between successive meridian passages of the vernal equinox.

sidereal month   The period of the Moon's revolution about the Earth with respect to the stars.

sidereal period   The orbital period of one object about another with respect to the stars.

sidereal time   Time reckoned by the location of the vernal equinox.

sidereal year   The orbital period of the Earth about the Sun with respect to the stars.

SMC   The Small Magellanic Cloud.

solar activity   Phenomena that occur in the solar atmosphere such as sunspots, plages, flares, and so forth.

solar constant   The average amount of energy received from the Sun per square centimeter per second, measured just above the Earth's atmosphere.

solar nebula   The cloud of gas and dust from which the Sun and solar system formed.

solar system   The Sun, planets, their satellites, asteroids, comets and related objects that orbit the Sun.

solar wind   A radial flow of particles (mostly electrons and protons) from the Sun.

solstice   Either of two points along the ecliptic at which the Sun reaches its maximum distance north or south of the celestial equator.

special theory of relativity   A description of mechanics and electromagnetic theory formulated by Albert Einstein, which explains that measurements of distance, time, and mass are affected by the observer's motion.

spectral class   A classification of stars according to the appearance of their spectra.

spectrogram   The photograph of a spectrum.

spectrograph   A device for photographing a spectrum.

spectroheliograph   A device for photographing the Sun in the monochromatic light of one spectral line.

spectroscope   A device for directly viewing a spectrum.

spectroscopic binary star   A binary star whose binary nature is deduced from the periodic Doppler shifting of lines in its spectrum.

spectroscopic parallax   The distance to a star derived by comparing its apparent magnitude to an absolute magnitude inferred from the star's spectrum.

spectroscopy   The study of spectra.

speed   The rate at which an object moves.

spicule   A narrow jet of rising gas in the solar chromosphere.

spiral arms   Lanes of interstellar gas, dust, and young stars that wind outward in a plane from the central regions of a galaxy.

spiral galaxy   A flattened, rotating galaxy with pinwheel-like spiral arms winding outward from the galaxy's nucleus.

spring tide   Ocean tides that occur at new-moon and full-moon phases.

standard time   Local mean solar time at a standard meridian adopted for convenience at surrounding geographical areas.

star   A self-luminous sphere of gas.

Stefan–Boltzmann law   A relationship between the temperature of a black body and the rate at which it radiates energy.

Stefan's law   *See* Stefan–Boltzmann law.

stellar evolution   The changes in size, luminosity, temperature, and so forth that occur as a star ages.

stellar model   The result of theoretical calculations that give details of physical conditions inside a star.

stratosphere   A layer in the atmosphere of a planet directly above the troposphere.

strong nuclear force   The force that binds protons and neutrons together in nuclei.

subduction zone   A location where colliding tectonic plates cause the Earth's crust to be pulled down into the mantle.

subdwarf   A star of lower luminosity than main-sequence stars of the same spectral type.

subgiant   A star whose luminosity is between that of main-sequence stars and normal giants of the same spectral type.

summer solstice   The point on the ecliptic where the Sun is farthest north of the celestial equator.

Sun   The star about which the Earth and other planets revolve.

sunspot   A temporary cool region in the solar photosphere.

sunspot cycle   The semiregular 11-year period with which the number of sunspots fluctuates.

supergiant   A star of very high luminosity.

superior conjunction   The configuration of a planet being behind the Sun as viewed from the Earth.

superior planet   A planet that is more distant from the Sun than the Earth is.

supernova   A stellar outburst during which a star suddenly increases its brightness roughly a millionfold.

surface gravity   The weight of a unit mass at the surface of an object such as a planet.

synchrotron radiation   The radiation emitted by charged particles moving through a magnetic field; nonthermal radiation.

synodic month   The period of revolution of the Moon with respect to the Sun; the length of one cycle of lunar phases.

synodic period   The interval between successive occurrences of the same configuration of a planet.

T Tauri stars   Young variable stars associated with interstellar matter that show erratic changes in luminosity.

T Tauri wind   A flow of particles away from a T Tauri star.

tail (of a comet)   Gas and dust particles from a comet's nucleus that have been swept away from the comet's head by the radiation pressure of sunlight and the solar wind.

tektites   Rounded glassy objects believed to have a meteoritic origin.

telescope   An instrument for viewing remote objects.

temperature (Celsius)   Temperature measured on a scale where water freezes at 0° and boils at 100°.

temperature (color)   The temperature of a star determined by comparing the intensity of starlight in two wavelength bands.

temperature (effective)   The temperature of a black body that would radiate the same total amount of energy that a particular star does.

temperature (excitation)   The temperature of a star determined from the strengths of various spectral lines that originate in atoms with different stages of excitation.

temperature (Fahrenheit)   Temperature measured on a scale where water freezes at 32° and boils at 212°.

temperature (Kelvin)   Absolute temperature measured in units (kelvins, K) equivalent to the degree Celsius.

temperature (kinetic)   Temperature directly related to the average speed of atoms or molecules in a substance.

terminator   The line dividing day and night on the surface of the Moon or a planet; the line of sunset or sunrise.

terrestrial planet   Any of the planets Mercury, Venus, Earth, or Mars, and sometimes also including the Galilean satellites and Pluto.

thermal energy   The energy associated with the motions of atoms or molecules in a substance.

thermal equilibrium   A balance between the input and outflow of heat in a system.

thermal radiation   The radiation naturally emitted by any object that is not at absolute zero.

thermodynamics   The branch of physics dealing with heat and the transfer of heat between bodies.

thermonuclear reaction   A reaction resulting from the high-speed collision of nuclear particles that are moving rapidly because they are at a high temperature.

thermosphere   A region in the Earth's atmosphere between the mesosphere and the exosphere.

threshold temperature   The temperature above which photons spontaneously produce particles and antiparticles of a particular type.

tidal force   A gravitational force whose strength and/or direction varies over a body and thus tends to deform the body.

total eclipse   A solar eclipse during which the Sun is completely hidden by the Moon, or a lunar eclipse during which the Moon is completely immersed in the Earth's umbra.

transit   The passage of a celestial body across the meridian; the passage of a small object in front of a larger object.

triple alpha process   A sequence of two thermonuclear reactions in which three helium nuclei combine to form one carbon nucleus.

triple point   The pressure and temperature at which a substance can exist simultaneously as a solid, liquid, and gas.

Trojan asteroid   One of several asteroids that share Jupiter's orbit about the Sun.

tropical year   The period of revolution of the Earth about the Sun with respect to the vernal equinox.

troposphere   The lowest level in the Earth's atmosphere.

turbulence   Random motions in a gas or liquid.

UBV system   A system of stellar magnitude involving measurements of starlight intensity in the ultraviolet, blue, and visible spectral regions.

ultraviolet radiation   Electromagnetic radiation of wavelengths shorter than those of visible light, but longer than those of X rays.

umbra   The central, completely dark portion of a shadow.

universal time   Local mean time at the prime meridian.

universal constant of gravitation (G)   The constant of proportionality in Newton's law of gravitation.

universe   All space, along with all the matter and radiation in space.

Van Allen belts   Two doughnut-shaped regions around the Earth where many charged particles (protons and electrons) are trapped by the Earth's magnetic field.

variable star   A star whose luminosity varies.

vernal equinox   The point on the ecliptic where the Sun crosses the celestial equator from south to north.

visual binary star   A binary star in which the two components can be resolved through a telescope.

VLBI   Very long baseline interferometry; a method of connecting widely separated radio telescopes to make observations of very high resolution.

watt   A unit of power; 10 ergs expended per second.

white dwarf   A low-mass star that has exhausted all its thermonuclear fuel and contracted to a size roughly equal to the size of the Earth.

Widmanstätten figures   Crystalline structure seen in certain types of meteorites.

Wien's law   A relationship between the temperature of a black body and the wavelength at which it emits the greatest intensity of radiation.

winter solstice   The point on the ecliptic where the Sun reaches its greatest distance south of the celestial equator.

Wolf–Rayet star   A class of very hot stars that eject shells of gas at high velocity.

X rays   Electromagnetic radiation whose wavelength is between that of ultraviolet light and gamma rays.

year (yr)   The period of revolution of the Earth about the Sun.

zap crater   A tiny glass-lined crater on a lunar rock presumed to be the result of the impact of a micrometeorite.

Zeeman effect   A splitting or broadening of spectral lines due to a magnetic field.

zenith   The point on the celestial sphere opposite to the direction of gravity.

zero-age main sequence   The main sequence of young stars that have just begun to burn hydrogen at their cores.

zodiac   A band of 12 constellations around the sky centered on the ecliptic.

Answers

Chapter 1

2. 3350 km   **3.** 8 minutes   **5.** 4×10^{36} times larger
6. 3×10^6 km   **7. a.** 140 cm   **b.** 86 m   **c.** 5.2 km
8. 8700 km   **9.** 4 km

Chapter 2

2. 9:04 PM   **7.** 18:00   **8.** September 22

Chapter 3

2. a. Waning crescent moon   **b.** Full moon   **c.** Waxing
crescent moon   **d.** At sunrise   **e.** At noon   **f.** At
sunrise   **8.** From the west   **10.** 50 arc seconds

Chapter 4

3. 5.2 square AU in 1984; 26.0 square AU in five years
4. Average distance = 100 AU; farthest distance = 200 AU
5. The star is 16 times more massive than the Sun   **7.** 100
times weaker   **8.** Yes. Sidereal period = synodic period = 2
years. Mars (sidereal period = 1.88 yr; synodic period = 2.14 yr)
9. 38,000 km

Chapter 5

3. The Palomar telescope collects a million times more light
than the human eye.   **7.** 9.5×10^{-18} ergs = 5.9×10^{-6} eV
8. 3×10^{11} Hz   **9.** At best, the Space Telescope will resolve
surface features as small as 50 km across on Jupiter's moons.
Only those features on our Moon larger than 110 km across can
be seen with the unaided human eye.

Chapter 6

8. 6×10^{26} g; 4 g/cm^3 **9.** $T_C = T_K = -40°$; $T_K = T_F = 574°$

Chapter 7

6. 3.4×10^{39} ergs   **7.** 4.8×10^{-4} cm   **9.** 10^{-4} Å
10. 30; CO_2 and O_2

Chapter 8

6. $(6.14)^4$ or about 1420 times weaker   **8.** $(1.87)^2$ or about
$3\frac{1}{2}$ times larger for Mercury

Chapter 9

4. Core = 30%; mantle = 69%; crust = 1%   **7.** About 220
million years ago   **9.** Atmospheric pressure = 1 millibar at an
altitude of about 55 km

Chapter 10

8. 1.5×10^5 kg/yr; 5×10^{17} years   **9.** 550,000 km

Chapter 11

8. Yes. Under favorable circumstances, the angular separation
between the Earth and Moon as seen from Mars is about
20 arc min.

Chapter 12

8. Outside the Sun, at a distance of about 7.8×10^6 km from
the Sun's center

Chapter 13

7. 2×10^{27} kg   **9.** 6×10^{10} yr, which is roughly 15 times
longer than the age of the solar system   **10.** 9 minutes

Chapter 14

5. Yes. At opposition, features larger than 120 km across
should be detectable.   **8.** The orbital period of particles in
the Keeler Gap is 13.8 hours, which is not a simple resonance
with any of Saturn's major satellites.   **9.** Gases with
molecular weights greater than 10 are retained.

Chapter 15

6. 1470 times fainter   **8.** 7000 km

Chapter 16

1. 678 years; 0.53 degrees per year   **9.** 50 km   **10.** In each case, the comet's lifetime is its orbital period multiplied by 100,000.   **a.** 354 years   **b.** 11,200 years   **c.** 354,000 years   **d.** 11.2 million years

Chapter 17

3. $7\frac{1}{2}$ times   **6.** 4102 Å   **7.** 2500 Å   **8.** The large, cool star is 6.25 times brighter than the Sun.

Chapter 18

1. 17   **8.** 100 magnitudes   **9.** New radius is twice the old radius; New luminosity is 64 times the old luminosity **10.** 1.5 $M_\odot$

Chapter 19

10. During 5 billion years, 2 percent of the Sun's hydrogen is converted into helium

Chapter 20

9. 3.6×10^4 atoms/cm^3 **11.** 10,000 times the volume of the Sun

Chapter 21

4. a. 750,000 years   **b.** 1 billion years   **5.** One tenth of the Sun's mass   **6.** 9.3×10^8 years

Chapter 22

4. 2×10^6 g/cm^3 **5.** The star's diameter is about one-tenth the Sun's diameter.

Chapter 23

8. The maximum correction is 10^{-4} sec for pulsars with periods of roughly 1 second.   **11.** It takes about 30 years for the edge of the Crab nebula to move outward by 1 arc sec.

Chapter 24

7. 2×10^{14} g/cm^3 **8.** About 10^8 $M_\odot$ for an average density of 1 g/cm^3 **9.** 2.9 $M_\odot$

Chapter 25

6. 2000 times   **9.** A 10% error in the distance would produce a 10% error in the mass; a 10% error in the orbital speed would produce a 20% error in the mass.   **10.** One supernova about every 800 years

Chapter 26

10. 2.5 Mpc   **11.** The distance would be in error by a factor of roughly 2.

Chapter 27

5. $0.52c$; 10^{10} ly ($H_0 = 15$ km/sec/10^6) ly   **6.** 20 AU **9. a.** $0.61c$ **b.** 1.2×10^{10} ly   **c.** -26 **d.** 1.6×10^{46} erg/sec

Chapter 28

3. 20 billion years; 15 billion years; 10 billion years **7.** radius = c/H_o

Chapter 29

9. The neutrino mass range is 25 to 100 eV

Illustration credits

Index

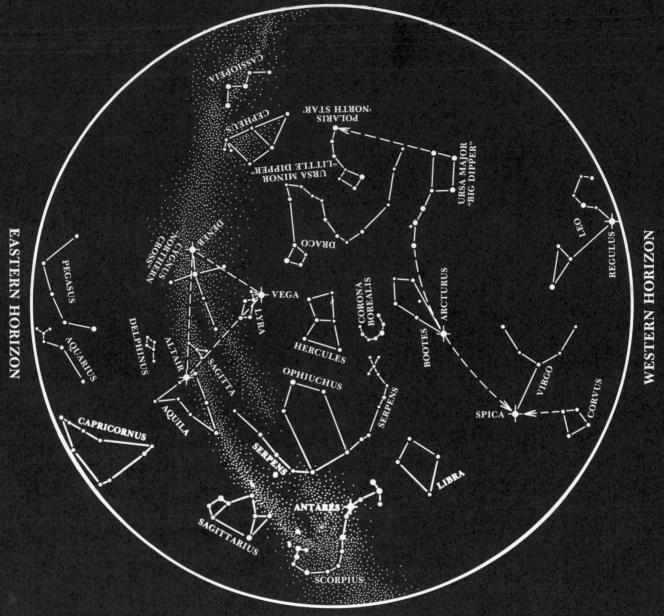

NORTHERN HORIZON

CASSIOPEIA

CEPHEUS

POLARIS "NORTH STAR"

URSA MAJOR "BIG DIPPER"

URSA MINOR "LITTLE DIPPER"

DENEB

CYGNUS "NORTHERN CROSS"

DRACO

VEGA

LYRA

CORONA BOREALIS

ARCTURUS

LEO

REGULUS

EASTERN HORIZON

WESTERN HORIZON

PEGASUS

AQUARIUS

DELPHINUS

ALTAIR

SAGITTA

AQUILA

HERCULES

OPHIUCHUS

SERPENS

BOOTES

VIRGO

CORVUS

SPICA

CAPRICORNUS

SERPENS

LIBRA

SAGITTARIUS

ANTARES

SCORPIUS

SOUTHERN HORIZON

THE NIGHT SKY IN SUMMER

Latitude of chart is 34°N, but it is practical throughout the continental United States.

To use: Hold chart vertically and turn it so the direction you are facing shows at the bottom.

Chart time (daylight saving time):
Mid–June . . . . Midnight
Mid–July . . . . 10 pm
Mid–August. . . . . 7 pm

Star Chart from *GRIFFITH OBSERVER*, Griffith Observatory, Los Angeles